HANDBUCH DER VIRUSFORSCHUNG

HERAUSGEGEBEN VON

PROF. Dr. R. DOERR UND PROF. Dr. C. HALLAUER

BASEL   BERN

II. ERGÄNZUNGSBAND

TECHNIC AND APPLICATION OF ROLLER TUBE CULTURES · TECHNIC AND APPLICATION OF DRYING OF VIRUSES IN THE FROZEN STATE · DIE AUFLICHT- UND DUNKELFELDMIKROSKOPIE · VARIATION IN INFLUENZA VIRUSES · IMMUNITY AND VACCINATION IN INFLUENZA · VIRUS PNEUMONIA AND PNEUMONITIS VIRUSES OF MAN AND ANIMALS · DIE HAEMAGGLUTINATION DURCH VIRUSARTEN · DIE ELEKTRONENMIKROSKOPIE

BEARBEITET VON

F. M. BURNET-MELBOURNE · M. D. EATON-BOSTON, MASS. · A. E. FELLER-CLEVELAND, O. · E. W. FLOSDORF-FOREST GROVE, PA. · TH. FRANCIS-ANN ARBOR, MICH. · C. HALLAUER-BERN · M. KAISER-WIEN · H. RUSKA-BERLIN-DAHLEM · P. VONWILLER-RHEINAU

MIT 187 ABBILDUNGEN IM TEXT

WIEN

SPRINGER-VERLAG

1950

ISBN 978-3-7091-5690-2   ISBN 978-3-7091-5688-9 (eBook)
DOI 10.1007/978-3-7091-5688-9

Inhaltsverzeichnis.

Seite

Technic and Application of Roller Tube Cultures. By A. E. FELLER, M. D., Cleveland, Ohio. With 2 figures 1

   I. Introduction 1

   II. Technic 2

      A. Assembly and care of roller tube cultures 2

      B. Modifications of technic 6

   III. Virus studies 7

   References 9

Technic and Application of Drying of Viruses in the Frozen State. By Earl W. FLOSDORF, Forest Grove, Pennsylvania. With 4 figures 11

Pre-freezing 11

Objectives in drying after freezing 11

Mechanism of freeze-drying 12

Freezing without drying 13

Low temperature without freezing 13

Results with freeze-dried viruses 13

Laboratory sources of low temperature for freezing and storage 15

Removing water vapor 16

Apparatus for laboratory research 19

Bibliography 22

Die Auflicht- und Dunkelfeldmikroskopie in der Virusforschung. Von M. KAISER, Wien und P. VONWILLER, Rheinau. Mit 15 Abbildungen 23

A. Einleitung 23

B. Die Auflichtapparatur 25

C. Erste eigene Versuche einer Auflichtbetrachtung virusbedingter Veränderungen im Gewebe 28

D. Weitere Versuche anderer Autoren über die Virusmikroskopie im Auflicht 32

E. Beobachtungen von virusbedingten Veränderungen im Dunkelfeld . . . 36

F. Weitere eigene Beobachtungen an vaccinierten Hornhäuten und Organen mit virusbedingten Veränderungen 40

G. Schlußbetrachtungen 44

Literaturübersicht 45

Variation in Influenza Viruses. By F. M. BURNET, Melbourne. With 1 figure 47

   I. The influenza group of viruses 47

   II. Variation as a necessary preliminary to laboratory study 48

      1. Adaptation to the allantoic cavity 49

      2. Adaptation to mice 51

Seite

III. Differences amongst influenza virus strains which have been adapted to normal laboratory passage ... 52

   1. Antigenic differences amongst strains of influenza virus ... 52
      a) The varying breadth of activity in human convalescent sera ... 53
      b) The significance of antigenic differences observed with human immune sera ... 55
      c) Antigenic differences as determined by neutralization with immune ferret sera ... 57
      d) The use of immune rabbit sera ... 57
      e) Cross immunity tests in mice ... 58
   2. Variations in the enzymic activity of different influenza virus strains ... 58

IV. Further modification of influenza viruses induced by laboratory manipulation ... 59

V. The general principles of virus variation as exemplified in influenza virus strains ... 61

VI. The significance of variation in influenza viruses in regard to the epidemiology of influenza ... 62

References ... 63

Immunity and Vaccination in Influenza. By Thomas Francis Jr., M. D., Ann Arbor, Michigan ... 66

  I. Epidemiologic concepts ... 66
 II. Resistance in experimental animals ... 67
III. Resistance in man ... 69
IV. Factors influencing immunity by vaccination ... 78

Bibliography ... 83

Virus Pneumonia and Pneumonitis Viruses of Man and Animals. By Monroe D. Eaton, Boston, Mass. With 2 figures ... 87

Introduction ... 87

Historical ... 87

  I. Early observations on pathology ... 87
 II. Attempts to determine etiology ... 88
III. Experiments in animals with known viruses ... 89

Virus pneumonia in man ... 91

  I. Differentiation from bacterial pneumonia ... 91

    Epidemiological differentiation ... 92
    Characteristic pathology ... 93
    Failure of response to chemotherapy ... 93

 II. Primary atypical pneumonia ... 93

    Nomenclature ... 93
    Distribution ... 94
    Epidemiology ... 95
    Clinical characteristics ... 97
    Pathology ... 99
    Serological reactions in primary atypical pneumonia ... 101

Inhaltsverzeichnis.   V

                                                              Seite

III. Etiology of primary atypical pneumonia   109

   Experimental transmission to man   109
   Experimental transmission to laboratory animals   112
Pneumonitis viruses of animals transmissible to man   126
   I. Agents of the psittacosis group isolated from human beings   126
   II. Other pneumonitis viruses of animals possibly infectious for man . .   131
Other pneumonitis agents of animals   135
   Diseases associated with coccobacilliform bodies   135
   Pleuropneumonia-like organisms   135
   The gray lung virus   136
   Hamster agent W 1   137
   Cotton rat agents W 2 and W 3   137
Pneumonitis agents of unknown origin   138
   Agents isolated in mice   138
   Agents isolated in guinea pigs   138
References   139

Die Haemagglutination durch Virusarten (Phaenomen von G. K. Hirst). Von
   Prof. Dr. C. Hallauer, Bern   141
   I. Die Entdeckung und die Phaenomenologie der virusbedingten Haemaggluti-
      nation   141
   II. Die Analyse des Phaenomens   144
      1. Die Resistenz des Haemagglutinins gegen inaktivierende Einflüsse im
         Vergleich zu den anderen Virusfunktionen   144
      2. Die Bildung des Haemagglutinins als Funktion der Virusvermehrung   149
      3. Die Beziehungen des Haemagglutinins zu den geformten Viruselementen   157
      4. Die Reaktion des Haemagglutinins mit der Erythrocytenoberfläche   161
      5. Die Fermenthypothese   181
      6. Das Verhalten des Virushaemagglutinins in vivo   184
      7. Die nichthaemagglutinierenden Virusarten   189
      8. Die Haemagglutinine und erythrocytentransformierenden Agenzien bak-
         terieller und pflanzlicher Provenienz   190
   III. Die Anwendungsmöglichkeiten   193
   IV. Die Technik   194
      1. Allgemeine Richtlinien   194
      2. Spezielle Methoden   202
         a) Methode von Hirst   202
         b) Methode von Burnet   203
         c) Methode von Whitman   204
         d) Methode von Salk   204
         e) Nachweis von Influenzaantikörpern im menschlichen Serum . . .   205
   Literatur   206
   Nachtrag bei der Korrektur   220

Die Elektronenmikroskopie in der Virusforschung. Von Helmut Ruska Berlin-
   Dahlem. Mit 163 Abbildungen   221
   I. Die Bedeutung der Elektronenmikroskopie für die Virusforschung . . .   221
      1. Die Elektronenmikroskopie in der Morphologie   223
      2. Das Verhältnis der Elektronenmikroskopie zu anderen Methoden der
         Virusforschung   227

Seite

a) Licht- und ultraviolettmikroskopische Methoden 227

    α) Hellfeldmikroskopie 227

    β) Dunkelfeld- und Ultramikroskopie 229

    γ) Fluoreszenzmikroskopie 230

    δ) Strömungsdoppelbrechung 230

b) Die Röntgenographie. 230

c) Nicht-optische Verfahren 231

    α) Ultrazentrifugierung 231

    β) Ultrafiltration 232

    γ) Viskositätsmessung 232

    δ) Statistische Ultramikrometrie 232

II. Die Grundlagen der Mikroskopie mit Elektronenstrahlen 233

  1. Die Elektronenstrahlen 233

  2. Die Elektronenlinsen 234

    a) Elektrische Felder und Linsen 234

    b) Magnetische Felder und Linsen 236

  3. Die Elektronenstrahlerzeuger 238

  4. Die Elektronen-Indikatoren 239

  5. Der Strahlengang im Elektronenmikroskop 239

  6. Die Eigenschaften des elektronenmikroskopischen Bildes 242

  7. Die physikalischen Grenzen der Elektronenmikroskopie 246

    a) Optische Grenzen 247

    b) Durch Objekt und Elektronenindikatoren bedingte Grenzen. . . 247

        α) Leistungsgrenze der Objektbelastung 248

        β) Dosisgrenze der Objektbelastung 249

  8. Die historische Entwicklung 249

III. Die Elektronenmikroskope 250

  1. Der Gesamtaufbau 250

  2. Die Mikroskopröhre 251

  3. Die elektrischen Anlagen 255

  4. Die Vakuumanlagen und Kühlwasserleitungen 257

  5. Die im Handel befindlichen Elektronenmikroskope. 258

IV. Die Präparationsmethoden 262

  1. Die Objekthalterung 263

    a) Objektnetze und Objektblenden 263

    b) Reinigung der Objektblenden 264

    c) Objektträgerfassungen 264

  2. Die Objektträgerfilme 265

    a) Die Herstellung von Kollodiumfilmen 265

    b) Die Berechnung der Dicke von Kollodiumfilmen 267

    c) Aluminiumoxydfilme 268

    d) Siliziumdioxyd- und Siliziummonoxydfilme 268

    e) Silikatglasfilme 269

    f) Berylliumfilme 269

  3. Die Beschickung der Filme mit Objekten 270

    a) Tropfenmethode 270

        α) Der Objektträger-Vibrator 271

        β) Die Gefrier-Vakuumtrockenkammer 271

    b) Niederschlagsmethode. 273

        α) Der elektrische Kernfäller 274

        β) Der Versuchsaufbau 275

Seite

c) Oberflächenmethoden 277

    α) Das Abklatschverfahren 278
    β) Das Abziehverfahren. 278
    γ) Das Filmbewuchsverfahren 279
    δ) Die Abdruckverfahren 280

4. Die Objektvorbereitung 283

  a) Verfahren zur Untersuchung von Zellen und Geweben. 283

    α) Schnitt-Schall-Verfahren 283
    β) Hochgeschwindigkeits-Mikrotome 284
    γ) Schnittverfahren für Bakterien 284

  b) Anreicherung und Reinigung von Virussuspensionen 285
  c) Virusadsorption an roten Blutkörperchen 287
  d) Verfahren zur Kontraständerung 288

    α) Färbemethoden 288
    β) Metallbedampfung (Schattenwurftechnik) 289

V. Die Auswertung von Beobachtungen und Aufnahmen. 291

1. Die Objektschädigungen durch Elektronenstrahlen 292
2. Die Vermessung der Objekte 293

  a) Messen und Meßgenauigkeit in der Bildebene. 293
  b) Messen und Meßgenauigkeit in Richtung des Strahlenganges. . . 295

    α) Messung aus dem Einzelbild 295
    β) Messung aus einem stereoskopischen Bildpaar 297

3. Die statistische Auswertung 298
4. Die Bestimmung der Teilchenzahl in dispersen Systemen. 302

  a) Bestimmung aus Filmen mit eingebetteten Objekten 302
  b) Bestimmung aus mittlerem Teilchengewicht und Teilchenkonzentration 302
  c) Bestimmung über die aerodisperse Zerteilung einer Suspension 302

    α) Die Bestimmung der Teilchenzahl einer Suspension unter Zuhilfenahme der Beziehung zwischen Fleckgröße und Tröpfchengröße 304
    β) Die Bestimmung der Teilchenzahl einer Suspension unter Verwendung eines vermeßbar kristallisierenden Salzzusatzes . . 305

  d) Bestimmung der Teilchenzahl in einem Gasvolumen 306

VI. Ergebnisse der Virusforschung 306

1. Protozoen 308
2. Spirochaeten und Leptospiren 308
3. Bakterien einschließlich Rickettsien 311

  a) Zellmembran 312
  b) Kapsel und Schleimsubstanz 315
  c) Geißeln 316
  d) Zellinhalt 317
  e) Verhalten gegenüber physikalischen und chemischen Einwirkungen 325
  f) Serologische Reaktionen 328

4. Bakteriophagen 329

  a) Größe und Form 329
  b) Wirkungsmechanismus und Vermehrung 335

5. Filtrierbare Mikroorganismen und „große Virusarten" 338
6. Newcastle Virus 344
7. Quaderförmige Virusarten 346

  a) Form, Größe und Struktur 347
  b) Vermehrungsweise 352
  c) Verhalten gegenüber physikalischen und chemischen Eingriffen 355

Seite

8. Varicellen- und Zostervirus 356
9. Influenzavirus 358
10. Mumpsvirus 364
11. Kaninchen-Papillomvirus 364
12. Pferde-Enzephalomyelitisvirus 365
13. Poliomyelitisvirus 366
14. Virus der Maul- und Klauenseuche 369
15. Das Agens des übertragbaren Milchdrüsentumors der Maus 369
16. Polyedervirus 370
17. Pflanzenpathogene Virusarten 376
    a) Stäbchenförmige Virusproteine 376
    b) Kugelförmige Virusproteine 379
    c) Der Aufbau von Viruskristallen und -gelen 379
    d) Physikalisch-chemische und serologische Reaktionen 383
    e) Die identische Reproduktion der Virusmoleküle 387
18. Die Systematik der Virusarten 389

Literatur 394

Literaturnachtrag 414

Nachtrag bei der Korrektur 416

Sachverzeichnis 418

Technic and Application of Roller Tube Cultures.

By A. E. FELLER, M. D.

Associate Professor of Preventive Medicine and Assistant Professor of Medicine.
The School of Medicine, Western Reserve University, Cleveland, Ohio.

Introduction.

The roller tube method of tissue culture permits the prolonged cultivation of tissue in the active state without explantation, and, when virus is added to the culture, offers a means of studying the virus-cell relationship for extended periods. The tissue may be readily observed *in situ*. The method employs tissue fragments which are fastened to the inner wall of a test tube by imbedding the tissue in a plasma coagulum. A small amount of nutrient material in the liquid state is added. The culture is then rotated continuously in the horizontal position, alternately bathing the tissue in nutrient fluid and exposing it to the gaseous content of the tube. Fresh nutrient fluid is added when desired, and the gaseous environment of the tissue can be controlled. The culture may be infected with virus either when it is originally assembled or at any subsequent time. Nutrient fluid or bits of tissue may be removed for examination at will. However, the apparatus for continuous rotation of the culture requires considerable care and attention, and meticulous aseptic technic is necessary to avoid bacterial contamination when changing nutrient fluid and gaseous content of the culture.

There is no certain explanation for the more prolonged activity of the tissues in roller tube cultures as contrasted with cultures by "classical" methods. It is possible that alternate exposure of the tissue to fluid and gaseous environment is favorable to the metabolism of the cells, or that the continuous distribution of metabolic and nutritive substances is advantageous.

The roller tube method of tissue culture is not a new technic, but its use for the cultivation of viruses is a relatively recent development. CARREL (1) suggested the method in 1913, LÖWENSTÄDT (2) employed it in 1925, and subsequently the technic was modified by CARREL (3), GEY (4), GEY and GEY (5), and LEWIS (6). GEY and BANG (7) used roller tube cultures for the propagation of the virus of lymphopathia venerea in 1939. In 1940, FELLER, ENDERS, and WELLER (8) reported the cultivation of vaccinia virus by the roller tube method and carried out studies on the coexistence of the virus and living cells over extended periods. The technic, however, has not been widely used, despite its apparent advantages.

In the present communication, the technic of roller tube tissue culture will be described and certain modifications of this method will be noted. In addition, a review of certain studies which have employed the roller tube method will be presented together with possible problems for which the technic appears to be a useful tool. Emphasis will be given to general methods and technics rather than to detailed descriptions, since it is likely that the roller tube method, when used by others, will be modified according to the particular purpose or problem.

Technic.

The technic employed by Feller, Enders and Weller (8) will be described first to illustrate how a roller tube culture may be assembled, and the care which is given the culture for the prolonged cultivation of the tissue. It is not believed that this technic is better than the several others which might be employed, but it is the technic with which the author is most familiar. Following the description of this technic, various modifications will be described.

A. Assembly and care of roller tube cultures.

Tissue. Minced chick embryos of 8 to 10 days incubation are an excellent source of tissue. The embryos from which the eyes have been removed are readily minced in a Petri dish with curved scissors. Pieces of tissue approximately a cubic millimeter in size are selected for cultivation. The eyes are removed before the embryos are minced because the dark pigment in them might interfere with subsequent observation of the growing tissue and the cones and rods of the retina might be confused with bacteria. Embryos older than 8 to 10 days are not recommended as they may have feathers and the liver and bile of the older embryos may have an inhibitory effect upon growth (9).

Plasma and serum. Cardiac puncture of adult chickens provides plasma when clotting is prevented with 0.2 per cent heparin, and serum when the blood is defibrinated with glass beads. Chickens are used because plasma and serum are homologous with the tissues being cultivated. Chicken plasma is especially desirable as a supporting medium for the tissues, since it clots rapidly and forms a tenacious coagulum.

Simms' solution. The formula and preparation of this solution as described by Sanders (10) follows: It is made up in two concentrated solutions, since, by making up relatively small quantities of concentrates, a large store of salt solution is insured.

	Final Concentration Gm. per liter	Concentrate, Gm. per liter
Solution A		
Sodium chloride	8.000	160.00
Potassium chloride	0.200	4.00
Calcium chloride ($CaCl_2 . 2H_2O$)	0.147	2.94
Magnesium chloride ($MgCl_2 . 6H_2O$)	0.203	4.06
Solution B		
Sodium bicarbonate	1.010	20.20
Disodium phosphate	0.213	4.26
Dextrose	1.000	20.00
Phenol red	0.050	1.00

Solution A is autoclaved. Solution B is filtered through a neutral sintered $^5/_3$ Jena glass filter and kept, stoppered, in the refrigerator. Fifty cc. of solution A is added to 900 cc. of double distilled water. The whole is then autoclaved. 50 cc. of solution B is added after autoclaving.

Filtration of mother solution B may be effected with a Berkefeld N filter. If a precipitate forms, it may be redissolved by bubbling carbon dioxide through the solution.

Embryonic extract. Finely minced chick embryos (eyes removed) are extracted for 20 minutes at 37⁰ C in an equal volume of Simms' solution. The supernatant fluid is carefully removed after centrifugation for 20 minutes at approximately 2500 R. P. M. Embryonic extract is stored at 4⁰ C and may be used for at least 10 days or possibly longer.

Care in removing the supernatant fluid after centrifugation is necessary to avoid the admixture of tissue cells with the embryonic extract. This factor is of importance to avoid the addition of viable cells to a growing culture since the embryonic extract is one of the ingredients of the nutrient fluid which is added daily.

Preparation of the culture. Five drops of plasma are added to a 20 × 150 mm. test tube. A capillary pipette of thin bore is used to spread the plasma evenly over the lower ²/₃ of the tube. A large bore capillary pipette slightly curved at the tip is used to pick up and deposit from 30 to 50 pieces of tissue. As the plasma clots, the tissue is distributed evenly throughout the plasma. Any excess plasma which collects in the butt of the tube after 10 to 15 minutes should be removed.

Fig. 1. Drawing of a roller tube culture.

1.6 ml. of nutrient fluid consisting of Simms' solution 7 parts, embryonic extract 2 parts, and chicken serum 1 part, are added. At this point, it is important to roll or rock the tube so that the entire plasma-tissue surface is wetted with the nutrient fluid. The tube is then closed with a one-hole rubber stopper fitted with a short piece of glass tubing. The glass tubing is closed with a rubber vaccine cap (fig. 1).

Incubation. The culture is incubated at 37⁰ C in the horizontal position in a rotating device which turns approximately 10 times per hour. The simplest device consists of a large wooden cylinder mounted on a shaft (fig. 2). Holes of the proper size to admit the cultures are bored into the wooden cylinder parallel with the shaft. The cylinder containing the cultures is then turned by an electric motor connected to the shaft by means of reduction gears. The cultures are continuously rotated except when they are receiving routine care.

Routine care of cultures. Each day the nutrient fluid is removed by inserting a capillary pipette through the small

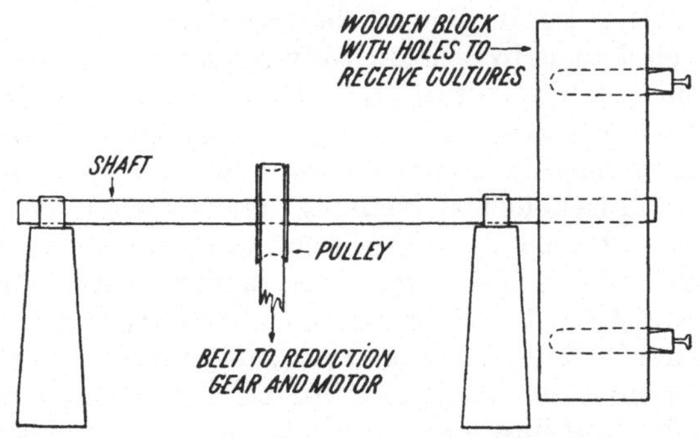

Fig. 2. Schematic drawing of a simple rotating device.

glass tube after removing the rubber vaccine cap. Approximately 20 ml. of steril air is then gently forced into the culture from a 20 ml. syringe equipped with a long needle, and filled with air drawn through sterile cotton. 1.6 ml. of nutrient fluid are added to the culture which is then re-sealed and returned to the rotating device.

It is usually necessary to change the nutrient fluid daily. In an actively growing culture the pH of the nutrient fluid is reduced from an initial of 7·8 to 7.0 or below in 24 hours. Retardation of growth and degenerative changes occur if the tissue is exposed to such an acid medium for long periods.

The plasma coagulum gradually becomes less dense and tends to peel away from the wall of the tube. It is then necessary to "patch" the plasma coagulum by removing the nutrient fluid completely and adding a few drops of plasma. Usually 3 to 5 drops of plasma are sufficient. The plasma is quickly distributed by a rotating and rocking motion. As the cultures become older the plasma clots poorly or not at all. A drop of embryonic extract will induce clotting in such instances. The addition of plasma for "patching" does not visibly increase or decrease the rate of growth.

Addition of virus. A simple means of adding virus to the culture without disturbing the quantitative relationship between the amount of tissue and the volume of nutrient fluid which is added daily, is to prepare the suspension of virus in SIMMS' solution which is then incorporated into the nutrient fluid in the manner noted above, except that the suspension of virus replaces an equivalent volume of SIMMS' solution. Successful infection with vaccinia virus may be accomplished either at the time the culture is assembled or may be delayed until tissue growth is well established.

Observation of culture. New cells appearing at the borders of the bits of tissue can be seen with the unaided eye, although the use of a hand lens or examination under the low power objective of a microscope is more satisfactory. For the latter purpose, the culture tube may be supported by two test tubes held parallel by adhesive tape, but separated from each other approximately one-half inch by means of two small corks interposed between the tubes at each end. The culture fits into the hollow between the two test tubes and is thus held motionless on the stage of the microscope. Separation of the two test tubes is necessary for the admission of sufficient light from below. Examination of the growing tissue with the high power objective is unsatisfactory due to the thickness of the wall of the test tube.

A permanent record of the day by day growth of a particular fragment of tissue may be obtained by placing a strip of translucent celluloid over the area and tracing the outline of the tissue on the celluloid with a sharp instrument. The celluloid strip is held firmly in place with rubber bands. A strong source of light beneath the culture and a hand lens held in place by a clamp are helpful. With a little practice, the same strip may be used for successive tracings and furnishes a reasonably accurate record of the growth arising from a single bit of tissue.

Morphological methods. Stained total mounts of tissue on cover slips may be obtained by a modification of the technic described by GEY (4, 5). Before assembling the culture, one or more cover slips (12 × 50 mm) are sealed to the wall of the tube by running melted 3.0 per cent agar in physiological saline beneath them. It is necessary to incubate such tubes with cotton stoppers in place for 24 hours at 37° C to drive off the water vapor which condenses on the cover slips and interferes with adhesion of the plasma coagulum. The culture is then assembled in the manner described above, except that the desired amount of tissue is placed on the cover slips. The cover slips may be removed for fixation and staining at any time during the course of cultivation without interfering seriously with the remaining tissue.

Removal of sheets of plasma with the enmeshed tissue from the tubes is difficult, particularly in young cultures, because the culture adheres firmly to the tube and is mangled by the manipulations required to remove it. Fixation *in situ* is more satisfactory, as the sheet of plasma and tissue may then be scraped or gently washed off. For morphological studies of chick embryonic tissue infected with vaccinia virus (8), ZENKERS fixative is useful.

Four stains are satisfactory for routine use: alum-hematoxylin and eosin; DOWNIE's modification of MANN's stain (11); GOODPASTURE's carbol-fuchsin methylene blue stain (12); and a modification of the acid-fuchsin stain described by BUDDINGH (13) in which LOEFFLER's methylene blue is used as a counter stain. The latter is the most satisfactory for staining inclusion bodies from vaccinia virus. It is probable that different methods of fixation and staining would have to be adopted or devised for tissues and viruses other than those noted above.

Titration of vaccinia virus. In the case of embryonic chick tissue infected with vaccinia virus (8), estimations of the amount of virus either in the tissue or in the nutrient fluid may be made by injecting 0.1 ml. of the material intradermally into the skin of a susceptible rabbit. The nutrient fluid is removed and diluted serially in broth. The tissue with the plasma coagulum is then scraped from the tube and ground in 1.6 ml. of infusion broth. Serial dilutions are made in broth. Control suspensions of tissue and nutrient fluid may be obtained from similar but unifected cultures.

General comments on technic. Several aspects of the technic require further comment because experience has indicated that attention to them will facilitate the cultivation of tissues in a viable state for prolonged periods.

Scrupulous aseptic technic is essential to avoid the loss of cultures by bacterial contamination. An adequate hood or a dust-free room in which to assemble the cultures, or to open them for the daily change of fluid and air, is recommended. A glass hood filled with steam, which is allowed to condense just before use, is satisfactory. A few days of practice with the technic before experiments are carried out is recommended.

Clean glassware must be used. If the test tubes are not carefully washed and then treated with acid cleaning solution for a period of 12 to 24 hours, difficulty in spreading plasma for the reception of the tissue will result and the plasma coagulum will not adhere properly to the glass.

In the case of chick embryonic tissues infected with vaccinia virus, the cultures may be left at ordinary room temperature in a test tube rack for one hour or more without apparent damage to the tissues. Longer periods are not recommended.

Proportions of embryonic extract in the nutrient fluid less than those recommended above may result in decreased growth and activity of the tissues.

A daily change of nutrient fluid and addition of sterile air is thought to be advantageous, although the interval may be lengthened to 48 hours provided the indicator in the nutrient fluid does not turn definitely yellow in color by this time.

A well built and reliable mechanical rotator for the cultures is recommended. The rotating devices described by OTT, TENNANT and LIEBOW (14) and by COMAN and STABLER (15) appear to fulfill these criteria. Time and expense involved in obtaining such apparatus will be repaid by trouble free operation 24 hours a day. Temporary failure of the rotating apparatus will not appreciably affect the cultures, but is undesirable. The author has not had experience with rates of rotation of the cultures appreciably different from the 10 rotations per hour recommended above, although speeds varying from 6 to 20 revolutions per hour have been employed by others.

B. Modifications of technic.

An exhaustive review of the various modifications of the technic described above will not be attempted, but brief mention of certain of them will be made to illustrate the versatility of the method.

The following tissues have been employed: human fibroblasts (7), brain (16), and placental tissue (17); chick embryo heart (8), brain (16), and membranes (18); rabbit testicle (19), liver (20), cardiac muscle (19), and leucocytes (19); mouse lung (21), and brain (16); rat brain (16); puppy brain (16); and Brown-Pearce carcinoma (22).

The observation that bits of cardiac muscle from the chick embryo beat rhythmically during 9 weeks of cultivation indicates that the tissues in roller tube cultures remain in the active state for long periods. Further indication is afforded by the observation that chick embryonic tissue cultivated for 37 days could be transplanted to new roller tubes and growth obtained (8).

When using certain of the tissues noted above, various workers have developed nutrient fluids appearing to be advantageous for that particular type of tissue. In many instances it appears that considerable study was given to the problem and it is suggested that the worker who is interested in cultivating any of the tissues noted above consult the indicated references. The recent report of the cultivation of rabbit liver cells indicates that an exhaustive study was required for the development of a nutrient fluid affording selective growth stimulation for these cells (20). On the contrary, it appears that certain other tissues may be kept in the viable state with relatively simple nutrient fluid. Chick embryo tissue remained viable for at least two weeks when nourished daily with Simms' solution only. However, the combination of mechanical exhaustion of air and the failure to change the nutrient fluids for 48 hours resulted in the apparent death of the cells in such cultures (8).

The frequency with which nutrient fluids have been changed has varied widely. With chick embryonic tissue a daily change of fluid appeared to be advantageous (8), but for the cultivation of human fibroblasts replenishment every 3 or 4 days, and occasionally every 5 days, was sufficient (7). This factor will probably depend upon the type of tissue being cultivated, the composition of the nutrient fluid, and the degree of tissue activity desired.

Observations on the gaseous environment of the tissue have varied widely. For the cultivation of rabbit leucocytes, testicle, or cardiac muscle, replenishment of the gaseous environment was limited to the exchange which occurred when the nutrient fluids were removed and replaced daily (19). On the other hand, an apparatus permitting continuous control of the atmosphere in roller tube culture has been described, and the growth of fibroblasts was found to be best with untreated air or with air containing 1%, 2% and 4% carbon dioxide (23, 24).

Although chicken plasma appears to have desirable properties for imbedding the tissue, rabbit plasma has been successfully used (20).

Various technics for the morphological or cytological studies of the tissues have been reported. Explants of tissue into lying drop preparations on cover slips have been used for microscopic study of living cells under oil immersion (7). Also, 0.3 per cent trypsin has been employed (7) for digestion of the substrate coagulum to facilitate removal of colonies for morphological study or subculture. A bottle with a hole in the side wall over which a cover glass is sealed in place has been described (25). The tissue is placed on the cover slip and may be readily observed during cultivation or fixed *in toto* for subsequent staining and morphological studies.

Virus studies.

The roller tube method has not, as yet, been widely used for the study of viruses. There are sufficient reports to indicate, however, that the method is a valuable tool for investigations of the growth and metabolism of viruses, the virus-cell relationship, the effects of sulfonamides or antibiotics, and mechanisms of immunity in virus diseases.

GEY and BANG (7) maintained the virus of lymphopathia venera in continuous roller tube cultures of a pure strain of human fibroblasts for more than seven months. In one particular culture, the presence of virus in cell-free tissue culture fluid was demonstrated after continuous cultivation for 89 days following infection of the culture with the virus. Although fluid from the cultures at first produced 100 % mortality when inoculated intracerebrally in mice, this power was eventually lost in subsequent cultures. The tissue culture fluid was also found to be a good antigen for the FREI test. The fluids which eventually lost virulence for mice were still active in the FREI test, thus indicating that the virus was present, but in an altered state so far as mouse virulence was concerned. Detailed morphological studies were also carried out and cyto-pathological changes believed to be specific were observed and studied in living cells infected with the virus. In one of their infected cultures, however, the "vesicular reaction" (interpreted as a specific cellular reaction to the virus) disappeared, the cells became granular for 21 days and then the "vesicular reaction" reappeared. For these reasons, they very properly warned against any simple interpretation of the significance of the "granules" within the vesicles. Of great interest was the observation that sulfanilamide could be used in low concentrations in the cultures. This drug was thought to have altered the course of the cellular reaction to the presence of the virus. In one set of cultures the cells in treated tubes were more granular than the controls, due, apparently, to the breakdown of "vesicles". A similar result was noted in another set of cultures with the additional observation that the cells in the treated tube appeared more healthy than those in the untreated tube. The cells in the treated cultures, however, were still cytologically abnormal and, aside from these changes, no definite evidence was presented to indicate that the virulence of the infection for the cultivated cells was decreased.

FELLER, ENDERS and WELLER (8) demonstrated the prolonged coexistence of vaccinia virus in high titer and living cells in roller tube cultures of chick embryonic tissues. The virus increased to maximal titer rapidly, and the quantity of virus in the tissue remained at or near maximum titer for approximately nine weeks. Considerable amounts were also present in the nutrient fluids removed daily, but there was always more virus in the tissue than in the fluid. Throughout the period of 9 weeks of continuous cultivation of the virus in a single roller tube culture, many, though not necessarily all, of the cells in the infected culture remained alive and retained the capacity to proliferate. The presence of living cells was demonstrated to be essential for the persistence of the virus. There was no discernible difference, in the gross, between the rate or amount of tissue growth in infected cultures and non-infected cultures. Cytoplasmic inclusion bodies were characteristic of tissues from vaccinia infected cultures, but inclusions were not seen in tissue from uninoculated cultures. Similar results were obtained when fragments of embryonic chick heart were employed rather than mixed chick embryonic tissue. Certain conditions of the experiments suggested, but did not prove, that virus was continually being propagated rather than simply preserved throughout the prolonged period of cultivation of the tissues.

Subsequently, Pearson and Enders (18) reported the prolonged cultivation of the Melbourne strain of influenza A virus in roller tube cultures of embryonic chick tissues. The titers of virus in tissue and nutrient fluid after initial increases remained almost constant during a period of 30 days continuous cultivation, provided the nutrient fluid contained serum and embryonic extract. As in the case of vaccinia virus propagated in embryonic chick tissue, more virus was present in the tissue than in the fluid. When only Tyrode solution or physiological saline solution was used as nutrient fluid, the titer of virus declined after one week. The influenza virus did not grow well in embryonic chick brain, heart, or skeletal muscle, but grew well in whole embryo, chorio-allantois, amnion, skin, lung and intestine. Growth was slow in yolk sac preparations.

The prolonged coexistence of living cells and virus in roller tube cultures (7, 8, 18) is in sharp contrast with available data on the survival of virus and tissue cells in suspended cell cultures. With the exception of Rous Sarcoma virus (26) and rabies virus (27), there has been no evidence in the literature until recently for survival beyond the 18th day in such cultures. Vaccinia virus has not been detected beyond the 12th day. In 1942, however, Enders and Florman (28) reinvestigated the problem of persistence of vaccinia virus and chick embryonic cells in suspended cell tissue cultures. Viable cells and vaccinia virus were demonstrated in cultures of Tyrode's solution and chick embryonic tissue for about 4 weeks following inoculation. Although this result was an improvement over those previously reported for the survival of vaccinia virus in suspended cell cultures, it was still in sharp contrast with the survival time of vaccinia virus in roller tube cultures, where survival of both cells and virus persisted much longer and multiplication of virus was greatly increased. In addition, fewer fragments of viable tissue could be recovered from suspended cell cultures infected with vaccinia virus than from uninfected cultures. In roller tube cultures (8) where conditions for cell growth and metabolism are more suitable, no definite evidence was detected of injury to the chick embryonic tissue beyond the changes seen in control cultures, even though the cells had been in contact with vaccinia virus for 9 weeks. Enders and Florman (28) also found that when 1.0 gram of tissue rather than only 0.1 gram was employed, neither virus nor living cells could be demonstrated after 15 days in suspended cell cultures.

These data (7, 8, 18, 28) indicate that the roller tube method of tissue culture is applicable for study of the problem of the prolonged coexistence of virus and living cells. Present technics of suspended cell cultures do not afford this opportunity for prolonged study. Available data (8), though not presented here, indicate that in many instances coexistence of viable cells and virus may be demonstrated for a longer period of time in plasma clot cultures than in suspended cell cultures, but not for extended periods as in the case of roller tube cultures.

Florman and Enders (19) have investigated the effect of homologous (rabbit) antiserum and complement on the multiplication of vaccinia virus in roller tube cultures of blood-mononuclear cells. The study was carried out with the general purpose of applying to the study of viruses the immunological principles which have proved so fruitful in researches with bacteria. The roller tube method was employed because it offered a means of maintaining and cultivating phagocytic cells over long periods of time. Their experiments showed that rabbit mononuclear blood cells not only survived in the roller tubes, but underwent cell-division and increase in numbers. The leucocytes actively ingested carmine particles, indicating that phagocytic cellular activity had not been suppressed. When infected with vaccinia virus, multiplication occured and virus could be recovered in considerable amounts from the nutrient fluids and cellular tissues for as long as 19 days after inoculation. These data indicated that the phagocytes did not destroy or inhibit the virus, but actually served as a medium in which the virus readily multiplied. The addition of specific homologous antiserum to infected cultures suppressed viral activity only temporarily. Furtermore, a mixture of virus and antiserum which did not produce a reaction in rabbit skin, produced infection of the mononuclear cells. The subsequent addition of antiserum did not eliminate the infection. Effectiveness of the antiserum was not enhanced by the addition of complement. Additional experiments, employing tissues consisting principally of fibroblats, demonstrated that the virus in such cultures increased more rapidly and reached a higher titer than it did in cultures of mononuclear cells. The impression was gained that complement and antiserum was more effective in temporarily suppressing viral activity in cultures of mononuclear cells than in cultures of fibroblasts. These important experiments of FLORMAN ans ENDERS (19) indicate that the roller tube technic is a valuable tool for further investigation of the problem of the mechanisms of immunity in virus diseases.

GEY, GEY, INUI and VEDDER (29) have recently reported the effects of crude and purified penicillin on continuous cultures of normal and malignant cells. Normal (rat) and tumor (human, rabbit, and rat) cells tolerated penicillin sodium in concentrations of 5000 Oxford units per cubic centimeter for long periods. Tumor cells were not more sensitive to penicillin than normal cells. These observations suggest that bacteria sensitive to penicillin could be eliminated from roller tube cultures without damage to the tissue cells or virus.

The present discussion has presented the technics and applications of the roller tube method of tissue culture for the study of viruses. In most instances, the possible applications of the method for further study are obvious. The detection of "masked" or altered virus from convalescent animals or in those where it has been impossible to demonstrate the causative agent has been noted (8). Simplification of the technic to provide a possible means for preparing vaccines has also been mentioned (8). The technic might also find application for the detection of certain viral or bacterial agents by virtue of the fact that tissues may be maintained in the active state for long periods.

References.

1. CARREL, A.: Neue Untersuchungen über das selbständige Leben der Gewebe und Organe. Berlin, Klinische Wochenschrift **50**; 1097—1101, 1913.
2. LÖWENSTÄDT, H.: Einige neue Hilfsmittel zur Anlegung von Gewebekulturen. Arch. F. Exp. Zellforsch. **1**; 251—256, 1925.
3. PARKER, R. C.: Methods of tissue culture. New York, Paul B. Hoeber, Inc., 118, 1938.
4. GEY, G. O.: Improved technic for massive tissue culture. Am J. Cancer, **17**; 752—756, 1933.
5. — and M. K. GEY: Maintenance of human normal cells and tumor cells in continous culture; preliminary report; cultivation of mesoblastic tumors and normal tissue and notes on methods of cultivation. Am. J. Cancer, **27**; 45—76, 1936.
6. LEWIS, W. H.: Carnegie Institution of Washington, Pub. No. 459. Contrib. Embryol. **25**; 161, 1935.

7. Gey, G. O., and F. B. Bang: Experimental studies on the cultural behavior and the infectivity of lymphopathia venera virus maintained in tissue culture. Johns Hopkins Hospital Bulletin 65; 393—417, 1939.

8. Feller, A. E., J. F. Enders and T. H. Weller: The prolonged coexistence of vaccinia virus in high titer and living cells in roller tube cultures of chick embryonic tissues. J. Exper. Med. 72; 367—388, 1940.

9. Plotz, H.: Personal communication.

10. Sanders, M.: Cultivation of the viruses: a critival review. Arch. Path. 28; 541—586, 1939.

11. Downie, A. W.: Study of lesions produced experimentally by cowpox virus. J. Path. & Bact. 48; 361—379, 1939.

12. Goodpasture, E. W.: Study of rabies, with reference to neural transmission of virus in rabbits, and structure and significance of negri bodies. Am. J. Path. 1; 547—582, 1925.

13. Buddingh, G. J.: Comparison of behavior of neurotesticular and dermal strain of vaccine virus in chorio-allantoic membrane of chick embryo- Am. J. Path. 12; 511—524, 1936.

14. Ott, L. H., R. Tennant and A. A. Liebow: An adaptable rotating device for roller tube tissue cultures. Science 91; 437—438, 1940.

15. Coman, D. R. and N. G. Stabler: An apparatus for roller tube tissue culture. Science 94; 569—570, 1941.

16. Hogue, M. J.: Tissue cultures of the brain (intercellular granules). J. Comp. Neurol. 85; 519—530, 1946.

17. Seegar, Jones, G. E., G. O. Gey and M. K. Gey: Hormone production by placental cells maintained in continuous culture. Johns Hopkins Hospital Bulletin 72; 26—38, 1943.

18. Pearson H. E. and J. F. Enders: Cultivation of influenza A virus in roller tubes. Proc. Soc. Exper. Biol. & Med. 48; 140—143, 1941.

19. Florman A. L. and J. F. Enders: The effect of homologous antiserum and complement on the multiplication of vaccinia virus in roller tube cultures of blood-mononuclear cells. J. Immunol. 43; 159—174, 1942.

20. Harris T. N. and S. Harris: The cultivation of mammalian liver cells in large numbers. Science 105; 75—76, 1947.

21. Feller, A. E.: Unpublished data.

22. Favorite, G. O. and F. S. Cheever: Observations on the Brown-Pearce carcinoma in roller tube tissue cultures. Cancer Research 1; 136—143, 1940.

23. Pfaf, G. H.: Control of pH in roller tube cultures. Proc. Soc. Exper. Biol. & Med. 62; 184—187, 1946.

24. — and G. S. Samuelsen: An apparatus for permitting control of the atmosphere in roller tube cultures. Proc. Soc. Exper. Biol. & Med. 62; 187—188, 1946.

25. Shaw, D. T., L. C. Kingsland and A. M. Brues: A roller bottle tissue culture system. Science 91; 148—149, 1940.

26. Carrel, A.: Conditions of reproduction in vitro of Rous virus. J. Exper. Med. 43; 647—668, 1926.

27. Webster, L. T. and A. D. Clow: Propagation of rabies virus in tissue culture. J. Exper. Med. 66; 125—131, 1937.

28. Enders, J. F., and A. L. Florman: Persistence of vaccine virus and chick. embryonic cells in suspended cell tissue cultures. Proc. Soc. Exper- Biol. & Med. 49, 153—155, 1942.

29. Gey, G. O., M. K. Gey, F. Inui, and H. Vedder: The effects of crude and purified penicillin on continuous cultures of normal and malignant cells. John Hopkins Hospital Bulletin 77, 116—131, 1945.

Technic and Application of Drying of Viruses in the Frozen State.

By EARL W. FLOSDORF, Forest Grove, Pennsylvania.

Pre-freezing.

Freezing quickly has been applied to various biological products like viruses to prevent alteration in chemical and biological characteristics before drying. No eutectic separation with resulting concentration of components occurs. This is in contrast with slow freezing, where two or more phases are obtained, one separating out first with solvent less rich in solute and the remaining liquid phase richer in solute. In the case of cellular materials, disruption of cell walls is avoided. Virus molecules, which frequently are large and relatively labile, are locked in place as a result of quickfreezing so that denaturation of protein and other changes do not occur. Viability is maintained.

In a word, quick-freezing is applied for the basic purpose of preventing chemical and physical change. Freezing more slowly may be utilized to effect changes where they are desired, as in concentrating penicillin, but it has not been applied to viruses.

The speed of freezing is not dependent entirely upon the temperature of the refrigerant or cold medium. The speed of freezing is a function of the rate of transfer of heat from the virus undergoing freezing to the refrigerant. The extent of surface exposed for transferring the heat from the one medium to the other, the type of the container, i. e. metal, glass or the like, are all important factors. The amount of agitation in the case of liquid preparations is important.

Objectives in drying after freezing.

Objectives involved in freeze-drying have been set forth in detail in the book "Freeze-Drying" (1). In summary, these may be set forth as follows. First, the temperature is below that at which many labile substances undergo chemical change. This applies to labile components in blood, to viruses and most forms of microorganisms, and to other biologicals and pharmaceuticals. Second, because of the low temperature, the loss of volatile constituents is minimal. This is particularly important in application to many foods like orange juice and pineapple juice. Third, since the product is frozen, there is no bubbling or foaming, which, in the case of drying some substances, would result in changes due to surface action such as the surface denaturation of proteins which occurs in drying their solutions even at low liquid-temperatures under vacuum.

Fourth, in most cases, the solute remains evenly dispersed and distributed without undergoing concentration, as frozen solvent sublimes, and the remaining dry residue emerges as a highly porous solid framework. It occupies essentially the same total space as the original solution did and the final residue is not the fine powder with which the chemist is mostly familiar. It consists of a friable, interlocking, and spongelike structure. As a result, solubility is extremely rapid and complete. For example, gelatin dried from a solution which must be prepared in the first place by boiling becomes instantly soluble in cold water. Fifth, since the molecules of solute are virtually "locked" in position in this way, the tendency for coagulation of even lyophobic sols is minimal. Even though the lipoidal constituents of dry blood plasma do not reconstitute perfectly after drying and do produce a slight degree of turbidity, there is far from complete coalescence. The particles are small enough to be safe for intravenous injection and do not cause capillary embolism.

Sixth, during drying the surface of the evaporating frozen ice layer gradually recedes to leave more and more of the highly porous residue of solute exposed. As a result, "case-hardening" never occurs. A far lower content of moisture may be obtained in the final product without use of excessively high final temperature. For this reason of lower moisture content, a greater degree of stability results than is the case after any other method of drying.

Seventh, bacteriological growth and enzymatic changes cannot take place under the frozen conditions of drying. This is important for foods as well as medical products used in parenteral injection. The final fully dried product likewise resists bacterial growth and enzymatic action. Eighth, because of the high vacuum used, in contrast with the degree of vacuum used in ordinary low-temperature liquid evaporation, the amount of oxygen present is so extremely small that even the most readily oxidizable constituents are protected.

Mechanism of freeze-drying.

A decade ago this process was largely a laboratory curiosity and viruses were kept in an ice-box or otherwise chilled to prevent thawing. As a result of recent advances (1) it has been recognized that by establishing proper conditions of vacuum for removal of water vapor and by rapid application of heat to the frozen virus, rather than keeping it in an ice-box, the process could be made more practical with improved products resulting. Even while being heated, viruses may be maintained at any desired range of temperatures such as —10⁰ C. or around —40⁰ C. simply by regulating the degree of vacuum and rate of evaporation.

There are two stages in drying by sublimation. In the first, ice is evaporated from a frozen mass. In the second, moisture is removed from the final dry solid to lower the residual content to a minimal level. During the first stage, depending on the particular product, some 98 to 99 per cent of all water is removed. In the second, the residual moisture content is reduced to 0.5 per cent of the final product or less, which represents final removal of 99.95 per cent of the original content of water (assuming 10 per cent solids originally). In the first stage, temperatures are well below 0⁰ C. Actual temperature varies with the product, as will be discussed later. Upon passing frcm the first stage to the second the temperature gradually rises and finally reaches that of the rocm or higher, depending upon whatever final ambient temperature is used.

Basically, during the first stage of drying a maximal rate of evaporation of the frozen product must be obtained. To achieve this, heat must be introduced into the frozen product as rapidly as possible without causing it to soften or melt. At the same time a maximal rate of flow away from the evaporating surface must be established. To accomplish this rapid flow adequate passageways must be provided for vapor, and this must then be condensed or evacuated efficiently.

Freezing without drying.

Oftentimes in carrying out an investigation with an unknown virus system, it has not been established whether there are labile components present or not. Manipulations may extend over a matter of days or even months. By utilizing quick-freezing and storage at low temperature at all times when not actually working with preparations, loss of possible labile components is avoided. Bacterial investigation of this type was encountered in studies with the antigenic structure of *H. pertussis* (2). Eventually it was found that agglutinogen is heat stable and freezing overnight and at other times is unnecessary. Of the two toxins present, one was found to be heat stable and the other extremely labile. By application of freezing during storage, it was reasonably certain that no possible labile components were being missed as a result of deterioration during the rather extended time required to conduct the investigation. Another advantage in freezing virus preparations is in elimination of bacterial and other contaminations. Use of preservatives frequently is undesirable because of adverse action on viruses.

Low temperature without freezing.

Cohn (3, 4) and his co-workers have separated various fractions from human blood plasma. These have been obtained in high degree of purity and as a result have been used for many specialized medical purposes. During World War II, fractions were obtained from plasma collected by the American Red Cross, for example, serum albumen in high concentration was used in the place of whole blood for osmotic effects, etc. In this work, Cohn avoided salting out, which is the classical procedure for separation of proteins. In this way, dialysis was avoided. Ethanol is used in various concentrations, and depending upon the relative solubility of various blood derivatives under conditions of varying salt content, pH and temperature, and with the use of relatively low temperature (0^0 to $—10^0$ C.) as well as other means the proteins are precipitated. By proper sequence of varying these conditions, separation into different fractions is accomplished (4). The Cohn system is, among other things, an application of low temperature in systems which do not actually freeze in order to carry out precipitations and other manipulations without chemical alterations such as denaturation of proteins which proceed at room temperature. Purified viral antibody preparations against measles, mumps, etc. have been obtained.

Results with freeze-dried viruses.

There is usually less activity after drying but the final desiccated virus is well stabilized and no further loss occurs during storage (5). Viruses are in general more difficult to dry than other substances. The temperature must be maintained lower than is frequently necessary with many other materials, preferably well below $—20^0$ C. (6).

Harris and Shackell presumably used freeze-drying in the preservation of rabic brains (7). Rivers and Ward have relied widely on freeze-drying in their work with an intradermal vaccine for Jennerian prophylaxis (8). Siedentopf and Green have reported great success in the preservation of modified canine distemper virus (ferret passage) (9). It is known that some substances after drying from the frozen state may not maintain activity in the presence of oxygen of the air, even though completely dry and hermetically sealed. Typical of such products are those high in lipoidal content. In such a case sealing either under original vacuum is necessary or under an inert gas such as nitrogen or argon. Siedentopf and Green report that such is the case with their distemper virus. It should be pointed out that release of original vacuum with dry air and subsequent evacuation or replacement of air with inert gas does not operate satisfactorily in all cases since oxygen may be adsorbed by the highly porous solid matter and cannot be readily swept out for removal. Also, for distemper virus the nitrogen must be purified by passage over hot copper (10).

Influenza virus may be successfully preserved (5) and freeze-drying is used widely for carrying various strains of the virus in many research laboratories. Hoffstadt and Tripi (11) report successful three-year preservation of Levaditi and Cutter strains of vaccinia, herpes simplex, laryngotracheitis of fowls and Rous sarcoma viruses. On the other hand, they found inconstant maintenance of viability of the virus of infectious myxomatosis of rabbits over the three-year periods; whether this was because of non-uniform residual content of moisture or other cause was not determined. Their culture of O A strain of Shope's fibroma did not survive the period. Other viruses which have been successfully kept are those causing hog cholera (12), rinderpest, ovine ecthyma (sheep scabs), yellow fever, Japanese encephalitis, and various fowl diseases such as laryngotracheitis and fowl pox prepared from chick embryos.

Libby (13) has used freeze-drying in immunochemical studies with tagged antigens involving radio-tracers, with tobacco mosaic virus tagged with radio-active phosphorus (P^{32}). Mice were injected with the tagged virus and twenty-four hours later the mice were killed in a dry-ice acetone bath and 3 to 4 mm. sagittal sections were prepared with a band saw. These sections after freeze-drying were impregnated with paraffin and 1 mm. sections were prepared with a microtome and placed in intimate contact with x-ray film for exposure for varying periods to obtain autoradiographs. In this way it is possible to locate regions and organs of greatest concentrations of radioactivity.

Wooley (14) has reported that lymphocytic choriomeningitis and St. Louis encephalitis after freeze-drying had been preserved for 378 and 833 days, respectively, tests for longer periods not having been made.

The viral vaccine for prevention of canine distemper has been produced in a small way commercially. Various fowl viruses are freeze-dried. Vaccine for the prevention of yellow fever in man has been successfully made on a rather large scale. In experimental work virus of influenza has been freeze-dried. Vaccine for intradermal injection in immunizing against small-pox by Jennerian prophylaxis (8) has all been distributed after freeze-drying. There is question about the efficacy of this method of immunization but this has no bearing on freeze-drying per se (relates to the nature of the material itself and the method of carrying out immunization).

Munce and Reichel (12) have shown that hog cholera virus of blood origin when desiccated under high vacuum and stored *in vacuo* in flame-sealed ampoules, remained infective after exposure to a temperature of 60° C. for 96 hours. At 37° C., the infectivity was maintained for 328 days. Phenolized liquid virus from the same mixture was non-infective after exposure to the higher temperature for only five hours. At 37° C. the period of infectivity of the dried preparation was approximately 23 times as long as for the phenolized liquid virus. After storage at 20⁰ C., the dried virus was still infective in the authors' last test which was conducted after a storage period of 1125 days which is 12 times as long as for the corresponding phenolized liquid virus.

It may be pointed out that hog cholera virus, as commercially available, consists of the phenolized, defibrinated blood of artificially infected swine which are undergoing an attack of acute hog cholera at the time their virus-laden blood is collected. This virus preparation is frequently refered to as "simultaneous hog cholera virus" because it is used simultaneously with anti-hog cholera serum in the immunization of swine against hog cholera. The animals would not survive the injection of the live virus were they not protected simultaneously with the serum. When used in conjunction with anti-hog cholera serum, the virus possessing high virulence and the proper antigenic properties stimulates a strong, active immunity of lasting duration. The product when dispensed as a liquid must be preserved with phenol in the amount of 0.5%. This agent exerts a viricidal action which gradually reduces the viability until it is no longer infective. For this reason, the United States Bureau of Animal Industry has assigned an expiration dating of only 90 days for the phenolized liquid from the date of the *production* (not sale) of the product. Accordingly, the results with the freeze-dried product is of particular importance, not only in extending the actual dating itself, but in assurance that the virus when used will have not undergone partial deterioration.

These authors have also reported that the freeze-dried samples maintained their infectivity for a longer period after freeze-drying when stored in flame-sealed ampoules than in rubber stoppered bottles, the containers having been evacuated in both cases.

Although unpreserved liquid virus maintains its infectivity for a longer period than phenolized liquid virus from the same mixture stored at the same temperature, Governmental regulations require the use of a suitable preservative for obvious reasons. Accordingly, the results of significance are those comparing the longevity of the freeze-dried product with the phenolized liquid virus.

One of the earliest applications of freeze-drying has been in control of Rinderpest (cattle plague) in Africa. The mortality for this disease is high, sometimes reaching 50 to 75%. In preparation of the virus the blood of infected animals is laked and centrifuged. This separates the leucocytes with the live virus is associated. Without drying the virus remains viable and effective as an immunizing agent for a matter of a few weeks only. The virus survives freeze-drying well and has its life extended to months and years. Cattle are immunized by the simultanoeus inoculation of living virus and immune serum, this being similar to the practice in immunizing swine against hog cholera.

Laboratory sources of low temperature for freezing and storage.

Before quick-freezing, dry-ice suspended in a bath of organic solvent such as ethanol, acetone or the like, is the simplest and most readily available refrigerant for very low temperature. Likewise for storage, dry-ice in large thermos bottles is simple and convenient. The ice-cube compartment of household refrigerators is suitable for storage of small quantities of virus and even for freezing where the amounts involved are small. For more rapid freezing than otherwise would be obtained by use of the ice-cube compartement, a bath of alcohol may be chilled to low temperature beforehand. Then by immersing a glass container of virus, rapid freezing may be obtained. Where larger amounts of material are involved, some of the currently available home freezers are suitable. The only technical skill required is good judgment in the matter of rotating the containers and other manipulations to achieve either rapid or slow transfer of heat depending upon whether quick or slow freezing is desired.

Removing water vapor.

In drying from the frozen state, the frozen virus is placed under high vacuum. A region of low vapor tension is established in the vacuum system to cause rapid sublimation so that the rapidity of evaporation results in sufficient cooling to keep the virus frozen until dried. It is necessary to supply heat to the frozen virus for compensation of the loss latent heat of sublimation. This heat must be supplied with sufficient rapidity to maintain rapid evaporation. At the same time, means must be established within the vacuum system for removing the water vapor as fast as it is produced.

To remove the water vapor under high vacuum, three general methods are used. These are condensation at low temperature, combination with desiccating substances and direct pumping (15, 1). Direct pumping finds greatest application in larger scale operation than is usually encountered in virus work except for commercial production of vaccines. The following is a summary of the three methods. The principles involved apply equally well in the smallest of equipment for laboratory research and in the largest of equipment used in industrial production of viral vaccines.

1. Condensation. At low temperature condensation is carried out with condensers chilled either with a dry-ice bath or with refrigerants such as Freon and ammonia. Originally, dry-ice was the most widely used refrigerant. However, because of cost it is now used only with products which have a high value, such as medicinals like viral preparations. The low temperature of dry-ice is not necessary. The proper refrigerant temperature to be used depends upon the heat-transfer conditions at the cold surface. There are four factors involved. The first is the noncondensable gases which contaminate the water vapor. The second is the heat-transfer characteristic of the condensing surface; e.g., a metallic surface has a better heat-transfer coefficient than glass or ice. The third is the thickness of the layer of the ice built up on the condenser which must be taken into account because of the poor thermal conductivity of ice. Fundamentally, it is the temperature of the surface of condensed ice which matters and not the temperature of refrigerant. The fourth factor is area of condensing surface. The area should be large to keep at a minimum the amount of heat to be transferres per unit area. The lower the amount of heat to be transferred per unit area, the less the differential in temperature required so the refrigerant can be warmer for maintenance of a given condensing surface temperature.

The noncondensable gases account for the differential ("split") between the temperature corresponding to the vapor pressure of ice at the total pressure in the system and the actual temperature of the surface of the ice on the condenser. This differential may also be expressed as the ratio of partial pressure of water vapor to the total pressure in the system and this has great influence on efficiency of condensation. This first condition then is controlled by using tight vacuum equipment and pumps of adequate capacity for noncondensables.

As soon as the process of condensation has gone on for a short period, the condensing surface is entirely covered with a coat of ice so that the true condensing surface becomes one of ice.

Accordingly, it becomes necessary to accept the loss of the poor heat-transferring ice surface. By use of condensers of very large surface, whereby the heat to be transferred per unit area is kept small, thickness of the ice layer is kept small, avoiding poor conductivity, so that both the third and fourth conditions are controlled. The condensed ice is spread thinly over a large area.

The net result of all these factors in combination may be expressed in terms of effective partial pressure of water vapor which can be established within the vacuum system by the condenser during drying. Assuming that the equipment is properly designed in order to avoid restrictive orifices in vacuum lines and elsewhere, the differential between this pressure and the vapor pressure of the evaporating viral preparation determines the theoretical maximal rate of flow of vapor. Vapor pressure of the virus at the temperature at which it should be dried, on this basis, determines the temperature at which the condenser should be maintained. Some viruses must be kept colder than others. In fig. 1 pressure differential is shown which is produced by various temperatures at the condensing surface in relation to three different temperatures at which viruses are kept during drying. This differential in pressure is the driving force and it will be observed that for each of the several temperatures of viruses, there is a minimal temperature of the condenser surface below which very little further increment is gained by further reduction in condenser temperature.

Further, as drying proceeds, it is limited in rate by diffusion of vapor through the interstices of the porous dry outer layer of virus, these acting as orifices. By laws of adiabatic gaseous flow

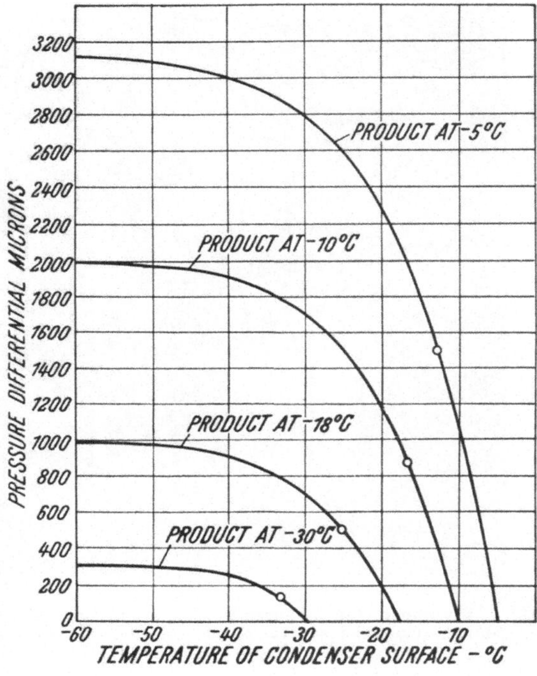

Fig. 1. Driving force in terms of differential in pressure produced by various condenser temperatures for products maintained at three different temperatures. "X" on each curve represents the temperature of the surface of condenser ice giving rise to the maximal differential of any value according to the Napier equation, assuming no restrictive orifices in pipe lines, etc., other than in interstices of the porous dry outer layer of product itself. This applies to the first stage of drying when ice is being sublimed.

through orifices (Napier equation), a differential between the vapor pressure at the surface of the condenser and that at the ice surface within the frozen viral preparation where the former is 55 % of the latter will result in the maximum rate of flow obtainable. This has been borne out experimentally with equipment similar to that illustrated in fig. 1 in which the temperature of the condenser surface may be varied from —10⁰ to —60⁰ C.

In drying by sublimation, pressures are well above one micron and the laws of flow of fluids apply to the movement of the water vapor. These laws as applied above are based upon the conditions of "screening" set up at an orifice when the downstream pressure is decreased below the limiting value. This screening accounts for the apparent discrepancy in not obtaining faster flow as a result of a greater differential in pressure. At pressures below a micron, the mean free path of the molecules approaches the diameter of the pipe line and the laws of flow of fluids no longer govern. Movement of the vapor is then controlled by the natural diffusion of molecules, but this condition is not reached in drying by sublimation.

This comparison of optimum differentials in pressures for maximum flow assumes little leakage and adequate pumping capacity to maintain low partial pressure of air. It further assumes no restrictions in vapor lines, necks of ampules, or elsewhere in the high-vaccum system. Otherwise, the temperature of the con-

Fig. 2. Equipment for commercial scale production at the rate of 60 l/day, using internal coil condenser (central black tank). May be used for drying either in bulk or in final ampules.
Courtesy of F. J. Stokes Machine Company.

densing surface would need to be reduced in order to produce a lower partial pressure of water vapor on the side toward the condenser of each restriction or orifice. Such reduction to less than 55 per cent of the pressure on the drying side of each orifice, by laws of adiabatic gaseous flow, can produce no further increment in rate of flow through the orifice. In other words, beyond a certain point, lowering of the temperature of the condenser cannot compensate for small vapor lines or restrictions.

2. **Desiccating substances.** Chemical desiccants such as phosphorus pentoxide and sulfuric acid have been suggested. These are impractical, however, on any scale of operation because of high cost, difficulty in using them, and other reasons.

Regenerable desiccants having a low enough vapor pressure have much in their favor under certain circumstances. The previous discussion of vapor pressure and condenser temperature applies also to the use of desiccants. Inasmuch as these desiccants are regenerated by heat, and since heat also is produced as a result of reaction of the desiccant with water vapor, a problem is introduced in maintaining properly low vapor pressure of the desiccant as drying proceeds. That is, the desiccants have a high thermal coefficient of change in vapor pressure. The heat of reaction is not readily conducted away in vacuum; the desiccants themselves are poor conductors.

The problem can be controlled, however, either by regulation of the quantity dried in relation to the amount of desiccant used (based on heat capacity of the entire bulk of desiccant), or by special means of cooling the desiccant. The desiccant can either be of the type which forms a fixed chemical hydrate, such as calcium sulfate, or of the physical type which combines with water by adsorption, such as silica gel or alumina. In the latter case, the problem of warming the desiccant during drying is magnified in that vapor pressure, even at fixed temperature, rises with each small increment of water vapor adsorbed.

However, specially prepared calcium sulfate (Drierite) in controlled amount without cooling has proved particularly satisfactory and is widely used, since it maintains a constant vapor pressure at any given temperature until saturated.

3. **Direct Pumping.** Direct pumping of the water vapor provides one of the simplest commercial procedures. Again the same pressures must be produced which are obtained with condensers and desiccants. In this case, the same means are used to pump out noncondensables as establishes the low partial pressure of water vapor. Either an ejector pump or an oil-sealed rotary pump may be used. In the latter case, the pump must be equipped with means for continuously removing water from the oil of the vacuum pump, such as by a centrifugal clarifier. In this way, even though the oil-sealed pump is carrying water vapor through it, freshly clarified oil from which all condensed water has been removed by centrifugation is returned to the high vacuum side of the pump and its efficiency is not impaired. A multistage steam ejector with interstage condenser is suitable and is widely used in large-scale operation.

Apparatus for laboratory research.

Widely used equipment for carrying on laboratory research was first described in 1935 (16). This equipment utilizes dry-ice for condensation as well as for prefreezing. The cost of dry-ice is not an important factor because of the

Fig. 3. Dry-ice equipment for virus research.
Courtesy of F. J. Stokes Machine Company.

small quantities used in ordinary laboratory research with viruses. However, in many cases it is not available readily in which case the cryochem type of equipment using a chemical desiccant is recommended (6).

In the 1935 type of equipment illustrated in fig. 3, the metal condenser "C" is contained within the large thermos bottle "D". The vacuum pump is shown at the left. Prefreezing in the various types of containers is carried out in the freezing bath "A". Containers are held horizontally and by gentle rocking back and forth a thin layer frozen on the inner wall of the container provides a large evaporating surface. These containers, Nos. 1, 2, 3, 4 and 5, are shown attached to the manifold "B" after freezing. Vacuum is quickly drawn by the pump to avoid thawing and then rapid sublimation takes place.

Temperature control of the virus is made simple and automatic because the rate of sublimation of water vapor is regulated by the surface conditions set up in freezing. That is, the virus is frozen in containers of proper size and shape, in such a way as to give the correct relationship of the evaporating surface of the frozen virus to the surface adjacent to the glass, through which atmospheric heat is transferred to the frozen solid. In other words, the evaporating surface must be large in relation to the surface of glass exposed to the atmosphere through which heat is being supplied. That is the reason for placing the bottles on their sides for freezing. Alternatively, by rotating the bottles during freezing in a properly shaped bath, the virus may be "shell frozen" around the entire inner periphery of the glass containers. Proper provision for the escape of water vapor from the container to the condenser is provided by having tubes of adequate diameter.

The virus in its containers is brought to a very low temperature for attaching to the manifold. Attachment of the containers to the manifold must be rapid. As indicated, all glass tubing connections must be of large diameter. Then there must be rapid establishment and maintenance of high vacuum throughout the entire apparatus. (The apparatus should be sufficiently tight for the vacuum pump to produce about 50 to 100 microns on "blank suction" without virus being dried — i. e., tested beforehand.)

Drying must be continued until the final moisture content has been reduced to less than 1.0% and preferably below 0.5%. There should be no bubbling or melting of the preparation during the early stages of drying or at any other time. As discussed above, the actual temperature necessary for a particular virus depends entirely upon the type of preparation. This must be determined experimentally.

During the early stages of drying, frost collects from the atmosphere on the outer walls of the containers of virus. After the ice within the frozen viral preparation has almost completely sublimed, the temperature automatically rises. At the same time the frost on the outside walls of the container will disappear and be replaced by liquid condensation. This indicates a temperature above 0^0 C. but below that of the room. Finally, the temperature of the virus reaches that of the room and all condensation on the outer walls of the container disappears. The total length of time to reach this point should be noted and then about half that number of hours additional should be allowed for reduction to proper final level of moisture content.

Usually in this type of work it is desirable to seal the exhaust tubes of the container under original vacuum by fusion of the glass with a flame. Pyrex glass is specified because of its low thermal coefficient of expansion. That is, it will withstand the temperature shock of sudden heating so that the glass may be drawn to a capillary and finally sealed by fusion. It is not possible to allow slow heating of the glass followed by slow cooling as is customary in glass blowing because of the large size of the glass tube holding vacuum; it would be "sucked in" to form a hole which would allow air to leak in.

Sealing under the original vacuum as just described provides a simple means of maintaining asepsis. Because of the shape of the exhaust tubes as illustrated, with everything used up to the manifold having been sterilized, it is not possible for any contamination to occur from the manifold itself. Therefore it need not be sterilized. After initial attachment of the sterile containers, the direction of flow of all air and vapors is outward from the containers into the manifold and this flow is rapid. Nothng can fall by gravity from the manifold into the con-

Fig. 4. Cryochem-type equipment for desiccating from the frozen state. A regenerable desiccant comprising mostly calcium sulfate (Drierite) with a small amount of silica gel (for a small amount of vapors other than water) contained in baskets within the white tank. The manifold is shown with outlets to which individual containers for drying are attached. At the end of drying, the glass tubes attaching the containers to these outlets are sealed by fusion with a flame.

Courtesy of F. J. Stokes Machine Company.

tainers. Finally, by sealing under original vacuum the viruses are obtained in final form without contamination.

Where dry-ice is not available, cryochem type of equipment mentioned above is used as illustrated in fig. 4. The specially prepared calcium sulfate is supported in special perforated trays containing the desiccant in a bed of about 2" or 3" thick in each tray. The trays must be supported with proper separations between them to allow easy access of water vapor to the desiccant. Theoretical capacity of the desiccant for water vapor is 6.6% by weight to form the hemi-hydrate. In practice, however, it is better to utilize only 6.0%. In a single batch not over one-quarter of this amount, i. e., 1.5% is utilized because of the temperature conditions within the desiccant, i. e., it is warmed as a result of heat of reaction with water vapor. However, four such batches of maximum capacity may be dried before regeneration is necessary.

For regeneration of the calcium sulfate, heating in an ordinary oven like a bacteriological dry wall sterilizing oven is carried out. The oven should be operated at a temperature of 180 to 200⁰ C. When the desiccant itself reaches 150⁰ C., regeneration is complete but it should be checked by moisture analysis to be certain that it is below about 0.2 or 0.3%. The desiccant is initially prepared from Gypsum in the same manner. It as well as the laboratory apparatus is available in the U. S. A. from the F. J. Stokes Machine Company, Philadelphia, 20.

With this apparatus the same type of manifold and containers may be used as was discussed with dry-ice equipment above. In the illustration, however, valves are shown on each outlet which facilitates operation.

Description and discussion of equipment used for industrial production of viral vaccines are beyond the scope of this book and will be found elsewhere (1).

Bibliography.

1. Flosdorf, E. W. 1949. Freeze-Drying. Reinhold Publishing Company, New York, N. Y. 281 pp.
2. — and A. C. Kimball. 1940. Separation of the phase I agglutinogen of H. pertussis from toxic components. The Jour. of Immunol., **39**, 475.
3. Cohn, E. J., J. L. Oncley, L. E. Strong, W. L. Hughes Jr. and S. H. Armstrong Jr. 1944. Chemical, clinical, and immunological studies on the products of human plasma fractionation. I. The characterization of the protein fractions of human plasma. The Jour. Clin. Invest., **23**, 417.
4. —, L. E. Strong, W. L. Hughes Jr., D. J. Mulford, J. N. Ashworth, M. Melin and H. L. Taylor. 1946. Preparation and properties of serum and plasma proteins. IV. A system for the separation into fractions of the protein and lipoprotein components of biological tissues and fluids. Jour. of the Amer. Chem. Soc. **68**, 459.
5. Scherp, H. W., E. W. Flosdorf and D. R. Shaw. 1938. Survival of the influenzal virus under various conditions. Journ. Immunol., **34**, 447.
6. Flosdorf, E. W. and S. Mudd. 1938. An improved procedure and apparatus for preservation of sera, micro-organisms and other substances — the cryochemprocess. The Journ. Immunol., **34**, 469.
7. Harris, D. L. and L. F. Shackell. 1911. Jour. Amer. Pub. Health Asso., **7**, 52.
8. Rivers, T. M. and S. M. Ward. 1935. Jour. Exper. Med., **62**, 549.
9. Siedentopf, H. A. and R. G. Green. 1942. Factors in the preservation of the distemper virus. Jour. Infectious Diseases, **71**, 253—259.
10. — U. S. Patent 2,380,339.
11. Hoffstadt, R. E. and H. B. Tripi. 1946. A study of the survival of certain strains of viruses after lyophilization and prolonged storage. Jour. of Infectious Diseases, **78**, 183—189.
12. Munce, T. W. and J. Reichel. 1943. The preservation of hog-cholera virus by desiccation under high vacuum. Amer. Jour. of Veterinary Research, **4**, 270—275.
13. Libby, R. L. 1947 (May). The use of tagged antigens in immuno-chemical studies. Transactions of The New York Academy of Sciences, P. 248.
14. Wooley, J. G. 1939. The preservation of lymphocytic choriomeningitis and St. Louis encephalitis viruses by freezing and drying *in vacuo*. Public Health Reports, **54**, 1077—79.
15. Flosdorf, E. W., L. W. Hull and S. Mudd, 1945. Drying by sublimation. The Jour. of Immunol., **50**, 21.
16. — and S. Mudd. 1935. Procedure and apparatus for preservation in "Lyophile" form of serum and other biological substances. The Jour. of Immunol., **29**, 389.

Die Auflicht- und Dunkelfeldmikroskopie in der Virusforschung.

Von **M. Kaiser**, Wien und **P. Vonwiller**, Rheinau.

A. Einleitung.

Das Erkennen krankhafter Prozesse im tierischen Gewebe ist durch die Benutzung mikroskopischer Betrachtung wesentlich gefördert worden. Insbesondere verdanken wir der Beobachtung mit starken Trockensystemen und mit Immersionslinsen einen tiefgehenden Einblick in die Natur pathologischer Prozesse. Die Undurchsichtigkeit mancher Objekte war aber immer ein Hindernis für ein optisches Erfassen tieferer Schichten des Substrates, zumal lebenden Gewebes, weil ein Durchleuchten von unten her nicht möglich war.

Wenn man nun heute das Gesamtgebiet der biologischen Mikroskopie überblickt — ganz besonders wenn man die Mikroskopie an höheren Organismen in Betracht zieht —, so stellt man fest, daß sich trotz eindringlicher Mahnrufe schon in früherer Zeit (z. B. Fischer, 1899) die klassische biologische Mikroskopie im ganzen hauptsächlich im Sinne der üblichen Fixierungs-, Schnitt- und Färbetechnik weiter entwickelt hat. Und so ist es im wesentlichen auch bis in die neueste Zeit hinein geblieben. Fragen wir nach den Gründen dieser bemerkenswerten Tatsache, so kann man namentlich zwei unterscheiden; 1. die Macht der Tradition, die sich ja auch auf imponierende Erfolge stützen konnte, und 2. das Fehlen einer allgemein verwendbaren, an jedem beliebigen lebenden Organ und mit jeder beliebigen Vergrößerung arbeitenden Vitalmikroskopie. Wir müssen uns jetzt die Frage vorlegen, ob dies auch heute noch zwangsmäßig so sein und bleiben müsse.

Zunächst wollen wir kurz die bisher in der Biologie bekannt gewordenen *Vitalmikroskopiermethoden* in Erinnerung rufen und uns fragen, welche von ihnen in der Virusforschung Anwendung gefunden haben. Bis in die neuere Zeit hinein war man in der ganzen Biologie mit ganz wenigen Ausnahmen immer sozusagen ausschließlich auf die Mikroskopie im durchfallenden Licht eingestellt. Das Objekt mußte zu diesem Zwecke durchsichtig sein. — Das ist ja nun bekanntlich, namentlich im Bereich höherer Organismen, nur ganz ausnahmsweise der Fall. Wollte man sich an wirklich lebendige Beobachtungsobjekte halten und nicht die das Leben schon vor der Beobachtung gewaltsam zerstörenden, eingreifenden Methoden der üblichen Gewebelehre wählen, so blieb man auf eine ganz kleine und inhomogene Gruppe von Natur aus durchsichtigen, lebenden Beobachtungsfeldern beschränkt.

In der Folge gelang es unter Zuhilfenahme eingreifender Operationen, eine weitere Gruppe von durchsichtigen Untersuchungsfeldern an höheren Organismen der vitalmikroskopischen Beobachtung zugänglich zu machen. Es sei an die Verfahren von Hirsch (19) und das Verfahren von Gramenitzki (11) erinnert.

Aber trotz alledem, wenn man die ganze Reihe der bisher einer eingehenden vitalmikroskopischen Beobachtung zugänglichen, durchsichtigen Beobachtungsfelder an höheren Organismen überblickt, so muß man doch sagen, daß diese Objekte immer noch recht seltene Ausnahmefälle darstellen und meist eine Zerstückelung der Objekte erfordern.

Will man nun den Notbehelf einer solchen Zerstückelung des lebenden Objekts und der anschließenden Untersuchung des so gewonnenen überlebenden Materials und ebenso den noch gefährlicheren Ausweg der „Verarbeitung" des Materials nach den künstlichen Methoden der traditionellen biologischen Mikroskopie vermeiden, so ergeben sich zwei Möglichkeiten des Vorgehens, welche die Beobachtung eines zweifellos normal *weiterlebenden Materials* ermöglichen. Entweder — da die Natur uns von sich aus nicht die genügende Anzahl und Auswahl lebendiger durchsichtiger Untersuchungsfelder zur Verfügung stellt — schreitet man dazu, künstlich solche herzustellen —, man verändert also das Objekt, läßt die übliche Mikroskopie unverändert, und hält sich an die Mikroskopie im durchfallenden Licht. Oder man muß eben das Objekt unberührt lassen, und eine ganz andere Technik einführen, d. h. man paßt Technik und Methodik dem lebenden, unveränderten Objekt an. Den ersteren Weg hat Clark (7) gewählt. Auf seine Versuche kann hier nicht eingegangen werden, es sei nur auf sie hingewiesen.

Der zweite Weg wurde von einer ganzen Reihe von Forschern eingeschlagen, mit recht verschiedenem Erfolg. Manche Autoren berichteten über gewisse Erfolge, die sich aber nicht verallgemeinern ließen, während andere bewußt danach strebten, eine allgemein anwendbare, auch den lebenden Menschen in den Bereich der Untersuchung mit einbeziehende Methodik zu entwickeln. Es muß vorausgeschickt werden, daß es ja schon seit geraumer Zeit in der Klinik eine weitgehend verwendete *Lebendmikroskopie* gab, die aber ausnahmslos im Bereich der schwachen Vergrößerungen steckengeblieben ist. Wir meinen die bekannten Verfahren der *Kapillaroskopie*, der *Spaltlampenmikroskopie* des lebenden Auges, ferner die *Kolpo-, Cysto-, Urethroskopie* und die von Lüscher entwickelte *Trommelfellmikroskopie*. Alle diese Verfahren arbeiten nur mit schwachen Vergrößerungen.

Es leuchtet ohne weiteres ein, daß sich für eine im *Virusgebiet* allgemein verwendbare Vitalmikroskopie nur eine Arbeitsweise als tauglich erweisen kann, welche es ermöglicht, mit allen beliebigen Vergrößerungen, an jedem beliebigen lebenden Organ, und sowohl an lebenden Tieren als auch an lebenden Pflanzen und an überlebenden Geweben zu arbeiten.

Zunächst möchten wir auf das von Knisely entwickelte Verfahren hinweisen. Ähnlich wie seinerzeit Barta (2), aber mit vervollkommneter Methodik, konnte Knisely, indem er sich lichtleitender Glas- oder Quarzstäbe bediente, deren spitzes Ende an das zu untersuchende Gewebe herangebracht oder in dasselbe eingeführt wurde, wohl jedes lebende Organ der Beobachtung zugänglich machen. Eine Übersicht über seine Arbeiten kann man der 1947 erschienenen Arbeit in der Zeitschrift „Science" (29) entnehmen. Es wäre denkbar, dieses Verfahren, das von Knisely zu gewissen Untersuchungen bei *Malaria* verwendet wurde, auch im Virusgebiet anzuwenden. Es scheinen zwar noch gewisse Einschränkungen mit Rücksicht auf die Anwendung stärkster Vergrößerungen zu bestehen (bis jetzt bis zu 600 ×), aber vermutlich ließe sich diese Grenze doch noch weiter hinausrücken.

Noch näher kommt dem gesteckten Ziel das von Ellinger und Hirt ausgearbeitete Verfahren der mikroskopischen *Beobachtung lebender Organe im Fluoreszenzlicht* (9). Es scheint sich vor allem bei Beobachtungen über den Durchtritt des Farbstoffes durch die Gefäßwand, z. B. bei Beobachtungen an der lebenden Niere, zu bewähren, kann aber offenbar auf jedes beliebige andere lebende Organ übertragen werden. Da es sich aber, abgesehen von der verwendeten Lichtart (ultraviolett) eng an das von uns ausgearbeitete Verfahren anlehnt, so genüge hier dieser Hinweis. Nach unserer Ansicht ist es nämlich näherliegend und technisch viel einfacher, wenn man mit „natürlichem Licht" arbeitet, außer wenn man eben gerade den Durchtritt von im ultravioletten Licht fluoreszierenden Farbstoffen genauer beobachten will. Es ist ja auch bekannt, daß viele lebende Strukturen von Natur aus gefärbt sind, wie z. B. Erythrozyten, Pigmentzellen, Fette usw., so daß eine künstliche Färbung mit fluoreszierenden Farben diesen Vorteil eventuell verdeckt, und daß ferner in vielen Fällen nicht Farben, sondern allerhand Lichtreflexe im lebenden Gewebe die Formen äußerst scharf erkennen lassen, wie z. B. markhaltige Nervenfasern, Fetttropfen usw.

Wir wiederholen deshalb noch einmal die Forderung, daß *für die Verwendung im Virusgebiet* nur eine Vitalmikroskopie taugen wird, welche ohne Unterschied alle lebenden Organe von Tieren und Pflanzen anzugehen ermöglicht, mit allen beliebigen Vergrößerungen und ohne Rücksicht auf die Größe des zu untersuchenden Organismus.

Wir haben danach getrachtet, ein solches Verfahren auszuarbeiten, und es hat gerade im Virusgebiet schon seine Brauchbarkeit, einerseits durch unsere eigenen Arbeiten KAISER und VONWILLER (23, 28), andererseits durch die Arbeiten von HIMMELWEIT (18), RACHEL E. HOFFSTADT und DOROTHY OMUNDSON (20, 21) erwiesen. Seither scheint sich allerdings niemand mehr damit beschäftigt zu haben, was sich wohl durch die der Forschung ungünstigen äußeren Umstände der vergangenen Jahre erklärt. Die Vorteile dieses Verfahrens sind aber so einleuchtend, daß die Weiterführung solcher Beobachtungen, gerade auch über das Virusgebiet wohl als eine Forderung der Zeit erscheint. An Pflanzen und Tieren sind weitgehende Möglichkeiten der Vitalmikroskopie am unverletzten lebenden Organismus vorhanden, und mittels kleiner bis zu sehr eingreifenden Operationen kann man sich jedes lebende Organ, so verborgen es auch liegen mag, so empfindlich es gegen Eingriffe sein mag, unter den schonendsten Bedingungen soweit freilegen, daß es einer eingehenden Vitalmikroskopie, mit jeder beliebigen Vergrößerung, zugänglich wird.

B. Die Auflichtapparatur.

Den Wünschen der Biologen ist die *optische Industrie* weitgehend entgegengekommen. Die Erfolge der Auflichtmikroskopie sind ihr nicht entgangen. Ausgehend von den Instrumenten der Metallurgen, den *Vertikalilluminatoren* oder *Opakilluminatoren* hat sie eine Reihe von Auflichtmikroskopen hergestellt, so die *Firma Leitz* ihren *Ultropak von Heine, Zeiß* seine *Epilampen* und *Reichert* seinen *Universalopakilluminator.*

Insbesondere hat die *Firma Leitz* dieses Problem gründlichst behandelt und in fortlaufender Arbeit eine Reihe von Verbesserungen der Auflichteinrichtungen geschaffen, die schließlich zu einem höchst geeigneten Instrument, dem Ultropak von Heine geführt haben, weshalb wir uns für die Verwendung dieser Einrichtung entschlossen haben. Ihr Wesen besteht darin, daß die Zuführung des Lichtes völlig getrennt von der Beobachtungsoptik mit Hilfe ringförmig um die Objektive angeordneter Kondensoren erfolgt, so daß Reflexionen im Objektiv nicht auftreten, und die Beobachtungen bei voller Objektivapertur vorgenommen werden können. Die von Leitz in das Ortholux eingebaute Niedervoltlampe (6 Volt, 5 Ampere), die für orientierende Beobachtungen ausgezeichnete Dienste leistet, und auch zum Aufsuchen der für das spätere Studium geeigneten Stellen sehr brauchbar ist, genügt für die von uns verfolgten Zwecke nicht, weshalb wir uns einer Kombination der zur Ultropak-Einrichtung, u. zw. des dazugehörigen Beleuchtungsansatzes und des Ortholux bedienten. Als Lichtquelle diente eine Bogenlampe, deren Licht wir durch eine mit 10% Kupfersulfatlösung gefüllten Küvette gehen ließen. Diese Untersuchungsanordnung ermöglichte photographische Aufnahmen und das Abtasten tieferer Schichten, für deren Durchleuchtung die Niedervoltlampe nicht mehr ausreichte. Für längere dauernde Beobachtung ist diese Lichtquelle wegen des die Augen stark in Anspruch nehmenden grellen Lichtes schwerer zu gebrauchen, auch wenn Dämpfungsfilter dazwischen geschaltet werden.

Wir bilden dieses Gerät der Firma Ernst Leitz und die von uns zuletzt gebrauchte Versuchsanordnung hier ab, weil sie uns ausgezeichnete Dienste geleistet hat und wir sie anderen Untersuchern

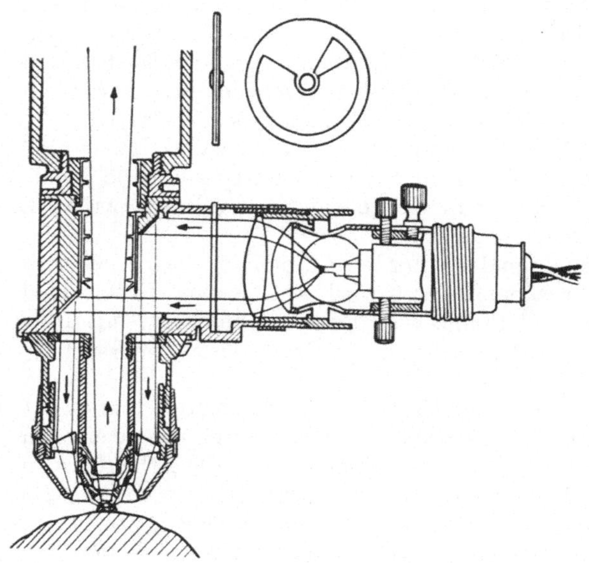

Abb. 1. Ultropakobjektiv mit seitlicher Beleuchtung. Die von der Beobachtungsoptik getrennte Zuführung des Lichtes ist durch Pfeile angedeutet. Oben Sektorenblende zur Lenkung des einfallenden Lichtes.

bestens anempfehlen können. Für die Betrachtung kleinerer oder überlebender Objekte diente uns, wie schon angedeutet, das binokulare Mikroskop der Firma Ernst Leitz, das Ortholux mit seiner Ultropakeinrichtung (30, 31). Das schwere

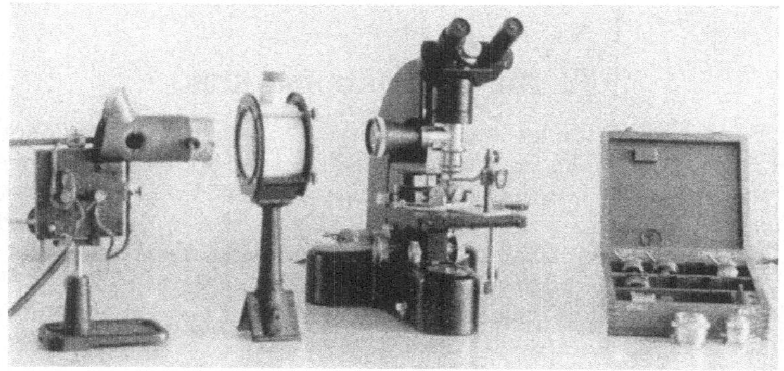

Abb. 2. Ortholux Ernst Leitz. Bogenlampe zur Beleuchtung. Kupfersulfatfilter. Objektivkasten (Liste Mikro D 7083).

und höchst solid gebaute Instrument ermöglicht ein ruhiges und sicheres Arbeiten. Sein vom Beobachter abgewandtes Stativ mit dem großen, mit Noniusteilung ausgestatteten, von unten her und vom Zahnbetrieb unabhängig verstellbaren Kreuztisch, ist für die Beobachtung dickerer Objekte im Auflicht von größter Bedeutung (Abb. 2).

Das zu untersuchende Objekt, ein Bulbus, ein geimpftes Ei, wird auf einer geeigneten Unterlage mit Hilfe der vorhandenen Klemmen fixiert und kann auf dem beweglichen Objekttisch verschoben werden. Für die Beobachtung un-

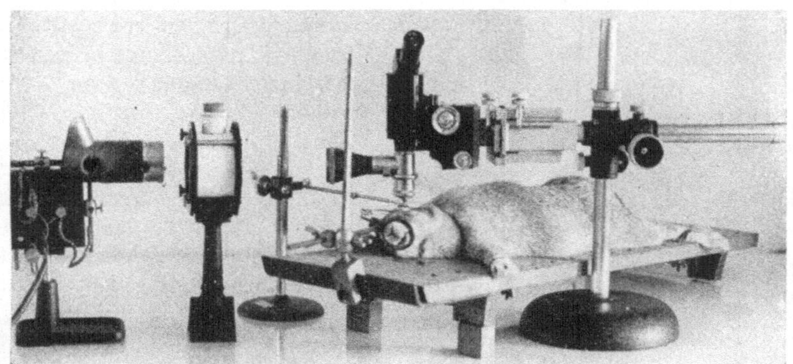

Abb. 3. Leitz Ultropakstativ U. S. II mit Kreuzschlitten, der eine Bewegung der Optik über dem Objekt gestattet. Dieses, ein mit Evipan narkotisiertes Kaninchen, ist auf einem Tierbrett befestigt. Der luxierte Bulbus wird mit einer Klammer gehalten und durch den Kompressor die Hornhaut eingeebnet. Bogenlicht mit Wärmefilter. Ultropakliste Ernst Leitz vom Oktober 1937, S. 51.

ebener Objekte ist eine Einebnung unerläßlich. Wir haben deshalb je nach der Art des zu beobachtenden Objektes entweder die Einrichtung des seitlich verschiebbaren Mikroskoptubus auf dem Stativ II oder III (Leitz Liste Mikro D

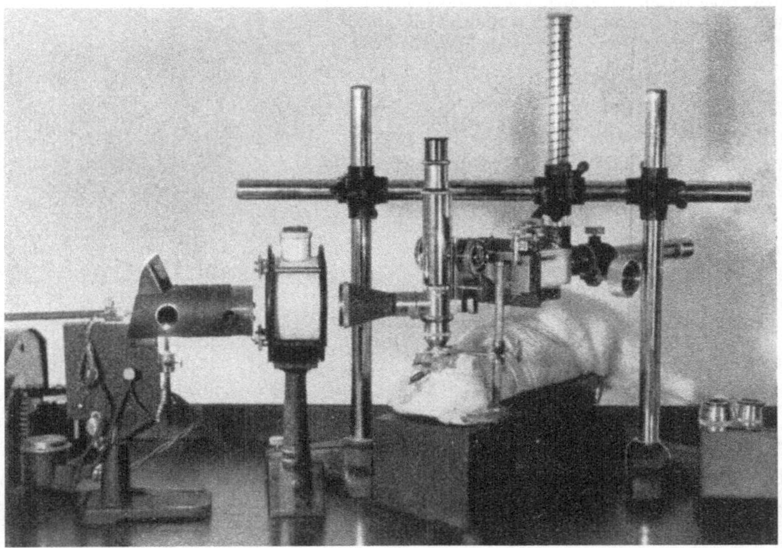

Abb. 4. Leitz Ultropakstativ U. S. III, stabiler als U. S. II. Kann auch an ganz großen Tieren angewendet werden. Beleuchtung und Bulbusfixierung wie bei Abb. 3. Ultropakliste Ernst Leitz, März 1934, S. 45.

Nr. 7184) mit Kreuzschlitten benützt und die Objekte (Bulbus) mit Hilfe eines Kompressors, den uns die Firma Reichert (Wien) beistellte, eingeebnet oder aber einen kleinen Kompressor auf den beweglichen Objekttisch aufgebaut, der mit ihm und mit dem Objekt verschiebbar ist. Ähnliche Kompressoren sind auch von der Firma Leitz zu beziehen (Abb. 3—5).

Mit Hilfe des Kreuzschlittens der US-Stative (30) kann der Mikroskoptubus über das zu betrachtende Objekt nach verschiedenen Richtungen bewegt werden, was am lebenden narkotisierten Tier notwendig ist. Die Betrachtung kleiner überlebender Objekte (Organteile, kleinere Tiere) erfordert Spezialeinrichtungen zum Fixieren solcher Geräte. Für die geeignete Einstellung von enukleierten Bulbi hat die *Firma Wild* (*Heerbrugg, Schweiz*) ein geeignetes Gerät hergestellt (Abb. 6), das eine Verstellung nach verschiedenen Richtungen mit Hilfe von Mikrometerschrauben

Abb. 5. Zwei Geräte zur Hornhaut- und sonstiger Objekteinebnung; das große von der Firma Reichert, Wien, das kleine in der eigenen Werkstatt hergestellt. Es ist ein mobiles Gerät, das auf den beweglichen Objekttisch aufmontiert und mit ihm bewegt werden kann, wenn kleine Objekte beobachtet werden.

Abb. 6. Halter für Bulben mit kardanischer Einstellung der Firma Wild, Heerbrugg, Schweiz.

erlaubt. Wird mit dem Ortholux gearbeitet, so muß auch die Einebnungsvorrichtung mobil gemacht werden. Für diesen Zweck haben wir das bereits erwähnte entsprechende Gerät in der eigenen Werkstätte herstellen lassen, das auf dem Objekttisch montiert und mit Hilfe von Stellschrauben festgehalten werden kann. Es hat sich bestens bewährt (Abb. 6).

C. Erste eigene Versuche einer Auflichtbetrachtung virusbedingter Veränderungen im Gewebe.

Unsere ersten Untersuchungen befaßten sich mit dem Studium vaccinaler Veränderungen am Kaninchenauge. Wir haben die von uns angewandte Technik im Zbl. Bakt. (28) veröffentlicht, so daß wir hier nur auf die wichtigsten Punkte dieser Technik einzugehen brauchen.

1. Die Betrachtung völlig ungefärbter Hornhautzellen erwies sich als unbefriedigend. Zur Unterscheidung krankhafter Veränderungen und ihrer Kontraste gegen die gesunden Gewebselemente ist eine Färbung förderlich, sie mußte deshalb bei dem zu untersuchenden Material angewendet werden, wozu sich eine Mischung von 2 % Phloxin und 2 % Wasserblau in wäßriger Lösung im Verhältnis 1 : 3 auch heute noch bewährt. Will man ohne Kontraste färben, so bewährt sich auch eine 1—2 % wäßrige Trypanblaulösung.

2. Es war ferner nötig, einen entsprechenden Untergrund unter dem zu beobachtenden Gewebe zu schaffen, einen Reflektor, der sich allenfalls auch am narkotisierten Tier anwenden ließ. Anfangs arbeiteten wir am lebenden Tier, das wir teils mit Äther, teils mit Evipan narkotisierten. Es erwies sich jedoch im Laufe

der Untersuchungen als völlig überflüssig, unsere Beobachtungen am lebenden Tier vorzunehmen, weil das *überlebende* Gewebe, das einer Betrachtung unmittelbar nach Enukleation des Bulbus unterzogen wurde, für unsere Studien ebenso verwendet werden konnte. Die Untersuchung am isolierten Bulbus gestattet allerdings nicht eine Verfolgung des vaccinalen Prozesses durch mehrere Tage. Hier hat es sich jedoch gezeigt, daß nach längstens 72 Stunden an der Hornhaut alle Stadien des Ablaufes bis zum Zerfall des Hornhautepithels wahrzunehmen waren, soweit sie *unter natürlichen Verhältnissen* vorkommen, so daß eine fortlaufende Beobachtung des Bulbus *am lebenden Tier* und dessen wiederholte Narkose entbehrlich wurden.

Der frisch enukleierte Bulbus wurde wie 1. c. beschrieben, vorerst in ein Holzklötzchen montiert und in geeigneter Weise festgehalten; später benützten wir das Gerät der Firma Wild. Aus dem Bulbus muß vorerst das Kammerwasser entfernt und durch fettfreie Vollmilch als geeignetem Reflektor ersetzt werden. Zwecks Einbringung der Milch, die mitunter Schwierigkeiten macht, erwies es sich in letzter Zeit als zweckmäßig, mit einer sehr feinen Nadel, die auf die Spritze angesteckt ist (Nr. 20), vom Äquator des Bulbus aus durch einen Muskel hindurch in die Kammer einzudringen und in die Nadel vor dem Einstechen in den Bulbus einen Mandrin einzuführen, der jedoch die Nadelspitze nicht überragen darf. So verhindert man eine Verstopfung der Kanüle beim Einstechen. Ist man in die Augenkammer eingedrungen, so wird die Spritze abmontiert, der Mandrin entfernt, und nach neuerlichem Montieren der Spritze das Kammerwasser abgesaugt, beim Kaninchen etwa 0,5 ccm. Daraufhin wird der Nadelkonus mit Milch gefüllt, die vollgefüllte Spritze angesetzt und die Milch unter schwachem Druck eingespritzt, worauf man die Spritze rasch entfernt. Die Einstichöffnung braucht nicht mehr abgedichtet zu werden, weil die Milch durch den langen Einstichkanal nicht ausfließt. *Die Einhaltung dieses Vorgehens ist dringend nötig*, weil man sonst Luftblasen in die Kammer bringt, welche die Beobachtung stören. Das Verfahren sollte am besten in einigen Blindversuchen geübt werden, weil es eine gewisse Übung erfordert. Ist die Füllung der Kammer einwandfrei gelungen, so ist das Kammerwasser vollkommen durch Milch ersetzt und störende Luftblasen sind nicht zu sehen.

Nun muß das Epithel der Hornhaut gefärbt werden. Der Bulbus wird mit einem Pean am Sehnervenstumpf gehalten, in ein Schälchen getaucht, das die Farblösung (einige ccm, s. o.) enthält. Es empfiehlt sich, die Farbe höchstens 2mal zu benützen.

Der etwa 3 Minuten lang (das muß ausprobiert werden) gefärbte Bulbus wird nunmehr mit steigenden Vergrößerungen betrachtet. Bei Verwendung eines Objektives UO 3,8 und 6,5 UO 11, (31) und bei Öl oder Wasserimmersion UO 11 bekommt man bereits ein völlig klares Übersichtsbild über die Beschaffenheit des Hornhautepithels und einen Einblick in die Ausbreitung der vaccinalen Reaktionen, was mit gewöhnlicher Lupenvergrößerung und mit der sonst üblichen Beleuchtung nicht möglich ist. Es zeigt sich folgendes Bild, das durch seine Klarheit überrascht (Abb. 7).

Von einem rein weißen Untergrund heben sich bei der normalen Kaninchenhornhaut gestochen scharf die blau gefärbten Hornhautzellen ab. In ihrer Größe beinahe gleich, reihen sie sich mosaikartig aneinander, mit ihrem blaßblauen strukturlosen, polygonalen Protoplasma und ihrem dunkler blaugefärbten, fast kreisrunden Kern. Auch im normalen Gewebe färbt sich eine große Zahl von Zellen, doch bleiben dazwischen große Partien ungefärbt, optisch leer, rein weiß, so daß der Kontrast zwischen gefärbten und ungefärbten Stellen ein sehr großer ist. Da sich nach der allgemeinen Auffassung nur abgestorbene Zellen diffus färben sollen, so ist man erstaunt über die große Anzahl dieser Zellen, die auch am normalen

Bulbus des lebenden Tieres in ganzen Partien gefärbt werden. Ob diese Zellen als völlig abgestorbene Zellen zu bezeichnen oder nur in ihrer Vitalität geschädigt sind, läßt sich wohl nicht ohne weiteres entscheiden. Jedenfalls stellen wir fest,

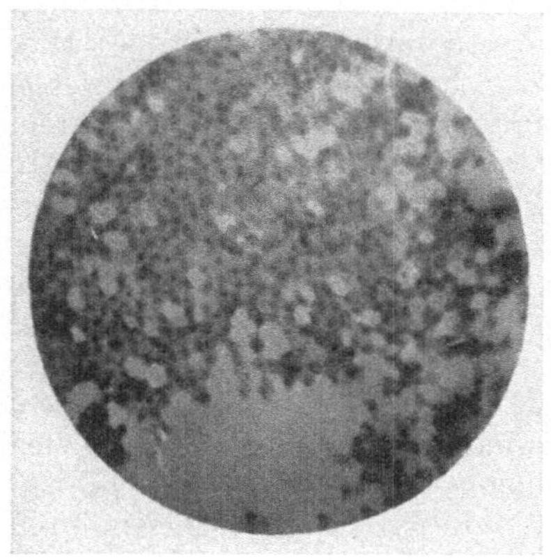

daß die Färbbarkeit der Hornhautzellen nicht mit allen Farbstoffen, die wir versucht haben, eine gleichartige ist, daß gewisse Farbstoffe sehr gut aufgenommen werden, andere weniger gut. Viele, minder gut gefärbte Zellen gehören den tieferen Gewebsschichten an. Zu den Farbstoffen, die sehr gut aufgenommen werden, gehören in erster Linie Phloxin und Wasserblau, die sehr schöne Bilder geben; Wasserblau gibt besser gefärbte als Phloxin, und mit einer Mischung beider Farben färben sich einzelne Zellen blau, andere rosa oder mischfarben, ohne daß ein Grund für dieses differente Verhalten ein und desselben normalen Gewebes angegeben werden kann.

Abb. 7. Normale Hornhaut im Auflicht. Leitz Objektiv UO 11, Ok. 8. Nur ein Teil des Hornhautepithels hat sich mit einem Gemisch von Phloxin gefärbt. Die Annahme der Farben ist verschieden. Photo Romeiskamera.

Erst die Verwendung von Objektiv UO 60, Okular 6—10 (Vergrößerung 450—750), ermöglicht es, die Struktur der Zellen besser zu studieren und Einzelheiten zu erkennen.

Vor allem fällt es im Vergleich zu den bekannten Hornhautschnittpräparaten auf, daß sowohl Protoplasma als auch Zellkerne, insbesondere aber ersteres, fast strukturlos sind, daß selbst Vergrößerungen bis zu 1500 nur wenig daran

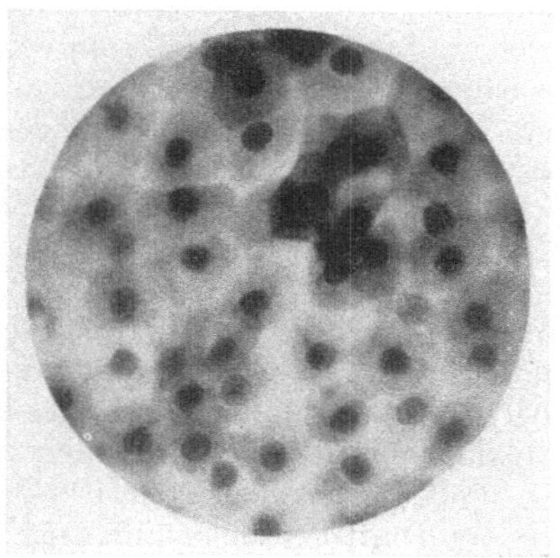

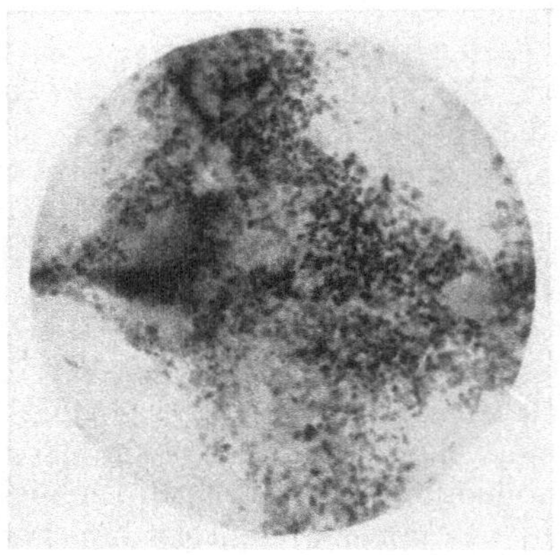

Abb. 8. Normale Hornhaut, gefärbt mit Phloxin Wasserblau. Leitz Objektiv UO 60 Oel. Ok. 8. Photo Romeiskamera.

Abb. 9. Vaccinierte Hornhaut. Leitz Objektiv UO 11 Oel, Ok. 8, Photo Miflex. Man sieht längst der Vaccinationsstreifen eine Verwirrung in dem Mosaik des normalen Hornhautepithels.

ändern, daß die Kerne höchstens eine ganz feinschaumige Struktur zeigen, jedoch niemals ein Netzwerk mit gröberen und feinen Maschen aufweisen, wie es in den fixierten Schnittpräparaten der Fall ist. Hier lernt man erst den Einfluß der Fixationsmittel kennen und wir glauben, daß gerade zum Studium dieses Gewebes der Vitalmikroskopie im Auflicht eine sehr große Rolle zukommt (Abb. 8).

Das normale Protoplasma der Hornhautzellen ist also homogen und der Kern zeigt bei nicht zu langer Einwirkung der Farbe höchstens eine feinschaumige Struktur, ist jedoch stärker gefärbt als das Protoplasma und weist nur selten ein deutliches, stärker gefärbtes Kernkörperchen auf.

Wird nun eine Hornhaut durch lineare Skarifikationen vacciniert — wir empfehlen, zuerst ein Übersichtsbild mit schwächeren Vergrößerungen zu betrachten (Abb. 9), so scheint die erste Folge dieses Eingriffes die Abstoßung des verletzten Epithels zu sein. Dementsprechend bildet sich an dieser Stelle ein seichter Graben. Das Virus diffundiert in die Nachbarschaft und verursacht die Proliferation des Gewebes, wie sie uns aus Schnitten bekannt ist. In der Daraufsicht sieht man den Graben, man sieht langkernige, spindelige Zellen mit Ausläufern (fixe Hornhautzellen) und zu beiden Seiten des Gewebes eine Anhäufung von Zellen, in welcher Einzelheiten meist nicht mehr zu erkennen sind. Diese Stellen sind es, die sich am intensivsten färben, bei denen auch der Reflektor nicht mehr durchscheint. Seitlich verflachen sich die Proliferationswülste immer mehr, und immer deutlicher kommt der helle Untergrund zum Vorschein, auf dem sich in immer dünner werdenden Schichten die einzelnen Zellen klarer und immer klarer abheben. Hier hat auch die Untersuchung auf Zellveränderungen einzusetzen. Dazu dient vorerst eine Betrachtung mit Objektiv UO 60, Okular 6—15, später mit Objektiv UO 100, Okular 6—15.

Wir konnten in den Jahren 1934 und 1935 hier nicht weiter vordringen. Wir haben jedoch einige histologische Veränderungen feststellen können, die dem vaccinalen Prozeß eigen sein dürften, und wollen sie, soweit wir sie heute noch für richtig halten, kurz anführen.

Hier ist vor allem die bekannte Erscheinung der *retikulierenden Degeneraion* zu erwähnen. Derartige Zellen sind deutlich wahrnehmbar bei allen von

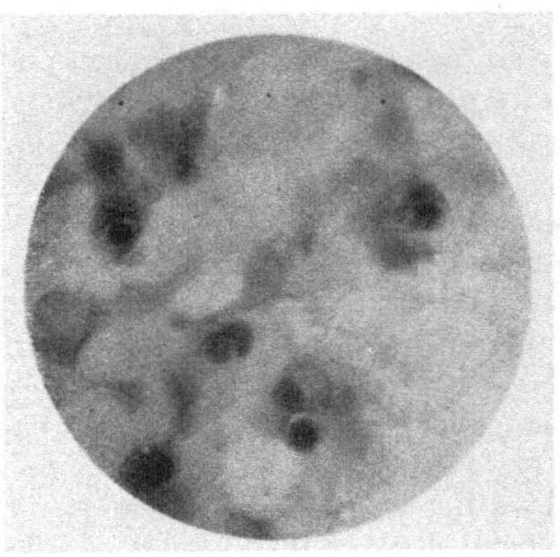

Abb. 10. Veränderungen des Hornhautepithels nach Vaccination mit Variolavaccine. GUARNIERIsche Körperchen. (Opt. Daten in Verlust geraten.)

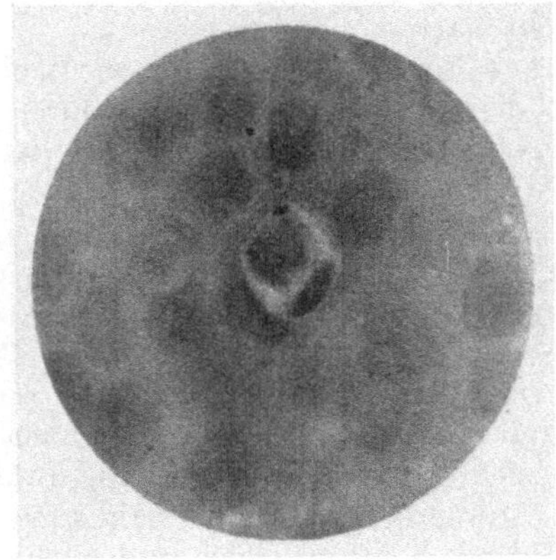

Abb. 11. GUARNIERIsches Körperchen. Leitz Objektiv UO 60 Oel, Ok. 8, Photo Miflex. Man sieht deutlich den an den Rand gedrückten Kern und ein GUARNIERIsches Körperchen.

uns am überlebenden Gewebe angewandten Färbungen. Wie ein Siegelring nimmt sich der Kern mit dem Protoplasma aus. Es ist durch eine Vacuole ganz an den Rand der Zelle gedrängt und zu einem schmalen Streifen rareficiert. An einer Stelle sitzt der Kern von sehr zarter, feinstschaumiger Struktur. Vielleicht erscheint eine disperse Struktur in der Daraufsicht in dieser Form. Solche

Zellen finden sich häufig, ohne daß man einen Grund für ihr Auftreten finden kann. Vielleicht ist es ein Symptom der Zellnekrobiose.

Eine weitere Erscheinung unserer damaligen Versuchsergebnisse waren *mehrkernige Zellen*, platte facettierte Zellen, sehr seltene Mitosen, ferner *Invaginationsformen*, die als Schachtelzellen beschrieben werden, große Zellen, die eine zweite enthalten, allenfalls noch eine dritte.

Schließlich müssen die s. z. schon von PFEIFFER (39) und JÜRGENS (22) gesehenen und erst durch GUARNIERI zu Ehren gekommenen (12) und nach ihm benannten *Guarnierikörperchen* (*G. K.*) erwähnt werden, welche heute als pathognomonisch für den vaccinalen Prozeß angesehen werden dürfen, wenn sie nicht zu sporadisch auftreten (Abb. 10).

Als Kernderivate können sie heute bekanntlich nicht mehr aufgefaßt werden. Über sie wird später noch zu berichten sein. Nach gewissen Versuchsergebnissen (27) schien es KAISER und GHERARDINI, als ob bei Verwendung von Phloxin und Wasserblau (damals hielten sie irrtümlich beide Farben für sauer) die G. K. stets eine Vorliebe für Phloxin hätten. Diese Anschauung kann nun, wie noch zu berichten sein wird, nicht mehr aufrechterhalten werden (Abb. 11).

Soweit waren die Beobachtungen in den Jahren 1934 und 1935 gediehen, als sich HIMMELWEIT (18) derselben Methodik zum Studium analoger Veränderungen im bebrüteten Hühnerei bediente.

D. Weitere Versuche anderer Autoren über die Virusmikroskopie im Auflicht.

HIMMELWEIT (18) weist zunächst auf die von BARNARD durch seine ultraviolett-mikroskopischen Beobachtungen erzielten Fortschritte in der Virusforschung hin und erinnert daran, daß BLAND und CATTI 1935 bei der Psittacosis mittels gewöhnlicher Dunkelfelduntersuchungen gute Erfahrungen gemacht hatten.

Am Beginn seiner eigenen Beobachtungen beschäftigte sich HIMMELWEIT mit den Veränderungen, die durch das verimpfte Vaccina- und Ektromelie-Virus an der *Allanto-Chorioidea* der sich entwickelnden Enten- und Hühnerembryonen bedingt sind.

Zunächst erfolgten diese Beobachtungen an Membranfragmenten, die sie den mit Virus infizierten Eihäuten entnommen hatten. Die Ergebnisse waren zwar ermutigend, aber die Präparate verschlechterten sich rasch. Infolgedessen ging HIMMELWEIT über zu Beobachtungen an der lebenden, infizierten Chorio-Allantoismembran in situ mittels des Ultropaks. Er brachte also das Ei mit seinem lebenden Inhalt nach Abhebung der Schale über dem Beobachtungsfeld unter das Mikroskop. Anfänglich erzielte HIMMELWEIT beim Versuch der Anwendung des Ultropak auch bei schwächeren Vergrößerungen nur ungenügend kontrastreiche und nicht hinreichend helle mikroskopische Bilder.

Es gelang ihm aber mittels einer von ihm eingeführten besonderen Beleuchtungsart bessere Bilder zu erzielen, welche vorher durch die Opaleszenz des Beobachtungsmaterials verschleiert geblieben waren. Er stellte fest, daß die Ultropakbeleuchtung bei stark opaleszierendem biologischem Material gewisse Vorteile bietet, wofür gerade lebende Zellen und Gewebe hervorstechende Beispiele darstellen. Solche Strukturen, die an der Oberfläche oder unmittelbar darunter gelegen sind, erleiden nach seinen Beobachtungen nur „wenig Interferenz von Seiten der darunterliegenden Strukturen". Das kommt nach HIMMELWEIT's Ansicht davon her, daß die tiefer gelegenen Strukturen verhältnismäßig wenig stark beleuchtet werden. Im Dunkelfeld und bei Beobachtungen im durchfallenden Licht seien die Bedingungen umgekehrt. .Bei diesen wird das Licht von unten her zugeführt und werden also die tieferen Schichten zuerst und mit größter Intensität beleuchtet. Eine besondere Schwierigkeit mußte erst noch überwunden werden, welche darin bestand, daß diejenige Portion der Allanto-Chorioidea, welche die zu untersuchenden krankhaften Veränderungen aufweist, in einer gewissen Tiefe liegt und Oberflächenunregelmäßigkeiten aufweisen kann. In der Folge hat HIMMELWEIT ein Verfahren ausgearbeitet, welches es ermöglicht, diese Oberfläche in einer geraden Ebene auszubreiten und zugleich den Abstand vom Mikroskopobjektiv zu vermindern, ohne die Zellen zu schädigen oder zu verlagern.

Die Impfung wurde im wesentlichen nach dem Verfahren von BURNET und GALLOWAY (6) und BURNET (4) vorgenommen. Zur Trennung der Chorio-Allantois von der Schalenmembran wurde eine kontinuierliche Saugwirkung mittels eines Gummihütchens ausgeübt, was die Trennung der Membran ohne Auftreten von Blutungen erleichtert. Dann wird, nachdem das infizierte Ei eine bestimmte Zeit lang bebrütet wurde, Schale und Schalenhaut über dem Beobachtungsfeld entfernt, wobei über dem Ansatz der Chorio-Allantois noch ein Schalenrand von 1—2 mm stehengelassen wird. Nun wird nach außen an diesen Schalenrand ein Wachsring aufgebaut (Paraffinwachs vom Schmelzpunkt 56° und Vaseline zu gleichen Teilen), 2 mm hoch und 5 mm dick, und sodann mit einem Deckglas der eröffnete Raum luftdicht abgeschlossen. Dann wird durch eine besondere Vorrichtung (eine an der Schale angekittete Glas- und Gummiröhre) Luft in den natürlichen Luftsack eingeblasen, bis sich die Allanto-Chorioidea an das Deckglas anlegt. Zum Entweichen der Luft über dem Beobachtungsfeld muß vorher noch ein kleines Loch in den Wachsrand gebohrt werden, das wieder verschlossen wird, sobald sich die Membran von unten an das Deckglas angelegt hat.

Enteneier eignen sich wegen der bedeutenderen Größe besser als Hühnereier. Die Impfung wurde gewöhnlich an Eiern vom 9. bis 15. Bebrütungstage vorgenommen. Jüngere Embryonen sind geeigneteres Beobachtungsmaterial, einmal wegen der länger möglichen Beobachtungszeit und dann auch, weil die Allanto-Chorioidea dünner und feiner ist. Später wird sie dicker und stärker opaleszierend und ergibt nicht mehr ebensogute mikroskopische Bilder.

Zwei Virusarten wurden verwendet: ein Hautvaccinestamm aus Kalbslymphe, und ein Ektromeliestamm (aus Mausleber isoliert), die sich bei den Untersuchungen als stabil erwiesen. Eine bestimmte Verdünnung des Infektionsmaterials muß eingehalten werden; zu konzentrierter Impfstoff erzeugt zusammenfließende, für die mikroskopische Untersuchung ungeeignete Veränderungen. Die Bebrütung erfolgt bei 37° C und wurde 1—7 Tage lang nach der Infektion angewendet.

Die optische Ausrüstung bestand im wesentlichen aus dem Ultropak von Heine (Leitz), wobei die Objektive von 22 bis 100 zur Verwendung kamen. Anstatt des zum Ultropak gelieferten Sammelkondensors wurde ein Langfocuskondensor, kombiniert mit einer Bogenlampe, einer Pointolit- oder einer Osirishochdruck-Quecksilberdampflampe, verwendet.

Bei schwachen Vergrößerungen erscheint die Chorio-Allantois als einförmig gelbes Feld, in welchem größere und kleinere Blutgefäße zu sehen sind. Die Blutbewegung in Arterien und Venen ist erkennbar. Bei stärkeren Vergrößerungen sieht man auch die oberste aus polygonalen Epithelzellen zusammengesetzte Ektodermschicht, wobei das Plasma beinahe strukturlos, und der Kern als rundes oder ovales Feld in seiner Mitte erscheint. Man sieht dank dieser Strukturarmut des Epithels die tiefer darunter gelegenen Blutgefäße. Diapedese kann häufig beobachtet werden — vielleicht infolge der intensiven Beleuchtung — fügt Himmelweit (18) bei. Die ausgewanderten Leukocyten können in den extravaskulären Strukturen erkannt werden.

Die von Himmelweit auf diese Weise — in vivo et in situ — beobachteten *pathologischen Veränderungen* zeigten zunächst bei schwachen Vergrößerungen folgende Eigentümlichkeiten: die Ektromelieknötchen schärfer begrenzt als diejenigen der Vaccine, und bei Ektromelie weniger ausgedehnte Infiltration mit Leukocyten; bei beiden Veränderungen ist vornehmlich das Reticuloendothel betroffen.

Bei *Ektromelie* war die erste sichtbare Veränderung die Anwesenheit von infiltrierenden Leukocyten, sodann die vermehrte Anzahl von Leukocyten in den Blutgefäßen der befallenen Gegend, ferner vermehrte Opacität in der infiltrierten Region. Der Grund liegt darin, daß Leucocyten opaker sind als Gewebezellen. Auch die Epithelzellen werden opaker.

Unmittelbar nach der Impfung mit Ektromelievirus können Einschlußkörperchen nicht beobachtet werden, aber schon nach 24 Stunden enthalten die Zellen eines oder mehrere, von welchen die kleinsten 1 bis 2 Mikron Durchmesser haben. Man sieht bis zu 5 bis 6 in einer Zelle oder noch mehr und von verschiedener Größe. Mit stärksten Vergrößerungen betrachtet, erweisen sich diese frühesten Einschlußkörperchen der Ektromelie als aus zahlreichen Granula zusammengesetzt, welche den Elementarkörperchen entsprechen. In älteren Läsionen können solche Einschlußkörperchen zu Riesenstrukturen zusammenfließen, welche gewöhnlich unregelmäßig umgrenzt sind, oft freiliegend zwischen Zellresten vorgefunden werden und bis zu 30 bis 50 Mikron im Durchmesser haben können. Scheinbar noch in lebende Zellen eingeschlossene und auch frei schwimmende, extrazellulär gelegene Einschluß- und Elementarkörperchen konnten beobachtet werden. Die Veränderungen auf der Allanto-Choroidea nach Verimpfung von *Vaccinavirus* zeigen eine Masse aus infizierten Zellen und infiltrierenden Leukocyten. Viele Zellen sind im Zerfall begriffen, weshalb starkes Opaleszieren auftritt und nur Bilder von geringem Kontrast und arm an Einzelheiten entstehen. Die entstandenen Herde sind weniger scharf begrenzt als bei Ektromelie. Im Randbezirk sind die Zellen noch nicht zersetzt, deutlich sichtbar und bilden ein ideales Material für die Untersuchung der Zellveränderungen unter der Einwirkung der Infektion. Die Epithelzellen der infizierten Allanto-Chorioidea zeigen die Einwirkungen der Infektion in charakteristischer Ausprägung. Viruskörperchen wurden zuerst am zweiten Tage nach der Infektion beobachtet. Die offenbar als *Paschenkörperchen* zu deutenden Granula zeigten in der Zelle andauernde *Brownsche Bewegung*. Mit fortschreitender Infektion vergrößern sich die Virusanhäufungen, dringen tiefer ins Cytoplasma ein und können sich mit benachbarten Anhäufungen verbinden. Zuletzt erscheint ein großer Teil oder das ganze Zellcytoplasma mit zitternden Viruskörperchen erfüllt. Die größten Körperchenanhäufungen zeigten 15 μ im Durchmesser. Vom dritten Tag an traten auch Kernveränderungen in den infizierten Zellen ein.

Himmelweit weist darauf hin, daß die Gewebekultur gewisse Nachteile aufweist, welche bei der Beobachtung in der Allanto-Chorioidea wegfallen. Abgesehen davon, daß es schwer ist, in der Gewebekultur genügend dünn ausgebreitete Zellen herzustellen, ist bei der Beobachtung in situ der Metabolismus durch das normale Blutgefäßsystem des Embryos erhalten. Tatsächlich stellt die lebende intakte Allantochorioidea eine „ideale Gewebekultur" vor, und die in der Membran auf diese Art beobachteten Veränderungen können als sehr ähnlich den Veränderungen bei der natürlichen Infektion betrachtet werden. Das gleiche Beobachtungsfeld kann tagelang unter fortlaufender Beobachtung verfolgt werden, ein Vorteil, der kaum auf andere Weise erreicht werden kann. Hinsichtlich der eingehenderen Angaben HIMMELWEITS über den genaueren Infektionsvorgang der einzelnen Zelle müssen wir auf das Original verweisen. Hingegen interessieren hier noch besonders HIMMELWEITS Angaben über die Einwirkung der Fixations- und Färbungsmethoden bei der Entstehung der G. K. Bei sorgfältiger Beobachtung könne man feststellen, daß sie aus granulärem Material bestehen. Schon nach TOPLEY und WILSON (44) stellten die G. K. wahrscheinlich intrazelluläre Mikrokolonien des Virus dar, und auf Grund der von HIMMELWEIT dargestellten Befunde könne man nicht länger daran zweifeln, daß die G. K. einfach eine Ansammlung von vaccinalen Elementarkörperchen darstellen.

Die HIMMELWEITsche Arbeit ist interessant als erster, an unsere gemeinsamen Untersuchungen anschließender, erfolgreicher Versuch, den Grundsatz der Beobachtung in vivo et situ auf ein neues, für die Virusforschung so besonders wichtiges Beobachtungsfeld, wie es die lebende Allanto-Chorioidea darstellt, zu übertragen.

Wir möchten uns aber erlauben, einige kritische Bemerkungen, speziell über die Beleuchtungsmethode beizufügen. Es ist HIMMELWEIT zwar gelungen, mit dem Ultropakverfahren an einem ausgesprochen schwierigen Untersuchungsfeld sehr klare Bilder zu erhalten, speziell auch Viruskolonien, in vivo et in situ, mit den dazu nötigen starken Vergrößerungen, zu beobachten. Es stellt also sein Verfahren einen Spezialfall der von einem von uns schon vor langen Jahren herausgearbeiteten „Parasitoskopie" dar. Damals wurde zum ersten Male der Beweis der Möglichkeit einer solchen an mit *Trypanosoma Lewisi* infizierten Ratten geführt (VONWILLER und DEMOLE, 1927).

Später gelang es in unseren gemeinsamen Untersuchungen, über deren Ergebnisse wir bereits berichtet haben, am lebenden mit Vaccine infizierten Kaninchenhornhautepithel [KAISER und VONWILLER (28)], die sich außerdem noch auf Untersuchungen am lebenden regenerierenden Froschhornhautephitel [VONWILLER und KOTLIAREWSKAYA (48)] stützen konnten, die Beobachtung bis zum Erscheinen von G. K. fortzuführen. Wir bedienten uns dabei noch eines anderen, wesentlichen Verfahrens, das von HIMMELWEIT nicht berücksichtigt wurde, und das in schwierigen Fällen, wie es z. B. die lebende menschliche Haut darstellt, zum Ziele führt, auch da, wo die übliche Ultropakbeobachtung bis dahin versagt hatte. Nach der Schilderung von HIMMELWEIT liegen nun an der lebenden Allanto-Chorioidea ganz ähnliche optische Verhältnisse vor, die wir von HIMMELWEIT abweichend, erklären möchten. HIMMELWEIT spricht nämlich von der Schwierigkeit des „opaleszierenden" Objektes und der Einwirkung der unter der zu beobachtenden Oberfläche liegenden tieferen, „opaleszierenden" Schicht auf das Beobachtungsfeld. Er überwand diese Schwierigkeit durch Zuführung stärkeren Lichtes, durch Verwendung einer stärkeren Lichtquelle, als sie gewöhnlich zur Ultropakbeleuchtung verwendet wird. Er äußert aber selbst Bedenken wegen der dadurch eventuell hervorgerufenen schädlichen Einwirkung auf das lebende Objekt, welche sich in der starken, von ihm beobachteten Diapedese äußert. Es könnte also durch die zu starke Beleuchtung, vielleicht auch die damit verbundene zu starke Erwärmung, das lebende Objekt geschädigt werden. Die Erwärmung kann allerdings durch ein entsprechendes Wärmefilter korrigiert werden.

Es gibt nun aber noch ein anderes, wahrscheinlich weniger schädliches Verfahren für derartige Untersuchungen, und das ist das Arbeiten mit *künstlichen Reflektoren*, über deren Anwendung auf unser Forschungsgebiet wir bereits berichtet haben. Grundsätzlich dürfte nun die Lage der Dinge in diesem Gebiete folgende sein: bei Objekten, denen ein natürlicher Reflektor abgeht, wählt man bei ungefärbtem Objekt, wie z. B. bei Beobachtungen an der lebenden menschlichen Haut (wobei wir wohl mit Recht annehmen, daß in der ungefärbten Allanto-Choroidea ähnliche Verhältnisse herrschen) einen spiegelnden Reflektor, kombiniert mit dem Opakilluminator; arbeitet man hingegen mit vitalgefärbten Objekten, wie z. B. in unseren Beobachtungen am lebenden oder überlebenden vitalgefärbten Kaninchenhornhautepithel, so wählt man einen diffusen Reflektor, kombiniert mit Ultropak oder Opakilluminator.

Nach unserer Ansicht bewirkt eben die unter dem Beobachtungsfeld liegende „opaleszierende" Schicht nicht deshalb eine Verschleierung des Bildes, weil sie „opalesziert", wie Himmelweit meint, sondern deshalb, weil sie zu viel Licht durchläßt, das sich ungenützt in der Tiefe verliert und nicht ausgenützt wird. Bei Anwendung eines künstlichen Reflektors hingegen erreichen wir, auch bei verhältnismäßig schwacher Lichtquelle, eine maximale Ausnützung des auffallenden Lichtes.

Die Himmelweitsche Methodik ist von Rachel E. Hoffstadt und Dorothy Omundson (20, 21) in manchen Punkten verbessert worden, die ein handliches Gerät für Zwecke der kontinuierlichen oder zeitweisen Beobachtung der Chorio-Allantois angegeben haben.

E. Beobachtungen von virusbedingten Veränderungen im Dunkelfeld.

Mit der Aufklärung der Genese der vaccinalen Veränderungen des geimpften Versuchstieres hat sich ferner K. B. E. Merling (33) beschäftigt, indem er ungefärbte Substrate in *Dunkelfeldbeleuchtung* studierte.

Diese Technik, welche eine gewisse Verwandtschaft mit der Auflichttechnik hat und ausgezeichnete Bilder gibt, ist folgende:

Mit einem Graefemesser, wie es für die Iridektomie benötigt wird, wurden in der Regel 3 Tage nach der Impfung dünne, tangentiale Abschabungen des Hornhautepithels vorgenommen, in einem Tropfen Tyrodelösung auf einen dünnen Objektträger ausgebreitet und mit einem Deckglas abgedeckt, indem man durch sanften Druck die überschüssige Flüssigkeit entfernte. Das Deckglas wird dann mit Paraffin dicht abgeschlossen, um ein Verdunsten der Flüssigkeit zu verhindern. So adjustierte Präparate wurden entweder bei Raumtemperatur oder bei 37°C durch mehrere Tage beobachtet.

Vaccinavirus kann frühestens 24 Stunden nach der Infektion erkannt werden. Infizierte Stellen sind leicht zu finden infolge ihrer Helligkeit, welche sich von der bläulichen, normalen Umgebung in den Zellen deutlich abhebt.

Eine nähere Betrachtung der Zellen zeigte folgende Bilder:

1) Die ganze Oberfläche der Zellen kann mit Elementarkörperchen (Paschenkörperchen) übersät sein, welche unbeweglich dem Gel des Zytoplasmas aufsitzen. Erst in einem späteren Stadium befinden sich die Elementarkörperchen infolge der allmählichen Verflüssigung des Zytoplasmas in Brown'scher Bewegung und es zeigen sich „Scheibchenformen" („discs").

2) Eine Häufung der P.K., die sich in rascher Brownscher Bewegung in der verflüssigten Zone des Zytoplasmas befinden, ähnelt dem nach der Färbung von Schnitten bekannten Bild: Die Haufen erinnern an Guarnierikörperchen, insbesondere, wenn die verflüssigte Zone durch einen Halo von der gelartigen Schichte abgegrenzt ist.

3) Die Einschlußkörperchen sind unregelmäßig und passen sich in ihrer Form der Gestalt der Zelle an.

4) Eine Zelle kann mehrere Einschlußkörperchen von verschiedener Größe enthalten, oder es kann sogar das ganze Zytoplasma von Viruselementen ausgefüllt sein, die sich in BROWNScher Bewegung befinden. Die kleinsten Elementarkörperchen enthalten ein einzelnes Viruselement. Im Mittel sind sie so groß wie der Kern und enthalten etwa 150 Viruselemente, sofern man sie zu zählen vermag.

5) In einer Zelle können die Viruselemente entweder der Oberfläche anhaften oder sie können sich in einem Einschlußkörperchen in lebhafter BROWN'scher Bewegung befinden.

6) Es wurden nicht nur einzelne Viruselemente beobachtet, sondern auch unter günstigen Bedingungen Scheibchen („discs") und Kugeln („spheres"). Die Scheibchen sind entweder bewegungslos im Zytoplasma oder beweglich in den Einschlüssen. Kugeln konnte man in den Einschlüssen und frei außerhalb der Zellen wahrnehmen.

7) Einschlußkörperchen wurden, wenn auch seltener, in den langen Zellen des der Cornea benachbarten Gewebes gefunden. Die einzigen Elemente, die mit Virus verwechselt werden konnten, sind kleinste Fettröpfchen, die jedoch bewegungslos sind. Auch Pigment kann Verwechslungen veranlassen. Auch dieses ist bewegungslos. Nicht unwichtig ist der Hinweis, daß die Elementarkörperchen anfänglich den Eindruck vollkommen gleicher Größe machen; es konnten jedoch später Viruselemente von verschiedener Gestalt gefunden werden, von den kleinsten angefangen bis zu den größeren Scheibchenformen. Die Größe der Einschlußkörperchen schwankte zwischen 0,5—15 µ, die der Kugeln zwischen 0,5 und 1,0 µ, die der Paschenkörperchen 80—120 mµ, der Scheibchen 150—200 mµ, wie es auch frühere Angaben bestätigen [K. B. E. MERLING (33—36)].

Der Lebenszyklus des Vaccinavirus ist nun folgender, wie er sich aus einer neuntägigen Beobachtung ergeben hat:

Es zeigt sich, daß Vaccinaviruselemente der Oberfläche der Zelle anzuhaften scheinen und daß sie dann das Zytoplasma verflüssigen, in das sie einsinken und BROWNsche Bewegung zeigen.

Die kleinen verflüssigten Zonen, die Viruselemente enthalten, vereinigen sich und bilden ein Guarnierikörperchen.

Der Entwicklungszyklus endet nun damit, daß die Zelle, welche Einschlußkörperchen enthält, schließlich platzt und daß die Viruselemente in die Nachbarschaft ausgestreut werden.

Die Einschlußkörperchen in den lebenden Zellen sind dieselben Gebilde, die wir später als Guarnierikörperchen in fixierten und gefärbten Präparaten sehen. Das zur Färbung von Schnitten benützte Haemotoxylin und das Eosin färben die gesamte Zelle und die Schrumpfung des coagulierten, flüssigen Anteils erzeugt das bekannte Bild des Halo um Einschlußkörperchen (Guarnierikörperchen).

Weitere Beobachtungen an lebenden Zellen wurden von K. B. E. MERLING (36) im gleichen Jahre durchgeführt. Die Technik war die bereits bewährte im Dunkelfeld. Die Entnahme der zu beobachtenden Objekte erfolgte am dritten Tag nach der Impfung, ihre Bebrütung bei Zimmertemperatur oder 37° C bis zu sechs Monaten. Die Schwierigkeit der Beobachtung von Objekten, die so lange fern von normalen Verhältnissen belassen werden, betont K. B. E. MERLING selbst, indem er auf die Gefahr der Austrocknung sowie der bakteriellen Infektion hinweist, denen er wohl noch *den unvermeidlichen Einfluß der Zellnekrobiose* zurechnen darf, der ein Gewebe unter unnatürlichen Verhältnissen kaum entzogen werden kann.

Diese Auffassung hat auch H. A. GINS (10) auf der zwölften Tagung der Ver-einigung für Mikrobiologie in Wien (1927) vertreten, indem er die in Zellexplantaten vorkommenden atypisch verlaufenden Zellteilungen als durch eine Reihe von Schädigungen bedingt ansah, die durch die Skarifikation der lebenden Hornhaut und die Entnahme des Gewebes vom lebenden Tier entstehen, wobei der Sauerstoff fehlt und die entstehenden Stoffwechselprodukte aus dem explantierten Gewebe nicht abtransportiert werden können.

Im allgemeinen wurden von K. B. E. Merling während der langen Beobachtungszeit *drei Typen von Veränderungen* wahrgenommen, deren Einteilung jedoch eine mehr weniger künstliche ist, weil die verschiedenen Formen auch in einer einzigen Probe festgestellt werden können.

Typus 1. Die Zellen enthalten ziemlich viel Virus entweder in Form von G. K. oder diffus verbreitet und erleiden keine weiteren Veränderungen als eine langsame Verflüssigung mit nachfolgender Degeneration, wobei das Virus in die verflüssigte Zone übergeht. Solche Fälle können auch in vier Wochen keine Veränderungen zeigen. Die im Laufe des Zellzerfalls auftretenden Körnchen kann man von Viruselementen leicht durch ihre unregelmäßige Gestalt und Größe und ein anderes Brechungsvermögen unterscheiden.

Zum Typus 2 führen einzelne dieser Formen. Dieser Typus zeigt Veränderungen, deren Entstehung längere Zeit erfordert. Hier kommt es zu Zellaustreibungen („Cellextrusions") von verschiedener Größe, die nicht nur von der Leibessubstanz der Zelle selbst, sondern auch von Guarnierikörperchen ausgehen können. Diese Austreibungen sind bedingt durch den Druck der Virusmassen auf die sie umgebende elastische Membran.

Der Typus 3 ist gekennzeichnet durch eine große Anzahl von Austreibungen, welche von der Mutterzelle abgeschnürt werden. Diese Zellen bezeichnet K. B. E. Merling als *Cysten*, die mit Virus vollgepfropft sind. Zu ihrer Bildung benötigt die Zelle einen Tag bis zu mehreren Wochen. Schließlich birst die Cyste und entleert bis zu 10 000 Viruselemente, darunter eine Anzahl von Scheibchen und Kugelformen.

Bei seinen Versuchen kam es K. B. E. Merling (33 bis 36) darauf an, nachzuweisen, ob für die Züchtung der zu untersuchenden Virusformen lebende Zellen notwendig sind, welche das nötige Nährmedium für das Virus abgeben, wie es von verschiedenen Forschern vorausgesetzt wird. Nach seiner Ansicht ist diese Forderung nicht unbedingt berechtigt, weil viele Virusformen durch Jahre hindurch bei niederer Temperatur erhalten bleiben können. Er meint deshalb, daß lebende Zellen ein oder mehrere Produkte des Zwischenstoffwechsels erzeugen, welche für das aktive Leben des Virus nötig sind.

In seinen ersten Untersuchungen hatte K. B. E. Merling den Entwicklungszyklus des Virus nachgewiesen. Es war nur noch zu untersuchen, wie lange sich das Virus, welches mit den es beherbergenden Zellen von seinem Wirt entfernt wird, in diesen Zellen lebend zu erhalten vermag. Es darf wohl angenommen werden, daß die Zellen selbst in wenigen Tagen absterben, wenn das Organ, dem sie entstammen, vom Wirtstier entfernt wird.

Die Frage, ob das Virus sich dann weiter zu erhalten vermag, konnte positiv beantwortet werden und es ist vorauszusetzen, daß die erforderlichen Stoffwechselprodukte in ausreichendem Maße in diesen Zellen vorhanden waren, um die Lebensfähigkeit des Virus zu sichern. Obwohl der anfängliche Impuls für die Vermehrung des Virus von den Zellen ausgehen kann, solange diese Zellen noch in Berührung mit dem Wirtstier sind, so zeigen die beschriebenen Versuche doch, daß dieser Impuls zur Vermehrung des Virus auch wirksam sein kann, wenn das infizierte Organ vom Wirtstier entfernt ist, und die das Virus enthaltenden Zellen bereits abgestorben sind.

Die 3 Typen zeigen Virus im *überlebenden* Gewebe, in dem die beschriebenen Zysten die Stelle der Guarnierikörperchen lebender Zellen vertreten, wobei es zu einer nachweisbaren Vermehrung der Viruselemente kommt, die durch Platzen der Zysten frei werden. In den beschriebenen Zysten kommt es somit zur Bildung von *Mikrokulturen des Vaccinavirus*, das auf Versuchstiere mit Erfolg verimpft werden kann. Insbesondere erweisen sich die Formen des Typus 3 als wirkliche Reinkulturen, ähnlich, wie wir sie von der Bakterienzüchtung her kennen.

Daß sich das Vaccinavirus fern vom Wirtstier in der Hornhaut enucleierter Bulben unter gewissen Voraussetzungen zu vermehren vermag, hatten schon ALDERSHOFF und BROER (1) und später (1913) E. STEINHARDT und A. LAMBERT (42) nachgewiesen, die in 3 Subkulturen dieses Virus auf das Kaninchen übertragen, und einen konfluierenden Ausschlag zu erzeugen vermochten.

Später ist auch im Hühnerembryonalgewebe, das in Hühnerplasma aufgeschwemmt war, eine Virusvermehrung beobachtet worden. Wenn die Forderung lebenden Gewebes immer wieder erhoben wird, so soll damit wohl in erster Linie der Gegensatz zu dem künstlich in der Nährbodenküche hergestellten Nährboden besonders betont werden.

Russische Autoren, insbesondere SILBER und WOSTRUCHOWA (40) haben versucht, das Vaccinavirus in Symbiose mit Hefezellen zu züchten. Wie verschiedene Nachprüfer finden konnten, mußten diese Versuche fehlschlagen. Es ist lediglich gelungen, das Virus eine gewisse Zeit durch die Kulturen fortzuschleppen, eine richtige, zahlenmäßig durch Titrierung nachgewiesene Vermehrung hat nicht stattgefunden, weil, wie wir es längst wissen, das Virus nur auf dem Gewebe gewisser Wirtstiere gedeiht, nicht aber auf Hefezellen. Wir glauben es nicht, daß sich außer diesen spezifischen Gewebsarten *völlig fremde* als Nährsubstrat eignen werden [vgl. C. HALLAUER (16)].

Trotzdem sind die Arbeiten von K. B. E. MERLING verdienstvoll, weil sie auch dem Vaccinebereiter einen Weg zeigen, der möglicherweise betreten werden könnte.

Es war bisher nicht bekannt, daß sich Vaccinavirus *monatelang* bei Bruttemperatur hält (vgl. Wärmeresistenz des Vaccinavirus). Es scheint nunmehr, daß dies innerhalb eines spezifischen Nährsubstrates möglich ist. Das soll ein Wink sein für die Herstellung von Trockenimpfstoffen, es wird aber auch zu versuchen sein, ob man nicht zerkleinertes Hautgewebe zur Anzucht von Vaccinavirus verwenden kann.

Von einem ähnlichen Gedanken scheinen H. B. MAITLAND (32) und A. W. LAING ausgegangen zu sein, die das Vaccinavirus mit zerkleinerter Niere in Tyrodelösung züchteten, indem sie annahmen, daß auf dem *überlebenden* Gewebe das Virus eine Grundlage für sein Wachstum finden werde.

Seine Beobachtungen hat K. B. E. MERLING im Jahre 1943 noch erweitert (35) indem er photographisch die Bildung von *Virusketten* nachwies und, was besonders wichtig ist, daß nach völliger Auflösung der wirtseigenen Zellen die Vermehrung in den Zellexplantaten weiter vor sich geht, so daß man annehmen darf, daß das Virus für die Erhaltung seines Lebens in vitro eigentlich nur der aus den abgestorbenen Zellen freigewordenen Stoffwechselprodukte bedarf.

Es war auch interessant zu erfahren, daß die in späten Stadien der Beobachtung gefundenen Cysten i. e. Viruskolonien eine Bewegung zeigen und die Fähigkeit, sich in zwei oder mehrere Tochtercysten zu teilen, die eine sehr dünne Membran besitzen, durch welche das Virus innerhalb der Cyste sehr gut zu sehen ist.

Das Vorkommen von Fissuren vor der Teilung war deutlich wahrzunehmen.

In das Gewebe aufgenommene, in den Zellen zur Vermehrung und Kolonien-
bildung gelangte Viruselemente werden schließlich ein Opfer der Phagocytose,
wenn sie von ihrer Bildungsstätte in Freiheit gelangen.

Über ihr weiteres Schicksal hat K. B. E. Merling Untersuchungen im Dunkelfeld angestellt (36), die zu demselben Ergebnis führten, wie seine Beobachtungen
an Hornhautzellen. Es ist überraschend, daß in den Leukocyten, denen man
eine Vernichtung, etwa in Form einer Verdauung der Viruselemente zuschreiben
möchte, eine ganz analoge Vermehrung zu beobachten war, wie in den primär
befallenen Gewebszellen.

Die im Dunkelfeld beobachteten Leukocyten, die auf gleiche Weise unter
dem Deckglas mit Paraffin abgeschlossen waren, lebten einige Tage und phago-
cytierten das Virus, wobei es nicht abstarb, sondern Kolonien bildete. Die Unterscheidung von Granula anderer Natur wird mit der Zeit nicht schwer, wenn
man auf Größe, Gestalt, erhöhte Helligkeit, Bewegung, Farbe achtgibt. Insbesondere zeichnen sich *Viruselemente* durch *größere Helligkeit* vor den dunkleren Leukocytengranula aus.

An den Leukocyten kann häufig ihre Freßtätigkeit beobachtet werden,
indem einzelne Viruselemente, aber auch kleine Cysten in kürzester Zeit von
ihnen verschlungen werden.

Wie analog sich sonderbarerweise zwei, phylo- und ontogenetisch differenten
Geweben angehörige Wirtszellen, dem Virus gegenüber verhalten, zeigten nicht
nur die in *Hornhautzellen, sondern auch in Leukocyten auftretende Bildung von
Kolonien und Cysten* und die virushältigen Austreibungen (Extrusionen) der
Leibessubstanz der weißen Blutkörperchen. Die Bildung von Guarnierikörper-
chen scheint in ihnen in gleicher Weise vor sich zu gehen wie in Hornhautzellen.

Vielleicht erlaubt auch dieser Befund die Hoffnung auf Gewinnung des
Virus auf einem in vitro gehaltenen Substrat.

F. Weitere eigene Beobachtungen an vaccinierten Horn-
häuten und Organen mit virusbedingten Veränderungen.

Um nun zu weiteren eigenen Versuchen überzugehen, so sei, anknüpfend
an die eigenen, im Abschnitt C beschriebenen Beobachtungen, bemerkt, daß
wir vorerst zu untersuchen trachteten, wie sich das implantierte Virus in den
Hornhautzellen weiterverbreitet. Hierbei bedienten wir uns mit Erfolg der
bereits erwähnten Züchtung des überlebenden Gewebes, die auch von Jean
v. Cooke und R. J. Blattner (8) zur Betrachtung der beimpften Chorio-Allantois
benützt worden ist. Beobachtet man eine Vaccinationswunde der Hornhaut,
so findet man, daß das Virus in die Nachbarschaft diffundiert und zuerst möglicherweise bereits durch die Impfung verletzte Zellen angreift, indem es in sie
eindringt. Hier versagt die lückenlose Verfolgung des vaccinalen Prozesses.
Man sieht aber gelegentlich bei stärkerer Vergrößerung einige wenige in die
Zellen eingedrungene Elementarkörperchen, dann mehrere gruppiert. Dabei
tauchen zum ersten Male neben den deutlich sichtbaren Elementarkörperchen
auch die von K. B. E. Merling (33—36) beobachteten *Scheibchenformen* („discs")
auf. Es ist schwer zu sagen, ob sie die Ausgangsformen für die Elementarkörperchen sind, in die sie dann zerfallen, oder ob ihnen im Entwicklungsprozeß
eine andere Rolle zukommt. Auch größere, kreisrunde Formen sieht man, wie
sie K. B. E. Merling beschrieben hat („spheres"). An ihnen sieht man, daß sie
ein rundes Konglomerat von Elementarkörperchen sind, die frei in ihrer Aufschwemmung liegen und vielleicht ein Ergebnis einer Agglutination sind, die

sie zusammenhält. Sie befinden sich in lebhafter Bewegung über das ganze Gesichtsfeld zerstreut und sind derart zahlreich, daß sie nicht einen zufälligen Befund darstellen können. Die Scheibchen sind nicht immer leicht zu differenzieren, weil Elementarkörperchen bei verschiedenen Kondensoreinstellungen ein ähnliches Bild zeigen können. Man kann sich aber auch aus einer anderen Darstellung der Elementarkörperchen ein Bild von ihrer Existenz machen. Wird ein Kaninchen intravenös mit 5 ccm einer verdünnten zentrifugierten Vaccine gespritzt, so ist es etwa am 5. Tage moribund. In diesem Zustand wird das Tier entblutet und die Gerinnung des Blutes durch Heparin oder Natriumzitrat verhindert. Läßt man nach der Weisung von HACKENTHAL (15) oder W. SMITH (41) das Blut in eine dünne Glasröhre fließen und sedimentieren, so setzt es sich in drei Schichten ab, deren mittlere aus einer schmalen Zone weißer Blutkörperchen besteht. Untersucht man sie im Dunkelfeld oder auch im Ultropak, so findet sich in dem einen von uns beobachteten Fall eine Gruppe von aneinandergebackenen weißen Blutkörperchen, die durch ihre leuchtend weiße Farbe im Dunkelfeld besonders ins Auge sticht. Sie ist bedingt durch eine Unzahl von Elementarkörperchen, die als leuchtend weiße Granula die Leukocyten erfüllen (Abb. 12).

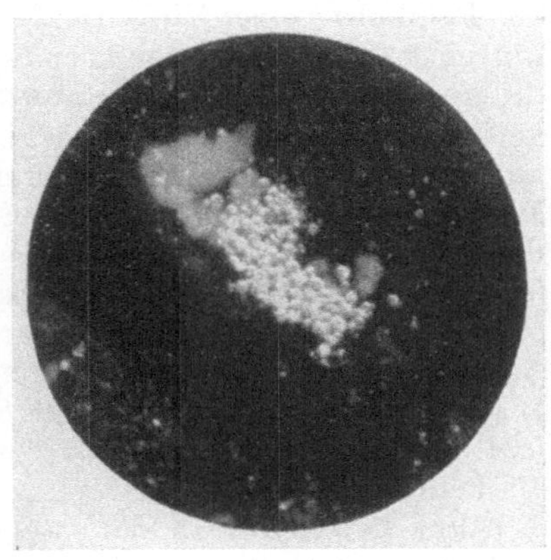

Abb. 12. Leukocyten aus Blut, überladen mit helleuchtenden Elementarkörperchen. Dunkelfeld Objektiv 40, Ok. 8, Zeiß. Photo Miflex.

Beobachtet man das gleiche Präparat im Auflicht bei starker Vergrößerung, so ergänzt sich das Bild in anderer Richtung. Um die von K. B. E. MERLING beschriebenen ,,cellextrusions" sichtbar zu machen, wurde auf die Wiedergabe der Elementarkörperchen verzichtet, die in einer anderen Ebene liegen dürften. Man sieht deutlich in dem ungemein plastischen Bild die Auswüchse der Leukocyten (Abb. 13).

Man hat sich daran gewöhnt, die Elementarkörperchen der Vaccina als vollkommen gleich große Körperchen zu beschreiben und diese gleichmäßige Größe als ein Charakteristikum derselben zu bezeichnen. Das ist nur bis zu einem gewissen Grade richtig. Jedenfalls muß festgestellt werden, daß diese gleichmäßige Größe nicht für alle Formen, die man sieht, zutrifft. Es gibt Übergänge. Das sind Elementarkörperchen, die etwa um die Hälfte größer sind als die Durchschnittsformen, sie heben sich als leuchtend weiße Körner von der Umgebung deutlich ab, und es gibt Formen, die kleiner sind als die Durchschnittselementarkörperchen in dieser völlig bakterienfreien Umgebung. Von diesem Vorkommen ungleich großer Formen von Elementarkörperchen kann

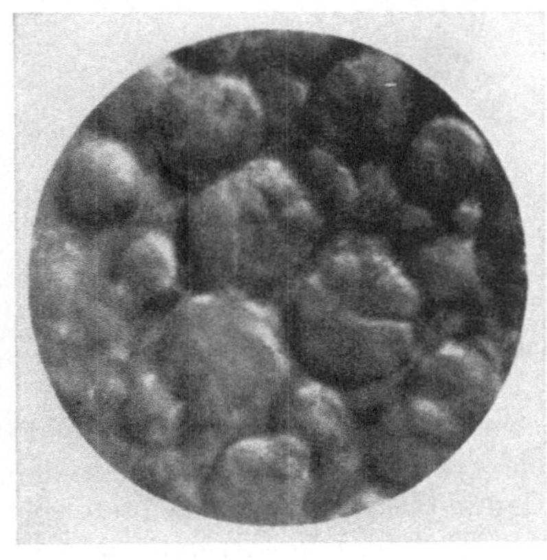

Abb. 13. Leukocyten der Abb. 13 im Auflicht. Leitz Objektiv UO 100 Oel, Ok. 10, Photo Miflex. Man sieht das ungemein plastische Bild, in dem in einzelnen Leukocyten die ,,Cellextrusions" von MERLING festzustellen sind. Elementarkörperchen verschwinden bei dieser Darstellung.

man sich auch überzeugen, wenn man am 6. Tage eine Vaccinapustel ansticht und den ausquellenden Tropfen getrübten Serums auf einem Objektträger auffängt und im Dunkelfeld oder im Auflicht betrachtet. Da sieht man das Gesichtsfeld übersät mit Elementarkörperchen in Brownscher Bewegung. Sie scheinen einander völlig gleich zu sein und man ist von dieser Gleichförmigkeit der Reinkultur einer Virusart überrascht. Unter ihnen findet sich jedoch immerhin eine Anzahl größerer korpuskulärer Elemente an Stellen, die dort gut zu differenzieren sind, wo die Brownsche Bewegung wegen des Mangels an Flüssigkeit aufhört. Dort sieht man größere und kleinere Formen, wobei man wohl annehmen darf, daß die kleineren allmählich größer werden. Man sieht aber auch seltene, ganz große Formen, die uns wieder zu unserem Ausgangspunkt, den weißen Blutkörperchen führen, die wir aus dem aufgefangenen Zitratblut isoliert haben. Hier sieht man neben den normal großen Elementarkörperchen eine Anzahl größerer korpuskularer Elemente, die wir bereits erwähnt haben. Ähnliche Formen hat auch Morosow (37) beschrieben, sie wurden auch im I. Band dieses Handbuches auf Seite 265 abgebildet. Im gesamten Gesichtsfeld sind Bakterien nicht nachzuweisen, bakteriologisch erwies sich das Blut als keimfrei.

Ähnliche, oder man darf wohl sagen, ganz gleiche Formen sind auch auf andere Art und Weise zu finden. Wenn man ein intravenös mit Vaccinavirus behandeltes Tier untersucht, so findet man mitunter seine Harnblase stark

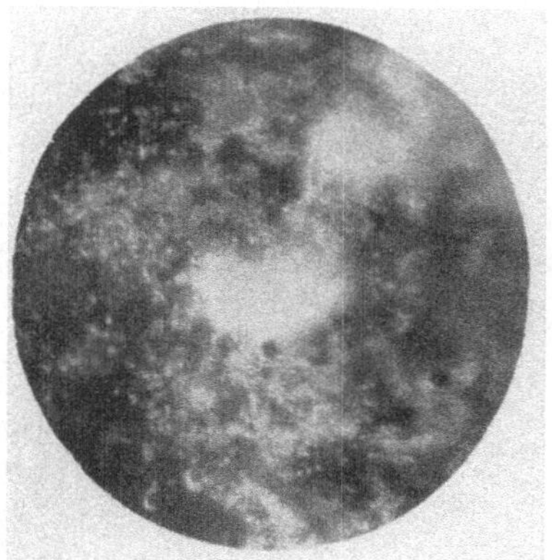

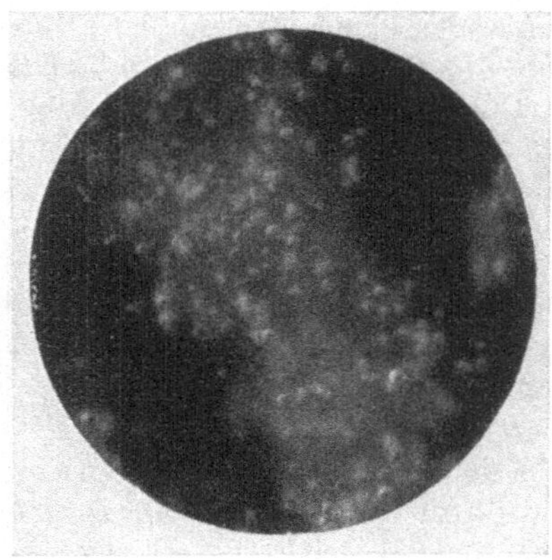

Abb. 14. Aufnahme von Harnsediment eines vacc. Kaninchens im *Dunkelfeld*. Obj. 90, Ok. 8, Zeiß, Leicaaufnahme. Man sieht das Gesichtsfeld übersät mit leuchtenden Elementarkörperchen.

Abb. 15. Dieselbe Aufnahme wie Abb. 15 *im Auflicht* mit Schwarzscheibe als Untergrund. Leitz Objektiv UO 100 Oel, Ok. 10, Leicaaufnahme. Im Gesichtsfeld zahlreiche Ringformen.

erweitert; sie ist mit trüben, jedoch bakterienfreiem Harn erfüllt. Im Dunkelfeld oder im Ultropak zeigen sich zahlreiche Schleimfäden, über und über mit Elementarkörperchen beladen, die auch das ganze Gesichtsfeld ausfüllen. Neben den Elementarkörperchen sind zahlreiche ähnliche Kugeln und Scheibchen zu finden, wie wir sie im Blute sahen. Sie mußten also trotz ihrer Größe durch die Niere hindurch, oder sie bilden sich auch im Harn, nachdem sie als Elementarkörperchen die Niere passiert haben. Wo sie zustande kommen, ist unsicher, jedenfalls sind sie im Blute zu finden. Entstehen sie dort, oder gibt es einen anderen Ort ihrer Entstehung? (Bild 14.)

Das eindrucksvollste Bild dieser Betrachtung bilden jedoch die weißen Blutkörperchen im auffallenden Licht. Man ist bestürzt über die eigene Unkenntnis des Aussehens dieser wundervollen Gebilde. Aus gefärbten Präparaten, aus

Nativpräparaten und auch vom Dunkelfeld her hat man das Bild eines flachen Scheibchens verschiedener Gestalt in Erinnerung, in dem sich granulierte hellere oder ungefärbte Zonen und ein dunkler verschieden gestalteter, gelappter oder ungelappter Kern abhebt. Das Auflicht gibt von der Beschaffenheit dieser Körperchen eine ganz andere Vorstellung. Von einer Ebene ist nichts mehr zu sehen. Am besten läßt sich das Gebilde mit einer Mondlandschaft vergleichen, in der es Randgebirge, Krater, Täler gibt und einige wenige ebene Stellen. Offenbar ist es der Kern, der in der Tiefe eines solchen Kraters sitzt, in dem er sich wie ein See ausnimmt mit glatter Oberfläche. Die ganze Landschaft ist bespickt mit leuchtenden weißen Körnchen, die man auch außerhalb der Leukocyten sieht, mit Elementarkörperchen, die entweder phagocytiert wurden oder den weißen Blutkörperchen nur aufgelagert sind. Untersucht man Schnittflächen von Organen im Auflicht bei starker Vergrößerung, so hat man Schwierigkeiten, Einzelheiten zu differenzieren und entsprechend zu deuten. Man fühlt sich den sich hier bietenden Bildern gegenüber wie ein Neuling, der zum ersten Male in ein Mikroskop blickt. Um klarer zu sehen, war es nötig, Farbstoffe zu Hilfe zu nehmen. Da zeigen sich Elementarkörperchen, die über die befallenen Zonen zerstreut sind. Sie nehmen weder unser Phloxin noch Wasserblau an, sie bleiben hell, scharf leuchtend und heben sich mit ihrer glänzend weißen Farbe von der gefärbten Umgebung und von dem dunkleren Granula der Leukocyten ab. Auch hier sieht man im gefärbten Präparat Zelleiber, Kerne, Elementarkörperchen, Kugeln und Scheibchen (Abb. 15).

Derartige Bilder kann man in allen Organen, die wir untersucht haben, finden: Milz, Lunge, Leber, Niere.

Es fehlt jedoch den Ansammlungen von Elementarkörperchen, die man sonst auch zerstreut über das ganze Gesichtsfeld sieht, ein fester Zusammenhang. Eine einwandfreie Aufnahme ist uns aus technischen Gründen vorläufig noch nicht gelungen.

Wie HIMMELWEIT (18) hervorhebt, kann man das Gebilde, das man gemeinhin als Guarnierikörperchen bezeichnet, in Organen im Auflicht nur selten sehen, es sei denn, daß man die als Cysten bezeichneten Ansammlungen von Virus als Guarnierikörperchen bezeichnen will. Der eine von uns (K.) hat es wohl einem besonderen Zufall zu verdanken, daß er in gefärbten Schnitten Viruskolonien in Hornhautzellen darstellen und mit Agfa-Kolorfilm mikrophotographieren konnte, wodurch ein weiterer unzweifelhafter Beweis für die Virusnatur der Guarnierischen Körperchen gegeben war. (27, dort auch Einzelheiten dieses Befundes nachzulesen.) Auf Grund unserer Befunde und der von HIMMELWEIT (18) erhobenen und auch von K. B. E. MERLING (33—36) bestätigten darf wohl mit größter Wahrscheinlichkeit angenommen werden, daß die in gefärbten Schnitten wahrgenommenen Guarnierischen Körperchen ein Koagulat sind, das durch die verschiedenen Fixierungsverfahren zustande kommt.

Auflichtbetrachtungen von Pustelausstrichen. Im Auflicht sieht man Elementarkörperchen in Massen plastisch erscheinen, es sind auch Scheibchen und Kugelformen zu sehen. Häufig beobachtet man Leukocyten, die mit Elementarkörperchen beladen sind. Der Kern zeigt im Auflicht keinerlei Unebenheiten und ist frei von Granula. Die Elementarkörperchen sind viel prominenter weiß, hie und da mit einem Stich ins Gelbe (möglicherweise hängt dies von der Beleuchtung ab), während die Granula dünkler sind.

Beobachtungen am Myxom. Im Dunkelfeld: verriebenes myxomatöses Hautmaterial zentrifugiert, zeigt dieselben Formen wie das vaccinale. Daneben kommt eine große Anzahl von Ringformen vor, die als Luftblasen verdächtig waren. Es wurde der Objektträger unter Vakuum gestellt und dann neuerdings mikroskopiert. Die Ringformen blieben bestehen, sie hätten verschwinden müssen, wenn sie Luftblasen gewesen wären. Es ist noch nicht sicher, ob es nicht Diffraktionserscheinungen sind, die auf die verschiedene Kondensorstellung zurückgeführt werden müssen. Um hier klarer zu sehen, wird man weitere Proben von infiziertem Gewebe im Dunkelfeld oder im Auflicht zu betrachten haben, Versuche, die noch nicht beendet wurden.

Objektträgerreinigung. Einen Hinweis darauf halten wir für dringend notwendig. Es ist wichtig, weniger Erfahrene darauf aufmerksam zu machen, wie leicht man „Elementarkörperchen" finden kann. Die Dunkelfeldbetrachtung oder auch die Auflichtbetrachtung eines völlig leeren Objektträgers liefert schon Beweise dafür. Ungereinigte Objektträger sind für Dunkelfeldbeobachtungen nicht zu brauchen; selbst die beste mechanische Reinigung genügt nicht; wir haben sie deshalb mit einer Schicht Kollodium überzogen, nachdem sie vorher mechanisch gereinigt wurden. Die Kollodiumhaut wird dann abgezogen und reißt korpuskulare Elemente mit sich. [Vgl. F. W. OELZE (38) und M. KAISER (24)$_3$.]

Diese Art der Reinigung hat sich auch für die Reinigung von Deckgläsern bewährt, die mit einer Pinzette in Kollodium eingetaucht werden. Auf diese Art bildet sich eine Kollodiumhaut auf beiden Seiten des Deckglases und auch über die Spitzen der Pinzette hinweg. Diese Haut läßt sich dann leicht abziehen und hinterläßt hochgradig gereinigte Flächen, auf denen korpuskulare Elemente, die Elementarkörperchen vortäuschen könnten, nur viel seltener zu finden sind.

G. Schlußbetrachtungen.

Wir sind uns dessen bewußt, keine wesentlich neuen Befunde für die Virusforschung mitgeteilt zu haben. Sie haben jedoch eigene, vor mehr als 10 Jahren vorgenommene, erstmalige Beobachtungen bestätigt und bringen auch anderen Forschern den Beweis, daß ihre Untersuchungsergebnisse von uns anerkannt werden.

Dabei zeigte es sich auch, daß die nach der MOROSOW-Methode gefärbten P. K., über deren Natur MOROSOW selbst nicht im klaren war (37), sich nunmehr als eine Entwicklungsform dieser Viruselemente erwiesen haben, die sowohl von K. B. E. MERLING (33—36), wie auch von uns gesehen wurden, also unzweifelhaft vorhanden sind und nicht als ein Kunstprodukt oder als eine Diffraktionserscheinung aufgefaßt werden dürfen. Es war ferner interessant, zu finden, daß die von einem von uns (K.) in Agfa-Koloraufnahmen nachgewiesene verschiedene Färbung der Guarnierischen Körperchen, sei es mit saueren oder mit basischen Farbstoffen, auch in überlebenden Hornhautzellen gelingt. So nahmen die Guarnierischen Körperchen bald das sauere Phloxin (meist ist es so) oder auch das basische Wasserblau an, oder aber sie färbten sich metachromatisch und wurden weinrot. Es ist also auch hier der Beweis erbracht, daß man bei einer Beurteilung der chemischen Zusammensetzung dieser Elemente recht vorsichtig sein muß und daß man aus der Färbung nicht immer einen Schluß auf die chemische Natur des gefärbten Substrates ziehen darf.

Unsere Versuche, die in der gegenwärtigen Zeit unter den denkbar schwierigsten Verhältnissen und technischen Widerwärtigkeiten durchgeführt wurden, haben aber über das rein Histologische hinaus den weiteren Beweis erbracht, daß der *Ultropak ein für die Virusforschung unentbehrliches Instrument* ist, auf dessen wundervolle Einrichtung wir jeden Arbeiter, der sich genußreiche Forscherfreuden verschaffen will, wärmstens aufmerksam machen.

Literaturübersicht.

1. ALDERSHOFF, H. et M. BROERS: Contribution à l'etude des corps intraépithéliaux de Guarnieri. Ann. de l'Inst. Pasteur **9**, 779, 1906.
2. BARTA, E.: Der Mikroilluminator, Z. f. wiss. Mikroskopie **52**, 276, 1935.
3. BEVERIDGE, W. I. B.: Simplified Techniques for inoculating chick embryos and a means of avoiding egg white in vaccines. Science Oct. **3**, 1947.
4. BURNET, F. M.: The growth of viruses on the chorioallantois of the chick embryo. In R. DOERR und C. HALLAUER Hdb. d. Virusforschung. Springer Wien I/1, 419, 1938.
5. —, D. LUSH and A. V. JACKSON: Propagation of herpes and pseudorabies viruses on chorioallantois. Austral. J. Exp. Biol. and Med. **17**, 35, 1939.
6. — and J. A. GALLOWAY: The propagation of the virus of vesicular stomatitis in the chorioallantoic membrane of the developing hen's egg. Brit. J. Exp. Path. **15**, 105, 1934.
7. CLARK, E. R. and E. L.: Observations on living preformed blood vessels as seen in an transparent chamber inserted in the rabbit's ear. Anat. Rec. **45**, 211, 1930 und **49**, 441, 1932.
8. COOKE, JEAN and R. J. BLATTNER: Vital staining of virus lesions on chorioallantoic membranes by trypan blue. Proc. Soc. Exp. Biol. and Med. **43**, 255, 1940.
9. ELLINGER, PH. und A. HIRT: Eine Methode zur Beobachtung lebender Organe mit stärksten Vergrößerungen im Luminiszenzlicht (Intravitalmikroskopie). In E. ABDERHALDEN Hdb. d. biol. Arbeitsmethoden. Urban-Schwarzenberg, Berlin, Wien, Abt. V. **2/2**, 1753, 1932.
10. GINS, H. A.: Absterbebedingungen an normalen Epithelzellen in Explantaten. Zbl. Bakt. Orig. I, **104**, Beiheft, 44, 1927.
11. GRAMENITZKI, M. J.: Episkopie, Diaskopie und Mikroskopie des schlagenden Herzens des Frosches. Archiv Biologitscheskich Nauk **34**, 617, 1934.
12. GUARNIERI, G.: Ricerche sulla patogenesi ed eziologia dell'infezione vaccinica e variolosa. Arch. per le Science med. **16**, 1892.
13. — Ulteriori recherche sulla eziologia et sulla patogenesi della infezione vaccinica. Clinica moderna 1897.
14. — Studi sulla struttura e sullo sviluppo dei parasiti dell'infezione vaccinica. Clinica moderna **7**, No. 34, 1902.
15. HACKENTHAL, H.: Über die Lokalisation des Kuhpockenvirus im Blute von mit Pockenlymphe infizierten Meerschweinchen. D. med. Wschr. **55**, 906, 1929.
16. HALLAUER, C.: Die Züchtung der Virusarten außerhalb ihrer Wirte. In R. DOERR und C. HALLAUER, Hdb. d. Virusforschung, Springer, Wien, **1**, 370, 1938.
17. HEINE, H.: Der Ultropak. Z. f. wiss. Mikroskopie **48**, 538, 1932.
18. HIMMELWEIT, F.: Observations on living vaccinia and ectromelia viruses by high power microscopy. Brit. J. Exp. Path. **20**, 108, 1938.
19. HIRSCH, G.: Die Lebendbeobachtung der Restitution des Sekretes im Pankreas. Z. f. Zellforschung **15**, 35, 1931.
20. HOFFSTADT, RACHEL E., and DOROTHY OMUNDSON: A convenient apparatus for the manipulation of eggs in the study of the chorio-allantoic membrane. Science **91**, 459, 1940.
21. — — Study of living virus of infections myxoma of rabbits by high power microscopy. J. Infect. Dis. **68**, 207, 1941.
22. JÜRGENS: Die aetiologische Begründung der Pockendiagnose. Deutsche med. Wschr. 30, 1904, 1636.
23. KAISER, M.: Die Färbemethoden der Viruselemente. In R. DOERR und C. HALLAUER, Hdb. d. Virusforschung, Springer, Wien, **1**, 251, 1938.
24. — Ein einfaches Verfahren zum Reinigen von Objektträgern und Deckgläsern. Zbl. Bakt. Orig. 147, 351, 1941.
25. — Viruskolonien in Hornhautzellen. Klin. Med. **3**, 405, 1948.
26. —, M. GHERARDINI: Über zwei Methoden zur Kontrastfärbung der GUARNIERIschen Körperchen, Zbl. Bakt. Orig. 131, 128, 1934.

46   M. KAISER und P. VONWILLER: Mikroskopie in der Virusforschung.

27. — — Ein Beitrag zur Kenntnis der GUARNIERIschen Körperchen nach Verimpfung „originärer Kuhpocken" auf die Kaninchenhornhaut. Arch. f. d. ges. Virusforschung **1**, 497, 1940.

28. — und P. VONWILLER: Weitere Beobachtungen an vaccinierten Hornhäuten. III. Mitt. Studien mit dem Ultropak von Ernst Leitz. Zbl. Bakt. O. **133**, 149, 1935.

29. KNISELY, M. H., E. H. BLOCH, TH. S. ELIOT and L. WARNER: Sludged Blood, Science **106**, 431, 1947.

30. LEITZ, E.: Der Ultropak für Auflicht-Beobachtungen bei jeder Vergrößerung. Druckschrift Mikro D Nr. 7358 b. Wetzlar, 1937.

31. — Leitz-Ortholux mit Aristophot. Druckschrift Mikro A Nr. 7880 b. Wetzlar, 1941.

32. MAITLAND, H. B. and LAING A. W.: On the function of tissue cells in media used for growing vaccinia virus. J. Path. and Bact. **53**, 419, 1941.

33. MERLING, K. B. E.: Observations on living vaccinia virus in the corneal cells of the rabbit. J. Path. and Bact. **50**, 279, 1940.

34. — Further observations on living vaccinia virus. J. Path. and Bact. **51**, 391, 1940.

35. — Ultramicroscopical observations on the morphology and development of vaccinia virus in vitro. Brit. J. Exp. Med. **24**, 240, 1943.

36. — Phagocytosis of vaccinia virus in vitro. J. Path. and Bact. **57**, 21, 1945.

37. MOROSOW, M. A.: Die Färbung der PASCHENschen Körperchen durch Versilberung. Zbl. Bakt. O. **100**, 385, 1926.

38. OELZE, F. W.: Methodik der Dunkelfeldmikroskopie in E. ABDERHALDEN Hdb. d. biol. Arbeitsmethoden. Urban-Schwarzenberg, Berlin, Wien. Abt. XII, 2/1, 446, 1939.

39. PFEIFFER, L.: Ein neuer Parasit der Pockenprozesse. Korr. Blätter d. ärztl. Vereins v. Thüringen 1887 und 1888. Cit. nach E. GILDEMEISTER in O. LENZ und H. A. GINS Hdb. d. Pockenbekämpfung und Impfung. R. Scholtz, Berlin, 1927, Seite 768.

40. SILBER und WOSTRUCHOWA: Über Züchtung der filtrierbaren Virusarten auf nicht pathogenen Mikroben. Zbl. Bakt. O. **129**, 389, 1933.

41. SMITH, W.: The distribution of virus and neutralizing antibodies in the blood and the pathological exudates of rabbits infected with vaccinia. Brit J. Exp. Med. **10**, 93, 1929.

42. STEINHARDT, E. I. C. and R. LAMBERT: Studies on the cultivation of the virus of vaccinia. J. Inf. Dis. **13**, 294, 1913.

43. — — Studies on the cultivation of the virus of vaccinia II. J. Inf. Dis. **14**, 87, 1914.

44. TOPLEY, W. W. C. and G. S. WILSON: The principles of bacteriology and immunity. Arnold, London, 1936, Seite 1476.

45. VONWILLER, P.: Neue Wege der Gewebelehre. Z. f. Anatomie und Entwicklungsgeschichte **76**, 498, 1925; **84**, 478, 1927; **95**, 512, 1931.

46. — Lebendige Gewebelehre. Eine Histophysiologie auf neuer Grundlage. Zollikofer, St. Gallen, 1945.

47. — und V. DEMOLE: Observation de trypanosomes vivants dans le sang du mammifère vivant. Cpt. rend. Soc. Anatomistes London **232**, 1927.

48. — und M. A. KOTLIAREWSKAYA: Die Mikroskopie des lebenden und regenerierenden Hornhautepithels in Hinsicht auf die mikroskopische Beobachtung seiner Erkrankungen. Trav. de l'Inst. de rech. phys. de Moscou **2**, 490, 1934.

49. ZEISS, C.: Auflicht-Geräte. Druckschrift Mikro 476/40.

Variation in Influenza Viruses.

By **F. M. BURNET**, Melbourne.

From the Walter & Eliza Hall Institute of Medical Research, Melbourne, Australia.

There are two aspects to any discussion of variation amongst the influenza viruses. On the one hand there are the directly observable changes in character that occur in the course of experimental manipulation and on the other there are the differences that are observed in the epidemiology of different outbreaks of virus influenza and the differences which are observed in strains as isolated. It is natural and probably legitimate to infer that the different antigenic types of influenza virus A that were isolated in Britain during the 1936—7 epidemic were variants from a common ancestral stock. But until a great deal more is known of the natural history of influenza it will be difficult to guess how far back that common ancestry might be. The related problems of how the influenza virus persists in interepidemic years and what is responsible for the genesis of a new epidemic, have not yet been solved. It is a possibility but no more than a possibility that as suggested by ANDREWES (1942) each new influenza epidemic represents the emergence of a more virulent mutant from the trickle of low grade infection that persists through interepidemic periods. Provided such a more active strain finds a population with low specific immunity, it will find opportunities both for spread and for the appearance of further mutants. An equally conceivable alternative is that epidemics arise by the simultaneous development into overt activity of many strains which have persisted in carriers or otherwise during the interepidemic period. On this view the epidemic is determined only by (a) the existence of a large enough susceptible population, and (b) conditions, social and meteorological, that are apt for its spread. Preliminary mutation would not be required nor would the existence of multiple strains within a single epidemic be valid evidence for the occurrence of contemporary variation.

Under the circumstances, we are compelled to treat the problem very much in the same way that a zoologist would use in discussing the systematics of mice or fruit-flies. Where there is sufficient resemblance in basic pattern of organization we must assume common descent, and the minor differences within the basic common pattern which occur in nature must be interpreted according to relevant circumstances and with particular attention to the phenomena of variation, hybridization etc that can be observed in controlled laboratory studies.

I. The influenza group of viruses.

It is universally agreed that influenza A and B viruses are sufficiently alike to be given the same name and a mere alphabetical differentiation. This is based primarily on the character of the human disease produced. Their behaviour in experimental animals is not very impressively similar. Ferret infections with A strains are usually fever producing while B infections are nearly always asymptomatic. A strains at least after a few passages through ferrets or chick embryos are readily adapted to mice, B strains only with the greatest difficulty. The real similarity emerges only in their behaviour in the chick embryo. Both are readily isolated from human material by amniotic inoculation; both can be adapted to multiply freely after injection into the allantoic cavity; both produce haemagglutination of approximately similar character. It has been suggested (BURNET, 1945) that those viruses should be included in the influenza group which multiply readily in the allantoic cavity giving fluids which agglutinate fowl erythrocytes — the agglutination being due to the virus particles themselves. On this criterion we must include in the group the following viruses and their minor variants: Influenza A and B, Swine influenza, Mumps, Newcastle Disease of Fowls and Fowl plague. It is possible even probable that other virus types may be found to fall into the group. The reality of this suggested relationship is strongly supported by the evidence that the same receptors in red cells are concerned with the haemagglutination of each of these viruses. BOVARNICK and de BURGH (1947) consider that the cell receptors for mumps virus differ from those reacting with influenza virus mainly because of the failure of sheep cells to be agglutinated by mumps virus. Work in this Institute (LIND and FAZEKAS DE ST. GROTH, unpublished) however strongly supports our contention that despite minor differences from one species of cell to another the same receptors are concerned. Sheep cells for instance, after minimal treatment with periodate are readily agglutinated by mumps virus. As far as information is available all of the viruses fall within the same range of size between 80 and 120 mμ in diameter. All three influenza virus types are of similar appearance in electronmicrographs, uniform spherical particles predominating. Newcastle disease virus differs in having a less well defined shape. (BANG 1946, 1947, CHUNHA et al 1947.)

The appearance seems to be very sensitive to physical conditions and the "tails" when present are by no means the well defined appearances shown by the larger bacterial viruses. There is certainly a clear difference in these respects from the influenza viruses and therefore a valid objection to the classification of NDV in the influenza group. It is to be hoped that electronmicrographs of the main variant types of influenza A as well as of mumps and fowl plague will clarify the discrepancy. The resemblance of NDV to influenza viruses can be seen also in the fact that the pathology in the chick embryo is almost identical with that of a fully egg adapted strain of influenza A and that on intranasal inoculation of infected embryo fluids in mice influenza-like consolidation is produced although the virus cannot be serially transmitted by this method.

Further discussion will be limited to the true influenza viruses including under that term A, B, Swine and their variants, but one can feel certain that relevant information may also be forthcoming from the variational behaviour of other viruses of the group. The extremely interesting range of natural disease produced by strains of NDV ranging from the very mild pneumoencephalitis of California to the extremely lethal East Indian disease, may be of special importance in this connection.

II. Variation as a necessary preliminary to laboratory study.

The agent of primary interest to the worker on influenza is the virus in its functional capacity as a human pathogen, and the major practical significance of variation is in regard to variations in virulence for human beings. But except for an occasional critical experiment in human volunteers all experimental work must of necessity be done with virus which rapidly changes its genetic character. For any type of quantitative study of influenza virus strains, they must be adapted to multiply regularly either in the lungs of mice or in the allantoic cavity of chick embryos. It is well known that no human strain yet described will produce lesions in mice on primary intranasal inoculation, and it is our contention (BURNET and BULL 1943) that human virus will not multiply in the allantoic cavity until a suitable mutant has appeared. This point may need qualification to provide for the occasional appearance as human pathogen of a strain which could multiply freely in the allantoic cavity.

It therefore seems to us to be of primary importance to analyse the nature of the changes in the genetic character of the virus that are undergone in the course of adaptation to laboratory passage. Influenza A strains have been more carefully studied in this regard than B strains.

1. Adaptation to the allantoic cavity.

The available evidence suggests that of the population of influenza A virus particles present in the mouth and throat secretions of a patient acutely ill with the disease, a majority are capable of initiating infection in the amniotic cavity of the chick embryo or in the nasal cavity of the ferret (HIRST 1947b). Occasionally mice inoculated intranasally with human material will be infected and develop a detectable antibody rise (FRANCIS and MAGILL, 1937, BURNET and FOLEY 1941), hamsters with greater regularity (TAYLOR 1940). Inoculations into the allantoic cavity usually give no evidence of infection but in a proportion sufficient multiplication occurs for passage to be possible with the eventual appearance of a variant capable of ready multiplication (THIGPEN and CROWLEY, 1943).

When human influenza A material is inoculated into the amniotic cavity by an appropriate method (BURNET 1943) multiplication is usually evidenced four to five days after the inoculation by the appearance of power to agglutinate red cells in the amniotic fluid. Characteristically fowl cells are either not agglutinated or agglutinated to very low titre while human and guinea pig cells are agglutinated to equal and much higher titres (BURNET and BULL, 1943). It is convenient to express the character of such fluids in the form of a ratio $\dfrac{\text{Titre with fowl cells}}{\text{Titre with guinea pig cells}}$, F/G. Typical F/G ratios for primary amniotic fluid virus are 0/40, 10/120, 20/320. In addition to the low titre, the type of agglutination of fowl cells differs from that given by active allantoic virus. There is very little cohesion between the cells and the pattern of agglutination at the bottom of the tube slips readily so that the picture by the time deposition is complete is little more than a central button of cells with a ragged granular edge.

When a primary fluid of this character is subinoculated undiluted to another embryo by the amniotic route, infection will be evident in two days and the fluid obtained will agglutinate fowl cells almost as actively as guinea pig cells; a typical F/G ratio would be 240/320 and the fowl cell agglutination is of typical active character. If instead of simple subinoculation a full titration of the fluid is made using both allantoic and amniotic routes of inoculation, a clear indication of the process at work can be obtained. Table 1 gives a schematic version of the type of result obtained — actual examples can be seen in the paper by BURNET and BULL (1943).

Table 1. *Analysis of a primary amniotic fluid culture of influenza virus A.*

	Dilutions of fluid F/G 10/320						
	10^{-2}	10^{-3}	10^{-4}	10^{-5}	10^{-6}	10^{-7}	10^{-8}
Amniotic route	D D D	D D D	D D O	D O D	O O O	O O –	– – –
Allantoic route	D D D	D D D	D D –	D – –	– – –	– – –	– – –

In the table, O represents a positive fluid with an F/G ratio $< \frac{1}{5}$, while D represents fluids with a higher ratio almost always close to unity. It will be seen that those dilutions which give D fluids in the allantoic cavity, will also infect by the allantoic route and here also give D fluids. The higher dilutions, 10^{-5}, 10^{-6}, 10^{-7}, however, behave like the original human material giving infection of O type only in the amnion and failing to infect the allantoic cavity.

If the same procedure is repeated with one of the O fluids from the higher dilutions, the same picture will be obtained of D fluids at low dilutions, O at higher. The process can be continued indefinitely, in our experience for at least twenty passages.

Results of this general character can be obtained with all A strains isolated by amniotic inoculation from human material. Minor modifications will be described later. The form taken by these results is probably typical of all virus mutations — it is only the sharp indication of change given by the haemagglutination results that makes this particular example so favourable a one. The initial human inoculum is a very small sample of the virus particles which have been responsible for the infection. In all probability they are of uniform O character, i. e. if present in sufficient concentration they would give a fluid agglutinating guinea pig or human but not fowl red cells. The particles infect the susceptible cells in contact with the amniotic fluid and multiply. Within the increasing population mutations occur as in all organisms at some characteristic rate. The O—D mutation that we are concerned with probably occurred about once in every 10^4 new virus particles produced with the strain BEL used in most of our work. Once the new type appeared, it has a potentially greater rate of multiplication than the original, but if as is usually the case, the D mutant appears after extensive O infection has taken place, opportunity for its multiplication will be reduced by the interference phenomenon of Henle and Henle (1944). In general, any embryo into which only O particles were inoculated will give a fluid of O character, but containing a proportion of D particles; any embryo receiving both O and D will give rise to a predominantly D fluid owing to the greater speed of multiplication of the D form even when initially it is associated with a large number of O particles.

For details of the behaviour of influenza A strains, the following references may be consulted: Burnet and Bull (1943), Burnet and Stone (1945), Anderson and Burnet (1947). Minor complications arise in different strains by the existence of intermediates ω and δ between O and D forms or by a lower advantage of growth rate of D over O. Amongst strains isolated in 1946, the O form grew fairly well in the allantoic cavity and by a little manipulation it was possible to select an O strain which would grow quite satisfactorily in that situation.

Although the details have not been fully worked out, our general experience has been that in ferrets O virus multiplies freely but gives rise to a proportion of D mutants.

Very little study of the O—D change has been made in other laboratories. HIRST (1947b) agrees that most A strains on primary isolation are in the O form but was unsuccessful in maintaining the O character by passage at limiting dilution. DUDGEON (personal communication) found that strains of influenza A isolated by the amniotic method in England during 1946—7 were initially in the O phase and this seems also to have been FRANCIS'experience (FRANCIS et al 1947) in America. On the other hand considerable numbers of primary isolations by the allantoic method have been reported (THIGPEN and CROWLEY 1943). This on our view would mean only that a sufficiently large inoculum had been used to ensure that a few D mutants were already present. (BURNET and STONE, 1945). All who have compared the two methods agree that the amniotic is far more sensitive (about 10^4 times according to Hirst 1947b).

Nothing closely corresponding to the O—D change has been observed with influenza B, although large numbers of strains have now been isolated by the amniotic method. Most strains as isolated show nearly equal fowl and guinea pig titres and fully adapted strains either have similar ratios or show a tendency to lower guinea pig cell titres. Growth occurs much more easily in the amniotic than in the allantoic cavity but it is usually a simple matter to adapt the strain to full allantoic growth. Both in A and B strains a virus adapted to the allantoic cavity has lost none of its ability to multiply in the amniotic cavity — as might be expected it actually multiplies much more freely than the original virus.

2. Adaptation to mice.

When influenza A virus isolated by amniotic inoculation and adapted to allantoic passage is inoculated intranasally in mice no lesions are produced unless a very large dose is given. The virus however multiplies freely and even minimal doses will produce solid immunity to a mouse-adapted strain of the same antigenic type. If passage is made at 2 or 3 day intervals there is almost always a rapid increase in lesion producing power. This has recently been studied in detail by HIRST (1947b). He finds that the virus content of successive passage lungs remains essentially constant when judged by power to infect chick embryos, although in the first passage or two no lesions are produced while in the later ones lesions can be produced in mice to approximately the limiting dilution capable of infecting the embryo. His interpretation is very similar to the one we have given for the O—D change viz. that the original "egg" virus multiplies readily but gives rise to mutants which are selectively favoured by the new environment and rapidly replace the original form. He finds further that on occasion at least the antigenic character of the mouse virulent form which develops is demonstrably different from that of the original "egg" virus. In our experience all mouse adapted strains have conformed to D phase character but HIRST mentions one in which a guinea pig cell titre very much higher than that for fowl cells developed during mouse passage.

Owing to the technical difficulties involved we have made few efforts to determine whether O virus can be adapted directly to mice. There is no doubt that fairly small doses of the strain Bel O will immunize mice (BURNET and BULL 1943) but no passage experiments were done with this strain. With a particularly stable and hence rather atypical O strain isolated in 1946 (IAN) a comparison of O and D phase material showed that the former failed to establish itself on passage while the D phase material showed normal adaptation to produce typical lesions. (BURNET unpublished, 1947).

The adaptation of influenza B to mice is apparently a most unpredictable affair. The classical LEE strain was isolated after a long series of "blind" ferret and mouse passages by Francis, but the strain TM obtained by Magill from the same epidemic was readily transferred from ferret to mice by usual procedures. Our own experience with B strains isolated in chick embryos (Burnet, Stone and McCrea 1946) which is paralleled by that of Dudgeon et al (1946) and of Hirst (1947a) is that we have failed completely to adapt them to mouse passage even when they have reached the stage of producing good allantoic cultures.

In summary, influenza virus A exists as a human pathogen in a particular form "O" which is incapable of growth either in the allantoic cavity or in the mouse lung. It regularly produces "D" mutants, sometimes directly, sometimes throug intermediate steps which grow readily in the allantoic cavity and with some further adaptation in the mouse lung. All comparative studies of influenza A strains in the literature have been made with D variants and their implications for the human pathogenic strains must be made with caution.

Althoug influenza B virus does not show the characteristic changes in haemagglutinating behaviour, human strains cannot be grown in the allantoic cavity until after passage by the amniotic method or in ferrets, and essentially the same principles apply as for influenza A.

III. Differences amongst influenza virus strains which have been adapted to normal laboratory passage.

In any collection of laboratory strains of influenza virus differences may be found in almost any of the observable characters of the virus but it is always very difficult to be sure whether the observed differences between strains have been inherent since isolation or have developed during laboratory manipulation. From the practical point of view antigenic differences are by far the most important and most of this section will be concerned with this aspect. Even more important epidemiologically may be intrinsic differences in human virulence and in capacity to give rise to variants of differing virulence. Unfortunately evidence bearing on this point can only be obtained from the very limited amount of work that has been done on experimental human infections. Other differences concern the ability of strains to agglutinate red cells from which "receptors" have been partially removed and the correlated activity in destroying mucoid inhibitors of haemagglutination. Minor differences amongst strains in regard to "toxicity" (Henle and Henle, 1946) and heat stability (Salk, 1946) will not be further discussed.

1. Antigenic differences amongst strains of influenza virus.

There is no doubt about the existence of significant antigenic differences amongst influenza A strains, but it is extremely difficult to provide a satisfactory account of the extent, type and significance of these differences. The difficulty resides principally in the fact that antigenic differences determined by one method may not be evident by another which appears to be equally valid. From the practical point of view the important antigenic differences are those which concern the human subject. Two strains are significantly different when an attack by one provides no immunity against infection by the other a few weeks afterwards, and when a vaccine prepared by one provides immunity against its own type, but not against the other. For obvious reasons these criteria are extremely difficult to apply in practice. A more practical approach is to determine the neutralizing power of a human convalescent serum against both the homologous type and other strains whose antigenic character is being studied.

The following account of the antigenic differences amongst influenza strains is therefore based primarily on the findings with human material.

As far as natural active immunity is concerned the only important finding is the sharp and apparently absolute distinction between influenza A and B. BURNET, STONE and ANDERSON (1946) have described two cases in which persons suffered consecutive B and A influenza at two to three weeks interval. Both attacks were typical and in one individual both strains of virus were isolated and identified. Serological changes in the patients were typical and corresponded to the type of virus isolated. No suggestion seems ever to have been made that there is any significant common factor between influenza viruses A and B and the epidemiological evidence speaks strongly for the same conclusion.

It is not easy to define influenza A virus. The evidence suggests rather strongly that within the group currently accepted as influenza A there are viruses which will actively immunize against their own subtype but not against others nominally similar. Our standard procedure in Melbourne for the identification of influenza A or B viruses has been to test newly isolated strains against paired human sera from known cases of influenza A and B, using the inhibition of haemagglutination as the actual test. Based on this experience we would define influenza A as comprising *those influenza virus strains which resemble the type strain of Smith Andrewes and Laidlaw W. S. in being inactivated to significantly high titre by a standard adult human convalescent serum chosen for its breadth of action and to a much lower titre by the "acute" serum taken from the same patient in the first days of his illness.*

On this criterion the standard American swine influenza virus, Swine 15, of Shope is an influenza A strain.

Similarly influenza B is defined as those strains which resemble FRANCIS' strain LEE in being inactivated by a similar type of human convalescent serum. No influenza strains have been described which fail to fall into type A or B by these criteria.

There are two rather distinct problems involved in the use of human immune sera to determine significant antigenic differences between influenza virus subtypes within the major A and B groups.

The first concerns the often times sharp differences in the specificity or conversely the "breadth" of different sera obtained from the same epidemic or after the same vaccination procedure.

The second is to know whether the antigenic differences observed, necessarily with partial or completely adapted laboratory strains, were present in the original human form of the virus strains concerned.

The two problems are necessarily interrelated but an attempt will be made to discuss them separately.

a) The varying breadth of activity in human convalescent sera.

There is general agreement that there are relatively large antigenic differences between the classical strains WS and Swine 15 and between either of these and the common type of influenza A virus isolated between 1934 and 1944 (including PR8, Melbourne, CHRISTIE, TALMEY, BEL and many others). The question of minor differences between the strains in this last group will be deferred. All present indications are that 1946—47 A strains from Australia and America will also be accepted as another major subtype but only preliminary information on the American work has reached us. The Australian strain CAM will be taken as provisionally representative of this group. Only on the basis of such recognized differences is it possible to discuss the breadth of action of human convalescent or post vaccination sera.

The fact that human convalescent sera from the same epidemic may differ in the range of strains neutralized was first suggested in 1935 by Andrewes, Laidlaw and Smith. They observed that when random human sera were studied for neutralizing actively against the strains WS and Swine, both adult and children's sera were effective against WS but that sera from children under ten years showed no inactivating power against Swine virus. The tests were made in mice against mouse adapted strains. It was not at that time recognized that an attack of influenza A in an adult produce an increase in antibody titre against Swine virus. This was shown conclusively in 1938 by Stuart-Harris, Andrewes and Smith, and has been abundantly confirmed since.

Burnet and Lush (1937) using strains of Melbourne, WS and Swine 15 adapted to produce foci on the chorioallantois found that amongst random human sera with moderate to high amounts of antibody against "Melbourne" which was taken as representative of strains recently active in Australia, those from adults usually had commensurate titres against WS and Swine, but children's sera showed much lower titres against the heterologous strains.

No direct confirmation of these findings in regard to the relation between age and breadth of response has been published, but using antihaemagglutinin titrations, Adams, Thigpen and Rickard (1944) found that convalescent children sera were in general relatively specific in action. Anderson (1947) has recently studied the breadth of antibody response in a group of adolescents convalescent from influenza A after a school epidemic. A wide variety of patterns of response was found but if attention is confined to the homologous strain CAM and the three "classical" strains, Melbourne, WS and Swine 15, there is a regularity in the results which can be illustrated in Table 2.

Table 2.

Patient	Antibody rise against strains			
	CAM	Melbourne	WS	Swine 15
TOW	32+	0	0	0
CAM	8+	16+	1	1
MEY	8	4	4+	0
IAN	32	5	12+	12+

Figures show the ratio $\dfrac{\text{convalescent serum titre}}{\text{acute serum titre}}$

O No neutralization shown by either serum.

The only two children under ten years in the series showed a specific response involving only the homologous strain CAM, but amongst the boys from fourteen to nineteen years who comprised the rest of the series there was no relation between age and the degree of specificity of response.

Bodily and Eaton (1942) studied the antibody in human convalescent serum using the strains Melbourne, PR8, WS and Swine and including both antihaemagglutination and mouse neutralization tests. They found that titres against Melbourne and PR8 were closely similar but that against WS and Swine, although some individuals gave high titres, most gave considerably lower ones and some with definite antibody rises against PR8 showed no increase against WS or Swine. In all probability the strains of influenza A current in California in 1940/41 were closely related to Melbourne and PR8. In examining sera from persons immunized with vaccines based respectively on PR8 and Melbourne, the same general findings were obtained but in addition a significant advantage in favour of the homologous strain was evident as between PR8 and Melbourne antibodies.

A study of the neutralizing power of human convalescent sera for different influenza A strains by FRIEDEWALD (1944) is of particular interest in showing that absorption methods can be used and that the homologous virus will absorb all antibody including that active against relatively distant strains such as WS and Swine. Absorption by heterologous strains removed the corresponding activity of the serum without much influence on its activity against homologous and more closely related strains. Haemagglutination methods were used in this work.

The only comparable studies with influenza B convalescent sera are those of BURNET and STONE (1946). Using a homologous strain and the standard LEE strain, antihaemagglutinin titrations indicated that the sera could be divided fairly sharply into specific and non specific, some neutralizing only the homologous strain and others showing equivalent titres with both. Had other subtypes of B been available it seems probable that different grades of "breadth" would have been observed quite comparable to the findings with influenza A.

On the basis of these findings we should consider that the classification of subtypes of A or B viruses should be based on their reaction with selected specific sera from each epidemic under consideration, just as the diagnosis of A or B should be based on the reactions with selected "broad" sera.

b) The significance of antigenic differences observed with human immune sera.

Using the human sera shown in Table 2 or others with eqivalent activity ANDERSON (1947), was able to distribute the other strains currently in use in Melbourne as shown in Table 3.

The position of the American 1947 strains is only provisional as they show some differences from CAM with certain sera. It is hoped to clarify the position by comparative studies with American convalescent sera from this epidemic.

Table 3. Classification of influenza A strains.

Type strains	CAM (homologous)	Melbourne	WS	Swine 15
Representatives	CAM, IAN (Aus 1946) MIL (Aus 1945) FMI (Am 1947)	Melbourne (Aus 1935) BEL, CKS (Aus 1942) GAT (Br 1936)	WS (Br 1935)	Swine 15 (Am 1931)

Place and year of origin is shown — Aus (Australia), Am (America), Br (Britain).

HIRST (1947a) has recently shown that apparently fortuitous new antigenic qualities can be introduced in the course of adaptation of influenza A strains to mice. This work was with ferret immune sera but it seems reasonably likely

that somewhat similar changes would be detectable with appropriate human
sera. Hirst claims that if strains isolated and maintained by chick embryo
methods only are used the strains from a given American epidemic are serologically uniform. This has been the impression obtained in Melbourne by Anderson
(1947) in working with strains from the rather scattered prevalence of influenza
A in 1945—46 and using specific human antisera. It is however strongly at
variance with the findings of Magill and Sugg (1944) on five human sera from the
1940—1 epidemic and strains of virus isolated from each of the patients providing
the serum. The virus strains were adapted to mice and complete cross neutralization
tests were made in these animals. Their tabulated results show very variable
degrees of neutralization against the different strains, the only consistent feature
being that the "acute" serum never showed any antibody against the infecting
strain. The general impression obtained is consistent with Hirst's view that
new minor antigenic qualities may appear on adaptation to mice and that human
antisera of varying breadth may show quite unpredictable activity against these
new antigenic components.

The question of the uniformity or otherwise of strains from a single epidemic
raised by Hirst was tested in some experiments with strains isolated in the
Melbourne 1945 epidemic of influenza B. None of these strains could be adapted
to mice and the allantoic fluids used should correspond to their normal antigenic
character. Two strictly homologous sera both non-specific in character were

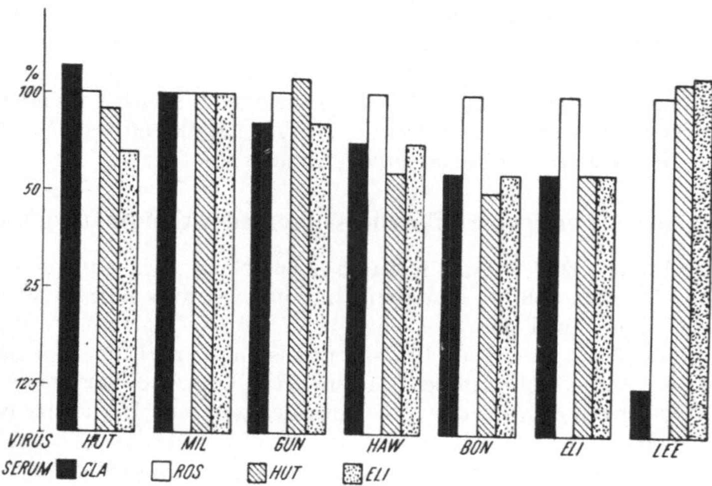

Fig, 1. Cross neutralization with human sera of influenza B strains. Two homologous serum-virus pairs
are included the other two sera being respectively a highly specific serum CLA and a broadly non-
specific one ROS. The titres are adjusted according to Hirst's method the titre shown by ROS
with each virus being taken as 100 %. The sharp difference between LEE and the other strains shown
by serum CLA should be noted.

used and two other sera, one "Ros" the most broadly active one available and
the other CLA a highly specific example. These were tested against five virus
strains from 1945, one BON from 1943 and LEE. In Figure 1 the antihaemagglutinin titres corrected by the methods introduced by Hirst (1943 a) are shown.
The sharp differentiation of LEE from the other strains by the specific serum
CLA is clearly shown. The differences amongst the other struins are slight but
at least BON and ELI approach a significant difference from the other three.

Much more work remains to be done in regard to the differences in subtypes from epidemic to epidemic. The matter is of the greatest practical importance as shown by the failure of a vaccine containing strains PR8 and Weiss to protect against the American 1947 epidemic (FRANCIS, SALK and QUILLIGAN, 1947).

c) Antigenic differences as determined by neutralization with immune ferret sera.

Ferret sera were adopted as standard reagents by SMITH and ANDREWES (1938) for use in mouse neutralization tests and by HIRST (1943 a and b) and FRIEDEWALD (1944) who used haemagglutination methods.

Both groups of workers agree that the strains WS, CHR, TAL, GAT and Swine are each almost unrelated to any of the others and that at least up to 1943 the other available strains of influenza A could be described as carrying one or more of the antigenic components present in these five type strains. The great majority of strains that were isolated during the period 1934—1942 in America and Australia were of intermediate type reacting particularly with CHR and TAL antisera. These include Melbourne PR8, Phila and all the 1940—41 American strains studied by HIRST as well as the Australian 1939 and 1942 strains.

HIRST (1947 a) has raised the important objection that mouse adaptation may seriously alter the antigenic quality of influenza A strains and has raised doubts as to the real status of the types obtained by SMITH and ANDREWES in the English epidemic of 1936—37. This objection is to a certain extent supported by the experience of BURNET and LUSH (1940). They found that the strain "Bundoora" at first gave highly specific immune ferret sera with very little action on Melbourne but after adaptation to mice the ferret immune sera were equally active on heterologous and homologous strains.

HIRST's conclusions are that using ferret sera influenza A strains of 1940—41 are uniform serologically as are 1943—4 strains but that there are significant minor differences between the types characteristic of the 2 epidemics. An addendum to his paper (HIRST, 1947b, p. 380) suggests that 1947 strains differ sharply. With influenza B strains his conclusions are the same as those of BURNET and STONE (1946) using human sera viz. that Lee (1940) differs sharply from the 1945 strains which amongst themselves are virtually uniform.

d) The use of immune rabbit sera.

If rabbits are injected intravenously with influenza viruses actively neutralizing antisera are obtained. According to MAGILL and FRANCIS (1938) the sera are relatively specific in differentiating strains, but the grouping obtained does not agree with that obtained from immune ferret sera. For instance, PR8, Phila and Melbourne fall into three distinct groups yet those who have worked with ferret antisera agree that all three are of intermediate type related to CHR and TAL. No other groups of workers have made any extensive use of rabbit sera and despite the general use of the rabbit in analogous work with bacteria it appears unlikely that the method is of any real value in connection with the problems of human influenza. It may however provide a valuable means of detecting antigenic changes occurring within a given strain under laboratory manipulation.

e) Cross immunity tests in mice.

It is a simple matter to immunize mice solidly against infection with a given strain of virus either by giving living or killed virus of the same strain by a route other than the intranasal, or by giving living unadapted virus intranasally. The immunity produced is less against heterologous strains, but a standard method for comparing the antigenic characters of an extensive series of strains is not available. The results obviously depend on many factors notably the dosage of the immunizing injection and the lesion producing power of the challenge strain in relation to its content of virus. In general, Smith and Andrewes (1938) found that the immunity response was broader than would be expected from the results with ferret serum neutralization tests.

2. Variations in the enzymic activity of different influenza virus strains.

As a development of Hirst's (1942, 1943) work on the nature of haemagglutination by influenza viruses we (Burnet, McCrea and Anderson, 1947, Burnet, 1948) have recently developed the conception that the initiation of infection by influenza viruses (and others including the viruses of mumps and Newcastle disease of fowls) depends on the action of an enzyme incorporated in the virus surface. The substrate of this enzyme can be any one of a number of glandular mucins and mucoids. Presumably there is some atomic configuration common to all substrate molecules but the nature of this common aspect is as yet unknown. The most direct means of demonstrating the enzymic activity is by the use as substrate of a highly purified mucoid from the contents of certain pseudomucinous cysts of the ovary. The mucoid is an extremely potent inhibitor of heated influenza B virus (Francis, 1947) being active to a dilution of 1 : 4 million. The inhibitory effect is rapidly and progressively destroyed by influenza viruses. The same enzymic action is responsible, as Hirst recognized, for the elution of virus from red cells to which it had been adsorbed and for the destruction of the cell "receptors".

Viruses differ very considerably in their enzymic activity, the strain Lee being in our experience more active than any of the influenza A strains that we have tested. This subject is under active study at present and it is not yet possible to make any comprehensive statement as to the range and significance of the differences between strains. Earlier work on the adsorption and elution of influenza and related viruses from red cells, however, gave some indication of the nature of the differences involved. If red cells are treated with a given strain of virus and maintained for some hours at 37^0 to allow complete elution of the virus they are found to be insusceptible to agglutination by the virus used. In most cases however, they are still agglutinated by some viruses of the group and not by others. If an extensive range of virus strains is studied in this way it will be found that they fall into a linear series, which we have called the receptor gradient. Cells rendered insusceptible to agglutination by a given virus are insusceptible to all viruses which precede it in the series but may be agglutinated by any of the viruses which follow it. With the strains we have used the order is Mumps, Newcastle disease, influenza A (Melbourne), A (WS), B Lee, A Bel, Fowl plague, Influenza A (Swine 15) and B (MIL), [Burnet McCrea and Stone, (1946)]. This series appearsto be determined primarily by differences in the readiness with which the nine virus strains are adsorbed to progressively modified substrate. All strains are adsorbed sufficiently firmly to the unmodified cell receptor to produce haemagglutination but as the substrate is progressively damaged by enzyme action either of the viruses or by the similar soluble enzymes produced by *Cl. welchii* or *V. cholerae* adsorption of viruses becomes successively too weak to allow visible haemagglutination. It has not yet been possible to determine whether enzymic activity runs parallel with adsorptive capacity. In general apparent enzymic activity on a purified mucoid shows approximately the same order of strains as the receptor gradient but this is most readily explicable on the assumption that enzyme activity can only be manifested after adsorption and will therefore naturally parallel adsorptive activity for the particular substrate under study.

Examination of O and D phases of the same influenza A strains has shown that with human cells the O phase precedes the D phase in the gradient, but the difference is not great. Using fowl cells a partial O phase (ω) virus giving a weak fowl cell agglutination comes first in the gradient. Minimal treatment of the cells by any of the enzymes will tender them inagglutinable by such virus.

There are hints that the specificity of adsorption to different types of mucopolysaccharide may be modified by passage in different hosts but no definite statement on the point is yet possible. There is now a large body of evidence from our laboratory (STONE, 1948a, b, FAZEKAS DE ST. GROTH, 1948) that the initiation of infection in the susceptible cells of the mouse lung or the chick embryo follows essentially the same pattern as the in vitro action on the red cell. It is not impossible that future work may indicate that differences in adsorptive power and enzymic activity on the primary cell substrate are closely correlated with virulence.

IV. Further modification of influenza viruses induced by laboratory manipulation.

In the first section of this review it has been stressed that considerable modification of the human pathogen is necessary before an influenza virus can be studied by the standard laboratory methods involving growth in the allantoic cavity or the mouse lung and the use of haemagglutination methods. But in addition a number of further modifications have been produced by deliberate laboratory manipulation, in general by the continued passage of the virus in some unusual environment.

Three topics call for discussion (1) changes produced in the course of prolonged passage on the chorioallantois (2) the development of neurotropic strains of virus and (3) the rather scanty evidence in regard to the changes in human pathogenicity associated with different forms of laboratory passage.

1. The strain Melbourne isolated by ferret inoculation in 1935 was transferred from third ferret passage material (a) to mouse passage, and (b) to continued passage on the chorioallantois of the chick embryo. The mouse strain Melbourne is similar to any other well adapted mouse strain of influenza A virus. The egg passage strain showed during 1935—6 a steady development of pathogenicity for the new host (BURNET, 1936). Discrete focal lesions gradually became larger and more numerous and at about the 50th passage death of the embryo occured. Shortly afterwards haemorrhagic brain lesions appeared and at full adaptation the embryo was killed within 40 hours with generalized haemorrhages throughout the body. The strain M. E. (Melbourne Egg) has been maintained at this level since. It has virtually no pathogenicity for mice or ferrets but immunizes solidly against the mouse passage strain. It is neither infective nor immunizing in human

beings inoculated intranasally. It grows irregularly in the allantoic cavity. It has very little power to agglutinate fowl cells but actively agglutinates guinea pig or human cells (Burnet and Bull, 1944).

Strain WS has also been adapted with considerable ease to produce similar lesions in the embryo (Burnet and Lush, 1938, Stuart-Harris, 1939). It differs from Melbourne however, in retaining relatively unaltered its mouse and ferret pathogenicity as well as its capacity to grow freely in the allantoic cavity and to show normal haemagglutination.

Several other strains Swine 15, BUR and R (Influenza A) and Lee B produced after repeated passage well defined foci on the chorioallantois but showed no sign of producing the fatal haemorrhagic lesions of the other two strains. It should be added that except for WS which is slightly active no influenza strains produce definite lesions on the chorioallantois on first inoculation on this site.

2. Stuart-Harris (1939) showed that a substrain of WS which had developed the power to produce "haemorrhagic encephalitis" in chick embryos could be further adapted to produce symptomatic infection in the mouse brain. Twelve rapid passages were necessary before deaths with symptoms of encephalitis occurred. With a few more passages deaths were consistently produced, usually in five or six days. Francis and Moore (1940) confirmed the production of a "neurotropic" variant from WS with relative ease and were also able to produce a similar variant from Melbourne but failed with other influenza A strains. Andrewes (1942) used the neurotropic variant in studying interference phenomena in tissue culture.

3. The human pathogenicity of laboratory strains of influenza virus is fairly obviously of a low order since laboratory infections are most uncommon. Most workers (Salk et al, 1944, Francis et al, 1944) who have studied the matter have used prolonged breathing of air containing atomized virus to produce infection. This must represent many thousands of times the dose received in natural contagion and capacity to produce symptoms and antibody under these circumstances can hardly be taken as an index of pathogenicity.

Our own experience in Melbourne using a single brief spraying of virus into nose and throat has been extensive but not productive of much useful information. Typical influenza was produced in two of three subjects with low antibody level and or sharp antibody rise in the third with virus which had had one ferret passage and one amniotic passage only (Burnet and Foley, 1940). It may have still contained a considerable proportion of O phase virus but at the time haemagglutination methods had not been developed.

Tests were made with influenza A virus after various numbers of amniotic passages, in volunteers chosen for low antibody content. A number showed minor symptoms and antibody rise but no typical clinical attacks were produced.

The B strain LEE similarly administered gave a very regular antibody rise in subjects with initially low antibody level (Bull and Burnet, 1943). Subsequent tests of the same subjects showed no further antibody rise, but it was noteworthy that minor symptoms were practically as common after the second as after the first inoculation.

It is highly desirable that human experiments should be made to determine whether the O phase of influenza A is as appears possible the only form with full human pathogenicity. All that can be said at present is that no investigator has ever demonstrated that laboratory material is capable of producing typical contagious influenza in human subjects. It is almost certain that the capacity to do so is rapidly lost on mouse or embryo passage.

One final point of interest concerns the fortuitous appearance of new characteristics not obviously related to the type of manipulation used. Previous mention has been made of the appearance of new antigenic characters on mouse passage and the inability of the "Melbourne-egg" strain to agglutinate fowl cells. In the course of amniotic passage of the influenza strain R (BURNET and FOLEY, 1941) mouse pathogenicity originally negligible rose to a moderately high level at the 30th passage and thereafter slowly fell. The capacity to produce distinct foci on the chorioallantois followed an approximately similar course. Another strain "Bundoora" tested at the same time failed to show such development of virulence for mice or to produce chorioallantoic foci.

V. The general principles of virus variation as exemplified in influenza virus strains.

At the present time there is in progress an interesting controversy in regard to the nature of "adaptive" variation in microorganisms. This is a problem of great practical importance in its bearing on the development of strains of bacteria resistant to sulphonamides or to penicillin. On the one hand American workers appear to have proved conclusively that the appearance of resistance to bacterial viruses in *B. coli* is essentially equivalent to a gene mutation in higher forms (LURIA and DELBRUCK, 1943, DEMEREC and FANO, 1945). Similarly DEMEREC (1945) working with staphylococci has provided almost equally cogent evidence that the development of resistance to penicillin is also mutational in character. This example is of particular interest as the development of substantial resistance to penicillin is only possible in a series of steps. Starting with a normally sensitive strain of staphylococci mutation to a first level of resistance occurs at an approximately measurable low rate per generation. Once such a mutation has occurred the experimental conditions ensure that it will have an overwhelming advantage over the original form and provide a population in which further mutations toward a higher level of penicillin tolerance can occur with a frequency similar to that of the first step. By continuing the process with gradually increasing concentrations of penicillin a very highly resistant form can be obtained.

On the other hand, HINSHELWOOD (1945) has concluded from his studies of the adaption of coliform organisms to growth in media containing sugars not normally fermentable, that the adaptive changes result from specific modification of all cells and not from the selection of appropriate variants.

Both groups of workers appear to have provided cogent arguments for their interpretation of the phenomena with which they are concerned. There is at present no clear indication of any generalized interpretation which will cover the general character of adaptive changes in unicellular microorganisms. The recent development of the concept of "phenocopies" i. e. changes in adult structure resembling in detail those produced by genetic mutation, but produced by environmental manipulation and not inheritable, perhaps suggests an approach toward the reconciliation of the two points of view.

Our own work with influenza viruses strongly supports the occurrence of changes analogous to gene mutation as opposed to any lamarckian interpretation. The fact that merely by passaging an O virus at limiting dilution it can be maintained indefinitely in this form is hardly open to any other interpretation. The question still remains however as to whether the phenotypic character of the D variant is a direct manifestation of genotypic change or whether the primary mutation is merely such as makes possible a subsequent adaptive change perhaps in enzyme character induced by some aspect of the cellular environment. With the development of concepts of the mucinolytic activity of a virus enzyme there is a considerable incentive to consider this possibility more closely. With further knowledge of the enzyme or enzymes concerned a direct attack on this problem may become possible. As far as the O—D change is concerned the one piece of evidence available speaks against a secondary adaptive process. It is possible to convert a O virus *in vitro* to a "phenocopy" of the D phase by simple exposure to mildly alkaline solutions (pH9) (Briody, unpublished). Since no adaptive or selective process is concerned in this phenomenon it is easy enough to conceive that the similar change observed *in vivo* is a direct result of genotypic change.

There is one characteristic of virus variation which we have not seen discussed but which has impressed us strongly on two occasions. This is the occasional inability to reproduce the course of variation in a given strain. Probably the two most striking changes in character that have been produced in influenza viruses are the development by chorioallantoic passage of the strain Melbourne-egg with its intense systemic action on the embryo and unusual haemagglutinating qualities (Burnet, 1936) and the development by Stuart-Harris (1939) of a neurotropic strain of WS. Recent endeavours in our laboratory to repeat these transformations have failed completely despite the closest possible adherence to the conditions formerly used. It seems that the potentiality to undergo a series of progressive mutations may be lost without any evident reason.

VI. The significance of variation in influenza viruses in regard to the epidemiology of influenza.

There are still too many phases of the natural history of influenza that have hardly been investigated, to allow more than speculation about the part played by virus variation in determining the intensity and distribution of influenza epidemics. One cannot doubt the importance of variation in this respect but a direct experimental or observational approach to the problem seems virtually impossible at present.

Some years ago Burnet and Clark (1942) working wholly at the epidemiological level attempted an interpretation of the major phases of the 1918—19 pandemic. The etiology of the pandemic is unknown but there is no reason to doubt that the essential infective agent was a virus or group of viruses closely similar to current strains of influenza A. It was suggested in the monograph mentioned that the pandemic was initiated in France by the appearance about April 1918 of strains with a qualitative difference in human virulence from those previously present. A second change (mutation) in the same direction of increased virulence took place about August possibly in more than one locality. There are hints derived from the fact that in some localities exposure to the spring epidemic protected highly against the severe autumn wave while in other areas no such protective effect was evident, that antigenic mutations also occurred during the progress of the pandemic.

All this however is speculation in the absence of the necessary facts. Shope (1944) has taken a completely different view in ascribing the severe autumn mave to the development of an effective synergism of the virus with strains of *Haemophilus influenza*. If Shope's identification of swine influenza with pandemic human influenza is correct there is much to be said in favour of this point of view.

In the event of a new pandemic visitation the possibility of antigenic variation would need be very closely considered since any attempt at control by vaccine prevention would be nullified by such an occurrence. It is of the greatest importance that every effort be made to determine what antigenic differences between strains are significant in the sense of requiring vaccines of different composition to provide effective protection against the natural disease produced by the types in question.

References.

1. ADAMS, J. M., M. P. THIGPEN and E. R. RICKARD: An Epidemic of Influenza A in Infants and Children. Clinical and Laboratory Investigations. J. Amer. med. Ass. **125,** 473 (1944).
2. ANDERSON, S. G.: Sporadic and Minor Epidemic Incidence of Influenza in Victoria, 1945—46. 2. The Breadth of Antibody Response following Influenza A Infection. Aust. J. exp. Biol. med. Sci. **25,** 243 (1947).
3. — and F. M. BURNET: Sporadic and Minor Epidemic Incidence of Influenza in Victoria, 1945—46. 1. Phase Behaviour of Influenza A Strains in relation to Epidemic Characteristics. Aust. J. exp. Biol. med. Sci. **25, 235** (1947).
4. ANDREWES, C. H., P. P. LAIDLAW and W. SMITH: Influenza: Observations on the Recovery of Virus from Man and on the Antibody Content of Human Sera. Brit. J. exp. Path. **16,** 566 (1935).
5. — Interference by One Virus with the Growth of Another in Tissue-culture. Brit. J. exp. Path. **23, 214** (1942).
6. — Thoughts on the Origin of Influenza Epidemics. Proc. Roy. Soc. Med. **36,** 1 (1942).
7. BANG, F. B.: Filamentous Forms of Newcastle Virus. Proc. Soc. exp. Biol. N. Y. **63,** 5 (1946).
8. — Newcastle Virus: Conversion of Spherical Forms to Filamentous Forms. Proc. Soc. exp. Biol. N. Y. **64,** 135 (1947).
9. BODILY, H. L. and M. D. EATON: Specificity of the Antibody Response of Human Beings to Strains of Influenza Virus. J. Immunol. **45,** 193 (1942).
10. BOVARNICK, M. and P. M. DE BURGH: Virus agglutination. Science, **105,** 550 (1947).
11. BULL, D. R. and F. M. BURNET: Experimental Immunization of Volunteers against Influenza Virus B. Med. J. Aust. **1,** 389 (1943).
12. BURNET, F. M.: The Specificity of Active Immunity in Mice against Influenza Virus. Brit. J. exp. Path. **19,** 388 (1938).
13. — and D. LUSH: Influenza Virus Strains Isolated from the Melbourne 1939 Epidemic. Aust. J. exp. Biol. med. Sci. **18,** 49 (1940).
14. — Influenza Virus Infections of the Chick Embryo Lung. Brit. J. exp. Path. **21,** 147 (1940).
15. — and M. FOLEY: The Results of Intranasal Inoculation of Modified and Unmodified Influenza Virus Strains in Human Volunteers. Med. J. Aust. **2,** 655 (1940).
16. — — Two Methods for the Detection of Influenza Virus in Human Throat Washings without the Use of Ferrets. Med. J. Aust. **1,** 68 (1941).
17. — — Influenza Virus Infections of the Chick Embryo by the Amniotic Route. 3. Changes in the Activity of Influenza Virus on Continued Amniotic Passage. Aust. J. exp. Biol. Med. Sci. **19,** 101 (1941).
18. — and E. CLARK: Influenza. Monograph of the Walter and Eliza Hall Institute, No. 4 MacMillan, Melbourne (1942).
19. — and D. R. BULL: Changes in Influenza Virus Associated with Adaption to Passage in Chick Embryos. Aust. J. exp. Biol. Med. Sci. **21,** 55 (1943).
20. — Immunization against Epidemic Influenza with Living Attenuated Virus. Med. J. Aust. **1, 385** (1943).
21. — and D. R. BULL: Reexamination of the Influenza Virus Strain "Melbourne Egg." Aust. J. exp. Biol. med. Sci. **22,** 173 (1944).

22. Burnet, F. M. and J. D. Stone: Further Studies of the O—D Change in Influenza Virus A. Aust. J. exp. Biol. med. Sci. **23**, 151 (1945).
23. — — The Significance of Primary Isolation of Influenza Virus by Inoculation of Mice or of the Allantoic Cavity of Chick Embryos. Aust. J. exp. Biol. med. Sci. **23**, 147 (1945).
24. — Haemagglutination by Mumps Virus: Relationship to Newcastle Disease and Influenza Viruses. Aust. J. Sci. **8**, 81 (1945).
25. — Influenza Virus on the Developing Egg. 1. Changes associated with the Development of an Egg Passage Strain of Virus. Brit. J. exp. Path. **17**, 282 (1936).
26. —, J. D. Stone and S. G. Anderson: An Epidemic of Influenza B in Australia. Lancet **1**, 807 (1946).
27. — — Serological response to Influenza B Infection in Human Beings: Differentiation of Specific and Nonspecific Type Reactions. Aust. J. exp. Biol. med. Sci. **24**, 207 (1946).
28. —, J. F. McCrea and J. D. Stone: Modification of Human Red Cells by Virus Action. 1. The Receptor Gradient for Virus Action in Human Red Cells. Brit. J. exp. Path. **27**, 228 (1946).
29. — — and S. G. Anderson: Mucin as Substrate of Enzyme Action by Viruses of the Mumps-Influenza Group. Nature, **160**, 404 (1947).
30. — The Initiation of Cellular Infection by Influenza and Related Viruses. Lancet **1**, 7 (1948).
31. Chunha, R., M. L. Weil, D. Beard, A. R. Taylor, D. G. Sharp and J. W. Beard: Purification and Characters of the Newcastle Disease Virus (California Strain). J. Immunol. **55**, 69 (1947).
32. Demerec, M.: Genetic Aspects of Changes in Staphylococcus Aureus Producing Strains Resistant to Various Concentrations of Penicillin. Ann. Missouri Bot. Garden. **32**, 131 (1945).
33. Dudgeon, J. A., C. H. Stuart-Harris, C. H. Andrewes, R. E. Glover and W. H. Bradley: Influenza B in 1945—46. Lancet **2**, 627 (1946).
34. Fazekas de St. Groth, S.: Destruction of Influenza Virus Receptors in the Mouse Lung by an Enzyme of V. Cholerae. Aust. J. exp. Biol. med. Sci. (in press).
35. Fano, V. and M. Demerec: Genetics, Physical Aspects. In Glasser, Medical Physics, Chicago (1944).
36. Francis, T. and T. P. Magill: Direct Transmission of Human Influenza Virus to Mice. Proc. Soc. exp. Biol. N. Y. **36**, 132 (1937).
37. — — Antigenic Differences in Strains of Epidemic Influenza Virus. II. Cross Immunization Tests in Mice. Brit. J. exp. Path. **19**, 284 (1938).
38. — and A. E. Moore: Study of the Neurotropic Tendency in Strains of the Virus of Epidemic Influenza. J. exp. Med. **72**, 717 (1940).
39. — A New Type of Virus from Epidemic Influenza. Science **92**, 405 (1940).
40. —, H. E. Pearson, J. E. Salk and P. N. Brown: Immunity in Human Subjects Artifically Infected with Influenza Virus. Am. J. Pub. Health. **34**, 317 (1944).
41. —, J. E. Salk, H. E. Pearson and P. N. Brown: Protective Effect of Vaccination against Induced Influenza A. Proc. Soc. exp. Biol. N. Y. **55**, 104 (1944).
42. — — and J. J. Quilligan: Experience with Vaccination against Influenza in the Spring of 1947. Am. J. Pub. Health. **37**, 1013 (1947).
43. — Apparent Serological Variation within a Strain of Influenza Virus. Proc. Soc. exp. Biol. N. Y. **65**, 143 (1947).
44. — Dissociation of Haemagglutinating and Antibody-measuring Capacities of Influenza Virus. J. exp. Med. **85**, 1 (1947).
45. Friedewald, W. F.: Qualitative Differences in the Antigenic Composition of Influenza A Virus Strains. J. exp. Med. **79**, 633 (1944).
46. Gordon, I.: Demonstration of Antigenic Differences Between Different Strains of Influenza B. J. Immunol. **44**, 231 (1942).
47. Henle, W. and G. Henle: Toxicity of Influenza Viruses. Science. **102**, 398 (1945).

48. HENLE, W. and G. HENLE: Interference between Inactive and Active Viruses of Influenza. 1. Am. J. med. Sci. **207,** 705 (1944).
49. HINSHELWOOD, C. N.: "The Chemical Kinetics of the Bacterial Cell." Oxford (1946).
50. HIRST, G. K.: Absorption of Influenza Haemagglutinins and Virus by Red Blood Cells. J. exp. Med. **76,** 195 (1942).
51. — Studies of Antigenic Differences Among Strains of Influenza A by Means of Red Cell Agglutination. J. exp. Med. **78,** 407 (1943a).
52. — Adsorption of Influenza Virus on Cells of Respiratory Tract. J. exp. Med. **78,** 99 (1943b).
53. — Studies on the Mechanism of Adaptation of Influenza Virus to Mice. J. exp. Med. **86,** 357 (1947a).
54. — Comparison of Influenza Virus Strains from Three Epidemics. J. exp. Med. **86,** 367 (1947b).
55. LURIA, S. E. and M. DELBRUCK: Mutations of Bacteria from Virus Sensitivity to Virus Resistance. Genetics. **28,** 491 (1943).
56. MAGILL, T. P. and T. FRANCIS: Antigenic Differences in Strains of Epidemic Influenza Virus. 1. Cross Neutralization Tests in Mice. Brit. J. exp. Path. **19,** 273 (1936).
57. — A Virus from Cases of Influenza-like Upper-respiratory Infection. Proc. Soc. exp. Biol. N. Y. **45,** 162 (1940).
58. — and J. Y. SUGG: The Significance of Antigenic Differences among Strains of the A Group of Influenza Viruses. J. exp. Med. **80,** 1 (1944).
59. SALK, J. E., H. E. PEARSON, P. N. BROWN and T. FRANCIS: Protective Effect of Vaccination Against Induced Influenza B. Proc. Soc. exp. Biol. N. Y. **55,** 106 (1944).
60. — Variation in Influenza Viruses. A Study of Heat Stability of the Red Cell Agglutinating Factor. Proc. Soc. exp. Biol. N. Y. **63,** 134 (1946).
61. SHOPE, R. E.: Old, Intermediate and Contemporary Contributions to our Knowledge of Pandemic Influenza. Medicine. **23,** 415 (1944).
62. SMITH, W. and C. H. ANDREWES: Serological Races of Influenza Virus. Brit. J. exp. Path. **19,** 293 (1938).
63. STONE, J. D.: Prevention of Virus Infection with Enzyme of V. cholerae. a) Studies with the Mumps-Influenza Group of Viruses in Chick Embryos. Aust. J. exp. Biol. med. Sci. (in press). b) Studies with Influenza Virus in Mice. Aust. J. exp. Biol. med. Sci. (in press).
64. STUART-HARRIS, C. H., C. H. ANDREWES and W. SMITH: Study of Epidemic Influenza with Special Reference to the 1936—37 Epidemic. Medical Research Council, Great Britain, Special Report Series No. 228 (1938).
65. — A Neurotropic Strain of Influenza Virus. Lancet. **1,** 497 (1939).
66. TAYLOR, R. M.: Detection of Human Influenza Virus in Throat Washings by Immunity Response in Syrian Hamster (Cricetus auratus). Proc. Soc. exp. Biol. N. Y. **43,** 541 (1940).
67. THIGPEN, M. P. and J. CROMLEY: Isolation of Influenza A by Intra-allanto c Inoculation of Untreated Throat Washings. Science. **98,** 516 (1943).

Immunity and Vaccination in Influenza.

By THOMAS FRANCIS, Jr., M. D.

Henry Sewall Professor of Epidemiology, School of Public Health, University of Michigan Ann Arbor, Michigan.

I. Epidemiologic Concepts.

For years efforts were made to settle the question of immunity to influenza by enumeration during an epidemic of the frequency of illness in members of a population who had suffered from influenza in earlier epidemics. The accumulated data were in most instances not susceptible to accurate analysis and highly contradictory opinions were derived. Difficulties have been constantly exaggerated by the lack of uniformity in clinical criteria for diagnosis of influenza and by the lack of diagnostic tests for confirmation or denial of the clinical impression. One conclusion was that influenza gave rise to no subsequent immunity but rather, rendered the subject more prone to attack. The frequent relapses were considered to be reinfections; the supposed second attacks in the same epidemic, infection of the same persons during different waves of the same pandemic, or the actual recurrence of epidemic waves were indicative of a lack of resistance. On the other hand, evidence of resistance is observed in the fact that during an epidemic, even with intimate exposure, a high proportion of the population escapes disease; in the observation made in stable populations, especially military units in 1918, indicating that persons affected during the mild spring wave escaped illness during the severe autumnal epidemic, or that those who took sick in the second wave were men who had joined the unit after the first wave; in the fact that the incidence of influenza is highest in children and lowest in the older ages. The lower incidence at older ages in 1918 was in reality considered by some to indicate a resistance carried over from influenza in 1889—90.

The essential conclusion warranted by the earlier epidemiologic studies is that immunity to influenza can result from the disease but that its duration is difficult to estimate. So long as influenza is considered an ill-defined miscellany of clinical symptoms, the problem of upper respiratory infection will remain confused. In this discussion the problem will be primarily considered in terms of the established viruses of influenza and of their common epidemic distribution, with full realization that the causative agent of certain outbreaks such as that of 1918 has not been established, though probably related to the viruses now recognized.

The absence of this information, in view of present knowledge of antigenic variation in types and strains of influenza virus, makes acceptance of earlier conclusions as to immunity difficult and much of this discussion will deal with information obtained since investigations with the viruses have been possible.

II. Resistance in Experimental Animals.

After Infection. The development of immunity after infection with influenza virus is readily demonstrable in the experimental animal. Shope (67) showed originally that *hogs* recovering from infection with swine influenza virus were staunchly immune to reinoculation with that virus or with a combination of the virus and H. *influenzae suis*. Moreover, he noted (Shope 68) that the recurrent epidemics seen each fall in the middle west of the United States involved essentially the young pigs while those that had been infected the previous year remained unaffected.

Mice may be resistant to reinoculation for 5 to 11 months after recovery from influenza virus infection (Francis 16) and can be shown to become resistant even after receiving virus in human throat washings or strains unadapted to mice which induce no gross pathologic lesions. And after infection with PR8 human-mouse passage strain of Type A it was early found that active immunity to swine influenza virus (S15) developed in mice even though neutralizing antibodies to the latter were absent or present in only small amounts (Francis and Shope 17). A similar result occured when swine virus was followed by the PR8 strain. Interestingly enough, immunity is to some extent proportionate to the size of the infecting dose or to the severity of infection. Burnet (Burnet and Clark 5) has stated that using egg-adapted Melbourne strain, non-virulent for mice, the degree of resistance is related to the dosage and that while 1000 egg-infective doses intranasally immunized against 100 MLD of the antigenically homologous strain as much as 100 million doses were required to give immunity to the heterologous W. S. or swine strains. Eaton and Pearson (13) have observed with mouse-unadapted ferret strain a difference in the size of dose required to give immunity to homologous *vis a vis* heterologous but the differences were of much smaller order. Intranasal inoculation was more effective than intraperitoneal.

The relation between clinical disease and immunity has been more effectively studied in the *ferret*. Smith, Andrewes and Laidlaw (73) promptly demonstrated that the ferret recovered from a first infection was for a time immune to reinfection with the same strain. But the period of immunity is transient, so that after two to three months reinoculation with the same strain gives rise to fever, clinical illness, and destruction of the respiratory epithelium. Nevertheless, the animal still retains a definite degree of immunity, illustrated by the absence of typical pulmonary lesions, by reduction in severity of illness and by the increased difficulty in recovering virus from the reincoulated animal (Francis and Magill 19). Moreover, the duration of immunity appears to be longer if resistance is tested by the less severe method of exposure to infected animals (Smith, Andrewes and Stuart-Harris 74). In this species, too, there are suggestions that the degree and duration of resistance are related to the severity of the primary disease (Francis 18). For instance, ferrets recovering from nearly fatal infection appear to withstand a second test more effectively than animals primarily undergoing mild subclinical disease. This may be related in part to the dosage of virus involved and partly to the extent of tissue injury and repair which ensues. And a solid immunity can be heightened and prolonged by repeated intranasal inoculations of active virus given at such intervals as to avoid evidence of clinical disease. Furthermore, subclinical infection of itself is capable of inciting immunity.

Concomitant with the development of active immunity in the experimental animal, antibodies, detectable by various serologic methods, to influenza virus appear and the general evidence causally links the two phenomena. This correlation has furnished tools applicable to the study of immunity induced by vaccination of experimental animals.

After Vaccination. Shope (67) first reported in 1932 that immunity to intranasal inoculation of swine influenza virus could be demonstrated in *swine* given intramuscularly active swine virus in the form of hog lung suspension. Later studies demonstrated that swine could be immunized, even subcutaneously, with that virus maintained by passage in ferrets or mice. Using the subcutaneous route mice and ferrets were rendered immune only when virus from the homologous species was used but intraperitoneally the species effect was unimportant (Shope 69).

Using strains of virus of human origin it was readily shown that subcutaneous inoculation of virus gave rise to much less effective immunity in mice than intraperitoneal injection and that repeated inoculations resulted not only in more solid immunity than did single injections [Francis and Magill (19); Andrewes and Smith (1); Shope (69)], but also in the development of antibodies to distantly related strains.

On the other hand, solid immunity to infectious inoculum is exceptional after vaccination of ferrets with human strains. Nevertheless, the ferrets exhibit a resistance similar to that of the animal losing immunity after infection in that the lungs remain normal (Shope 69) and when tested by contact exposure, the effectiveness of the resistance is more apparent (74). Moreover, after vaccination antibodies may be present in a titer comparable to that observed in the recently recovered, immune animal but in the partially immune state they mount rapidly to new heights when subjected to infection, whereas the fully immune animal does not exhibit this secondary antibody response (Francis 18). It seems quite clear that some additional protective mechanism is at play in the early post-infectious period which diminishes sharply with time. The fact that the vaccinated animal or that with waning immunity after infection commonly responds with more rapid onset of illness and exaggerated antibody production strongly suggests an accelerated reaction of resistance.

In recent years most studies have been concerned with virus grown in fertile eggs and chick embryo or the extra-embryonic fluids have furnished the major source of material for immunization procedures. Certain difficulties have been observed in that antigenic deviation of strains takes place rather rapidly in egg passage and the pathogenicity of strains becomes modified in egg passage. Furthermore, there has been an increasing tendency to substitute estimations of antibody response for measurements of immunity. With egg material the degree of immunity elicited in other species appears to be less marked than when homologous animal passage material is employed, as seen in the example of vaccination of swine by Beard and his associates who failed to demonstrate distinct immunity to swine virus after subcutaneous vaccination (Mc Lean, Beard and Beard 49) in sharp contrast to the earlier clear cut results of Shope (67). Moreover, with egg passage material Sugg and Magill (83) have failed, according to their one report, to obtain cross protection after infection of ferrets with closely related strains, again in contrast to the marked cross protection earlier demonstrated between swine and human strains. All these results strongly suggest a narrowing in the virulence for other species of strains maintained steadily in egg passage and of a limited effect in those animals upon specific tissues, including those producing antibodies.

There has been a continued resistance to the idea that immunity can be properly engendered by vaccination with a non-infectious preparation of a virus.

In the case of influenza virus studies with a variety of inactivating procedures such as formalin, ultraviolet light, detergents, have shown that virus rendered non-infectious still possesses an effective immunizing property. SMITH, ANDREWES and LAIDLAW (75) first demonstrated the retention of this effect in filtrates of mouse lung suspension when small concentrations of formalin were employed but in most instances a definite reduction in immunizing capacity of the inactivated material has been observed, approaching a ten-fold reduction. However, with virus in allantoic fluid gentle procedures of inactivation appear at times to exert a negligible effect upon antigenicity. The quantitative level of antibody response alone is not acceptable evidence that antigenicity remains unimpaired since, as will be seen, they are not proportionately parallel.

III. Resistance in Man.

General Data. That resistance to influenza does develop in man seems evident from the increased incidence in younger ages, from the fact that every one does not take sick even under similar conditions of intimate exposure, from the observation that individuals may go through one epidemic only to take sick during a later one, thus showing a selective resistance. Moreover, the extensive unsuccessful attempts of ROSENAU and his associates (59, 48, 60) during the epidemic of 1918 to induce influenza in over 100 human volunteers strongly suggests that a resistant group was being tested. How long this resistance remains in the individual, or in the group, under natural conditions, is not known. We have repeatedly commented that when careful histories are taken a considerable interval is usually noted between attacks of similar clinical character. Nevertheless, the lack of cross immunity between Types A and B virus and the variation in strains obviously interferes with any simple effort to interpret resistance in a general population on the basis of history. There is, however, a lack of data concerning the frequency by repeated attacks due to the same agent. There is evidence of the same individuals undergoing more than one attack with essentially the same virus under laboratory conditions. Thus, there is no direct evidence of the duration of immunity in man after naturally acquired infection.

The relative similarity of influenza in man and in the experimental animal also strongly suggests that there is at least a period of firm immunity after the disease which may continue as a comparatively long period of increased resistance under natural conditions of exposure. The parallel between the development of circulating antibody and the development of resistance with recovery suggests that the heightened antibody level in a population for several years after an epidemic may reflect a continued resistance. But in a disease which has essentially a necrotizing action upon a superficial epithelium it seems probable that the effective factors of resistance are those available at the portal of entry where the virus may reach cells remote from the antibody in the blood and that the antibody in the respiratory secretions together with the susceptibility of the cells are the deciding factors. The latter influence may well be related to the seasoning observed in closely associated groups and to a lessening of incidence at older ages.

Experimental Infection. The first significant effort to test susceptibility and resistance with influenza virus in man were those of SMORODINTSEFF and his associates (76) who subjected 72 volunteers to test by inhalation of mouse passage Typei A nfluenza virus. Of these, 20 per cent became ill with clinical influenza; all the latter were individuals with low antibody titers whereas none of those with high antibody titers took sick.

The possibility of immunization by modified infection has many attractive features. It represents a principle which has been most effective in prophylaxis against smallpox and yellow fever viruses. It offers the advantage that if given by the natural respiratory route virus multiplication could take place and increase the antigenic dosage furnished the subject by the wider distribution of virus. Furthermore it could simulate the actual disease process so as to affect susceptible cells and obtain any additional benefit of a refractory stage or of interference. The intranasal route also offers a simple channel for administration.

A number of studies have demonstrated that spraying of influenza virus cultivated in the egg is capable of inducing a clinical infection and that after the experimental disease a sharp rise in circulating antibodies takes place. On the other hand, a relatively large series of subjects given tissue culture fluid of the PR8 strain of virus by instillation into the nose or by packs soaked in virus have exhibited no signs of disease and extremely irregular antibody responses (Francis, unpublished). Similarly Henle et al have noted the limited effect gained by instillation of virus which caused clinical illness when sprayed (38). Chalkina (10) reported rather uniform antibody responses with minimal clinical symptoms but these are not in keeping with the experience of Burnet and Foley (7), Burnet and Lush (6) and those of Francis (20), who failed to obtain significant antibody rises with strains which did not give rise to signs of illness. The majority of the studies have been concerned essentially with the antibody response. But one series tested the clinical reactions of a group when sprayed with Type B virus four months after an initial spraying with the same virus (26). A third of the resprayed individuals again had illness of the same severity as in first test even though their antibody titers remained well above the original level. The majority did, however, exhibit a reduced response but evidence of staunch immunity after the four months interval was not pronounced. In one other series, however, spraying with Type A virus two months after a first exposure elicited no illness. Henle, et al (38) observed a second clinical attack in 5 of 9 subjects sprayed 9 months after a first experimental infection with Type A virus. Bull and Burnet (4) have reported that subclinical infection with attenuated Type B virus gave rise to considerable resistance to reinoculation but most of the data were serologic and not fully controlled. Nevertheless, Burnet and his associates (8, 9, 51) have repeatedly reported that mild infection of this sort stimulates antibody production in persons with low titer and suggest that they may be the ones who are most likely to take sick in epidemic times.

Recently the effect of repeated spraying with concentrated inactive virus was studied in children (Quilligan and Francis 53). Repeated sprays resulted in somewhat higher circulating antibody titers than did a single spray and the levels were better sustained. But a single subcutaneous infection containing one-eighth as much virus as was given intranasally gave higher and more persistent titers. Henle, Henle and Stokes (37) also noted that subcutaneous vaccination was more effective antigenically than intranasal infection with mild virus.

In experimental animals there is evidence that mild intranasal infection can give rise to a more effective resistance than much larger doses by other routes. But up to the present the results obtained with intranasal virus in man have not yielded results of sufficient uniformity or promise to indicate its practical application in prophylaxis. Nevertheless, the possibility remains that active virus, stabilized at the proper level of infectivity and pathogenicity and given by the respiratory route, can be utilized in the prevention of influenza. In the experimental animal it is clear that immunity can be increased and extended by repeated exposures to virulent virus which does not in the resistant animal give rise to clinical disease.

Immunity from Vaccination. The initial studies of vaccination of human subjects with influenza virus were undertaken to determine whether in man the administration of active virus by the subcutaneous or intracutaneous routes would incite infection and to gain evidence of the degree of antibody response which could be elicited. Utilizing the PR8 strain of Type A virus it was found that virus propagated in chick embryo-Tyrode's culture given by either route did result in the production of neutralizing antibodies to titers comparable with those observed in convalescent patients (FRANCIS and MAGILL 24, 25). The peak was reached in the second week and persisted at a relatively high level during the observation period of six months. There was little evidence that additional doses after the peak was reached had any significant influence in the adult subjects. And even at that time the suggestion appeared that those with the higher titers initially did not exhibit as great increments of increase as those with lower titers.

STOKES, CHENOWETH, WALTZ, GLADEN and SHAW (79) prepared Berkefeld filtrates of swine and Type A human virus from 10 per cent mouse lung suspensions and then inoculated members of a children's institution with three doses intramuscularly, while retaining an uninoculated group of approximately twice the size as controls. During an outbreak of respiratory disease in February, 1936, they recorded an incidence of 12.5 and 12.4 per cent of febrile illness in the controls and in the swine vaccinated groups, respectively, but only 2.7 per cent in those receiving the PR8 vaccine although only 31 per cent of this latter group showed an increase of antibodies after vaccination. The nature of the disease was not clearly established but it did occur in a season which was subsequently shown to have been largely occupied by influenza B. Nevertheless, certain serologic tests suggested that some cases of influenza A were occurring. The incidence of afebrile disease was uninfluenced.

The following year (STOKES, MC GUINNESS, LANGNER and SHAW 80) an expanded study in similar institutions was carried out with culture virus of the PR8 strain. In this instance some strains of influenza virus, Type A, were recovered and a reduction of febrile illness was observed varying from one-third to three-fifths as much in the vaccinated as in the controls. There was, therefore, an indication that, with active virus, vaccination in children had been influential in reducing the incidence of influenza and the suggestion was made that some of the differences in effect observed in the different institutions might be related to the length of interval between inoculation and appearance of disease.

In the same winter SMITH, ANDREWES and STUART-HARRIS (74) attempted a controlled prophylactic study of subcutaneous vaccination in military forces in England employing filtrates of 10 per cent suspension of the W. S. strain from mouse lung, inactivated by 1 : 2000 formalin. The disease occured before vaccination was fully carried out and the incidence was low. No effect was noted. TAYLOR and DREGUSS (84) observed no significant effect in their study with the same vaccine. It is of interest, however, that Type A virus was recovered from vaccinated individuals of both series and attention was drawn to the fact that the new strain differed from the W. S. of the vaccine. A polyvalent vaccine similarly prepared was employed in 1938—39 but no evidence of protection was observed in the group under study (STUART-HARRIS 81, 82).

The next extensive studies were those begun by HORSFALL, LENNETTE, RICKARD and HIRST (44) with material prepared from chick embryo tissue previously inoculated with PR8 strain of Type A virus and a strain of canine distemper virus. The minced embryo suspension was inactivated with 1 : 4400 formaldehyde and 1.0 cc doses were given subcutaneously to human subjects in a number of institutions. From 30 to 60 per cent of the populations, totalling some 16,000, were vaccinated; the remainders served as controls. Although a difference in total incidence among the vaccinated and controls was observed during an epidemic period, there was a significant reduction in only two of ten groups; in two other groups the incidence was higher among the vaccinated. Some of the ineffectiveness was attributed to the lack of potency of one large batch of the vaccine. Brown, Eaton, et al (3), using a similar preparation reported an inconstant but final reduction in incidence from 25 per cent in controls to 13 per cent in the vaccinated. Dalldorf, .Whitney and Ruskin (12), in a limited study of the same material noted no difference. Siegel, Muckenfuss, et al (70), employed different vaccine preparations through three successive outbreaks of influenza A in 1937, 1939, and 1941 but observed no difference between vaccinated and unvaccinated groups.

At this stage, then, there was little consistent evidence from field trials that subcutaneous vaccination was effective in protection against influenza in epidemic times, even though the studies had adequately shown that vaccination with various materials could induce an increase in antibody titer comparable to that observed after infection. The persistence of good levels of antibody for a period of several months was the rule. And continued emphasis was placed upon an apparent correlation between the height of the level of antibodies and resistance to the disease. The only conclusions which could be drawn were, therefore, that if influenza was not prevented in the vaccinated individuals it was because a) the vaccine was not of sufficient potency to excite antibody levels to uniformly adequate heights; b) the strains in the vaccine were not sufficiently similar to the epidemic ones; c) vaccination and the antibodies measured in the blood were not capable of controlling influenza.

There were obvious reasons for suggesting that the materials employed in the vaccination studies were not particularly potent in virus content. Eaton and Martin (14) noted that the titers obtained with the complex influenza-distemper vaccine were not as high as those in convalescent patients. Quantitative determinations of antibody levels in an attempt to appraise the nature of clinical illness occurring in the study groups had not been followed extensively until the studies of Rickard, Horsfall, Hirst and Lennette (57). These in the end were difficult of interpretation because a large number of cases with high titers who showed no increase in convalescence were classified as a different disease, influenza Y. This conclusion was dependent to some extent upon Horsfall and Rickard's conclusion (45) that the serologic response of the patient convalescent from influenza A was uniformly effective against all strains of Type A. Hence, if no increase was observed the causative agent must be of another type.

It had been demonstrated, however, in mice vaccinated intraperitoneally with active virus, that there was a progressive increase in the active immunity obtained as the amount of virus in the inoculum increased up to the point where maximal infecting doses were resisted (Francis 18). Subsequent to the foregoing vaccination studies in man, Hirst, Rickard, Whitman and Horsfall (39) carried out a study comparing the effect of different preparations containing different amounts of active or inactive virus obtained from infected chick embryo or from allantoic fluid. The two most concentrated vaccines were prepared by high speed centrifugation. Using adequate numbers of human subjects, and measuring their antibody levels by the Hirst anti-hemagglutination technic, they found with active influenza virus that the antibody titers attained increased as the amount of virus in the inoculum increased. The increase was not a direct proportion, however, as shown in Table 1.

Table 1.

Vaccine 56 contained the equivalent of 0.05 cc of PR 8 allantoic fluid;
„   55   „   „   „   „ 0.5 cc „   „   „   „
„   57   „   „   „   „ 5.0 cc „   „   „   „
„   64   „   „   „   „ 24.0 cc „   „   „   „
                                    and 10.0 cc of W. S.   „   „

An increase of 10 times the dose between vaccines 56 and 55 gave but a 20 per cent increase in titer, and to obtain levels in the same range as those observed in patients convalescent from influenza A, twice as high as that reached with unconcentrated fluid, ten times as much virus was required (vaccine 57) as was present in straight allantoic fluid (vaccine 55). A sixfold increase in the amount of virus administered beyond that quantity resulted in no significant enhancement of the titer. These results clearly indicated that unconcentrated, infected allantoic fluid definitely stimulated antibody production but that there was a sharp increase in the effectiveness as a concentration approximately ten times greater was approached, beyond which there was little additional gain. In the case of Type B virus a concentration only four times greater than that of straight allantoic fluid was required. One other feature of interest was that the titers fell rapidly in six weeks after vaccination, a point at variance with other studies. One can ask whether the process of concentration of itself influenced the antigenic potency by removal of stabilizing substances. Addition of the distemper virus had no influence. Moreover, when unconcentrated allantoic fluid was inactivated the antigenicity in terms of anti-hemagglutination was not influenced.

In contrast a series of preparations of formalinized allantoic fluid tested in 1941 showed sharp deterioration of immunizing potency for mice even though antibody was still elicited in man (26). Inactivation of mouse lung virus has usually resulted in a definite decline of immunizing potency to about $1/10$ the original even with mild procedures such as unltraviolet light or soaps. SMITH, ANDREWES and LAIDLAW (75) had stated that certain formalinized preparations appeared to be as effective as active ones, but they were not thoroughly titrated.

In 1942 HIRST, RICKARD and WHITMAN (40), and HARE, McCLELLAND and MORGAN (33) demonstrated that virus could be concentrated from allantoic fluid by freezing and then thawing at a low temperature. In the latter state virus precipitated and separated from other constituents of the fluid. Both groups of investigators (HIRST, et al (40); HARE, MORGAN, JACKSON and STAMATIS 34) demonstrated that after inactivation a concentrated vaccine induced better antibody responses than had been obtained with unconcentrated vaccine and as good, at least, as with active centrifuged material.

The Commission on Influenza of the Board for the Investigation and Control of Influenza and other Epidemic Diseases in the Army proceeded at this stage to untertake studies, in 1942—43, of the effect of concentrated vaccine containing influenza viruses of both Types A and B, inactivated with formalin. The studies were devised so that alternate numbers of persons would receive vaccine or a control inoculation. The records of vaccination were kept separate so as to avoid reference to them until the study was completed. Determinations of the antibody levels induced by the vaccine were made and close, continued clinical observation was kept of the illnesses in the experimental groups. In addition, material for virus study and for serologic determinations was obtained from cases of respiratory disease.

For the study at Cornell University Medical College allantoic vaccine concentrated by freezing was employed. Difficulties in the preparation were presented by the problem of maintaining and rendering large amounts bacteriologically sterile.

For the study in Michigan a process was employed which takes advantage of the adsorption of influenza virus to the erythrocytes of the chick. McCLELLAND and HARE (47) in their original paper suggested this as a method of concentration since the virus could be detected adsorbed to the red blood cells. By this means the major portion of the virus can be removed by the erythrocytes from the allantoic fluid and most of the normal protein remains in the fluid. HIRST (41) had shown that the virus eluted from the erythrocytes readily at room temperature or higher. The agglutination is carried out in the cold and the supernatant fluid is removed; to the mass of agglutinated red cells is added $^1/_{10}$ the original volume of physiological NaCl solution and then the material is placed at 23° to 37° C for 1 to 2 hours during which time virus elutes from the red cells (FRANCIS and SALK 27). The vaccine was then constituted so that 1.0 cc contained the PR8, Type A, strain obtained from 5.0 cc of the allantoic fluid and the Lee, Type B, strain obtained from 5.0 cc of fluid. Virus was inactivated by formalin 1 : 2000 and a mild bacteriostatic was added.

Because of the preceding two year cycle of influenza A observed in the United States the winter of 1942 was expected to present an outbreak of that disease. It did not occur. Late in that season a mild, unsuspected prevalence of influenza B was detected in March 1943, largely by serological tests. No influenza A appeared in the study group of approximately 8,000. It had been possible, however, to demonstrate that excellent antibody responses had occured to both types of virus and that in four months a decline of about one-third from the peak titer had taken place (SALK, PEARSON, BROWN, SMYTH and FRANCIS 63).

Since it had been found, however, that infection could be induced experimentally without serious risk it seemed desirable to test the efficacy of vaccination by that procedure. HENLE, HENLE and STOKES (37) had reported that a mixture of allantoic fluids containing PR8, W. S. and Melbourne strains of Type A virus inactivated by formalin, given four months earlier, or one of PR8 alone given two and one-half weeks earlier, protected all but one of 44 children against inhalation of a recently isolated strain cultivated in eggs (F—99). Ten of 28 controls took sick.

On the other hand, our studies of resistance to Type B influenza virus had shown that four months after receiving a spray of this virus in allantoic fluid a majority had symptoms when sprayed again with the same virus, and one-third had just as severe illness the second time even though their antibody levels were markedly elevated above the original titers. Nevertheless, a group of 66 vaccinated subjects was selected and sprayed with a strain of Type A virus relatively close antigenically but not identical. Of 36 unvaccinated controls half developed fever of 100° F or more and symptoms of influenza; of 28 vaccinated 4½ months before, 32 per cent had similar experience; among the 38 vaccinated two weeks prior to testing, 6 or 16 per cent, had fevers of 100°: but none higher — in sharp contrast to the other groups (FRANCIS, SALK, PEARSON and BROWN 21, 22).

The results in a group of 96 tested by inhalation of test B virus (SALK, PEARSON, BROWN and FRANCIS (61, 62) were that 11, or 41 per cent of 27 unvaccinated took sick; but of the 79 vaccinated either 4½ months, 4 weeks, or at both times before testing, only 8, or 10 per cent, had fevers of 100° F and none reached 101°. These results indicated a more effective influence of subcutaneous vaccination than was observed against influenza A, although the B test may have been somewhat less severe. In comparison with the resistance exhibitel four months after actual intranasal infection the effect of subcutaneous vaccination was much more impressive.

In 1943 a more extensive investigation was made by the Commission on Influenza in a series of Army units in colleges throughout the United States with vaccine prepared by adsorption and elution but incorporating equal parts of PR8 and Weiss strains as the A component. The study comprised six different groups of investigators working in nine different universities with an effort to maintain comparable conditions throughout. In all but one area alternate men of each unit served as vaccinated and controls; the same lots of vaccine were employed throughout; the same basic plan of clinical observation and handling and of etiologic studies was maintained (11). There were 6263 vaccinated and 6211 inoculated controls. Most of the vaccination was done in late October and early November and was soon subjected to test by an epidemic of *Type A influenza*. In the total group an incidence of 2.2 per cent of hospitalized cases was observed in the vaccinated and 7.1 per cent in the control group, with a consistent and significant reduction in all but the one study in California where a number of deviations in the pattern of the study occurred and where EATON and MEIKLEJOHN (11e) attributed the lack of effect to the occurrence of an antigenically divergent strain of virus. In five of the nine units the incidence in the controls was 3.5 to 6 times as great as in the vaccinated, an effect which may be considered minimal since the frequency of illness in completely unvaccinated units was greater than that among controls of the vaccinated groups, thus suggesting a reduction in risk among the unvaccinated through the reducing of susceptibility in the group as a whole.

The results clearly demonstrated that a consistent and pronounced lowering in incidence of clinical influenza A was attained by the subcutaneous vaccination with inactive influenza virus.

In two locations (HALE and MCKEE 11d; HIRST, PLUMMER and FRIEDEWALD 11f) because the epidemic began about the time vaccination was done the curves of incidence of disease in vaccinated and controls could be followed. In the first week no differences were observed but after 6 to 7 days they diverged sharply as the incidence in the vaccinated decreased; this indicated that the prophylactic effect of vaccine began at a time when circulating antibodies are ordinarily beginning to rise.

In 1945 by virtue of the uniform vaccination of the entire personnel of the U. S. Army, and the occurrence of an epidemic of *influenza B* it was possible through the Commission on Influenza to gain information of the effect of the same type of vaccine against that disease. There were, at the University of Michigan, 109 cases in the unvaccinated naval unit of 1100 men per cent and only 7 in the 600 vaccinated men of the Army unit 1.1 per cent living and being observed under almost identical conditions (FRANCIS, SALK and BRACE 28). At Yale University with similar circumstances and numbers there were three cases 0.5 per cent among 550 vaccinated Army personnel and 132, or 12.5 per cent among 1050 unvaccinated naval students (HIRST, VILCHES, ROGERS and ROBBINS 42).

Table 2.

Location	No.	Vaccinated		No.	Unvaccinated	
		Cases	Incidence %		Cases	Incidence %
Michigan	600	7	1.15	1100	109	9.9
Yale	550	3	0.5	1050	132	12.5
Alabama	30	2	6.7	95	18	19.0
Washington	360	7	1.94	4280	352	8.23

A small group at the University of Alabama also showed a reduced incidence in the vaccinated (Friedman 32). Norwood and Sachs (52) observed a sharp reduction in an industrial plant in Washington.

Although these studies were not made up of alternate controls within the same units the groups were so similar in all other respects as to make them readily comparable. The difference in incidence in the two groups certainly appears to be the effect of vaccination. Further support for this conclusion is found in the fact that the vaccinated Army in the same geographic areas had a sharply lower incidence of influenza during the epidemic period than did naval personnel. The greater effect observed against influenza B than in the earlier studies against influenza A is in keeping with the results noted with experimental infection after vaccination and also with the readier immunizing effect of Type B virus in mice. In addition, the fact that the vaccinated units were completely vaccinated and the controls were entirely unvaccinated may play a role in enhancing differences in mass resistance of the two groups. This influence was exhibited despite the fact that distinct differences could be demonstrated in the serological character of the epidemic strains from that of the Lee strain in the vaccine.

1947. With a lapse of three years since the influenza A of 1943 an outbreak of that disease was anticipated in the winter of 1946—47. At the end of October, 1946 a vaccination study was again instituted at the University of Michigan, where 10,328 persons received eluate vaccine containing the same strains as in previous years; 7615 were unvaccinated. When influenza occurred during March 1947 no evidence of protective effect was demonstrated. The incidence in vaccinated subjects was 7.19 per cent; in the controls 8.09 per cent (Francis, Salk and Quilligan 29). Although the outbreak did not begin until four months after vaccination the evidence was clear that the antibody titers of the vaccinated individuals when measured against the vaccine strains, remained about the same level as that observed two weeks after vaccination. On the other hand, the titers of the vaccinated were no higher than those of the controls when tested against epidemic strains.

Further evidence of the inefficacy of the vaccine was observed in a high frequency of the disease in vaccinated groups even though comparable numbers of controls were not available (Sigel et al 71). Data from the Army as a whole yielded no evidence of efficacy (66). A controlled study by Fowle and Weightman (15) noted an incidence of 7.05 per cent in 1250 vaccinated and of 7.3 per cent in 794 unvaccinated. Loosli, Schoenberger and Barnett (46), observed the same incidence, 9.5 per cent, in 790 vaccinated and 1230 unvaccinated comprising a test of three different preparations of vaccine. Van Ravenswaay (56) observed 20.2 per cent incidence in 237 vaccinated, 27.8 per cent in 284 unvaccinated.

The absence of prophylactic effect was so clear-cut as to differ sharply from, and to enhance the significance of, the results of 1943 and 1945. Studies from numerous laboratories pointed out clearly the serologic difference of the epidemic strains from the PR8 and Weiss of Type A incorporated in the vaccine (72, 55, 71, 29, 46). Antibodies to the epidemic strains were not generally induced by vaccination or only to low levels although excellent responses to the vaccine strains were demonstrable. There was no significant difference in mean titers to the 1947 strains among vaccinated and unvaccinated persons in the acute stage of the disease and the antibody increase observed in convalescence was essentially the same in the two groups (FRANCIS, SALK and QUILLIGAN 29). Furthermore many sera from the 1943 epidemic which showed a marked rise to the PR8 strain failed to show an antibody increase to the 1947 strains. That the strains were of Type A was shown by the fact that the majority of convalescent patients exhibited an antibody rise to PR8 or other A strains; this was demonstrable by neutralization, hemagglutination-inhibition, or complement fixation tests. That vaccinated individuals showed less rise to the PR8 than to 1947 strains after infection is to be expected because of their high post-vaccination titers to that strain.

Although the interval between vaccination and epidemic occurrence was greater in 1947 than in 1943 and 1945 it does not seem likely that the time interval is the essential feature in the lack of effect of vaccination upon disease incidence but rather that the antigenic difference in the infecting strains was sufficiently great to escape the influence of antibody developed against the vaccine strains. This experience has been most profitable for it has furnished information upon certain problems which have been up to the present largely speculative.

In order to meet the antigenic variant the Commission on Influenza recommended that a representative of the United States 1947 strains, now designated A prime, be incorporated in the Army vaccine in place of the Weiss strain. The difficulties involved have clearly dampened the idea that a newly isolated strain can be put into mass vaccine production immediately, at least with present methods. It has been found that these strains are less well stabilized than the other standard strains, undergoing rapid variation in eggs, more sensitive to formalin, slower to adaptation in allantoic sac. Moreover they appear to be somewhat less effective antigens than PR8 and Lee under comparable conditions. A controlled study of the effect of the new combination has been carried out only with difficulty since the winter of 1947—48 provided little epidemic influenza. Nevertheless, one study conducted by SALK (64) for the Influenza Commission has compared the incidence of disease in groups receiving vaccine with or without A prime virus at a time when A prime influenza again arose in the Spring of 1948. A clearly significant reduction occurred in those receiving the new vaccine even though the overall incidence of the disease was low.

Passive Immunity. The prophylactic studies which have been discussed are those which tend to induce active immunity by modified infection with active virus or by vaccination with active and inactive virus. But if the superficial cells lining the air passages represent the vulnerable tissues it might be possible to protect them by applying antibodies at the surface where the virus alights. In other words the virus-inactivating effect of the respiratory secretions might be augmented by antibodies introduced directly into the respiratory tract. SMORODINTSEFF and SHISKINA (77) reported in 1938 that when immune serum was given to mice intranasally before the virus infection or even for a period thereafter it was much more effective in protection than a much larger dose given by other routes. This has been repeatedly confirmed but it is important to note

that with spraying of serum it has been necessary to use excessive amounts of
serum over long periods of time.

Smorodintseff, Gulmow and Tshalkina (78) reported that administration
of approximately 2.0 cc of immune animal sera by respiratory spray greatly
reduced the incidence of natural infection from 82 per 1,000 to 8 per 1,000 in
the treated. But using serum of a foreign species introduces the serious problem
of sensitization.

In one study by the Commission on Influenza, as yet unpublished, efforts
were made to prevent artificial infection with sprayed virus by large amounts
of inhaled human convalescent serum. No preventive effect was demonstrated.

This procedure must then await further investigation of its attractive and
promising possibilities.

IV. Factors Influencing Immunity by Vaccination.

Strain Differences. The experience of 1947 has apparently established the
fact that a single vaccination with certain strains of Type A virus is incapable
of eliciting antibodies and furnishing protection against certain other strains
of the same type. The problem remains to ascertain how similar strains must
be, to be encompassed within the influence of others. It has been repeatedly
suggested that the failure of earlier vaccination studies was related to differences
in the antigenic make-up of the vaccine and epidemic strains. Eaton and Meiklejohn (11e) interpreted the exceptional lack of effect of vaccine in California
in 1943 to be due to the fact that antibodies to the current Olson strain were
not adequately induced by the vaccine. And Salk, Menke and Francis (11g)
observed that the antibody response to the PR8 strain was not as full as to the
Weiss strain. On the other hand many of the strains of Type B influenza virus
encountered in 1945 were readily shown to differ considerably from the vaccine
strain, Lee, but the effect of vaccination upon the disease was pronounced
Francis, Salk and Brace 28).

It is difficult to transfer the meaning of strain differences as determined
in serological tests with animal sera to man, at least to adults, who presumably
have had repeated experiences with the disease. For just as the animal with
repeated inoculations develops a broader resistance and broader antibody effect
so the adult in response to infection gives rise to antibodies which overlap strains
which are quite different antigenically. This may be in part due to a genetic
heritage. Rickard, Thigpen and Adams (58) have observed the development
of antibodies to swine virus in the serum of infants undergoing their first infection with influenza A but Hare and Riehm (35) have pointed out that this
is less commonly observed in the first decade than later. But even in the experimental animal, vaccination tends to elicit a more strain-specific response than
infection does. Bodily and Eaton (2) showed the same effect in man, as did
Hare (36). This was more clearly seen in 1947 (Francis, Salk and Quilligan
(29) when vaccination with PR8 and Weiss strains induced little antibody production to the epidemic strains (now termed A prime) but the patients convalescent from A prime infection developed good titers against the PR8 strain.
It is conceivable that passage of strains for vaccine in eggs tends to maintain
a sharper strain specificity and that the breadth of response to these preparations
is limited in scope. Otherwise vaccination of the adult might well be expected
to give rise consistently to polyvalent antibody.

The fact that repeated inoculations of animals with a given strain increases
the capacity of their sera to neutralize other strains as well as their resistance

naturally suggests the same probability in man. On the contrary, however, in adults the common experience is that with reasonable amounts of virus in the vaccinating dose the complete response is obtained with a single injection and repeated injections at intervals of a few days or a week have no added effect. Revaccination after four months when the titer is declining has neither an accelerating or booster effect, nor does it strikingly affect the titers except of those who had failed to respond originally (FRANCIS, SALK, PEARSON and BROWN 23). Further in 1946, adults who had been vaccinated one year earlier showed no higher final titer to revaccination than those vaccinated for the first time (FRANCIS, SALK and QUILLIGAN 29). Hence the usual experience with human individuals is that a ceiling on antibody titer appears to exist and that this is not exceeded as a result of repeated inoculations even at intervals which in the experimental animal would give rise to an excess secondary response. Man behaves like a resistant animal serologically.

It is still conceivable that while the titer to the homologous strain is not appreciably heightened repeated inoculations could cause an increasingly greater antibody titer to heterologous strains. This possibility was studied in young children most of whom had been born since influenza A occurred in the area (QUILLIGAN, MINUSE and FRANCIS 54). They received repeated large or small doses of the PR8 strain subcutaneously at two week or two day intervals and the titers of antibody against different Type A and swine strains were measured. Interestingly enough the development of antibodies to the heterologous strains was entirely parallel to that for the homologous although the levels arrived at depended upon the closeness of relationship to the PR8. In children of four years or older the full effect was attained with a single large dose; but in the younger children the highest titer was not reached until after three large doses. When smaller amounts of virus were given, the peak was reached at all ages only after three doses. Additional inoculations caused no further increase, but rather a tendency to decline. It was of interest to note, however, that in contrast to the lack of antibodies to A prime strains induced in adults by standard vaccine, a number of the children receiving repeated doses of PR8 vaccine did develop antibodies to A prime strains. The results then failed to show that a progressive increase in heterologous antibody occurred even in inexperienced children except as it paralleled the development of homologous antibody. But they did demonstrate that children respond more slowly to the stimulus of vaccine than do adults.

It is clear that the factor of antigenic variation is one requiring continued consideration. The search is for strains of broad antigenic capacity which may be incorporated in vaccine and a continued study of the character of strains in circulation. It is not an imponderable task for the polyvalent reactivity of the human individual may be a helpful influence. This part of the problem is a whole field of investigation to which viruses, or most other infectious agents, have only become susceptible and constant thought must be given to the importance of variations other than antigenic constitution.

Relations of Resistance to Antibodies. Since the recognition that animals recovering from influenza virus infection develop antibodies, measurable by a variety of methods, and become resistant, the two phenomena have been linked as they have in most infectious diseases. And since man recovering from disease also develops antibodies it is assumed that he has acquired resistance, a probability supported by epidemiologic evidence. It is equally true, however, that the presence of antibodies in the blood is no guarantee of immunity for in man and experimental animals infection occurs even though considerable levels of antibody to the same virus are found in the blood. Apart from the problem of strain variation, induced infection with a single strain has clearly demonstrated that at all ranges of antibody level some individuals are resistant while others are susceptible. If only 30 to 40 percent of those with little or no antibody in the blood take sick when exposed to a standard dose of virus there is strong basis for presumption that other factors may be involved in determining who is to take sick.

The animal recovered from infection will in the resistant state accept a second exposure without further antibody rise while somewhat later without much change in antibody titer it develops clinical disease and its antibody level rises very sharply to exaggerated heights. But the resistant animal's immunity is reinforced by the new exposure so as to be more durable even without further accession of antibody level. In the animal after infection a given level of antibody has a greater correlation with staunch immunity than after vaccination. The crossing of immunity between strains such as PR8 and swine, after recovery, is not closely related to the level measured by neutralizing antibodies. We have but recently demonstrated the presence of well-marked resistance of mice who possessed little antibody to the heterologous strain as measured by agglutination-inhibition. It is quite clear, therefore, that immunity and antibody are evidence of experience with influenza virus but they are not causally interrelated quantitatively in a parallel manner.

On the positive side is the trend constantly observed under conditions of natural or induced infection: that the incidence of influenza in those with high antibody titers is consistently less than in those with low titers. In a review of the data from numerous studies in animals the same trend exists. Moreover, the trend exists in vaccinated or unvaccinated parts of the same human population (63). The effect of vaccine has been observed to begin about one week after inoculation when antibodies are developing. Hence there is consistent support for the interpretation that protective antibody is related to resistance in influenza as in other diseases.

There is no reason to postulate an independent, *specific* cellular immunity operating in the absence of antibody. Nevertheless, there is excellent evidence to support the probability that modification in the behavior and character of the susceptible cells plays a role in modifying susceptibility or in enhancing the efficiency of antibody. Following the concept that the effectiveness of circulating antibody in virus infections is dependent upon the mechanism of the disease process it is believed that circulating antibody exerts its prophylactic influence by becoming available in the respiratory secretions about the susceptible epithelial cells which the virus attacks. With recovery from disease or after proper vaccination the amount of antibody in the secretions increases. These factors may strongly influence whether antibody of the blood is, or is not associated with resistance (Francis 30, 31). They may also explain the difficulty, because of physiologic variations in the hosts, of establishing a numerical proportion between circulating antibody and immunity. There is no well defined level which separates the immune from the susceptible. But the trend, most readily measurable in the blood, still indicates, in general, an inverse relationship between circulating antibody level and resistance.

Influence of Dosage. As has earlier been mentioned the size of the dose of virus by intranasal infection or by other routes has an influence upon the degree of immunity and upon the broadening of the resistance. But in man there is little precise information except as it applies to antibody titers. Hirst, et al (39) recorded the small enhancement in antibody titer attained in man as the amount of virus in different preparations of vaccine was markedly increased. McClean, Beard, Taylor, Sharp and Beard (50) observed in swine that a 10—50 fold increase in the amount of virus must be administered to obtain a twofold increase in antibody. Salk (65) found with a single centrifuged preparation that 0.01 mgm of protein was less effective than 0.05 mgm but that there was no significant difference beween the effect of 0.05 and 0.50 mgm. When a dose of 2.0 mgm was used however the titer was but two to three times as high as that obtained with 0.05 or one-fortieth the dose. Thus in terms of antibody titer there is a broad zone of dosage with which differences are not observed.

In this respect the question of intracutaneous versus subcutaneous route is of interest. The original investigation of this point by Francis and Magill (25) indicated no striking difference in results even though smaller doses were given intracutaneously. This may again reflect the wide range of dosage which can give rise to similar antibody levels or may be indicative of a greater efficiency of the intracutaneous route. But this needs more complete comparative study.

It had been suggested that large doses could be used to advantage but this did not consider the pharmacologic toxic effect which appears clearly to be a property of the virus and limits the amount which can be used (Salk 65). The virus contents of the vaccines which have been recently used approach the practicable limit. In the studies of Salk the 2.0 mgm dosage was entirely too toxic. In children much smaller doses must be used. Nevertheless, there is a need for hesitancy in dropping the dose because a "sufficient" level of antibody can be reached with a smaller quantity. In animals larger doses give more resistance; the antibody titer of man is not a reliable index and until more complete information as to the efficacy of influenza vaccine is obtained it seems unwise to jeopardize the demonstration of a prophylactic influence by use of too little virus.

Duration of Effect. The duration of immunity following infection or vaccination is not established. Experimental infection of ferrets and man has demonstrated that a second exposure after 4—5 months may cause clinical disease though modified in character. Nevertheless, the possibility remains that milder tests by contact will in man, as in the ferret, demonstrate a longer period of firm resistance. It was suggested by Hirst, Rickard and Friedewald (43) in an analysis of the results of the 1943 studies, that the effect of vaccination in reducing the incidence of influenza among the vaccinated as compared with the controls had declined sharply in 6—7 weeks after vaccination. This interpretation seems to be extremely tenuous in that it suggests that the susceptibility of the vaccinated has declined while the resistance of the unvaccinated has increased sufficiently to reduce the attack rate to a level approximating that in the vaccinated. The real change was a decline in incidence in controls rather than an increase in the vaccinated. They also suggest that the lack of significant lowering of incidence in California (11 e) was due to the interval between vaccination and major prevalence. There are no other data to support this assumption. At the same time they present data to indicate a 35 per cent reduction in disease one year after vaccination of an institutional population. And contrary to their earlier report of rapid declines in antibody titer in three months they noted that at the end of a year the levels were still considerably elevated.

Frankly, the duration of resistance after vaccination of man is not known nor has it been possible to determine it adequately for the problem is still that of how much immunity is evoked by vaccination at any interval. Salk, Pearson, Brown, Smyth and Francis (63) reported observations during an epidemic of influenza A in an institution one year after vaccination. Among 1319 persons in unvaccinated wards an incidence of 12.4 per cent was recorded; among 1916 persons in partially vaccinated wards the incidence was 1.9 per cent. As mentioned earlier their antibody levels four months after vaccination had declined one-third from the peak level and after one year to one-half the peak but still five to eight times higher than before vaccination. In 1946 a group of veterans vaccinated in 1945 were found still to have titers five to six times higher than the rest of the population (Francis, Salk, Quilligan 29). In serological terms, therefore, the effect of vaccination is readily demonstrable for more than a year. The crucial question is, of course: Is this paralleled by resistance? Since evidence suggests that the higher antibody levels, whether derived from vaccination or natural infection, are associated with a decreased incidence of disease, the persistence of high titers for a year, at least, after vaccination, would indicate that an increased resistance was also retained. Thus the weight of evidence at present favors the conclusion that a beneficial effect of proper vaccination persists for several months to a year. There is as yet no evidence to warrant the idea that the effect is for only a few weeks. In the individual the result cannot be predicted but since influenza is essentially an infection of the group it seems likely that the production and retention of increased antibody titers in the population group will serve for mass protection for an extended period. If this be reinforced by natural exposure a firm effect may result.

This again constitutes a problem for further study and reliable data. And these features clearly indicate that the problem of immunity and vaccination in influenza must be considered one for continuing investigation. It would be extremely short-sighted to assume that the problem was near completion. The very questions of how little material can be effectively used and what may be the best preparations are but touched. One might say that materials of proper antigenicity and stability should be effective but up to the present many different types of preparation have not had satisfactory field trials for prophylactic effect. There has been difference of opinion as to the time at which vaccination should be done. Some estimates of the probabilities can be made by observing the time of recurrences. Despite opinion to the contrary it has not proved possible to vaccinate a general population adequately after the onset of an epidemic and for that reason the early autumn is the time of choice with the intent of preceding an outbreak. Vaccination of half a population, especially the youngest half, would raise the antibody level, and presumably the resistance, more completely than an epidemic comparable in intensity to those of recent years. But its application beyond special groups is at the present a matter for administrative decision.

It is an interesting commentary that since evidence has been obtained that vaccination can prevent influenza many objections have been raised. One is that the epidemics are too mild to warrant the expenditure of the necessary funds and effort. Another that the frequency of systemic reactions is too high; this statement is made without much thought of the evidence or of the preparations employed. The reactions resemble those occurring after typhoid vaccine but in recent years the frequency of systemic febrile reactions has been well below one per cent and of little importance in collegiate or industrial groups where absence from work from the disease is an important matter. The possibility of sensitization to the vaccinating material has been considered but up to the present has not been of significance. Nevertheless, inoculation of naturally sensitive persons must be done with caution or avoided.

Consequently, it can be said that subcutaneous vaccination with the concentrated materials employed has been shown to be of benefit in the prevention of epidemic influenza A or B. The lack of effect in 1947 emphasizes the other positive results and indicates the need for continued investigation and flexibility in the improvement of a procedure which will progressively establish its value.

Bibliography.

1. ANDREWES, C. H., and W. SMITH: Influenza: further experiments on the active immunization of mice. Brit. J. Exp. Path. **18**, 43, 1937.
2. BODILY, H. C., and M. D. EATON: Specificity of the antibody response of human beings to strains of influenza virus. J. Immunol. **45**, 193, 1942.
3. BROWN, J. W., and M. D. EATON, G. MEIKLEJOHN, J. B. LAGEN and W. J. KERR: An epidemic of influenza. Results of prophylactic inoculation of a complex influenza A-distemper vaccine. J. Clin. Invest. **20**, 663, 1941.
4. BULL, D. R., and F. M. BURNET: Experimental immunization of volunteers against influenza virus B. Med. J. Austral. **1**, 389, 1943.
5. BURNET, F. M., and E. CLARK: Influenza. Monograph Walter and Eliza Hall Institute, No. 4, 23, 1942.
6. — and D. LUSH: Influenza virus on the developing egg: The antibodies of experimental and human sera. Brit. J. Exp. Path. **19**, 17, 1938.
7. — and M. FOLEY: The results of intranasal inoculation of modified and unmodified influenza virus strains in human volunteers. Med. J. Austral. **2**, 655, 1940.
8. — Influenza B: II. Immunization of human volunteers with living attenuated virus. Med. J. Austral. **1**, 673, 1942.
9. — Immunization against epidemic influenza with living attenuated virus. Med. J. Austral. **1**, 385, 1943.
10. CHALKINA, O. M.: Immunological changes in the blood of men, vaccinated against the virus of epidemic influenza. Arch. d. Sci. Biol. **52**, 126, 1938.
11. (a) Commission on Influenza, United States Army. A clinical evaluation of vaccination against epidemic influenza. Preliminary report. J. A. M. A. **124**, 982, 1944.
    (b) FRANCIS, T., Jr.: The development of the 1943 vaccination study of the Commission on Influenza. Amer. Jour. Hyg. **42**, 1, 1945.
    (c) RICKARD, E. R., M. THIGPEN and J. H. CROWLEY: Vaccination against influenza at the University of Minnesota. Amer. Jour. Hyg. **42**, 12, 1945.
    (d) HALE, W. M., and A. P. McKEE: The value of influenza vaccination when done at the beginning of an epidemic. Amer. Jour. Hyg. **42**, 21, 1945.
    (e) EATON, M. D., and G. MEIKLEJOHN: Vaccination against influenza: a study in California during the epidemic of 1943—44. Amer. Jour. Hyg. **42**, 28, 1945.
    (f) HIRST, G. K., N. PLUMMER and W. F. FRIEDEWALD: Human immunity following vaccination with formalinized influenza virus. Amer. Jour. Hyg. **42**, 45, 1945.
    (g) SALK, J. E., W. J. MENKE, Jr. and T. FRANCIS, Jr.: A clinical, epidemiological and immunological evaluation of vaccination against epidemic influenza. Amer. Jour. Hyg. **42**, 57, 1945.
    (h) MAGILL, T. P., N. PLUMMER, G. SMILLIE and J. Y. SUGG: An evalution of vaccination against influenza. Amer. Jour. Hyg. **42**, 94, 1945.
12. DALLDORF, G., E. WHITNEY and A. RUSKIN: Controlled clinical test of influenza A vaccine. J. A. M. A. **116**, 2574, 1941.
13. EATON, M. D., and H. E. PEARSON: Quantitative aspects of homologous and heterologous active immunity to strains of the virus of epidemic influenza. J. Exp. Med. **72**, 635, 1940.
14. — and W. P. MARTIN: Analysis of serological reactions after vaccination and infection with the virus of influenza A. Am. J. Hyg. **36**, 255, 1942.
15. FOWLE, L. P., and J. WEIGHTMAN: Report of effectiveness of influenza virus vaccine A and B at Bucknell University, 1946—47. J. Lancet (Minneapolis) **67**, 388, 1947.

16. FRANCIS, T., Jr.: The immunology of epidemic influenza. Amer. Jour. Hyg. **28**, 63, 1938.
17. — and R. E. SHOPE: Neutralization tests with sera of convalescent or immunized animals and the viruses of swine and human influenza. J. Exp. Med. **63**, 645, 1936.
18. — Quantitative relationships between the immunizing dose of epidemic influenza virus and the resultant immunity. J. Exp. Med. **69**, 283, 1939.
19. — and T. P. MAGILL: Immunological studies with the virus of influenza. J. Exp. Med. **62**, 505, 1935.
20. — Intranasal inoculation of human individuals with the virus of epidemic influenza. Proc. Soc. Exp. Biol. and Med. **43**, 337, 1940.
21. —, J. E. SALK, H. E. PEARSON and P. N. BROWN: Protective effect of vaccination against induced influenza A. Proc. Soc. Exp. Biol. and Med. **55**, 104, 1944.
22. — — — — Protective effect of vaccination against induced influenza A. J. Clin. Invest. **24**, 536, 1945.
23. —, H. E. PEARSON, J. E. SALK and P. N. BROWN: Immunity in human subjects artificially infected with influenza virus Type B. Am. J. Pub. Health. **34**, 317, 1944.
24. — and T. P. MAGILL: Vaccination of human subjects with virus of human influenza. Proc. Soc. Exp. Biol. and Med. **33**, 604, 1936.
25. — — The antibody response of human subjects vaccinated with the virus of human influenza. J. Exp. Med. **65**, 251, 1937.
26. —, H. E. PEARSON, E. R. SULLIVAN and P. N. BROWN: The effect of subcutaneous vaccination with influenza virus upon the virus-inactivating capacity of nasal secretions. Am. J. Hyg. **37**, 294, 1943.
27. — and J. E. SALK: A simplified procedure for the concentration and purification of influenza virus. Science **96**, 499, 1942.
28. — — and W. M. BRACE: The protective effect of vaccination against epidemic influenza B. J. A. M. A. **131**, 275, 1946.
29. — — and J. J. QUILLIGAN, Jr.: Experience with vaccination against influenza in the spring of 1947. Am. J. Pub. Health. **37**, 1013, 1947.
30. — Factors conditioning resistance to epidemic influenza. The Harvey Lectures. New York, N. Y. Series **37**, 69, 1941—42.
31. — Mechanisms of infection and immunity in virus diseases of man. Bact. Rev. **11**, 147, 1947.
32. FRIEDMAN, L. L.: The value of influenza virus vaccine, Types A and B. Southern Med. J. **39**, 809, 1946.
33. HARE, R., L. McCLELLAND and J. MORGAN: A method for concentration of influenza virus. Canadian J. Pub. Health. **33**, 325, 1942.
34. — J. MORGAN, J. JACKSON and D. STAMATIS: Immunization against influenza A. Canadian J. Pub. Health. **34**, 353, 1943.
35. — and W. C. RIEHM: Long term variations in the titer of neutralizing antibody for influenza virus in the sera of adults and children. J. Immunol. **40**, 253, 1941.
36. — Active Immunity to Influenza Virus in the Mouse. J. Immunol. **40**, 267, 1941.
37. HENLE, W., G. HENLE and J. STOKES, Jr.: Demonstration of the efficiency of vaccination against influenza Type A by experimental infection of human beings. J. Immunol. **46**, 163, 1943.
38. — — — and E. R. MARIS: Experimental exposure of human subjects to viruses of influenza. J. Immunol. **52**, 145, 1946.
39. HIRST, G. K., E. R. RICKARD, L. WHITMAN and F. L. HORSFALL, Jr.: Antibody response of human beings following vaccination with influenza viruses. J. Exp. Med. **75**, 495, 1942.
40. — — — A new method for concentrating influenza virus from allantoic fluid. Proc. Soc. Exp. Biol. and Med. **50**, 129, 1942.
41. — Adsorption of influenza hemagglutinins and virus by red blood cells. J. Exp. Med. **76**, 195, 1942.

42. HIRST, G. K., A. VILCHES, O. ROGERS and C. L. ROBBINS: The effect of vaccination on the incidence of influenza B. Am. J. Hyg. **45**, 96, 1947.

43. —, E. R. RICKARD and W. F. FRIEDEWALD: Studies in human immunization against influenza; duration of immunity induced by inactive virus. J. Exp. Med. **80**, 265, 1944.

44. HORSFALL, F. L., Jr., E. H. LENNETTE, E. R. RICKARD and G. K. HIRST: Studies on efficacy of complex vaccine against influenza A. Pub. Health Rep. 56, 1863, 1941.

45. — and E. R. RICKARD: Neütralizing antibodies in human serum after influenza A. The lack of strain specificity in the immunological response. J. Exp. Med. **74**, 433, 1941.

46. LOOSLI, C. G., J. SCHÖNBERGER and G. BARNETT: Results of vaccination against influenza during the epidemic of 1947. J. Lab. and Clin. Med. **33**, 789, 1948.

47. McCLELLAND, L., and R. HARE: Adsorption of influenza virus by red cells and a new in vitro method of measuring antibodies for influenza virus. Canad. J. Pub. Health. **32**, 530, 1941.

48. McCoY, G. W., and W. DE RICHEY: II. series of experiments at San Francisco, November and December, 1918. Ibid, p. 42.

49. McLEAN, I. W. Jr., D. BEARD and J. W. BEARD: Studies on the immunization of swine against infection with the swine influenza virus. J. Immunol. **56**, 109, 1947.

50. — — A. R. TAYLOR, D. G. SHARP and J. W. BEARD: The relation of antibody response in swine to dose of the swine influenza virus inactivated with formalin and with ultraviolet light. J. Immunol. **51**, 65, 1945.

51. MAWSON, J., and C. SWAN: Intranasal vaccination of humans with living attenuated influenza virus strains. Med. J. Austral. **1**, 394, 1943.

52. NORWOOD, W. D., and R. R. SACHS: The protective effect of vaccination against epidemic influenza B in an industrial plant. Indust. Med. **16**, 1, 1947.

53. QUILLIGAN, J. J. Jr, and T. FRANCIS, Jr.: Serological response to intranasal administration of inactive influenza virus in children. J. Clin. Invest. **26**, 1079, 1947.

54. —, E. MINUSE and T. FRANCIS, Jr.: Homologous and heterologous antibody response of infants and children to multiple infections of a single strain of influenza virus. J. Clin. Invest. **27**, No. 5, 572, 1948.

55. RASMUSSEN, A. F., Jr., J. C. STOKES and J. E. SMADEL: The army experience with influenza 1946—47. Am. J. Hyg. **47**, 142, 1948.

56. RAVENSWAAY, VAN, A. C.: Prophylactic use of influenza virus vaccine. J. A. M. A. **136**, 435, 1948.

57. RICKARD, E. R., F. L. HORSFALL, Jr., G. K. HIRST and E. H. LENNETTE: The correlation between neutralizing antibodies in serum against influenza viruses and susceptibility to influenza in man. Pub. Health Rep. **56**, 1819, 1941.

58. —, M. P. THIGPEN and J. M. ADAMS: Antibody response to strains of influenza A and swine influenza viruses in the serums of infants experiencing their first infection with influenza A. J. Inf. Dis. **76**, 203, 1945.

59. ROSENAU, M. J., W. J. KEEGAN, J. GOLDBERGER and G. C. LAKE: I. Series of experiments at Boston, November and December, 1918. Hygiene Laboratory, USPHS Bull. No. 123, 5, 1921.

60. — —, DE W. RICHEY, J. GOLDBERGER, G. W. McCoY, J. P. LEAKE and G. C. LAKE: Ibid. p. 54.

61. SALK, J. E., H. E. PEARSON, P. N. BROWN and T. FRANCIS, Jr.: Protectiv effect of vaccination against induced influenza B. Proc. Soc. Exp. Biol. and Med. **55**, 106, 1944.

62. — — — — Protective effect of vaccination against induced influenza B. J. Clin. Invest. **24**, 547, 1945.

63. — — —, C. J. SMYTH and T. FRANCIS, Jr.: Immunization against influenza with observations during an epidemic of influenza A one year after vaccination. Am. J. Hyg. **42**, 307, 1945.

64. Salk, J. E. unpublished data.
65. — Reactions to concentrated influenza virus vaccines. J. Immunol. **58**, 369, 1948.
66. Sartwell, P. E. and A. P. Long: The army experience with influenza, 1946—47. Am. J. Hyg. **47**, 135, 1948.
67. Shope, R. E.: Studies on immunity to swine influenza. J. Exp. Med. **56**, 575, 1932.
68. — Swine influenza III. Filtration experiments and etiology. J. Exp. Med. **54**, 373, 1931.
69. — Immunization experiments with swine influenza virus. J. Exp. Med. **64**, 47, 1936.
70. Siegel, M., R. S. Muckenfuss, M. Schaeffer, H. L. Wilcox and A. G. Leider: Studies in active immunization against epidemic influenza and pneumococcus pneumonia at Letchworth Village. III. The results of active immunization against epidemic influenza from 1937 to 1940. Am. J. Hyg. **35**, 55, 1942. IV. Results in an epidemic of influenza A in 1940—41. Ibid. p. 186.
71. Sigel, M. M., F. W. Shaffer, M. W. Kirber, A. B. Light and W. Henle: Influenza A in a vaccinated population. J. A. M. A. **136**, 437, 1948.
72. Smadel, J. E.: Research in virus diseases. Bull. u. S. Army Med. Dept. **7**, 795, 1947.
73. Smith, W., C. H. Andrewes and P. P. Laidlaw: A virus obtained from influenza patients. Lancet. **2**, 66, 1933.
74. — — and C. H. Stuart-Harris: Studies on immunization of ferrets and mice. Medical Research Council. Special Report Series, No. 228. Section VI 125, 1938.
75. — — and P. P. Laidlaw: Influenza: Experiments on the immunization of ferrets and mice. Brit. J. Exp. Path. **16**, 291, 1935.
76. Smorodintseff, A. A., M. D. Tushinsky, A. I. Drobyshevskaya, A. A. Korovin and A. I. Osetroff: Investigation of volunteers infected with the influenza virus. Am. J. Med. Sci. **194**, 159, 1937.
77. — and O. I. Shishkina: Arch. Biol. nauk. (russ.) **52**, 132, 1938.
78. —, A. G. Gulmow and O. M. Tsalkina: Über die spezifische Prophylaxe der epidemischen Grippe durch Inhalation antigrippischen Serums. Ztschr. f. klin. Med. **138**, 755, 1940.
79. Stokes, J., Jr., A. D. Chenoweth, A. D. Waltz, R. G. Gladen and D. Shaw: Results of immunization by means of active virus of human influenza. J. Clin. Invest. **16**, 237, 1937.
80. —. A. C. McGuinness, P. H. Langner, Jr. and D. Shaw: Vaccination against epidemic influenza with active virus of human influenza. Am. J. Med. Sci. **194**, 757, 1937.
81. Stuart-Harris, C. H., W. Smith and C. H. Andrewes: The influenza epidemic of January-March 1939. Lancet **1**, 205, 1940.
82. — Influenza epidemics and the influenza viruses. Brit. Med. J. **1**, 209 and 251, 1945.
83. Sugg, J. Y., and T. P. Magill: Significance of antigenic differences among strains of influenza A virus in reinfection of ferrets. Proc. Soc. Exp. Biol. and Med. **63**, 1, 1946.
84. Taylor, R. M., and M. Dreguss: Experiment in immunization against influenza with formaldehyde-inactivated virus. Am. J. Hyg. **31**, 31, 1940.

Virus Pneumonia and Pneumonitis Viruses of Man and Animals.

By MONROE D. EATON.

Associate Professor of Bacteriology and Immunology, Harvard Medical School, Boston, Mass.

Formerly Director of the Virus Laboratory California State Dept. of Public Health.

Introduction.

A wide variety of viruses may cause pulmonary infections in man and in animals. Knowledge of virus pneumonia has developed not only from study of the human infections but also from the experimental production of this disease with various known viruses in animals, and the discovery of new pneumotropic agents in experimental animals during attempts to transmit the agents of human respiratory infections to them.

Since the fundamental tissue reactions to many different viruses are quite similar, virus pneumonia in man must be considered as a clinical and pathological syndrome rather than a disease entity. Epidemiological, clinical and pathological data may serve for a rather incomplete differentiation of various forms. Pneumonia produced by viruses in animals has many of the characteristics of the disease seen in man. The lethality, cellular composition, distribution, and rate of development of the lesions can sometimes be used as criteria for differentiating one virus infection from another. Certain identification can be made, however, only by isolation of the causative virus and by neutralization or cross immunity tests.

It is by no means certain that the tissue reactions seen in virus pneumonia can be produced only by viruses and not by other agents. The possibility that some of the features in the pathology of virus pneumonia may also be duplicated by physical or chemical influences, toxins, or even bacteria has been cited from time to time by various authors.

Historical.

I. Early observations on pathology.

One of the early descriptions of what is now known to represent a virus infection of the lungs was given by BARTELS in 1860. The pulmonary complications in an epidemic of measles were studied by this investigator. Psittacosis which produces a pneumonitis in man was recognized by RITTER and others about 1880 as a disease entity, then refered to as "pneumo-typhus". An attempt to classify various forms of bronchopneumonia according to clinical and pathological data was made by DELAFIELD in 1884. The acute catarrhal bronchitis of adults and children and the acute idiopathic bronchopneumonia described by him seem to have many resemblances to virus pneumonia. At about this time there were also a number of German publications describing a similar form of bronchopneumonia in children after diphtheria, measles, and whooping cough. Many cases of pneumonia following measles occurred in the United States during the War of 1812 an the Civil War. MacCallum in 1919 studied sections from the lungs of three cases from the Civil War. These specimens had been preserved in alcohol at the Army Medical Museum for over 50 years. The type of pneumonia was similar to that seen by the same author in 1918—19.

The etiology of the great influenza pandemics of 1889—90 and 1918—19 is unknown but it is probable that a virus was the causative agent. In the epidemic of 1889—90 Leichtenstern while noting the almost constant association of Pfeiffer's bacillus with the influenzal pneumonia, described the pathology of what is now considered a virus pneumonia. The disease process produced a suppurative exudate in the bronchi and round cell infiltration of the peribronchial connective tissue. It then extended into the surrounding alveoli with mononuclear infiltration of the lumina and septa.

The term acute interstitial bronchopneumonia was introduced by MacCallum in 1918 to denote the disease process found in the lungs of patients with pneumonia following measles, and later this form of pneumonia was recognized as the fundamental anatomic lesion in some cases of influenzal pneumonia. Descriptions of what might be considered other virus infections of the lungs are to be found in the years immediately preceeding the influenza pandemic of 1918. MacCallum noted cases of interstitial pneumonia without history of measles in United States Army camps during the winter of 1917—18. There were also other pulmonary infections in which the clinical and pathological findings suggested a viral etiology as noted in a recent review by Dingle and Finland. Among them were the purulent bronchitis found by Hammond in British troops in northern France during 1916 and 1917 and the central pneumonia demonstrated in roentgenographic studies by Hesse.

In the pandemic of 1918 very similar pathologic lesions were described by a number of investigators. The earliest acute pneumonic lesion of that form of influenza was probably a hemorrhagic and oedematous lobular consolidation with foci of necrosis. The pulmonary lesions following measles did not resemble the early acute hemorrhagic oedematous lesion of influenza but were rather similar to the interstitial changes seen in patients who survived the earlier stages of influenza. McCordock and Muckenfuss in reviewing the subject note that the interstitial infiltration in post-measles pneumonia was often more marked, and that collections of large mononuclear cells were found in the alveoli more frequently than in influenza.

II. Attempts to determine etiology.

The possibility that a toxic injury to the lung predisposed to secondary bacterial infection in influenzal pneumonia was considered by Leichtenstern and later by Goodpasture and Burnet. The latter investigators in autopsies performed during the 1918—19 epidemic found that in certain very early cases bacteria of any kind were scarce or not present in significant numbers. This led them to believe that "notwithstanding the demonstration of influenza bacilli in pure culture in the lung in all but one instance the primary injury is due to a very potent toxic agent elaborated in and disseminared through the larger air passages."

The almost invariable presence of secondary infection in the pneumonias associated with influenza and measles probably obscured the changes which in the light of more recent knowledge could be ascribed to a virus infection. There is still some uncertainty as to whether the interstitial pneumonia described by MacCallum and others was of bacterial or viral origin. MacCallum believed that the interstitial process in post-measles pnemonia was caused by the hemolytic streptococcus although similar changes associated with the Pfeiffer's bacillus were later described by him both in cases of influenzal pneumonia and in post-measles pneumonia. He noted that streptococci were much more common in the bronchi and lymphatics than in the alveoli despite infiltration of the air sacs with mononuclear cells. Lobar pneumonia caused by the pneumococcus was seen after measles as well as after influenza. In the cases of influenza there was no general agreement as to the predominant bacterium found in the lungs and although a viral etiology was suspected the experimental evidence for this was inconclusive.

Early animal experiments attempted to prove the role of bacteria in the forms of pneumonia seen after measles and influenza. BLAKE and CECIL in an extensive series of observations with monkeys succeeded in producing various forms of respiratory disease and pneumonia with H. influenzae, hemolytic streptococcus and pneumococcus. Ten of 27 monkeys inoculated with fluid cultures of H. influenzae developed a bronchopneumonia with hemorrhage, oedema, purulent bronchiolitis and peribronchial consolidation. The predominant cells in the exudates were apparently polymorphonuclear leucocytes. These changes may have resulted at least in part from a bacterial toxin since FILDES and McINTOSH were able to produce oedema and hemorrhages in the lungs with filtrates from broth cultures of H. influenzae; BLAKE and CECIL also produced in monkeys a disease resembling acute interstitial bronchopneumonia with a strain of hemolytic streptococcus given in small amounts while larger inocula of the same organism produced confluent lobular pneumonia. That toxic gases such as chlorine and phosgene may produce hemorrhagic oedematous and necrotic lesions of the lungs resembling acute fulminating influenzal pneumonia was shown by WINTERNITZ and his coworkers.

III. Experiments in animals with known viruses.

A clarification of the nature of the changes produced in pulmonary tissue as a result of virus infection came from the experimental production of pneumonia in animals by intranasal or intratracheal inoculation. In several of these early experiments strains of vaccine virus were used but not all of them produced lung lesions. Bosc in 1904 abtained a proliferative reaction with mononuclear exudate in the lungs of three rabbits among eight inoculated intratracheally with vaccine virus. His observations were later repeated by other investigators but the nature of the reaction obtained by ARMSTRONG and LILLIE and later by McCORDOCK and his associates seems to have been slightly different in that oedema and coagulation necrosis were very prominent. Bronchopneumonia associated with small pox in which Guarnieri bodies were found in the lungs was described in 1909 and later investigators have noted the interstitial reaction in bronchopneumonia following generalized vaccinia in human beings.

Psittacosis was recognized as a virus disease in 1930 as a result of the work of a number of investigators. RIVERS and BERRY produced by intratracheal inoculation an experimental psittacosis pneumonia in monkeys and rabbits which resembled the pneumonia seen in man. The experimental pneumonia was not associated with bacteria. In monkeys it began around the large bronchi and vessels near the hilum and spread towards the periphery along the alveolar walls which were greatly thickened by cellular proliferation and infiltration. Necrosis, hemorrhage and serous exudation were observed in some animals.

The bacterial component in experimental virus pneumonia. In 1931 Shope demonstrated that Berkefeld filtrates of infectious material from experimental cases of swine influenza administered intranasally to swine induced a mild disease with minimal pulmonary lesions. These were characterized by damage to the bronchial epithelium, peribronchial infiltration with round cells, and scant patchy lobular atelectasis. Simultaneous infection with Hemophilus influenzae suis resulted in a severe disease resembling the human influenza seen in 1918. The H. influenzae suis was incapable by itself of producing the disease. Thus it was shown for the first time that a virus and bacterium can act synergistically in the production of a pneumonia resembling the acute interstitial bronchopneumonias of man.

The observations of Shope on the mild filtrate disease supplemented the work of other investigators working at about the same time with other viruses in defining the nature of the pulmonary lesions produced by filtrable agents. McCordock and Muckenfuss called attention to the resemblance of vaccine virus pneumonia in rabbits to human influenza and reported that the lobular or bronchopneumonic component could be superimposed by injecting streptococci or staphylococci into the lungs of rabbits after intranasal inoculation with dilute vaccine virus.

Another example of a naturally occurring disease of animals in which the primary cause is a filtrable virus with secondary pulmonary complications due to bacteria was found in distemper of dogs. Dunkin and Laidlaw in 1926—28 previous to the observations of Shope on swine influenza had shown that distemper of dogs and of ferrets was transmitted by a filtrable virus despite the fact that previous observers had found a rather constant association of Br. bronchiseptica with this disease. The observations of Laidlaw and Dunkin indicated that in natural distemper in which pneumonia is not uncommon, Br. bronchiseptica though readily cultivable from the lung can rarely be cultivated from the blood thus indicating that the organism is a secondary invader possibly implicated in the production of the pneumonia.

Transmission of influenza to experimental animals. In 1933 Smith, Andrewes and Laidlaw transmitted human influenza to ferrets by inoculating filtrates of pharyngeal washings intranasally. These investigators also found that swine influenza was transmissible to ferrets and demonstrated a partial immunological relationship between the viruses from man and swine. Shope in 1934 observed that ferrets lightly anesthetized with ether and inoculated intranasally with swine influenza virus developed an extensive bloody oedematous lobar pneumonia four to five days after infection. In contrast to influenza in swine the ferret disease was not modified in character when cultures of H. influenzae suis were inoculated together with virus. Francis and others at about this time observed that adaptation of the human virus to ferrets using anaesthesia at the time of inoculation, led to the production of pulmonary consolidation. Later the viruses of human and swine influenza were adapted to mice by Andrewes, Laidlaw and Smith, by Francis and by Shope with the production of a uniformly fatal virus pneumonia. The human virus required several preliminary passages in ferrets before being adapted to mice but the swine agent was transmitted directly to mice. H. influenza suis did not modify the disease in mice.

Recognition of the character of virus pneumonia. From the experimental work on human and swine influenza a great advance in the knowledge of virus pneumonia was accomplished. New techniques which were somewhat more reliable and simpler than those previously used were developed. The production with influenza and other viruses of pulmonary lesions free of bacteria demonstrated that these agents by themselves can, under certain conditions, produce a pneumonia. The characteristic pathology of virus pneumonia as seen in experimental animals was recognized and it was found that there were differences but at the same time many similarities between pneumonias produced by various viruses. Finally the relationship of bacteria to virus infections of the respiratory tract was defined. In some virus diseases the pneumonia could be attributed to secondary infection with a variety of organisms. In certain hosts a synergistic action of a virus and a specific organism was observed, while in other species the same bacterium failed to augment the pathogenic effect of the virus in the lungs.

Virus pneumonia in man.
I. Differentiation from bacterial pneumonia.

In secondary bronchopneumonia as a complication of surgical operations acute illnesses, cardiac failure, chronic degenerative diseases, neoplasms, and as a terminal event in other fatal illnesses, histological sections usually reveal bacteria in considerable numbers either as one species or as mixtures of several kinds of organisms. Secondary bronchopneumonia can usually be recognized as a bacterial infection when it follows such diseases as diphteria, scarlet fever, or typhoid, the specific causative organism being found in the sputum or lungs. In many cases it is very difficult to draw a line between those bacterial pneumonias which result from physiological debilitation and those infections of the lungs in which a virus may be the primary incitant and the bacterial infection secondary. This has been emphasized for influenza and measles in the previous section.

Primary bronchopneumonia may be caused by the pneumococcus, the tubercle bacillus, streptococci, staphylococci, FRIEDLANDER's bacillus, and Bacterium tularense. Mycotic infections of the lungs also result in a primary bronchopneumonia or pneumonitis. Important among the latter is coccidiomycosis which was common among troops at airfields and training camps in the western United States during the last World War. Pulmonary histoplasmosis may also be considered a primary bronchopneumonia. Pneumonitis occurs in toxoplasmosis. In all such infections the diagnosis can be established by culture of the causative organisms from the sputum or by their demonstration in significant numbers in the pulmonary tissue.

The separation and identification of the various forms of bacterial bronchopneumonia leaves an ill-defined group of so-called primary bronchopneumonias in which no bacterial etiology has been demonstrated. Many of these illnesses are mild in character and the pneumonia is recognized only by x-ray, but some may be severe or fatal. On clinical basis alone some of these are very difficult to differentiate from the known bacterial pneumonias, and the failure to find a causative organism is merely an indication of the possible virus etiology of the disease.

The clinical characteristics of virus pneumonia are those of an acute infection of gradual onset in which constitutional symptoms such as headache, fever malaise, and chilliness predominate over respiratory symptoms during the early stages. Dry irritating cough may develop soon after onset in some forms of the disease but is said to be less frequent in the pneumonitis of Q fever. In a small proportion of the cases, dyspnea, cyanosis, and circulatory collapse may occur, but in most instances and particularly in the disease known as primary atypical pneumonia, the illness is of mild or moderate severity. Virus pneumonia caused by the psittacosis group is usually more severe than other forms of the disease. Rarely the moderately severe or severe illness may be complicated by pleural effusion, bronchiectasis, or signs suggestive of involvement of the central nervous system.

The absence of a significant increase in polymorphonuclear leucocytes may be indicative of a virus infection. The incubation period is generally considered to be longer in virus pneumonias than in bacterial pneumonias. The pulmonary lesions seen by x-ray are frequently diffuse in character and difficult to detect by physical examination. There is often incomplete consolidation in the affected areas of the lung and the lesions may wander from one lobe to another during the febrile period. Bloody sputum is rarely seen in virus infections of the lungs and involvement of the pleural surfaces is minimal or absent. Clinical differentiation of the virus pneumonias from one another is at present impossible.

Epidemiological differentiation.

The recognition of pneumonias associated with epidemic influenza caused by viruses type A or B is based in part on epidemiological considerations. A widespread explosive outbreak of respiratory disease with the clinical characteristics of influenza is accompanied by an increase in the number of bronchopneumonias. There is at present some doubt as to the relative importance of the influenza virus itself and of secondary bacterial infection in the causation of influenzal pneumonia in man, but it is generally assumed that the virus acts as a potentiating factor for the production of bacterial bronchopneumonia. Influenzal pneumonias seldom occur as isolated or sporadic cases during interepidemic periods.

Primary atypical pneumonia may occur sporadically or in small local outbreaks which are not necessarily associated with an increase in the rate of upper respiratory illnesses. Sometimes the incidence of primary atypical pneumonia, of bronchitis resembling atypical pneumonia, and of acute upper respiratory disease, may rise simultaneously suggesting that the same infectious agent may be causing both the pneumonia and the milder respiratory illnesses. In general the attack rate in such outbreks rises more gradually than in epidemics of influenza and a high proportion of the population may have pulmonary involvement. The occurrence of atypical pneumonia either in epidemic or sporadic form is not closely related to the season. Outbreaks have occurred during the summer when influenza and other respiratory illnesses are rare.

Two other diseases, namely Q fever and psittacosis, differ from influenza and primary atypical pneumonia in that they are not usually contracted from human sources and are not primarily diseases of man. Q fever was originally described as an occupational disease among persons in contact with cattle. The causative agent, Rickettsia burnetti (or diaporica), is found in certain species of ticks. Recently the rickettsiae of Q fever have been found in milk. Explosive outbreaks of pneumonitis caused by the rickettsiae of Q fever have been described, and these probably resulted from the simultaneous exposure of a number of persons to a highly infectious source of the agent. Sporadic cases are not uncommon.

Persons infected with viruses of the psittacosis group frequently give histories of contact with psittacine birds, pigeons, chickens or ducks, but such histories cannot always be elicited.

Characteristic pathology.

The nature of the pulmonary lesion called acute interstitial pneumonitis has been discussed in a previous section, and will be described in detail for primary atypical pneumonia. This type of lesion differentiates virus pneumonias from bacterial pneumonia. There is nothing about the pathology of virus pneumonias, however, which would enable one to diagnose the causative virus from the gross or microscopic changes in the lungs or other tissues. The pneumonias caused by influenza, psittacosis, Q fever, or the agent of primary atypical pneumonia are all very similar in man. Variations in the degree of proliferation of the alveolar epithelium, in the proportion of polymorphonuclear to mononuclear cells in the alveolar exudate, and in the amount of perivascular or peribronchial infiltration do occur but it is uncertain whether or not these are significantly related to etiology.

A virus pneumonia of infants described by ADAMS and GREEN may be differentiated from other forms of the disease by the presence of cytoplasmic inclusion bodies both in the cells of the pulmonary epithelium and those of the oropharynx. Inclusion bodies have also been seen in a number of cases of pertussis by McCORDOCK and SMITH, but here the position was intranuclear. The significance of these intranuclear inclusion bodies in pertussis has never been accurately determined. Cytoplasmic inclusion bodies in various organs have been described in other diseases of children. The presence of inclusion bodies in the lungs cannot therefore be considered as pathognomonic of a particular form of virus pneumonia.

Failure of response to chemotherapy.

In the treatment of virus pneumonias sulfonamides have generally proved to be of little value. Possible exceptions are the secondary bacterial pneumonias following influenza A or B. Penicillin has definite inhibitory action on viruses of the psittacosis group, and in large doses is effective against experimental pneumonia caused by this group of agents. There have been reports of the successful treatment with penicillin of psittacosis in man, but there is no evidence that this drug is effective against primary atypical pneumonia. Streptomycin has recently been tried with some success in experimental infection of guinea pigs with the rickettsiae of Q fever, but there are at present no reports of its use in the human disease. This antibiotic is inactive against experimental infections with the viruses of influenza, psittacosis, and primary atypical pneumonia[1].

II. Primary atypical pneumonia.

Nomenclature.

There is much confusion in the nomenclature of a disease which is gradually becoming, but is not yet universally, recognized as a specific etiologic entity. The term atypical pneumonia was first used commonly in the last decade as

---

[1] Since this article was sent to press experimental and clinical evidence for effectiveness of a new antibiotic, Aureomycin, against primary atypical pneumonia has been published:

SCHOENBACH, E. B., and M. S. BRYER: Jour. Am. Med. Assn. **139**, 275, 1949. KNEELAND, Y. Jr., H. M. ROSE, and C. D. GIBSON: Am. Jour. Med. Sci. **6**, 41, 1949. FINLAND. M. H. S. COLLINS, and E. B. WELLS: New England Jour. Med.,**240**, 241, 1949. MEIKLEJOHN, G. and R. I. SHRAGG: Jour. Am. Med. Assn. **140**, 391, 1949. EATON, M. D.: Jour. Clinical investigation, **28**, (No. 4) 1949.

a general name to designate a form of pneumonia which was not the typical lobar pneumonia caused by the pneumococcus. In this sense it may include a variety of primary bronchopneumonias either bacterial or viral in origin. The disease under consideration was recognized as a rather characteristic syndrome among the respiratory diseases beginning about 1935 and was seen with increasing frequency through 1945. This syndrome has been variously designated as acute interstitial pneumonitis or pneumonia, acute influenzal pneumonitis, bronchopneumonia, variety X, disseminated focal pneumonia, atypical-bronchopneumonia, virus pneumonia, and viroid.

During the recent World War when this disease was prevalent among military personnel, the term "Primary Atypical Pneumonia, Etiology Unknown" was adopted by the Surgeon General's office of the United States Army. This term was preferred because of the possibility that agents other than a filtrable virus may cause a similar disease entity and because of doubt as to whether one or several diseases were included.

There is now an increasing body of evidence to indicate that the disease prevalent during the years 1939 to 1945 was caused by a virus. Therefore we have less reason to consider primary atypical pneumonia a syndrome of which there may be many causes, and it would seem justifiable to use the term to designate a specific etiologic entity among the various forms of virus pneumonia. Continuation of the suffix "etiology unknown" is now misleading because at least two extensive experimental studies have demonstrated that the causative agent is a filtrable virus and there is good evidence that this virus has been propagated in the laboratory.

Distribution.

It is, of course, impossible to know whether or not some of the diseases mentioned in the previous historical section were identical with the entity now called primary atypical pneumonia. Many cases of this disease were probably not recognized as pneumonia at all because of the difficulty of eliciting physical signs and because roentgenograms were seldom made. Several authors have regarded primary atypical pneumonia as a new disease.

Primary atypical pneumonia became prominent about 1938, but cases of what may have been the same disease were described by Bowen as occurring among white troops in Hawaii in 1932 and 1933. About the same time Gallagher described cases of what appears to be a similar form of bronchopneumonia which he observed in a boy's school. Numerous other descriptions of the disease were given in the United States, England, and the European Continent. Cases seem to have been particularly noticed in schools or colleges before the war and this emphasized the susceptibility of young adults to this disease.

Several reasons may be given for the apparent increase in the incidence of primary atypical pneumonia during the last 12 years.

(1) The increase may be real and a result of the natural fluctuation in incidence common for many infectious diseases.

(2) The diagnosis has been made more frequently because of interest in the disease and use of x-ray.

(3) A relative diminution of the incidence of pneumococcal and other bacterial pneumonias as a result of chemotherapy has emphasized unusual forms of pneumonia which were not previously studied.

During the second World War the incidence among military personnel was greater than that of all other forms of pneumonia combined. In some hospitals, the cases of primary atypical pneumonia outnumbered pneumococcal pneumonia by as much as 3 to 1. Since 1945 the incidence of the disease appears to have decreased but considerable numbers of cases are still seen in various parts of the world. Recently the disease has been described in Australia and interest in it continues in England and various European countries.

Epidemiology.

The disease called primary atypical pneumonia may occur as sporadic cases, in family or institutional outbreaks, or as small epidemics. Transmission of the disease from one member of a family to another has been repeatedly observed and transmission from patients in a hospital to medical personnel is not an infrequent occurrence. The interval between exposure and the onset of the disease in such sequences of cases has been ten days to three weeks. Frequently mild and severe illnesses have occured in the same series. Presumably the causative agent is transmitted from person to person by droplet infection or contact. No reservoir other than man has been found.

Between 1938 and 1941 rather sharp epidemics of the disease were frequently observed in schools or colleges. The data presented in Table I indicate an increasing incidence of the disease. The length of such outbreaks varied from one to six months. During such epidemic periods, primary atypical pneumonia has often affected between one and five per cent of the population. During the war, in 1941—45, the disease was very prevalent in various Army camps in the United States. Between March 1942 and May 1943 the number of cases was approximately 26,000. Special studies on the clinical characteristics of the disease and its epidemiology and etiology were undertaken by the Army Commission on Acute Respiratory Diseases, first at Camp Claiborne, Louisiana (DINGLE et al 1944) and later at Fort Bragg, North Carolina.

Table 1. *Incidence of Primary Atypical Pneumonia Among School and College Students*

Academic Year	Totals Reported By:		
	GALLAGHER	SMILEY et al	MURRAY
1935—36	6	*	5
1936—37	16	*	25
1937—38	4	59	35
1938—39	8	27	81
1939—40	37	*	*

* No cases recorded by the authors.

The epidemic at Camp Claiborne, Louisiana started in June 1941, reached its peak during the summer and autumn, and declined during the winter months of 1942. In late July, hospital admissions for atypical pneumonia reached an incidence of 88 per 100,000 population per week. The disease remained endemic for a long period of time after the first sharp outbreak. The attack rate was three times higher among hospital personnel than the rate for the remaining troops. Focal concentrations of cases in companies or regiments were rare but the disease was widespread and occurred in nearly every organization during the endemic and epidemic periods. A limited amount of evidence indicated that the atypical pneumonia at Camp Claiborne was transmissible from patient to patient by contact. Evidence of contaminated water, food or milk as possible factors in the occurrence of the disease was not found. There was no evidence for insect transmission or non-human reservoirs. Transmission from inapparent infections or from persons having the symptoms of acute respiratory disease without pneumonia was considered likely. During the summer of 1941 there seemed to be a relation between the incidence of atypical pneumonia and that of other acute disease of the respiratory tract. The admission rate for acute respiratory disease was approximately 200 per 100,000 per week in June, reached a peak of 500 per 100,000 per week in July and was not associated with the existence of any other recognized epidemic disease. The symptomatology and course of illness exhibited by many of the patients with acute respiratory infections were indistinguishable from those of atypical pneumonia.

At Fort Bragg, North Carolina, where further studies were conducted by the Commission on Acute Respiratory Diseases, atypical pneumonia was endemic throughout the war. In the winter of 1942—43 the incidence was highest during the months of December, January, and February, but there was no definite epidemic peak. Both atypical pneumonia and acute upper respiratory infections were four to ten times as frequent among new recruits entering the camp as among seasoned troops who had been in the Army for some time. The incidence of atypical pneumonia varied from 86 to 400 per 100,000 per week among new recruits and from 7 to 98 per 100,000 per week among seasoned troops. Approximately a 10 to 1 ratio of respiratory admissions to cases of atypical pneumonia was maintained both among new recruits and among seasoned men. Separate small epidemics of respiratory disease and atypical pnemonia corresponded each month with the time of arrival of new recruits.

The description of an outbreak of atypical pneumonia in a relatively isolated small civilian community as presented by Breslow is of interest because this is one of the few recorded epidemics in which a probable etiologic relationship to a virus propagated in the laboratory was demonstrated. An explosive outbreak of respiratory disease associated with atypical pneumonia occurred in Kasson, Minnesota (population 1,230) in October 1942. The epidemic started in the only grade school of the town and spread to households of the school children. There were 191 cases of respiratory disease with or without pneumonia. The symptoms were almost identical with those described by other investigators. Of the total of 191 cases 31 were classed as severe, 47 as moderate, and 113 as mild. The "severe" cases of this outbreak would, however, correspond to a fairly mild form of the disease as commonly described, and the "mild" cases generally had coryza and cough with little or no fever and were ill for only a few days. Of the 26 patients examined by a physician during the acute stages of the illness only 12 had physical signs of pulmonary involvement. In the grade school the attack rate was 63 per cent. Among 230 persons exposed only in the households of the children there were 41 secondary cases, an attack rate of 18 per cent. The incubation period was estimated to be 8 to 21 days. The epidemic period extended from September 29 to October 26 but nearly half of the cases occurred during the week of October 6 to 12. The available evidence suggested respiratory spread. Because numerous pigeons were present in and around the schoolhouse the possibility of infection by one of the psittacosis group of viruses (ornithosis) was considered but tests on sera and sputum from patients failed to substantiate this possibility. The positive results of neutralization tests with a virus isolated by Eaton et al in California from patients with atypical pneumonia and other serological studies will be discussed in a later section.

These and other observations have indicated that in relatively well defined epidemics of acute respiratory disease associated with atypical pneumonia almost the entire group of cases may be caused by a single virus. In sporadic cases of the disease, on the other hand, the etiology tends to be more diverse because several forms of virus pneumonia and even bacterial or mycotic pneumonias may be confused clinically with primary atypical pneumonia.

Very little is known about immunity and susceptibility to primary atypical pneumonia. No differences in incidence with respect to race, color or sex have been observed. The disease apparently may affect any age group, but the majority of outbreaks have been described in children, adolescents or young adults. There is some evidence for a diminished susceptibility of those over 30 years of age. The incidence of pneumonia is higher in those heavily exposed to the disease. The general population is apparently not highly susceptible to clinically recognizable infection although in some epidemics the evidence points to a rather high incidence of non-pneumonic acute respiratory disease caused by the same agent. There is no evidence from epidemiological results or the studies on neutralizing antibodies to be presented later that immunity to the disease is more lasting than to other respiratory infections. Re-infection in a partially immune individual would be more likely to result in uncomplicated acute respiratory infection rather than in the production of pneumonia.

Clinical characteristics.

Since there are so many detailed descriptions in the literature of the clinical signs and symptoms of primary atypical pneumonia, another extended presentation of this data would seem superfluous. A summary of the observations of several investigators is presented in tables 2 and 3. The onset in primary aty-

Table 2. *Clinical data on atypical pneumonia as reported by several investigators.*

	Per Cent of Cases as Reported By:			
	CALLAGHER (1941)	DINGLE et al. (1944)	MURRAY (1940)	CURNEN et al. (1945)
Onset:				
Sudden	45	26	27	27
Gradual (cold or sore throat)	55	74	73	73
Malaise	96	77	70	61
Headache	78	78	53	65
Chill and rigor	—	13	13	—
Cough	77	99	84	98
Coryza	—	41	20	59
Sore Throat	8	36	38	30
Sputum	54	64	33	82
Substernal discomfort	12	26	15	24
Pleural pain	—	18	3	—
Pulmonary consolidation detectable:				
By percussion	—	41	50	54
By abnormal breath & voice sounds	14	4	15	70
Rales or ronchi	83	93	—	93
	Number of cases studied			
	87	69	132	106

Table 3. *Range of Pulse Temperature, Respiration, and Leucocyte Count in Primary Atypical Pneumonia.*

	Gallagher (1941)	Dingle and al. (1944)	Murray (1940)	Meiklejohn et al. (1945)
Duration of fever (days)	2—18	2—45	1—7	7—24
Temperature °F maximum	101—105	99—105	101—105	101—105
Pulse rate during fever	90—136	75—125	90—110	90—150
Respiration rate .	20—38	18—32	normal	20—50
Leucocyte count .	6000—23400	5000—13000	5000—12000*	5000—37000*

* Late rise in leucocyte count noted in some cases.

pical pneumonia is usually gradual with premonitory symptoms of coryza or pharyngitis. Fever is almost uniformly present although a few cases have been described in which pulmonary consolidation was found by x-ray in the absence of known rise in temperature. Systemic symptoms of an acute infection are usually present but definite chill with rigor is rare. The cough develops during the early phases of the disease but seldom becomes very productive. Sputum is generally not bloody although traces of blood are seen in some cases.

The physical signs of pulmonary consolidation are generally rather difficult to elicit and become evident only after the first appearance of shadows by x-ray. Rales or ronchi are usually found over the consolidated portion of the lung and persist after the roentgenogram has become negative. The maximum temperature varies from 99 to 105° F; the fever is intermittent in character, and persists from 2 days to two weeks. In the average case the maximum temperature is about 101 to 103° F and the duration of fever, one week. The pulse is relatively slow in proportion to the fever and the rate of respiration is normal except in the most severe cases. The leucocyte count is variable but a marked rise is

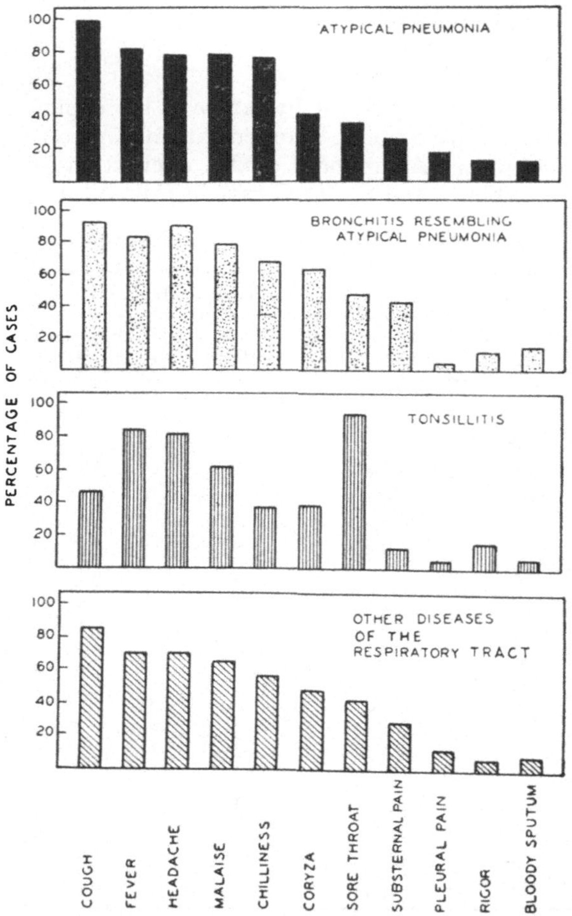

Fig. 1. Comparison of clinical characteristics of primary atypical pneumonia with certain other diseases of the respiratory tract. Data collected by Dingle et al. from cases at Camp Claiborne, Louisiana, 1941—42. Reprinted from War Medicine **3**, 223, 1943.

infrequent except in a few of the more severe cases late in the disease. From the data presented in Figure 1 it may be seen that the clinical symptoms of primary atypical pneumonia differ very little from those of bronchitis resembling atypical pneumonia and other acute diseases of the respiratory tract.

Pathology.

A virus has been isolated from only two fatal cases of atypical pneumonia. The pathology of one of these has been studied by two independent groups of investigators and will be reported in detail.

The patient, a male aged 26, was admitted to the hospital on November 11, 1942, with symptoms and clinical findings characteristic of primary atypical pneumonia and died six days later. At autopsy performed one hour after death the essential pathologic changes were limited to the lungs. The right lung weighed 1,085 gms., and the left lung 945 gms. The bronchi contained abundant yellowish, purulent, frothy exudate. The pleural surfaces were smooth. On section both lungs were found to be extensively infiltrated with whitish peribronchiolar nodules up to one centimeter in diameter and by confluent nodules measuring up to four centimeters in diameter. The nodular areas could be seen as pink or grey dry granular foci of irregular size and shape which stood up above the cut surface of the lung. The pneumonic process extended from apex to base involving both lobes.

Post-mortem bacteriological cultures of the lung tissue which had been collected with aseptic precautions showed no growth after 72 hours. Specimens of tissue were sealed in glass ampules and stored in dry ice immediately after collection. These were used for the isolation of virus to be described later.

Microscopic examination of the lung tissue showed acute bilateral interstitial pneumonitis with almost uniform ulceration of the epithelium in the bronchioles. Practically all combinations of cellular, fibrinous, and serous exudate could be found in the various alveoli, the exudate in some appearing fresh while in others it seemed to be undergoing resolution. Some of the alveoli contained patches of polymorphonuclear exudate which seemed to be derived by direct extension from involved bronchioles. There were certain areas in which necrosis of the alveolar walls had occurred with predominent polymorphonuclear exudate and partial liquefaction of the pneumonic focus. Fibrin was present in considerable amount and often tightly adherent to the walls of the alveolus. In other areas invasion by young fibroblasts was evident. In many alveoli sparse mononuclear exudate was found.

In addition to the frequent desquamation, a marked morphological change had taken place in the epithelial lining of a few of the bronchi. The epithelium had changed wholly or in part to a stratified squamous type, with loss of cilia. Mitotic figures were extremely numerous in these areas. Acute bronchiolitis with marked dilatation was widespread. The thick collars of round cells around the bronchioles extended widely into alveolar walls, thickening them markedly.

Numerous tissue sections stained for bacteria revealed none, even in the bronchiolar lumens. In several other cases cultures of the lungs have shown by ordinary post mortem bacteriologic methods no significant bacteria and none have been seen in appropriately stained tissue sections. In others of the cases reported by GOLDEN examination of sections stained for bacteria revealed occasional organisms in five of thirteen instances in the areas of acute interstitial pneumonitis. These were most often confined to the lumina of bronchioles. The organism usually found was a gram positive encapsulated diplococcus or streptococcus. These bacteriologic studies indicate that in many cases of primary atypical pneumonia it is impossible to account for any part of the pneumonic process by bacterial infection.

The second case from which a virus was isolated was more acute and fulminating than the one just described. The patient, a male, 24 years of age, was admitted to the hospital on March 21, 1944, in an extremely critical condition and gave a history of an illness of only one day's duration. X-ray showed an extensive pneumonia involving all lobes and tending to be denser in the central portions. The patient died one day after admission to the hospital. Postmortem examination showed almost complete hemorrhagic and edematous consolidation of all lobes of the lungs. The right lung weighed 1500 grams and the left 1200 grams. On section of the completely consolidated lung, large amounts of blood poured from the cut surface, which was somewhat irregular in countour but without distinct foci. In those lobes not completely consolidated the transition from the central pneumonic area to normal lung tissue was fairly sharp. Watery whitish mucoid material was present in the trachea and larger bronchi. The mediastinal lymph nodes were only slightly enlarged. Evidence of acute myocardial involvement was also found in this case.

Microscopic examinations of sections of the lung revealed much less cellular reaction than in the first case, probably because of the acute nature of the infection. Edema fluid and erythrocytes were seen in the alveoli with sparse mononuclear exudate in some areas. Peribronchial, perivascular, or septal infiltration were not prominent features in these sections of pulmonary tissue. Examination of sections and impression smears revealed no bacteria. Staphylococcus aureus was cultured from material which had been taken with sterile precautions and immediately frozen in dry ice for purposes of virus isolation.

Comparison of these two cases can be made with descriptions of the pathologic anatomy of atypical pneumonia of undetermined etiology published by a number of investigators. In the group of cases studied by Golden emphasis is put on the bronchiolar lesions and extension to adjacent alveoli. The degree of hemorrhage and edema in the pulmonary tissue varies tremendously. In some cases the tissue surrounding the infiltrated bronchioles is spongy and normal in appearance or atelectatic. In other cases marked hemorrhage and edema were found. The mucous membranes of the larger bronchi of the bronchial tree was usually normal or only slightly edematous. Ulceration of the bronchioles appeared quite early, being present in the lungs of patients who were ill for only a few days before death. Fragmentation of the smooth muscle fibers of the bronchiolar walls and of elastic tissue in the alveolar walls was observed.

Golden suggests that the primary lesion is in the bronchiolar mucosa and that the presence of abundant polymorphonuclear leucocytes in the bronchioles represents a response to the necrosis and desquamation of cells in the absence of significant numbers of bacteria. The round cell interstitial exudate would correspond to the known leucocytic responses to viruses in general. In some of these cases secondary bacterial infection did occur and in these it was possible to find areas of acute interstitial pneumonitis adjacent to zones of typical bronchopneumonia, lobar pneumonia, or pulmonary abscesses.

In a few cases lesions of the hemorrhagic type were found in the brain, and more rarely, small zones of necrosis and perivascular infiltration or cuffing with glial cells were observed.

The desquamation of bronchiolar mucosa is not emphasized in the two fatal cases described by LONGCOPE, both of which had chronic rheumatic heart disease in addition to atypical pneumonia. In these lungs the lesions were characterized by peribronchial infiltration with monunuclear cells and exudate of the same type in the alveoli. In the case described by KNEELAND several different stages of the pneumonic process were found in various parts of the lung. The earliest phase was characterized by a hemorrhagic exudate which later was succeeded by a mononuclear exudate with marked tendency to fibrosis. Lesions resembling acute periarteriitis nodosa were seen in medium sized branches of the pulmonary artery. The single fatal case occurring at Camp Claiborne as described by DINGLE et al had firm dark hemorrhagic consolidation of portions of several lobes of the lungs with plugging of the small bronche. Microscopically the involved areas of the lungs showed congestion and monocytic infiltration, and in the alveoli many large actively phagocytic monocytes, a moderate number of eosinophiles, scant neutrohils, fibrin edema fluid and erythrocytes. The walls of the bronchi showed an acute inflammatory process with extensive ulceration of the bronchial mucosa and thick polymorphonuclear exdate in the lumina. No bacteria were found.

Bronchoscopy was performed on several of the patients at Camp Claiborne. Mild to moderate acute inflammation and exudate were observed in one or more portions of the lower bronchial tree. The trachea was usually normal. In stained smears of the exudate obtained from twenty patients by bronchoscopy no bacteria could be seen in twelve specimens while in the remainder organisms were usually rare. The exudates contained variable proportions of polymorphonuclear leucocytes, lymphocytes, large mononuclear cells and epithelial cells.

Serological reactions in primary atypical pneumonia.

Cold Agglutination. In 1943 PETERSON, HAM, and FINLAND observed that sera collected late in the illness of during convalescence from patients with primary atypical pneumonia possessed the property of agglutinating type 0 human erythrocytes at 2 to 4⁰ Centigrade. Later FINLAND et al presented the results of tests for cold isohemagglutinins in 200 cases with characteristic clinical and x-ray findings of atypical pneumonia.

The method of performing the test described by FINLAND et al is identical or very similar to that used by other workers. The collection blood is best done by venupuncture with precautions as to sterility. Clotting should occur at room temperature or 37⁰ C and the serum should be separated by centrifugation without placing the specimen in the ice box. Adsorption of cold agglutinins by the erythrocytes will occur at low temperature and subsequent centrifugation can result in partial or complete removal of the agglutinins from the serum. Oxalated plasma may be used in place of serum. Prolonged storage of serum results in diminished titer of agglutinins.

The erythrocytes to be used in the test are obtained from group 0 donors as oxalated or citrated blood and washed three times. A two per cent suspension in physiologic saline is prepared. Cells stored in the original plasma are more satisfactory than those stored in saline. In the series done by FINLAND and his collaborators almost all of the tests were carried out with cells 2 to 4 days old. In sera with low titers the use of cells 2 to 4 days old gave significantly higher titers than did fresh or 1-day old cells. Cells more than 4 days old tended to behave like fresh cells. Use of 2 to 4 day old cells also minimized the differences in titer obtained with a given serum and cells from different donors.

Serial two-fold dilutions of serum or plasma in saline are made in 0.5 ml amounts in 100 × 11 mm test tubes. The first dilution varies in different studies from 1 : 2 to 1 : 10. Equal volumes of the 2 per cent erythrocyte suspension are added, the tubes shaken and stored in the refrigerator overnight. Other workers have kept the cell and serum mixtures in the ice box for shorter periods, and in some studies the test has been read after one hour. Accelerated sedimentation and coarse agglutination are interpreted as a positive test. After the readings are made in the cold the racks of tubes are placed in the water bath or incubator for 2 hours. All positive agglutination should be completely reversible at 37⁰ Centigrade.

Cold agglutinins rarely appear before the seventh day after onset of atypical pneumonia. The most frequent time of first appearance is between the tenth and twentieth days. The titers usually reach a maximum between the eleventh and twenty-fifth days. In a number of cases the maximum titers were not attained until the fifth week or later. The value of the maximum titer varies tremendously. In a few individuals the titer may not rise above 20 or 40 at any stage of the disease but in the majority of cases the titers are between 80 and 640. Maximum titers as high as 10,000 have been recorded. The first significant drop from the maximum titer occurs generally between the third and fifth weeks and a drop below the significant level is found between the fourth and sixth weeks. Rarely the cold agglutinins may persist for 2 months or more. Reappearance of cold agglutinins in two patients with recurrent attacks of atypical pneumonia has been reported by Finland. In one of these an interval of ten months separated the two attacks while in the other case the interval was six months.

It is the impression of a number of workers that the titer of cold agglutinins is related in a general way to the severity of the symptoms, to the extent of the pulmonary lesion, and to the duration and height of the fever. In a group of the mildest cases the incidence of titers over 20 may be 20 per cent or less while among the severe cases more than 85 per cent may have cold agglutinins. The height to which the cold agglutinin titer rises is not related to the age of the patient, the administration of sulfonamides, or the season.

A considerable variation is found in the reports of different investigators as to the proportion of cases of primary atypical pneumonia with significant titers of cold hemagglutinins. A summary is presented in Table 4. While the failure of certain investigators to find significant titers of cold hemagglutinins in their cases have admittedly been due to prolonged storage of the sera with consequent loss of titer, this factor cannot be considered the cause in other studies. Cressy of the Commission on Acute Resopiratry Diseases at Fort Bragg, North Carolina, when using 2 per cent erythrocyte suspensions for the test obtained only a few positive results. Use of a 0.2 per cent suspension made the test more sensitive but even then less than half of his cases had significant titers. Sensitivity of the test also varied with the age of the erythrocyte suspension.

Cold agglutination in other diseases. The only infectious disease other than atypical pneumonia in which cold agglutinins have been found with any regularity is trypanosomiasis. Occasional instances in which cold agglutination titers up to 40 have been reported include a wide variety of diseases, notably in blood dyscrasias, liver diseases, and in allergic and peripheral vascular disorders. Among infections of the respiratory tract, atypical pneumonia is the only disease in which cold hemagglutination is a frequent phenomenon. They occur at titers of 80 or over in less than one per cent of bacterial pneumonias, and of cases of influenza. Lower titers have been reported with some frequency in cases of upper respiratory infection by certain workers. Cressy found titers between 8 and 32 in 31 of 94 cases diagnosed as exudative pharyngitis or tonsillitis (Table 5), and Favour reported titers between 10 and 80 in 35 of a group of 50 cases diagnosed as tracheobronchitis, colds, or influenza. These results are in disagreement with the observations of other investigators and may have been due to the use of erythrocyte suspensions of 0.2 and 0.5 per cent respectively in contrast to the 2 per cent suspensions employed by others.

Table 4. *Maximum Titers of Cold Agglutinins in Primary Atypical Pneumonia.*

	FINLAND	MEIKLEJOHN	CRESSY*
	per cent of cases		
Titer <10.....................	22.5	16.0	45.0
Titer 10—80...................	32.0	49.0	29.0
Titer 160 or over	45.0	35.0	26.0
Number tested................	200	74	93

* Using 0,2 % cells (Commission on Acute Respiratory Diseases).

Table 5. *Maximum Titer of Cold Agglutinins In Other Diseases and Normal Persons*

	FINLAND	MEIKLEJOHN	CRESSY*
	per cent of cases		
Titer < 10	94.0	96.5	68.0
Titer 10—40...................	5.0	3.5	31.2
Titer 80 or over	1.0	0.0	0.8
Number tested................	809	138	121

* Using 0,2 % cells (Commission of Acute Respiratory Diseases).

Cold agglutination is apparently uncommon in virus pneumonia of the psittacosis group and Q fever. In recent outbreaks of the latter disease sera from a number of patients giving positive results for Q fever by complement fixation or agglutination showed no significant titers of cold agglutinins with one exception. Atypical pneumonia could not be excluded in this case.

Characteristics of the cold agglutinin. The aggluntination of group 0 human erythrocytes which takes place in the cold is dispersed as the temperature is raised above 5⁰ C. The temperature at which the reversal occurs with any given serum depends on the titer and intensity of the cold agglutination reaction in that blood. In most instances the agglutination is completely dispersed at 20 to 25⁰ C but in sera of high titer, incubation at 37⁰ C for 30 minutes or longer may be required. In certain sera with titers of over 1,000 at 5⁰ C, agglutination at low dilutions may be obtained by incubating at 20⁰ C.

Finland describes two cases of primary atypical pneumonia; one with hemolytic anemia, in which cold agglutinins could not be demonstated with the cells of other donors (isohemagglutinins) but agglutination did occur in the cold with the patient's own cells (autohemagglutinins). It is possible that the presence of a high titer of cold autohemagglutinins might result in agglutination and injury of red cells within a patient's own circulation and thus in the development of embolic or hemolytic reactions. Since the cold agglutinins are generally more active at temperatures lower than 37^0 C it is possible that they have their greatest effect either in peripheral vascular areas or in those organs which have a relatively static circulation. It may therefore be of considerable practical impotance to take measures to prevent chilling of patients with primary atypical pneumonia and other conditions in which cold agglutinins are found. Hemogobinemia and hemoglobinuria have been produced by exposure to cold of the limbs of patients with primary atypical pneumonia or certain forms of hemolytic anemia.

Cold isohemagglutinins are unaffected by heating to 56^0 C for 30 minutes. Partial destruction occurs at 60^0 to 62^0 C or higher. The agglutinins may be partially or completely removed from serum by adsorption on charcoal, norit permutite, and kaolin, and by filtration through Seitz EK, Berkefeld or Mandler filters. Alundum, silica, or casein do not adsorb the agglutinins.

After storage at refrigerator temperatures for several months, a large proportion of sera show significant deterioration in cold agglutinin titer. The rate of loss in agglutinins varies markedly with different sera. In about half of the sera, with titers of 20 to 80 which were stored at 5 to 10^0 C for several months, the titer may drop below 10. In sera with titers of 80 to 640 about 30 per cent may lose their cold agglutinins. When sera are kept at room temperature deterioration is two or three times as rapid as in the refrigerator.

The cold agglutinins are independent of the isohemagglutinins of the blood groups which are active at 37^0 C as well as at 5^0 C. Adsorption of the blood group agglutinins from sera containing moderate or high titers of cold agglutinins left the latter in undiminished titer.

Cold agglutinins for the erythrocytes of several species of animals exist in normal human sera. Some of these agglutination reactions are reversed at 37^0 C. They are not usually removed to a significant degree by absorption with human type 0 cells at 4^0 C. This indicates the non-identity of these cold agglutinins for animal cells with the human cold isohemagglutinins. In certain cases of atypical pneumonia there may be an increase in the titer of reversible cold agglutinins for sheep or rhesus monkey erythrocytes but this increase does not necessarily parallel that for human erythrocytes.

In certain cases of atypical pneumonia an increase of titer associated with illness has been observed by agglutination tests at 37^0 C with monkey, sheep or hen erythrocytes. Here again there is no relation to the increase in titer of cold isohemagglutinins.

Cold isohemagglutinins may be partially or completely removed from human sera by adsorption at 4^0 C with the erythrocytes or lung tissus of certain species of animals. Finland found that adsorption with rabbit erythrocytes removed the largest amount of cold agglutinins for human cells while guinea pig cells were less effective. Thomas et al. observed that absorption with mouse lung removed the cold isohemagglutinins from the convalescent serum of a patient with primary atypical pneumonia, and the writer has made similar observations with the lungs of normal hamsters. The extent of this phenomenon has not been thoroughly explored and it is therefore probably not limited to these particular kinds of erythrocytes or tissues.

The observations just recorded suggest that an antigen related to the one concerned in cold isohemagglutination is present in the erythrocytes and tissues of the lower animals. The possibility that a similar antigen is present in bacteria or in the virus causing atypical pneumonia should be considered.

Non-specific Complement Fixation. The puzzling phenomenon of complement fixation with a wide variety of dissimilar antigens has been observed in atypical pneumonia by several investigators. EATON and COREY did complement fixations with antigens derived from several members of the psittacosis group and sera from over 70 patients with atypical pneumonia. They observed positive reactions of a group specific character in 10 to 15 per cent. Four cases of atypical pneumonia also showed an increase in complement fixation titer with normal yolk sac antigen. Although the remaining positive sera had no increase in titer with the normal yolk sac antigen the specificity of the positive results with infected yolk sac is open to question in some cases because of the failure to isolate psittacosis-like viruses from the sputum of these patients and because of subsequent observations that similar complement fixation reactions were obtained with tissues infected with other agents.

THOMAS and his associates later did studies with various dissimilar antigens. They found increases in antibody titer of two-fold or greater for mouse lung infected with pneumonia virus of mice in 25 of the 35 sera obtained from patients patients with atypical pneumonia and tested by complement fixation. Some of these sera also showed increases in antibody titer for normal mouse lung, normal yolk sac and mouse lung infected with the viruses of influenza or the psittacosis group. In some cases an increase in titer was obtained with the infected mouse lung antigen, but not with the normal mouse lung antigen. Similar changes were observed with antigens prepared from normal human, guinea pig, rabbit and rat lung. These sera failed to hemolyze sheep cells when tested by the usual method and did not fix complement with psittacosis virus antigen prepared from tissue culture or the soluble antigen for lymphocytic choriomeningitis. Sera from patients with other repiratory diseases did not exhibit this property. Adsorption of the sera with a suspension of normal mouse lungs or heating at 65⁰ C for 30 minutes abolished the capacity of the atypical pneumonia sera to fix complement with the different antigens. The component of the antigens responsible for the non-specific complement fixation was removed from the supernatant fluid by centrifugation at 12,000 RPM for 30 minutes. The antigen deteriorated quite rapidly on standing.

Unpublished observation by the author made independently and concurrently with the above studies confirmed the existence of non-specific complement fixing antigens in the lungs of cotton rats and hamsters infected with a virus isolated form normal cotton rats or with a virus of the psittacosis group. The fixation of complement with serum from patients with atypical pneumonia and these antigens was much stronger at 0⁰ C for 18 hours than at the usual incubation period of one or two hours at 37⁰ C. The antibody was not destroyed by heating at 60⁰ C for 15 minutes although higher temperatures did reduce the titer. Complement fixation was much stronger with suspensions of infected hamster lung than with comparable preparations of normal hamster lung. The antigen could be sedimented by centrifugation at 5,000 RPM for one hour, and in contrast to the preparations of Thomas et al, was stable in the ice box for at least two weeks. Increase in titer of non-specific complement fixation were usually associated with the appearance of cold agglutinins in the sera. It has also been observed that transitory positive Wasserman or Kahn reactions occur more frequently in patients with primary atypical pneumonia than in comparable control groups of patients with other acute respiratory infections (HEGGLIN and others).

The inter-relation of these non-specific complement fixation reactions in primary atypical pneumonia has not been satisfactorily worked out but it has been shown by adsorption experiments that reagin or antibody is not the same as that responsible for the cold agglutination reaction. Adsorption of sera positive for both cold agglutinin and complement fixation with group 0 human cells in the cold removes the cold agglutinin but fails to reduce the titer of complement fixation with mouse lung or hamster lung. The equivalence of normal and infected lung in the complement fixation reaction is open to question. The writer has observed that absorption with a suspension of normal hamster lung failed to remove the ability of a serum to fix complement with hamster lung infected with viruses from animals. This and the observation by several workers that infected tissue gives stronger reactions than normal tissue indicates the possibility of an antigen resulting from tissue damage. Various infections may cause an increase in the amount of this non-specific "tissue-damage" antigen.

Agglutination of an indifferent Streptococcus. Serums from certain cases of atypical pneumonia develop agglutinins for a particular strain of indifferent or non-hemolytic streptococcus (Streptococcus M. G. or strain 344) isolated by Thomas and his associates at the Rockefeller Institute Hospital from the lungs of a patient who died of atypical pneumonia. The relations of this organism to the etiology of the disease is at present uncertain but the previously reported observations that the lungs in certain fatal cases are bacteriologically sterile seem to exclude the streptococcus as the primary cause in at least certain forms of the disease. Further discussion of this organism will be found in the section on etiology of atypical pneumonia. The indifferent streptococcus might in some cases produce secondary infection and a more severe illness but in certain respects the serological reactions with this organism are analogous to the Weil-Felix reaction with strains of proteus bacilli in rickettsial diseases.

The method of performing the agglutination is as follows. Bacterial suspensions for agglutination tests with streptococcus MG are prepared from 18 hour broth cultures. The bacterial cells are washed three times, suspended in saline, heated to 65⁰ C for one hour and preserved by adding merthiolate in a final concentration of 1 : 10,000. Serums for test are used without inactivation at 56⁰ and are set up in final dilutions from 1 : 10 to 1 : 320. Dilutions lower than 1 : 10 of normal human or rabbit serum often give non-specific agglutination of streptococcus MG. Equal volumes of serum dilutions and streptococcal suspension are mixed and placed in the water bath for 2 hours at 37⁰ C and then in the ice box at 4⁰ C for 18 hours. An additional incubation period of 2 hours at 37⁰ C before reading the test is used by many investigators. One modification consists in incubating for 18 hours at 37⁰ C. Non-specific agglutination of streptococcus MG may occur at 4⁰ C, but is reversible at 37⁰ C. The agglutination appears as flocculant clumps similar to that seen with type specific pneumococci and varies from solid disc formation (4+) to granules just visible to the unaided eye (1+).

A summary of the results of three groups of workers is presented in Tab. 6. It is generally agreed that the development of agglutinins for the indifferent streptococcus and the occurence of rises in titer or of increases in the intensity of the agglutination during the course of the disease or in convalescence is limited almost entirely to cases of atypical pneumonia. In a group of 99 patients with atypical pneumonia from whom two or more samples of serum were obtained during illness a two-fold increase in the titer of the convalescent serum was demonstrated in 47 patients, a fourfold increase in 25, an eight-fold increase in 13, a sixteen-fold increase in 2, and a 64-fold increase in one (Thomas et al). Agglutinins often appear during the second week of the illness and are most frequently found in the fourth or fifth week.

Table 6. *Results of Streptococcal Agglutination in Primary Atypical Pneumonia and Other Acute Infectious Diseases as Reported by Several Investigators.*

	Finland et al.			Thomas et al.			Meiklejohn & Hanford	
	No. tested	% Titer		No. tested	% Titer		No. tested	% Titer
		10+	20+*		10+	20+*		10+
Primary Atypical Pneumonia	78	89	77	193	67.0	39.0	156	47
Other Acute Infectious Diseases	104	·40	13	120	6.0	1.5	128	1.6

* 10 + or 20 + mean percentages of total group with titers of 10 or more and 20 or more respectively.

Agglutinins are found in the sera of normal individuals and in other diseases, but less frequently and in lower titers than in atypical pneumonia. An increase in titer of streptococcal agglutinins has been reported by Finland in a few cases diagnosed as pneumococcal lobar pneumonia. High titers have been observed occasionally in hemolytic streptococcal infections, or rheumatic fever and subacute bacterial endocarditis, but there is no correlation between the titers of antistreptolysin and agglutinins for the indifferent streptococcus.

Eaton and van Herick observed in the sera of five patients with upper respiratory infections a rise in titer of streptococcal agglutinins from less than 10 to 10 or from 10 to 20. Two of these patients' sera had an increase in antibodies for influenza A. Thomas and his associates noted a particularly high incidence (13.9 per cent) of titers of 10 or over in a group of 36 cases of acute respiratory infection without pneumonia. The specificity of the test appears to vary somewhat with different techniques since the incidence of titers of 20 or over in cases other than atypical pneumonia was 1.5 and 1.6 per cent as observed by two workers, while the third reported a figure of 13 (Table 6).

As is true of cold hemagglutinins, antibodies against streptococcus MG do not become demonstrable in all cases of primary atypical pneumonia and the development of streptococcal antibodies seems to be related to the severity of the illness. There is a positive correlation between the height of the antibody titer, the duration of fever, and x-ray, and physical signs of pneumonia.

Relation of Streptococcal Agglutinins to other serological reactions. Since cold agglutination tests can be done satisfactorily only with fresh serum, a negative cold agglutination test on serum that has been stored has less significance than on a specimen tested soon after collection. Streptococcus agglutination tests on sera from patients with atypical pneumonia have been done successfully, however, with sera kept in the ice box for as long as four years. In the series studied by Eaton and van Herick, over 40 per cent of the 62 specimens collected in 1941 or 1942 gave agglutination of the indifferent streptococcus when tested three to four years later, but none had cold agglutinins. It appears therefore that the cold agglutinin is much more labile to storage at 4° C than is the streptococcal agglutinin.

An analysis of the relative frequency of occurence of the two agglutination reactions as reported by three investigators is presented in Table 7. A considerable discrepancy between the results in various laboratories will be noted. Two factors in these discrepancies may be (1) differences in the severity

Table 7. *Relation of Cold Agglutination to Streptococcal Agglutination in Atypical Pneumonia.*

| Cold aggl. | Investigator | Agglutination of Streptococcus MG | | | |
| | | Positive | | Negative | |
		No.	per cent*	No.	per cent**
Pos.	CURNEN	13	18	3	4
,,	MEIKLEJOHN	36	41.5	23	26
,,	FLORMAN**	11	**30.5**	9	**25**
,,	Total ...	60	(av.) 31	35	(av.) 18
Neg.	CURNEN	19	27	36	51
,,	MEIKLEJOHN	3	3.5	25	29
,,	FLORMAN**	6	**16.5**	10	**28**
,,	Total ...	28	(av.) 14.5	71	(av.) 36.5

* Per cent of total group tested by investigator named.
** "Doubtful" results counted as negative.

of the illnesses selected for study, and (2) deterioration of cold agglutinin in some serum specimens as a result of variations in the period of storage. The averages for the three sets of results show that in the largest group of sera (36.5 per cent of 194 cases) both reactions were negative. The next largest group consisted of those in which both reactions were positive (31 per cent). The cold agglutination was positive and the streptococcal agglutination negative in 18 per cent. The remaining group (14.5 per of the total) where the cold agglutination was negative and the streptococcal agglutination positive, may be too large because of the possibility that some of the negative results for cold agglutinins may have been due to prolonged storage of the sera. There is therefore some statistical correlation between the appearance of cold agglutinins and streptococcal agglutinins in atypical pneumonia.

Absorption tests reported by THOMAS indicate that the two agglutinins are distinct. Group 0 human cells fail to absorb the streptococcal agglutinin while adsorption with streptococcal suspensions removes the bacterial agglutinin but leaves the cold agglutinin in undiminished titer. Absorption with mouse lung failed to remove the streptococcal agglutinins, and the results of the reverse absorption procedure were also negative.

Streptococcus MG is a single serologic type of non-hemolytic streptococcus which previously was undifferentiated. In analogy to type specific pneumococci, it elaborates capsular polysaccharide which gives precipitin and flocculation tests as well as agglutination. The antibodies against streptococcus MG which develop during primary atypical pneumonia appear to be specifically directed against this microorganism. They usually do not react with other streptococci, either hemolytic or non-hemolytic, nor with a wide variety of other bacterial species. The antibodies produced in primary atypical pneumonia apparently react not only with the capsular polysaccharide but also with a somatic antigen in the rough form of the streptococcus. Since these two antigens differ chemically and are immunologically distinct, HORSFALL and his associates conclude that the streptococcal agglutinins are formed as a result of stimulation by streptococcus MG and are not the resolt uf an accidental immunologic relationship between one antigenic fraction of this microorganim and some other infectious agent. THOMAS observed an antigenic relationship between streptococcus MG and strept. salivarius Type I which is apparently due to antigenic similarities in the capsular polysaccharides of these two organisms. Among 99 sera from patients with atypical pneumonia that agglutinated streptococcus MG six also agglutinated Strept. salivarius Type I and an increase in titer of four-fold or more for the latter organism was demonstrated. in three of the six instances. This indicated a slight non-specificity of the test which may not have any practical importance.

III. Etiology of primary atypical pneumonia.

Many attempts have been made to propagate in the laboratory the infectious agent or agents which cause atypical pneumonia. In most instances the viruses obtained by passage in laboratory animals were either known previously or represented newly discovered infectious agents of animals. These will be discussed in the subsequent section on pneumonitis viruses of animals. Only in one or two instances was evidence for a causal relationship to atypical pneumonia obtained by the demonstration in convalescent sera of neutralizing antibodes for the virus propagated in the laboratory. Several agents at first believed to have come from patients with atypical pneumonia were later isolated from the laboratory animals used in the study. In the remaining the origin of the agent was uncertain because neither neutralization by convalescent serum nor isolation from the laboratory animals themselves was accomplished. The results of experimental transmission of the disease to human volunteers seem to have deminstrated quite conclusively that a filtrable agent is the cause of primary atypical pneumonia. Although these studies were done after most of the laboratory studies with animals, they will be reported first because they establish the nature of the agent involved.

Experimental transmission to man.

The first attempt to transmit atypical pneumonia to human volunteers was made by VANCE and his associates in 1943. They failed to transmit infection with 20 to 40 cc of Berkefeld V or N filtrates of sputum or nasal washings taken from patients with atypical pneumonia and inoculated intranasally into seven persons.

During the second World War an extensive series of experiments was carried out by the Commission on Acute Respiratory Diseases (1946). The subjects were conscientious objectors who volunteered after the nature of the investigation had been completely explained to them. The first experiment was done in 1943 with twelve men whose throats were sprayed with a pool of untreated and unfiltered throat washings and sputa from seven atypical pneumonia patients. The inoculum was sprayed from an atomizer and nebulizer into the nose and pharynx of each volunteer, synchronizing the spray with inhalation so that a part of the material was presumably aspirated. Ten of the twelve developed respiratory infections which varied in their clinical characteristics, severity and course. Two men had mild afebrile upper respiratory infections similar to those that had developed in three uninoculated individuals of same group before the experiment. In the remaining eight cases fever developed from approximately 7 to 22 days after inoculation. Five had auscultory evidence of pneumonia and in three of these roentgenograms showed minimal patchy infiltration in one or both lower lobes; illnesses consistent clinically with moderately severe atypical pneumonia. Three of the eight patients showed no evidence of pneumonia. The sera in three cases of experimentally induced atypical pneumonia showed cold agglutinins in titers of 64, 128, and 256 respectively during the course of the illness. None of the patients developed agglutinins for the indifferent streptococcus or antibodies for influenza viruses. Because three of the fifteen originally selected developed mild respiratory disease in a five-day period of observation prior to the date of inoculation and were then excluded, there remained a possibility that the infections observed in the test subjects were spontaneous.

In the second experiment done in June and July 1944, thirty-six volonteers were placed in strict isolation for a period of three weeks before inoculation. Complete examinations including roentgenographic electrocardiographic and laboratory studies were done and the men were observed on alternate days for symptoms and signs of respiratory infection.

The material used for inoculation had been collected from patients during the two months before the experiment. Immediately after collection the specimens were placed in Pyrex test tubes sealed with rubber stoppers, frozen in a bath of alcohol and solid carbon dioxide and then stored at —70° C. Two days before inoculation the frozen specimens were thawed rapidly in a 37° C water bath and pooled in a proportion of one part sputum to five parts of throat washing. After thorough mixing in a Waring Blendor the material was divided into thirds and treated as indicated below. The three varieties of inoculum were then frozen at —70° in rubber stoppered Pyrex Erlenmeyer flasks until use.

Three groups of twelve persons each were inoculated with three kinds of material as follows:

1) Untreated throat washings and sputa pooled from seven cases of primary atypical pneumonia.

2) The same material filtered through sintered glass or Seitz filters.

3) The same material autoclaved.

Each man recieved a total inoculum of 10 cubic centimeters adminstered in three equally divided doses during one day. Part of the inoculation on a given subject were done with a handspray, and part with a power spray. The equipment was sterilized between each set of inoculations. Isolation precautions were taken in an attempt to prevent cross-infection or extraneous exposure, and a seperate room was used for each group.

The results are shown in Table 8. Ten cases of atypical pneumonia developed after eight to fourteen days among the thirty-six persons inoculated. In addition there were fifteen cases of minor repiratory illness. In both the cases of minor respiratory illness and those of atypical pneumonia, symptoms and signs appeared from one to twelve days after inoculation but the date of onset of the atypical pneumonia was defined as the day of development of persistent symptoms and signs which resulted in pneumonia. The first symptoms appeared 2 to 7 days before the latter. The incubation period was shortest in those receiving untreated material. Whether or not these results indicate the presence of multiple agents of respiratory disease in the inoculum or were merely due to variations from mild to severe in a single etiologic entity cannot be determined by experiments of this nature. There was no change in the bacterial flora of the nose and throat after inoculation or during illness and no serological evidence of infection with a virus of the psittacosis-lymphogranuloma group, or with influenza A or B.

In this experiment atypical pneumonia and minor respiratory illness occurred in volunteers who recieved the autoclaved throat washings and sputum as well as in the others. This was interpreted as indicating cross infection. In addition an attendant who assisted in inoculating twelve volunteers with untreated inoculum also developed atypical pneumonia, and a laboratory worker who assisted in preparing the inoculum developed atypical pneumonia 12 days after the start of the experiment. Possible explanations advanced by DINGLE (1945) for the occurrence of atypical pneumonia in three persons receiving autoclaved material were airborne infection in the hallway on the day of inoculation, contamination of the inner surface of the air pump used for the power spray or non-specific evocation of a latent virus in the respiratory tract.

The study was repeated in August of 1944 with increased precautions including performance of the inoculations out-of-doors, use of a tank of nitrogen as a source of pressure for the power spray, allowance of an interval of four days between inoculation of the autoclaved and filtered matcrial, and an additional interval of 8 days before administration of the untreated inoculum, and segregation of the three groups of men by construction of partitions in the hallways. The inoculum consisted of pooled throat whashings and sputum collected from patients who developed primary aytpical pneumonia during the second experiment and represented therefore a second serial passage.

The results of the experiment are presented in Table 9. In this series three each of the groups of twelve men receiving filtered inoculum or untreated inoculum developed atypical pneumonia and five in each group had minor respiratory illness. Among the eighteen persons receiving autoclaved material none developed pneumonia but there was one case of minor respiratory illness with an incubation period of 19 days. This patient is said to have broken isolation. The onset of illness was earlier and the disease somewhat more severe in those receiving untreated material as compared with those receiving filtrates.

The results of the experiments done by the Commission on Acute Respiratory Diseases indicate that a filtrable agent is implicated in the infection. The clinical types of the disease produced in about three-fourths of those inoculated varied from the mildest repiratory illness to severe atypical pneumonia indistinguishable clinically from the disaeses occurring by spontaneous infection. The results suggest but do not prove that non-pneumonic infection is an important manifestation of the disease. The possibility that some of the milder forms were caused by another agent cannot be excluded.

Table 8. *Production of primary atypical pneumonia and minor respiratory illness in human volunteers inoculated with untreated, filtered, and autoclaved sputa and throat washings from hospitalized cases.*

Inoculum	No. Men	Result		
		AP	MRI	No Illness
Untreated	12	3	9	0
Filtered	12	4	5	3
Autoclaved	12	3	1	8
Total ...	36	10	15	11

AP—Primary atypical pneumonia.
MRI—Minor respiratory illness of undifferentiated type.

Table 9. *Production of primary atypical pneumonia and minor respiratory illness in human volunteers inoculated with untreated, filtered, and autoclaved sputa and throat washings from experimentally produced cases of atypical pneumonia.*

Inoculum	No. Men	Result		
		AP	MRI	No Illness
Untreated	12	3	5	4
Filtered	12	3	5	4
Autoclaved	18	0	1	17
Total ...	42	6	11	25

AP—Primary atypical pneumonia.
MRI—Minor respiratory illness of undifferentiated type.

(Reprinted from the article by J. H. Dingle, Bulletin of the New Aork Academy of Medicine, second series, Yol. 21, page 235; 1945.)

Evidence for the identity of the experimental and natural diseases is also available from the results of cold agglutination tests. Among the 16 cases of experimental atypical pneumonia in the second and third studies, 13 had a four-fold or greather increase in the titer of cold agglutinins between the acute phase and convalescent serum specimens, the maximum titers being 16 to 8192 in these cases. Eleven of the 26 patients showing minor respiratory illnesses without pneumonia developed cold agglutination titers, the increase in 9 cases being only four fold. Four additional cases showed a two fold increase in cold agglutinin titer. The proportion of patients developing agglutinins for streptococcus MG was much lower than in many surveys of the naturally occurring disease. Only 2 of the 16 experimental subjects with atypical pneumonia developed these streptococcal agglutinins.

Experimental transmission to laboratory animals.

A number of the viruses discovered during attempts to transmit atypical pneumonia to laboratory animals are described in two later sections comprising (1) pneumonitis viruses of animals where the agents were subsequently isoladet from animals not inoculated with human materials, and (2) pneumonitis viruses of unknown origin or unclassified where the agents were not actually shown to have come from the animals but where no evidence of etiologic relation to the human disease is available.

Experiments with the Mongoose. In 1939 Weir and Horsfall inoculated sputa and throat washings from patients with acute pneumonitis intranasally into the wild mongoose (Herpestes griseus) and reported the production of pulmonary lesions. The agent producing the lesions was said to be neutralizable by convalescent serum. Although the work was not continued and efforts to repeat it four years later by other investigators were unsuccessful it is still possible that the strains of virus studied by Weir and Horsfall were different from those causing later outbreaks of atypical pneumonia. Because of the fact that evidence for the development of neutralizing antibodies in the patients was presented, the agent transmissible in the mongoose is included as one of the possible causes of pneumonitis in man.

Pulmonary lesions were obtained by serial intranasal passage of lungs suspension from mongooses inoculated with throat washing from four different patients with acute pneumonitis. One of these throat washings had been filtered before the initiation of the serial passage and the others were used unfiltered. Two distinct types of lesions were observed. They consisted of circumscribed areas of consolidation and generalized pulmonary hyperemia. The areas of consolidation were plum colored and of gelatinous consistency varying in size from one millimeter or more in diameter to large lesions filling an entire lobe. Lesions appeared in 37 to 74 per cent of the total number of animals used in each of the four series of passages. Passages in a comparable number of normal animals in four different series resulted in the production of pulmonary lesions in 3 of 11 animals in one series but in the remaining three series of passages of normal mongoose lung using a total of 48 animals no pulmonary lesions were seen.

Bacteria were stated to be absent or present only in small numbers except in the case of one mongoose in the control passages which developed pulmonary consolidation. A heavy growth of gram positive cocci was obtained from the lung of this animal. The agent producing the pulmonary consolidation in the passages initiated from throat washings was filtrable through Berkefeld V and N candles and was not inactivated by glycerin or by freezing and drying in vacuo.

Serial passages on the chorioallantoic membrane of the chick embryo were initiated with filtered throat washings and with a suspension of infected mongoose lung. These chick embryo passages also produced pulmonary lesions in the mongoose. The virus propagated in mongoose lung was not pathogenic for ferrets of mice by serial passage. The material obtained by egg-membrane passages failed to produce lesions in mice, deer mice, moles, woodchucks, oppossums or monkeys on serial intranasal passage. Known viruses such as influenza. A and the pneumonia virus of mice failed to produce pulmonary lesions and were not propagated by serial intranasal passage in the mongoose.

Virus neutralization tests were done with specimens of serum taken early in the disease and during convalescence from patients with primary atypical pneumonia. The sera were mixed with 5 per cent suspension of infected mongoose lungs incubated for 30 minutes at room temperature and inoculated intranasally into two mongooses. Two to five serial intranasal passages were then done from the lungs of the first passage animals. With the acute phase sera from four patients apparently no neutralization was obtained since pulmonary lesions were seen in 25 to 71 per cent of the animals. In the animals receiving infected mongoose lung and convalescent serum and those in subsequent passages the incidence of pulmonary lesions varied from 0 to 25 per cent. No cross neutralization between the mongoose agent and influenza A could be demonstrated. The agent in mongoose lungs was also neutralized by the serum of mongooses recovered from pulmonary infection.

In the summer of 1943 DAMMIN and WELLER did further experiments in the mongoose with material from cases of atypical pneumonia. These specimens had been stored at — 20⁰ C for a number of days before use and also at — 70⁰ C for variable periods of time. Animals were also inoculated intranasally with suspensions of mongoose lung used in the studies of WEIR and HORSFALL, with an agent isolated in cotton rats by EATON et al (to be described later), with influenza viruses A and B, and with the meningopneumonitis virus (a member of the psittacosis group).

Of the total of 265 mongooses used in passages initiated with throat washings from 7 patients, sputa from 6 patiente and lungs from two fatal cases of atypical pneumonia, none developed gross or microscopic evidence of the presence of a transmissable agent. In one animal a pneumonic process consisting of a small area with polymorphonuclear leukocytes and round cells distributed through the interstitial tissue and alveoli was seen. No subsequent lesions appeared in the two later passages. In one other experiment, the first passage from a sample of human lung, gross lesions were present in the lungs of four experimental animals. In three of these the changes were found to be due atelectasis and congestion and in the fourth there was a moderate inflammatory reaction with interstitial infiltration of polymorphonuclear leukocytes and large monocytes. Later passages were negative.

Passages in the mongooses initiated with two specimens of the mongoose agent of Weir and Horsfall resulted in no gross pulmonary lesions In two lungs, scattered areas of minimal interstitial infiltration with polymorphonuclear leukocytes and lymphocytes, moderate bronchiolitis and bronchitis were seen by microscopic examination but no lesions appeared on subsequent passage. No pneumonic process was seen in animals inoculated with the cotton rat agent or with the viruses of influenza A or B.

Definite pneumonia was produced in the mongoose with the meningopneumonitis virus, but without definite signs of illness. The lungs of animals sacrificed on the fourth day presented numerous irregular areas of reddish gray consolidation in several lobes which on microscopic examination showed the cellular changes characteristic of a virus pneumonia. Meningopneumonitis virus was recovered by intranasal inoculation of mice.

Experiments with Cotton Rats. In 1941 the writer observed that suspensions of a specimen of lung tissue from a patient dying with atypical pneumonia when inoculated by the intranasal route into cotton rats of the species *Sigmodon hispidus tremicus* produced non-bacterial pulmonary lesions in 21 of 28 animals. Attempts to reproduce the lesions on the 2nd or 3rd passage from the animals receiving the original human material were unsuccessful. Later in 1941 J. C. Talbot in the same laboratory inoculated cotton rats with sputum from patients with atypical pneumonia and observed pulmonary lesions in a portion of the animals.

More extensive studies were then done with both eastern cotton rats (*Sigmodon hispidus hispidus*) and western cotton rats (*Sigmodon hispidus eremicus*), both of which are about equally susceptible. The results published by Eaton et al in 1942 and 1944 indicated that sputum taken early in the disease often produced lung lesions rather consistently when several cotton rats were inoculated with the same specimen. Specimens from a total of 153 cases of primary atypical pneumonia were studied. With 38 specimens of unfiltered sputum and 3 specimens of lung tissue non-bacterial pulmonary lesions were obtained in cotton rats after primary intranasal inoculation. Specimens from 9 of these cases produced pulmonary lesions in 25 to 33 per cent of the cotton rats inoculated, specimens from 18 patients produced lesions in 35 to 40 per cent, and from the remaining 14 patients in 60 to 100 per cent of the cotton rats. Of the total of 190 cotton rats used for these 41 positive specimens, 83 animals developed pulmonary lesions.

Control series of cotton rats of both species were inoculated with material presumed to be free of any agent pathogenic for cotton rats. These materials included throat washings from patients with diseases other than atypical pneumonia, sputa from atypical pneumonia heated to 60° C for one hour, broth, suspensions of normal chick embryo tissue, and normal cotton rat lung. The incidence of pulmonary lesions in a total of 136 animals in this control series was 4.2 per cent or about one sixth of that seen in all animals inoculated with sputa from cases of atypical pneumonia.

These observations indicated that an agent capable of producing pulmonary lesions in 25 to 100 per cent of the cotton rats inoculated was present in certain samples of sputum from patients with atypical pneumonia. This agent was apparently absent from other samples of sputum particularly those taken late in the disease, or those kept at room temperature or in the ice box for some time before inoculation. It was also absent from materials obtained from other sources and from heated sputum. Further analysis of the results showed that specimens obtained early in the disease were particularly likely to produce pulmonary lesions in cotton rats. A summary of these data is presented in Table 10. It will be seen that 52.5 per cent of the specimens collected between the third and fifth day after onset produced significant pulmonary lesions in cotton rats while only 7.1 per cent of those collected between the tenth and twenty-fifth day of illness produced lesions. There was also a significant but less definite relation to the subsequent development of cold agglutinins in the patient's serum. Pulmonary

Table 10. *Relation of Time of Collection of Sputum Specimens to Results of Primary Inoculation of Cotton Rats*

Kind of Specimen	Days After Onset	Number of Positive	Specimens Negative	Per cent Positive*
Sputum	3—5	21	19	52.5
	6—9	11	29	27.5
	10—25	2	26	7.1
	Unknown	4	16	20.0
Lung	5—9	1	6	16.6
	10—25	2	6	33.3
Total specimens		41	102	

* Positive specimen produces pulmonary lesions in 25 to 100 per cent of the group of cotton rats inoculated.

lesions were obtained in cotton rats with the sputum of 10 of 18 patients in whom the presence of cold agglutinins was subsequently demonstrated while positive results were obtained in only 3 of 16 patients who did not develop cold agglutinins in their serum.

The gross appearance of the pulmonary lesions in cotton rats inoculated with the various specimens was of uniform character. These lesions were found most frequently near the hilum and varied in width from 2 to 10 millimeters. Sometimes several small irregular patches were scattered throughout one or more lobes. The lesions reached a maximum intensity between the seventh and fourteenth days after inoculation, were grayish-red or gray and usually level with the surrounding tissue which was pink and normal in appearance or only slightly hyperemic. On microscopic examination pathological changes resembling those of a virus pneumonia were seen. Perivascular and peribronchial infiltration with lymphocytic cells was a prominent feature and this sometimes occurred in the absence of definite alveolar reaction. Atelectasis infiltration of the alveolar walls, and alveolar exudate consisting of mononuclear cells and a very few polymorphonuclear leucocytes were seen in the consolidated areas. Frequently the alveolar walls contained pathological cells of polyhedral form with large pale — staining nuclei. Microphotographs of sections of the lung of a cotton rat inoculated with sputum are shown in Figure 2. No inclusion bodies, elementary bodies, rickettsiae or visible microorganisms were seen in sections or impression smears stained by the methods of Gram, Giemsa, or Macchiavello. Cultures on blood agar and horse serum broth were negative.

Bacterial pulmonary infection were seen very infrequently in cotton rats inoculated with unfiltered human material. In the series of experiments conducted by Eaton et al sputum from seven patients and lung tissue from one patient produced pneumococcal pneumonia in cotton rats which was fatal for all animals within 72 hours. In three instances gram-negative bacilli probably of the hemophilus group produced edematous red pulmonary consolidation. The pneumonia produced by the bycteria was definitely suppurative in contrast to the virus-like lesions described in the previous paragraph.

At about the same time as the above studies in California independent observations with eastern cotton rats were carried on by Johnson in Alabama who inoculated these animals with throat washings and sputum specimens from six cases of atypical pneumonia. The specimens from 4 patients produced an acute hemorrhagic type of pneumonia which appeared 9 or more days after inoculation. In some instances lesions were obtained in cotton rats on further intranasal passage of lung suspensions but little or no increase in pathogenicity was apparent on passage. In other passage series the lesions failed to persist with any degree of constancy. Acute splenitis and variable degrees of hepatic necrosis were also observed by Johnson and in occasional instances intracytoplasmic inclusion bodies in sections of the lung were reported.

Horsfall and his associates in 1943 observed that one of several sputum specimens studied by them quite consistently produced pulmonary lesions in cotton rats after primary intranasal inoculation but not after serial passage. Lung consolidation was obtained in 25 of 32 cotton rats recieving this specimen. Neutralization tests were done by mixing a 20 per cent suspension of the sputum with human sera and inoculating a group of 3 or 4 cotton with each sputum-serum mixture. With the sputum and acute-phase serum from 6 patiens with atypical pneumonia 68 per cent of 25 inoculated rats developed pulmonary consolidation, whereas no lesions were obtained in an equal number of cotton rats inoculated with sputum and convalescent serum from the same 6 patients. In similar tests with acute-phase and convalescent serum from 5 other patients no pulmonary lesions were obtained. This indicates neutralization by specimens taken both early and late in the disease in the latter 5 cases. These observations suggested that some patients with atypical pneumonia developed during their illness antibodies which neutralized the infective agent in sputum demonstrable by inoculation of cotton rats. The paper cited contains other observations on the development in the inoculated animals of neutralizing antibodies to the pneumonia virus of mice. Since these observations now appear to be unrelated to the production of pulmonary consolidation as discussed above, they will be covered in a subsequent section.

It is evident from the above data that three different investigators have observed the presence of a non-bacterial agent in sputum or lung from cases

---

Fig. 2. Section of lung from a cotton rat inoculated intranasally with sputum from a patient with primary atypical pneumonia and sacrificed ten days later. Stain, hematoxylin and phloxine.

a) Low power X 15. In several places the alveolar walls are thickened and merge with patchy areas of consolidation. Round dark spots represent perivascular and peribronchial infiltration.

b) Higher magnification (X 185) showing infiltration with lymphocytic cells around bronchioles. Exudate in bronchial lumen consists of polymorphonuclear leucocytes and lymphocytes.

c) Same at high magnification (X 500) Alveolar exudate consists of mononuclear cells and a very few polymorphonuclear leucocytes. Many pathological cells of polyhedral form with large pale-staining nuclei are also to be seen.

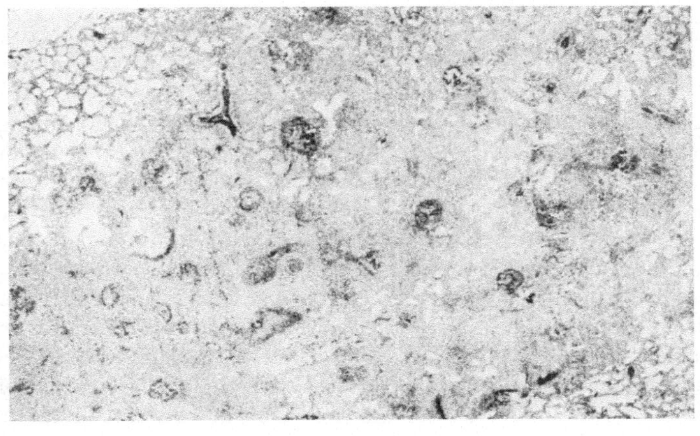

a

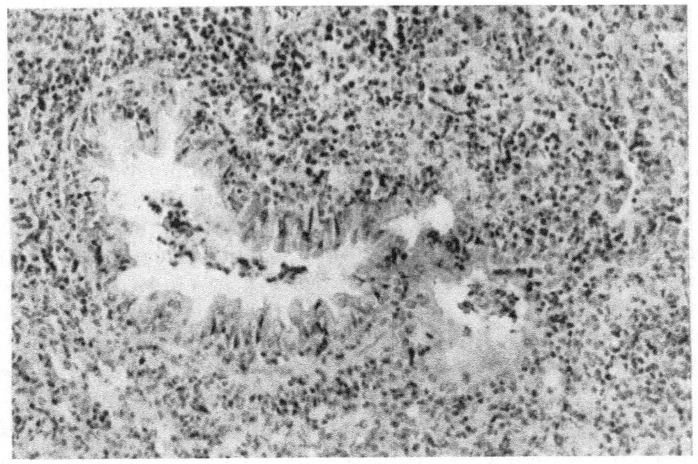

b

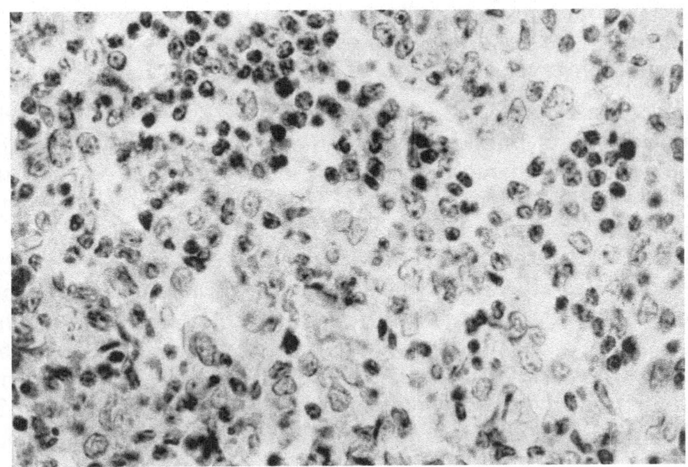

c

of atypical pneumonia which produced pulmonary lesions in cotton rats. The studies so far presented do not serve to elucidate the nature of the agent whether organism, virus, or toxin. Attempts to adapt the agent or agents to cotton rats by serial intranasal passages of suspension of lung led to confusing and contradictory results because of contamination with latent respiratory agents already present in the animals. Similar difficulties had been encountered previously by other investigators during attempts to adapt viruses in human throat washings or sputum to mice by serial intranasal passage of lung suspensions. In these and subsequent investigations with other animals this procedure has led almost invariably to contamination or overgrowth with animal viruses. The one important exception is the adaptation of influenza virus to ferrets and mice by this method. As influenza has a relatively short incubation period in both of these species it seems likely that it overgrows or interferes with the multiplication of the latent animal viruses. Unfortunately this was not true of the cotton rat agents. The virus present in the human material aparently had a longer incubation period than that of the latent cotton rat agents with the result that the human strain became contaminated with the cotton rat viruses and was probably displaced by it.

Contaminating viruses. The repeated appearance of a contaminating virus on serial passages of cotton rat lungs was quite evident from the work of Eaton, Meiklejohn and van Herik (1944). In a total of 52 cases studied by them several intranasal passages were carried out. Sputum from 24 patients failed to produce lung lesions on primary inoculation and "blind" passages from the lungs of the first passage animals were uniformly negative. In 19 cases lesions were obtained in one or more cotton rats on primary inoculation but subsequent passage were either negative or lesions appeared in only the first few passages. Continued serial passage in nine cases resulted in the establishment of an agent in cotton rats which consistently produced lung lesions and in four cases this result was repeated starting with the original material. The agents in cotton rats which were responsible for the production of lesions in the later passages were not, however, identical with the virus in the human material. Definite evidence of neutralization by convalescent human serum could not be obtained and an identical agent was isolated from normal cotton rat lungs. This agent was designated W_2. In two or three other passage series an unrelated agent W_3 was found. These two viruses will be described in greater detail in the section on animal viruses. The successful passage of the human atypical pneumonia virus in chicken embryos is described in the next section.

Toxic agents in lung tissue. Unpublished observations by the writer suggest that certain samples of human lung tissue contain a toxic agent which produces an acute hemorrhagic edematous reaction in the lungs of cotton rats resulting in death of the animals in 2 to 5 days. This shorter incubation period and the high fatality seem to differentiate this reaction from the slowly developed nonfatal interstitial infiltration observed with specimens of sputum and other samples of lung tissue as described above. The acute toxic reaction in cotton rat lungs has been obtained with only two samples of human lung tissue, one of which was collected from a fatal case in 1941[1] and the other in 1945. Both samples of lung contained staphylococci in small numbers but broth cultures of the organism failed to produce the pulmonary reaction in cotton rats. The significance of this acute edematous reaction resulting from the inoculation of human

---

[1] The case referred to as Bl in Journ. Exp. Med. **79**, 649, 1944.

lung is at present uncertain. The following summary serves to differentiate this reaction (agent X) from the lesions produced by the atypical pneumonia virus as found in sputum or as proagated in chick embryo.

	Atypical pneumonia virus	Agent X
Type of pulmonary lesion in cotton rat	Interstitial	Hemorrhagic&edematous
Incubation period	8 to 14 days	2 to 5 days
Fatality	5 per cent	30 to 60 per cent
Passage in chick embryos ...	successful	unsuccessful
Homologous immunity in recovered animals	definite	irregular

Agent X was destroyed by heating to 56⁰ C for 30 minutes, by 0.1 per cent formal dehyde at 5⁰ C for 4 days but was not inactivated at 37⁰ C for 4 hours. The human lung suspension incubated at 20⁰ C with 500 units of penicillin per cubic centimeter still produced pulmonary edema and death in cotton rats. The agent was filtrable with diminished activity through a Seitz E. K. filter. The apparent neutralization of this agent by certain rabbit sera and the production of immunity to it by inoculation of cotton rats with certain strains of cotton rat virus are at present unexplained phenomena.

Experiments with chick embryos. After the unsuccessful attempts to adapt the virus of atypical pneumonia to cotton rats and similar difficulties with contaminating viruses on passage in hamsters, adaptation to chick embryos was attempted as a possible means of propagating the virus. Suspensions of sputums were filtered through collodion membranes of 600 millimicrons average pore diameter and the sterile filtrates were inoculated into the amnion of 11 or 12 day old chick embryos. On serial passage in chick embryos the adaptation of the virus from human material was accomplished in 5 of 10 trials. One strain was isolated from a bacteriologically sterile sample of human lung tissue. As the chick embyros showed no significant pathology it was nesecsary to inoculate Syrian hamsters (Cricetus auratus) and cotton rats intranasally with suspensions of chick embryo tissues to demonstrate the virus. This procedure produced in 25 to 75 per cent of the animals pulmonary lesions which appeared on gross and microscopic examination to be identical with those resulting from inoculation of sputum. The differences between the plan originally used for passage in cotton rats and hamsters and the experimental plan for the chick embryo passages together with the results are shown in Table 11. As indicated in the previous section the "orthodox" method shown at the top of the table was unsuccessful due to the appearance of contaminating viruses in the cotton rats and hamsters. The chick embryo procedure resulted in the propagation of a filtrable virus which was neutralizable by convalescent serum.

In the first series of experiments the 11 day old chick embryos were inoculated into the amniotic sac through a 5 mm. opening in the shell according to the method of Burnet. In later experiments a modification of the stab method described by Hirst was used with equal success. The egg was transilluminated from below and the air sac elevated with the long axis at an angle of about 45⁰ to the horizontal, then rotated so that the embryo could be located in the angle formed by the inner air sac membrane and the shell. Inoculation was done through a hole in the air sac with a $1\text{-}^1/_4$ inch 24 gauge long bevel needle attached to a syringe of one or two cubic centimeters capacity. By a short quick motion the embryo was gently impaled on the point of the needle and 0.1 cc of the inoculum injected. If the latter operation was performed so that only the point of the needle entered the embryo most of the inoculum passed through the open part of the bevel into the amniotic cavity. The embryos were then incubated at 36.5^0 C for an additional period of 5 to 7 days and the amniotic membranes, lungs and trachea were collected, ground in broth containing 50 units of penicillin per cubic centimeter and stored in

Table 11. *Methods of Isolating the Virus of Atypical Pneumonia*

	Primary inoculation	Early passages	Late passages
Unfiltered sputum	⟶ Cotton rats	⟶ Cotton rats	⟶ Cotton rats
	Hamsters	⟶ Hamsters	⟶ Hamsters
	"Specific" lesions	Small irregular lesions Questionable neutralization	"Non-specific" lesions No neutralization

Filtered sputum ⟶ chick embryos ⟶ chick embryos

hamsters   cotton rats

"specific" lesions on primary inoculation: virus in chick embryo material is neutralizable.

(Reprinted from California and Western Medicine, Vol. 63, No. 3, 1945.)

sealed glass ampules at -70^0C in a solid carbon dioxide ice box. For intranasal tests in hamsters or cotton rats aliquote samples representing a pool of 6 to 20 embryos were used. Growth of the virus in individual embryos was found to be irregular, some developing eggs containing much more of the agent than others.

Cotton rats and hamsters were inoculated intranasally under ether anaesthesia with the 20 per cent suspensions of chick embryo tissue. It was found that a more effective aspiration of the inoculum could be obtained by inducing a second brief period of light anaesthesia after the inoculation. Cotton rats inoculated with chick embryo tissue were autopsied 10 to 11 days after inoculation while hamsters were sacrificed at 8 days. Animals were killed with chloroform and autopsied as soon as respiration had ceased. The pulmonary lesions were graded immediately after removing the lungs, each lobe being examined carefully. Immediate examination is essential because lungs of animals which appear normal on gross examination immediately after autopsy frequently develop, after standing for 15 minutes or more, areas of collapse or local congestion which can be mistaken for pulmonary lesions. The lesions characteristic of the chick embryo passage virus resembled those seen after inoculation of human material. They occured as irregular but well demarcated grayish-red patches of consolidation 2 mm to 10 mm in width often near the hilum or sometimes scattered through one or more lobes. The lesions produced in hamsters were very similar to those seen in cotton rats. The histopathology of the lesions in both species of animal was almost identical to that shown in Figure 2[1].

Using this method of adaptation to chick embryos with intranasal subinoculation of hamsters or cotton rats to demonstrate the presence of the virus, five seperate strains have been isolated. The first strain was isolated in December 1942 from a bacteriologically sterile sample of human lung tissue which was passed to chick embryos without filtration. This strain has been carried for approximately fifty passages in chick embryos. These passage experiments were successfully repeated, and the same strain was reisolated from the original sample of human lung. The pathology of this case has been presented in a previous section. Three strains were isolated from suspensions of sputum filtered through collodion membranes of 600 millimicrons average pore diameter but two of these strains were subsequently lost. A fifth strain was isolated from a filtered suspension of human lung in 1944.

In other experiments filtered throat washings from two patients with an influenza-like illness and filtered sputum from one patient with type A influenza were inoculated into the amnion of chick embryos and serial passages were carried out. Several blind serial passages starting with broth were also done in the same manner on two different occasions. None of the material from the embryos in these control passages produced significant pulmonary lesions in cotton rats or hamsters.

Cross immunity tests by reinoculation of recovered hamsters or cotton rats indicated the identity of the agent present in human lung or sputum with the filtrable agent carried in chick embryos. Animals immunized with chick embryo material by repeated intranasal inoculation did not develop pneumonia after intranasal inoculation of a sample of sputum which produced definite signs of pulmonary infections in the controls. Hamters immunized by intranasal inoculation with suspensions of human lung were immune to the same strain which had been adapted to chick embryos.

The virus propagated in chick embryos grew best after amniotic inoculation and was present in highest concentration in the lungs and amniotic membrane. Chorionic and allantoic passages were apparently unsuccessful. Yolk sac inoculation and passage resulted in survival of the virus through four passages in the yolk sac but the degree of multiplication was not great enough to produce pulmonary lesions in hamsters, when suspensions of yolk sac were inoculated intranasally. There were marked variations in the incidence of pulmonary lesions produced by different lots of chick embryo material. Only about one third of all the lots of suspension of chick embryo, lung, trachea, and amniotic membrane prepared from the various passages produced pulmonary lesions in a high proportion of the animals inoculated. Of the total of 970 animals inoculated with infected tissues 572 or 59 per cent developed pulmonary lesions. Only 8 or 2.24 per cent of 375 animals inoculated concurrently with suspensions of lungs, trachea and amniotic membranes of normal 19 day old chick embryos developed pulmonary lesions. The virus was infectious for

---

[1] Other studies not reported in detail had shown that hamsters as well as cotton rats develop pulmonary lesions after inoculation of sputum or suspensions of lung tissue from patients with atypical pneumonia.

chick embryos at a dilution of 10^{-4} and produced pulmonary lesions in hamsters only at a dilution of 1:500 or 1:2500 of wet chick embryo tissue.

Chick embryo suspensions could be stored in sealed glass tubes at -70^0 C for as long as 14 months without complete loss of activity but storage in rubber-stoppered tubes resulted in inactivation within one to seven months probably from access of CO_2 gas to the virus suspensions. The virus remained active for 2 to 3 days at 5^0 C. Preparation kept at 20^0 C for 4 to 6 hours showed marked diminution or complete loss of activity. This high lability of the virus may account for some of the difficulties experienced by certain observers in isolating or working with the agent. Material dried from the frozen state showed marked loss of activity and in only one or two instances could the virus be recovered by passage in chick embryos. Filtration experiments using graded collodion membranes showed that the virus was apparently retained by filters of 300 and 336 millimicrons average pore diameter but passed filters of 400 millimicrons indicating a particle size of about 180 to 250 millimicrons. Because of the irregularities of propagation of the virus in chick embryos and other uncertainties of the filtration technique these results can be considered to establish the maximum size of the virus but not necessarily its true size. The agent also passed a Berkefeld N filter.

Growth of the virus in chick embryos is inhibited by aureomycin inoculated into the yolk sac in two doses of one milligram each 2 and 48 hours after amniotic infection. Development of pulmorary lesions in cotton rats is also prevented by intraperitoneal injection of 1 to 2 milligrams daily and the antibiotic is effective when the treatment is delayed for 4 days or more after the virus is inoculated intranasally. (Eaton, M. D. Jour. Clinical Investigation 28 No. 4, 1949 and other observations to be published.)

These properties of the filtrable agent propagated in chick embryos should be compared with the findings of others who worked with the mongoose, and with human volunteers. Important differences of the mongoose agent were its apparent stability to drying and its ability to grow on the chorioallantoic membrane. The agent inoculated into human volunteers was unfortunately not studied for its ability to grow in chick embryos, and no inoculations of human volunteers have yet been done with the virus propagated in chick embryos. It is therefore uncertain whether the same agents were involved in the two series of experiments but there is a strong indication from serological data that this was the case. Increases in neutralizing antibody titer have been observed in persons working in the laboratory with the chick embryo virus[1].

The recent observations on effectiveness of aureomycin in the treatment of atypical pneumonia in human beings also suggest that the filterable agent causing the human disease and the virus propogated in the laboratory are related or identical. The only other agents smaller than bacteria at present known to be inhibited by aureomycin are the rickettsiae and viruses of the psittacosis-lymphogranuloma group and there is no evidence that atypical pneumonia is caused by agents belonging to either of these classifications.

Antigenic relationship of the several strains isolated was shown by re-inoculation experiments in cotton rats and hamsters. The strains were also neutralized by the serum of recovered cotton rats and hamsters and by the serum of rabbits repeatedly inoculated with chick embryo material. In certain instances incomplete reciprocal cross neutralization with immune rabbit sera

---

[1] Eaton — unpublished observations.

was found[1]. This indicates possible antigenic differences among the several strains. Hamsters and cotton rats were also immunized against pulmonary lesions by intraperitoneal inoculation of active or formalinized chick embryo material. After infection the duration of immunity in hamsters was apparently more than five months and in cotton rats less than three months. Immunization with three different infectious agents isolated from hamsters or cotton rats produced immunity against the homologous animal viruses but not against the agents propagated in chick embryos. Further details as to the identity of the several pneumonitis agents are given in the section on animal viruses.

These observations established that the pulmonary lesions produced in cotton rats or hamsters by inoculation of infected chick embryo tissue were due to an agent propagated in the embryos and not to mobilization of a latent virus in the animalis by some unknown factor in the inoculum. These facts were substantiated by the unpublished observations of three independent investigators who worked with the strains isolated by EATON et al.

Neutralization results with human serum. Neutralization tests with sera from patients in the acute and convalescent stages of atypical pneumonia were done by mixing these sera with suspensions of the infected chick embryo tissues and inoculating groups of cotton rats or hamsters intranasally with the mixtures. Because of the instability of the virus at room temperature it was necessary to limit the period of incubation to 20 minutes at 20^0 C and to inoculate the animals as rapidly as possible. When large series of tests were set up a portion of the tubes were kept in the ice box after the preliminary incubation period until the time for inoculation. Suspensions of virus containing 25 to 50 per cent normal horse serum remained active for longer periods of time at room temperature than did plain broth suspensions. Consequently dilutions of the human serum were made in normal horse serum. To the serum was added an equal volume of virus suspension obtained by pooling lungs trachea and amniotic membrane pooled from 10 to 20 chick embryos. Preliminary tests were done on various lots of 20 per cent suspensions of chick embryo material and lots which produced on preliminary test pulmonary lesions in over half of the number of animals inoculated were selected for use in neutralization tests. Acute-phase and convalescent serum specimens from the same individual were, with a very few exceptions, tested at the same time. Four fold dilutions of the acute-phase and convalescent serum starting with serum undiluted or diluted 1 in 2 depending on the quantity available were tested in parallel and when no change in titer was found, additional tests were done with 2-fold dilutions. Virus controls were done with mixtures of virus and normal horse serum, and controls for intercurrent repiratory disease in the animals by inoculation of a mixture of normal horse serum and suspension of lungs, trachea, and amniotic membrane from normal chick embryos.

Paired acute-phase and convalescent serum specimens from 213 persons with various acute respiratory infections were titrated for neutralizing antibodies against the atypical pneumonia virus (EATON and VAN HERICK, 1947). Among 84 cases of broncho-pneumonia, apparently non-bacterial in etiology and resembling clininally primary atypical pneumonia, increases in neutralizing antibody titer of 4 to 64 fold were found in 52 cases or 62 per cent of the total. In addition there were 10 patients in whose serum a two-fold increase in neutralization titer occurred during illness. Neutralization tests were done on

---

[1] Ibid. (Footnote p. 71).

77 patients with acute upper respiratory infection other than influenza A or B. Of these 15 or 19.5 per cent had increases of neutralizing antibodies amounting to 4-fold or greater. Among the cases of atypical pneumonia and acute respiratory infection with increases in neutralizing antibodies for the virus propagated in chick embryos, 84 per cent of the acute phase sera had titers of 4 or less while 80 per cent of the convalescent sera had titers of 8 to 256.

Neutralization tests were also done with acute-phase and convalescent sera from 12 patients with pneumococcal pneumonia, 8 patients with pneumonia associated with influenza A, 27 cases of influenza A without pneumonia and 5 cases of pneumonia caused by psittacosis-like viruses. There were no significant changes in antibody titer for the atypical pneumonia virus in this group except in one case of influenza A where a four-fold increase in titer was observed and 2 cases of bacterial pneumonia in which two-fold increases were found. These small and infrequent antibody increases in other diseases might be interpreted as anamnestic reactions or they may have resulted from dual infections with two viruses or with a virus and bacterium. Interpreted provisionally the results indicated that antibody responses to the atypical pneumonia virus of four-fold or greater signified probable infection with this virus and were designated as positive neutralization. An antibody increase of two-fold, although suggestive was not considered to be positive.

An analysis of the relation of virus neutralization to other serological reactions was made. The association of streptococcal agglutination and cold agglutination in primary atypical pneumonia has been discussed previously. The data presented in Table 12 indicate that those patients with atypical pneumonia who develop agglutinins for the indifferent streptococcus MG and cold isohemagglutinins almost invariably also have a significant increase in titer of virus neutralizing antibodies. It is not known at present whether this is due to association of the streptococcus MG with the virus in the causation of primary atypical pneumonia or to production of agglutinins for the streptococcus as a result of the virus infection alone. Since a rise in neutralizing antibodies occurs in most instances where development of streptococcal and cold agglutinins has accompanied primary atypical pneumonia these simpler tests when applied to a group of cases in an epidemic might if positive in a high proportion be provisionally considered indicative of the virus etiology even when no neutralization data were available. An analogy exists in the interpretation of the Weil-Felix reaction as indicative of typhus infection. On the other hand negative agglutination reactions in an individual case do not rule out infection with the atypical pneumonia virus. It will be seen in Table 12 that in nine cases of atypical pneumonia with four-fold or greater increases in virus neutralizing antibodies both cold agglutination and streptococcal agglutination were negative and in 24 of the other cases where a four-fold or greater increase in neutralization occurred either the streptococcal agglutination or the cold agglutination was negative.

By means of the neutralization test limited etiological and epidemiological studies were made on atypical pneumonia in various parts of thr United States. In four localized epidemics of acute respiratory illness associated with pneumonia occurring in 1941 and 1942 a high proportion of the cases showed positive neutralization with significant increase in titer of antibodies for the atypical pneumonia virus (Table 13). One of these outbreaks at Kasson, Minnesota has already been described in detail.

The outbreak of atypical pneumonia first studied in detail by Eaton et al occured at the University of California in August 1941 and subsequent months

Table 12. *Relation between streptococcal agglutination, and virus neutralization*

Streptococcal agglutination	Cold agglutination	Number	Response in virus-neutralizing antibodies		
			No. increase	Increase of 2-fold	Increase of 4-fold or more
Positive*	Positive*	14	0	1	13
Positive*	Negative	4	1	0	3
Positive*	No test	7	1	0.	6
Totals		25	2	1	22
Negative	Positive*	14	4	2	8
Negative	Negative	22	8	5	9
Negative	No test	23	8	2	13
Totals		59	20	9	30

* Increase in titer.

(Reprinted from American Journal of Hygiene, 45, 82, 1947.)

when there was a marked increase in the incidence of this disease among students with continued prevalence but gradual decrease in incidence throughout 1942. Positive neutralization tests were obtained on the sera of 6 cases occurring in 1942. In 28 of 41 cases occurring in 1941 and 1942 positive agglutination

Table 13. *Primary Atypical Pneumonia and Undifferentiated Upper Respiratory Infections Epidemic and Sporadic in Various Population Groups*

Place	Year	Virus neutralization in				Agglutination streptococcus no. 344 in		
		Primary Atypical Pneumonia		Upper Respiratory Infections		Primary Atipical Pneumonia and Upper Respiratory Infections		
		Number Negative	Number Positive	Number Negative	Number Positive	Number Negative	Number Positive	% Pos.
University of California .	1941—42*	0	6	6	2	80**	46	58
Kasson, Minn..........	1942*	1	4	0	3	17**	12**	41
Barnes G. H., Wash. ...	1942*	1	3	2	1	21	2	9
Fort Baker, Calif.	1942*	1	2	1	0	12	1	8
Camp Adair, Ore.......	1942*	—	—	5	0	15***	0	0
Santa Anita, Calif......	1943*	—	—	6	0	15***	1****	7
Camp Kearns, Utah ...	1943*	3	0	4	1	13**	0	0
Camp Roberts, Calif. ..	1942*	—	—	4	1	6***	0	0
Camp Roberts, Calif. ..	1944	0	3	2	1	8**	0	0
Camp Roberts, Calif. ..	1945	0	3	—	—	3	3	50
Santa Ana AAB, Calif. .	1943	4	4	—	—	37	0	20
Fort Bragg, N. C.......	1943—44	6	7	2	0	14	1	7

* Epidemic respiratory disease.
** Includes both primary atypical pneumonia and upper respiratory infections.
*** Includes only upper respiratory infections.
**** Upper respiratory infection with no increase in virus-neutralizing antibodies.

(Reprinted from American Journal of Hygiene, 45, 82, 1947.)

of streptococcus M. G. was demonstrated. Several other local epidemics gave negative serological tests both for the atypical pneumonia and for influenza A and B. These date are summarized in Table 13. In addition atypical pneumonia and acute upper respiratory infections with positive virus neutralization were found to occur sporadically from 1941 through 1945 in serval regions of the United States. Increases in neutralizing antibody titer were found in 7 of 13 samples of serum submitted by the Commission on Acute Respiratory Diseases from Fort Bragg, North Carolina. The presence of these positive neutralization tests and the appearance of cold agglutinins in others of the Fort Bragg cases suggest that the disease studied in human volunteers may have been the same as that furnishing the virus propagated in chick embryos. There is however no direct evidence for this. Additional positive neutralization tests not included in Table 13 have been obtained from cases in the eastern United States and from cases occurring in Norway[1] in 1946—47. Serum obtained by Dr. Gordon Meiklejohn from natives on the island of Okinawa in East Asia also showed low titers of neutralizing antibodies. These data suggest a possible world-wide distribution of the agents producing neutralizing antibodies to the virus of atypical pneumonia propagated in chick embryos.

Etiologic relation of the Indifferent Streptococcus M. G. to Primary Atypical Pneumonia. Data on the serological reactions with the streptococcus isolated by Thomas et al have been given in a previous section. The organism in pure culture is not pathogenic for experimental animals. The present discussion is limited to the work of investigators who have studied the distribution of this organism in normal persons and in patients with atipical pneumiona. Thomas and his associates found the organism in the respiratory tract or the respiratory secretions of about 25 per cent of a group of persons with acute respiratory disease other than atypical pneumonia. In primary atypical pneumonia this bacterium was recovered from sputum or throat swabs in 55 per cent of their patients and from the lungs of 6 out of 8 fatal cases of atypical pneumonia. In the studies with human volunteers done by the Commission on Acute Respiratory Diseases, intensive efforts were made to recover streptococcus M. G. from each patient. Among the 16 individuals who developed pneumonia the organism was not recovered from 7 either before or after inoculation, and in 2 other cases it disappeared during the period of the experiment. The streptococcus M. G. was found in the throats of about 25 per cent of the human volunteers who did not develop primary pneumonia. These results and the pathological data previously presented where no organisms could be seen in the lungs indicate that primary atypical pneumonia may occur in the absence of streptococcus M.G. from the respiratory tract. The development of agglutinins for this organism in the more severe forms of atypical pneumonia suggests a possible specific synergistic relation of this bacterium and the virus in such cases. Proof of this relation is at present not available.

Pneumonitis viruses of animals transmissible to man.

1. Agents of the psittacosis group isolated from human beings.

The psittacosis-lymphogranuloma group of viruses contains a large number of related infectious agents widely distributed in nature. This group now includes the viruses of psittacosis, ornithosis, lymphogranuloma venereum, and pneumonitis of mice, of man, of hamsters, and of cats. The agents of trachoma

---

[1] These specimes were sent to us by Dr. Bjorn Knutsen.

and inclusion blennorrhea are also included because of morphological and serological relationships to the other viruses. This group of agents has been placed in a family designated *Chlamydozoaceae* by MOSHKOVSKY and this classification is being adopted in the new edition of BERGEY's Manual.

The viruses of psittacosis and ornithosis found in parrots, pigeons, chickens, and other birds are occasionally transmitted to human beings. In such cases of virus pneumonia the disease is usually severe and the patient gives a history of contact with birds. In many cases where ornithosis is transmitted from pigeons or chickens to human beings there is evidence of massive infection by inhalation of dried finely divided feces while the person was cleaning the cages. Since psittacosis and ornithosis are adequately discussed by other authors this section will be limited to those viruses of the psittacosis-lymphogranuloma group which are not know to be of avian origin. Some of these agents are closely related to psittacosis or ornithosis. Other members of the group which are not known to be pathogenic for man come from animals such as cats or rodents. Some of the agents to be described may have been transmitted from animals to man but epidemiological evidence of this is lacking.

In 1940 EATON, BECK, and PEARSON isolated a strain of psittacosis-like virus from the lungs of two fatal cases and the sputum of two non-fatal cases of atypical bronchopneumonia. At the time of these studies the disease was called atypical pneumonia because of its resemblance to other cases described in the literature. It was later recognized, however, that the disease differed both clinically and etiologically from the atypical pneumonia caused by the unrelated agent transmissible to chick embryos, cotton rats and hamsters. For this reason the term atypical pneumonia is now considered a misnomer when applied to disease caused by a virus of the psittacosis group and use of the term, virus pneumonia or pneumonitis of the psittacosis group would appear to be a better designation.

The virus which was called the S-F strain was isolated by direct intranasal inoculation of mice. Intraperitoneal injection of the same human material into mice gave negative results. The virus had a relatively high virulence for mice by intranasal or intracerebral inoculation but did not kill after intraperitoneal inoculation. Its virulence for ricebirds, pigeons, and parakeets was relatively low. A carrier state in surviving ricebirds, pigeons, or mice inoculated intraperitoneally was rarely found. The S-F strain formed elementary bodies and was antigenically related to the viruses of meningopneumonitis and psittacosis by complement fixation and active immunity tests in mice. It could be differentiated from these viruses and from ornithosis by appropriate cross immunity tests in mice immunized by the intraperitoneal route and challenged by the intracerebral route. HILLEMAN has shown that this strain differs from the viruses of ornithosis, meningopneumonitis, mouse pneumonitis and lymphogranuloma venereum by neutralization tests with chicken serum. In 1942 another psittacosis-like virus apparently identical with the S-F strain was isolated from the lung of a fatal case of atypical pneumonia.

Among the six naturally occuring infections presumably due to psittacosis-like strain S-F, four occurred in nurses in attendance on the two original patients. This indicates a high person-to-person communicability. Neither of the two original patients gave a definite history of contact with birds. Five of the 6 cases terminated fatally, an exceedingly high mortality. In addition there were two laboratory infections with the S-F strain. This agent was isolated from the sputum of these patients and the development of psittacosis complement fixing antibodies in their serum was demonstrated. Strains of the S-F type may represent viruses which have become adapted to human beings as a result of several direct human transmissions, losing in the process some of their pathogenicity for mice and birds and part of the antigenic components of classical psittacosis strains.

Another virus of unknown origin was isolated by Olson and Larson from the blood, sputum, lung, and spleen of fatal human cases of pneumonitis that occurred in the Bayou region of southwestern Louisiana during the winter and spring of 1942-43. In this small epidemic there were 19 human cases of the disease with 8 deaths. The initial case occurred in the wife of a trapper, three nurses contracted secondary infections from her, and one of these, a fatal case, gave rise to a chain of other cases apparently representing five or six successive human-to-human passages with an average incubatiod period of 10 to 19 days. Elementary bodies were seen in the alveolar cells, pneumonic exudate, and in Kupfer cells of the liver in two fatal cases. The virus which infected both mice and guinea pigs after intraperitoneal injection and was apparently more virulent for these animals than the S-F strain or other human strains of psittacosis or ornithosis. Mice died after intracerebral, intranasal, intraperitoneal, intramuscular, or subcutaneous inoculation. The fatal infection by the latter two routes also differentiated this Louisiana pneumonitis strain from other, members of the psittacosis group.

A psittacosis-like virus resembling in certain respects the Louisiana strain was isolated by Zichis and Shauhgnessy in Chicago during the winter of 1944 from two fatal cases of pneumonia. This has been designated the Illinois strain. The virus was obtained by inoculating mice by the intranasal route with a suspension of lung tissue. It killed mice when injected by the intracerebral, intraperitoneal, intranasal and subcutaneous routes with the production of characteristic elementary bodies. Immunological comparison with psittacosis, ornithosis, meningopneumonitis, and S-F strains showed that the Illinois strain was not antigenically identical with any of these viruses, but was related to them. Neutralization tests in mice with chicken serum done by St. John and Gordon revealed that the Illinois strain was not neutralized by antiserum against the meningopneumonitis virus. This antiserum did neutralize the ornithosis strain.

The meningopneumonitis virus was isolated by Francis and Magill from ferrets inoculated with throat washings from patients with a disease resembling influenza. The virus is identical in its biological properties and antigenic composition with the ornithosis strain isolated by Meyer. It is undertain whether the meningopneumonitis strain came from ferrets, from avian contacts of the ferrets in nature, or from human beings. A case of laboratory infection with this virus is described by Meiklejohn, Beck and Eaton. The meningopneumonitis virus was isolated from the sputum of the patient and an increase in complement-fixing antibodies to the psittacosis group was demonstrated in her serum. One other human laboratory infection has occured with this virus. In both cases the illness was much milder with much less pulmonary involvement than that seen in infection with psittacosis, ornithosis, or other strains pathogenic for man.

During the studies on atypical pneumonia done by the Commission on Acute Respiratory Diseases at Camp Claiborne, Louisiana the meningopneumonitis virus or an ornithosis strain was isolated eight times from patients without definite evidence of relation to their disease. The isolations were done in the laboratory of DR. THOMAS FRANCIS [1] by repeated passage of sputum through mice or by passage of sputum through Java ricebirds to mice. Control passages with normal mouse lung were consistently negative. In none of these cases was an increase in complement-fixing antibodies for the psittacosis group demonstrated. The significance of these observations in relation to the etiology of the pneumonia or upper respiratory infection is not clear and the possible existence of latent respiratory infection or saprophytic existence of the virus in the human respiratory tract was considered.

The laboratory differentiation of atypical pneumonia from virus pneumonia of the psittacosis group can now be accomplished by appropriate serological tests. (EATON 1945). In cases of primary atypical pneumonia increases in neutralizing antibodies for the virus propagated in chick embryos and no significant development of specific complement-fixing antibodies for agents of the psittacosis group are found. Generally the patients with atypical pneumonia develop cold agglutinins, streptococcal agglutinins or both. In cases of virus pneumonia of the psittacosis group the agent is usually quite readily isolated from sputum by intranasal or intraper itoneal inoculation of mice and increases in complement-fixing antibodies for the psittacosis group are demonstrable. In such cases there is no development of neutralizing antibodies for the atypical pneumonia virus and cold agglutinins or streptococcal agglutinins are usually absent.

Other agents of the psittacosis group. Other viruses of the psittacosis group isolated from animals but without definite relation to human infection include two different agents obtained from mice, a virus of pneumonitis of cats, and a virus isolated from Syrian hamsters and wild ferrets. Complement fixing antibodies to most of these agents can be demonstrated in certain human sera particularly from persons infected with one of the psittacosis-like agents, but because of the group reactive character of the antigens from the various members of this class of agents, the mere presence of complement fixing antibodies for one of these animal viruses is not necessarily significant of infection with that virus. The possibility of inapparent infection with one of the psittacosis-lymphogranuloma groups of viruses seems quite possible in view of the presence of high titers of complement fixing antibodies to these agents in certain persons with no recent history of virus pneumonia and because of the tendency to latency of these viruses in animalis or birds.

A virus latent in certain mouse stocks has been encountered by several investigators. The virus of mouse pneumonitis described by NIGG and EATON produces focal or patchy gray lung lesions, kills more rapidly than influenza virus after serial passage, and tends to produce systemic infection or carrier states. It forms elementary bodies and fixes complement with antiserum against several members of the psittacosis-lymphogranuloma group. Like certain other members of the psittacosis-lymphogranuloma group the virus of mouse pneumonitis is inhibited by penicillin and sulfonamides. It is distinguished from other members of the group by its failure to produce fatal infection in mice either by the intracerebral or intraperitoneal routes. Other strains identical in their properties and antigenically related to the mouse pneumonitis agent of NIGG have been isolated in various parts of the United States and in Germany. A virus apparently closely related but not antigenically identical to this agent was isolated in Buenos Aires by DR. R. M. TAYLOR [2] on several occasions. One

---

[1] FRANCIS, T. Jr., Appendix I in *Dingle* et al., Am. Jour. Hygiene, **39**, 177, 1944.
[2] See article by NIGG and EATON.

of his strains was found in hamsters and another apparently came from wild ferrets. These strains of virus introduce a hazard in working with other pneumotropic viruses in mice because of the possibility of their acting as contaminants. Certain instances have been recorded of the overgrowth and replacement of influenza virus during passages in mice by the elementary-body agent of mouse pneumonitis. This virus is not to be confused with the unrelated pneumonia virus of mice (PVM) isolated by Horsfall and Hahn which is described later. These mouse pneumonitis agents are not related by cross immunity or cross neutralization tests to the psittacosis, ornithosis, meningopneumonitis, or S-F strains.

De Burgh and his associates in Australia reported the isolation of an agent which like the mouse pneumonitis virus of nigg was a natural inhabitant of the respiratory tract of mice. The virus differed from nigg's agent, however, and was closely similar to psittacosis in the following properties (1) fatal infection by the intracerebral and intraperitoneal routes, (2) pathogenicity for guinea pigs, (3) easy cultivation in the allantois of chick embryos, and (4) resistance to the action of sulfanilamide. The importance of this virus lies in the fact that its close resemblance to psittacosis virus might easily lead to confusion and error.

In 1942 Baker isolated from cats ill or dying with pneumonitis another elementary body agent which is closely similar to the Nigg mouse pneumonitis virus in producing pneumonia inoculated intranasally. One report indicated that fatal infection can be established by the intracerebral or intraperitoneal routes using very large doses but other investigators have obtained negative results on attempted passage of the agent in brain or spleen by the respective routes. The virus is also highly infectious for kittens either by inoculation or by contact with infected animals.

A summary of the important differential properties of the agents just described is presented in Table 14. All of the agents isolated from human infec-

Table 14. *Differentiation of Pneumonitis Viruses of the Psittacosis Group*

	Lethal infection in mice			Inhibited by Sulfonamides
	I. N.	I. C.	I. P.	
Human pneumonitis Agents .	+	+	— or*	0
Ornithosis and Meningo- pneumonitis...............	+	+	±	0
de Burgh Mouse Virus	+	+	+	0
Nigg Mouse Pneumonitis & Hamster Pneumonitis	+	0	0	+
Feline Pneumonitis	+	0	0	0
Lymphogranuloma venereum .	+	+	0	+

* S-F strain does not produce lethal infection by the intraperitoneal route; Illinois and Louisiana strains do.

tions are lethal for mice by the intracerebral route and resistant to sulfonamides [1] but are variable in their infectivity for mice by the intraperitoneal route. The feline pneumonitis virus is not lethal for mice by the intracerebral or intraperitoneal routes in moderate doses and is not inhibited by sulfonamides ex-

---

[1] An exception are two strains of psittacosis inhibited by sulfadiazine.

cept when these drugs are given in very large doses in the treatment of respiratory infections with this virus in rats or hamsters. As indicated in the table the virus of lymphogranuloma venereum can also be differentiated from the others on the basis of the four properties mentioned. Other means of differentiation include ability to grow on the allantoic membranne or in tissue culture, degree of resistance to penicillin, host range in other animals or birds, and the degree of chronicity or carrier state after infection. All grow in the lungs of mice and in the yolk sacs of chick embryos.

II. Other pneumonitis viruses of animals possibly infectious for man.

One other virus isolated from animals is considered potentially infectious for man because of the frequent finding of neutralizing antibodies for this agent in human serum. This is the pneumonia virus of mice isolated by HORSFALL and HAHN. A virus of feline pneumonia which is also unrelated to members of the psittacosis group was isolated by BLAKE et al and some evidence of neutralization by human serum was obtained. It is uncertain at present whether either of these agents actually produces active respiratory infection in human beings or whether they produce a subclinical syndrome without obvious signs of infection.

The Pneumonia Virus of Mice. This agent hereinafter refered to as PVM was isolated in 1939 by HORSFALL and HAHN during attempts to adapt agents to mice from nasopharyngeal washings of patients with several different acute diseases of the repiratory tract. Its widespread distribution in various species of animals particularly in the respiratory tracts of small rodents used in the laboratory has led to much confusion and error in attempts to isolate agents from human beings by intranasal passage of lung suspensions or other materials. Strains identical with or related to the original strain were isolated from apparently healthy mice and have been obtained in several other laboratories from mice, hamsters, and cotton rats.

The virus is not carried by all stocks of mice. In the studies of HORSFALL and HAHN is was obtained only from mice supplied by three breeders and not from those obtained from five other sources. These authors reported that the virus appeared to be relatively avirulent or latent in the mouse lungs but increased gradually in virulence on intranasal passage at intervals of 7 to 9 days. In the original studies no disease was observed in ferrets, rabbits, guinea pigs, monkeys, moles, deer mice, skunks, woodchucks, oppossums or Syrian hamsters after intranasal inoculation of infected mouse lung suspensions but later studies have revealed evidence of infection in some of these species. The failure to infect cotton rats or hamsters at certain times of the year is attributable to naturally acquired immunity in these animals. The virus was not infectious for mice by routes other than the intranasal.

The virus was readily differentiated from mouse adapted influenza strains by cross-immunity or cross neutralization tests but the character of the pulmonary lesions seen in mice resembles grossly those produced by the influenza virus. Neutralization by specific antiserum and by some normal mouse serums is readily demonstrable. The agent was extremely labile and suspensions in saline or broth became inactivated within a few hours at room temperature. Suspensions could be stabilized by addition of normal horse serum. Ultrafiltration results indicated an infectious particle diameter of 100 to 150 millimicrons although later studies have indicated a smaller diameter of the virus particle itself.

In 1940 Pearson and Eaton isolated from the lungs of Syrian hamsters (Cricetus auratus) a virus which produced fatal pulmonary consolidation in these animals, and which, on the basis of cross immunity tests was shown to be antigenically related to the pneumonia virus of mice. Subsequent observations showed that cotton rats may carry the same agent. The strains of PVM isolated from hamster and cotton rat lungs were usually directly transmissable to mice by intranasal inoculation. This property is useful in differentiating PVM in these animals from other respiratory agents which are' not transmissible to mice. The virus isolated from hamsters produced no immunity in these animals to the agent of atypical pneumonia propagated in chick embryos.

In cotton rats and hamsters the pneumonia due to PVM often appeared suddenly without previous passage of lung suspensions and could be induced by intranasal inoculation of a variety of materials including saline, broth, or suspensions of chick embryo tissues. The behavier of the virus in these animals suggested that the agent was present in the upper respiratory tract during epizootics of the disease. The process of intranasal inoculation under ether anaesthesia probably resulted in aspiration of the virus into the lungs with production of pneumonia. The pneumonia in cotton rats was usually fatal after a period of 4 to 10 days, while in hamsters the disease tended to be less severe.

During epizootics of infection with PVM in hamsters a large proportion of animals inoculated intranasally with non-infectious suspensions develop pulmonary consolidation and at such times it becomes impossible to use these animals for the demonstration of other respiratory agents from human or other sources. The epizoctic in the colony of hamsters proceeds without definite evidence of illness in the animals except those inoculated intranasally and no pneumonia is found on autopsy of uninoculated animals. The epizootic subsides after a period of 3 to 4 weeks when pulmonary lesions are no longer produced by intranasal inoculation of non-infectious material. At this time the animals have neutralizing antibodies for PVM in their serum and are solidly immune to intranasal inoculation with lung suspensions of this virus. Animals can be immunized against infection by intraperitoneal inoculation of 10 % lung suspensions of active virus. This procedure is useful in controlling intercurrent respiratory infection with PVM in stocks of animals being used for intranasal passage of atypical pneumonia or other viruses.

Horsfall and Hahn found that the sera of 22 out of a group of 67 apparently healthy human beings contained sufficient antibodies to neutralize 100 or more lethal doses of PVM, and some sera even neutralized at a dilution of 1 : 50. There was no correlation with the ability to neutralize influenza A. Eaton, van Herick and Meiklejohn (1945) also observed the neutralization of PVM in hamsters and mice by human sera but found no correlation of this neutralization with the ability of the sera to neutralize the atypical pneumonia virus. No direct or convincing evidence of an increase in neutralizing antibodies for PVM has been found in any known respiratory infection in human beings despite the frequent presence of such antibodies in human sera. Until such evidence is found it must be assumed that the antibodies result from inapparent infection, possibly early in life, whenever exposure to animals carrying the virus takes place.

It appears from the work of Eaton and van Herick (1944) and Horsfall and Curnen (1946) that several species of laboratory animals have antibodies to the pneumonia virus of mice on arrival at the laboratory or develop these antibodies after being kept in the laboratory for some time. Antibodies to PVM have been found in apparently normal mice, hamsters, cotton rats, rabbits, monkeys, chimpanzees and guinea pigs but not in chicken or wild mongoose sera. HORSFALL and CURNEN reported that intranasal inoculation of broth, normal chick embryo material, or sputum into cotton rats, hamsters or mongooses stimulated the production of antibodies to PVM. Intraperitoneal or intravenous injection of various materials into rabbits also caused the production of neutralizing antibodies to PVM irrespective of the kind of material inoculated. Such development of antibody occurred on 10 to 50 per cent of the animals. The incidence of antibodies in uninoculated control animals was lower.

Marked seasonal variations in infection with PVM in hamsters and cotton rats have been observed. Evidence for this is found in the previously cited experiments demonstrating pneumonia after intranasal inoculation of noninfectious material and in measurements of antibodies. The incidence of infection in cotton rats and hamsters seems to be highest in winter and spring, and lowest in the summer. Cotton rats or hamsters having neutralizing antibodies to PVM were resistant to intranasal infection with this virus while those with no antibodies were fully susceptible.

The pneumonia virus of mice, like influenza, mumps, and certain other viruses will cause agglutination of erythrocytes. MIILS and DOCHEZ discovered that suspensions of lungs from mice infected with this virus could agglutinate mouse erythrocytes. The agglutination ocurred only with the supernatant of lung suspensions heated to 80^0 C for five minutes and in this respect it differed from the erythrocyte agglutinins of influenza and mumps viruses which are relatively heat-labile. The agglutination was specifically inhibited by antiserum to PVM. CURNEN and HORSFALL studied this phenomenon in some detail and concluded that the component in lung tissue causing the agglutination was the virus particle itself rather than some other substance present in infected lungs. In mice killed at various times after inoculation the titer of virus infectivity and hemagglutination increased and decreased in parallel. It was observed that consolidated lungs of mice in the frech unheated state showed agglutination of erythrocytes exuding from their cut surfaces whereas similar hemagglutination with normal mouse lungs did not occur. Once such infected lungs were ground the capacity to produce hemagglutination became masked and was not manifested until the suspension was heated. Presumably the virus combined with a thermolabile substance in the tissue to form a complex which would not agglutinate. This was destroyed on heating with subsequent liberation of the virus particle in a state capable of agglutinating erythrocytes. The minimum estimated size of the infectious particles of PVM (ie: complex of virus and tissue substance) is 100 to 150 millimicrons. This is several times greater than the estimate of size of the hemagglutinating component released by heat which was placed at 40 millimicrons. Free infectious virus in an undenatured state and of a similar particle size was obtained by saline extraction (without grinding) of slices of mouse lung. It was also found that unheated suspensions of infected lung failed to fix complement with specific antisera while heated suspensions did react in complement fixation tests.

Later studies by VOLKERT and HORSFALL indicated that the lungs of mice, hamsters, cotton rats, rabbits, guinea pigs and man contained a component capable of combining with the pneumonia virus of mice. The virus-combining component was absent from kidney, brain, liver, spleen, and muscle, and was not found in whole chick embryo or in chick embryo lungs. Their results suggested a correlation between the amount of virus combining component in the lungs of mammals and the degree of susceptibility to infection. The virus receptor was destroyed by trypsin. Dissociation of the hemagglutinating factor (or virus) from the lung was accomplished by treatment with antibody or alkali, heating or digestion with trypsin. These authors also observed that PVM showed increased lability to heat in concentrated lung suspensions and in the presence of 0.005 M glutathione. Addition of iodoacetamide which combines with sulfhydryl groups prevented the deleterious effect of lung tissue or glutathione. The change in the virus produced by glutathione and lung was reversible after short periods but later became irreversible. The authors conclude that sulfhydryl groups produce adverse effects on PVM.

The observations on attachment of PVM to a receptor in lung tissue are analogous to observations of a similar nature with the influenza virus. These and other studies strongly suggest that the first phase in a virus infection of the respiratory tract is specific adsorption of virus particles from the inspired air to receptors on the surfaces of the cells of susceptible tissues probably the columnar epithelium of the bronchi or bronchioles or the respiratory epithelium of the alveolar ducts.

Horsfall and McCarty observed that certain polysaccharide preparations derived from various bacterial species as well as blood group A specific substance and agar when given intranasally were capable of lessening the severity of infection with PVM and inhibited multiplication of the virus in the lungs of mice as measured by the hemagglutination test. With Friedlander Type B polysaccharide virus multiplication was stopped or retarded when the carbohydrate was given as long as 4 days after intranasal inoculation or, when marked multiplication of virus had occurred. When the various substances were given intranasally 2 days before the virus, the observed titer obtained by inoculation of serial dilutions was reduced on the average 78-fold as compared with the controls. The materials when inoculated intraperitoneally had no effect on pneumonia due to PVM. They had no effect on the multiplication of influenza virus in mice.

The mechanism of the action of the various polysaccharide preparations was studied. The active proparations did not cause agglutination of erythrocytes or inhibit agglutination of erythrocytes by PVM. The did not prevent combination of PVM with lung tissue particles. From these experiments it seemed improbable that the effect was due to modification or blockade of virus receptors. The authors suggested that the active preparations may exert their modifying effect by competing with PVM for some intracellular system essential to multiplication of the virus.

Virus of Feline Pneumonia (Blake et al). The virus isolated by Blake, Howard and Tatlock from cats is apparently unrelated to the agent of the psittacosis-lymphogranuloma group isolated by Baker. Blake's cat virus does not form elementary bodies and is not infectious for mice and in these respects differs from Baker's agent. An outbreak of acute respiratory infection occurred in 4 out of the 8 members of a family and concurrently a highly fatal respiratory illness was seen in 10 of the 12 cats owned by them. The infection in the cats may or may not have been the same as the feline catarrhal distempers previously described by other investigators. On serial passage in cats the causative agent was established as a filtrable virus. The pneumonia produced in the animals was similar to other virus pneumonias and bacteria were found infrequently. Necrosis of the bronchiolar and alveolar epithelium and infiltration with large mononuclear cells were seen in sections. Kittens appeard to be more susceptible to the virus than full grown cats. The association of the human and feline illnesses, and the results of a preliminary neutralization test with human serum sugested but did not establish that the human infections may have come from the cats.

Other pneumonitis agents of animals.

The number of filterable infectious agents causing respiratory disease in each species of animal used for laboratory experiments is probably as great as the frequency of similar agents in human beings. Although some of these animal agents such as the pneumonia virus of mice have been studied in detail, very little is known about the nature or distribution of most of them. The possible relationship of similar agents isolated in different laboratories has not been adequately determined. The importance of viruses carried by animals as possible contaminants in experimental transmission of disease from human beings cannot be overemphasized and a more detailed knowledge of these agents would seem justified solely on the grounds of being better able to avoid them. Others are of interest because of their unique characteristics.

Diseases associated with coccobacilliform bodies.

NELSON has described a highly fatal infectious catarrh of mice characterized by rhinitis, otitis media, and pneumonia. It is transmissible by contact or by intranasal inoculation and is very chronic in nature, the pneumonia sometimes persisting for a year before death of the animal. Small gram negative coccobacilliform bodies 300 to 400 millimicrons in size and regularly found in the nasal an middle ear exudates of infected mice were considered the cause of the disease. The agent can be grown in chick embryo tissue culture but has not been cultivated on artificial media free of living cells. Its staining characteristics differentiate the coccobacilliform body from the agents of the psittacosis-lymphogranuloma group, the former staining blue by the Macchiavello method while most of the Lille-Coles-Levinthal bodies stain red. There is no immunological relationship by complement fixation between the two groups of agents.

The agent of mouse catarrh is frequently found in various stocks of apparently normal mice and produces a massive gray focal pneumonia when animals are infected by the intranasal route. Since the pulmonary lesions develope very slowly the agent does not often cause confusion with other viruses having a shorter incubation period in mice. When, however, pulmonary lesions of greater than two weeks duration are under study, the pneumonia produced by the agent of mouse catarrh is easily confused with that caused by other viruses which also tend to cause chronic infections; for example the psittacosis group. Complete resolution of the pulmonary lesions in infected mice is rare, but a relative immunity to respiratory infection can apparently be produced by intraperitoneal inoculation of the agent in large amounts. Infected mice do no develop circulating antibodies demonstrable by the ordinary neutralization procedures.

Infectious catarrh of albino rate as described by NELSON is a relatively benign infection associated with the presence of coccobacilliform bodies and is similar to mouse catarrh. Similar coccobacilliform bodies have been described in fowl coryza (NELSON, 1936).

Pleuropneumonia-like organisms.

HORSFALL and HAHN isolated with equal ease pleuropneumonia-like organisms from the lungs of normal mice, and from the lungs of mice infected with the virus of influenza or the pneumonia virus of mice. These organisms did not produce any evidence of pulmonary consolidation when inoculated by themselves nor did they increase the extent or severity of the pulmonary lesions

when inoculated with the pneumonia virus of mice. The strains studied by
HORSFALL and HAHN were apparently carried as saprophytes in the respiratory
tracts of mice.

EDWARD obtained a strain of pleuropneumonia-like organisms which was
definitely pathogenic for mice by the intranasal route and produced experimental pneumonia in a large percentage of these animals. The histological
changes in the lungs were a combination of inflammatory exudation and localized reticulum-cell hyperplasia. After intraperitoneal or intravenous inoculation microscopic focal infiltrations of the liver and lesions in the lungs were
seen. The organisms did not produce a neurotoxin and did not cause polyarthritis in mice as do certain other strains. Other pleuropneumonia-like organisms infectious for mice by the intranasal route have been recorded by
HERZBERG and by GROSS.

A pleuropneumonia-like organism pathogenic for chick embryos and cotton
rats was isolated by VAN HERICK and EATON. This organism appeared as a contaminant during serial passages of one strain of the atypical pneumonia virus
in chick embryos but was not obtained in other serial passages or in a repetition
of the passages with the same strain of virus. It elaborated a substance which,
after intranasal instillation, caused edematous pulmonary consolidation and
death in cotton rats. This lung poison or toxin was not inactivated by heating
to 90° C for 30 minutes and could be separated from the organisms by Seitz filtration of broth cultures or of suspensions of chick embryo material. The presence of a contaminant in the chick embryo passages was first suspected because of the production in cotton rats of the acutely fatal pneumonia which
differed in its characteristics from the lung lesions produced by the atypical
pneumonia virus. The pleuropneumonia-like organism or its toxin did not
produce significant pulmonary lesions after intranasal inoculation into mice
or hamsters. Another interesting property of this organism was the agglutination of the erythrocytes of various species of animals. This hemagglutinin
seemed to be associated with the bodies of the pleuropneumonia-like organisms
and its action was specifically inhibited by hyperimmune rabbit serum. Inhibition of the hemagglutination was also observed with a high proportion of
the hens at the hatchery which furnished the eggs from which the organism
was isolated.

The gray lung virus.

A filterable agent pathogenic for mice and other rodents was isolated by
ANDREWES and GLOVER apparently from mice during the course of two series
of intranasal lung passages. One of these passage series had been started with
material from calf pneumonia and the other with a specimen from a case of
infantile diarrhea. The lesions appeared six to eight days after inoculation
and persisted for at least six months. The virus was strictly pneumotropic
and was transmissible by contact. Dense perivascular cuffs of rather large
mononuclear cells were a striking and regular feature in the lungs of infected
animals. These changes were associated with diffuse edema, swelling and proliferation of cells in the alveolar walls.

The gray lung virus was retained by collodion membranes of average pore
diameter smaller than 450 millimicrons but passed membranes of this size or
larger. Elementary bodies characteristic of the mouse pneumonitis virus (*Nigg*)
or the coccobacilliform bodies of mouse catarrh were not found. The virus
was also differentiated from other known pneumotropic agents of mice.

It multiplied in the lungs of rats, hamsters, and voles (microtus arvensis) but did not produce conspicuous pulmonary lesions in these animals. In cotton rats scattered or diffuse gray lesions resembling those in the mouse were seen after intranasal inoculation. Neither active immunity nor the production of antibodies could be demonstrated.

Hamster agent W1.

A passage experiment in apparently normal hamsters conducted in the laboratory of the author by W. VAN HERICK resulted in the isolation of an unidentified agent which produced pneumonia in these animals. This agent is referred to as W1 in the publications of EATON et al. The agent W1 also produced pulmonary lesions in cotton rats and mice after intranasal inoculation. The pulmonary lesions were characterized by a marked tendency to chronicity and those in the mice appeared after an incubation period of a month or more. No bacteria, coccobacilliform bodies or elementary bodies could be found in the lungs of infected animals. The agent was not neutralized by the serum of animals with chronic infections or by serum from rabbits repeatedly inoculated with infected hamster lung suspensions. In certain respects agent W1 resembled the gray lung virus described by ANDREWES and GLOVER but it was apparently more virulent for cotton rats and hamsters and had a longer incubation period in mice.

Cotton rat agents W2 and W3.

The agent designated W2 was isolated about fifteen times during the studies of EATON et al on atypical pneumonia. It was obtained both in passage series initiated with material from atypical pneumonia and with normal rat lung, appearing after three to six intranasal passages in cotton rats and increasing in virulence with subsequent passage. The agent was filtrable with some difficulty through Berkefeld V filters and was sedimented at 5,000 RPM in an angle centrifuge, observations which indicate a relatively large particle size although attempts to demonstrate elementary bodies were unsuccessful. The agent W2 is apparently carried by both of the species *Sigmodon hispidus eremicus* and *Sigmodon hispidus hispidus*. The pulmonary lesions are usually non-fatal and may readily be mistaken on superficial gross examination for the lesions produced by the atypical pneumonia virus but they differ from those of the latter virus in being more purulent in character. After adaptation to cotton rats the virus W2 is transmissible to hamsters and microtus species with the production of pulmonary lesions but causes no detectable disease in mice, rabbits, or guinea pigs.

Animals recovered from pulmonary infection with the agent W2 in two to three weeks and were then solidly immune to reinoculation by the intranasal route. It produced neither signs of infection nor immunity when inoculated by the intracerebral or intraperitoneal route into cotton rats. Animals immunized with the agent W2 were not immune to intranasal inoculation with the atypical pneumonia virus propagated in chick embryos or to the agent W3 described below. Serum from cotton rats, hamsters, or rabbits repeatedly inoculated with the agent W2 gave definite neutralization but the virus was not significantly neutralized by human serum from normal persons or from patients with primary atypical pneumonia or other acute respiratory diseases.

The other agent also isolated from cotton rats and designated W3 was not related immunologically to W2 and differed in several other properties. It was isolated only twice during the many attempts to adapt a virus to cotton rats from human material. Unlike W2, the agent W3 was not infectious for hamsters or other species of rodents. The lesions in cotton rats developed between the eighth and fifteenth day after intranasal inoculation and disappeared after the third week. Recovered animals were solidly immune to reinoculation by the intranasal route. Agent W3 was differentiated by crossimmunity tests from the atypical pneumonia virus and the pneumonia virus of mice. No or ganisms or elementary bodies were seen in the lungs of infected animals. A few neutralization tests with human serum gave negative results.

Pneumonitis agents of unknown origin.

Agents isolated in mice.

Some of the earliest work on the etiology of atypical pneumonia was done by Stokes Kenney and Shaw in 1938, who observed in one ferret inoculated intranasally with throat washings coma, labored respiration, and high fever. Passage to other ferrets failed but intracerebral inoculation of mice with material of the first ferret passage resulted in slowly developing paresis of the hind legs and intranasal inoculation of mice produced pneumonia and involvement of the central nervous system. Intracerebral inoculation of guinea pigs produced symptoms of infection of the central nervous system similar to those seen in mice and a bluish-gray firm patchy bronchopneumonia. The agent passed a Berkefeld V filter. Successive passages in mice and guinea pigs seemed to result in decreased virulence and the agent was eventually lost in the passage series in both species of animal. It could not be recovered from frozen or glycerinated material that had been stored for seven months. Because of the loss of this virus it is impossible to state whether it was of human or animal origin. The incomplete data available suggest that these authors may have had a strain of lymphocytic choriomeningitis virus but if this were the case it is difficult to explain the loss of the agent on continued animal passage. A strain of the virus of lymphocytic choriomeningitis which produced pneumonia in mice after intranasal inoculation was isolated from mouse lungs by W. P. Martin working in the writer's laboratory in 1941[1].

Agents isolated in guinea pigs.

Rose and Molloy isolated 10 strains of a readily transmissible pneumotropic agent in recently weaned guinea pigs inoculated with throat washings, blood, sputum, lung or spleen of patients with atypical pneumonia. The agent produced a disseminated bronchopneumonia in guinea pigs, the lesions becoming manifest within 12 to 20 days and showing complete resolutions after 45 to 60 days. The agent was transmissible only by the intranasal route of inoculation. It passed a Berkefeld V filter but was retained by filters of smaller porosity.

Identity of the various strains of the guinea pig agent was demonstrated by crossimmunity tests in recovered animals. Neutralizing antibodies could not be found either in the serum of recovered animals or in serum of patients convalescent from atypical pneumonia. The guinea pig agent was not transmissible in chick embryos or in tissue culture but it produced pneumonia regularly after intranasal inoculation of cotton rats. Guinea pigs and cotton rats repeatedly injected by the intranasal route with human material were said

---

[1] Unpublished observations.

to develop partial or complete immunity to strains of the agent passed in cotton rats. Although this evidence was suggestive the relationship of the guinea pig agent to human infection was not definitely established and ROSE has more recently advanced the opinion that no causal association exists [1].

EATON et al (1944) observed that guinea pigs frequently developed pulmonary lesions after intranasal inoculation of broth or other non-infectious materials. One pneumotropic agent isolated by intranasal passage of lung suspensions in these animals was found to be very similar in its properties to the hamster agent W1. The relation of this virus isolated from guinea pigs and of the normal hamster agent to the virus described by ROSE and MOLLOY has not yet been investigated.

References.

ADAMS, J. M., R. G. GREEN, C. A. EVANS and N. BEACH: Jour. Pediat. **20,** 405 1942.

ANDREWES, C. H., and R. E. GLOVER: Brit. Jour. Exp. Path. **24,** 379, 1945.

—, P. P. LAIDLAW and W. SMITH: Brit. Jour. Exp. Path. **16,** 566, 1945.

ARMSTRONG, C., and R. D. LILLIE: U. S. Public Health Reports, **44,** 2635, 1929.

BAKER, J. A: Science, **96,** 475, 1942; Jour. Exp. Med. **79,** 159, 1944.

BARTELS: Virchow's Arch. f. Path. Anat. **21,** 65, 1861.

BLAKE, F. G., and R. L. CECIL: Jour. Exp. Med. **32,** 401, 1920.

—, M. E. HOWARD and H. TATLOCK: Yale Jour. Biol. and Med. **15,** 139, 1942.

BOSC, F. J.: Centralblatt f. Bakt. Pt. 1. Orig. **36,** 487, 630, 1904; **37,** 39, 195, 1904.

BOWEN, A.: Am. Jour. Roentgenol. **34,** 168, 1935.

BRESLOW, L.: Jour. Clinical Investigation. **24,** 775, 1945.

DE BURGH, P., A. V. JACKSON and S. E. WILLIAMS: Australian Jour. Exp. Biol. and Med. Science. **23,** 107, 1945.

BURNET, F. M.: Brit. Jour. Exp. Path. **21,** 147, 1940; Australian Jour. Exp. Biol. and Med. Science. **18,** 333, 1940.

Commission on Acute Respiratory Diseases., Am Jour. Public Health. **34,** 335, 1944; Am. Jour. Med. Science. **208,** 742, 1944; Bull. Johns Hopkins Hospital. **79,** 97, 1946.

CURNEN, E. C., G. S. MIRICK, J. E. ZIEGLER Jr., L. THOMAS and F. L. HORSFALL Jr.: Jour. Clin. Investigation. **24,** 209, 1945.

DAMMIN, G. J., and T. H. WELLER: Jour. Immunol. **50,** 107, 1945.

DINGLE, J. H., and M. FINLAND: New England Jour. Med. **227,** 342, 378, 1942.

—, et al (Commission on Acute Respiratory Diseases), Am. Jour. Hygiene. **39,** 67, 197, 269, 1944.

— Bull. New York Academy Med. **21,** 235, 1945.

EATON, M. D., M. D. BECK and H. E. PEARSON: Jour. Exp. Med. **73,** 641 1941.

— and M. COREY: Proc. Soc. Exp. Biol. and Med. **51,** 165, 1942.

—, G. MEIKLEJOHN, W. VAN HERICK and J. C. TALBOT: Science **96,** 518, 1942.

— — — and M. COREY: Jour. Exp. Med. **79,** 649, 1944; **82,** 317, 329, 1945.

— and W. VAN HERICK: Proc. Soc. Exp. Biol. and Med. **57,** 84, 1944.

— Proc. Soc. Exp. Biol. and Med. **60,** 231, 1945.

— and W. VAN HERICK: Am. Jour. Hygeine **45,** 82, 1947.

EDWARD, D. G.: ff., Jour. Path. and Bact. **50,** 409, 1940.

FAVOUR, C. B.: Jour. Clinical Investigation, **23,** 891, 1944.

FILDES, P., and J. MCINTOSH: Brit. Jour. Exp. Path. **1,** 119, 159, 1920.

FINLAND, M., et al: Jour. Clinical Investigation, **24,** 451, 1944.

FLORMAN, A. L. and A. B. WEISS: Jour. Lab. and Clin. Med. **30,** 902, 1945.

FRANCIS, T.: Jr. Science, **80,** 475, 1934.

GALLAGHER, J. R.: Yale Jour. Biol. and Med. **7,** 23, 1934; **13,** 663, 769, 1941.

GOLDEN,A.: Arch. of Path. **38,** 187, 1944.

---

[1] Unpublished statement made at the symposium on primary atypical pneumonia, Society of American Bacteriologists, General Meeting, May 1944.

Goodpasture, E. W., and F. L. Burnett: U. S. Naval Med. Bull. **13**, 177, 1919.
Gross, W. O.: Centralblatt f. Bakt. Pt. 1, Orig. **149**, 366, 1943.
Hammond, J. A. B., W. Rolland and T. H. G. Shore: Lancet **2**, 41, 1917.
Hegglin, R.: Helvet. Med. Acta. **7**, 497, 1941; Schweiz. med. Wchschr. **71**, 578, 1941.
Herzberg, K.: Centralbaltt Bakt. Pt. 1 Orig. **149**, 362, 1943.
Hesse, W.: Munch. Med. Wchschr. **65**, 1125, 1918.
Hilleman, W. R.: Jour. Infect. Disease **76**, 96, 1945.
Hirst, G. K.: Jour. Immunol. **45**, 293, 1942.
Horsfall, F. L. Jr., and R. Hahn: Jour. Exp. Med. **71**, 391, 1940.
— et al., Science **97**, 289, 1943.
— and E. C. Curnen: Jour. Exp. Med. **83**, 25, 43, 105, 1946.
— New York State Jour. Med. **46**, 1810, 1945; Ann. Internal Med. **27**, 275, 1947.
— and M. McCarty: Jour. Exp. Med. **85**, 623, 1947.
Kneeland, Y. Jr., and H. F. Smetana: Bull. Johns Hopkins Hospital **67**, 229, 1940.
Leichtenstern, O.: in Nothnagel's Encyclopedia of Practical Medicine, W. B. Saunders and Co., New York, 1905.
Longcope, W. T.: Bull. Johns Hopkins Hospital **67**, 268, 1940.
McCallum: Monographs of the Rockefeller Institute for Medical Research, New York, No. 10, 1919; Jour. Am. Med. Assn. **70**, 1153, 1918; **72**, 720, 1919.
McCordock, H. A., and R. S. Muckenfuss: Am. Jour. Path. **7**, 552, 1931; **8**, 63, 1932; **9**, 221, 1933.
— and M. G. Smith: Am. Jour. Diseases Children, **47**, 771, 1933.
Meiklejohn, G.: Proc. Soc. Exp. Biol. and Med. **54**, 181, 1943.
— and V. L. Hanford: Proc. Soc. Exp. Biol. and Med. **57**, 356, 1944.
—, M. D. Beck and M. D. Eaton: Jour. Clinical Investigation, **23**, 167, 1944.
—, M. D. Eaton and W. van Herick: Jour. Clinical Investigation **24**, 241, 1945.
Meyer, K. F.: Schweiz. Med. Wchschr. **71**, 1377, 1941.
Mills, K. and A. R. Dochez: Proc. Soc. Exp. Biol. and Med. **57**, 140, 1944; **60**, 141, 1945.
Moskovsky, C. D.: Uspeki. Sowiem. biol. **19**, 1 (1945).
Murray, M. E.: New England Jour. Med. **222**, 565, 1940.
Nelson, J. B.: Jour. Exp. Med. **63**, 515, 1936; **65**, 833, 843, 851, 1937; **72**, 645, 655, 1940.
Nigg, C.: Science, **93**, 49, 1942.
— and M. D. Eaton: Jour. Exp. Med., **79**, 497, 1944.
Olson, J. B., and C. L. Larson: U. S. Public Health Reports **60**, 1488, 1945.
Pearson, H. E., and M. D. Eaton: Proc. Soc. Exp. Biol. and Med. **45**, 677, 1940.
Peterson, O. L., T. H. Ham and M. Finland: Science **97**, 167, 1943.
Ritter, J.: Dtsch. arch. klin. med. **25**, 53, 1880.
Rivers, T. M., and G. P. Berry: Jour. Exp. Med. **54**, 105, 119, 129, 1931.
Rose, H. M., and E. Molloy: Science, **98**, 112, 1943.
Shope, R. E.: Jour. Exp. Med. **54**, 361, 373, 1931; **62**, 561, 1935.
Smiley, D. F., E. D. Showacre, W. F. Lee and A. W. Ferris: Jour. Am. Med. Assn. **112**, 1901, 1939.
Smith, W., C. H. Andrewes and P. P. Laidlaw: Lancet, **2**, 66, 1933.
St. John, E. and F. B. Gordon: Jour. Infections Diseases, **80**, 297, 1947.
Stokes, J. Jr., A. S. Kenney and D. R. Shaw: Transactions and Studies, College of Physicians Philadelphia, **6**, 329, 1939.
Thomas, L., G. S. Mirick, E. C. Curnen, J. E. Zeigler Jr. and F. L. Horsfall Jr.: Jour. Clinical Investigation, **24**, 227, 1945.
— — — — — Proc. Soc. Exp. Biol. and Med. **52**, 121, 1943.
Vance, D. H., T. Scott and H. C. Mason: Science, **98**, 412, 1943.
van Herick, W., and M. D. Eaton: Jour. Bact. **50**, 47, 1945.
Volkert, M., and F. L. Horsfall Jr.: Jour. Exp. Med. **86**, 383, 1947.
Weir, J. M., and F. L. Horsfall Jr.: Jour. Exp. Med. **72**, 595, 1940.
Winternitz, M. C., I. M. Wason and F. P. McNamara: The Pathology of Influenza, Yale University Press, New Haven, 1920.
Zichis, J., and Shaugnessy: Jour. Bact. **51**, 55, 1946.

Die Haemagglutination durch Virusarten (Phaenomen von G. K. Hirst).

Von Professor Dr. C. HALLAUER, Bern.

I. Die Entdeckung und die Phaenomenologie der virusbedingten Haemagglutination.

Bei der Gewinnung von Influenzavirus aus befruchteten Hühnereiern machten HIRST (191) sowie MC CLELLAND und HARE (246) erstmalig und gleichzeitig (1941) die Beobachtung, daß embryonales Hühnerblut, welches aus zufällig verletzten Gefäßen in die virushaltige Allantoisflüssigkeit einströmte, nahezu augenblicklich agglutiniert wurde. Wie diese Autoren zeigten, kann dieser Vorgang auch in vitro nachgewiesen werden, wobei die Möglichkeit gegeben ist, das haemagglutinierende Prinzip zu titrieren. Daß die haemagglutinierende Wirkung als eine spezifische Virusfunktion zu betrachten ist, ergab sich nicht nur aus der Beobachtung, daß auch ein Influenzavirus anderer Provenienz (Mäuselunge) dieselbe haemagglutinierende Eigenschaft aufweist, sondern vor allem aus der Feststellung, daß zwischen der Infektiosität und der agglutinierenden Aktivität eine weitgehende quantitative Parallelität besteht, und daß die Haemagglutination von Influenzavirus A und B durch korrespondierende Immunseren in typspezifischer Weise und nach Maßgabe des Antikörpergehaltes aufgehoben bzw. gehemmt werden kann. Der Haemagglutinationstiter kann demnach — zumindest unter Verwendung von frischen, bzw. gut konservierten Virusproben — als Maßstab für die Virusinfektiosität, und der Hemmungstiter des Immunserums für die Wertbemessung virusneutralisierender Antikörper verwendet werden. Schließlich lieferten Ultrazentrifugierungsversuche den stringenten Nachweis, daß das Haemagglutinin in der Sedimentierungsfraktion der Viruspartikel von 80—100 mμ angereichert wird und von diesen Viruselementen nicht abzutrennen ist. Dessen ungeachtet ist die haemagglutinierende Eigenschaft nicht unbedingt an das infektiöse Vermögen gebunden, da schonend inaktivierte Virusproben einen kaum verminderten Haemagglutinationstiter aufweisen können, und — umgekehrt — liegen Beobachtungen vor, wonach Influenzavirus in bestimmten Gewebesuspensionen trotz hoher Infektiosität kein haemagglutinierendes Vermögen besitzt. Eine solche „Dissociation" von infektiöser und haemagglutinierender Qualität ist jedoch im einen Fall lediglich auf die unterschiedliche Stabilität dieser Virusfunktionen gegenüber äußeren Einflüssen zurückzuführen und im anderen dadurch verursacht, daß das Haemagglutinin durch gewebliche Hemmstoffe reversibel inaktiviert ist, bzw. maskiert wird.

Wie schon sehr frühzeitig nachgewiesen wurde, erstreckt sich die haemagglutinierende Wirkung von Influenzavirus nicht nur auf Hühnererythrocyten, sondern auch auf die Blutkörperchen anderer Vogelarten, sowie zahlreicher Säugetier- und Kaltblüterspecies, wobei jedem Virustypus und selbst jedem Stamm eine bestimmte Anzahl agglutinabler Erythrocytenspecies zugeordnet ist.

Eine Korrelation zwischen der Virusempfänglichkeit und der Agglutinierbarkeit der Erythrocyten einer Tierspecies besteht demnach nicht; vielmehr erinnert die anscheinend regellose Verteilung der agglutinablen Substanz im Tierreich an die selektiven Wirkungen der sogenannten Phytagglutinine (Ricin, Abrin, Phasin usw.), deren haemagglutinierende Wirkungsbreite ebenso unterschiedlich und ausgedehnt und in ihrer „Specifität" ungeklärt geblieben ist [Doerr (95)]. In welcher Weise das Influenzavirus mit der Erythrocytenoberfläche reagiert, wurde eingehend analysiert und führte zum Ergebnis, daß der Vorgang in zwei distinkten, zeitlich sich folgenden Stufen abläuft, nämlich 1. der *Adsorptionsphase*, in welcher das Virus, bzw. das Haemagglutinin nahezu quantitativ an die Erythrocyten gebunden wird. Die Raschheit, mit der diese Bindung zustande kommt, ist bei höherer Temperatur (37⁰ C) derartig groß, daß eine starke Bindungsaffinität zwischen den Reaktionskomponenten angenommen werden muß. Der Adsorptionsvorgang läßt sich sowohl mit aktivem als inaktivem Virus erzielen und führt in beiden Fällen zum sichtbaren Phaenomen der Haemagglutination. Eigenartiger und für die Virushaemagglutination charakteristischer ist nun aber 2. die *Elutionsphase*, in welcher das Virus von der Zelloberfläche wieder spontan abdissociiert, wobei die Agglutination rückgängig gemacht wird und die Erythrocytensuspension ihre ursprüngliche Stabilität zurückgewinnt. Diese Virusliberierung, die sich bemerkenswerterweise nur unter Verwendung von aktivem Virus erzielen läßt, erreicht bei 37⁰ C schon nach wenigen Stunden ihr Maximum und liefert ein Virus, das in qualitativer Hinsicht völlig unverändert erscheint, d. h. die ursprüngliche Infektiosität, Toxicität und haemagglutinierende Aktivität aufweist. Die Nutzanwendung dieser reversiblen Bindung von Influenzavirus an Hühnererythrocyten als einfache Methode der Virusreinigung und Konzentrierung ist damit gegeben. Im Gegensatz zum Virus zeigen nun aber die Erythrocyten nach der Virus-Adsorption und Elution charakteristische Veränderungen, nämlich 1. ein refraktäres Verhalten, bzw. eine Inagglutinabilität gegenüber der erneuten Zufuhr desselben Virus (oder eines Virus mit ähnlicher „Haemagglutininspecifität") und 2. eine unspezifische Panagglutinabilität gegenüber zahlreichen tierischen Normalseren, d. h. eine Veränderung, die mit dem von Thomsen und Friedenreich beschriebenen Phaenomen in auffallender Analogie steht. Schließlich wurde festgestellt, daß der Adsorptions- und Elutionsprozeß auch in Gegenwart haemoglobinfreier Erythrocytenstromata stattfindet.

Über die stoffliche Natur des Haemagglutinins und der reagierenden Zellkomponente liegen bisher nur unzureichende Informationen vor. Hirst machte schon frühzeitig darauf aufmerksam, daß die Reaktionskinetik des Adsorptions- und Elutionsprozesses alle Kriterien eines *fermentativen Vorganges* aufweist, nämlich die rasche Bindung an das Zellsubstrat, die Inaktivierung, bzw. Zerstörung des Zellreceptors und die nachfolgende Abspaltung des unveränderten, zu neuer Umsetzung befähigten Virus. Die proteide Natur des Virushaemagglutinins, die auf Grund des Verhaltens gegen eiweißdenaturierende Einflüsse (Erwärmen, Formalin) und proteolytische Fermente mit Wahrscheinlichkeit angenommen werden kann, würde sich ebenfalls in diese Konzeption einfügen lassen. Die Fermenthypothese erhielt aber namentlich durch die Arbeiten der Burnet'schen Schule von einer ganz anderen Seite her erheblichen Auftrieb, nämlich durch die überraschende Feststellung, daß Kulturfiltrate von Cl. Welchii (Toxin A) und vor allem von V. cholerae Erythrocytenveränderungen (Zerstörung des Zellreceptors, Panagglutinabilität) bewirken, welche von den virusinduzierten nicht zu unterscheiden sind. Fernerhin reagieren diese Agenzien — konform dem Virus — mit den Erythrocyten nach dem Vorgang der Adsorption und der nachfolgenden spontanen Elution. Sowohl für das Toxin von Cl. Welchii als auch für Choleravibrionen war nun bekannt, daß diesselben Lecithinase enthalten, so daß die Annahme nahelag, daß die erythrocytenverändernde Wirkung dieser Substanzen — und vice versa auch des Virus — durch dieses Ferment bewirkt würde. Untersuchungen über die biochemische Natur des Zellreceptors wiesen nun aber in eine etwas andere Richtung. Wie HIRST zeigte, ist die mit dem Virus reagierende Komponente der Erythrocyten weitgehend stabil gegenüber thermischen Einflüssen, Verschiebungen des pH nach der sauren und alkalischen Seite, sowie gegenüber den meisten Oxydationsmitteln, dagegen in hohem Grade empfindlich gegenüber Trypsin und Kaliumperjodat, ein Befund, der schon vermuten ließ, daß das Substrat, auf welches das Virus einwirkt, glycoproteide Beschaffenheit haben könnte. Diese Vermutung wurde fernerhin durch den Nachweis verstärkt, daß Erythrocytenextrakte, die auf Grund ihrer stark hemmenden Wirkung auf die Virushaemagglutination offensichtlich den Zellreceptor in gelöster Form enthielten, durch einen hohen Polysaccharidgehalt charakterisiert waren. Schließlich wurde von BURNET und seinen Mitarbeitern nachgewiesen, daß zahlreiche körpereigene Stoffe von mucoproteiner Natur, wie z. B. die gereinigte Blutgruppensubstanz O, das Mucin der menschlichen Magenschleimhaut, bzw. der Cervix uteri, sowie der schleimige Cysteninhalt von pseudomucinösen Ovarialcysten stark hemmende Effekte auf die Virushaemagglutination ausüben, und daß die antagonistische Wirkung dieser Substanzen durch die Bebrütung mit aktivem Virus, bzw. der erwähnten bakteriellen Kulturfiltrate zerstört wird. Die Schlußfolgerung schien daher erlaubt, daß der Zellreceptor ein Mucoprotein und das Virushaemagglutinin — mutatis mutandum — eine Mucinase darstellt. Der Identifizierung des Zellreceptors mit einem mucinartigen Körper steht nun einstweilen nichts entgegen, dagegen erscheint die Hypothese von der enzymatischen Zerstörung der bindenden Receptoren agglutinabler Erythrocyten noch keineswegs endgültig gesichert [Doerr (97)].

Schon kurze Zeit nach der Entdeckung des HIRST'schen Phaenomens wurde die Möglichkeit erwogen, daß die Infektion von virusempfindlichen Epithelzellen durch ähnliche Vorgänge wie bei der Haemagglutination eingeleitet und bestimmt werden könnte. In dieser Richtung angestellte Versuche, in welchen die Adsorption und Elution von Influenzavirus in der Frettchen- und Mäuselunge sowie im Cavum Allantois befruchteter Hühnereier geprüft wurde, führten zu Ergebnissen, die sich — allerdings nur teilweise — mit den in vitro erhobenen Befunden deckten. In excidierten Lungenpraeparationen und in der (mit Formalin devitalisierten) Allantois verlief die Reaktion zwischen Virus und Epithelzellen grundsätzlich wie im Reagenzglasversuch mit Erythrocyten, d. h. das Virus wurde innert kürzester Zeit rasch und vollständig adsorbiert und hierauf in verlangsamtem Tempo eluiert, wobei das Vermögen der Zellen zur Virusbindung auch hier — zumindest vorübergehend — verloren ging. In Gegenwart von voll vitalen Geweben (in situ belassene Lungen noch lebender Frettchen und Mäuse, Allantois lebender Hühnerembryonen) konnte jedoch nur die adsorptive Phase beobachtet werden und fehlte eine sich anschließende Elution vollständig, wohl ein Hinweis dafür, daß in diesem Fall die Virusbindung fester ist, und zwar als Vorbedingung für die nun einsetzende Virusvermehrung. Das hauptsächlichste Ergebnis dieser Versuche kann demnach darin erblickt werden, daß auch die virusempfindlichen Gewebe mit Receptoren ausgestattet sind, die durch den Kontakt mit Virus abgesättigt, inaktiviert oder zerstört werden. Ob sich die bei Virusinfektionen so häufigen Interferenzphaenomene durch die Zerstörung oder Blockade solcher Receptoren in befriedigender Weise erklären lassen, erscheint jedoch nach den bereits in dieser Richtung unternommenen Untersuchungen sehr fragwürdig. Die Analogie zwischen den Vorgängen bei der Haemagglutination und der Zellinfektion ist demnach begrenzt, schon deshalb, weil die Virusvermehrung im einen Fall ausgeschlossen ist, im andern dagegen das weitere Schicksal des Virus bestimmt.

Eine generelle Bedeutung erlangte das Phaenomen von Hirst durch den Nachweis, daß auch andere Virusarten haemagglutinierende Wirkung entfalten. Ein solches Vermögen wurde bisher nachgewiesen für das Virus der *Pseudopest der Hühner* (38, 161), *klassischen Hühnerpest* (239, 161), *Mäusepneumonie* (261), *Parotitis epidemica* (233), *Vaccine* (264), *Variola* (266), *Ektromelie* (58, 160) und für bestimmte *murine Poliomyelitisstämme* (160, 163). Haemagglutinierende Eigenschaften wurden fernerhin bei verschiedenen Species von *Rickettsien* (267, 160) und bei einem *Pleuropneumonie-artigen Erreger* (190) festgestellt. Schließlich fehlt es in der ältern und neuern Literatur nicht an Beobachtungen, wonach auch *bakterielle Erreger* Haemagglutinations-Reaktionen auslösen, die möglicherweise in den Rahmen des Hirst'schen Phaenomens eingeordnet werden können. Schon aus den bisherigen Untersuchungen dürfte hervorgehen, daß die Haemagglutinine verschiedener Virusarten sich in mehrfacher Hinsicht voneinander unterscheiden, nämlich nicht nur dadurch, daß der Agglutinationstypus (Aspekt des Agglutinates) und das „Agglutinationsspektrum" (Anzahl und Artzugehörigkeit agglutinabler Erythrocytenspecies) für jede Virusart mehr oder weniger charakteristisch sind, sondern auch durch größere Differenzen, wie die bei einigen Virusarten nicht nachgewiesene (spontane) Eluierbarkeit des haemagglutinierenden Prinzips (Vaccine-, Ektromelie-, Mäusepneumonievirus) oder das Vorliegen bestimmter Virushaemagglutinine (Vaccine-, Ektromelievirus) in einer gelösten, d. h. nicht an die Viruspartikel gebundenen Form. Trotz dieser sicher bemerkenswerten Unterschiede wäre es wohl verfrüht, die Virusarten nach den Besonderheiten ihrer Haemagglutinine schon jetzt zu gruppieren, schon in Hinsicht auf die immer noch bescheidene Anzahl von Virusarten, bei welchen bisher ein haemagglutinierendes Vermögen nachgewiesen werden konnte und schließlich auch auf Grund der Unkenntnis, welche Bedeutung bestimmten Vorgängen im Rahmen des Gesamtphaenomens beigemessen werden muß.

II. Die Analyse des Phaenomens.

1. Die Resistenz des Haemagglutinins gegen inaktivierende Einflüsse im Vergleich zu den andern Virusfunktionen.

Temperatur. Bei niedern Temperaturen ($\lesssim +4^0$ C) bleibt der Haemagglutinintiter von Influenza-, Hühnerpest- und — in vermindertem Grade — auch von Mumpsvirus während mehrerer Monate nahezu unverändert, sofern die Virusproben in Form von Allantois- bzw. Amnionflüssigkeit konserviert werden. Dagegen wird die Stabilität des Influenza-Haemagglutinins durch das Einfrieren bei tieferen Temperaturen (—70⁰ C) deutlich herabgesetzt [Miller und Stanley (259), Stanley (308)][1] und durch wiederholtes Einfrieren und Auftauen anscheinend gänzlich zerstört [Bieling und Oelrichs (25)]. Die Trocknung der Virusproben im gefrorenen Zustand nach dem Verfahren von Flosdorf führt ebenfalls zu erheblichen Haemagglutininverlusten [Stanley (308), Pearson (270)] und bietet demnach als Konservierungsverfahren keinerlei Vorteile.

Auch bei höheren Temperaturen von 18, 25 und 37⁰ C besitzt das Haemagglutinin von Influenza- und Hühnerpestvirus noch eine bemerkenswerte Stabilität,

---

[1] Möglicherweise gilt diese Feststellung nicht für sämtliche Influenzastämme, da nach Hirst (200) das Einfrieren der menschlichen Rachenspülflüssigkeit bei — 72⁰ C die Infektiosität von A-Stämmen während fünf Jahren konserviert, B-Stämme dagegen schon innert vier Monaten gänzlich inaktiviert.

indem der Titerabfall in derartigen Proben nur langsam fortschreitet [STANLEY (308), SALK (285), HALLAUER (164)]. Die kritische Temperaturzone, innerhalb welcher die Haemagglutinine der meisten Virusarten verändert und in mehr oder weniger kurzer Zeit inaktiviert werden, liegt in einem Bereich von 50—60⁰ C. Eine höhere Thermoresistenz wurde bisher nur für das Virus der Mäusepneumonie nachgewiesen, dessen Haemagglutinin das Erhitzen auf 70—80⁰ C während kürzerer Zeit verträgt [MILLS und DOCHEZ (261, 262), CURNEN und HORSFALL (81), CURNEN, PICKELS und HORSFALL (83)]. Genaue Angaben über die Thermoresistenz der Haemagglutinine verschiedener Virusarten sind schon deshalb nicht möglich, weil diese Eigenschaft bei den verschiedenen Typen und Stämmen einer Virusgruppe erheblich variiert, ein Befund, der sowohl für die verschiedenen Stämme des Influenzavirus [HIRST (202), SALK (284), BRIODY (35), JONES (210)] als auch des Hühnerpestvirus [HALLAUER (164)] erhoben werden konnte. Außerdem ist die Hitzeempfindlichkeit des Haemagglutinins erwartungsgemäß von Milieufaktoren — wie der Gegenwart bestimmter Ionen (35), von Formalin (285), Glutathion (330) usw. — weitgehend abhängig.

Sämtliche der bisher nachgewiesenen Virushaemagglutinine sind gegenüber thermischen Einflüssen stabiler als die infektiositätsbestimmenden Strukturen, so daß die Möglichkeit besteht, die Infektiosität vollständig aufzuheben, ohne hierbei das haemagglutinierende Vermögen zu beeinträchtigen. Eine solche „Dissociation" tritt beim Influenzavirus schon spontan in Erscheinung, wenn Virusproben über einige Zeit bei Zimmer- oder Bruttemperatur belassen werden [HIRST (194)] oder kann durch eine zeitlich befristete Erhitzung auf 52—56⁰ C künstlich erzeugt werden [HIRST (194, 195), BRIODY (35)]. Ein schonend inaktiviertes Virus kann nun zweifellos in seiner haemagglutinierenden Wirkung vollständig unverändert sein [HIRST (195)]. Immerhin scheint dies nicht die Regel zu sein, da nach HIRST (202) und BRIODY (35) erhitzte Influenza- bzw. Hühnerpestvirusstämme wohl in kaum verändertem Grade haemagglutinieren, auch in unvermindertem Maße von den Erythrocyten adsorbiert werden, zur spontanen Elution dagegen nicht mehr befähigt sind. Nach der Fermenthypothese würde dieser Befund dahin zu interpretieren sein, daß erhitzte Haemagglutinine der enzymatischen Aktivität beraubt sind, so daß ein Abbau, bzw. eine Zerstörung des Zellreceptors nicht mehr zustande kommt. Mit dieser Vorstellung steht nun auch die Beobachtung im Einklang, wonach das Haemagglutinin von hitzeinaktiviertem Influenzavirus in auffallend hohem Grade von Normalseren (und andern Substanzen, die den Zellreceptor in gelöster Form enthalten) gehemmt wird [FRANCIS (125)], ein Befund, der ebenfalls in der Hauptsache darauf zurückgeführt werden konnte, daß das inaktivierte Virus — im Gegensatz zum aktiven — die hemmende Receptorsubstanz nicht abzubauen vermag [HIRST (202), BURNET (51)]. Schließlich ist, wie BRIODY (35) feststellte, auch der Agglutinationstypus in charakteristischer Art verändert, da die mit erhitztem Virus erzeugten Agglutinate stabiler sind und daher durch Aufschütteln nicht zu einer homogenen Suspension dispergiert werden können, ein Verhalten, das dieser Autor — wohl mit Recht — im Sinne einer festeren Bindung des Virus an den Receptor deutet.

Eine ausgesprochene Zunahme des Haemagglutinintiters beobachtete HIRST (202) bei der Erhitzung (56⁰ C während 60 Min.) von Proben eines auf der Ultrazentrifuge gereinigten und konzentrierten Influenzavirus. Da beim Ultrazentrifugieren stets mit der Bildung von Virusaggregaten zu rechnen ist, und der erhöhte Agglutinationstiter allmählich wieder auf den Ausgangswert abfiel, bezieht HIRST diesen paradoxen Effekt auf einen vorübergehend erhöhten Dispersitätsgrad der haemagglutinierenden Viruspartikel. Daß ein solcher Effekt auch schon bei niederer Temperatur (+4⁰ C) auftreten kann, wurde von KNIGHT (221) festgestellt. Proben eines aus Mäuselungen isolierten und hochgereinigten (von inertem Material bzw. natürlichen Hemmstoffen befreiten) Influenzavirus A zeigten während einer 16tägigen Lagerung in Phosphatpuffer eine Aktivitätszunahme von 5000 auf 18000 agglutinierenden Einheiten. Eine ähnliche Aktivierung des Haemagglutinins beobachteten Anderson (1) beim Virus der Newcastle-Krankheit der Hühner und Lind (235) beim Mumpsvirus in Virus-Erythrocytengemischen, die über 1—2 Stunden bei 37° C bebrütet wurden[1], fernerhin Beveridge und Lind (22) bei einem bestimmten Stamm von Mumpsvirus in formolisierter Allantoisflüssigkeit nach mehrwöchiger Lagerung bei + 4° C. Es sind demnach wohl nicht nur thermische, sondern auch andere Einflüsse (Formalin, Wegfall von Hemmstoffen, Kontakt mit Erythrocyten usw.), die eine solche desaggregierende Wirkung besitzen. Möglicherweise sind derartige Aktivitätsschwankungen nicht nur abhängig vom Dispersitätsgrad der Viruselemente, sondern von der jeweiligen Zustandsform der Viruspartikel selbst. So stellte Bang (14, 17) in elektronenoptischen Untersuchungen fest, daß das Virus der Newcastle-Krankheit in der Allantoisflüssigkeit und in elektrolytfreier, gereinigter Form sphärische Partikel aufweist, die jedoch durch den Zusatz kleiner Salzmengen (0,07—0,15 m NaCl) in Faden- oder geschwänzte Formen übergeführt werden konnten. Durch schonendes Erhitzen, partielle Inaktivierung mit Formaldehyd oder Senfgas wurde diese Transformation verhindert.

Henle und Henle (179) dehnten ihre vergleichenden Stabilitätsstudien auch auf die *toxischen Eigenschaften* des Influenzavirus aus und fanden, daß die Thermoresistenz des toxischen Agens verschiedener Stämme deutlich höher ist als diejenige der infektiösen Aktivität, dagegen ebenso deutlich niederer als die des Haemagglutinins. McKee und Hale (250) prüften beim Influenzavirus A (PR 8) vergleichsweise die Thermostabilität des haemagglutinierenden und *antigenen Vermögens* mit dem Ergebnis, daß das Haemagglutinin bei der verwendeten Temperatur (57° C während 70 Minuten) größtenteils inaktiviert, die antikörperbindende Wirkung dagegen kaum abgeschwächt wurde. Ebenso vermochten sich Beveridge und Lind (22) nicht davon zu überzeugen, daß die immunisierende Potenz von konserviertem Mumpsvirus von der Höhe des jeweils noch vorhandenen Haemagglutinintiters abhängig ist. Nach Mills und Dochez (262) und Curnen und Horsfall (81) besitzt das auf 70—80° C erhitzte Haemagglutinin des Mäusepneumonievirus ein kaum vermindertes immunisierendes Vermögen und reagiert mit Immunsera nicht nur im Hemmungs- und Neutralisationstest, sondern auch — im Gegensatz zum aktiven, unerhitzten Virus — im Komplementbindungsversuch. Möglicherweise sind die immunisierenden und komplementbindenden Antigene auch dieser Virusart noch thermoresistenter als das Haemagglutinin, da die antigenen Funktionen selbst bei Kochtemperaturen (100° C während 10—15 Minuten) oft nicht aufgehoben werden (81). Schließlich stellte Lind (235) fest, daß die Erhitzung auf 55° C wohl die Infektiosität und das Haemagglutinin des Mumpsvirus zerstört, das komplementbindende Antigen dagegen intakt beläßt.

Formol, Formalin, bzw. Formaldehyd beeinflußt das Haemagglutinin und die übrigen Virusfunktionen in ähnlicher Weise wie die Erwärmung. Formalinkonzentrationen von 0,05—0,75% schädigen das Haemagglutinin von Influenza-,

---

[1] Nach Anderson (1) und Lind (235) ist die Aktivierung an die Gegenwart von Erythrocyten gebunden. Hanig (165) konnte nun tatsächlich beobachten, daß Virusaggregate von Influenzavirus (PR 8) während des Adsorptions- und Elutionsvorganges an Erythrocyten desaggregiert werden, so daß das Eluat eine höhere Aktivität zeigt als das Ausgangsvirus.

Hühnerpest-, Mumps- und Vaccinevirus bei $+4^0$ C über mehrere Wochen nicht [Hirst (195), Stanley (308), Hallauer (164), Dinter et al. (94), Beveridge und Lind (21, 22), Nagler (265)], wirken dagegen — wie dies beim Influenzavirus nachgewiesen wurde — beim Einfrieren oder Trocknen in gefrorenem Zustand deletär [Stanley (308)]. Während die Infektiosität bereits durch niedere Formalinkonzentrationen rasch und vollständig aufgehoben wird, erfolgt die Inaktivierung des Haemagglutinins erst in Gegenwart größerer Formolmengen (0,5—1%, und auch dann nur langsam [Hirst (195), McLean et al. (251), Florman und Weiss (123), Hallauer (164), Nagler (256)]. Für das Influenzavirus wurden die quantitativen Bedingungen für die Dissociation ermittelt [Hirst (195)]; eine Formalinkonzentration von 0,5% bewirkte nach 24stündiger Einwirkung den Verlust der Infektiosität, ließ dagegen den Haemagglutinintiter unverändert (Virus B, Stamm Lee) oder senkte denselben um die Hälfte (Virus A, Stamm PR 8). Die Haemagglutinine verschiedener Virusstämme besitzen demnach auch gegenüber Formalin eine unterschiedliche Resistenz. Die adsorptiven und eluierenden Eigenschaften von formalinisiertem Influenzavirus untersuchte Hirst (195); Formalinkonzentrationen von 0,1—0,5% verzögerten den Adsorptions- und Elutionsvorgang beträchtlich, noch höhere Formolkonzentrationen von 1—2% wirkten — in völliger Analogie zum thermisch beeinflußten Haemagglutinin — „dissociierend", d. h. erlaubten zwar eine rasche Adsorption, verhinderten jedoch die Elution. Die Toxicität des Influenzavirus nimmt nach Henle und Henle (179) hinsichtlich der Empfindlichkeit gegen Formol eine ähnliche Mittelstellung zwischen Infektiosität und haemagglutinierender Aktivität ein, wie dies schon bei der Einwirkung erhöhter Temperaturen beobachtet wurde. Schließlich stellte Stanley (308) fest, daß der Haemagglutiningehalt in formolisierten Impfstoffproben von Influenzavirus im allgemeinen der immunisierenden Potenz entspricht und zumindest als Indikator eines undenaturierten Antigenbestandes bewertet werden kann.

Ultraviolettes Licht. Die Resistenzprüfungen gegen ultraviolettes Licht führten grundsätzlich zu denselben Ergebnissen. Wiederum zeigte sich, daß die infektiöse Qualität des Influenzavirus schon bei weit kürzerer Exposition aufgehoben werden kann, als die haemagglutinierende Wirkung [McLean et al. (251), Stanley (308)]. Typen- und Stammesunterschiede konnten ebenfalls nachgewiesen werden [Henle und Henle (180)]. Vergleichende Resistenzprüfungen zwischen dem Haemagglutinin und den übrigen Funktionen des Influenzavirus wurden von Henle und Henle (180) angestellt und ergaben die folgende Skala: Infektiosität < Toxicität < interferierendes Vermögen < Haemagglutinin < komplementbindendes Antigen (600-S-Komponente) < komplementbindendes Antigen (30-S-Komponente). Das immunisierende Vermögen wurde beim PR-8-Stamm bereits vor, beim Lee-Stamm gleichzeitig mit der Abnahme des Haemagglutiningehaltes abgeschwächt, konnte jedoch noch in Proben nachgewiesen werden, die keine haemagglutinierende Wirkung mehr erkennen ließen. Beim strahlenresistenteren Influenzavirus B (Lee) wurde außerdem beobachtet, daß der Haemagglutinintiter während der ersten 1—2 Stunden der Belichtung beträchtlich (um das 3—4fache des Ausgangswertes) anstieg.

Alkohol, Äther, Sublimat, Senfgas Perjodat. Daß *Methylalkohol* in der Kälte und bei neutraler Reaktion weder die infektiösen noch haemagglutinierenden und immunisierenden Eigenschaften beeinträchtigt und daher als ein nicht denaturierendes Fällungsmittel für die Virusreinigung und -konzentrierung verwendet werden kann, wurde von Cox und van der Scheer (78) nachgewiesen. Demgegenüber beobachteten Beveridge und Lind (21), daß das Ausschütteln von Mumpsvirus mit *Äther* einen starken Verlust an Haemagglutinin verursacht.

Nach Klein und Perez (217) wird die Infektiosität von Influenzavirus durch *Sublimat* in der Verdünnung 1 : 10.000 aufgehoben, das Haemagglutinin dagegen nicht alteriert. Denselben dissociierenden Effekt beobachteten Rose und Geelhorn (280) bei der Behandlung von Influenzavirus mit *Senfgas* (Dichlordiaethylsulfid, bzw. Dichlordiaethylamin). Fazekas de St. Groth und Graham (109) untersuchten die Wirkung von Perjodat (KJO_4) auf verschiedene Influenzavirusstämme und stellten hierbei fest, daß diese Substanz in einer Konzentration von m/100—m/400 die Infektiosität vernichtet, das haemagglutinierende Vermögen dagegen nicht beeinträchtigt, beim Lee-Stamm sogar steigert. De facto wurde aber auch das Haemagglutinin verändert, da dasselbe wohl in unverminderter Stärke an Erythrocyten adsorbiert, jedoch nicht mehr eluiert wurde. Bei noch höheren Perjodatkonzentrationen wurde auch die Adsorption des Haemagglutinins und zugleich das antigene Vermögen aufgehoben.

Wasserstoffionenkonzentration. Miller (257) prüfte vergleichsweise die Stabilität bei drei gereinigten und konzentrierten Influenzavirusstämmen (PR 8, Lee, Shope) in Pufferlösungen verschiedener Wasserstoffionenkonzentration. Beide Virusqualitäten zeigten die größte Stabilität bei pH 7, 0—8,0 (Lee, Shope), bzw. pH 7,0 (PR 8) und verloren ihre Wirksamkeit auf der sauren Seite dieses Optimums rascher als auf der alkalischen. Das Haemagglutinin erwies sich durchwegs als stabiler und konnte im alkalischen Gebiet von der Infektiosität „dissociiert" werden. Anscheinend ist auch das Haemagglutinin des Vaccine- und Ektromelievirus in einem größeren pH-Bereich stabil, da nach Burnet und Stone (63) eine Inaktivierung erst bei einem pH < 5,0 eintritt. Für das Mumpsvirus ermittelten Weil, D. Beard und J. W. Beard (332) eine optimale Stabilität zwischen pH 5,65—7,9. Die wohl größte Stabilität gegenüber Veränderungen der Wasserstoffionenkonzentration zeigt das Haemagglutinin des Virus der Mäusepneumonie, dessen Aktivität erst bei einem pH 3,0, bzw. > 11,0 zerstört wird [Curnen und Horsfall (81, 82)]. Im alkalischen Gebiet geht ebenfalls die Infektiosität früher und rascher verloren als die haemagglutinierende Eigenschaft.

Dialyse. W. Henle, G. Henle, Stokes und Maris (188) dialysierten influenzavirushaltige Allantoisflüssigkeit gegen gepufferte Kochsalzlösung bei pH 7,0 und konservierten das Dialysat bei — 10°C, mit dem Ergebnis, daß der Haemagglutinintiter während mehr als 18 Monaten unverändert blieb. In den Versuchen von Lauffer und Stanley (229), die gereinigtes Influenzavirus A (PR 8) während drei Tagen bei + 4°C gegen destilliertes Wasser dialysierten, wurde dagegen sowohl die haemagglutinierende wie die infektiöse Aktivität ganz erheblich vermindert, ein Befund, welcher mit der von Knight (218) festgestellten geringen Stabilität von Influenzavirus in elektrolytarmen Medien übereinstimmt. Nach Curnen und Horsfall (81, 82) ist das Pneumonievirus der Mäuse sowohl in gebundener als in freier Form nicht dialysabel. Die Dialyse bewirkte den Verlust der Infektiosität, ließ dagegen die haemagglutinierende Aktivität unverändert, die sich ausschließlich nur im wasserunlöslichen Praecipitat vorfand.

Fermente. Die enzymatische Wirkung von Influenzavirus B (Lee) gegenüber dem „Francis-Inhibitor" (vgl. unten) wird durch Trypsin ebenso aufgehoben wie durch Erhitzung [Stone (317)]. Stone vermochte auch nachzuweisen, daß der Haemagglutinintiter unter dem Einfluß einer milden tryptischen Verdauung ansteigt. Dieselbe Empfindlichkeit gegen Trypsin und Chymotrypsin wiesen Curnen und Horsfall (81, 82) für das Virushaemagglutinin der Mäusepneumonie nach. Nach Stone (311) wird das Haemagglutinin von Vaccine- und Ektromelievirus sowohl durch das α-Toxin von Cl. Welchii als auch durch

Kobragift rasch und vollständig zerstört. Da beide Gifte Lecithinasen vom Typ C, bzw. A enthalten, wird der inaktivierende Effekt als Lecithinasewirkung gedeutet.

Insgesamt lieferten diese Stabilitätsstudien recht wertvolle Ergebnisse: *1. Ergeben sich gewisse Anhaltspunkte für die stoffliche Natur des haemagglutinierenden Agens. Das Verhalten gegen Erhitzung, Formol, ultraviolettes Licht, Alkohol, Dialyse und proteolytische Fermente ist mit der Annahme einer proteiden Beschaffenheit des Haemagglutinins von Influenza-, Hühnerpest- und Mumpsvirus zumindest kompatibel. Dem Haemagglutinin des Vaccine- und Ectromelievirus müßte der Charakter eines Phospholipoids (eventuell in Verbindung mit einem Protein) zugeschrieben werden, falls der inaktivierende Einfluß von Lecithinasen sichergestellt würde; 2. besitzen einige Virushaemagglutinine (Influenza, Hühnerpest, Mumps) offensichtlich zwei Funktionen, nämlich, einerseits das Vermögen, sich mit dem Receptor unter dem Phaenomen der Haemagglutination zu verbinden und anderseits die — enzymatisch gedeutete — Wirkung, den Receptor zu inaktivieren, bzw. zu zerstören und hierbei sich wieder abzuspalten. Beide Aktivitäten sind in unterschiedlichem Maße gegenüber inaktivierenden Einflüssen stabil und können daher durch eine geeignete Erhitzung, Formolisierung oder Behandlung mit Perjodat voneinander „dissociiert" werden, und 3. bestehen bemerkenswerte Resistenzunterschiede zwischen dem haemagglutinierenden Agens und den übrigen Virusfunktionen. Die Haemagglutinine sind durchwegs stabiler als das infektiöse Vermögen, so daß durch zahlreiche Mittel physikalischer und chemischer Art (Temperatur, Ultraviolettbestrahlung, Dialyse, alkalische Reaktion, Formol, Sublimat, Senfgas) eine „Dissociation" zu erreichen ist, d. h. einerseits die Infektiosität aufgehoben und anderseits die haemagglutinierende Wirkung erhalten werden kann. Demgegenüber scheint erwiesen, daß die haemagglutinierende Eigenschaft früher erlischt als die antigene Qualität. Ob diese Unterschiede jedoch zur Annahme verschiedener Strukturen berechtigen, bleibt fraglich, da der Haemagglutininnachweis eine relativ wenig empfindliche Methode darstellt, und auch Virusantigene, welche der haemagglutinierenden Wirkung beraubt sind, zur Bildung von Antihaemagglutininen Anlaß geben. Die Unversehrtheit des Haemagglutinins ist jedenfalls ein geeigneter Indikator für einen intakten, wenig geschädigten Antigenbestand.*

2. Die Bildung des Haemagglutinins als Funktion der Virusvermehrung.

Zeitliche Beziehungen. HENLE und HENLE (170, 187, 172, 178, 179) untersuchten bei einer größeren Anzahl von Influenzavirusstämmen den Infektionsverlauf im Allantoissack von Hühnerembryonen, wobei für die verschiedenen Virusfunktionen (Infektiosität, Toxicität, Haemagglutinin, komplementbindendes Antigen) das zeitliche Verhältnis hinsichtlich der ersten Nachweisbarkeit, der maximalen Ausbildung und der Persistenz vergleichsweise ermittelt wurde. Die erzielten Versuchsergebnisse lassen sich dahin zusammenfassen, daß für jede Virusqualität eine charakteristische Entwicklungskurve nachgewiesen werden konnte; die Infektiosität erreichte bereits nach *24—48* Stunden ihr Maximum und fiel hierauf rasch innert 1—2 Tagen auf ein Minimum ab, die haemagglutinierende Eigenschaft war deutlich später nachweisbar und zeigte einen ebenfalls verspäteten Gipfelpunkt nach 24—*48* Stunden, persistierte jedoch in nahezu unvermindertem Grade während 96 Stunden, die toxische Wirkung war erst zwischen 48—72 Stunden voll ausgebildet und hatte einen flüchtigen Bestand, das komplementbindende Antigen stellte diejenige Komponente dar, die am spätesten auftrat, aber auch — ebenso wie das Haemagglutinin — am längsten nachweisbar blieb. Die von HENLE nachgewiesene zeitliche Verschiebung zwischen der Virusvermehrung und Haemagglutininbildung wurde nicht nur für das Influenzavirus bestätigt [McLean et al. (255), v. Magnus (243)] u. a., sondern auch beim Mumpsvirus [Weil, D. Beard und J. W. Beard (332), Ginsberg, Goebel und Horsfall (145)] und dem Virus der Newcastle-Krankheit der Hühner [Bang (16)] festgestellt. Eine Ausnahme von dieser Regel macht bisher nur das *Virus der Mäusepneumonie*, da von Curnen und Horsfall (81) nachgewiesen wurde, daß der Infektiositäts- *und* Haemagglutinintiter in der infizierten Mäuselunge sich zeitlich konform bewegen, d. h. gleichzeitig nach 6—7 Tagen den Höhepunkt und nach dem 10. Tag minimale Werte erreichen. Fernerhin ist bemerkenswert, daß sämtliche Faktoren (Größe der Infektionsdosis, Inokulationsmodus, Bebrütungstemperatur, Alter und Individualität der Embryonen), die erfahrungsgemäß das Ausmaß und die Geschwindigkeit der Virusvermehrung beeinflussen können, einen meist gleichsinnigen, d. h. hemmenden oder fördernden (eventuell auch nicht nachweisbaren) Effekt auch auf die Haemagglutininbildung ausüben, ein Befund, der ebenfalls nicht nur für das Influenzavirus (192, 170, 178, 295, 255, 256), sondern auch beim Mumpsvirus (332) und dem Virus der Newcastle-Krankheit (16) erhoben werden konnte. Die Tatsache, daß die Konzentration des Haemagglutinins zu einem Zeitpunkt am größten ist, in welchem der Höchsttiter der Infektiosität bereits überschritten ist und auch dann noch beträchtlich bleibt, wenn die infizierende Eigenschaft einen minimalen Grad erreicht hat, könnte am einfachsten dahin interpretiert werden, daß einerseits das verspätete Auftreten des Haemagglutinins nur dadurch vorgetäuscht wird, daß die Haemagglutininbestimmung eine weit weniger empfindliche Methode darstellt als die Infektiositätsprüfung[1] (vgl. S. 153) und daß anderseits die längere Persistenz des Haemagglutinins dadurch bedingt ist, daß im Haemagglutinationsversuch nicht nur aktives, sondern auch inaktiviertes Virus gemessen wird. Anderseits wäre aber auch die Deutung zulässig, daß die Haemagglutininproduktion einen selbständigen Vorgang darstellt, welcher mit der Infektiosität nur teilweise koordiniert ist.

Quantitative Beziehungen. Bei der Beurteilung des quantitativen Verhältnisses der haemagglutinierenden und infektiösen Aktivität ist zu berücksichtigen, daß im Haemagglutinationsversuch eine *neue* Virusqualität gemessen wird, die sich mit der Infektiosität nur in bedingtem Maß vergleichen läßt. Fernerhin wird bei der Infektiositätsprüfung nur die Konzentration der aktiven, vermehrungsfähigen Viruselemente bestimmt, während im Agglutinationstest die Höhe des Titers außerdem von der Konzentration der jeweilig vorhandenen inaktiven Viruspartikel abhängig ist, so daß die Verwendung von partiell inaktivierten Virusproben notwendigerweise zu unrichtigen Vergleichswerten führen muß. Schließlich sind bei Haemagglutininen, die nicht an die Viruspartikel gebunden sind (Vaccine-, Ektromelievirus), strengere quantitative Beziehungen zur Infektiosität möglicherweise überhaupt nicht vorhanden[2]. In jedem Fall sind nur Titerwerte vergleichbar, die nach einer einheitlichen Bewertungsmethode (Bestimmung des 50% Infektiositäts-, bzw. Agglutinationstiters nach Reed und Muench) ermittelt werden.

---

[1] Fernerhin ist mit der Möglichkeit zu rechnen, daß das Haemagglutinin durch normalerweise in der Allantoisflüssigkeit vorkommende Hemmstoffe (Beveridge und Lind (21), Rawlinson (cit. 21), Svedmyr (302)) zumindest temporär maskiert wird.

[2] Beim Vaccinevirus stehen — nach den Untersuchungen von Stone und Burnet (318) — Haemagglutinin- und Infektiositätstiter in einem sehr wechselnden Verhältnis, je nachdem das Untersuchungsmaterial den oberflächlichen oder tiefern Schichten der Chorionallantois, bzw. der Kaninchenhaut entstammt.

Hirst (194) untersuchte schon in seiner ersten ausführlichen Arbeit, in welcher die Brauchbarkeit des Haemagglutinationstestes für die Virusauswertung ermittelt werden sollte, die quantitativen Beziehungen zwischen Agglutinations- und Infektiositätstiter bei einer Reihe von Influenzavirusstämmen und gelangte hierbei zu den folgenden Schlußfolgerungen: 1. In vollinfektiösen Proben ein und desselben Virusstammes besteht zwischen den beiden Titerwerten über einen großen Bereich von Viruskonzentrationen ein weitgehend konstantes Verhältnis; 2. diese Relation ist für jeden Virusstamm mehr oder weniger fixiert und charakteristisch und daher unterschiedlich, wenn verschiedene Virusstämme miteinander verglichen werden. Diese Feststellungen von Hirst sind bisher — abgesehen von einigen Einschränkungen (vgl. unten) — unbestritten geblieben und haben anscheinend für alle Virusarten Gültigkeit, deren Haemagglutinin von den Viruselementen nicht getrennt werden kann. Das von Hirst berechnete Verhältnis

$$\frac{\text{(log.) Infektiositätstiter}}{\text{(log.) Agglutinationstiter}}$$

ist anschaulich, weil es die Anzahl der minimalen Letalitäts-, bzw. Infektiositätsdosen angibt, die in 1 agglutinierenden Einheit (höchste Virusverdünnung, die zu einer $++$Reaktion führt) enthalten sind. In diesem Wert würde demnach nicht nur zum Ausdruck kommen, in welchem quantitativen Verhältnis die infizierende zur agglutinierenden Einheit steht, sondern auch, welche Viruskonzentration im Minimum notwendig ist, damit der Agglutinationstest überhaupt anzeigt. In Tabelle 1 sind die für eine Reihe von Virusarten festgestellten maximalen Titerwerte nach der beschriebenen Art in Relation gesetzt, wobei nur diejenigen Versuche berücksichtigt wurden, in welchen der Agglutinations- und Infektiositätstiter nach einheitlichen Gesichtspunkten ermittelt wurde. Unter A sind die Titerwerte nativer (meist der Allantoisflüssigkeit entstammender) Virusproben, unter B diejenigen von künstlich (durch Ultrazentrifugieren) gereinigter und konzentrierter Viruspräparationen angeführt. Das wiedergegebene Untersuchungsmaterial ist zweifellos zu unvollständig und zu spärlich, um bindende und weittragende Schlüsse abzuleiten, immerhin illustrativ genug, um die Schwierigkeiten aufzuzeigen, die einer quantitativen Erfassung des in Frage stehenden Verhältnisses entgegenstehen.

Zunächst ist ohne weiteres ersichtlich, daß die *Haemagglutininkonzentration*, welche die verschiedenen Influenzastämme und auch das Mumpsvirus in der Allantoisflüssigkeit von Hühnerembryonen im Maximum erreichen, nicht so unterschiedlich ($10^{2,1}$—$10^{3,1}$) ist, als man dies im Hinblick auf die z. T. recht wechselnde Infektiosität ($10^{6,2}$—$10^{10,2}$) erwarten könnte. Noch geringfügiger sind die Unterschiede der Haemagglutininkonzentration bei gereinigten und konzentrierten Virusproben, wo sich die Haemagglutinintiter in einem Bereich von log. 6,2—6,8 bewegen, die Infektiositätstiter dagegen zwischen log. 11,6—13,7 variieren. Hirst (194) stellte diese Erscheinung bei verschiedenen Influenzavirusstämmen ebenfalls fest, und ebenso Hallauer (164) bei 13 Hühnerpeststämmen des Typus B, die — trotz beträchtlicher Infektiositätsunterschiede — in Allantoispassagen stets annähernd denselben Haemagglutinationstiter (log. 2,7—3,0) ergaben. Man erhält somit den Eindruck, daß die Haemagglutininbildung im Hühnerei einem Maximum zustrebt, das für zahlreiche Virusstämme unterschiedlicher Infektiosität und antigener Qualität annähernd dasselbe ist. Dieses Phaenomen ist — im Falle seiner Bestätigung — zweifellos überraschend und würde wohl darauf hinweisen, daß die Bildung des Haemagglutinins nicht

Tab. 1. *Verhältnis des Haemagglutinationstiters zum Infektiositäts-, bzw. Letalitätstiter,*
A. Natives, ungereinigtes Virus.

Virus-stämme	Autor	Aggluti-nations-titer ++ 50 % (log.)	Infektiositäts-, bzw. Letalitätstiter		Verhältniszahl [1]	
			L.D. 50 % Maus (log.)	I.D. 50 % Ei (log.)	L.D. 50 Maus	I.D. 50 Ei
Influenza A						
PR 8	Hirst (194)	3,1	6,7		$10^{3,6}$	
	Taylor et al. (322)	2,7		8,0		$10^{5,3}$
WS	Hirst (194)	2,1	6 5		$10^{4,4}$	
	Henle (187)	2,7	6,3		$10^{3,6}$	
F 99	Henle (179)	2,4		10,2		$10^{7,8}$
NY 43	Hirst (199)	2,5		8,2		$10^{5,7}$
Rhodes	Wang (74)	3,1		7,8		$10^{4,7}$
Influenza B						
	Hirst (155)	2,4	4.0		$10^{1,6}$	
Lee	Sharp et al. (291)	2,1		6 2		$10^{4,1}$
	Mc Lean et al. (255)	2,3		7,1		$10^{4,8}$
Influenza						
Swine	Hirst (194)	2,1	4 2		$10^{2,1}$	
	Taylor et al. (323)	2,7		7,5		$10^{4,8}$
Mumps						
	Ginsberg et al. (145)	2,7		7,3		$10^{4,6}$
	Weil et al. (332)	1,6–2,1		8,5–9,5		$10^{6,8}$–$10^{7,4}$
Pseudopest						
N.D.V.	Bang (16)	2,9–3 8		8,5–9,3		$10^{5,5}$–$10^{5,6}$
	Florman (120)	2,71		10,0		$10^{7,3}$

B. Gereinigtes, konzentriertes Virus.

Virus-stämme	Autor	Aggluti-nations-titer ++ 50 % (log.)	L.D. 50 % Maus (log.)	I.D. 50 % Ei (log.)	L.D. 50 Maus	I.D. 50 Ei
Influenza A						
	Taylor et al. (321)	6,8		13,4		$10^{6,6}$
PR 8	Stanley (306, 307)	6,2	10,0	13,0	$10^{3,8}$	$10^{6,8}$
	Knight (221)	$_2${ 6,6		13,4		$10^{6,8}$
		6,4	10,4	13,3	$10^{4,0}$	$10^{6,9}$
Influenza B						
Lee	Sharp et al. (291)	6,4		11,6		$10^{5,2}$
	Taylor et al. (321)	6 3		12,3		$10^{6,0}$
Influenza						
Swine	Taylor et al. (323, 324)	6,3		12,7		$10^{6,4}$
	Mc Lean et al. (251)	6,5		13,3		$10^{6,8}$
Mumps	Weil et al. (333)	6,2		12,7		$10^{6,5}$
Pseudopest N.D.V.	Cunha et al. (80)	6,3		13,7		$10^{7,4}$

[1] $\dfrac{\text{Letalitäts-, bzw. Infektiositätstiter}}{\text{Agglutinationstiter}}$ = Konzentration von L.D. 50, bzw. I.D. 50 in 1 Agglutinationseinheit.

[2] Aus Mäuselunge.

nur eine Funktion der Virusvermehrung ist, sondern auch — hiervon unabhängig — von seiten des Wirtes reguliert wird. Jedenfalls wäre die Vorstellung zulässig, daß das Haemagglutinin aus der „Normalkomponente" (vgl. S. 158, 173) der Allantoisflüssigkeit aufgebaut wird, die bekanntlich während der Virusvermehrung verbraucht [GARD und v. MAGNUS (141, 142), SVEDMYR (302)] und auch als konstanter Bestandteil gereinigter Viruspartikel nachgewiesen werden konnte [KNIGHT (219, 222)].

Nach HIRST läßt sich zwischen haemagglutinierender und infizierender Aktivität *keine* Proportionalität nachweisen, falls verschiedene Influenzastämme in dieser Hinsicht miteinander verglichen werden. Beide Viruseigenschaften verhalten sich hierbei wie zwei voneinander unabhängige Variable. Die Gültigkeit dieser Feststellung kann aus Tab. 1 erschlossen werden, aus welcher ersichtlich ist, daß gleichen Agglutinationstitern sehr unterschiedliche Infektiositätstiter zugeordnet sind, ein Verhalten, das in den variablen Verhältniswerten deutlich zum Ausdruck kommt. Geht man von der Voraussetzung aus, daß die Größendifferenz zwischen Agglutinations- und Infektiositätstiter für einen bestimmten, volladaptierten Virusstamm mehr oder weniger fixiert ist, so erhält man für die in Tab. 1 (A) angeführten Virusarten und -stämme — nach der Größe der Verhältniszahl (Anzahl M. L. D., bzw. M. I. D./1 aggl. E.) geordnet — die nachstehende Reihenfolge:

$10^{1,6}$	M. L. D. (Maus)	Influenza;	Lee
$10^{2,6}$	,, ,,	,,	Swine
$10^{3,6}$	,, ,,	,,	PR 8, W. S.
$10^{4,1}$	M. I. D. (Hühnerembryo)	,,	Lee
$10^{4,7}$	,, ,,	,,	Rhodes
$10^{4,8}$	,, ,,	,,	Swine
$10^{5,3}$	,, ,,	,,	PR 8
$10^{5,7}$	,, ,,	,,	NY 43
$10^{7,8}$	,, ,,	,,	F 99
$10^{4,6}$—$10^{7,4}$	,, ,,	Mumps	
$10^{5,5}$—$10^{7,3}$	,, ,,	Pseudopest;	N. D. V.

Das Verhältnis der agglutinierenden zur infizierenden Einheit wäre demnach sehr variabel; je nach dem Virusstamm würden 10'000 bis >1 Million M. I. D. erforderlich sein, damit der Schwellenwert für die Haemagglutination erreicht wird. Ob sich allerdings diese berechneten Werte mit den Erfahrungstatsachen decken, kann in Ermangelung ausreichender Angaben einstweilen nicht entschieden werden. Gesichert ist wohl nur das Faktum, daß der Haemagglutinationstest für den Virusnachweis eine viel weniger empfindliche Methode darstellt als die Infektiositätsprüfung, da Viruskonzentrationen $< 10^3$—10^4 M. I. D. im Test nach HIRST in der Regel nicht nachgewiesen werden können. Eine bemerkenswerte Ausnahme hiervon macht nur das Virus der Mäusepneumonie, bei welchem die Haemagglutinations- und Infektiositätstiter in der gleichen Größenordnung liegen.

Die aprioristische Erwartung, daß das Titerverhältnis bei nativen *und* gereinigten, bzw. konzentrierten Virusproben derselben Stämme (PR 8, Lee, Swine) annähernd gleich sein würde, trifft de facto nicht zu. In den gereinigten Viruskonzentraten sind die Titerabstände durchwegs um etwa eine Zehnerpotenz größer als beim Ausgangsvirus[1]. Mc Lean, Taylor und Mitarbeiter (255, 323) haben eine solche Diskrepanz bereits registriert, indem sie feststellten, daß die für Viruskonzentrate (Influenza B, Lee) kalkulierten Gewichtsmengen an Virus zwischen 61 und 5,4 mg (bezogen auf das Volum des Ausgangsmateriales) variieren, je nachdem diese Berechnung auf Grund der nachgewiesenen infektiösen oder haemagglutinierenden Aktivität vorgenommen wird. Nach Mc Lean und Mitarbeitern dürfte nur der kleinere, d. h. der auf der haemagglutinierenden Aktivität basierende Wert den tatsächlichen Verhältnissen entsprechen. Fernerhin wurde von Taylor und Mitarbeitern beobachtet, daß in den Viruskonzentraten wohl die bei den Ausgangsstämmen (Influenza B, Lee bzw. Swine) bestehenden Infektiositätsunterschiede zum Ausdruck kommen, die ursprünglichen Differenzen im haemagglutinierenden Vermögen dagegen nicht mehr nachgewiesen werden können. Fernerhin ist auch für ein und denselben Virusstamm das quantitative Verhältnis zwischen Haemagglutininkonzentration und Infektiositätstiter je nach der Provenienz des Virus unterschiedlich. So vermochte Knight (221) in überzeugender Art nachzuweisen, daß hochgereinigte Präparationen des Influenzavirus PR 8 je nach dem Ausgangsmaterial — Allantoisflüssigkeit, bzw. Mäuselunge — eine verschiedene haemagglutinierende Aktivität — 40'000, bzw. 30'000 E/mg Protein — besitzen, obschon die Infektiosität für den Hühnerembryo bei beiden Fraktionen dieselbe (10—^{14,0}/g Protein) war. In diesen Feststellungen dürfte die Problematik der quantitativen Beziehungen zwischen Infektiosität und haemagglutinierender Wirkung klar zum Ausdruck kommen. Eine Klärung dieser Frage ist wohl erst zu erwarten, wenn die Beziehungen zwischen Masse und Aktivität für beide Virusfunktionen ermittelt sind.

Die Annahme einer „gesetzmäßigen" quantitativen Beziehung zwischen der haemagglutinierenden und infektiösen Eigenschaft stützt sich demnach in der Hauptsache auf den von Hirst geführten Nachweis, daß das Verhältnis von Infektiositäts- und Agglutinationstiter für einen bestimmten Virusstamm über einen größeren Bereich von Viruskonzentrationen annähernd konstant bleibt, ein Befund, der bisher in eindeutiger Weise nicht widerlegt, sondern im Gegenteil wiederholt bestätigt worden ist. Immerhin ist die Konstanz dieses Verhältnisses nicht so groß, als wie man dies aus den Protokollen von Hirst annehmen könnte, da die für den PR 8-Stamm berechneten Verhältniszahlen zwischen $10^{3,3}$ und $10^{3,9}$ [2] (=2000—8000 M. L. D. für die Maus) oscillieren, also Schwankungen aufweisen, die bei der Trägheit, mit welcher der Haemagglutinationstest reagiert, wohl eine Bedeutung haben könnten. Wie aus den Protokollen von Bang (13) berechnet werden kann, sind die Differenzen der Verhältniszahlen beim Virus der Newcastle

---

[1] Welche Gründe hierfür maßgebend sind, ist nicht klar ersichtlich. Berechnungsgemäß müßte angenommen werden, daß die Verluste während des Reinigungsprozesses beim Haemagglutinin größer sind als diejenigen des infektiösen Vermögens, eine Annahme, die auch mit der praktischen Erfahrung, wonach beim Ultrazentrifugieren größere Haemagglutininverluste eintreten, in Einklang steht. Wodurch nun aber diese Einbuße größtenteils bedingt ist, steht zur Diskussion. Die Bildung von Virusaggregaten ist sicher zuzugeben, anderseits ist nicht ohne weiteres verständlich, daß hierdurch nur die haemagglutinierende Aktivität beeinträchtigt, die Infektiosität dagegen anscheinend nicht vermindert wird. Befriedigender ist die Vorstellung, daß der Haemagglutinintiter durch einen — aus dem Wirtsgewebe stammenden — Hemmstoff gesenkt wird, der nach den Untersuchungen von Knight (219—223) und Miller (256) auch in gereinigten Viruskonzentraten regelmäßig nachzuweisen ist (vgl. S. 158), und nach Hardy und Horsfall (166) ausschließlich nur haemagglutininbindende Wirkung aufweist (vgl. S. 174).

[2] Hirst gibt versehentlich eine Amplitude von $10^{3,5}$ bis $10^{3,9}$ an.

Krankheit ($10^{4.7}$,—$10^{6.1}$, entsprechend 50'000—1'000'000 M. L. D.[50] für den Hühnerembryo) noch bedeutend größer. Davon abgesehen, steht wohl außer Zweifel, daß eine derartige Proportionalität zwischen den Titerwerten nur in Virusproben nachweisbar ist, die in der sog. logarithmischen Phase der Virusvermehrung, d. h. in einem Zeitpunkt entnommen sind, in welchem sowohl das Haemagglutinin wie auch die Infektiosität maximale Werte erreichen. In der initialen Phase der Virusvermehrung ist das Verhältnis der Titerwerte eindeutig zugunsten der Infektiosität verschoben, so daß die Verhältniszahlen zu groß werden, und in der terminalen Phase tritt eine gegenteilige Verschiebung in Erscheinung [HENLE und HENLE (172, 170)]. Unter Berücksichtigung der unterschiedlichen zeitlichen Verhältnisse hinsichtlich Ausbildung und Persistenz dieser Virusfunktionen (vgl. S. 149), sind diese Abweichungen ohne weiteres verständlich. Von größerem Interesse sind dagegen Beobachtungen, wonach durch die Größe der infizierenden Virusdosis das quantitative Verhältnis zwischen Haemagglutinin und Infektiosität weitgehend verändert werden kann. Werden nämlich große Dosen von Influenzavirus in den Allantoissack von Hühnerembryonen verimpft, so wird — wahrscheinlich durch einen Autointerferenzvorgang — die Infektiosität viel stärker gehemmt als die Bildung des Haemagglutinins [HENLE und HENLE (179, 170), v. MAGNUS (243, 244), VIEUCHANGE, REINIÉ und SAUTTER (328)]. Durch eine geeignete Passagetechnik ist es sogar v. MAGNUS gelungen, die Infektiosität auf ein Minimum zu senken, ohne daß hierdurch die Haemagglutininbildung in nennenswerter Weise eingeschränkt wurde. v. MAGNUS neigt daher zur Vorstellung, daß diese „Dissociation" auf eine atypische Virusvermehrung hinweist, bei welcher eine Virusvariante gebildet wird, die zwar haemagglutinierende und immunisierende Eigenschaft besitzt, dagegen kaum mehr infektiös ist. Tatsächlich konnte diese Annahme durch spätere Versuche von GARD und v. MAGNUS (141, 142) noch wesentlich gestützt werden.

Schließlich hat die von HIRST postulierte quantitative Relation höchstenfalls Gültigkeit für Virusstämme, die an eine Wirtsspezies vollständig angepaßt sind. Bei der Adaptierung von Influenzavirus an andere Wirte ((Hühnerembryo, Maus) wird dieses Verhältnis zumindest temporär gestört, bzw. verändert, allerdings nicht nur in quantitativer, sondern auch in qualitativer Hinsicht. Die erste Beobachtung dieser Art machten BURNET, BEVERIDGE, BULL und CLARK (71) und BURNET und BULL (59); bei frisch vom Menschen in der Amnionflüssigkeit von Hühnerembryonen isolierten Influenza-A-Stämmen konnte ein sehr eigenartiger Agglutinationstypus nachgewiesen werden, der hauptsächlich dadurch charakterisiert ist, daß der Titer gegenüber Hühnererythrocyten extrem niedrig ist, wogegen andere Erythrocytenspecies (Meerschweinchen, Mensch) in einer der Norm entsprechenden Titerhöhe agglutiniert werden. Die Infektiosität dieser „O-Stämme" erreicht im Amnion einen hohen Grad, ist dagegen nur schwach ausgeprägt bei der Überimpfung auf die Allantois oder die weiße Maus. Schließlich ist die O-Phase durch ihre große Unbeständigkeit gekennzeichnet, da in der Regel schon eine Allantoispassage ausreicht, um eiadaptierte „D-Stämme" zu gewinnen. Daß auch ein vollständig an den Hühnerembryo adaptierter Influenzastamm gelegentlich wieder in die O-Phase zurückschlagen und hierbei das haemagglutinierende Vermögen gegenüber Hühnererythrocyten nahezu vollständig einbüßen kann, wurde ebenfalls von BURNET und BULL (60) bei einem bestimmten Influenzastamm („Melbourne egg") nachgewiesen. Von nicht geringem Interesse ist die Feststellung, daß die haemagglutinierende Aktivität von Influenzavirus bei der Anpassung von Eipassage-Stämmen an die weiße Maus in ähnlicher Art abgewandelt wird. So zeigte ein von HIRST (199) untersuchter Influenzavirusstamm (N. Y. 43) von typischem D-Charakter während 30 Anpassungspassagen auf der weißen Maus ein agglutinierendes Verhalten, das der O-Form von Burnet weitgehend entsprach, d. h. der Agglutinationstiter gegenüber Hühnererythrocyten fiel auf ein Minimum ab, erfuhr dagegen eine Zunahme gegen Meerschweinchenerythrocyten, die Infektiosität für den Hühnerembryo hielt sich während der ganzen Passagedauer auf einer — gegenüber dem Ausgangsvirus — kaum verminderten Höhe. In welchem Grade das quantitative Verhältnis zwischen Haemagglutinin -und Infektiositätstiter in Abhängigkeit von der Adaption eines Virusstammes an einen Wirt wechselt, geht besonders deutlich aus den Untersuchungen von Wang (74) hervor, der dieses Verhalten mit einem mäuseadaptierten, bzw. nichtadaptierten Stamm desselben Influenzavirus (Stamm Rhodes) vergleichsweise prüfte. Der adaptierte Stamm zeichnete sich durch einen hohen Infektiositäts- und Agglutinationstiter aus, während der nichtangepaßte Stamm zwar ebenfalls einen beträchtlichen Infektiositätsgrad erreichte, jedoch nur spärlichste Mengen an Haemagglutinin (für Hühnererythrocyten) erzeugte. Die Befunde von Hirst und Wang werden durch die Untersuchungen von Friedewald und Hook (136) über das Verhalten von Influenzastämmen bei der Anpassung an den Hamster in wertvoller Weise ergänzt. Eipassagestämme von Influenzavirus A (PR 8, Ga 45) bewirkten während den ersten 6 Passagen nur latente Infektionen, wobei jeweils ein beträchtlich hoher Infektiositätstiter erreicht wurde, irgendwelche Haemagglutinine jedoch nicht nachgewiesen werden konnten. Von der 7. Passage an wurde dagegen eine plötzliche Veränderung festgestellt, und zwar durch den gleichzeitigen Nachweis einer hohen Pathogenität, eines beträchtlichen Haemagglutinintiters für Menschen- und Meerschweinchen-Erythrocyten und einer bedeutenden Zunahme des komplementbindenden Antigens. Haemagglutinine gegenüber Hühnererythrocyten waren dagegen erst von der 11. Passage an nachweisbar. Soweit stimmen die Versuche von Friedewald und Hook mit den Befunden von Hirst und Wang überein. Friedewald und Hook vermochten nun aber fernerhin nachzuweisen, daß das stark verspätete Erscheinen der Haemagglutinine für Hühnererythrocyten nur vorgetäuscht, d. h. auf einen in der Hamsterlunge vorkommenden Inhibitor zu beziehen ist. Schließlich ist bemerkenswert, daß diese Veränderungen beim Influenzavirus B (Lee), welches schon in der Ausgangspassage für den Hamster pathogen war, überhaupt nicht beobachtet werden konnten. Wenn nun auch die Bedeutung des O-D-Phasenwechsels noch keineswegs klargestellt ist (vgl. S. 164), so scheint aus diesen Untersuchungen doch hervorzugehen, daß während der Anpassungsperiode einer Virusart an einen neuen Wirt die Bildung des Haemagglutinins in einem weit höheren Grade gehemmt sein kann als das infektiöse Vermögen.

Die Frage der quantitativen Beziehungen zwischen haemagglutinierender und infizierender Virusaktivität ist demnach noch keineswegs abgeklärt. Die bestehende Unsicherheit ist zweifellos darin begründet, daß zwei Vorgänge in Relation gesetzt werden, deren wohl unterschiedliche Dynamik einen derartigen Vergleich kaum zuläßt. Schon jetzt ist nicht zu übersehen, daß die Haemagglutininbildung bei verschiedenen Virusstämmen und Arten in sehr unterschiedlichem Grade vom Ausmaß der Virusvermehrung abhängig ist und daß anderseits die im Maximum erreichte Haemagglutininkonzentration bei verschiedenen Virusstämmen unterschiedlicher Infektiosität in derselben Größenordnung liegen kann. Auch für ein und denselben Virusstamm ist das Verhältnis der Titerwerte nicht konstant, sondern variiert in Abhängigkeit von der Phase des Infektionsverlaufes, der Virusprovenienz, der Virusreinigung und von Interferenz- und Adaptierungsvorgängen. Der Schlüssel zur Aufklärung dieser Diskrepanzen könnte darin gefunden werden, daß die Haemagglutination eben doch andere Zustandsformen des Virus mißt als die Infektiositätsprüfung, nämlich, einerseits auch inaktives (nicht infektiöses) Virus und anderseits

*nur diejenige Quote von infektiösem Virus, die in „nicht-gebundener, freier Form"
vorliegt. Aus den Untersuchungen von* HARDY *und* HORSFALL *(166) scheint jedenfalls hervorzugehen, daß das Influenzavirus in der Allantoisflüssigkeit in zwei, im
Gleichgewicht stehenden Formen unterschiedlicher Aktivität vorkommt, nämlich als
„gebundenes" und als „freies" Virus. Beide Formen sind gleichermaßen infektiös,
aber nur das freie Virus haemagglutiniert, während das an den Inhibitor der Allantoisflüssigkeit (vgl. S. 174, 183) gebundene Virus hierzu nicht befähigt ist. Da das quantitative Verhältnis von freiem zu gebundenem Virus nicht konstant, sondern je nach der
Virusart, dem Virusstamm, der Phase der Virusvermehrung, der Virusreinigung
und dem Grad der Anpassung unterschiedlich sein dürfte, wäre in befriedigender
Weise erklärt, weshalb zwischen Haemagglutinations- und Infektiositätstiter in zahlreichen Fällen keine quantitative Parallelität erwartet werden kann.*

3. Die Beziehungen des Haemagglutinins zu den geformten Viruselementen.

Aus zahlreichen Untersuchungen, in welchen das Verhalten von *Influenzavirusstämmen* im Gravitationsfeld der Ultrazentrifuge (20—27'000 U/Min.) geprüft wurde, schien mit Eindeutigkeit hervorzugehen, daß das Haemagglutinin
mit derselben Geschwindigkeit (S 600—800) sedimentiert wie die als Träger der
Infektiosität angenommenen Viruspartikel und im Sediment nahezu quantitativ
nachgewiesen werden kann. Damit schien der Nachweis gesichert, daß das Haemagglutinin bei sämtlichen geprüften Influenzastämmen (PR 8, W. S., F 12, F 99,
Lee, Swine) an Viruspartikel von zirka 100 mμ gebunden und von denselben
nicht abzutrennen ist (137, 322, 138, 290, 291, 324, 323, 306, 307, 310, 321).
Dieselbe intime Association zwischen Haemagglutinin und geformten Viruselementen wurde auch für das *Mumpsvirus* (21, 235, 333, 140, 183), *Virus der
Newcastle-Krankheit der Hühner* (12, 80, 17) und der *Mäusepneumonie* (83) nachgewiesen. Dagegen besteht eine solche Bindung beim *Vaccine-* und *Ektromelievirus* offensichtlich nicht, da durch Ultrazentrifugieren die Viruspartikel zwar
abgeschleudert werden, das Haemagglutinin jedoch in der überstehenden Flüssigkeit verbleibt, so daß auf diese Weise die infektiöse von der agglutinierenden
Eigenschaft dissociiert werden kann (45, 318, 58). Ob diese Unterschiede hinsichtlich der Bindung, bzw. Liberierung des Haemagglutinins von so grund-
sätzlicher Bedeutung sind, daß eine Klassifikation der Virushaemagglutinine auf
Grund dieser Eigenschaft gerechtfertigt wäre [BURNET und BOAKE (58)], erscheint jedoch fragwürdig. Einesteils sind die derzeitigen Kenntnisse über die
löslichen Haemagglutinine des Vaccine- und Ektromelievirus — erstaunlicher-
und bedauerlicherweise — noch sehr spärlich, und anderseits ist es anscheinend
BOURDILLON (32) in jüngster Zeit gelungen, mit Hilfe organischer Solventien
das Haemagglutinin des Influenzavirus von den Viruselementen zu isolieren[1].
Die Annahme, daß das Haemagglutinin als integrierende Komponente der
Viruspartikel von zirka 100 mμ mit einer Sedimentationskonstante von S 600—800
im Gravitationsfeld wandert, wurde durch eine eingehende Analyse der physikalischen und biologischen Eigenschaften der übrigen Sedimentierungsfraktionen
gesichert. Die Notwendigkeit derartiger Untersuchungen ergab sich schon aus
der Tatsache, daß das Virusausgangsmaterial (Allantoisflüssigkeit, Gewebe-
extrakt) ein polydisperses Gemisch hochmolekularer Substanzen darstellt, deren

---

[1] Das Haemagglutinin von BOURDILLON ist nun allerdings von besonderer Art
und wird vor allem durch Normalsera in hohem Grade gehemmt, ein Verhalten,
das an die von BURNET und STONE (63, 312) beschriebenen unspezifischen Lipoid-
Haemagglutinine erinnert.

Sedimentierungskonstanten von S 1000—30 und Partikelgrößen von 100—10 mμ variieren und die erfahrungsgemäß auch in wechselnder Menge in die Viruskonzentrate übergehen können. Fernerhin lagen Beobachtungen vor, wonach den leichteren, langsamer sedimentierenden Fraktionen eine bedeutende biologische Aktivität (Infektiosität, haemagglutinierendes Vermögen) zukommen würde [Chambers et al. (72, 73), Friedewald und Pickels (138), Gard und v. Magnus (141, 142)]. Die Abtrennung der schweren Fraktion (S 600—800) von den leichteren Komponenten (S 30, bzw. 170[1]) kann auf verschiedene Art erreicht werden, nämlich durch Adsorption und Elution an Hühnererythrocyten (322, 222, 258), Ultrazentrifugieren mit der Separatorzelle (307, 229, 228, 258) oder durch elektrophoretische Fraktionierung (260). Die Ergebnisse derartiger Versuche sind in mehrfacher Hinsicht aufschlußreich. Lauffer und Miller (228) prüften die Sedimentierungsgeschwindigkeit der *schweren Komponente* der Influenzastämme PR 8 und F 12 im unterschiedlich hohen Gravitationsfeld und stellten — in Übereinstimmung mit den früheren Befunden von Friedewald und Pickels (137) — fest, daß die Sedimentierungsrate für das Haemagglutinin, die infektiöse Aktivität (für die Maus und den Hühnerembryo) und den N-Gehalt dieselbe ist. Von Stanley und Mitarbeitern (307, 229) wurde die *leichte Fraktion* (S 30 bzw. 170) der Influenzastämme F 12, Lee und PR 8 isoliert und auf physikalische und biologische Eigenschaften näher untersucht. Der Anteil dieser Komponente in den Viruskonzentraten schwankte zwischen 10—30%, je nach den untersuchten Virusstämmen (F 12 > Lee > PR 8). Die infektiöse und haemagglutinierende Aktivität dieser Fraktion war 100—10'000mal geringer als diejenige der schweren Komponente. Nach Stanley dürfte dieser geringe Grad von Aktivität mit Wahrscheinlichkeit auf eine nicht vollständige Abtrennung der aktiven schweren Komponente zurückzuführen sein, so daß die Fraktion S 30, bzw. S 170 de facto als inaktiv zu betrachten wäre. In physikalischer Hinsicht zeichnete sich diese leichte Fraktion durch eine hohe Viskosität aus und verhielt sich elektrophoretisch wie ein saures Protein mit einem isoelektrischen Punkt bei pH 2,2—2,4. Stanley und Mitarbeiter betrachten nun — wie schon erwähnt — diese Fraktion nicht als ein Virusderivat, sondern identifizieren dieselbe mit der von Knight (219, 221) aus normaler und virusinfizierter Allantoisflüssigkeit, bzw. Mäuselunge isolierten hochmolekularen Eiweissubstanz, der sogenannten *Normalkomponente*. Die Übereinstimmung dieser Substanzen nach physikalischen und elektrochemischen Eigenschaften ist nun tatsächlich eine vollkommene. Die Normalkomponente von Knight verhält sich im Gravitationsfeld der Ultrazentrifuge als polydispers, mit einer dominierenden Sedimentierungskonstante von S 30, bzw. S 170, ist elektronenoptisch durch sphärische Partikel von zirka 40 mμ charakterisiert, besitzt stark viskose Eigenschaft und hat den isoelektrischen Punkt bei pH 2,3. Die Normalkomponente zeigt kein haemagglutinierendes Vermögen und wird auch von Hühnererythrocyten nicht adsorbiert. Dessen ungeachtet ergeben sich zwischen Normalkomponente und Virus bzw. Haemagglutinin in mehrfacher Hinsicht bedeutungsvolle Beziehungen. So weisen mehrere Beobachtungen darauf hin, daß die Aktivität des Haemagglutinins in Anwesenheit der Normalkomponente gehemmt ist. Knight (221, 223) selbst stellte fest, daß die Eliminierung von Normalprotein aus Viruskonzentraten einen bedeutenden Anstieg des Haemagglutinintiters zur Folge hat, und Miller (256) beobachtete, daß das Haemagglutinationsvermögen von Viruskonzentraten vom Ausmaß der Verunreinigung mit Normalprotein abhängig ist. Den direkten Nachweis der hemmenden Wirkung der nativen Allantoisflüssig-

---

[1] Der Wert S 30 ergibt auf Viskosität korrigiert (229) S 170.

keit, bzw. einer aus dieser durch Ultrazentrifugieren gewonnenen Fraktion auf die Haemagglutination von Influenzavirus in vitro leistete SVEDMYR (302)[1] und bestätigte hierdurch frühere Beobachtungen von BEVERIDGE und LIND (21) und LIND (235), wonach normale Allantoisflüssigkeit die Haemagglutination von Mumpsvirus hemmt. In diesem Zusammenhang ist die Feststellung von KNIGHT (223) von Interesse, wonach das Normalprotein als charakteristisches Kohlehydrat Glucosamin, also einen Baustein von Mucoproteinen enthält, die ihrerseits als spezifische Hemmstoffe der Virushaemagglutination zu gelten haben (vgl. S. 172).

Ebenso bemerkenswert sind die von KNIGHT (219, 222) festgestellten *serologischen Beziehungen*, die zwischen Normalkomponente und Virushaemagglutinin bestehen. Gegen die Normalkomponente gerichtete Immunsera präcipitieren nicht nur das normale Protein, sondern auch hochgereinigte Präparationen von Influenzavirus (PR 8 und Lee) und hemmen in beträchtlichem Grade die Haemagglutination dieser Virusstämme, besitzen jedoch keine virusneutralisierende, bzw. antiinfektiöse Wirkung. Umgekehrt präcipitiert ein Influenzaimmunserum (anti-PR 8) nicht nur das zugehörige Virus, sondern auch die Normalkomponente. Beide Immunsera sind nun aber unterschiedlich hinsichtlich ihrer Spezifität und Wirkungsbreite. Das Immunserum gegen das Normalprotein ist artspezifisch, d. h. reagiert — je nach der Provenienz (Allantoisflüssigkeit, Mäuselunge) des zur Immunisierung verwendeten Normalantigens — nur mit den Antigenen (Normalkomponente, Virus) entsprechender Herkunft; die Typenzugehörigkeit des Influenzavirus spielt bei diesen Seren weder im Präcipitationsnoch Agglutinationshemmungsversuch eine Rolle, ebenso ist keine antiinfektiöse Wirkung vorhanden. Demgegenüber besitzen Influenzaimmunsera artspezifischen Charakter im Präcipitationsversuch und typenspezifische Eigenschaft im Hemmungs- und Neutralisationstest. Dieser Sachverhalt entspricht demnach weitgehend den Verhältnissen, die bei den Virusarten der Geflügeltumoren und der Hühnerleukämie nachgewiesen wurden, und KNIGHT (222) zieht denn auch dieselbe Schlußfolgerung, nämlich, daß das Normalprotein in hochgereinigten Virusproben nicht mehr als eine abtrennbare Verunreinigung, sondern — mit Wahrscheinlichkeit — als eine dem Influenzavirus inkorporierte Komponente zu betrachten ist[2].

Die wichtige Frage, ob zwischen dem Normalprotein und dem Virus, bzw. Haemagglutinin genetische Beziehungen bestehen, bleibt dagegen vorläufig unbeantwortet. Immerhin liegen Beobachtungen vor, die auf einen derartigen Zusammenhang hinweisen könnten. Nach KNIGHT (221) und SVEDMYR (303) nimmt das Normalprotein in der Allantoisflüssigkeit befruchteter, aber nicht infizierter Eier mit zunehmendem Alter der Embryonen mengenmäßig zu, ein Vorgang, der nicht nur physikalisch, sondern auch biologisch (im Agglutinationshemmungsversuch) festgestellt werden kann. In der mit Influenzavirus (PR 8) infizierten Allantoisflüssigkeit ist dagegen das Normalprotein in Form der leichten Komponente (S 30, bzw. S 170) anscheinend nur während der Inkubationsperiode nachweisbar [MILLER (256), SVEDMYR (303), GARD und v. MAGNUS (141, 142)], erscheint jedoch während der Virusvermehrung in zunehmender Menge in der schweren Fraktion (S 600—800) z. T. als Verunreinigung, z. T. als Virusbestandteil [KNIGHT (221, 222, 223)]. Die Annahme liegt daher nahe, daß das Normalprotein in den Vorgang der Virussynthese einbezogen wird und möglicherweise auch am Aufbau des Haemagglutinins beteiligt ist. Die letztgenannte Möglichkeit würde jedoch zur Vorstellung nötigen, daß aus einem

---

[1] Vgl. auch HARDY und HORSFALL (166) S. 174, 183.

[2] HARDY und HORSFALL (166) schließen sich dieser Auffassung allerdings nicht an, sondern sie deuten diese Association als spezifische Verbindung zwischen Virus und Inhibitor der Allantoisflüssigkeit, bzw. Normalkomponente.

Anti-Haemagglutinin ein Haemagglutinin gebildet wird, ein Widerspruch, der übrigens auch in der biologischen Aktivität der Normalkomponente (agglutinationshemmende Wirkung des Antigens und Antihaemagglutininfunktion des Antikörpers) zum Ausdruck kommt.

Weniger abgeklärt ist die Bedeutung einer *mittelschweren Fraktion (S 460—650)*, die bei den Influenzastämmen PR 8 (138, 141, 142) und Lee (291, 258) wiederholt nachgewiesen worden ist. Die Aktivität dieser Fraktion ist durch eine geringfügige, bzw. überhaupt nicht nachweisbare Infektiosität einerseits und ein oft beträchtliches haemagglutinierendes und interferierendes Vermögen anderseits gekennzeichnet. Nach Friedewald und Pickels (138) und Sharp und *Mitarbeitern* (291) wird diese Fraktion durch degradierte, inaktive Viruspartikel repräsentiert, während Gard und v. Magnus (141, 142) in dieser Komponente eine vermehrungsfähige, atypische Virusvariante bzw. einen Virusvorläufer erblicken. Miller (258) bestätigte das Vorkommen dieser Fraktion beim Influenzavirus Lee, deutete dieselbe jedoch — auf Grund eingehender physikalischer Studien — als Verunreinigung durch ein Normalprotein von außergewöhnlicher Partikelgröße.

An welche Sedimentierungsfraktion das *komplementbindende* Antigen des Influenzavirus gebunden ist, wurde durch die Untersuchungen von Friedewald und Pickels (132, 138) und Henle und *Mitarbeitern* (187) abgeklärt. Der größte Anteil dieses Antigens wurde in der schweren Fraktion von S 600—800 nachgewiesen und scheint demnach ebenso wie das Haemagglutinin an die Viruspartikel von zirka 100 mμ gebunden zu sein; eine kleine Quote erwies sich als nicht sedimentierbar und liegt offensichtlich in Form eines löslichen Antigens („soluble substance") in den leichtesten Fraktionen (S 30) vor. Beide Fraktionen unterscheiden sich hinsichtlich Adsorption und Elution an Hühnererythrocyten, Grad der Spezifität und antigener Qualität (132, 187, 342). Für das *Mumpsvirus* stellten Beveridge und Lind (21) und Lind (235) gleichfalls fest, daß das komplementbindende Antigen auf der Ultrazentrifuge in gleichem Grade sedimentiert wie das Haemagglutinin, ließen jedoch die Frage offen, ob nicht ein gewisser Anteil des Antigens in der überstehenden Flüssigkeit zurückbleibt. Nach Henle, Henle und Harris (183) liegen ähnliche Verhältnisse wie beim Influenzavirus vor; ein Großteil des komplementbindenden Antigens bildet wie das Haemagglutinin einen Bestandteil der Viruspartikel, eine kleinere Quote findet sich in den leichteren Fraktionen und ist auch serologisch unterschiedlich.

Die Analyse mit Hilfe der Ultrazentrifuge und der Elektrophorese führte demnach, zusammengefaßt, zu den folgenden Ergebnissen:

1. Die in der schweren Fraktion ausgeschleuderten Viruspartikel sind nicht nur die Träger der Infektiosität, sondern auch des Haemagglutinins und eines großen Anteils des komplementbindenden Antigens. Dieser Tatbestand wurde mit Sicherheit für das Influenzavirus und mit großer Wahrscheinlichkeit auch für das Mumpsvirus, Virus der Pseudopest der Hühner und der Mäusepneumonie nachgewiesen. Beim Vaccine- und Ektromelievirus konnte dagegen eine derartige Association zwischen Haemagglutinin und infektiösen Viruselementen bisher nicht nachgewiesen werden, da diese Komponenten durch Ultrazentrifugieren voneinander getrennt werden können.

2. Ob das Haemagglutinin des Influenzavirus in leichteren Fraktionen als selbständige, nichtinfektiöse Virusvariante vorkommt, ist nicht genügend sichergestellt.

3. In sämtlichen Fraktionen ist ein aus dem Wirtsgewebe stammendes hochmolekulares Protein von saurem Charakter, großer Viskosität und sehr unterschiedlicher Dispersität nachzuweisen. Diese „Normalkomponente" zeigt Beziehungen zum Haemagglutinin, nämlich einerseits als Hemmstoff und anderseits als Antigen, indem Antikörper gebildet werden, welche die haemagglutinierende Wirkung aufheben, jedoch keine virusneutralisierende Wirkung entfalten.

4. Eine Quote des komplementbindenden Antigens des Influenzavirus findet sich in den leichtesten Fraktionen.

4. Die Reaktion des Haemagglutinins mit der Erythrocytenoberfläche.

Agglutinationstypus. Läßt man ein haemagglutinierendes Virus auf geeignete Erythrocyten einwirken, so kann die Verklumpung der Blutzellen schon nach kurzer Zeit festgestellt werden; auf dem Objektträger zeigt sich meist schon nach 1 Minute ein zunächst körniges, dann fetziges Agglutinat und im Sedimentierungsröhrchen sind die ersten Anzeichen der Agglutination — je nach Virusart, Erythrocytenspezies und Bebrütungstemperatur — bereits nach 10—20 Minuten deutlich sichtbar. Die virusbedingte Haemagglutination findet bei jeder Temperatur ($+4^0$, 25^0, 37^0 C) statt, wird jedoch bei höheren Temperaturen wesentlich beschleunigt. In dieser Hinsicht besteht zwischen der Haemagglutination durch Virus oder Antikörper kein wesentlicher Unterschied. Charakteristischer für die virusbedingte Haemagglutination ist der makroskopische Aspekt des im Sedimentierungsröhrchen gebildeten Agglutinates, die lockere Beschaffenheit desselben und vor allem die spontane Reversibilität der Zellveränderung. Das im Teströhrchen feststellbare Agglutinat besteht — je nach der Stärke der Reaktion — aus einem flach ausgebreiteten Sediment mit gezacktem oder ausgefranstem Rand, bzw. einem fein granulierten Wandbeschlag, der sich entweder in fetzigen Membranen loslöst und dann ebenfalls sedimentiert oder an der Glaswand adhärent bleibt. Verschiedene Virusarten zeigen Unterschiede im Agglutinationstypus, die jedoch im allgemeinen nicht sehr bedeutend sind. Agglutinate sehr ähnlicher Art bilden nun aber auch „Agglutinine", die in der normalen Amnionflüssigkeit von Hühnerembryonen (77) und im Normalmäusegehirn (164) nachgewiesen wurden, fernerhin auch Rhesusantikörper, gewisse Phytagglutinine usw., so daß von einem „virusspecifischen" Agglutinationstypus nicht die Rede sein kann. Daß die durch Virus bewirkten Agglutinate nicht fest, sondern nur locker zusammengehalten sind, zeigt sich in ihrer leichten Aufschüttelbarkeit, wobei sich allerdings aus den homogenisierten Suspensionen erneut — und meist in noch kürzerer Zeit — wieder Zellaggregate abscheiden. Auch dieses Merkmal kann nun aber nicht als typisch für die Haemagglutination durch Virus bezeichnet werden, da einige Virusarten (Vaccine-, Ektromelie-, Poliomyelitisvirus) nach Art von haemagglutinierenden Antikörpern stark kohärente Agglutinate bilden, und ein solcher Agglutinationstypus auch bei Virusarten (Influenza, Pseudopest) zu beobachten ist, die vorgängig durch Erhitzen inaktiviert worden sind [BRIODY (35)]. Immerhin zeigt die Art des Agglutinationstypus (lockere oder feste Beschaffenheit der Zellaggregate) eine wichtige Viruseigenschaft an, nämlich die Befähigung oder das Unvermögen zur sekundären Elution und damit auch, ob die Agglutination reversibel oder irreversibel ist.

Nach LOWELL und BUCKINGHAM (238) unterscheidet sich die Virus- von der Immunhaemagglutination durch ihr entgegengesetztes Verhalten gegenüber geringen und hohen Salzkonzentrationen; die Virushaemagglutinine werden in elektrolytarmen Medien frühzeitiger (0,03 m NaCl) gehemmt als die haemagglutinierenden Antikörper, sind jedoch im salzreichen Milieu (0,58 m NaCl), in welchem die Antikörperwirkung bereits aufgehoben ist, noch vollständig aktiv.

Ein *Virushaemolysin* wurde von MORGAN, ENDERS und WAGLEY (263) bei mehreren Eipassagestämmen von *Mumpsvirus* nachgewiesen; bei 37^0 C war dieses Haemolysin gegenüber Hühner-, Hammel- und (in geringerem Grade) Menschenerythrocyten wirksam, d. h. beeinflußte also dieselben Zellspecies wie das Haemagglutinin. Auch hinsichtlich der Adsorption und Elution an Erythrocyten und der spezifischen Hemmbarkeit durch Immunsera bestand zwischen Haemagglutinin und Haemolysin eine weitgehende Parallelität. Beim Influenzavirus vermochten diese Autoren ein Haemolysin nicht festzustellen; nach STONE

(313) werden dagegen mit Influenzavirus vorbehandelte Erythrocyten in Gegenwart von Komplement haemolysiert.

Agglutinationsspektrum. Wie schon McCelland und Hare (246) in ihrer ersten Mitteilung feststellten, vermag Influenzavirus nicht nur die Erythrocyten des Hühnerembryos, des Kückens und des erwachsenen Huhnes zu agglutinieren, sondern auch die Blutzellen einiger Säugetiere. Untersuchungen dieser Art wurden in der Folge nicht nur auf verschiedene Virusstämme, sondern auch auf eine größere Anzahl von Erythrocytenspecies verschiedener Tierklassen (Vogelarten, Säugetiere, Amphibien und Reptilien) ausgedehnt, und zwar nicht nur für das *Influenzavirus* (76, 196, 326, 25, 276), sondern auch für das *Virus der Newcastle-Krankheit* (38, 76), zahlreiche *Hühnerpeststämme vom Typus A und B* (161, 164), *Mumpsvirus* (21, 33), *Virus der Mäusepneumonie* (261, 262), *Vaccinevirus* (264, 76), *Ektromelievirus* (58), *murine Poliomyelitisvirusstämme* (163), *peripneumonie-artige Erreger* (190) und *Rickettsien* (267). Die hierbei erzielten Ergebnisse sind in Tabelle 2 zusammengestellt.

Selbst bei der Unvollständigkeit der vorliegenden Resultate ist ohne weiteres ersichtlich, daß die Anzahl und die Artzugehörigkeit agglutinabler Erythrocyten für die verschiedenen Virusarten und -stämme sehr unterschiedlich sind. Berücksichtigt man nur diejenigen Erythrocytenspecies, die von den meisten Stämmen einer Virusart regelmäßig und in höherem Titer agglutiniert werden, so ergibt sich die folgende Anordnung: *Hühnerpest* (Huhn, Mensch, Meerschweinchen, Ratte, Maus, Kaltblüter), *Influenzavirus* (Huhn, Mensch, Meerschweinchen, Kaltblüter), *Mumpsvirus* (Huhn, Meerschweinchen, Hammel, Mensch), *Virus der Mäusepneumonie* (Maus, Hamster), *Ektromelievirus* (Huhn, Maus), *Vaccinevirus* (Huhn), *Poliomyelitisvirus* (Hammel). Auffallend sind die Unterschiede im „Agglutinationsspektrum" bei den verschiedenen Virusstämmen einer Art. Beim Influenzavirus sind dem Stamm Lee zahlreiche, den A-Stämmen mäßig viele und dem porcinen Stamm nur wenige agglutinable Erythrocytenspecies zugeordnet, wobei man nicht den Eindruck hat, daß dieses unterschiedliche Verhalten durch die immunologische Typenzugehörigkeit bestimmt wird. Daß die immunologische Specifität das „Agglutinationsspektrum" nicht beeinflußt, konnte Hallauer mit Eindeutigkeit beim Hühnerpestvirus (15 Stämme) nachweisen; zwischen den immunologisch distinkten A- und B-Typen bestanden im haemagglutinierenden Vermögen gegen verschiedene Erythrocytenarten keine wesentlichen Unterschiede, gelegentliche Abweichungen erwiesen sich stets als Stammeseigentümlichkeiten. Fernerhin lehrt schon ein flüchtiger Blick auf die Tabelle, daß zwischen der Agglutinierbarkeit der Erythrocyten verschiedener Tierspecies und deren Empfänglichkeit für eine bestimmte Virusart — zumindest in der Großzahl der Fälle — keine Korrelation bestehen kann; die jeweils agglutinable Substanz scheint vielmehr in regelloser Art im Tierreich verbreitet zu sein, ein Verhalten, das an die ebenso unterschiedliche Wirkung der Phytagglutinine und der von Gujot (155) beschriebenen haemagglutinierenden Aktivität verschiedener Stämme von E. Coli erinnert.

Die Angaben über das „Agglutinationsspektrum" einer Virusart sind häufig unzuverlässig, wechselnd und widersprechend, was nicht erstaunlich ist, wenn man berücksichtigt, daß das Ergebnis des Haemagglutinationsversuches von zahlreichen Variabeln (Qualität der Erythrocyten, Aktivität des Virus, Beschaffenheit des Reaktionsmilieus) abhängig ist. Daß die *Agglutinabilität verschiedener Blutproben* derselben Spezies sehr unterschiedlich ist, ist durch die Erfahrung hinlänglich belegt. Miller und Stanley (259) beobachteten beim Influenzavirus Titerschwankungen von 36—38% unter Verwendung von Hühnererythrocyten verschiedener Provenienz. Differenzen ähnlicher Größe sind beim

Tab. 2. *Haemagglutinierendes Verhalten von Virusarten gegenüber verschiedenen Erythrocytenspecies.*

	Influenza					Geflügelpest		Mumps	Mäuse-pneumonie	Vaccine	Ektromelie	Poliomyelitis SK/MM	Pleuro-pneumonie	Ricketts. orient.
	A			B		A	B (NDV)							
	PR 8	WS	Melb.	Lee	Swine	3 Stämme	11 Stämme							
I. Vogelarten Huhn	++	++	++	++	++	++	++	++	-	++**	++**	-	+	++**
Gans	(+)													
Truthuhn	+													
Ente	+	+	+	+	+		+			-,+**	-			+
Taube	+								+	-,+**	-,+**			
Kanarien	(+)									-				
Sperling				++	++		-,++			+				
II. Säugetiere Mensch	++	++	++	++	-	++	++	+	-	-	-	-,+***	+	-
Affe	-			++		-	-			-	-			
Pferd	-	-	-	(+)	-	+	-			-		-	-	
Esel		-	-	(+)	-		-			-				
Rind	-	-	-	++	-	(+)	++			-	-			
Hammel	-	-,++*	+	++	-,++*	++	++	++	-	-		++	(+)	-
Ziege		-	+	++	-		++							
Schwein	-	-	-	++	-	(+),+	(+)					-		
Dachs	(+)													
Oppossum		-	-	++	-		(+)							
Hamster									++				+	
Hund										-				
Katze		-	-	++	-		-		-	-				
Frettchen	+	+	+	+	-		-			-				
Kaninchen	-	+	+	+	+	(+)	(+)		(+)	-	-	-	(+)	
Meerschw.	++	+	+	++	-	++	++	++	-	-	-	-,+***	+	-
Ratte (alb.)	(+)	+	+	+	-	++	++		-	-	-	-,+***		
Ratte (Cotton)	-					(+)	-		-	-		-	+	
Maus	(+)	+	+	+	-	++	++		++	++	+	-		
III. Amphibien u. Reptilien Frosch		++	++	++	++		++			-				
Schildkröte							-							
Eidechse		++	++	++	++		++							
Schlange		++	++	++	++		++							

Legende: ++ Regelmäßige Agglutination von höherem Titer.
 + Meist vorhandene Agglutination.
 (+) Inkonstante Agglutination von niederem Titer.

* In Phosphatpuffer oder mit Perjodat vorbehandelte Erythrocyten.
** Nur bestimmte Blutproben.
*** Unter Glucosezusatz.

Hühnerpest-, bzw. Poliomyelitisvirus nachzuweisen, wenn der Haemagglutinationstest mit verschiedenen Hühner-, bzw. Hammelerythrocyten durchgeführt wird [Hallauer (161, 164)]. In noch höherem Maße wird die Haemagglutination des Vaccine- und Ektromelievirus [Nagler (264), Clark und Nagler (76), Burnet und Boake (58)] und anscheinend auch von Rickettsien [O'Connor (267)] von der Qualität der Hühnererythrocyten beeinflußt, da nach den Untersuchungen von Clark und Nagler nur rund 50% aller Hühner Erythrocyten liefern, die gegenüber Vaccinevirus vollagglutinabel sind. Diese Differenzen treten besonders dann hervor, wenn Virusproben von geringer haemagglutinierender Aktivität verwendet werden. Fernerhin bestimmt auch das *Alter des blutspendenden Tieres*, in welchem Grad Erythrocyten gegenüber bestimmten Virusarten agglutinabel sind. Für *Influenzavirus* sind die Erythrocyten des Hühnerembryos empfindlicher als die Blutzellen des erwachsenen Huhnes; von Burnet und Bull (60) wurde ein durch fortgesetzte Eipassagen modifizierter Influenzastamm („Melbourne egg") beschrieben, der nur noch embryonale Erythrocyten agglutinierte. Umgekehrt liegen die Verhältnisse beim *Vaccine-* und *Ektromelievirus*; nach Clark und Nagler und Burnet und Boake sind embryonale Hühnererythrocyten gegenüber diesen Virusarten vollständig unempfindlich und können agglutinable Erythrocyten erst von Kücken im Alter von 2—4 Wochen gewonnen werden. Demgegenüber agglutiniert das *Variolavirus*, wie North (266) feststellte, die Blutzellen sowohl des Hühnerembryos als auch des erwachsenen Tieres, wenn auch in ziemlich schwachem Grade.

Daß aus den agglutinierenden Eigenschaften *eines* Virusstammes nicht ohne weiteres auf ein ähnliches Verhalten anderer Stämme geschlossen werden darf, wurde bereits erwähnt. Aber auch für *denselben Virusstamm* variiert das „Agglutinationsspektrum" in Abhängigkeit zahlreicher Faktoren. Von Twyble und Mason (326) wurde der Einfluß der *Virusprovenienz* (Allantois, bzw. Mäuselunge) für den Influenzastamm PR 8 nachgewiesen. Bekanntlich wird die haemagglutinierende Qualität von Influenzavirus auch in der *Phase der Anpassung* an neue Wirte in charakteristischer Weise verändert. So stellten Burnet und Bull (59) fest, daß menschliche Virusstämme, die in der Amnionflüssigkeit von Hühnerembryonen isoliert werden, zunächst in der O-Phase vorliegen, die dadurch gekennzeichnet ist, daß derartige Stämme — im Gegensatz zu den vollangepaßten D-Formen — zwar die Erythrocyten von Säugetieren (Meerschweinchen, Mensch) in normaler Titerhöhe agglutinieren, jedoch gegenüber Hühnererythrocyten nur schwach oder überhaupt nicht wirksam sind. Ähnliche Beobachtungen machten Hirst (199) und Wang (74) bei der Anpassung von Eipassagestämmen an die weiße Maus. Das Hauptcharakteristikum dieser O-Stämme liegt nun aber in der zweifellos veränderten Infektiosität und Pathogenität, wogegen der besondere Agglutinationstypus möglicherweise nur als „Hemmungsphänomen" zu betrachten ist. Mehrere Beobachtungen deuten auf diese Möglichkeit hin. So gelang Magill und Sugg (241) der Nachweis, daß ein O-Stamm sich in vitro wie ein D-Stamm verhält, wenn der Haemagglutinationstest mit Hühnererythrocyten nicht in Kochsalz-, sondern in Phosphatpufferlösung von pH 5,6 angesetzt wird; die Pufferung des Reaktionsmilieus genügte demnach, um die Agglutinationshemmung gegenüber Hühnererythrocyten vollständig aufzuheben. Ebenso manifestieren intermediäre O-D-Stämme (δ bzw. ω) ein unvermindertes Agglutinationsvermögen gegenüber Hühnererythrocyten, wenn das Virus schonend erhitzt(Briody[1]), der Agglutinationstest in der Kälte bei 0^0 C ausgeführt wird (Stone[1]) oder die Erythrocyten mit Perjodat vorbehandelt werden [Fazekas de St. Groth (108)].

---

[1] Cit. nach Fazekas de St. Groth (108).

Die *Inagglutinabilität von Hammelerythrocyten gegenüber Influenzavirus* ist ebenfalls keine absolute, sondern kann auf mehrfache Art beseitigt werden, nämlich, durch die Ausführung der Haemagglutination in der Kälte [CLARK und NAGLER (76)], geeignete Pufferung des Reaktionsmilieus [MAGILL und SUGG (240)], schonende Hitzeinaktivierung des Virus [GOTTSCHALK und LIND (150)] oder durch die Vorbehandlung der Erythrocyten mit Kaliumperjodat [FAZEKAS DE ST. GROTH (108)]. Desgleichen beobachtete HALLAUER (163), daß die murinen Poliomyelitisstämme SK und MM Hammelerythrocyten nur unter Glucosezusatz in einem konstant hohen Titer agglutinieren, und daß auch andere Erythrocytenspezies (Mensch, Meerschweinchen, Ratte) für diese Stämme agglutinabel werden, sofern die Blutzellen nicht in Kochsalz-, sondern in Glucoselösung suspendiert werden.

Reaktionskinetik (Adsorption und Elution). Nach den ersten Beobachtungen von HIRST (191, 195) und MC CLELLAND und HARE (246) reagiert das Influenzavirus mit Hühnererythrocyten in zwei zeitlich sich folgenden Vorgängen, nämlich, 1. der Bindung, die sich im Phänomen der Haemagglutination kundgibt (Adsorptionsphase) und 2. der Wiederabspaltung, bzw. Liberierung des Virus von der Erythrocytenoberfläche, wobei die Erythrocytensuspension wieder eine homogendisperse Beschaffenheit annimmt (Elutionsphase). Der Reaktionsvorgang ist demnach durch eine spontane Reversibilität ausgezeichnet. In den Adsorptions- und Elutionsprozeß werden beim *Influenzavirus* sämtliche Virusfunktionen einbezogen, nämlich nicht nur das Haemagglutinin, sondern auch die Infektiosität (195), Toxicität (178), ein Anteil des komplementbindenden Antigens (132, 187) und die immunisierende Qualität (127), woraus schon hervorgeht, daß die geformten Viruselemente in Reaktion treten[1]. Grundsätzlich ebenso verhalten sich das *Mumpsvirus* (21, 235), die verschiedenen Typen des *Geflügelpestvirus* (1, 161) und wahrscheinlich auch bestimmte murine Stämme von *Poliomyelitisvirus* (163). Beim Virus der *Mäusepneumonie* bildet zwar das Haemagglutinin ebenfalls eine Komponente der Viruspartikel, so daß durch den Kontakt mit Erythrocyten sowohl die haemagglutinierenden als auch infektiösen Eigenschaften gebunden werden; diese Adsorption ist jedoch fest und von keiner spontanen Elution gefolgt (81, 82). In ähnlicher Weise reagieren Virusarten mit ,,selbständigem", d. h. nicht an die Viruselemente gebundenem Haemagglutinin, nämlich das *Vaccine-* und *Ektromelievirus* (264, 58, 63); die Adsorption erstreckt sich hier nur auf das Haemagglutinin und ist gleichfalls irreversibel. Sämtliche Virushaemagglutinine werden demnach — erwartungsgemäß — von agglutinablen Erythrocyten adsorbiert, sind jedoch nicht durchwegs imstande, wiederum von denselben abzudissociieren. Hieraus ergibt sich, daß die Bedingungen für die Adsorption nicht dieselben sind wie für die Elution, so daß eine gesonderte Betrachtung dieser Vorgänge gerechtfertigt erscheint.

Der vorwiegend physikalische Charakter des *Adsorptionsvorganges* zeigt sich in seiner relativ geringen Abhängigkeit von der biologischen Qualität der Reaktionskomponenten, der herrschenden Temperatur und den Milieufaktoren.

Sämtliche *Virusarten* und *-stämme* erreichen — unter optimalen Bedingungen — annähernd dasselbe Bindungsmaximum (90 bis 99 % Adsorption), allerdings — je nach der Stärke der Bindungsaffinität — mit unterschiedlicher Geschwindigkeit. Die infektiöse Aktivität des Virus spielt für den Bindungsvorgang keine Rolle; durch Erhitzung (195, 35, 81), Formolisierung (195), Ultraviolettbestrahlung (180) wird — sofern das Haemagglutinin nicht geschädigt ist — die Adsorption nicht beeinträchtigt, sondern im Gegenteil verstärkt. Der Einfluß der *Erythrocyten* auf den Bindungsvorgang ist mehrfacher Art. Erythrocyten verschiedener Artzugehörigkeit besitzen —

---

[1] Von HEINMETS (169) wurde die Adsorption der Viruspartikel an die Erythrocytenoberfläche auch elektronenoptisch nachgewiesen.

wie schon aus dem wechselnden Agglutinationstiter geschlossen werden kann — ein
unterschiedliches Bindungsvermögen [vgl. Lind (235)]. Fernerhin wird der Grad
der Virusadsorption durch die Konzentration der Zellsuspension bestimmt [Hirst
(195)]. Formolisierte Erythrocyten besitzen nach Flick (118) dieselbe Bindungs-
kapazität wie native Zellen. Auch die milde Vorbehandlung der Erythrocyten mit
verdünntem Perjodat, einem zellreceptorzerstörenden Agens [Hirst (201)], beeinträchtigt, wie Fazekas de St. Groth (108) feststellte, die Adsorption nicht, sondern
bewirkt im Gegenteil eine festere Bindung (vgl. S. 182). Stromata sind für den Bindungsvorgang ebenso geeignet (195, 201, 81) und verlieren die adsorptive Eigenschaft
auch durch die kurzfristige Erhitzung auf 100° C nicht (195, 201). Immerhin ist die
Unversehrtheit · der Zellstruktur für den Adsorptionsvorgang wesentlich, da zur
Erzielung desselben Bindungseffektes weit größere Mengen von Blutzellschatten
erforderlich sind. Nach Hirst (201) ist die Gewinnungsart der Stromata nicht gleichgültig. Erfolgt die Haemolyse durch Aqua dest., so ist das Adsorptionsvermögen
stark vermindert, nach Saponinhaemolyse dagegen, welche die Zellform weitgehend
intakt beläßt, noch voll erhalten. Die Phase der Adsorption ist an bestimmte *Temperaturen* nicht gebunden; in der Kälte ($+4°$C) erfolgt die Virusbindung zwar erheblich
langsamer, aber meist vollständiger, wie in der Wärme (37° C), die lediglich beschleunigend wirkt (195). Den Effekt der *Salzkonzentration* auf die Adsorption von Influenzavirus an Menschen- und Hühnererythrocyten, bzw. des Mäusepneumonievirus an
Mäuseerythrocyten untersuchten Lowell und Buckingham (238) und Davenport
und Horsfall (84). Die erzielten Versuchsergebnisse sind von besonderem Interesse,
weil sich die geprüften Virusarten nicht gleichartig verhielten. Während das Influenzavirus in einem Konzentrationsbereich von 0,6 bis 0,03 m NaCl gleich stark agglutinierte
und adsorbiert wurde und dieses Vermögen erst bei einer Verminderung der Elektrolyte
auf 0,003 bis 0,001 m NaCl verlorenging, wurde die Adsorption des Pneumonievirus
schon bei einer Salzkonzentration von 0,01 m NaCl stark herabgesetzt, dagegen im
salzreichen Medium (0,73 m NaCl) um das 2- bis 4fache gesteigert. Welche Art von
Elektrolyten verwendet wurde, erwies sich als irrelevant.

Der *Elutionsprozeß* kontrastiert in zahlreicher Hinsicht von der Adsorptionsphase und imponiert zufolge seiner großen Abhängigkeit von der Aktivität der
Reaktionskomponenten und von Milieufaktoren als biologischer Vorgang.
Zunächst steht außer Zweifel, daß das Phänomen der Elution nur mit aktiven
(infektiösen) Virusproben auszulösen ist; durch Hitze oder Formolisierung
inaktivierte Virusarten (mit noch erhaltenem haemagglutinierendem Vermögen)
werden nur adsorbiert, aber nicht eluiert [Hirst (195, 202), Briody (35)]. Der
Grad der Viruseluierung hängt nach Briody von der Viruskonzentration ab
und ist am größten, d. h. — wie dieser Autor annimmt — komplett, wenn die
Erythrocyten nahezu oder völlig mit Virus „gesättigt" sind. Auch unter dieser
Voraussetzung ist nun aber das Elutionsvermögen bei den verschiedenen Virusarten und -stämmen unterschiedlich. Derartige Differenzen wurden bei verschiedenen *Influenzavirusstämmen* nachgewiesen. So beobachtete Hirst (195),
daß beim Stamm Lee 100 %, beim Stamm PR 8 dagegen nur 25 % des adsorbierten
Virus in das Eluat übergehen, und Briody (35) wies für den Stamm WSE nach,
daß nur 5 % des gebundenen Virus liberiert wird. Nach Anderson (1) und
Lind (235) ist auch die Elution beim *Virus der Newcastle-Krankheit der Hühner*
und *Mumpsvirus* nicht vollständig, da stets eine Quote von zirka 10 % „nicht-
eluierender Viruspartikel" an der Erythrocytenoberfläche haften bleibt, die
hiedurch gegenüber virusneutralisierenden Seren sensibilisiert wird (55). Sehr
beachtenswert sind fernerhin die Versuche von Björkman und Horsfall (28),
denen es gelang, durch die einmalige Vorbehandlung der Influenzastämme PR 8
und Lee mit Lanthan[1], bzw. Ultraviolettbestrahlung eine Dauermodifikation

---

[1] Nach Hammarsten und Teorell reagiert Lanthan mit Nucleinsäuren unter
Bildung eines unlöslichen Komplexes.

zu erzeugen, die sich gegenüber dem Ausgangsvirus nur in einer Hinsicht unterschied, nämlich im stark verlangsamten und verminderten Elutionsvermögen. Diese Veränderung war namentlich bei dem sonst stark dissociierenden Lee-Virus ausgeprägt. BJÖRKMAN und HORSFALL interpretieren ihre Befunde dahin, daß der Lee-Stamm eine heterogene Population von dissociierenden und nicht-dissociierenden Partikeln darstellt, die mit den angewandten Mitteln zugunsten der letzteren selektioniert wird.

Die Ausschaltung der Elutionsphase kann nicht nur durch die Inaktivierung des Virus, sondern auch durch eine partielle Inaktivierung des Zellreceptors erreicht werden. Wie nämlich FAZEKAS DE ST. GROTH (108) nachwies, wird Influenzavirus von schonend mit Perjodat vorbehandelten Erythrocyten wohl in unvermindertem Grade adsorbiert, aber — trotz vollerhaltener Virusaktivität — nicht mehr eluiert. FAZEKAS DE ST. GROTH deutet diesen Befund im Sinne der Fermenttheorie (vgl. S. 182), nämlich dahin, daß das Receptorsubstrat durch Perjodat soweit verändert wird, daß ein enzymatischer Abbau — und damit auch die Liberierung von Virus — nicht mehr möglich ist.

Die Elutionsphase ist nun auch in hohem Grade *temperaturgebunden*, indem die Virusabspaltung erst bei Temperaturen von $\gtrless 37^0$ C stark beschleunigt und vollständig ist. Ebenso ist der bestimmende Einfluß des *Reaktionsmilieus* (Salzkonzentration, Gegenwart bestimmter Elektrolyte, Wasserstoffionenkonzentration) unverkennbar. Nach DAVENPORT und HORSFALL (84) und BJÖRKMAN und HORSFALL (28) wird das Elutionsvermögen von Influenzavirus im elektrolytarmen Medium stark gehemmt, im salzreichen Milieu (3 % NaCl) dagegen bedeutend gesteigert. Wie BRIODY (35) zeigte, ist auch die Art der vorhandenen Elektrolyte keineswegs gleichgültig; elutionsfördernde Wirkung besitzen Calcium- und Magnesiumsalze, wogegen Anticalciumionen (Oxalat, Citrat, Natriumhexametaphosphat) hemmen. Fernerhin ist die Elutionsrate im alkalischen Milieu (pH 9,0) stark erhöht. In wohl noch höherem Grade wird das dissociierende Verhalten des Mäusepneumonievirus durch die Salzkonzentration bestimmt, allerdings in einer — im Vergleich zum Influenzavirus — völlig gegensätzlichen Weise. Bekanntlich zeigt dieses Virus in Gegenwart von Elektrolyten, gleichgültig welcher Art, nicht die geringste spontane Elution. Werden dagegen die Elektrolyte durch die Verwendung einer isotonischen Glucoselösung weitgehend ausgeschaltet, so wird — wie DAVENPORT und HORSFALL (84) feststellten — in kürzester Zeit eine völlige Dissociation erreicht, die durch einen geringen Salzzusatz (0,15 m NaCl) sofort wieder rückgängig gemacht werden kann.

Das Elutionsvermögen ist daher nicht nur von Viruseigenschaften, sondern auch in hohem Grade von Milieuverhältnissen abhängig. Bei schwer dissociierenden Influenzastämmen kann bei alkalischer Reaktion oder in Gegenwart bestimmter Elektrolyte die Elutionsrate beträchtlich gesteigert werden [BRIODY (35)] und beim Mäusepneumonievirus genügt die Erhitzung (261, 81), Alkalisierung (82) oder die Eliminierung der Salze (84), damit die Dissociation zustande kommt. Eine festere Bindung an die Erythrocyten scheint nur für das Vaccine- und Ektromelievirus nachgewiesen, bei welchen die Dissociation bisher nur unter Verwendung von Immunseren gelungen ist [BURNET und STONE (63)]. Angaben über eine *vollständige* Eluierung des adsorbierten Virus bedürfen noch der Verifizierung, was am besten durch die elektronenoptische Prüfung eluierter Blutzellen geschehen könnte. Jedenfalls ist die quantitative Virusbestimmung im Eluat für die Entscheidung dieser Frage nicht ausreichend, da mit Potenzierungseffekten im Sinne von DOERR und *Mitarbeitern* (97, 98) zu rechnen ist. Tatsächlich wurde auch mehrfach beobachtet, daß die im Eluat aufgefundene Virusmenge größer ist, als wie berechnungsgemäß erwartet werden konnte

(1, 235, 165, 163). Anderseits ist der serologische Nachweis des noch adsorbierten Virus zweifellos möglich (47, 55, 21, 1, 108), dürfte jedoch als quantitative Methóde und für die Bestimmung kleiner Virusmengen nicht geeignet sein.

Daß das Virus, bzw. Haemagglutinin durch den Adsorptions- und Elutionsvorgang keine qualitative Veränderung erleidet, wurde zuerst von Hirst (195) festgestellt und seither wiederholt bestätigt.

Veränderung der Erythrocyten. Wie Hirst (195) gleichfalls feststellte, sind Erythrocyten nach der Adsorption und Elution von *Influenzavirus* gegenüber der erneuten Zufuhr desselben Virus refraktär, d. h. inagglutinabel und außerstande, Virus zu binden. Dasselbe Verhalten wurde von Burnet für Blutzellen nachgewiesen, die mit dem *Virus der Newcastle-Krankheit* (42, 43) oder dem *Mumpsvirus* (44) vorbehandelt waren. Die Annahme schien daher gerechtfertigt, daß die Erythrocyten unter der Einwirkung dieser Virusarten den virusbindenden Zellreceptor, und damit auch ihre ursprüngliche Reaktivität, verlieren. Bei den spontan nicht eluierenden Virusarten konnte ein derartiger Receptorverlust nicht nachgewiesen werden. Nach Burnet und Stone (63) erlangen mit *Vaccine-* oder *Ektromelievirus* vorbehandelte Hühnererythrocyten durch die Stabilisierung mit Immunserum ihre Agglutinabilität in vollem Umfange zurück. Noch eindeutiger tritt dieses Verhalten beim *Virus der Mäusepneumonie* zutage, das nach Davenport und Horsfall (84) mit denselben Mäuseerythrocyten beliebig oft reagiert, wenn durch den Wechsel der Elektrolytkonzentration alternierend die Association, bzw. Dissociation begünstigt wird. Die Frage der *Spezifität der Zellveränderung*, d. h. ob den verschiedenen Virusarten und -stämmen dieselben oder unterschiedliche Zellreceptoren zugeordnet sind, wurde durch die Untersuchungen von Hirst (195), Burnet und Mitarbeitern (42, 43, 44, 69, 248, 313) und Florman (121) teilweise aufgeklärt. Nach der ersten Beobachtung von Hirst (195) sind Hühnererythrocyten, die mit den Influenzastämmen PR 8 oder Lee vorbehandelt werden, sowohl für den homologen als auch heterologen Stamm inagglutinabel, ein Verhalten, das auf einen gemeinsamen Receptor schließen ließ. Burnet und Mitarbeiter (69) prüften eine größere Anzahl von Influenzastämmen und ebenso auch das Newcastle- und Mumpsvirus auf einen receptorverändernden Einfluß auf Hühner- und Menschenerythrocyten und gelangten zu den folgenden Ergebnissen: Sämtliche Influenzavirusstämme bewirken die Inagglutinabilität der Erythrocyten gegen Mumps- und Pseudopestvirus, wogegen mit Mumps- oder Newcastlevirus vorbehandelte Erythrocyten nur gegen Mumps-, bzw. Newcastle- und Mumpsvirus refraktär sind, von sämtlichen Influenzastämmen jedoch noch agglutiniert werden. Einseitige Beziehungen ähnlicher Art zeigten die verschiedenen Influenzavirusstämme untereinander, da die erythrocytenverändernde Wirkung für jeden Stamm in unterschiedlichem Grade (z. B. porcines Virus > Virus B > Virus A) ausgeprägt war. Diese Verhältnisse können nach Burnet, McCrea und Stone (69) dadurch veranschaulicht werden, daß die geprüften Virusstämme in einer Skala („Receptor gradient") angeordnet werden, die ihrer receptorverändernden Wirkung entspricht:

Mumps < Newcastle < Influenza A (Großteil der Stämme) < Influenza B <
< Influenza Schwein.

Die Stellung einer Virusart in dieser Skala ist dadurch gekennzeichnet, daß die — mit diesem Virus — vorbehandelten Erythrocyten nicht nur für den homologen Stamm, sondern für alle vorangehenden Virusstämme inagglutinabel, für alle nachfolgenden Stämme dagegen noch agglutinabel sind. Burnet und Mitarbeiter halten diese Anordnung für weitgehend fixiert, was allerdings nur der Fall ist, wenn dieselbe Erythrocytenspezies (Hühner-, bzw. Menschenerythrocyten) verwendet wird (248, 313). Immerhin zeigten sich wiederholt

Unstimmigkeiten und Diskrepanzen [vgl. STONE (313)], die nach FLORMAN (121) auch stets zu erwarten sind, falls derartige Versuche nicht nach quantitativen Gesichtspunkten durchgeführt werden. Wie FLORMAN nachwies, ist der Versuchsausfall in hohem Grade abhängig sowohl von der Intensität der Vorbehandlung der Erythrocyten („mit Virus gesättigte, bzw. ungesättigte Zellen") als auch von der Viruskonzentration, die zur Prüfung der Erythrocyten auf Agglutinabilität verwendet wird. Im übrigen vermochte FLORMAN die Befunde der BURNETschen Schule zu bestätigen.

Die Bedeutung der Receptorskala von BURNET ist nicht ausreichend klargestellt. Immerhin dürften die folgenden Schlüsse erlaubt sein: 1. Sämtliche Virusarten und -stämme dieser Gruppe greifen am selben Zellreceptor an. Ein von STONE (313) durchgeführter „Blockierungsversuch" scheint hiefür beweisend: Erythrocyten, die mit einem bestimmten Virusstamm „beladen" sind, besitzen nicht das geringste Adsorptionsvermögen für alle übrigen Stämme, woraus gleichzeitig hervorgeht, daß in der Adsorptionsphase ein „Gradient" nicht vorkommt; 2. die Receptorveränderung ist für alle Virusstämme dieser Gruppe[1] in qualitativer Hinsicht wahrscheinlich dieselbe. Jedenfalls läßt sich eine spezifische Wirkung in immunologischem Sinne bei den verschiedenen Stämmen nicht nachweisen; 3. der „Gradient" kommt dadurch zustande, daß die verschiedenen Virusstämme den Zellreceptor in unterschiedlichem Grade „zerstören", wobei nur Virusstämme mit starken Adsorptionsaffinitäten mit den verbleibenden Substratmengen noch zu reagieren vermögen.

Das *Vaccine-* und *Ektromelievirus* lassen sich dagegen, wie BURNET und BOAKE (58) feststellten, in diese Receptorskala nicht einreihen, da gegen Pseudopest- und Influenzavirus inagglutinabel gewordene Erythrocyten von diesen Virusarten noch in ungehemmtem Grad agglutiniert werden.

Bei virusvorbehandelten Erythrocyten sind nun auch eine Reihe *serologisch nachweisbarer Veränderungen* festgestellt worden. Nach der Beobachtung von BURNET, MCCREA und STONE (69) erwerben menschliche Erythrocyten der Gruppe 0 durch die Vorbehandlung mit Pseudopest- oder Influenzavirus eine beträchtliche *Panagglutinabilität* gegenüber menschlichen und tierischen Normalseren. Die verschiedenen Virusstämme bewirken diese Veränderungen nicht in gleichem Grade, sondern nach Maßgabe ihrer Stellung im „Gradient", bzw. ihrer receptorzerstörenden Aktivität. Fernerhin zeigen virusmodifizierte Erythrocyten ein verändertes Verhalten gegenüber bestimmten *Immunseren.* So wurde von BURNET und ANDERSON (55) und STONE (313) beobachtet, daß mit Newcastle- oder Schweineinfluenzavirus vorbehandelte Menschenerythrocyten von menschlichen Seren *infektiöser Mononucleose* in auffallend hohem Titer agglutiniert werden[2]. Von Interesse ist fernerhin die Feststellung von CHU und COOMBS (75) und STONE (313), wonach die Verwendung von virusmodifizierten Rh_1-Erythrocyten den Nachweis *inkompletter Rhesusantikörper* erlaubt. Der Agglutinationstiter wechselte je nach der zur Vorbehandlung verwendeten Virusart (Newcastle > Schweineinfluenza > Influenza B > Influenza A).

Die virusinducierte Panagglutinabilität der Erythrocyten steht nun, wie schon von BURNET und Mitarbeitern (69) mit Recht hervorgehoben wurde, in auffallen-

---

[1] Möglicherweise mit Ausnahme des Mumpsvirus, dem anscheinend doch eine Sonderstellung zukommt [BOVARNICK und DE BURGH (33), STONE (313), GINSBERG, GOEBEL und HORSFALL (144, 146)].

[2] Nach EVANS und CURNEN (105) ist dieser Effekt für Mononucleosesera nicht spezifisch, d. h. unabhängig vom heterophilen Antikörper oder dem hypothetischen Erreger dieser Affektion. Von gleicher Wirksamkeit erwiesen sich auch bestimmte menschliche Rekonvalescentensera und Kaninchenimmunsera gegen Influenzavirus.

der Analogie zum Phänomen von Thomsen und Friedenreich (130), welches bekanntlich dadurch gekennzeichnet ist, daß Erythrocyten unter der Einwirkung bakterieller Kulturfiltrate panagglutinable Eigenschaften erhalten. Die Übereinstimmung der Erythrocytenveränderung durch Virus einerseits und durch das transformierende Agens von Thomsen und Friedenreich anderseits ist nun tatsächlich weitgehend und überraschend. So vermochten Burnet und seine Schüler nachzuweisen, daß Kulturfiltrate von Choleravibrionen Erythrocyten in gleicher Art modifizieren wie Influenza- und Pseudopestvirus, nämlich im Sinne einer Agglutinabilität gegenüber Normalsera (313), Mononucleosesera (313), inkomplette Rhesusantikörper (274) und Inagglutinabilität gegenüber einer Reihe von Virusarten [Mumps, Pseudopest, Influenza (69, 131)]. Von Friedenreich wird nun angenommen, daß das erythrocytentransformierende Agens von Thomsen ein Ferment ist, das den Umbau eines im Erythrocytenstroma bereits vorhandenen Antigens in den panagglutinablen „Receptor T" bewirkt, der mit den im Normalserum natürlicherweise vorhandenen T-Agglutininen in Reaktion treten kann. Die tatsächliche Existenz eines solchen T-Antigens wurde im Immunisierungsversuch nachgewiesen [Hallauer (158), Burnet und Anderson (54)], wobei nun wiederum bemerkenswert ist, daß die erzeugten Immunagglutinine auch die mit Virus vorbehandelten Erythrocyten in hohem Titer agglutinieren (54). Leider ist die umgekehrte Prüfung, nämlich welche qualitative Veränderung der Antigenbestand des Erythrocytenstroma unter der Einwirkung von Virus erfährt, bisher nicht durchgeführt worden. Ein derartiger Versuch könnte möglicherweise — je nach der Anzahl der gebildeten „Extraagglutinine" — darüber Aufschluß geben, ob nur ein oder mehrere Zellreceptoren umgeformt werden. Aus dem refraktären Verhalten von mit Virus, bzw. bakteriellen Kulturfiltraten vorbehandelten Erythrocyten gegenüber der nochmaligen Zufuhr von Virus muß ja nicht notwendigerweise auf einen weitgehenden Abbau, bzw. eine Zerstörung des Zellreceptors geschlossen werden, sondern diese Unempfindlichkeit ließe sich ebensogut — im Sinne von Friedenreich — mit einer Spezifitätsveränderung erklären. Ob die nachgewiesenen Zellveränderungen, d. h. die erworbene Inagglutinabilität gegen Virus, bzw. Agglutinabilität gegen Antikörper als Manifestationen eines einheitlichen Vorganges (Modifikation desselben Receptors durch dasselbe Agens) gedeutet werden können [vgl. Stone (313)], ist jedoch nicht entschieden.

Andere Antigene des Erythrocytenstroma werden durch Virus nicht alteriert; so bleibt die serologische Spezifität der Isoagglutinogene menschlicher Erythrocyten intakt, und Hammelerythrocyten werden — auch nach der Einwirkung von Poliomyelitisvirus — von Immunhaemolysinen gelöst [Hallauer (164)].

Morphologische Veränderungen des Erythrocytenstroma konnte Heinmets (169) in elektronenoptischen Aufnahmen nicht wahrnehmen, wobei allerdings zu berücksichtigen ist, daß an Stromata, die für die Untersuchung im Elektronenmikroskop einzig geeignet sind, feinere Strukturveränderungen kaum feststellbar sind. Die *osmotische Resistenz* bei virusvorbehandelten Erythrocyten ist bisher nicht geprüft worden, dagegen wurde festgestellt, daß derartige Zellsuspensionen unter bestimmten Bedingungen haemolysieren. So wurde von Stone (313) beobachtet, daß Erythrocyten nach der Vorbehandlung mit Influenzavirus, bzw. Cholerakulturfiltrat in Gegenwart von frischem (komplementhaltigem) Meerschweinchenserum lysiert werden. Nach Beveridge und Lind (21) ist auch die Elution von Mumpsvirus häufig von Haemolyse begleitet, ein Befund, der nun aber nicht notwendigerweise eine verminderte Zellresistenz anzeigt, da dem Mumpsvirus selbst Haemolysinwirkung zukommt (263). Bemerkenswert sind Untersuchungen von Hanig (165) über das *elektrokinetische Potential* von menschlichen Erythrocyten in Gegenwart von Influenzavirus; bei der Adsorption von

Virus konnte ein scharfer Abfall der elektrophoretischen Mobilität festgestellt werden, die auch nach der Viruseluierung ihren niederen Wert beibehielt.

In welcher Art und durch welche Kräfte die *Agglutinate* zusammengehalten werden, ist ungewiß. A priori sind zwei Möglichkeiten in Betracht zu ziehen, nämlich 1. daß die an der Zelloberfläche adsorbierten Viruspartikel selbst als verbindende Brücken fungieren, oder 2. daß die Oberflächen der Blutzellen durch adsorbiertes Virus sensibilisiert und infolgedessen sekundär agglutiniert werden. Nach den bisherigen Beobachtungen besitzt die erste Annahme größere Wahrscheinlichkeit. Auch HEINMETS (169) neigt auf Grund seiner elektronenoptischen Studien eher der Ansicht zu, daß die Viruspartikel die Bindeglieder zwischen den Erythrocyten darstellen, und zwar vor allem aus quantitativen Gründen. Einen Sensibilisierungseffekt hält HEINMETS schon deshalb für wenig wahrscheinlich, weil im elektronenoptischen Bild zahlreiche *nichtagglutinierte* Blutzellen nachgewiesen werden konnten, obschon diese Zellen einen größeren „Besatz" von adsorbierten Viruspartikeln aufwiesen; bei niedrigeren Viruskonzentrationen waren derartige isoliert liegende Zellen sogar in der Überzahl. HEINMETS nimmt daher an, daß zur Bildung von Agglutinaten unverhältnismäßig große Mengen adsorbierter Viruspartikel erforderlich sind, eine Annahme, die nun aber mit den Berechnungen anderer Autoren in Widerspruch steht. So veranschlagen FRIEDEWALD und PICKELS (138) die Menge der Viruspartikel, die bei den Influenzastämmen PR 8 und Lee in der Titerendverdünnung enthalten ist, auf 45—92 Millionen und das Verhältnis der Anzahl der Viruspartikel zur Anzahl der Erythrocyten auf rund 1. Nach ähnlichen Berechnungen von HANIG (165) wäre nun allerdings die Anzahl der Viruspartikel, die für die vollständige Absättigung der Zellreceptoren notwendig ist, wesentlich größer, da HANIG die Zahl der auf 1 Erythrocyten entfallenden Viruspartikel (Stamm PR 8) auf 298 einschätzt, wobei rund 1:80 der Zelloberfläche mit Virus besetzt wäre.

Substrat des Zellreceptors. Die physikalischen und chemischen Eigenschaften des Zellreceptors können nach den folgenden Prinzipien ermittelt werden: 1. durch die Isolierung des Receptors; 2. durch die Prüfung des Receptors in situ auf sein Verhalten gegenüber physikalischen und chemischen Einflüssen, und 3. durch die Ermittlung receptorähnlicher Substanzen anderer biologischer Provenienz. Sämtliche drei Wege sind beschritten worden und führten auch zu weitgehend konvergierenden Ergebnissen.

Die Abtrennung des Zellreceptors von den Erythrocyten erstrebten FRIEDEWALD, MILLER und WHATLEY (139), BOVARNICK und DE BURGH (33, 88) und WOOLLEY (345). Die Vollkommenheit der eingeschlagenen Technik ist unterschiedlich. FRIEDEWALD und Mitarbeiter begnügten sich, gewaschene Blutzellen auf mechanischem Wege zu detritieren und in Kochsalzlösung zu extrahieren, ein Verfahren, das wohl die völlige Beseitigung von Stromata nicht erlaubt. BOVARNICK, DE BURGH und Mitarbeiter extrahierten die Elininfraktion der Stromata mit organischen Solventien und erhielten schließlich eine wasserlösliche Fraktion von hoher Wirksamkeit. WOOLLEY gelangte zu seinem Präparat durch die Vorbehandlung der Stromata mit Alkohol und Chloroform und Fällung der wäßrigen Extrakte mit Alkohol. Die von FRIEDEWALD und BOVARNICK und DE BURGH erzielten Versuchsergebnisse sind weitgehend übereinstimmend. Die hergestellten Extrakte hemmten noch in hohen Verdünnungen, bzw. kleinen Substanzmengen — $1,1\,\gamma/cc$ des bestgereinigten Präparates (88) — die Haemagglutination von Influenza- und Mumpsvirus. Die „Specifität" dieser Hemmung ergab sich aus der Feststellung, daß nur Extrakte aus denjenigen Erythrocytenspecies wirksam waren, die auch den Zellreceptor für eine bestimmte Virusart enthielten. So hemmten die aus Menschen- und Hühnererythrocyten hergestellten

Präparate sowohl Influenza- als auch Mumpsvirus, während Extrakte aus Schaferythrocyten ausschließlich nur gegen Mumpsvirus wirksam waren, ein Verhalten, das dem Agglutinationsspektrum dieser Virusarten entspricht. Ebenso lieferten virusvorbehandelte Erythrocyten — erwartungsgemäß — keine wirksamen Extrakte. Fernerhin wurde der Nachweis geleistet, daß aktives Influenzavirus mit dem extrahierten Hemmstoff ebenso reagiert wie mit dem Zellreceptor der Erythrocyten, d. h. denselben bei höherer Temperatur (20—37° C) in zunehmendem Grade inaktiviert und hierbei wieder liberiert wird. Beziehungen zu den menschlichen Gruppensubstanzen A und B und dem Rhesusantigen bestanden nicht. Die chemische Analyse des gereinigten Präparates von DE BURGH, YU, HOWE und BOVARNICK (88) ergab ein wasserlösliches, hochmolekulares Polysaccharid, eventuell in Verbindung mit Eiweiß- und lipoiden Substanzen. Die Versuche von WOOLLEY sind weniger eindeutig, führten jedoch — nach Ansicht des Autors — zu einem ähnlichen Ergebnis.

HIRST (201) suchte die Natur des Zellreceptors an intakten Erythrocyten durch eine Reihe von Resistenzprüfungen gegen physikalische und chemische Einflüsse zu ermitteln, wobei die Unversehrtheit oder Schädigung des Receptors jeweils im Adsorptions- und Elutionsversuch mit den Influenzastämmen PR 8 und Lee geprüft wurde. Das Resultat dieser Versuche läßt sich dahin zusammenfassen, daß der Receptor gegenüber Erhitzung und alkalischer Reaktion eine beträchtliche Resistenz aufwies, dagegen durch Perjodat (als einziges der geprüften Oxydationsmittel) und Trypsin rasch zerstört wurde. Auf Grund dieser Eigenschaften hält HIRST die Annahme für wahrscheinlich, daß der Receptor einen Polysaccharid-Protein-Komplex darstellen dürfte.

Untersuchungen an nicht zellgebundenen Receptoren wurden durch die Beobachtung von HIRST (194) und namentlich von FRANCIS (125) ermöglicht, wonach die meisten menschlichen und tierischen Normalsera einen thermostabilen Hemmstoff („FRANCIS-Inhibitor“) gegen das Influenzahaemagglutinin enthalten, dessen Aktivität besonders gegenüber dem schonend erhitzten Lee-Stamm hervortritt. Die Vermutung, daß dieser im Normalserum vorkommende Inhibitor mit dem Zellreceptor der Erythrocyten identisch sein könnte, wurde durch die gleichzeitigen Untersuchungen von BURNET und seinen Mitarbeitern (68, 4, 249) und HIRST (201, 202) verifiziert. Die Gleichartigkeit von Zellreceptor und Seruminhibitor kommt in den folgenden gemeinsamen Merkmalen zum Ausdruck: 1. der relativ hohen Resistenz gegenüber der Erhitzung und Verschiebung der Wasserstoffionenkonzentration nach der sauren und alkalischen Seite (201, 249); 2. der großen Empfindlichkeit gegenüber Perjodat und proteolytischen Fermenten (201, 249), und 3. der Zerstörung durch Virus oder Cholerakulturfiltrat (201, 202, 4). McCREA (249) ist es auch gelungen, den Seruminhibitor zu reinigen und zu konzentrieren; das hierbei gewonnene Präparat zeigte ebenfalls — mit Ausnahme einer größeren Labilität — die bereits erwähnten Eigenschaften. Die Annahme scheint daher gerechtfertigt, daß auch der Seruminhibitor ein Glycoprotein, bzw. ein Mucopolysaccharid darstellt (68, 201, 249).

Wie BURNET (68, 49, 50, 51, 52, 53) zeigte, finden sich mucinartige Substanzen von hemmender Wirkung auf Virushaemagglutinine nicht nur in den Erythrocyten und im normalen Serum, sondern auch in den meisten Schleimdrüsensekreten des Menschen, in welchen wahres Mucin nachgewiesen werden kann[1], nämlich in besonders reichlichen Mengen in Ovarialcysten und im Schleim der Cervix uteri, aber auch im Magenschleim, Nasensekret und wahrscheinlich auch in der Tränenflüssigkeit (57). Von hoher hemmender Aktivität erwiesen sich auch

---

[1] Eine einfache Nachweismethode für Mucin wurde von BURNET (48, 53) angegeben.

die von MORGAN und WADDELL aus pseudomucinösen Ovarialcysten gewonnenen, gereinigten Blutgruppensubstanzen 0 und A. Die Verteilung dieser Hemmstoffe im menschlichen Organismus erinnert an das Vorkommen löslicher Blutgruppenstoffe, läßt jedoch keine Identifizierung dieser Substanzen zu[1]. Einige natürlich vorkommende Mucine, wie die des menschlichen Speichels, des Schweinemagens und der Cervix uteri des Rindes verhielten sich völlig inaktiv (36, 7), ein Hinweis dafür, daß die verschiedenen Mucine je nach ihrer Provenienz keineswegs gleichartig und daher auch vorgreifende, verallgemeinernde Schlüsse nicht zulässig sind.

Schließlich wurden auch im *Hühnerei* Hemmstoffe mit den Eigenschaften des Inhibitors von FRANCIS, bzw. des Zellrezeptors nachgewiesen. Nach LANNI und BEARD (226, 227) findet sich im *Eiklar* ein hochwirksamer Hemmstoff für das Haemagglutinin des Influenzavirus, welcher durch Perjodat und aktives Virus inaktiviert wird. Wie die ergänzenden Untersuchungen von HARDY und HORSFALL (166) zeigten, besitzt dieser Hemmstoff auch die übrigen Merkmale des Inhibitors von FRANCIS (Thermostabilität, Resistenz bei alkalischer Reaktion, Empfindlichkeit gegen proteolytische Fermente). Das gereinigte Präparat vermochte die Haemagglutination von erhitztem Virus noch in einer Menge von 0,5 γ/1 A. E. zu hemmen und zeigte eine positive BIURET- und MOLISCH-Reaktion. Aus dem gesamten Verhalten konnte wiederum auf das Vorliegen eines Mucopolysaccharides geschlossen werden[2].

In den *extraembryonalen Flüssigkeiten des Hühnerembryos (Allantois, Amnion)* sind ähnliche Hemmstoffe für die Haemagglutinine des Mumps- und Influenzavirus von BEVERIDGE und LIND (21), LIND (235), SVEDMYR (300, 303) und BIELING und OELRICHS (26) nachgewiesen worden. KNIGHT (219, 223) und GARD und v. MAGNUS (144) haben diesen Inhibitor in der Sedimentierungsfraktion von 170 bis 220 S isoliert und als „Normalprotein" von hoher Viskosität und einem beträchtlichen Gehalt an Glucosamin (223) charakterisiert. Von besonderem Interesse ist die Feststellung, daß der Inhibitor im normalen, nichtinfizierten Ei während der Bebrütungszeit (7. bis 14. Tag) an Konzentration erheblich zunimmt (21, 256, 300, 166), in der infizierten Allantoisflüssigkeit dagegen mit dem Beginn der Virusvermehrung verschwindet, bzw. sich dem Nachweis entzieht (144, 300). Die naheliegende Vermutung, daß der Hemmstoff der Allantoisflüssigkeit mit dem in Bildung begriffenen Virus in vivo in ähnlicher Weise reagiert, wie dies bereits in vitro für andere wirtseigene Inhibitoren (vgl. oben) nachgewiesen wurde, konnte von HARDY und HORSFALL (166) und SVED-

---

[1] Den Angaben von BURNET (68, 49, 53) und GREEN und WOOLLEY (152), wonach der gereinigten Gruppensubstanz 0 und A virusbindende Eigenschaft zukommt, stehen die Befunde von FRIEDEWALD, MILLER und WHATLEY (139) sowie von DE BURGH, YU, HOWE und BOVARNICK (88) entgegen, welche die Unabhängigkeit ihrer aus Erythrocyten gewonnenen Hemmstoffe von A- und B-Substanzen nachzuweisen vermochten. Ebenso inaktiv verhielten sich aus Schweinemagen isolierte Gruppenstoffe. Divergenzen dieser Art sind wohl nicht nur durch die unterschiedliche Reinheit, sondern auch durch die verschiedenartige Provenienz der Präparate bedingt. Eine Identität der Gruppensubstanzen mit den virusbindenden Hemmstoffen ist schon deshalb nicht anzunehmen, weil die serologische Spezifität der Isoagglutinogene durch Influenzavirus nicht verändert wird (68).

[2] Die Feststellung, wonach im Hühnereiklar anscheinend dieselben mucinartigen Hemmstoffe nachzuweisen sind wie in den Sekreten menschlicher Schleimdrüsen, ist auch von Interesse in Hinsicht auf das in denselben Substraten vorkommende antibakterielle *Lysozym* von FLEMING, ein Ferment, das nach den Untersuchungen von FEINER, MEYER und STEINBERG (110) ebenfalls hochpolymerisierte Mucopolysaccharide angreift.

MYR (303) in überzeugender Art verifiziert werden. Wie Hardy und Horsfall in Reagensglasversuchen feststellten, reagiert das Influenzavirus mit dem Inhibitor der normalen Allantoisflüssigkeit grundsätzlich ebenso wie mit dem Zellreceptor, d. h. nach dem Vorgang der Adsorption und Elution, wobei auch der Inhibitor inaktiviert wird. Eine vollständige Reversibilität der Reaktion konnte nun aber in vitro niemals festgestellt werden, da stets nur eine Quote des gebundenen Virus wieder in Freiheit gesetzt wurde. Dieses Verhalten ist nach Hardy und Horsfall insofern gesetzmäßig, als in jedem Virus-Inhibitor-Gemisch ein ganz bestimmtes Gleichgewicht zwischen gebundenem und freiem Virus erreicht wird, das nur durch den weiteren Zusatz von Virus, bzw. Inhibitor verändert werden kann. Hardy und Horsfall erblicken in dieser unvollständig ablaufenden, einem Gleichgewicht zustrebenden Reaktion einen auffallenden Gegensatz zur vollständigen Viruselution aus Erythrocyten. Ob eine solche Gegensätzlichkeit besteht, ist nun allerdings fraglich, da auch die Dissociation von Virus und Zellreceptor häufig — möglicherweise stets — nur unvollständig ist (vgl. S. 166, 184). Nach Hardy und Horsfall wird der Inhibitor weder in vivo noch in vitro zerstört, sondern nur soweit alteriert, daß seine hemmende Wirkung nicht mehr nachgewiesen werden kann, eine Auffassung, die wohl auch für den Zellreceptor Gültigkeit haben dürfte (vgl. S. 170). In der mit Influenzavirus infizierten Allantoisflüssigkeit würde demnach ein Gleichgewicht $(V+J \rightleftharpoons VJ)$ zwischen drei Komponenten bestehen, nämlich 1. dem freien, haemagglutinierenden Virus (V); 2. dem nicht mehr bindungsfähigen Inhibitor (J) und 3. dem an den Inhibitor gebundenen Virus (VJ)[1].

Gegenüber inaktivierenden Einflüssen (Hitze, Alkali, Perjodat, proteolytische Fermente) verhält sich der Hemmstoff der Allantoisflüssigkeit wie die übrigen Inhibitoren von mucopolysaccharider Beschaffenheit.

Inhibitoren. Eine einheitliche Gruppe von Inhibitoren bilden die bereits erwähnten *Mucopolysaccharide*, die im Tierreich anscheinend weitverbreitet und bisher in normalen Seren, in den Schleimdrüsensekreten des Menschen, im Eiklar und in der Allantois- und Amnionflüssigkeit von Hühnerembryonen nachgewiesen worden sind. In ihren physikalischen und chemischen Eigenschaften (Stabilität gegen Hitze und alkalische Reaktion, Empfindlichkeit gegen Perjodat und proteolytische Fermente) und ihrem biologischen Verhalten gegenüber aktivem Virus (Adsorptionsvermögen, Inaktivierbarkeit) zeigen diese Inhibitoren alle Merkmale des Zellreceptors (für Virusarten der Influenza-, Mumps-, Hühnerpestgruppe). Die Hemmung der Haemagglutination beruht daher auf der Konkurrenz zweier ähnlicher Substrate (Inhibitor, Zellreceptor) um dasselbe Agens (Virus).

Dieser Gruppe von Inhibitoren am nächsten stehen einige *Hemmstoffe*, die in *virusinfizierten Geweben* nachgewiesen wurden. So beobachtete Hirst (194, 196), daß das aus der *Frettchenlunge* gewonnene *Influenzavirus* — trotz hoher Infektiosität — nicht haemagglutiniert. Daß diese Reaktionslosigkeit auf einen in der Frettchenlunge vorkommenden Hemmstoff zu beziehen war, ergab sich durch den Nachweis der stark virusbindenden, bzw. haemagglutininhemmenden Wirkung von Suspensionen normaler Lunge. Die von Hirst vermutete Gleichartigkeit dieses Inhibitors mit dem Receptor der Erythrocyten konnte jedoch in eindeutiger Weise nicht bewiesen werden. Von Friedewald, Miller und Whatley (139) wurden ähnliche Hemmstoffe für das Haemagglutinin des

---

[1] Die Association von Virus und Normalprotein wurde bereits von Knight (vgl. S. 159) festgestellt, jedoch nicht im Sinne eines Virus-Inhibitor-Komplexes gedeutet.

Influenzavirus in der Lunge des Menschen und des Meerschweinchens nachgewiesen, die sich — im Gegensatz zu den übrigen aus Leber, Niere und Muskel gewonnenen Gewebsextrakten — als thermostabil erwiesen und daher möglicherweise mit dem Seruminhibitor von FRANCIS, bzw. dem Zellreceptor identisch sind. Ähnliche Verhältnisse bestehen beim *Virus der Mäusepneumonie.* Bekanntlich liegt dieses Virus in der (zermörserten) Mäuselunge stets in der „gebundenen Form" vor, d. h. ermangelt jeglicher haemagglutinierender Aktivität. Daß das Haemagglutinin de facto aber nur maskiert, bzw. durch seine Bindung an eine gewebliche Komponente reaktionsunfähig ist, zeigt der Erfolg von Dissociationsverfahren. Die Liberierung des Haemagglutinins kann erreicht werden: 1. durch Mittel, die den geweblichen Receptor inaktivieren, bzw. zerstören, nämlich Erhitzung auf 70 bis 80⁰ C (261, 262, 81), Alkalisierung (82); oder proteolytische Fermente (331); 2. durch den spezifischen Antikörper, dessen virusbindende Kraft offensichtlich stärker ist als diejenige des Inhibitors, und nachfolgende Erhitzung (331), und 3. durch Elution in elektrolytarmen, gepufferten Medien (84), das einzige Verfahren, das sämtliche Komponenten (Haemagglutinin, Infektiosität, geweblicher Receptor) intakt beläßt und ein vollaktives „freies" Virus zu gewinnen erlaubt. Die Realität des Dissociationsvorganges konnte durch Filtrations- und Ultrazentrifugierungsversuche sichergestellt werden, in welchen für das gebundene und freie Virus sehr unterschiedliche Größendimensionen (140 und 40 mμ) ermittelt wurden (83). Der gewebliche Hemmstoff, bzw. Receptor konnte ausschließlich nur im Lungengewebe virusempfänglicher Tierspezies (Maus, Hamster, Baumwollratte) nachgewiesen werden (81, 82, 331) und ist mit großer Wahrscheinlichkeit mit dem Receptor agglutinabler Erythrocyten (Maus, Hamster) identisch. Beide Receptoren gehen mit dem freien Virus eine stabile Verbindung ein, die nur mit Hilfe der bereits erwähnten Dissociationsverfahren gesprengt werden kann, und werden auch durch die Bindung von Virus in keiner Weise verändert, sondern behalten — bei Anwendung eines schonenden Dissociationsmodus — ihre ursprüngliche Reaktionsfähigkeit bei (84). Die Labilität des Inhibitors, bzw. Receptors gegenüber höheren Temperaturen und Alkali sowie seine leichte Zerstörbarkeit durch proteolytische Fermente (Trypsin, Chymotrypsin), scheinen auf seine proteide Natur hinzuweisen (331). Von Interesse ist die Feststellung von DAVENPORT und HORSFALL (84), wonach die Bindungskräfte zwischen Pneumonievirus und Receptor anscheinend nur durch die Konzentration der Elektrolyte und Wasserstoffionen des Reaktionsmilieus gesteuert werden, so daß der Virus-Receptor-Komplex seinem Verhalten nach als ein schwaches Salz zu betrachten wäre. Die Gegensätzlichkeit zu den beim Influenza-, Mumps- und Hühnerpestvirus festgestellten Verhältnissen ist offensichtlich; die stoffliche Natur des Inhibitors und die Kinetik des Bindungsvorganges sind unterschiedlich und eine Übereinstimmung ergibt sich nur in einer Hinsicht, daß nämlich auch der Inhibitor des Mäusepneumonievirus mit dem Zellreceptor identisch ist.

Von einer größeren Anzahl *wirtseigener Hemmstoffe* (Normalsera, Organextrakte) steht lediglich fest, daß in ihrer Gegenwart die haemagglutinierende Wirkung von Virus (Influenza, Mumps) aufgehoben ist; die Natur und die Wirkungsart dieser Inhibitoren sind jedoch größtenteils unbekannt.

In *Normalseren* zahlreicher Tierspecies (Kaninchen, Frettchen, Meerschweinchen Maus etc.) und auch des Menschen finden sich — außer dem Inhibitor von FRANCIS — *thermolabile Stoffe* von haemagglutininhemmender Wirkung [HIRST (194), BURNET und McCREA (62), McCREA (247) u. a.]. Im Gegensatz zu den mucinartigen Inhibitoren werden diese Hemmstoffe bei Temperaturen von 62⁰ C meist inaktiviert, von Perjodat wenig geschädigt und von Erythrocyten adsorbiert [McCREA (249)]. Bei der

Serumfraktionierung kann der thermolabile Hemmstoff im γ-Globulin nachgewiesen werden [McCrea (247, 62)].

Charakteristisch ist fernerhin, daß die hemmende Wirkung gegenüber verschiedenen Virusstämmen unterschiedlich groß und vor allem von der Individualität der verwendeten Erythrocytenproben in hohem Grade abhängig ist (62, 247, 6, 2, 319). Schließlich ist bemerkenswert, daß derartige Normalseren gelegentlich auch antiinfektiöse Wirkung in vivo entfalten [Burnet und McCrea (62), Hirst (200)]. Friedewald, Miller und Whatley (139) prüften den hemmenden Einfluß von *Organextrakten* von Mensch und Kaninchen auf die Haemagglutination von Influenza- und Mumpsvirus. Sämtliche Extrakte (aus Lunge, Leber, Niere, Muskel) hemmten, z. T. in erstaunlich hoher Verdünnung, ohne daß hierfür eine Antikörperwirkung verantwortlich gemacht werden konnte. Mit Ausnahme des Lungenextraktes büßten diese Hemmstoffe schon bei mäßig hohen Temperaturen (56⁰ C) einen Großteil ihrer Wirksamkeit ein. Beziehungen zum Zellreceptor sind nach Friedewald und Mitarbeitern wahrscheinlich, aber nicht sichergestellt. Untersuchungen ähnlicher Art führten Bieling und Oelrichs (26) mit Zellaufschwemmungen und Extrakten aus Organen verschiedener Vogelarten und Säugetiere durch, wobei die hemmende Wirkung gegenüber Influenzavirus geprüft wurde. Sämtliche Suspensionen und Extrakte erwiesen sich als hemmend, aber in unterschiedlichem Ausmaß, nämlich am schwächsten Lungenextrakte (Maus, Mensch), am stärksten Extrakte aus Mäuse- und Menschenniere und besonders hochgradig der Pferdeleberextrakt. Der Mangel an Spezifität ist damit wohl hinreichend dokumentiert. Die Versuchsergebnisse werden von Bieling und Oelrichs dahin interpretiert, daß die Hemmstoffe die haemagglutinierende Wirkung des Virus schädigen, dessen Bindung an die Erythrocyten jedoch nicht verhindern, eine Deutung, die wohl kaum befriedigen kann [vgl. Doerr (96)].

Bei der Prüfung von virushaltigen Gewebssuspensionen ist zu berücksichtigen, daß das haemagglutinierende Vermögen durch *reduzierende oder autolytische Vorgänge* beeinträchtigt oder gehemmt werden kann. Beobachtungen dieser Art wurden beim Mäusepneumonievirus und bestimmten murinen Poliomyelitisstämmen gemacht.

Nach Volkert und Horsfall (330) verliert das sonst so stabile Haemagglutinin des Pneumonievirus in dichten Mäuselungensuspensionen rasch an Stabilität, so daß sein Nachweis schon nach kurzfristiger Lagerung nicht mehr gelingt, obschon die Infektiosität noch vollständig erhalten ist. Dieser paradoxe Effekt konnte auf den Einfluß reducierender Sulfhydrylgruppen zurückgeführt werden, nämlich einerseits durch den Nachweis, daß der Zusatz von Glutathion zu vollaktivem Haemagglutinin dieselbe Thermolabilität bewirkt und anderseits durch die Feststellung, daß die Haemagglutininaktivität durch die spontane Oxydation, bzw. durch den Zusatz von Jodacetamid wieder hergestellt werden kann.

Einen ähnlichen Hemmungseffekt beobachtete Hallauer (163) in poliomyelitis-virushaltigen Mäusegehirnsuspensionen, in welchen schon unmittelbar nach der Herstellung ein auffallendes Mißverhältnis zwischen niedrigem Haemagglutinin- und hohem Infektiositätstiter festgestellt werden kann. In kurzfristig gelagerten Proben (24 bis 48 Stunden, + 4⁰ C) sinkt der Haemagglutiningehalt rasch auf den Nullwert ab, ohne daß merkliche Infektiositätsverluste eintreten. Daß das Haemagglutinin in derartigen Proben nicht zerstört, sondern gehemmt ist, zeigte der reaktivierende Einfluß von Glucose, durch deren Zusatz nicht nur der ursprüngliche Haemagglutinintiter wiederhergestellt, sondern um das Mehrfache gesteigert wurde. Die reducierenden Eigenschaften der Glucose sind für den reaktivierenden Vorgang nicht maßgebend, da sich andere Reduktionsmittel (Glutathion, Methionin, Cystein) als hierzu ungeeignet erwiesen.

Die Möglichkeit der *spezifischen Hemmung der Haemagglutination durch Antikörper* ist im Serum und in Organextrakten stets vorhanden und — in geweblichen Extrakten — auch dann nicht auszuschließen, wenn der Infektiositätsnachweis geleistet werden kann.

NAGLER (264) vermochte in der vom Kalb gewonnenen *Dermovaccine* trotz hoher Infektiosität keine Haemagglutinine nachzuweisen, während derselbe von der Eimembran gewonnene Virusstamm stets haemagglutinierte. Nach STONE und BURNET (318) ist die Inaktivität der Kälberlymphe mit Wahrscheinlichkeit auf die Gegenwart *virusneutralisierender Antikörper* zurückzuführen, eine Annahme, die schon deshalb wahrscheinlich ist, da der Antikörpernachweis — wie OERSKOV und ANDERSEN feststellten — in den Pusteln des Rindes bereits vom zweiten Tage an gelingt. Möglicherweise ist auch der von LEVENS und ENDERS (233) erhobene Befund, wonach die mit *Mumpsvirus* infizierte Parotisflüssigkeit des Affen — im Gegensatz zum eiadaptierten Virus — kein haemagglutinierendes Vermögen besitzt, ebenfalls auf eine immunologische Hemmung zu beziehen.

In den extraembryonalen Flüssigkeiten des Hühnerembryos (Allantois, Amnion) spielen Antikörper als haemagglutininhemmende Faktoren keine Rolle, wohl aber im *Dotter*, in welchen die Antikörper des Huhnes übergehen und angereichert werden [HALLAUER (159), BIELING und OELRICHS (26)].

Von besonderem Interesse sind Beobachtungen, wonach eine Reihe von *Polysacchariden höherer Pflanzen und bakterieller Kapselsubstanzen* die Virushaemagglutination antagonistisch beeinflussen, da die Vermutung naheliegt, daß eine solche Hemmung durch receptorähnliche Substanzen bedingt sein könnte. GREEN und WOOLLEY (152) prüften eine größere Anzahl pflanzlicher Polysaccharide auf ihre haemagglutininhemmende Wirkung gegenüber Influenzavirus; als wirksam erwiesen sich Leinsamenschleim, Akazien- und Myrrhengummi, Zitronenpektin und — in besonders hohem Grade — *Apfelpektin*. Welche Strukturen für die Aktivität maßgebend sind, konnte nicht ermittelt werden. Ebenso ist der Mechanismus, welcher der Hemmung zugrunde liegt, nicht hinreichend abgeklärt. Nach GREEN und WOOLLEY reagiert Apfelpektin sowohl mit dem Virus als auch mit den Erythrocyten. Mit Sicherheit ist nun aber nur nachgewiesen, daß das Pektin an die Erythrocytenoberfläche adsorbiert wird und in höheren Konzentrationen — wie auch die meisten übrigen Polysaccharide dieser Art — haemagglutinierende Wirkung besitzt. Auffallend und ungeklärt ist fernerhin, daß der hemmende Effekt von Pektin an die Gegenwart von Calciumionen gebunden ist (345). WOOLLEY (345) bemühte sich in einer eigenartigen Versuchsanordnung — Messung des antagonistischen Verhältnisses von Pektin und gereinigter Receptorsubstanz — den Nachweis zu leisten, daß beide Substanzen zufolge ihrer strukturellen Ähnlichkeit um das Haemagglutinin (als Enzym) konkurrieren, und stellte hierbei fest, daß die hemmende Wirkung von 0,75 mg Pektin bereits durch 0,00004 γ (!) gereinigter Receptorsubstanz aufgehoben wird. Selbst wenn man sich der Meinung von WOOLLEY anschließt wonach das Pektin mit dem Virus grundsätzlich ebenso reagiert wie die Rezeptorsubstanz, wird man aus diesem Ergebnis kaum auf eine nahe Verwandtschaft dieser Inhibitoren schließen können. Wahrscheinlicher ist wohl die Annahme, daß die hemmende Wirkung des Pektins auf einer Veränderung der Zelloberfläche oder des Zellreceptors selbst beruht. Ein derartiger Hemmungsmechanismus scheint nun auch für die von GINSBERG, GOEBEL und HORSFALL (144, 146) beschriebenen *Polysaccharide* aus *bakteriellen Kapselsubstanzen* nachgewiesen. Nach den Untersuchungen dieser Autoren sind Hühnererythrocyten, die mit dem Polysaccharid aus FRIEDLÄNDER-Bazillen des Typus B vorbehandelt werden, vollständig inagglutinabel gegenüber Mumps- und Influenzavirus B und nur schwach agglutinabel gegenüber Influenzavirus A und dem Newcastle-Virus der Pseudopest. Bemerkenswerterweise konnte nun aber nur für das Mumpsvirus nachgewiesen werden, daß der Zellrezeptor „blockiert", d. h. auch der Adsorptions- und Elutionsvorgang aufgehoben ist. Das Polysaccharid hemmte daher die verschiedenen Virusarten in ungleicher Art und vermochte die haemagglutinierende Wirkung von Mäusepneumonievirus — wie Horsfall und McCarty (208) an vorbehandelten Mäuseerythrocyten feststellten — überhaupt nicht zu beeinflussen. Der Ausfall des Hemmungsversuches erwies sich auch abhängig von der Art der verwendeten Erythrocytenspezies; vorbehandelte Menschenerythrocyten wurden vom Influenzavirus A und B in unverminderter Stärke agglutiniert. Das hemmende Prinzip wurde nur in der Kapselsubstanz B und deren Derivate (durch Alkali degradiertes, bzw. mit Perjodat oxydiertes Polysaccharid) nachgewiesen, fehlte jedoch in der Kapselsubstanz von Friedländer-Bazillen des Typus A und C. Eine direkte Wirkung auf das Haemagglutinin des Mumpsvirus vermochten Ginsberg, Goebel und Horsfall mit Polysaccharidkonzentrationen, die für die Erythrocytenveränderung ausreichten, nicht festzustellen.

Nach Green (151) hemmt *Tanninsäure* in kleinsten Mengen (5 bis 20 γ/1 cc) die Haemagglutination von Influenzavirus und wirkt in etwas größerer Konzentration (45 γ/1 cc) selbst haemagglutinierend, in noch größerer Konzentration (0,2 mg/1 cc) virusinaktivierend. Für den in vitro beobachteten Hemmungseffekt entscheidend ist wohl auch hier die Erythrocytenveränderung[1].

Spezifische Hemmung durch Immunserum. Hirst (191, 194) und McClelland und Hare (246) stellten schon in ihren ersten Arbeiten fest, daß die Haemagglutination der Influenzavirustypen A und B durch den Zusatz korrespondierender Immunsera in typspezifischer Weise aufgehoben, bzw. gehemmt wird, und daß der Agglutinationshemmungstiter dem in vivo ermittelten Neutralisationsvermögen eines Immunserums weitgehend entspricht und somit eine Antikörpertitration in vitro möglich ist. Dieser Befund wurde nicht nur für Influenzavirus (39, 71, 30, 148, 186), sondern auch für das Newcastle-Virus (38), klassische Hühnerpestvirus (239, 159), Mumpsvirus (233, 21), Vaccinevirus (264, 265), Ektromelievirus (58) und das Virus der Mäusepneumonie (81) bestätigt. Fernerhin erlaubt die Hemmungsreaktion — unter Verwendung bekannter Testsera — Virusarten serologisch zu differenzieren (38, 45, 58, 159, 266) und selbst feinere antigene Unterschiede bei Virusstämmen derselben Art aufzudecken (194, 148, 61, 71, 209, 272, 197, 147, 133).

Mit der zunehmenden Anwendung der Haemagglutinationshemmungsreaktion wurde offensichtlich, daß dieser Test von zahlreichen *Variablen* abhängig ist, deren Nichtbeachtung die Zuverlässigkeit der Antikörperauswertung und der serologischen Antigenanalyse stark beeinträchtigt. Alle drei Reaktionskomponenten (Serum, Virus, Erythrocyten) sind, wie nachfolgend gezeigt werden soll, als Variable zu betrachten. So ist die virusbindende Potenz des *Immunserums* nicht nur abhängig von der Konzentration spezifischer Antikörper, sondern auch vom wechselnden Gehalt an thermolabilen „Normalglobulinen" (vgl. S. 175) und thermostabilen Inhibitoren receptorähnlicher Qualität (vgl. S.172—174). Die Möglichkeit der Interferenz zwischen Antikörper und Inhibitoren ist daher gegeben, wobei nun allerdings zu berücksichtigen ist, daß die Bindungsaffinität des Antikörpers ungleich größer ist als diejenige des Inhibitors. Die unspezifische Hemmung macht sich daher vor allem bei Seren von niederem Antikörpergehalt störend bemerkbar, kann dagegen in hochwertigen Immunseren durch die vorgängige Inaktivierung des Serums, Verwendung selektionierter Erythrocyten, bestimmter Virusstämme und die Anwendung einer besonderen Auswertungstechnik (vgl. S. 199) weitgehend ausgeschaltet oder zumindest

---

[1] Nach Olitsky und Cox verbindet sich Tanninsäure mit Proteinen und verleiht denselben eine hohe Resistenz gegenüber proteolytischen Fermenten (cit. nach Green).

von der spezifischen Hemmung differenziert werden [vgl. ANDERSON, BURNET und STONE (6)].

Fernerhin wird der Ausfall des Hemmungstestes durch die *individuelle Qualität der Erythrocyten* in hohem Grade beeinflußt. Nach HIRST und PICKELS (203), STUART-HARRIS (319) und HIRST (197) schwankt der Hemmungstiter desselben Serums gegenüber dem gleichen Testvirus erheblich, wenn verschiedene Proben von Hühnererythrocyten verwendet werden, und sind daher reproduzierbare Werte nur mit Erythrocyten derselben Provenienz zu erreichen. Da im direkten Agglutinationsversuch die individuelle Qualität der Erythrocyten die Höhe des Haemagglutinationstiters zwar ebenfalls beeinflußt (259), aber in einem weit geringeren Grade, liegt die Annahme nahe, daß ein derartiger Effekt nur in Gegenwart von Serum in prononcierter Form hervortritt. Von ANDERSON, BURNET und STONE (6) und ANDERSON (2, 4) wurde nun auch tatsächlich nachgewiesen, daß die unspezifische Hemmung von Normalserum und — in vermindertem Maße — von Immunseren je nach der Individualität der verwendeten Erythrocytenproben ungleich groß ist. Mit der Großzahl der Erythrocyten konnte ein hoher unspezifischer Hemmungstiter nachgewiesen werden, und nur wenige Erythrocytenproben — bemerkenswerterweise diejenigen, die im direkten Agglutinationsversuch mit dem Virus am stärksten reagierten — ergaben minimale unspezifische Titer. Nach BURNET und Mitarbeitern ist daher die Auswahl geeigneter Erythrocyten für die Spezifität des Hemmungstestes von ausschlaggebender Bedeutung, da nur die Verwendung von Erythrocyten großer „Avidität" (Bindungsaffinität für Virus) die weitgehende Ausschaltung unspezifischer Reaktionen erlaubt. Bei frisch isolierten Virusstämmen, die erfahrungsgemäß besonders leicht in unspezifischer Weise gehemmt werden (vgl. unten), erscheint eine solche Selektion der Erythrocyten unerläßlich (6, 2, 320, 99). Den Einfluß der *Zelldichte* auf den Hemmungstiter bestimmten HIRST (194) und SALK (282). Nach HIRST nimmt der Hemmungstiter in absteigenden Erythrocytenkonzentrationen ab, ein Befund, der mit der Feststellung von SALK, wonach das Verhältnis von Haemagglutinintiter zur Dichte der Zellsuspension invers proportional ist, in Einklang steht. Wie sich SALK fernerhin überzeugte, ist der Hemmungstiter eines Serums stets derselbe, wenn zwischen Viruskonzentration und Zelldichte konstante Proportionen eingehalten werden.

Die größte Variable stellt zweifellos das *Virus* selbst dar. So machte HIRST (197) die überraschende Beobachtung, daß der gleiche Influenzastamm (PR 8) derselben Eipassage und in konstanter Gebrauchsdosis (vier agglutinierende Einheiten) — je nach seiner Provenienz (verschiedene Allantoisflüssigkeiten) — vom gleichen Immunserum in sehr verschiedenem Titer gehemmt wurde. Ähnliche Unterschiede in der Hemmbarkeit durch Immunserum zeigten Influenza-A-Stämme, die zu gleicher Zeit aus einer geschlossenen menschlichen Gemeinschaft isoliert wurden. Da antigene Differenzen unter den gegebenen Verhältnissen kaum anzunehmen waren, ist HIRST geneigt, die beobachteten Divergenzen auf Unterschiede in der „*Avidität*" der verschiedenen Virusproben gegenüber dem Zellreceptor zurückzuführen, wobei ein „avides Virus" dadurch gekennzeichnet wäre, daß zu seiner Hemmung eine relativ große Serummenge erforderlich ist. HIRST vermochte fernerhin in überzeugender Art nachzuweisen, daß derartige Aviditätsunterschiede auch bei der Antigenanalyse im gekreuzten Hemmungstest eine namhafte Rolle spielen und für paradoxe Befunde (z. B. stärkere Hemmung von heterologen Stämmen) verantwortlich sind. Durch die Einführung eines ad hoc zu bestimmenden „Aviditätsfaktors" gelingt es nach HIRST, diese Variabilität zu korrigieren und als Fehlerquelle auszuschalten. Ein sehr unregelmäßiges Verhalten im Hemmungstest zeigen nun bekanntlich auch *frisch isolierte Influenzavirusstämme*, für die im allgemeinen charakteristisch ist, daß sie mit heterologen Immunseren und Normalseren leicht reagieren, d. h. in unspezifischer Weise gehemmt werden [Anderson Burnet und Stone (6), Anderson (2), Francis, Salk und Brace (128), Dudgeon, Stuart-Harris, Glover, Andrewes und Bradley (100)]. Nach der Konzeption von Hirst würden derartige Stämme eine schwache Avidität aufweisen. Immerhin wurde auch das Gegenteil beobachtet; so verfügten Stuart-Harris und Miller (320) über recente Influenza-A-Stämme, die im allgemeinen durch die homologen menschlichen Rekonvalescentensera weniger leicht gehemmt wurden, als stammverwandte alte Passagestämme. Fernerhin bestand zwischen dem Hemmungstiter in vitro und dem in vivo ermittelten Neutralisationswert häufig keine Übereinstimmung.

Der von Hirst eingeführte Aviditätsbegriff charakterisiert den Tatbestand, gibt hierfür jedoch keine Erklärung. Hirst hält es für möglich, daß der Dispersitätsgrad der Virussuspension, bzw. das Ausmaß der Bildung von Virusaggregaten in der Allantoisflüssigkeit [vgl. Chambers und Henle (72)] die Avidität bestimmen könnte. Wahrscheinlicher erscheint die Annahme, daß die von Hardy und Horsfall (166) für das Influenzavirus der Allantoisflüssigkeit beschriebenen Zustandsformen (freies Virus, gebundenes Virus, alterierter Inhibitor) eine Rolle spielen (vgl. S. 174, 183), da der von diesen Autoren angenommene Gleichgewichtszustand zwischen diesen Komponenten durch den im Hemmungstest erfolgenden Zusatz von Seruminhibitor zwangsläufig und in unterschiedlichem Grade verschoben werden müßte.

Hirst (199, 200) untersuchte mit Hilfe des Hemmungstestes auch die *antigene Variabilität* von Influenzavirusstämmen und vermochte hierbei festzustellen, daß menschliche Stämme, die im Hühnerei isoliert und ausschließlich nur in Eipassagen weitergeführt werden, ihre ursprüngliche Spezifität unverändert beibehalten, während dieselben Stämme nach der Isolierung und Passage in Frettchen und Mäusen serologisch nachweisbare Antigenveränderungen erleiden. Die Vorgeschichte eines Virusstammes ist demnach bei der Antigenanalyse im Haemagglutininhemmungsversuch keineswegs gleichgültig, da das serologische Verhalten von Frettchen- und Mäusepassagestämmen keine Rückschlüsse auf die ursprüngliche Spezifität gestattet. Für epidemiologische Studien sind nur Eipassagestämme geeignet, die nun auch nach Hirst (200) für jede Epidemie eine große Homogenität aufweisen und selbst dann — wie Hirst an A- und B-Stämmen aus den Jahren 1940/41, 1943/44 und 1945/46 nachwies — keine größeren Abweichungen zeigen, wenn Stämme verschiedener Epidemien miteinander verglichen werden.

Zusammenfassend ergibt sich die Feststellung, daß der Haemagglutininhemmungstest ein kompliziertes Reaktionssystem darstellt, an welchem nicht nur die drei Hauptkomponenten (Virus, Immunserum, Erythrocyten) beteiligt sind, sondern außerdem noch eine Reihe von Inhibitoren, wie das „unspezifische Serumglobulin" und vor allem die im Serum und in der Allantoisflüssigkeit vorhandenen Hemmstoffe receptorähnlicher Natur. Bis zu einem gewissen Grade kann der Reaktionsablauf in „spezifischer" Richtung gelenkt werden, nämlich, einerseits durch die Einschränkung, bzw. Ausschaltung des Inhibitoreneinflusses (Inaktivierung und Gewinnung des Immunserums von bestimmten Tierspezies[1]), und anderseits durch

---

[1] Für den Hemmungstest sind *Hühnerimmunsera* in hohem Grade geeignet, da derartige Sera nicht nur leicht zu gewinnen sind und eine hohe Wertigkeit und Spezifität aufweisen, sondern auch durch den Mangel an Antikörpern gegenüber Hühnereiweiß und unspezifischen Hemmstoffen ausgezeichnet sind [Hudson, Sigel und Markham (209), Hallauer (161, 164)].

*die Auswahl von „aviden" Erythrocyten, Virusstämmen mit hohem Agglutinations-
titer und hochwertigen Immunseren. Für die serologische Differenzierung von
Virusstämmen besteht außerdem die Notwendigkeit, Aviditätsunterschiede zu
korrigieren und ausschließlich nur im Hühnerei isolierte und passierte Stämme zu
verwenden (vgl. das Kapitel über Technik S. 201).*

5. Die Fermenthypothese.

HIRST (195) stellte als erster die Fermentnatur des Haemagglutinins zur
Diskussion, indem er darauf hinwies, daß die Reaktion zwischen *Influenzavirus*
und Erythrocyten insofern die Kriterien eines katalytischen Vorganges aufweist, als das Haemagglutinin nach Art eines Enzyms an den Zellreceptor gebunden wird, den Receptor zerstört und hierbei wieder in unveränderter und unver-
brauchter Form in Freiheit gesetzt wird. Die Rezeptorveränderung bei gleichzeitiger Viruselution ist nun auch, wie von BURNET und Mitarbeitern (42, 43,
44, 69) gezeigt wurde, charakteristisch für das *Mumps-* und *Pseudopestvirus*,
so daß auch für diese Virusarten enzymatisch wirksame Haemagglutinine zu
postulieren wären. Dagegen hätte die Fermenthypothese keine Geltung für das
Vaccine-, Ektromelie- und *Mäusepneumonievirus*, bei welchen eine spontane
Dissociation des Haemagglutinins nicht zu beobachten ist, und auch — wie dies
mit Sicherheit für das Pneumonievirus feststeht — der Receptor nicht alteriert
wird.

Die Schwierigkeiten, den enzymatischen Charakter von Virushaemagglutininen
in eindeutiger Art nachzuweisen, sind nun allerdings beträchtlich und wohl
auch solange nicht zu meistern, als die üblichen Methoden der fermentchemischen
Analyse nicht angewendet werden können. Die Isolierung des Haemagglutinins
ist bekanntlich bisher nicht gelungen und seine proteide Beschaffenheit[1] läßt sich
lediglich auf Grund des inaktivierenden Effektes eiweißdenaturierender Mittel
(Hitze, Formol, proteolytische Fermente) vermuten. Ebenso ist nicht bekannt,
an welche chemischen Strukturen der Zellreceptor in den ermittelten Mucopolysacchariden gebunden ist, so daß eine chemische Analyse der Receptorveränderung und auftretender Spaltprodukte einstweilen nicht möglich ist. Die
Beweisführung mußte sich daher auf die Anwendung biologischer Methoden
beschränken und führte dessenungeachtet zu Ergebnissen, die in ihrer Gesamtheit die enzymatische Hypothese als wohl fundiert erscheinen lassen.

Zunächst ist die Feststellung von Interesse, daß die *„enzymatische Reaktion"*
an bestimmte *Bedingungen* geknüpft ist, nämlich an die vollerhaltene Virusaktivität einerseits und den intakten Receptor anderseits, und sich außerdem
in hohem Grade als abhängig erweist von der Temperatur und den vorhandenen
Milieuverhältnissen. Nur aktive, noch infektiöse Virusproben verändern den
Zellreceptor der Erythrocyten oder inaktivieren den mucopolysacchariden
Inhibitor unter gleichzeitiger Wiederabspaltung von Virus, bzw. Haemagglutinin
(195, 69, 201, 202, 4, 51, 227, 166), und der Ablauf dieser Reaktion ist an höhere
Temperaturen (195, 69, 35, 51), höhere Salzkonzentrationen (84, 28), Anwesenheit bestimmter Ionen (35, 51) und bestimmte Wasserstoffionenkonzentrationen
(35, 51) weitgehend gebunden. Bei partiell inaktivierten Virusproben ist dagegen

---

[1] Im Gegensatz zum Influenza-, Mumps-, Geflügelpest- und Pneumonievirus
dürfte für das Haemagglutinin des Vaccine- und Ektromelievirus ein P-haltiges
Lipoid als Träger der agglutinierenden Aktivität in Frage kommen [BURNET und
STONE (63)], wobei nahe Beziehungen zu einer Reihe haemagglutinierender „Normal-
lipoide" nachgewiesen werden konnten [STONE (312)].

wohl die Bindungsaffinität zum Zellreceptor, bzw. Inhibitor noch vollerhalten, so daß auch das Phänomen der Haemagglutination, bzw. der Haemagglutininhemmung in Erscheinung tritt, die „enzymatische Reaktion" (Receptorinaktivierung, Viruselution) kommt jedoch nicht zustande (195, 202, 35, 4, 49, 226, 166, 303). Derselbe Blockierungseffekt wird erreicht durch die Vorbehandlung des Erythrocytenreceptors, bzw. des mucoiden Inhibitors mit partiell inaktivierenden Dosen von Perjodat; in diesem Fall wird auch vollaktives Virus sowohl von den Erythrocyten als auch vom Inhibitor in irreversibler Art gebunden (108, 50), und verhält sich gegenüber dem Receptor inaktiv, obschon sämtliche Virusfunktionen erhalten sind. Die Interpretation dieses Funktionsausfalles ist — im Sinne der Fermenthypothese — einfach: im einen Fall wird das labile Ferment bei der Virusinaktivierung zerstört, im andern ist das modifizierte Receptorsubstrat der fermentativen Spaltung nicht mehr zugänglich. Schwer verständlich ist jedoch die Dissociation zwischen dem Bindungsvorgang, bzw. der Haemagglutination und der „enzymatischen Reaktion", die sowohl von seiten des Virus als auch des Receptors nachgewiesen ist. Jedenfalls wäre man zur Vorstellung genötigt, daß das Virushaemagglutinin nicht einheitlich ist, sondern aus einem thermostabilen, bindungsfähigen und haemagglutinierenden Anteil und einer „fermentierenden" Komponente bestehen würde, der man ebenfalls bindende Eigenschaften zubilligen müßte [vgl. ANDERSON (3, 4) und FAZEKAS DE ST. GROTH (108)].

Die Annahme von HIRST, daß der Erythrocytenreceptor unter der Einwirkung von Virus „enzymatisch abgebaut, bzw. zerstört" wird, ergab sich aus der Gesamtheit des beobachteten Reaktionsverlaufes und konnte zunächst nur als Arbeitshypothese gewertet werden. An dieser Sachlage hätte sich wohl nichts geändert, wenn nicht von BURNET und Mitarbeitern (69) das Phänomen von Thomsen und Friedenreich (vgl. S. 170) bei virusvorbehandelten Erythrocyten nachgewiesen und zugleich der überraschende Befund erhoben worden wäre, daß mit bestimmten bakteriellen Kulturfiltraten prinzipiell dieselben Erythrocytenveränderungen erzeugt werden können, wie mit Influenza-, Mumps- und Pseudopestvirus. Als wirksam erwiesen sich Kulturfiltrate von V. cholerae (313, 66) und das α-Toxin von Cl. Welchii Typ A (248)[1]. Das meist verwendete Cholerakulturfiltrat wurde nicht nur in Analogie zum Virus von Erythrocyten adsorbiert und eluiert, sondern bewirkte hierbei die Inagglutinabilität gegenüber Virus (69, 313), Panagglutinabilität gegenüber Normalseren (69, 313) und Agglutinabilität gegenüber inkompletten Rhesusantikörpern (274). Daß de facto derselbe Receptor beeinflußt wird wie durch das Virus, ergab sich im gekreuzten Adsorptions- und Blockierungsversuch (313) sowie aus der Feststellung, daß festgebundenes Virus durch das Filtrat vom Zellreceptor abgespalten werden kann (1, 35), und daß das bakterielle Agens auch mucoide Inhibitoren zu inaktivieren vermag (68, 4, 49, 51, 52)[2, 3]. Fernerhin zeigt das gereinigte Präparat gegenüber Erwärmung, Formol und Trypsin ein ähnliches Verhalten wie das Virushaemagglutinin und ist in seiner Wirksamkeit auch von ähnlichen Faktoren abhängig wie das Virus (66). Die Übereinstimmung zwischen dem aus Choleravibrionen gewonnenen Agens und dem Virushaemagglutinin ist demnach —

---

[1] Ebenso, wenn auch in vermindertem Grade, *Blutegelextrakte* [BRIODY (35)].

[2] Nach WHITTEN [cit. nach BURNET (52)] wird auch das *gonadotrope Hormon* sowohl vom Influenzavirus (Lee) als auch Cholerakulturfiltrat rasch und vollständig inaktiviert.

[3] Dieselbe Wirkung zeigt nach SVEDMYR (303) das Filtrat von CL. WELCHII, das ebenfalls den Inhibitor inaktiviert und das an den Inhibitor gebundene Virus liberiert.

mit Ausnahme des haemagglutinierenden Vermögens — eine nahezu vollständige. BURNET und STONE (66) bezeichneten dieses Agens als *„receptor-destroying enzyme"* (RDE). Die ursprüngliche Vermutung, daß das receptorzerstörende Agens durch eine *Lecithinase* repräsentiert wird, konnte weder für das α-Toxin von CL. WELCHII noch für das Cholerakulturfiltrat bestätigt werden (248). Mit dem Nachweis, daß das gereinigte Cholerakulturfiltrat — ebenso wie das Virus — mucinartige Substanzen angreift und inaktiviert, wurde beiden der Charakter einer *Mucinase* zugewiesen, die mit einer ebenfalls von Choleravibrionen produzierten „epitheldesquamierenden" Mucinase jedoch nicht identisch ist. Beide Mucinasen wirken nur auf das echte Mucin der Schleimdrüsen und unterscheiden sich hierdurch von der Hyaluronidase (65, 48, 68, 52). Das große Interesse, das diesen Untersuchungen zukommt, soll nicht bestritten werden, immerhin wurde der angestrebte Beweis, wonach das bakterielle Agens — und eo ipso auch das Virushaemagglutinin — ein Enzym ist, in stringenter Art nicht geleistet; jedenfalls ist es bisher nicht gelungen, die receptortransformierende Mucinase in fermentchemischem Sinne zu identifizieren.

Ob der Zellreceptor unter der Einwirkung von Virus, bzw. bakteriellem Kulturfiltrat abgebaut, zerstört oder transformiert wird, ist ungewiß. Auch kann das Ausmaß der Receptorveränderung an Erythrocyten nur auf indirekte Art, nämlich durch die Prüfung der noch vorhandenen Reaktionsfähigkeit gegenüber den Virusstämmen des „Receptorgradients" (vgl. S. 168), abgeschätzt werden. Dagegen besteht die Möglichkeit, die „enzymatische Aktivität" in *gelösten Receptorsubstanzen* (Serum, gereinigtes Mucin, Eiklar, normale Allantoisflüssigkeit, Erythrocytenextrakt) zeitlich zu verfolgen und auch zu messen (202, 4, 49, 51, 227, 166). Immerhin erlaubt auch diese Methode einstweilen nur die Feststellung, daß der Inhibitor eine progressive Einbuße an Wirksamkeit erleidet und sich schließlich dem Nachweis entzieht; durch welche Vorgänge der Inhibitor inaktiviert wird, ist auch in diesem Fall unbekannt. Die Beobachtung von BURNET (49), wonach ein Mucinsubstrat vor und nach der Inaktivierung dieselbe Viskosität aufweist, würde vermuten lassen, daß der Receptor wohl kaum in tiefgreifender Art verändert wird. Der Einfluß der Virus-, bzw. Inhibitorkonzentration auf den Reaktionsablauf wurde nur unzureichend untersucht, so daß weittragende Schlüsse hinsichtlich der Reaktionskinetik nicht möglich sind. BURNET (51) gelangte unter Verwendung eines gereinigten Mucopolysaccharides und bestimmter Influenzavirusstämme zu den folgenden Feststellungen: 1. Die Geschwindigkeit der Reaktion ist während der ganzen Meßperiode konstant und der Viruskonzentration proportional; 2. kleine Substratmengen beschleunigen, große verlangsamen den Reaktionsablauf, und 3. die Inaktivierung des Inhibitors schreitet in „logarithmischem Rhythmus" bis zur Vollständigkeit fort. Mit anderen Virusstämmen konnte eine derartige gleichmäßig und vollständig verlaufende Inaktivierung nicht beobachtet werden. Von besonderem Interesse sind die Untersuchungen von HARDY und HORSFALL (166), die nicht nur die Abnahme des Inhibitors (Allantoisflüssigkeit), sondern auch die Zunahme des sich wiederabspaltenden Virus (Influenzastamm PR 8) im Reaktionsgemisch fortlaufend bestimmten. Die Geschwindigkeit der Inhibitorinaktivierung, bzw. der koordinierten Virusliberierung erwies sich ebenfalls abhängig von der Virusausgangskonzentration, wobei die Reaktion stets einem Maximum zustrebte, in welchem der Inhibitor komplett inaktiviert war und das abdissoziierte Virus den Höchsttiter erreicht hatte. Die Viruselution war nun aber niemals vollständig, sondern eine mehr oder weniger große Quote (50 bis 100 %) des Virus blieb mit dem Inhibitor verbunden. Die Menge des freien, bzw. gebundenen Virus stand jeweils in direkter Relation zur gesamten Viruskonzentration, bzw. zum

Verhältnis $\dfrac{\text{Viruskonzentration}}{\text{Inhibitorkonzentration}}$, so daß die Gleichgewichtsgleichung $V + J \rightleftarrows VJ$ angenommen werden konnte. Hardy und Horsfall hatten nun keineswegs die Absicht oder den Gedanken, ihre Versuchsergebnisse im Sinne der Fermenthypothese zu deuten. Nichtsdestoweniger zeigt das nachgewiesene Reaktionsgeschehen eine auffallende Ähnlichkeit mit der von Henri[1] ermittelten Reaktionsgleichung der fermentativen Rohrzuckerspaltung, bei welcher die Zerfallsgeschwindigkeit in jedem Augenblick der Konzentration der Saccharose-Invertin-Verbindung proportional ist, und die Konzentration dieser Verbindung in jedem Augenblick bestimmt wird durch die Konzentration des Fermentes, der Saccharose und der ebenfalls an das Ferment bindungsfähigen Spaltprodukte.

Die unvollständige Dissociation zwischen Virus und Inhibitor kontrastiert nach Hardy und Horsfall mit der meist vollständigen Viruselution aus Erythrocyten. Ob eine solche Gegensätzlichkeit besteht, ist nun aber fraglich, da auch die Reaktion mit Erythrocyten in zahlreichen Fällen nur unvollständig abläuft und eine „vollständige" Viruselution sehr leicht durch Potenzierungseffekte im Sinne von Doerr und Mitarbeitern (97, 98) vorgetäuscht werden kann. Die Unerschöpflichkeit der Haemagglutininwirkung, die von Hirst (195) mit dem Influenzavirusstamm Lee demonstriert wurde, steht demnach nicht völlig außer Zweifel. Eine enzymatische Reaktion wäre aber auch dann nicht ausgeschlossen, wenn nachgewiesen würde, daß die Reaktion de facto unvollständig ist und daher nicht ad infinitum fortgesetzt werden kann.

Zugunsten der Fermenthypothese könnte schließlich auch angeführt werden, daß kleine Haemagglutininmengen relativ große Inhibitorkonzentrationen umzusetzen vermögen; so wurde von Lanni und Beard (227) berechnet, daß eine haemagglutinierende Einheit (entsprechend $10^{-7.6}$ g Virus) eine Inhibitormenge inaktiviert, die zur Hemmung von 100 haemagglutinierenden Einheiten ausreicht. Nach Lanni und Beard erlaubt diese Disproportion keine stöchiometrische Erklärung des Reaktionsablaufes.

6. Das Verhalten des Virushaemagglutinins in vivo.

Reaktion mit den Wirtszellen. Die Vermutung von Hirst (195), daß die Haemagglutination im Infektionsvorgang ihr Gegenstück finden könnte, hätte — im Falle der Bestätigung — dem Phänomen eine höchst bedeutungsvolle Ausweitung verliehen. Die engen Beziehungen zwischen haemagglutinierender und infektiöser Funktion einerseits und der Nachweis von receptorähnlichen Substanzen außerhalb von Erythrocyten schienen a priori die Annahme zuzulassen, daß die erste Phase der Infektion durch einen ähnlichen „enzymatischen" Bindungsvorgang zwischen Virus und geweblichem Receptor gekennzeichnet sein könnte. Fernerhin erschien es nicht ausgeschlossen, daß die im Wirtsorganismus nachgewiesenen „Interferenzphänomene" zwischen immunologisch gleich- oder verschiedenartigen Virusinfektionen [vgl. Ziegler und Horsfall (347), Delbrück (89), Henle und Henle (172) Vilches und Hirst (329)] auf einer Konkurrenz um Zellreceptoren beruhen könnten, in Analogie zum refraktären Verhalten virusvorbehandelter Erythrocyten gegenüber dem nochmaligen Kontakt mit Virus oder zur Haemagglutinationshemmung durch Inhibitoren von receptorartigem Charakter.

Den *Bindungsvorgang* zwischen Influenzavirus und den Epithelien der Frettchenlunge, Mäuselunge und der Allantoismembran von Hühnerembryonen

---

[1] V. Henri, Lois générales de l'action des diastases, Paris 1903. Vgl. auch Michaelis und Menten, Biochem. Z. **49**, 333 (1913).

analysierten HIRST (196), FAZEKAS DE ST. GROTH (106) und STONE (314, 315) und gelangten hierbei zu konkordanten Ergebnissen. Von den Zellen der excidierten Lungenpräparate, bzw. der in situ belassenen (leicht formolisierten) Allantoismembran wird Influenzavirus rasch und nahezu vollständig adsorbiert und nach einiger Zeit auch wieder eluiert, wobei die gleichen Stammeseigentümlichkeiten wie bei der Reaktion mit Erythrocyten zutage treten. Partiell inaktiviertes Virus wird nur adsorbiert, jedoch nicht mehr liberiert. Die Übereinstimmung mit dem Adsorptions- und Elutionsvorgang an Erythrocyten war demnach eine vollkommene. In vivo, d. h. mit vollvitalem, zur Virusvermehrung befähigtem Gewebe konnte dagegen *nur* die Adsorptionsphase beobachtet werden und freies Virus war erst in der Phase der Virusvermehrung nachweisbar. Anzeichen einer „enzymatischen Reaktion" waren demnach nicht vorhanden. Daß enzymatische Vorgänge in der ersten Phase der Zellinfektion kaum eine Rolle spielen dürften, scheint nun auch aus den Untersuchungen von FAZEKAS DE ST. GROTH und GRAHAM (109) hervorzugehen, in welchen der Nachweis geleistet wurde, daß mit Perjodat modifizierte Epithelzellen (Mäuselunge, Allantoismembran) als Receptoren geeignet sind, d. h. die Zellinfektion in ungestörter Art vermitteln. Unter diesen Umständen erscheint die Hypothese, wonach die „fermentative Receptorenstörung" eine notwendige Phase für die Zellinfektion darstellt, wenig begründet.

Im Gegensatz hierzu stehen nun aber die Versuchsergebnisse, die STONE (314, 315) und FAZEKAS DE ST. GROTH (106, 107) mit dem receptorzerstörenden Ferment (RDE) von Choleravibrionen in vivo erzielten, allerdings nur, wenn man sich der von diesen Autoren gegebenen Interpretation anschließt. Unter künstlichen Versuchsbedingungen (excidierte Lungen, formolisierte Allantoismembran) hatte das RDE-Präparat denselben Effekt wie gegenüber Erythrocyten, d. h. veränderte die Zellreceptoren derart, daß eine nachfolgende Adsorption von Virus nicht mehr möglich war, und vermochte außerdem das bereits an den Zellreceptor gebundene Virus wieder abzuspalten. Ähnliche in vivo durchgeführte Versuche ergaben eine weniger eindrucksvolle Receptorzerstörung und nötigten zur Annahme, daß der gewebliche Receptor relativ rasch „regeneriert" wird [FAZEKAS DE ST. GROTH (107)]. Den stringenten Beweis für die receptorzerstörende Wirkung von RDE hält STONE dadurch geleistet, daß die Vorbehandlung der Allantoismembran mit RDE die nachfolgende Infektion mit verschiedenen Influenzavirusstämmen in einem höheren Prozentsatz verhinderte oder zumindest verzögerte. Die für die Schutzwirkung erforderliche RDE-Menge war nun aber unverhältnismäßig groß und der Grad des verliehenen Schutzes variierte für die verschiedenen Virusstämme beträchtlich, und zwar unabhängig von deren Stellung im „Gradient". Schließlich wurde in diesen Interferenzversuchen eine Receptorveränderung de facto überhaupt nicht nachgewiesen, sondern deren Bestehen lediglich aus dem refraktären Verhalten vorbehandelter Rezeptoren in künstlichen Präparationen abgeleitet, ein Vorgehen, das nun aber nicht zulässig ist, da anscheinend — wie übrigens STONE und FAZEKAS DE ST. GROTH selbst feststellten — nur Gewebezellen von geschädigter Vitalität mit dem Virus nach Art der Erythrocyten reagieren. FLORMAN (121) vermochte nun auch in eindeutiger Art nachzuweisen, daß aus den festgestellten Interferenzen an Erythrocyten keineswegs auf ein ähnliches Verhalten von interferierenden Virusinfektionen im Wirtsorganismus geschlossen werden kann. Beiden Phänomenen liegt nach FLORMAN nicht derselbe Mechanismus zugrunde. Die grundsätzliche Verschiedenartigkeit der Interferenz im Haemagglutinations-, bzw. Infektionsversuch ergibt sich fernerhin aus den Untersuchungen von HENLE und Mitarbeitern (172, 173, 174, 175, 184, 185, 180), aus welchen hervorgeht:

1. daß das interferierende Vermögen von Influenzavirus gegenüber der Zweit-
infektion schon durch kurze U.-V.-Bestrahlung (3 bis 5 Minuten) verloren geht,
während die Adsorption von Virus an Erythrocyten oder Allantoiszellen erst
bei einer weit längeren Expositionszeit (60 bis 120 Minuten) aufgehoben wird (180);
2. daß resistent gewordene Wirtszellen noch weiterhin Virus zu adsorbieren
vermögen, und daher der Zellreceptor nicht blockiert sein kann (184), und 3. daß
das interferierende (homologe) Virus auch dann noch einen inhibierenden Einfluß
ausübt, wenn die Adsorption des erstinfizierenden Virus an die Zellreceptoren
schon längst stattgefunden hat (185). Nach Henle und Mitarbeitern (185, 184)
dürfte die Viruskonkurrenz in vivo einen ähnlichen Mechanismus aufweisen
wie die von Delbrück (89) analysierten Interferenzphänomene bei Bakteriophagen.

Einige *Polysaccharide pflanzlicher und bakterieller* Provenienz hemmen Virusarten sowohl in vitro als auch in vivo. Ein solches Verhalten wurde von Green
und Woolley (152, 346) für das Apfelpektin gegenüber Influenzavirus und von
Ginsberg, Goebel und Horsfall (144, 146) für das Kapselpolysaccharid von
Friedländer-Bazillen des Typus B gegenüber Mumpsvirus nachgewiesen. Auch
in diesem Fall wird jedoch — wie von Ginsberg und Mitarbeitern (146) für das
Friedländer-Polysaccharid festgestellt wurde — die Haemagglutination auf eine
andere Art gehemmt als die Infektion; im einen Fall (Haemagglutination) ist
die Virusadsorption an den Zellreceptor verhindert, im andern (Infektion) spielt
eine derartige Receptorblockade keine Rolle, sondern interferiert das Polysaccharid in der Phase der Virusvermehrung. Die Polysaccharidhemmung in
vivo (Hühnerei) dürfte nach den Untersuchungen von Ginsberg und Horsfall
(142) auch einen ähnlichen Mechanismus aufweisen wie die eigentlichen Interferenzphänomene, bei welchen ja ebenfalls das inhibierende Virus nicht notwendigerweise infektiöse Qualität haben muß. Im übrigen sind die erwähnten
Beispiele von Polysacchariden, die zugleich in vitro und in vivo hemmen, als
Ausnahmen zu betrachten. So hemmt die Kapselsubstanz von Friedländer-
Bazillen des Typus B zwar die Infektion des Pneumonievirus in der Mäuselunge,
ohne jedoch die haemagglutinierenden Eigenschaften dieses Virus aufzuheben
[Horsfall und McCarty (208)]. Denselben Effekt besitzen die Polysaccharide
von Friedländer-Bazillen des Typus A und C gegenüber Mumpsvirus (144, 146)
Fernerhin kann durch chemische Eingriffe die hemmende Wirkung der Kapselsubstanz B gegenüber Mumpsvirus in einseitiger Weise verändert werden, derart,
daß im einen Fall (Perjodat, Ribonucleinsäure) nur der haemagglutininhemmende
Effekt, im andern (Alkali) nur das interferierende Vermögen unterbunden wird
(146). Eine irgendwelche Parallelität zwischen der antagonistischen Wirksamkeit
dieser Inhibitoren im Haemagglutinations- und Interferenzversuch besteht
demnach nicht.

*Zusammenfassend dürfte feststehen, daß die Zellen virusempfänglicher Gewebe
nur unter künstlichen Versuchsbedingungen, unter welchen eine Virusvermehrung
nicht stattfindet, nach Art von Erythrocyten mit dem Virus, bzw. Haemagglutinin
reagieren. In der ersten Phase der Infektion wird das Virus — erwartungsgemäß —
von der Wirtszelle adsorbiert, wobei die virusbindenden Zellreceptoren möglicherweise mit denjenigen der Erythrocyten identisch sind. Eine „enzymatische Reaktion"
als notwendige Vorbedingung für die Zellinfektion konnte jedoch nicht beobachtet
werden. Ebenso fehlen sämtliche Anzeichen dafür, daß die im Wirtsorganismus
beobachteten Interferenzphänomene durch die Blockierung oder den Verlust von
Zellreceptoren verursacht sind. Das Phänomen von Hirst ist daher — nach den
bisher vorliegenden Untersuchungen — kein geeignetes Reaktionsmodell für das
Verständnis des Infektionsvorganges.*

Reaktion mit Erythrocyten. Die Frage der intravitalen Haemagglutination wurde bisher weder gestellt noch untersucht. Bei der *Hühnerpest* wäre diese Möglichkeit gegeben, da das Virus in der Blutbahn eine hohe Konzentration erreicht und die Erythrocyten des Wirtes reaktionsfähig sind. Von DOERR und Mitarbeitern (97, 98) wurde auch nachgewiesen, daß der größte Teil des zirkulierenden Virus an den roten Blutzellen haftet, so daß die Reaktion zwischen Virus und Zellen wohl auch im kreisenden Blut stattfindet. Es ist daher auch keineswegs ausgeschlossen, daß intravasale Haemagglutinate gebildet werden, die nun aber wegen der hohen Körpertemperatur des Wirtes nur eine sehr lockere und unbeständige Beschaffenheit haben dürften. An Körperstellen, die der Abkühlung ausgesetzt sind (Kamm und Kehllappen) könnten jedoch Haemagglutinationseffekte auch klinisch hervortreten, und es wäre daher sehr wohl möglich, daß das immer noch ungeklärte Symptom der „Cyanophilie" bei der Hühnerpest eine lokale Haemagglutination anzeigt. Allerdings ist es HALLAUER (164) nicht gelungen, durch die intravenöse Einverleibung von größeren Mengen (10 cc) eisgekühlter, influenzavirushaltiger Allantoisflüssigkeit dieses Symptom bei Hühnern künstlich zu erzeugen; derartige Injektionen wurden im allgemeinen anstandslos vertragen, bewirkten jedoch in Einzelfällen Schocktodesfälle, deren Genese indessen nicht geklärt werden konnte. In Extravasaten, Blutungen in Trans- und Exsudate, wurde jedoch die Virushaemagglutination — wie schon die Entdeckung des HIRSTschen Phänomens zeigt — wiederholt festgestellt, und zwar anscheinend auch bei der *lymphocytären Choriomeningitis der Mäuse* [WENNER (338)], bei welcher zwar bekannt ist, daß das Virus an den Erythrocyten des Wirtes zum großen Teil adsorbiert ist [SHWARTZMAN (293)], ein Haemagglutinin in vitro jedoch bisher nicht nachgewiesen werden konnte [HALLAUER (164)].

Von BURNET und ANDERSON (54) wurde die Frage geprüft, ob die Erythrocytenveränderung von THOMSEN und FRIEDENREICH, die in vitro mit bakteriellen Kulturfiltraten und Virusarten erzeugt werden kann und zur Bildung eines neuen Antigens („T-Receptor") führt, auch in vivo zustande kommt und zur Bildung von Autoantikörpern („T-Agglutininen") Anlaß gibt. Das Versuchsergebnis war jedoch — in Übereinstimmung mit früheren Befunden von HALLAUER (158) — negativ, da Meerschweinchen auf die intracardiale Einverleibung von Virus, „RDE" und transformierten arteigenen Erythrocyten mit der Bildung von Autoantikörpern nicht reagierten. Ebensowenig vermochten LIND und McARTHUR (236) den Nachweis zu leisten, daß die natürlichen T-Agglutinine des menschlichen Serums durch die Infektion mit Influenza- und Mumpsvirus eine Zunahme erfahren; der in Rekonvaleszentenseren von atypischer Pneumonie nachgewiesene erhöhte Titer konnte — in Analogie mit den Kälteagglutininen — ebenfalls nicht auf einen Autoimmunisierungseffekt bezogen werden.

Antigenes Vermögen. Immunsera gegen haemagglutinierende Virusarten enthalten neben virusneutralisierenden ausnahmslos auch agglutinationshemmende Antikörper. Das Haemagglutinin als integrierende Komponente der geformten Viruselemente besitzt demnach vollantigene Wirksamkeit. Über die antigene Qualität der „freien Haemagglutinine" des Vaccine- und Ektromelievirus liegen anscheinend noch keine Untersuchungen vor. Nur unzureichend geprüft sind auch die Beziehungen des Haemagglutinins zu anderen immunisierenden Antigenen, und ebenso ist die Frage nicht abgeklärt, ob der virusneutralisierende mit dem agglutinationshemmenden Antikörper identisch ist, oder ob zwei verschiedene Antikörperqualitäten angenommen werden müssen.

Nach STANLEY (308) ist der Haemagglutinationstiter eines Influenzaimpfstoffes ein praktisch brauchbarer, wenn auch grober Indikator für die immunisierende Potenz. Für die Bewertung der immunisierenden Wirksamkeit von frischeren Virusproben und Impfstoffen vor und unmittelbar nach der Inaktivierung ist die Haemagglutinationsprüfung wohl tatsächlich geeignet, da zumindest die Konzentration des Antigens und das Ausmaß der Inaktivierung, bzw. Denaturierung abgeschätzt werden kann. Angaben, wonach die immunisierende Qualität von der Höhe des Haemagglutinationstiters weitgehend unabhängig oder auch dann noch erhalten ist, wenn Haemagglutinine nicht mehr nachgewiesen werden können, beziehen sich entweder auf Virusstämme verschiedener Provenienz (124) oder auf längere Zeit gelagerte Virus- und Impfstoffproben (22, 25), bei welchen der ursprünglich vorhandene Haemagglutiningehalt nicht in Rechnung gestellt wird. Henle und Henle (180) prüften nun vergleichsweise die Resistenz des immunisierenden Antigens und des Haemagglutinins von Influenzavirus gegenüber der U.-V.-Bestrahlung und stellten hierbei — zumindest für einen Virusstamm — fest, daß das immunisierende Vermögen schon deutlich abnimmt, noch bevor das Haemagglutinin verändert wird, jedoch im Zeitpunkt der völligen Haemagglutinininaktivierung noch immer vorhanden ist. Die Unabhängigkeit des Haemagglutinins von den übrigen antigenen Funktionen ist hierdurch nicht bewiesen, da einerseits der direkte Agglutinationsversuch die progressive Veränderung (Verlust des Elutionsvermögens), die das Haemagglutinin während der Inaktivierung erleidet, nicht festzustellen erlaubt, und anderseits auch das „völlig inaktivierte" Haemagglutinin noch antigene Wirkung haben könnte. Daß diese Vermutung begründet ist, geht aus den Untersuchungen von McKee und Hale (250) hervor, die nachzuweisen vermochten, daß erhitztes Influenzavirus trotz stärkster Einbuße an haemagglutinierender Eigenschaft noch erhebliche Mengen von virusneutralisierenden und agglutinationshemmenden Antikörpern zu binden vermag, und fernerhin aus der Feststellung, daß auch das weitgehend inaktivierte Haemagglutinin noch agglutinationshemmende Antikörper erzeugen kann (81, 22).

Eine Differenzierung der gebildeten *Antikörperqualitäten* versuchte Friedewald (133) im Adsorptionsversuch mit konzentriertem Influenzavirus, wobei sich ergab, daß das Bindungsvermögen des Virus für die geprüften Antikörper unterschiedlich ist, nämlich in der Reihenfolge: virusneutralisierender — agglutinationshemmender — komplementbindender Antikörper abnimmt. Schließlich sei daran erinnert, daß Knight (219) durch die Immunisierung von Kaninchen mit dem „Normalprotein" der Allantoisflüssigkeit einen Antikörper gewinnen konnte, welcher die Haemagglutination von Influenzavirus — in nicht typspezifischer Art — aufhob, jedoch keine virusneutralisierende Eigenschaft aufwies.

Welche Bedeutung dem Haemagglutinin als immunisierendes Antigen zukommt, dürfte aus den vorliegenden Untersuchungen mit Gewißheit nicht zu entscheiden sein. Gesichert ist wohl lediglich die Tatsache, daß auch ein partiell, möglicherweise selbst völlig inaktiviertes Haemagglutinin, dem keine fermentative Wirkung mehr zugeschrieben werden kann, noch vollantigene Wirksamkeit besitzt, d. h. zur Bildung von virusneutralisierenden und agglutinationshemmenden Antikörpern Anlaß gibt. Fernerhin liegt zur Zeit kein zwingender Grund vor, für den virusneutralisierenden und agglutinationshemmenden Antikörper verschiedene Wirkungsqualitäten anzunehmen. Beide Antikörperqualitäten sind in jedem Immunserum vorhanden, zeigen weitgehend parallelgehende Titer und wirken wohl auch auf dieselbe Weise, nämlich durch die Absättigung derjenigen Gruppen, die mit den Zellreceptoren in Reaktion treten können.

7. Die nichthaemagglutinierenden Virusarten.

Es darf wohl mit Sicherheit angenommen werden, daß seit der Entdeckung von HIRST sehr zahlreiche Virusarten auf haemagglutinierende Eigenschaften geprüft worden sind. Mit wenigen Ausnahmen liegen über diese offenbar negativ ausgefallenen Versuche keine Berichte vor.

Nach ANDREWES und GLOVER (10) vermag das von diesen Autoren entdeckte *Virus der grauen Mäuselunge* („grey lung virus") Mäuseerythrocyten nicht zu agglutinieren und unterscheidet sich auch hierdurch vom Pneumonievirus (PVM). Auf der Chorionallantoismembran gezüchtetes *Kaninchenfibromvirus* zeigt, wie SMITH (297) feststellte, sowohl im nativen als auch erhitzten Zustand kein haemagglutinierendes Vermögen gegen die Erythrocyten von Hühnern, Enten, Mäusen, Meerschweinchen, Kaninchen, Pferden und Schafen. Ein völlig negatives Ergebnis ergaben auch die Versuche von HALLAUER (160) und SKINNER (296) mit *Maul-* und *Klauenseuchevirus*, wobei die Typen A, B und C aus Aphthen von Rindern und Meerschweinchen gegen eine größere Anzahl von Erythrocyten (Huhn, Mensch, Pferd, Rind, Schaf, Schwein, Kaninchen, Meerschweinchen, Ratte, Maus) im direkten Agglutinations- oder Adsorptionsversuch geprüft wurden. Auch die Erhitzung des Virusmaterials zwecks Eliminierung eventuell vorhandener geweblicher Hemmstoffe zeitigte keinen Erfolg. Auch das von SKINNER (296) untersuchte Virus der *Vesicularstomatitis* (Stamm Indianapolis) zeigte in Form von Meerschweinchenblaseninhalt, Eimembran und Amnionflüssigkeit höchstenfalls unspezifische Reaktionen mit Hühner- und Mäuseerythrocyten. Zweifelhafte Resultate erzielte HALLAUER (164) mit *Gelbfiebervirus;* sämtliche der geprüften Stämme (17 D, neurotroper und pantroper Mäusestamm) agglutinierten in Form von Suspensionen infizierter Mäuseleber und -milz, Eimembran und Allantoisflüssigkeit ausschließlich Hühnererythrocyten in zum Teil erheblich hohem Titer. Die Agglutination wurde jedoch — zumindest beim neurotropen und 17-D-Stamm — durch Kaninchennormalseren ebenso gehemmt, wie durch Immunsera von Affen und Kaninchen, und nur der pantrope Stamm „Grandpa" zeigte Anzeichen eines spezifischen Verhaltens. Ungeklärt blieben auch die haemagglutinierenden Eigenschaften des *Virus der infektiösen Anämie der Pferde* (160), da der mit gereinigtem und konzentriertem Pferdevirus, bzw. Mäuseleber- und -milz beobachtete Haemagglutinationseffekt gegen Hühner-, Mäuse- und Rattenerythrocyten in Ermangelung eines Immunserums serologisch nicht verifiziert werden konnte. Durchwegs negativ verhielten sich, wie HALLAUER (162) feststellte, eine größere Anzahl von *Bakteriophagen*.

Kaum zufällig und in hohem Grade auffallend ist nun aber die Tatsache, daß *neurotrope Virusarten* anscheinend mit Erythrocyten nicht zu reagieren vermögen, d. h. das HIRSTsche Phänomen — mit einer möglichen Ausnahme (vgl. unten) — nicht ergeben. Von HALLAUER (160) wurden die folgenden Virusarten und -stämme gegen eine größere Anzahl von Vogel-, Säugetier- und Kaltblütererythrocyten mit durchwegs negativem Resultat geprüft: *Lyssavirus* (zwei fixierte Stämme), *Poliomyelitisvirus* (humane und murine Stämme vom Typus Lansing), *Virus der Mäuseencephalomyelitis Theiler* (GD VII, FA), *Encephalitis St. Louis, equinen Encephalomyelitis* (Western, Eastern), *lymphocytaeren Choriomeningitis*. Sämtliche dieser Virusarten wurden — je nach Möglichkeit — in verschiedenen Ausgangsmaterialien (Gehirn und Rückenmark von Mäusen, Meerschweinchen, Kaninchen, Menschen; Hühnereikulturen) untersucht und nur Virusproben von hohem Infektiositätstiter ($> 10^{-6}$ M. I. D.) verwendet. Zur Ausschaltung eventuell vorhandener Hemmstoffe wurden die Gehirnsuspensionen auf unterschiedlich hohe Temperatur erhitzt, autolysiert, entfettet und zum Teil

(Poliomyelitis) auch durch die Behandlung mit Alkohol in der Kälte, bzw. Ammon-sulfatfällung gereinigt. Alle diese Prozeduren vermochten das negative Resultat nicht zu verändern. Die zunächst möglich erscheinende Annahme, daß die Haemagglutination durch gewebliche Inhibitoren oder Antikörper gehemmt werden könnte, wurde schon dadurch unwahrscheinlich, als eine hemmende Wirkung von Normalgehirnsuspensionen auf haemagglutinierendes Virus nicht festgestellt werden konnte, und durch die Züchtung von Virus (equine Encephalo-myelitis) im Hühnerembryo derartige Hemmstoffe und Antikörper ohnehin in Wegfall gekommen wären.

Der einzige Erfolg, den diese Untersuchungen zeitigten, war die Feststellung, daß die *murinen Poliomyelitisvirusstämme SK und MM*[1] Hammelerythrocyten agglutinieren (160), ein Effekt, der durch Glucosezusatz erheblich verstärkt wird. In glucosehaltigen Medien werden von diesen Stämmen auch Meerschwein-chen-, Ratten- und Menschenerythrocyten agglutiniert. Außerdem besitzt die Glucose die Eigenschaft, durch Lagerung inaktiv gewordenes Haemagglutinin zu reaktivieren (163). Daß das haemagglutinierende Vermögen dem Virus selbst zukommt und nicht durch irgendwelche Begleitstoffe verursacht wird, erscheint durch den Nachweis gesichert: 1. daß auch Eipassagestämme dieselbe Wirkung zeigen; 2. daß die infektiösen und haemagglutinierenden Eigenschaften des Virus durch Hammelerythrocyten adsorbiert und wiederum von denselben abgespalten werden können, und 3. daß die Agglutination durch Kaninchenimmunsera (Anti-SK, bzw. MM) und Affenhyperimmunserum (Aycock) in hohem Titer gehemmt, durch andere Normal- und Immunsera (Lansing „low" und „high", MEFH, Theiler GD VII und FA) dagegen nicht beeinflußt wird.

Die auffallende Gegensätzlichkeit im haemagglutinierenden Verhalten zwischen den Stämmen SK und MM einerseits und den übrigen murinen Stämmen[2] (Lansing, MEFH, Theiler GD VII und FA) anderseits legte die Vermutung nahe, daß hierfür die Unterschiedlichkeit in der peripheren Haftfähigkeit, bzw. der Grad der neurotropen Qualität ausschlaggebend sein könnte, eine Annahme, die jedoch bisher nicht ausreichend gestützt werden konnte.

8. Die Haemagglutinine und erythrocytentransformierenden Agenzien bakterieller und pflanzlicher Provenienz.

In der älteren und neueren Literatur findet sich eine größere Reihe von Mitteilungen, wonach auch Bakteriensuspensionen, bakteriellen Extrakten, Kulturfiltraten und Toxinen sowie Samenextrakten aus höheren Pflanzen haemagglutinierende oder erythrocytenverändernde Eigenschaften zukommen[3]. Offensichtlich sind zwei distinkte Phänomene beobachtet worden: 1. die *Haem-agglutination durch Suspensionen und Extrakte von Bakterien und pflanzlichen Samen.* Das agglutinierende Agens ist zellgebunden und kann nur künstlich liberiert

---

[1] Sämtliche Stämme wurden mir freundlicherweise von Herrn Prof. Cl. Junge-blut zur Verfügung gestellt.

[2] Aus einer kürzlich von Bremer und Mutsaars (34) veröffentlichten Mitteilung würde hervorgehen, daß der von diesen Autoren verwendete Lansing-Stamm ebenfalls Hammelerythrocyten zu agglutinieren vermag.

[3] Erwähnenswert ist die haemagglutinierende Wirkung von *Schlangengiften*, die von Flexner und Noguchi (117) in den Toxinen von Colubriden und Viperiden nachgewiesen worden sind. Von diesen Haemagglutininen ist lediglich bekannt, daß sie mit den Haemolysinen nicht identisch und gegenüber zahlreichen Säugetier-erythrocyten wirksam sind.

werden. Die Agglutination erfolgt rasch, meist schon nach wenigen Minuten nach dem Kontakt, und 2. die meist *ohne Agglutination* einhergehende *Erythrocytenveränderung durch bakterielle Kulturfiltrate und Toxine.* Das umformende Agens wird während des Wachstums der Mikroben in zunehmendem Maße gebildet, bzw. in die Kulturmedien abgesondert und zeigt fermentähnliche Wirkung. Die Transformation der Erythrocyten erreicht erst nach einiger Zeit ihr Maximum und äußert sich in einer veränderten Agglutinabilität. Immerhin ist eine derartige Abgrenzung nicht in jedem Falle möglich.

Haemagglutination durch Bakterien und Phytagglutinine. Die agglutinierenden Eigenschaften bestimmter Stämme von E. COLI wurden von GUJOT (155), FUKUHARA (140), AYNAUD (12), ROSENTHAL (281) und KAUFFMANN (213) festgestellt und näher untersucht. Das agglutinierende Vermögen findet sich bei haemolysierenden Stämmen der 0-Gruppe 4 und 6 am häufigsten, kommt jedoch auch bei anhaemolytischen Stämmen vor (213). Der agglutinierende Effekt ist an die erhaltene Vitalität der Mikroben nicht gebunden. Durch alkoholische Extraktion gelingt es, das Agens von den Zelleibern zu isolieren (140). Das Agglutinin ist mäßig thermostabil und verhältnismäßig resistent gegenüber Formol. Agglutiniert werden zahlreiche Erythrocyten in unterschiedlichem Grade, ebenso auch Leukocyten, Thrombocyten, Spermatozoen, Sporen von Schimmelpilzen und Pollen (281). Eine Hemmung durch Immunserum wurde bisher nicht beobachtet (140). Ein ähnliches Verhalten konnte GRIFFITTS (154) für das Haemagglutinin von *Shigella alcalescens* nachweisen, dessen Haemagglutinationsspektrum allerdings kleiner ist. Auch in diesem Fall konnte das agglutinierende Agens nur auf dem Wege der Zellaufschließung liberiert werden. Die haemagglutinierenden Eigenschaften von *Staphylococcen* gegenüber Kaninchenerythrocyten wurden bereits von KRAUS und LUDWIG (224) festgestellt. Die agglutinierende Wirkung erwies sich als thermolabil und konnte durch Immunserum in spezifischer Weise aufgehoben werden. Mit *Pneumococcen*suspensionen und — einwandfreier — mit den Organen (Milz, Leber, Lunge) von infizierten Mäusen erzielte HALLAUER (160) eine Agglutination von Hühnererythrocyten, die im äußeren Aspekt mit der Virushaemagglutination große Ähnlichkeit hatte. Das agglutinierende Prinzip erwies sich als thermolabil und von der Typenzugehörigkeit unabhängig, da *ein* typspezifisches Immunserum das agglutinierende Vermögen sämtlicher Pneumococcentypen aufzuheben vermochte. Wie LÖFFLER (237) in späteren Versuchen feststellte, muß jedoch die Spezifität der Serumhemmung bezweifelt werden, und zeigt die Reaktion bei höheren Temperaturen (37° C) Besonderheiten, die keine Analogie zur virusbedingten Agglutination zulassen. Nach DINTER (92, 93) und SCHELLNER und SEYERL (304) agglutinieren bestimmte Stämme von *Rotlaufbazillen (Erysipelothrix rhusiopathiae)* Hühnererythrocyten, wobei die Agglutination in spezifischer Weise durch Immunseren gehemmt werden kann. Das agglutinierende Agens war zellgebunden und thermolabil. Zwischen dem Agglutinationsvermögen und der Virulenz, bzw. immunisierenden Wirkung der Stämme konnten keine direkten Beziehungen nachgewiesen werden. Die von KEOGH und Mitarbeitern (214, 215) untersuchte Haemagglutination durch *haemophile Bakterien* (H. pertussis, H. parapertussis, H. bronchisepticus, H. influenzae) bietet besonderes Interesse, da die Analogie zum HIRSTschen Phänomen in mehrfacher Hinsicht zutage tritt. Frisch isolierte Stämme von H. pertussis bilden in geeigneten Nährböden (116) ein thermolabiles Haemagglutinin, welches in das Kulturfiltrat übergeht und gegenüber einer größeren Anzahl von Erythrocytenspecies wirksam ist. Der Grad der Haemagglutininbildung entspricht weitgehend der Virulenz und dem immunisierenden Vermögen der Stämme und ist für diese Eigenschaften ein zuverlässigerer Indikator als der Nachweis der Phase I. Das Haemagglutinin besitzt vollantigene und immunisierende Wirksamkeit und der antiinfektiöse Schutzwert von Immunseren läßt sich durch die Bestimmung des haemagglutininhemmenden Titers eruieren. Fernerhin wurde der Nachweis geleistet, daß Erythrocyten, die mit Extrakten aus H. influenzae vorbehandelt wurden, gegenüber dem Pertussishaemagglutinin und Influenzavirus inagglutinabel sind. Die von KOBERT und Mitarbeitern entdeckten und von LANDSTEINER und RAUBITSCHEK näher untersuchten *Phytagglutinine*[1] (Ricin, Abrin, Crotin, Phasin etc.) gehören ihrer Wirkung nach ebenfalls in diese Gruppe. Das breite und für jede Art unterschiedliche Agglutinationsspektrum, das sich nicht nur auf Erythrocyten, sondern auch auf Leukocyten und Organzellen erstreckt, der rasch einsetzende Agglutinationseffekt, das Vermögen zur Adsorption, die Eluierbarkeit, die antigene Wirksamkeit und die spezifische Hemmbarkeit durch Immunsera charakterisieren auch die pflanzlichen Agglutinine. Fernerhin reagieren Phytagglutinine auch mit den zugehörigen Seren agglutinabler Erythrocyten und stellen mit Wahrscheinlichkeit Proteine von eventuell fermentativer Wirkung dar [WIENHAUS (343)]. *Die Haemagglutination durch bakterielle Mikroben und Samenextrakte zeigt demnach zumindest phänologische Ähnlichkeiten und — in Einzelfällen (haemophile Bakterien) — selbst darüber hinausgehende Analogien mit dem HIRSTschen Phänomen, und zwar namentlich zum haemagglutinierenden Verhalten bestimmter Virusarten (Vaccine, Ektromelie, Mäusepneumonie). Dagegen wurden die spontane Reversibilität der Reaktion und die Wiederabspaltung des agglutinierenden Agens unter gleichzeitiger „Desensibilisierung" der Erythrocyten, d. h. Vorgänge, die für die Haemagglutinine anderer Virusarten (Influenza, Mumps, Geflügelpest) so charakteristisch sind, anscheinend nicht beobachtet.*

Zellveränderungen durch bakterielle Kulturfiltrate und Toxine. Wie THOMSEN und FRIEDENREICH (130) feststellten, enthalten *Kulturfiltrate* bestimmter *Corynebakterien*, *Vibrionen* und *Spirillen* ein enzymartiges Agens, welches Erythrocyten im Sinne einer entstehenden Panagglutinabilität gegenüber Normalseren umzuformen vermag. Von DAVIDSOHN und TOHARSKY (85, 86) wurde fernerhin nachgewiesen, daß in Kulturfiltraten bestimmter Corynebakterien ein transformierendes Agens — anscheinend anderer Art — vorhanden ist, welches Normalseren panagglutinierende Eigenschaften verleiht. Beziehungen dieser Befunde zur Virushaemagglutination ergaben sich erst durch die Untersuchungen von BURNET und Mitarbeitern (69, 313), aus welchen mit Eindeutigkeit hervorging, daß Cholerakulturfiltrate prinzipiell dieselben Erythrocytenveränderungen bewirken, wie das Influenza-, Mumps- und Geflügelpestvirus (vgl S. 170, 182). Annähernd dieselbe Wirkung zeigte auch das *α-Toxin von* CL. WELCHII (69, 248, 302). Der virusähnliche Effekt dieser Agenzien bezieht sich anscheinend nur auf die übereinstimmende Erythrocytenveränderung, da eine gleichzeitige Haemagglutination nicht nachzuweisen ist. Immerhin ist dieser Gegensatz nicht vollständig, da einerseits Keime, die als Spender wirksamer Kulturfiltrate in Frage kommen, selbst haemagglutinierende Eigenschaften besitzen, wie dies sowohl für Corynebakterien (130, 131) als auch Vibrionen (224, 140, 154) festgestellt ist, und anderseits auch in transformierenden Kulturfiltraten von CL. WELCHII Agglutinine gegen Erythrocyten (302) oder Leukocyten (334) nachgewiesen worden sind. Außerdem wurden Haemagglutinine im *Toxin von Cl. botulinum Typ A* von LAMANNA (225) und Leukagglutinine in *Staphylo- und Streptococcentoxinen* von WELD und MITCHELL (335) aufgefunden.

---

[1] Vgl. die zusammenfassende Darstellung von JAKOBY im Handb. d. path. Mikroorg. v. KOLLE, KRAUS und UHLENHUTH III, 1, 109 (1930).

III. Die Anwendungsmöglichkeiten.

Die praktische Auswirkung des Haemagglutinationsphänomens ergab sich schon aus den ersten Arbeiten von HIRST (191) und McCLELLAND und HARE (246), in welchen auf die Möglichkeit der quantitativen Titration von Virus und Antikörper in vitro und der Reinigung und Konzentrierung von Virus auf dem Wege der Adsorption und Elution an Erythrocyten hingewiesen wurde. Seither wurde diese Untersuchungstechnik nicht nur wesentlich erweitert, sondern auch auf ihre Leistungsfähigkeit eingehend geprüft, und ist hierdurch zum obligaten Rüstzeug der Virusforschung geworden.

Virusnachweis im Haemagglutinationstest. Die Vorteile des direkten Agglutinationstestes liegen bekanntlich in der Einfachheit, Raschheit und Ökonomie, mit welcher der Nachweis von Virus geführt und der Virusgehalt abgeschätzt werden kann. Anderseits ist die Leistungsfähigkeit des Testes dadurch begrenzt, daß erst größere Virusmengen nachgewiesen werden können, die Unterscheidung von aktivem und partiell inaktiviertem Virus nicht möglich ist, und daher — und auch aus anderen Gründen (vgl. S. 154—157) — der Haemagglutinationstiter dem Infektiositätstiter nicht immer entspricht. Die Virusauswertung im Tierversuch wird daher durch den Haemagglutinationstest nicht ersetzt, jedoch insofern in vorteilhafter Weise ergänzt, als die *Virustitration im Hühnerei* durchgeführt werden kann, wobei der qualitative Haemagglutininnachweis — als Indikator für vorhandenes Virus — den 50%igen Infektiositätstiter festzustellen erlaubt. Nach HIRST (192) bietet diese Art der Infektiositätsbestimmung — gegenüber dem Mäuseversuch — den Vorteil der größeren Empfindlichkeit, der abgekürzten Versuchszeit und der möglichen Prüfung von nicht mäuseadaptierten Influenzastämmen. Bei der *Isolierung menschlicher Virusstämme (Influenza, Mumps) im Hühnerei,* leistet der Haemagglutinationstest ebenfalls vorzügliche Dienste, nicht nur zur raschen Eruierung „angegangener" Stämme, sondern auch — beim Influenzavirus — für die Feststellung der O-D-Phase, die schon Anhaltspunkte für die Typenzugehörigkeit gibt (193, 24, 71, 79, 325, 198, 287, 180, 102, 181, 234, 216). Ebenso kann der Haemagglutinationstest als rascher und einfacher Indikator der *Virusvermehrung im Gewebsexplantat* [WELLER und ENDERS (336)] oder bei der Prüfung chemotherapeutischer Stoffe im Hühnerei [GREEN, RASMUSSEN und SMADEL (153)] verwendet werden. Schließlich läßt sich auch die *Qualität von Impfstoffen* nach dem Ausfall des Agglutinationstestes abschätzen (31, 283, 308), so daß bekanntlich Influenzaimpfstoffe bestimmte Mindesttiter an agglutinierenden Einheiten aufweisen sollten (11).

Antikörpernachweis im Hemmungstest. Wie HIRST (192) zeigte, kann auch die *Antikörpertitration* (gegen Influenzavirus) *im Hühnerei* rascher und zuverlässiger durchgeführt werden als im Mäuseversuch, da der Agglutinationstest gestattet, das nichtneutralisierte, bzw. vermehrungsfähige Virus mit ausreichender Genauigkeit zu erfassen. Die vorzügliche Eignung dieser Methode wurde von BURNET und BEVERIDGE (56) und BLASKOVIC und SALK (29) bestätigt. Für die approximative Auswertung von Seren ist jedoch der in vitro durchgeführte *Agglutinationshemmungstest* gleichfalls geeignet und bei Massenuntersuchungen auch wohl einzig möglich. Ausgedehntere Untersuchungen über die *Zunahme des Antikörpertiters* und die *Qualität des gebildeten Antikörpers in Rekonvaleszentenseren* von Influenzafällen sowie *epidemiologische Studien* über die Fluktuation des Antikörperspiegels, Stammes- und Typenspezifität des Antikörpers konnten erst unter Verwendung des Hemmungstestes realisiert werden (194, 71, 61, 30, 319, 79, 122, 287, 100, 64, 305, 211, 212, 230, 276). Eine mindestens ebenso große Bedeutung erlangte der Hemmungstest bei der *Prüfung der immunisierenden Wirkung von aktivem Influenzavirus im Menschenversuch* (40, 37, 245, 186, 129, 273) oder von *Influenzavirusimpfstoffen im Tierversuch* (283, 29), bzw. *an Menschen*, wobei der Antikörpertiter den wegweisenden Indikator für die beste Herstellungsart und Qualität des Impfstoffes (207, 186, 134, 205, 251, 253, 254, 31, 283, 204, 177, 275) darstellte und auch die optimalsten Immunisierungsbedingungen hinsichtlich der Impfstoffkonzentration (207, 168, 251, 189, 308, 286), Anzahl und Intervall der Injektionen (20, 251, 252, 254, 275) und der Applikationsart (273, 337) für die Höhe und Dauer des Impfschutzes (207, 205, 251, 189, 288) zu ermitteln erlaubte. Selbstverständlich kann dieses Auswertungsprinzip auch für andere haemagglutinierende Virusarten angewendet werden; so führte Nagler (265) vergleichende Untersuchungen über das immunisierende Vermögen von Dermo- und Eihautvaccinevirus durch, und Hallauer (161) prüfte mit dem Hemmungstest Hühnerpestimpfstoffe an Hühnern. Schließlich benützte Fenner (112, 113, 114) den Hemmungstest für den Nachweis von Antihaemagglutininen gegenüber Vaccine- und Ektromelievirus im Mäuseserum als Indikator für das Bestehen einer klinisch latenten Ektromelie.

Virusdifferenzierung im Hemmungstest. Unter Verwendung von Testsera gelingt die serologische Differenzierung von Virusarten — Newcastle-Influenzavirus (38), Vaccine-Ektromelievirus (45, 58, 46), Vaccine-Variolavirus (266) —, von Virustypen und Subtypen beim Influenzavirus (194, 148, 61, 71, 209, 147, 272, 197), bzw. Geflügelpestvirus (239, 161, 90) und die Identifizierung von frisch isolierten Influenzastämmen (67, 64, 128, 100, 320, 99, 212, 200).

Konzentrierung und Reinigung von Virus im Adsorptions-Elutionsverfahren. Francis und Salk (76) sowie Hare, Mackenzie, McClelland und Curl (167) verwendeten dieses Prinzip für die Herstellung von konzentrierten Influenzaimpfstoffen. Nach Stanley (255) bietet nun allerdings dieses Verfahren — im Vergleich zur Viruskonzentrierung mit Hilfe der Sharpless-Zentrifuge — den Nachteil großer Virusverluste und der relativ geringgradigen Reinigung. Taylor, Sharp, Beard und Mitarbeiter (322, 231, 19) überzeugten sich ebenfalls davon, daß Influenzavirus nicht mit ausreichender Selektion an Hühnererythrocyten adsorbiert und eluiert wird und daß erhebliche Mengen von inertem Protein im Eluat vorhanden sind. Immerhin dürfte das Adsorptions-Elutions-Verfahren in Kombination mit der Ultrazentrifugierungsmethode für die Herstellung hochgereinigter Präparationen leistungsfähig sein [Knight (124)]. Hallauer (161, 163) erzielte fernerhin durch die Adsorption und Elution von Pseudopestvirus an Hühnererythrocyten, bzw. von Poliomyelitisvirus an Hammelblutkörperchen Viruskonzentrate von hoher Aktivität (10^{12} M. I. D./1 cc).

IV. Die Technik.
1. Allgemeine Richtlinien.

Agglutinationstest. Der Haemagglutininnachweis gelingt in einfacher Weise durch die Vermischung absteigender, halbschlächtiger Virusverdünnungen mit gleichen Teilen einer Suspension von ausgewaschenen Testerythrocyten, und zwar entweder auf dem Objektträger oder — in der Mehrzahl der Fälle — im Sedimentierungsröhrchen. Die *Objektträgermethode* (246, 94, 271) entspricht in jeder Hinsicht dem bei der Blutgruppenbestimmung geübten Verfahren und ist als rasch orientierender qualitativer Test zweifellos geeignet, da die Agglutination schon nach kürzester Zeit deutlich hervortritt. Genauere Titerbestimmungen sind jedoch kaum möglich. Bei der *Röhrchenmethode* läßt man das Gemisch —

ohne zu zentrifugieren — sedimentieren, wobei der Grad der Agglutination entweder nach dem gebildeten Zellsediment (71, 30, 272, 282, 341, 327) oder nach der Sedimentierungsgeschwindigkeit, bzw. Dichte der überstehenden Zellsuspension (194, 203, 259, 278) beurteilt wird. Die Röhrchenmethode ist durch die Untersuchungen von HIRST (194, 203), BURNET (71, 38), MILLER und STANLEY (259), SALK (282) und WHITMAN (341) weitgehend standardisiert worden und erlaubt die Ermittlung reproduzierbarer und selbst — zumindest innerhalb einer Methode — vergleichbarer Titerwerte.

Für die *praktische Durchführung* des Haemagglutinationstestes im Sedimentierungsverfahren können die folgenden Richtlinien aufgestellt werden:

a) *Materialien.* Für die Ausführung des Testes eignen sich *Glasröhrchen* (80 bis 85 mm lang, 8 bis 10 mm innerer Durchmesser), wie solche für die KAHNsche Flockungsreaktion verwendet werden, wobei darauf zu achten ist, daß der Röhrchenboden eine *gleichmäßige Rundung* aufweist. Für andere Zwecke sollten die Röhrchen nicht verwendet werden. Ebenso ist eine sorgfältige Reinigung notwendig, da Spuren von Säure und Alkali zu Fehlresultaten Anlaß geben können. Die Montierung der Röhrchen erfolgt auf *Holz-* oder *Metallgestellen* mit mindestens zwölf Bohrungen. Der Gestellboden sollte völlig plan (ohne Eindellungen), bzw. gelocht sein, damit das Zellsediment entweder von der Seite oder von unten (mit Hilfe eines Spiegels) betrachtet werden kann.

b) *Erythrocyten.* Die Entnahme des Blutes von Tier und Mensch erfolgt durch Herz- oder Venenpunktion in 3,8-%-Natriumzitratlösung (1 bis 2 cc auf 10 cc Blut). Die Erythrocyten werden dreimal in physiologischer Kochsalzlösung ausgewaschen. Die *Konservierung* der Blutzellen geschieht bei $+2$ bis 4^0 C in Form des konzentrierten Erythrocytensedimentes oder einer 10-%-Stammsuspension. Wie lange gelagerte Erythrocytenproben verwendet werden können, hängt ab von der Erythrocytenspezies und von den Ansprüchen, die man an die Genauigkeit und Reproduzierbarkeit der Titerbestimmung stellt. Hühnererythrocyten sind für routinemäßige Untersuchungen bis zum achten Tag gebrauchsfähig, für exaktere Versuche sollten jedoch auch in diesem Falle nur frischgewonnene oder nicht länger als vier Tage gelagerte Erythrocytenproben verwendet werden. Die Haltbarkeit von Säugetiererythrocyten ist im allgemeinen beschränkter, kann jedoch notwendigenfalls durch konservierende Flüssigkeiten [z. B. Alsever-Lösung für Menschenerythrocyten (341)] beträchtlich (um mehrere Wochen) verlängert werden. Die *Dichte der Erythrocytensuspension* ist für den quantitativen Haemagglutininnachweis nicht gleichgültig, da die Höhe des Agglutinationstiters zur Zellkonzentration in umgekehrt proportionalem Verhältnis steht (194, 282, 341). Durch die Verminderung der Zelldichte auf 0,25 % (entsprechend einer Endkonzentration von 0,125 %) kann daher die Empfindlichkeit des Haemagglutinintestes — wie das Verfahren von SALK (282) zeigt — gesteigert werden. Für die Auswertungsmethoden nach WHITMAN (341), BURNET (71) und HIRST (194, 203) erweisen sich dagegen Erythrocytensuspensionen von 1 bis 2 %, bzw. Endkonzentrationen von 0,5, 0,66, 0,75 und 1,0 % als optimal geeignet. Sofern eine genaue Einstellung der Zelldichte erwünscht ist, müssen die gewaschenen Erythrocytensuspensionen in kalibrierten Röhrchen bei gleicher Umdrehungszahl (1500, bzw. 1800 U) und Zeitdauer (10 Minuten) zentrifugiert und hierauf das Sediment auf das gewünschte Volumen verdünnt werden, wobei die Dichte der hergestellten Suspensionen noch durch Trübungsmessungen kontrolliert werden kann. Die Wahl der *Erythrocytenspecies* richtet sich nach der Virusart. Hühnererythrocyten sind nicht nur für den Nachweis der meisten Virushaemagglutinine (Influenza, Mumps, Geflügelpest, Vaccine, Ektromelie) geeignet, sondern auch gegenüber anderen Erythrocyten zu bevorzugen, wegen ihres raschen Sedimentierungsvermögens und ihrer großen Stabilität. In Ermangelung von Hühnererythrocyten oder für den Nachweis der O-Phase können beim Influenza- und Mumpsvirus auch Menschenerythrocyten der Gruppe 0, bzw. Meerschweinchenerythrocyten verwendet werden. Für das Mäusepneumonievirus, bzw. die haemagglutinierenden Poliomyelitisstämme kommen nur Mäuse- oder Hamstererythrocyten, bzw. Hammelblutkörperchen in Betracht. Die Agglutinabilität *individueller Blutproben* gegenüber haemagglutinierenden Virusarten (Influenza, Mumps, Geflügelpest) ist zweifellos unterschiedlich (259, 122) und beeinflußt die Höhe des Agglutinationstiters. Für vergleichende quantitative Untersuchungen ist daher die Mitführung eines an hochempfindlichen Erythrocyten geeichten Standardvirus — zum Ausgleich derartiger Unterschiede — empfehlenswert (259). In extremer Form tritt die individuelle Variabilität von Hühnererythrocyten hinsichtlich der Empfindlichkeit gegenüber den Haemagglutininen des Vaccine- und Ektromelievirus hervor, so daß für den Agglutinationstest nur selektionierte Erythrocytenproben von hoher oder zumindest ausreichender Empfindlichkeit verwendet werden müssen (264, 76, 58).

c) *Volumen der Reaktionskomponenten.* Virusverdünnungen und Zellsuspensionen werden meistens zu gleichen Teilen angesetzt; in welchen Volumina (0,25 + 0,25; 0,5 + 0,5; 1,0 + 1,0) dies geschieht, ist für die Höhe des Agglutinationstiters gleichgültig. Kleinere Volumina können bei sämtlichen Methoden, bei welchen das Erythrocytensediment als Indikator dient, angewendet werden, wobei die Einsparung an Material und die kürzere Sedimentierungszeit, bzw. raschere Ablesungsmöglichkeit einen Vorteil darstellt. Ein größeres Reaktionsvolumen (2 cc) ist dagegen notwendig, wenn der Grad der Agglutination nach der optischen Dichte der überstehenden Suspension beurteilt wird. Dieselben Volumina wie im Hemmungsversuch (vgl. unten) verwendet die Burnetsche Schule, nämlich 0,25 cc Virusverdünnung, 0,25 cc Erythrocytensuspension und 0,25 cc Kochsalzlösung.

d) *Reaktionstemperatur.* Mit steigender Temperatur (4° — 25° — 37° C) wird der Agglutinationsvorgang erheblich beschleunigt, gleichzeitig aber auch die Wiederabspaltung des Virus von den Erythrocyten begünstigt, so daß die Agglutinate bei höheren Temperaturen (37° C) sehr unbeständig und daher Titerbestimmungen kaum möglich sind. Die Bebrütung der Gemische bei 37° C bietet daher nur einen Vorteil (Ermittlung höherer Titer) für den Haemagglutininnachweis bei nichteluierenden Virusarten, wie dem Vaccine- und Ektromelievirus (264, 265, 58, 63) und auch für Mumpsvirus, wenn nicht Hühner-, sondern Menschenerythrocyten verwendet werden (235). Das Stehenlassen der Gemische bei Zimmertemperatur (20 bis 25° C) erlaubt bei allen haemagglutinierenden Virusarten, den Titer in eindeutiger Art festzustellen, falls die Ablesung innerhalb eines bestimmten Zeitintervalles — meist nach 60 bis 120 Minuten — vorgenommen wird. Ist eine Titerbestimmung im vorgeschriebenen Zeitpunkt nicht möglich, oder werden Virusarten mit hohem Elutionsvermögen — z. B. Pseudopest-, bzw. Newcastle-Virus — geprüft, so empfiehlt sich die Anstellung des Haemagglutinationstestes bei + 4° C, da bei niederer Temperatur die Agglutinate über längere Zeit stabil sind (38, 120). Nach Salk (282) ist der in der Kälte durchgeführte Agglutinationstest (für Influenzavirus) überhaupt am empfindlichsten und daher für den Nachweis kleinster Haemagglutininmengen besonders geeignet.

e) *Wasserstoffionenkonzentration des Reaktionsmilieus.* Wie Miller und Stanley (259) feststellten, sind die Titerwerte von Influenzavirus in einem Bereich von p_H 6,0 bis 8,0 nahezu dieselben. Die Verwendung einer 0,1-m-Phosphatpufferlösung von p_H 7,0 an Stelle der gewöhnlichen Kochsalzlösung als Verdünnungsmittel bietet daher keine nennenswerten Vorteile.

f) Ablesung (Kriterien für die Bestimmung des Agglutinationstiters). Die Einschätzung des „Endpunktes", bzw. des Haemagglutinationstiters[1] erfolgt bei den verschiedenen Auswertungsverfahren nach unterschiedlichen Gesichtspunkten. Nach dem Vorgehen von BURNET (71) und WHITMAN (341, 327) wird der Titer durch die seitliche Betrachtung des Erythrocytensedimentes bestimmt, wobei der Endtiter für dasjenige Röhrchen angenommen wird, in welchem eben noch eine „+-Reaktion", bzw. partielle Agglutination nachzuweisen ist. Wenn nun auch der Aspekt dieser Grenzreaktion spezifiziert wird („nichtagglutiniertes linsenförmiges Zentrum, umgeben von einem Randbeschlag agglutinierter Zellen"), so ist doch die Ermittlung dieses Endpunktes vom subjektiven Ermessen des Beobachters abhängig. Im Verfahren von SALK (282) dient die Art des Zellsedimentes zwar ebenfalls zur Ermittlung des Titers; da die Ablesung jedoch von unten erfolgt, ist das Aussehen des Agglutinates einförmiger und erlaubt nun auch eine ausreichend scharfe Abgrenzung zwischen der „+-Reaktion", bzw. totalen Agglutination und der „O-Reaktion", bei welcher eine Agglutination überhaupt nicht feststellbar ist. Die gelegentlich nachweisbare „±-Übergangsreaktion" wird nicht berücksichtigt, sondern das letzte Röhrchen mit „+-Reaktion" repräsentiert den Titer. Bei der Auswertung nach HIRST (194, 203), bei welcher der Grad der Agglutination nicht nach dem Erythrocytensediment, sondern nach der Sedimentierungsgeschwindigkeit, bzw. der in einem bestimmten Zeitpunkt vorhandenen Zelldichte in der mittleren Schicht des Gemisches beurteilt wird, repräsentiert die „++-Reaktion", bzw. die 50-%-Agglutination den Titer. Die Ermittlung dieses Agglutinationsgrades erfolgt durch die vergleichende Trübungsmessung mit einer Reihe von Erythrocytensuspensionen bekannter Zellkonzentration (194), oder — in noch objektiverer Art — durch die direkte Dichtigkeitsbestimmung mit einem photoelektrischen Turbidometer (203).

g) Standardisierung. Die meist verwendeten Auswertungsverfahren (vgl. S. 202) sind hinsichtlich der Dichte der Erythrocytensuspension, Reaktionstemperatur, Bebrütungsdauer und Art der Endpunktbestimmung weitgehend standardisiert. Jede Methode ist nun aber mit zwei Variabeln, nämlich der wechselnden Zimmertemperatur und der unterschiedlichen individuellen Qualität der Erythrocyten behaftet, die bei nicht gleichzeitig durchgeführten Untersuchungen die Reproduzierbarkeit und Vergleichbarkeit der Titerwerte beeinträchtigen. Zur Kontrolle dieser „Temperatur- und Erythrocytenfehler" dient die Mittitration eines Standardvirus, dessen haemagglutinierende Aktivität auf die Temperatur von 25 ° C korrigiert und gegenüber hochempfindlichen Erythrocyten geeicht ist. Das Verhältnis des theoretischen zum jeweils ermittelten Titer des Standardvirus stellt den Korrektionsfaktor dar, der — multipliziert mit dem festgestellten Titer der zu untersuchenden Virusprobe — beide Fehler auszugleichen erlaubt (259). Demgegenüber sind die mit *verschiedenen* Methoden ermittelten Titerwerte und agglutinierenden Einheiten schon auf Grund des ungleich angenommenen Endpunktes höchstenfalls nur annähernd übereinstimmend und vergleichbar. Eine Einigung auf *eine* möglichst einfache und

---

[1] Anmerkung zur Nomenklatur: *Endpunkt* = höchste Verdünnung des Ausgangsvirus, die noch eine partielle, bzw. komplette Agglutination ergibt. — *Haemagglutinationstiter* = Reziproker Wert der Virusendverdünnung (unter Berücksichtigung des Reaktionsvolumens) im Endpunkt. Gibt zugleich den Verdünnungsfaktor wieder, mit welchem die ursprüngliche Haemagglutininkonzentration auf eine agglutinierende Einheit reduziert wird. — *Haemagglutinierende Einheit* = Haemagglutininkonzentration im Teströhrchen des Haemagglutinationstiters.

zugleich hinreichend genaue Auswertungsmethode wäre daher wünschenswert [vgl. WHITMAN (341)]. Für streng quantitative Auswertungen und bei Verwendung eines konzentrierten Virus müssen die Virusverdünnungen unter beständigem Pipettenwechsel durchgeführt werden.

h) Fehlerquellen. Hochgradig *bakteriell verunreinigte* Allantoisflüssigkeiten — z. B. nach der Verimpfung von nicht vorbehandelter menschlicher Rachenspülflüssigkeit — können Hühnererythrocyten agglutinieren und hierdurch eine Virushaemagglutination vortäuschen [SALK, MENKE und FRANCIS (287), FLORMAN (119)]. Fernerhin isolierten VAN HERICK und EATON (190) — bei der Eipassage eines Virus der atypischen Pneumonie — von der Amnionmembran einen *pleuropneumonieartigen Erreger*, der selbst haemagglutinierende Eigenschaften aufwies und anscheinend als latent infizierender Keim gelegentlich in Hühnerembryonen vorkommt. Weiterhin muß stets mit der Möglichkeit gerechnet werden, daß die Flüssigkeiten und Gewebe des Wirtes, aus welchen das Virusmaterial bezogen wird, schon *normalerweise Agglutinine* enthalten können, deren Unterscheidung von Virushaemagglutininen oft keineswegs einfach ist. So finden sich im *Hühnereiklar* unbefruchteter Eier und im *Albuminsack* sowie — vom elften Brütungstag an —, auch in der *Amnionflüssigkeit* von Hühnerembryonen thermolabile Haemagglutinine, die gegenüber einer Reihe von Säugetiererythrocyten (Maus, Meerschweinchen, Kaninchen, Pferd, Mensch usw.) wirksam sind und anscheinend nicht als „natürliche Antikörper" qualifiziert werden können (77). Fernerhin sind im *Dotter* und auch in sämtlichen *Geweben des Hühnerembryos* mit Regelmäßigkeit *Haemagglutinine gegenüber Kaninchenerythrocyten* nachzuweisen (164), die wohl als passiv übertragene Antikörper aufzufassen sind (159). Nach BURNET und STONE (63, 312) besitzen auch sämtliche *Gewebslipoide* haemagglutinierende Eigenschaften, die sich von denjenigen des Vaccine- und Ektromelievirus nur durch die leichte Hemmbarkeit durch Normalseren unterscheiden lassen. Derartige unspezifische Lipoidhaemagglutinine sind nicht nur in wässerigen Emulsionen alkoholischer Extrakte nachzuweisen, sondern können auch in gelagerten Gewebssuspensionen (z. B. Chorionallantois) manifest werden (63) und sind sehr wahrscheinlich auch für die haemagglutinierende Wirkung des Normalmäusegehirns gegenüber Hühner- und Mäuseerythrocyten verantwortlich (164). Pseudoagglutinationen sind fernerhin möglich durch die Verwendung von ungenügend zentrifugierten Gewebssuspensionen oder durch die Anwesenheit von Kupferionen im destillierten Wasser, das zur Herstellung der Kochsalzlösung benützt wird (289). Bei sämtlichen Versuchen, bei welchen der Haemagglutinationstest den Nachweis der haemagglutinierenden Eigenschaft einer Virusart erbringen soll, oder die einzige Methode des Virusnachweises darstellt, ist demnach Vorsicht geboten. Der virusbedingte Charakter des Testes kann in diesen Fällen nur dadurch erbracht werden, daß 1. „Normalagglutinine", bzw. haemagglutinierende Antikörper mit Sicherheit ausgeschlossen werden, und 2. die Spezifität des Testes im Hemmungsversuch mit Normal- und Immunseren verifiziert wird.

Hemmungstest. Der Haemagglutinationshemmungstest stellt ein Dreikomponentensystem dar, in welchem — in den meisten Fällen — der Antikörper bis zur Titergrenze verdünnt, während die Virus- und Erythrocytenmenge konstant gehalten wird. Die Reaktionsverhältnisse werden dadurch erheblich kompliziert, daß Immunsera nicht nur spezifische Antihaemagglutinine, sondern auch hemmende „Normalglobuline" und vor allem Hemmstoffe receptorartiger Natur enthalten (vgl. S. 172, 175). Hierdurch ergibt sich eine Summation der hemmenden Wirkung, die namentlich dann zur Geltung kommt, wenn sowohl die virusbindende Kapazität des Antikörpers als auch die Bindungsaffinität des Virus zu den Erythrocyten von nur geringem Grade sind, so daß sich die Konkurrenz des receptorartigen Inhibitors (um das Virus, bzw. die Erythrocyten) in vollem Umfang auswirken kann. Durch eine Reihe von Maßnahmen ist es nun möglich, der Hemmungsreaktion eine „spezifische Richtung" zu geben und die „unspezifische Hemmung" weitgehend auszuschalten oder zumindest zu kontrollieren. Die Zuverlässigkeit des Hemmungstestes für die Wertbestimmung virusneutralisierender Antikörper ist von einer Reihe von Faktoren abhängig, deren Einfluß aus den folgenden Richtlinien zu ersehen ist.

a) Sera. Die *Inaktivierung der Sera* bei 56° bis 62° C während 30 Minuten erlaubt die Eliminierung eventuell vorhandener natürlicher Haemolysine und eines Großteiles thermolabiler Hemmstoffe (194, 62, 247). Enthält das inaktivierte Serum noch natürliche oder durch die Immunisierung ausgelöste Haemagglutinine, so ist die Vorverdünnung des Serums bis zur Unwirksamkeit dieser Antikörper oder die vorgängige Beseitigung dieser Haemagglutinine auf dem Wege der Adsorption notwendig. Die Ausschaltung des thermostabilen receptorartigen Inhibitors kann ebenfalls durch eine geeignete Vorverdünnung des Serums erreicht werden, ein Prinzip, das in bestimmten Auswertungsverfahren (272, 273), in welchen eine konstante Menge Immunserum (1 : 100) gegenüber fallenden Virusverdünnungen titriert wird, zur Anwendung kommt. Dieser „umgekehrte Hemmungstest" ist jedoch für den Nachweis kleiner Antikörpermengen ungeeignet und bietet auch Nachteile anderer Art [vgl. WHITMAN (341)]. Der störende Einfluß der „unspezifischen Hemmung" kann auch im üblichen Hemmungstest durch andere Maßnahmen — Verwendung „avider" Erythrocyten (vgl. unten) — auf ein Minimum beschränkt werden. Bei schwachen Immunseren mit niedrigem Hemmungstiter kann zur Entscheidung der Frage, ob die Hemmung als spezifisch oder unspezifisch zu werten ist, das von BURNET und Mitarbeitern (64, 6) empfohlene Vorgehen der mehrmaligen Ablesung bei wiederholt aufgeschüttelten Reaktionsgemischen angewendet werden. Bleibt hierbei der Hemmungstiter konstant, so liegt mit Wahrscheinlichkeit eine unspezifische Hemmung vor, steigt dagegen der Titer merklich an, so sind hemmende Antikörper anzunehmen.

b) Testvirus. Nach HIRST (194) ist der Hemmungstiter eines Serums umgekehrt proportional der im Gemisch vorhandenen Virusmenge. Die Größe der verwendeten Virusgebrauchsdosis ist demnach keineswegs gleichgültig; dieselbe sollte hinreichend groß sein, d. h. in dem für den Hemmungstest verwendeten Reaktionsvolumen in Abwesenheit von Serum noch eine deutlich feststellbare Agglutination ergeben, anderseits aber doch eine Grenzdosis darstellen, damit die Empfindlichkeit des Testes nicht beeinträchtigt wird. Die Bestimmung der Virusgebrauchsdosis ergibt sich aus der Vortitration des Virus, in welcher der Endpunkt mit möglichster Genauigkeit — eventuell durch die Interpolation (194, 203, 305) oder durch die Ermittlung mittlerer Titer aus zwei gegeneinander verschobenen Verdünnungsreihen (282, 300) — ermittelt werden sollte. Als Gebrauchsdosis dient ein Multiplum der im Endpunkt ermittelten Viruskonzentration, nämlich 4 bis 5 agglutinierende Einheiten[1] (203, 341, 327, 71) oder — zur Steigerung der Empfindlichkeit des Testes — gelegentlich auch nur 1 bis 2 Einheiten (282, 6). Für vergleichende Untersuchungen zu unterschiedlicher Zeit müssen Testvirusproben zur Verfügung stehen, die einen ausreichend hohen (Titer > 160) und möglichst konstanten Titerwert aufweisen. Die Konservierung des Virus (in Form von Allantoisflüssigkeit, bzw. eines gereinigten und konzentrierten Viruseluates) geschieht am besten bei + 4° C (259, 282), eventuell — zur Vermeidung des Ausfallens von Uraten — unter Zusatz von 10 % sterilem Glyzerin (6). Derartige Proben zeigen in der Regel während Monaten einen

---

[1] Bezogen auf das im Hemmungstest verwendete Volumen.

nahezu unveränderten Titer, dessen Konstanz aber doch durch periodische
Kontrollen festzustellen ist. Die Prüfung der Gebrauchsdosis erfolgt mit Hilfe
eines Standardimmunserums, dessen Mitführung in den jeweiligen Hemmungs-
testen auch den Ausgleich von „Temperatur- und Erythrocytenfehlern" erlaubt.
Die von Hirst (197) nachgewiesene unterschiedliche „Avidität" der Virusstämme gegenüber Erythrocyten muß wohl nur bei der serologischen Differenzierung frischisolierter Virusstämme in vollem Umfange berücksichtigt werden.

  c) *Erythrocyten.* Die Abhängigkeit des Hemmungstiters von der Zelldichte
wurde von Hirst (194), Salk (282) und Whitman (341) untersucht. Nach
Salk nimmt der Hemmungstiter proportional der Zellverminderung ab, ein
Befund, der mit der Feststellung in Einklang steht, wonach verdünnte Erythrocytensuspensionen kleine Mengen von freiem Haemagglutinin empfindlicher
anzeigen. In unterneutralisierten Gemischen mit einem stärkeren Überschuß
an freiem Virus hat dagegen, wie Whitman (341) feststellte, die von Salk ermittelte Relation keine Gültigkeit. Für die Praxis dürfte sich die Schlußfolgerung ergeben, daß im Hemmungstest dieselbe *mittlere Erythrocytenkonzentration*
(0,125 bis 0,75 % im Reaktionsvolumen), die schon im Agglutinationsversuch
Anwendung findet, von optimaler Eignung ist, da eine extreme Senkung, bzw.
Steigerung des Hemmungstiters kaum erwünscht erscheint. Die *individuelle
Qualität der Erythrocyten* spielt im Hemmungstest eine ungleich größere Rolle
wie im direkten Agglutinationsversuch (vgl. S. 179), da es gelingt, durch die
Auswahl bestimmter Erythrocyten von hoher „Avidität", bzw. starkem Virus-
bindungsvermögen die unspezifische Hemmung auf ein Minimum zu reduzieren
[Anderson, Burnet und Stone (6)]. Die Ermittlung geeigneter Blutspender
ist daher von Bedeutung, namentlich für vergleichende quantitative Studien
und die serologische Differenzierung von Virusstämmen. Die Wahl der *Erythrocytenspezies* entspricht in den meisten Fällen der Erythrocytenart (vorzugs-
weise Hühnererythrocyten), die auch im Agglutinationstest verwendet wird.
Eine Ausnahme von dieser Regel bildet die Verwendung von Meerschweinchen-
erythrocyten im Hemmungstest mit Mumpsvirus, da nach Beveridge und
Lind (21) nur diese Zellart in Gegenwart von Serum einen scharf begrenzten
Endpunkt festzustellen erlaubt. Aus den Untersuchungen von Whitman (341),
welcher die Agglutinations- und Hemmungstiter der Influenzatypen A und B
unter Verwendung von Hühner-, bzw. Menschenerythrocyten vergleichsweise
prüfte, scheint hervorzugehen, daß menschliche Blutzellen durch kleinere Antikörpermengen „geschützt" werden und daher schwächere Indikatoren für den
Nachweis von freiem, ungebundenem Virus darstellen als Hühnererythrocyten.

  d) *Herstellungsart der Gemische.* Nach Hirst (194) erfolgt die Absättigung
des Haemagglutinins durch den Antikörper derartig rasch, daß eine Vorbebrütung
der Virus-Serum-Gemische nicht notwendig ist. Salk (282) zieht aus dieser
Feststellung die Nutzanwendung, indem die Verdünnungen des Immunserums
in der Erythrocytensuspension hergestellt und das Virus als letzte Komponente
zugesetzt werden, ein Verfahren, das die Verringerung des Gesamtvolumens und
damit auch eine Abkürzung der Sedimentierungszeit bewirkt. Immerhin besteht
kein Zweifel darüber, daß durch eine längere Inkubation der Virus-Serum-
Gemische bei Zimmertemperatur, bzw. 37° C vor dem Zusatz der Testerythrocyten der Hemmungstiter gesteigert wird (282), so daß die Vorbebrütung —
namentlich für den Nachweis kleiner Antikörpermengen — von Vorteil sein
kann. In einigen Auswertungsverfahren — Hemmungstest mit Influenzavirus
(272), Vaccinevirus (264, 265) und Pseudopestvirus (120) — ist die zweizeitige
Herstellung der Gemische auch beibehalten worden. In jedem Fall müssen die
Virus-Serum-Gemische vor dem Zusatz der Testblutkörperchen durchgeschüttelt

werden, und ebenso wichtig ist die rasche Vermischung der Erythrocyten mit den übrigen Reaktionskomponenten, die am besten durch die Verwendung einer automatischen Abfüllspritze erreicht wird.

e) Endpunktbestimmung und Standardisierung der Titerwerte. Die Bestimmung des Endpunktes erfolgt — wie im direkten Agglutinationsversuch — nach unterschiedlichen Kriterien, nämlich, nach der partiellen Agglutination des Zellsedimentes (71, 21, 120), der 50-%-Abnahme der Zelldichte der überstehenden Suspensionen (194, 203) oder der totalen Hemmung (282, 122), welche durch die letzte Serumverdünnung zustande kommen. Damit die ermittelten Serumtiter, bzw. „Serumeinheiten" — wenigstens innerhalb *einer* Methode — nach einem einheitlichen Maßstab gemessen und auch zu unterschiedlichen Versuchszeiten miteinander verglichen werden können, ist die Verwendung eines *Standardimmunserums,* das am besten in lyophiler Form konserviert wird, erforderlich. Das Standardserum (1 : 100) erlaubt die Ermittlung oder die Kontrolle der bereits bestimmten Virustestdosis und der eruierte Endpunkt gibt an, welche Antikörpermenge der Virusgebrauchsdosis äquivalent ist. Die Stärke eines unbekannten Serums kommt dann im Verhältnis $\dfrac{\text{Titer von Serum x}}{\text{Titer von Serum Standard}}$ als Fraktion oder Multiplum der Potenz des Standardserums zum Ausdruck (38). Fernerhin gestattet die jeweilige Mittitration des Standardserums die Variabilität der Versuchsbedingungen (Virustiter, unterschiedliche Affinität individueller Erythrocyten, schwankende Zimmertemperatur) zu kontrollieren, bzw. auszugleichen (207, 134, 188). Ein Standardserum von internationaler Wertigkeit, mit welchem die nach verschiedenen Verfahren ermittelten Serumtiter in eine vergleichbare Übereinstimmung gebracht werden könnten, steht jedoch zur Zeit noch nicht zur Verfügung [vgl. WHITMAN (341)].

f) Serologische Differenzierung der Virusantigene. Wie bereits erwähnt (vgl. S. 178), können im Hemmungstest — mit Hilfe von Testseren — Virusarten, Virustypen und Subtypen, bzw. antigene Stammesunterschiede differenziert werden. Die Versuchsanordnung ist dieselbe wie bei der Antikörperauswertung, indem meistens zunehmende Verdünnungen der Testseren gegen eine konstante Virusgebrauchsdosis titriert werden. Immerhin ist auch hier das umgekehrte Vorgehen, nämlich die Prüfung einer konstanten Testserumdosis (1 : 100) gegenüber ansteigenden Virusverdünnungen möglich, (71, 272). Der Nachweis feinerer antigener Unterschiede ist nur dann zu erbringen, wenn die Versuchsbedingungen in jeder Hinsicht konstant gehalten werden. Die *Serumverdünnungen* sollten nicht einzeln, sondern im Gesamtvolumen der Kochsalzlösung, die zur Beschickung sämtlicher Teströhrchen ausreicht, ausgeführt werden. Die variable Bindungsfähigkeit individueller Erythrocyten ist dadurch auszuschalten, daß der gesamte Versuch zu gleicher Zeit mit *derselben Probe von frisch gewonnenen Testzellen möglichst hoher Avidität* (6) ausgeführt wird. Die *sorgfältige Einstellung der Virustestdosis* ist notwendig, erlaubt jedoch noch nicht, eventuell vorhandene „*Aviditätsunterschiede"* (vgl. S. 179) als Fehlerquellen zu eliminieren. Die unterschiedliche Avidität von Virushaemagglutininen kommt nach HIRST (197) in einer Reihe serologisch paradoxer Effekte zum Ausdruck, nämlich: 1. daß die gleiche Virustestdosis desselben Virusstammes und derselben Eipassage von einer ungleich großen Serummenge gehemmt wird, wenn verschiedene Viruspräparationen, bzw. Allantoisflüssigkeiten geprüft werden; 2. daß ein gegebenes Immunserum den heterologen Virusstamm stärker hemmt als den homologen und 3. daß die antigene Differenz zweier Virusstämme im gekreuzten Hemmungsversuch in verschiedenem Grade hervortritt, je nachdem ein Virus X mit einem Serum Anti-Y, bzw. ein Virus Y mit einem Serum

Anti-X geprüft wird. Im gekreuzten Hemmungsversuch können derartige, nicht immunologisch bedingte Differenzen bei verschiedenen Viruspräparationen desselben Stammes, bzw. einer Reihe von Virusstämmen durch die jeweilige Ermittlung des Verhältnisses $\dfrac{\text{heterologer Serumtiter}}{\text{homologer Serumtiter}}$ leicht festgestellt und durch einen ad hoc berechneten Korrektionsfaktor zum Ausgleich gebracht werden (197). Schließlich sollten die *Antezedenzien der zu prüfenden Virusstämme* (Art der Isolierung, Passagewirte) bekannt sein, da nach Hirst (199, 200) nur die im Ei isolierten und passierten Influenzastämme ihre ursprüngliche antigene Qualität behalten, während im Frettchen isolierte und in Mäusen passierte Stämme antigene Abwandlungen erleiden.

2. Spezielle Methoden.

Die verschiedenen Auswertungsverfahren, von denen hier nur die wichtigsten angeführt werden können, lassen sich nach der Art der Ablesung, bzw. der End-punkt- und Titerbestimmung in zwei Gruppen einteilen, nämlich: 1. in Verfahren, bei welchen der Agglutinationseffekt nach der Sedimentierungsgeschwindigkeit der Erythrocyten, bzw. der in einem bestimmten Zeitpunkt feststellbaren *Zell-dichte in der überstehenden Suspension* beurteilt wird [Hirst und Mitarbeiter (194, 207, 203)], und 2. in Methoden, bei welchen der Agglutinationsgrad nach dem *Aspekt des Zellsedimentes* eingeschätzt wird, wobei die Betrachtung des Agglutinates von der Seite her [Burnet und Mitarbeiter (71, 61), Whitman (341, 327)] oder von unten [Salk (282)] geschieht.

a) Methode von Hirst.

Für routinemäßige Untersuchungen dürfte das erste von Hirst (194, 207) angegebene Verfahren ausreichend genau sein.

Agglutinationstest: Je 1 cc halbschlächtiger Virusverdünnungen (1 : 10 bis 1 : 5120) werden mit je 1 cc einer 2-%-Erythrocytensuspension versetzt. Als Kontrolle dient 1 cc Erythrocytensuspension + 1 cc Kochsalzlösung. Die Gemische werden durchgeschüttelt und bleiben hierauf während 60 Minuten bei Zimmertemperatur. Die Ermittlung des Endpunktes (= 33 bis 55 % Agglu-tination) erfolgt durch den turbidometrischen Vergleich der Teströhrchen mit einer Reihe von frisch hergestellten Erythrocytenstandardsuspensionen bekannter Zelldichte (1,0, 0,75, 0,5, 0,3 %, bzw. — wenn eine feinere Einstellung erwünscht ist — 1,0, 0,87, 0,75, 0,67, 0,5, 0,37, 0,25 %) im Komparator. Der Endpunkt (+ +-Reaktion) wird für dasjenige Röhrchen angenommen, dessen Zelldichte zwischen den Standardsuspensionen von 0,75 und 0,5 %, bzw. 0,67 und 0,5 % liegt. Eventuell muß dieser Endpunkt auf dem Wege der Interpolation ermittelt werden.

Hemmungstest: Zu je 0,5 cc ebenfalls halbschlächtig hergestellten Serumver-dünnungen (1 : 16 bis 1 : 8192) werden je 0,5 cc einer konstanten Virustestdosis von vier agglutinierenden Einheiten (vierfache Viruskonzentration im Endpunkt) zugesetzt, die Gemische hierauf durchgeschüttelt und mit je 1,0 cc einer 2%igen Erythrocytensuspension mit einer automatischen Mischspritze versetzt. Mit-geführt werden drei Kontrollen, in welchen die drei Komponenten des Gemisches (1,0 cc Erythrocytensuspension + 1,0 cc NaCl-Lösung; 0,5 cc höchste Serum-konzentration + 1,5 cc NaCl-Lösung; 0,5 cc Virusgebrauchsdosis + 1,5 cc NaCl-Lösung) geprüft werden. Nach 75 Minuten langem Stehen erfolgt die Ablesung des Endpunktes (33 bis 50 % Agglutination) auf dieselbe Art wie im direkten Agglutinationsversuch.

Hirst und Pickels (203) haben den Agglutinations- und Hemmungstest noch dadurch weiter verfeinert und „objektiviert", daß die Bestimmung der 50-%-Agglutination nicht durch den makroskopischen Vergleich der Zelldichte mit Standardsuspensionen, sondern mit Hilfe eines photoelektrischen Trübungsmessers vorgenommen wird. Die verwendeten Reaktionsvolumina sind dieselben, dagegen ist die Konzentration der Zellsuspension auf 1,5 % (= 0,75 % Endkonzentration) reduziert, und erfolgt die Ablesung nach genau 75 Minuten. Durch die Mitführung eines Standardvirus im Agglutinationstest, bzw. eines Standardimmunserums im Hemmungsversuch können „Temperatur- und Erythrocytenfehler" korrigiert werden.

b) Methode von Burnet.

Agglutinationstest: Für den Haemagglutininnachweis wird ein Gemisch, bestehend aus drei Komponenten, hergestellt: je 0,25 cc halbschlächtiger Virusverdünnungen + je 0,25 cc Kochsalzlösung + je 0,25 cc einer 2%igen Erythrocytensuspension. Die Gemische werden durchgeschüttelt und verbleiben bis zur vollständigen Sedimentierung während 60 bis 120 Minuten (je nach der verwendeten Erythrocytenart) bei Zimmertemperatur oder, falls eine Ablesung in diesem Termin nicht möglich ist, bei $+4^0$ C. Die Ablesung erfolgt durch die seitliche Betrachtung des Agglutinates von bloßem Auge oder mit der Lupe, wobei die folgenden Abstufungen im Agglutinationsgrad festgestellt werden können: 1. ein klumpiges, grobgranuliertes Sediment, das sich über den ganzen Boden des Röhrchens ausbreitet (+++, maximale Agglutination), 2. ein feiner granuliertes Bodensediment mit einem überstehenden Randbeschlag, der sich teils in Fetzen ablöst, teils adhärent bleibt (++, komplette oder fast komplette Agglutination), 3. ein linsenförmiges Sediment, umgeben von einem granulierten Randbeschlag (+, inkomplette, partielle Agglutination) und 4. ein gerade noch angedeuteter Randbeschlag (±, Spur Agglutination). Die negative Reaktion (im Röhrchen der Erythrocytenkontrolle) präsentiert sich als scharf begrenztes, linsenförmiges Sediment ohne jeden Randbeschlag. Als Endpunkt wird die +-Reaktion, bzw. partielle Agglutination bewertet.

Hemmungstest: Die Volumina werden nicht verändert: Zu je 0,25 cc steigender Verdünnungen des Immunserums werden je 0,25 cc der Virustestdosis von fünf agglutinierenden Einheiten und — nach Durchmischen — noch je 0,25 cc der 2-%-Erythrocytensuspension zugesetzt; die Gemische nochmals aufgeschüttelt und während $1\frac{3}{4}$ Stunden bei $+4^0$ C, bzw. 60 Minuten bei Zimmertemperatur belassen. Der Endpunkt wird wiederum durch die +-Agglutination repräsentiert.

Die Burnetsche Methode erlaubt *Modifikationen* ohne wesentliche Verschiebung der Auswertungsresultate. Hallauer bevorzugt — für routinemäßige, rasch orientierende Versuche — das folgende Verfahren:

Für den *direkten Haemagglutininnachweis* werden halbschlächtige Virusverdünnungen (1 : 10 bis 1 : 5120) mit einer 1-%-Erythrocytensuspension zu gleichen Teilen (0,25 cc + 0,25 cc, bzw. 0,5 cc + 0,5 cc) versetzt, und die Gemische während 60 Minuten (Hühnererythrocyten), bzw. 90 bis 120 Minuten (kernlose Erythrocyten) bei Zimmertemperatur belassen. Die letzte Virusverdünnung, die noch einen deutlich wahrnehmbaren Randbeschlag zeigt, wird als Endpunkt (++-Reaktion) angenommen. Für den *Hemmungsversuch* werden die Gemische wie folgt hergestellt: 0,25 cc Immunserumverdünnungen (1 : 10 bis 1 : 5120) + + 0,25 cc Virustestdosis (vier agglutinierende Dosen = vierfache Viruskonzentration im Endpunkt) + 0,5 cc einer 1-%-Erythrocytensuspension. Der Endpunkt, der meistens schon nach 30 bis 45 Minuten Stehen bei Zimmertemperatur bestimmt

werden kann, entspricht der letzten Serumverdünnung, die eben noch eine
vollständige Hemmung ergibt.

Die Methode von Burnet bietet den Vorteil der Anwendbarkeit für sämtliche
haemagglutinierende Virusarten und die Möglichkeit, das Agglutinat in qualitativer Hinsicht zu bewerten. Die Bestimmung des Endpunktes unterliegt
jedoch — in wechselndem Grade — dem subjektiven Ermessen des Beobachters,
so daß die in verschiedenen Laboratorien erzielten Auswertungsresultate nur
annähernd vergleichbar sind.

c) Methode von Whitman.

Der von Whitman (341) standardisierte Hemmungstest ist vom *U. S. Army
Medical Department* (327) für die Auswertung menschlicher Influenzaseren über-
nommen worden und entspricht zum Teil dem Burnetschen Verfahren. Die
Ermittlung der Virustestdosis erfolgt durch die Vortitration des Virus (halbschlächtige Virusverdünnungen 1 : 10 bis 1 : 5120 in je 0,5 cc) gegenüber einer
1-%-Suspension von Menschenerythrocyten der Gruppe 0 (je 0,25 cc). Nach
zweistündigem Stehen der Gemische bei Zimmertemperatur wird der Endpunkt
im Röhrchen mit partieller Agglutination ermittelt. Die Virusgebrauchsdosis
beträgt vier agglutinierende Einheiten in 0,25 cc (= achtfach ermittelte End-
punktdosis). Für den Hemmungstest werden je 0,25 cc halbschlächtiger Serumverdünnungen (1 : 10 bis 1 : 4096) + 0,25 cc Virustestdosis + 0,25 cc 1-%-Ery-
throcytensuspension angesetzt, und die Gemische während zwei Stunden bei
Zimmertemperatur belassen. Der Endpunkt stellt diejenige Serumverdünnung
dar, die eben noch eine vollständige Hemmung bewirkt. Jedes Versuchsserum
und ebenso ein Standardimmunserum werden in zwei Versuchsreihen geführt,
wovon die eine mit Influenzavirus Typ A, die andere mit Virus Typ B geprüft
werden.

d) Methode von Salk.

Das von Salk (282) vorgeschlagene Verfahren bietet die Vorteile der Einfachheit, Ökonomie, hohen Empfindlichkeit und ausreichenden Genauigkeit
und scheint sich aus diesen Gründen in zunehmendem Maße einzubürgern. Die
Titration des Haemagglutinins geschieht in halbschlächtigen Virusverdünnungen
1 : 10 bis 1 : 5120 in einem Volumen von 0,5 cc Kochsalzlösung, zu welchen je
0,5 cc einer 0,25-%-Hühnererythrocytensuspension zugesetzt werden. Ein Teströhrchen enthält 0,5 cc Zellsuspension + 0,5 cc Kochsalzlösung zur Kontrolle
der Erythrocyten. Die durchgeschüttelten Gemische werden bei Zimmertemperatur bis zur vollständigen Sedimentierung der Erythrocyten, d. h. 90 bis
120 Minuten stehen gelassen. Die Kontrolle des Zellsedimentes erfolgt von
unten und erlaubt die *scharfe Differenzierung* zwischen der +-Reaktion, bzw.
vollständigen Agglutination (lachsfarbenes, über den ganzen Röhrchenboden
ausgebreitetes Sediment) und der 0-Reaktion, die der negativen Erythrocytenkontrolle entspricht (zentrales, scharfbegrenztes, linsenförmiges, dunkelrotes
Sediment). Die höchste Virusverdünnung, die eben noch eine +-Reaktion ergibt,
repräsentiert den Endpunkt. Zur genauen Ermittlung der Virustestdosis für
den Hemmungstest wird die Virustitration in zwei — gegeneinander um eine
eindrittelschlächtige Verdünnung verschobenen — Reihen (Beginn mit der
Virusverdünnung 1 : 20, bzw. 1 : 30) unter jeweiligem Pipettenwechsel durchgeführt und der Mittelwert der beiden Endpunkte bestimmt.

Im *Hemmungstest* werden die Reaktionsvolumina dadurch nicht verändert,
daß die halbschlächtigen Verdünnungen des Immunserums (1 : 32 bis 1 : 16000

Endverdünnung) in der Erythrocytensuspension (je 0,5 cc) ausgeführt werden, wobei das erste Röhrchen mit einer 0,5-%, das zweite bis zehnte Röhrchen mit einer 0,25-%-Zellsuspension beschickt werden. Die nach dem Aufschütteln der Gemische zugesetzte Virustestdosis beträgt zwei agglutinierende Einheiten in 0,5 cc, so daß im schließlichen Reaktionsvolumen von 1,0 cc eine Einheit enthalten ist. Die Kontrollen bestehen aus: 1. einer positiven Viruskontrolle (0,5 Virustestdosis + 0,5 cc einer 0,25-%-Erythrocytensuspension), 2. einer negativen Serumkontrolle (0,5 cc Serum 1 : 8, bzw. 1 : 16 + 0,5 cc Zellsuspension) und 3. einer negativen Erythrocytenkontrolle (0,5 cc Erythrocytensuspension + + 0,5 cc Kochsalzlösung). Nach eineinhalb bis zweistündigem Stehen der Gemische bei Zimmertemperatur wird die Ablesung in derselben Weise wie im direkten Agglutinationsversuch vorgenommen. Der Endpunkt ist wiederum scharf begrenzt und wird durch die höchste Serumverdünnung repräsentiert, die eben noch eine vollständige Hemmung (0-Reaktion) bewirkt.

e) Nachweis von Influenza-Antikörpern im menschlichen Serum.

Zur Serodiagnose der epidemischen Influenza dient der Antikörpernachweis in zwei Serumproben desselben Patienten, von denen die eine (Serum I) in der akuten fieberhaften Phase der Infektion, die andere (Serum II) 10 bis 12 Tage später in der Rekonvaleszenz entnommen wird. Ebenso läßt sich die immunisierende Wirkung einer Schutzimpfung durch die Bestimmung des Antikörpergehaltes im prä- und postvaccinalen Serum abschätzen. Für derartige Prüfungen sind sämtliche der angeführten Hemmungsteste geeignet, wobei nun aber die Auswertungsmethode von SALK wegen der technischen Einfachheit, mit welcher ein scharfbegrenzter Serumtiter ermittelt werden kann, besonders für Massenuntersuchungen den Vorzug verdient.

Nach einer von LÉPINE, SAUTTER und REINIÉ (232) angegebenen Modifikation der SALKschen Methode empfiehlt sich das folgende Vorgehen:

1. *Bestimmung der Virustestdosis:* Je 0,5 cc halbschlächtiger Verdünnungen (1 : 40 bis 1 : 5120) der Virustypen A und B werden mit je 0,5 cc einer 0,5-%-Suspension von Hühnererythrocyten versetzt. Die Bestimmung des Endpunktes (letzte Virusverdünnung mit +-Reaktion nach SALK) erfolgt nach eineinhalbstündigem Stehen der Gemische bei Zimmertemperatur. Die Gebrauchsdosis beträgt drei agglutinierende Einheiten in 0,5 cc.

2. *Bestimmung des Serumtiters:* Sämtliche Patientensera werden inaktiviert (30 Minuten, 56° C). Die Gemische bestehen aus:

0,5 cc halbschlächtiger Serumverdünnungen 1 : 40 bis 1 : 5120
+ 0,5 cc Gebrauchsdosis des Testvirus A, bzw. B
+ 0,5 cc einer 0,5-%-Hühnererythrocytensuspension.

Als Kontrollen dienen:

0,5 cc Virustestdosis + 0,5 cc Erythrocytensuspension + 0,5 cc NaCl
0,5 cc Serum 1 : 40 + 0,5 cc Erythrocytensuspension + 0,5 cc NaCl
0,5 cc Erythrocytensuspension + 1,0 cc NaCl.

Zur Spezifitätskontrolle des Testvirus A und B werden entsprechende Verdünnungsreihen mit Standardimmunseren bei jedem Versuch mitgeführt. Für die Untersuchung von zwei Serumproben (I und II) desselben Patienten sind acht Reihen zu je acht Röhrchen (ohne Kontrollen) erforderlich:

1. Patientenserum   I   + Testvirus A
2. Patientenserum   I   + Testvirus B
3. Patientenserum   II   + Testvirus A
4. Patientenserum   II   + Testvirus B
5. Standardserum Anti-A + Testvirus A
6. Standardserum Anti-A + Testvirus B
7. Standardserum Anti-B + Testvirus A
8. Standardserum Anti-B + Testvirus B.

In sämtlichen Gemischen wird der Endpunkt (höchste Serumverdünnung mit totaler Hemmung) nach eineinhalb Stunden Stehen bei Zimmertemperatur ermittelt. Für den Nachweis kleiner Antikörpermengen eignet sich die von Anderson, Burnet und Stone (6) vorgenommene Modifikation des Salkschen Hemmungstestes. Die weitgehende Ausschaltung der „unspezifischen Serumhemmung" wird durch die Verwendung *selektionierter Erythrocyten* von hoher Agglutinabilität und geringer unspezifischer Hemmbarkeit erreicht. Die Unterscheidung zwischen spezifischer, bzw. unspezifischer Hemmung kann durch die zweimalige Sedimentierung der Gemische, bzw. Endpunktbestimmung (Anstieg, bzw. Konstanz des Hemmungstiters) getroffen werden (vgl. S. 199).

Die *Bewertung des Versuchsausfalles* beschränkt sich meistens auf die Feststellung, ob und in welchem Grade der Antikörpergehalt des Serums II — im Vergleich zum Serum I — angestiegen, und gegenüber welchem Virustypus eine solche Zunahme erfolgt ist. Eine vierfache Steigerung des Hemmungstiters liegt außerhalb der Fehlergrenze und kann für den Antikörperanstieg als beweisend angesehen werden.

Literatur.

1. Anderson: The reaction between red cells and viruses of the influenza group. Studies with Newcastle disease virus. Austr. J. exp. Biol. a. Med. Sci 25, 163 (1947).
2. — Sporadic and minor epidemic incidence of influenza A in Victoria 1945—46. II. The breadth of antibody response following influenza A infection. Austr. J. exp. Biol. a. Med. Sci 25, 243 (1947).
3. — The mechanism of the union between Influenza virus and susceptible cells. Austr. J. exp. Biol. a. Med. Sci 10, 83 (1947).
4. — Mucins and mucoids in relation to influenza virus action. I. Inactivation by RDE and by viruses of the influenza group of the serum inhibitor of haemagglutination. Austr. J. exp. Biol. a. Med. Sci 26, 347 (1948).
5. Anderson and Burnet: Sporadic and minor epidemic incidence of influenza A in Victoria 1945—46. I. Phase behaviour of influenza A-strains in relation to epidemic characteristics. Austr. J. exp. Biol. a. Med. Sci 25, 235 (1947).
6. Anderson, Burnet and Stone: A modified Salk test for in vitro titration of influenza antibodies. Austr. J. exp. Biol. a. Med. Sci 24, 269 (1946).
7. Anderson, Burnet, Fazekas de St. Groth, McCrea and Stone: Mucins and mucoids in relation to influenza virus action. VI. General Discussion. Austr. J. exp. Biol. a. Med. Sci 26, 403 (1948).
8. Andrewes: Interference by one virus with the growth of another in tissue culture. Brit. J. exp. Path. 23, 214 (1942).
9. Andrewes and Elford: Infectious ectromelia: experiments on interference and immunization. Brit. J. exp. Path. 28, 278 (1947).
10. Andrewes and Glover: Grey lung virus: an agent pathogenic for mice and other rodents. Brit. J. exp. Path. 26, 379 (1945).

11. ARMY SERVICE FORCES. Office Surgeon General Washington 25, 2. IV. 1945: Recommendations for the preparation of influenza virus vaccine, typus A and B, for the U. S. Army.
12. AYNAUD: Action des microbes sur les globulins. C. R. Soc. Biol. Paris 70, 54 (1911).
13. BANG: Filamentous forms of Newcastle virus. Proc. Soc. exp. Biol. a. Med. 63, 5 (1946).
14. — Newcastle virus: Conversion of spherical forms to filamentous forms. Proc. Soc. exp. Biol. a. Med. 64, 135 (1947).
15. — Studies on Newcastle disease virus. I. An evaluation of the method of titration. J. exp. Med. 88, 233 (1948).
16. — Studies on Newcastle disease virus. II. Behaviour of the virus in the embryo. J. exp. Med. 88, 241 (1948).
17. — Studies on Newcastle disease virus. III. Characters of the virus itself with particular reference to electron microscopy. J. exp. Med. 88, 251 (1948).
18. BAYLOR and CLARK: Electron microscope studies of the „interference phenomenon" between bacterial viruses of the escherichia coli group. J. Bact. 53, 49 (1947).
19. BEARD: Purification of animal viruses. J. Imm. 58, 49 (1948).
20. BEVERIDGE: Lack of increase in antibody after second injection of influenza virus in man. Austr. J. exp. Biol. a. Med. Sci 22, 301 (1944).
21. BEVERIDGE and LIND: „Mumps". II. Virushaemagglutination and serological reactions. Austr. J. exp. Biol. a. Med. Sci 24, 127 (1946).
22. — „Mumps". III. Immunization experiments in man. Austr. J. exp. Biol. a. Med. Sci 25, 337 (1947).
23. BEVERIDGE, LIND and ANDERSON: „Mumps". Isolation and cultivation of the virus in the chick embryo. Austr. J. exp. Biol. a. Med. Sci 24, 15 (1946).
24. BEVERIDGE, BURNET and WILLIAMS: The isolation of influenza A and B viruses by chick embryo inoculations. Austr. J. exp. Biol. a. Med. Sci 22, 1 (1944).
25. BIELING und OELRICHS: Untersuchungen über die Haemagglutination durch Grippevirus. Z. Hyg. 127, 535 (1948).
26. — Untersuchungen über die Hemmungswirkung von Organzellen und Extrakten sowie von Allantoisflüssigkeit auf die haemagglutinierende Wirkung des Grippevirus. Z. Hyg. 127, 545 (1948).
27. — Über Hemmungsstoffe gegen Grippevirus und über Grippeimmunität der Nachkommen grippeimmunisierter Eltern. Z. Hyg. 127, 592 (1948).
28. BJÖRKMAN and HORSFALL: The production of a persistent alteration in influenza virus by lanthanum or ultraviolet irradiation. J. exp. Med. 88, 445 (1948).
29. BLASKOVIC and SALK: A method applicable to the standardization of influenza virus vaccines. Proc. Soc. exp. Biol. a. Med. 65, 352 (1947).
30. BODILY and EATON: Specificity of the antibody response of human beings to strains of influenza virus. J. Imm. 45, 193 (1942).
31. BODILY, COREY and EATON: Precipitation and concentration of influenza virus with alum. Proc. Soc. exp. Biol. a. Med. 52, 165 (1943).
32. BOURDILLON: On the haemagglutinins of beans and influenza virus. Experientia 5, 122 (1949).
33. BOVARNICK and DE BURGH: Virus Haemagglutination. Science 105, 550 (1947).
34. BREMER et MUTSAARS: Agglutination des globules rouges de mouton par le virus de la poliomyélite (souche Lansing). C. R. Soc. Biol. 142, 1194 (1948).
35. BRIODY: Characterization of the „enzymic" action of influenza virus on human red cells. J. Imm. 59, 115, (1948).
36. BRIODY and HANIG: Lack of action of influenza virus upon mucin of human or swine origin. Proc. Soc. exp. Biol. a. Med. 67, 485 (1948).
37. BULL and BURNET: Experimental Immunization of volunteers against influenza virus B. Med. J. Austr. 1, 389 (1943).
38. BURNET: The affinity of Newcastle Disease virus to the influenza virus group. Austr. J. exp. Biol. 20, 81 (1942).

39. Burnet: Influenza Virus B: I. Observations on growth in chick embryos and on the occurence of antibodies in Australian Serum. Med. J. Austral. 1, 671 (1942).
40. — Influenza Virus B: II. Immunization of human volunteers with living attenuated virus. Med. J. Austr. 1, 673 (1942).
41. — Characteristics of the influenza virus-antibody reaction as tested by the method of allantoic inoculation. Austr. J. exp. Biol. a. Med. Sci 21, 231 (1943).
42. — Partial dissociation of haemagglutinin and infectiosity of Newcastle disease virus. Austr. J. exp. Biol. a. Med. Sci 23, 178 (1945).
43. — Experiments on the elution of Newcastle disease virus and influenza virus from fowl red cells. Austr. J. exp. Biol. a. Med. Sci 23, 180 (1945).
44. — Hemagglutination by mumps virus: relationship to Newcastle disease and influenza viruses. Austr. J. Sci 8, 81 (1945).
45. — An unsuspected relationship between the viruses of vaccinia and infectious ectromelia of mice. Nature 155, 543, (1945).
46. — Vaccinia haemagglutinin. Nature 158, 119 (1946).
47. — Modification of human red cells by virus action. III. A sensitive test for Mumps antibody in human serum by the agglutination of human red cells coated with a virus antigen. Brit. J. exp. Path. 27, 244 (1946).
48. — The mucinase of V. cholerae. Austr. J. exp. Biol. a. Med. Sci 26, 71 (1948).
49. — Mucins and mucoids in relation to influenza virus action. III. Inhibition of virus hemagglutination by glandular mucins. Austr. J. exp. Biol. a. Med. Sci 26, 371 (1948).
50. — Mucins and mucoids in relation to influenza virus action. IV. Inhibition by purified mucoid of infection and haemagglutination with the virus strain WSE. Austr. J. exp. Biol. a. Med. Sci 26, 381 (1948).
51. — Mucins and mucoids in relation to influenza virus action. V. The destruction of "Francis Inhibitor" activity in a purified mucoid by virus action. Austr. J. exp. Biol. a. Med. Sci 26, 389 (1948).
52. — Mucins and mucoids in Medicine. Roy. Melbourne Hospital Clinical Reports. Centenary volume Aug. 1948.
53. — The initiation of cellular infection by influenza and related viruses. Lancet 1948, I, 7.
54. Burnet and Anderson: The T-Antigen of guinea pig and human red cells. Austr. J. exp. Biol. a. Med. Sci 25, 213 (1947).
55. — Modification of human red cells by virus action. II. Agglutination of modified human red cells by sera from cases of infectious mononucleosis. Brit. J. exp. Path. 27, 236 (1946).
56. Burnet and Beveridge: Titration of antibody against influenza viruses by allantoic inoculation of the developing chick embryo. Austr. J. exp. Biol. a. Med. Sci 21, 71 (1943).
57. — The action of human tears on influenza virus. Austr. J. exp. Biol. a. Med. Sci 23, 186 (1945).
58. Burnet and Boake: The relationship between the virus of infectious ectromelia of mice and vaccinia. J. Imm. 53, 1 (1946).
59. Burnet and Bull: Changes in influenza virus associated with adaptation to passage in chick embryos. Austr. J. exp. Biol. a. Med. Sci 21, 55 (1943).
60. — Re-examination of the influenza virus strain "Melbourne egg". Austr. J. exp. Biol. a. Med. Sci 22, 173 (1944).
61. Burnet and Clark: "Influenza". A survey of the last 50 years in the light of modern work on the virus of epidemic influenza. Monographs from the Walter and Eliza Hall Institute of Research in Pathology and Medicine, Melbourne No 4 (1942).
62. Burnet and McCrea: Inhibitory and inactivating action of normal ferret sera against an influenza virus strain. Austr. J. exp. Biol. a. Med. Sci 24, 277 (1946).
63. Burnet and Stone: The haemagglutinins of vaccinia and ectromelia viruses. Austr. J. exp. Biol. a. Med. Sci 24, 1 (1946).

64. BURNET and STONE: Serological response to influenza B infection in human beings. Differentiation of specific and non-specific type reactions. Austr. J. exp. Biol. a. Med. Sci **24**, 207 (1946).

65. — Desquamation of intestinal epithelium in vitro by V. cholerae filtrates. Characterization of Mucinase and tissue disintegrating enzymes. Austr. J. exp. Biol. a. Med. Sci **25**, 219 (1947).

66. — The receptor-destroying enzyme of V. cholerae. Austr. J. exp. Biol. a. Med. Sci **25**, 227 (1947).

67. BURNET, BEVERIDGE and BULL: Study of a strain of influenza B-virus isolated by chick-embryo inoculation. Austr. J. exp. Biol. a. Med. Sci **22**, 9 (1944).

68. BURNET, McCREA and ANDERSON: Mucin as substrate of enzyme action by viruses of the mumps-influenza group. Nature **160**, 404 (1947).

69. BURNET, McCREA and STONE: Modification of human red cells by virus action. I. The receptor gradient for virus action in human red cells. Brit. J. exp. Path. **27**, 228 (1946).

70. BURNET, STONE and ANDERSON: An epidemic of Influenza B in Australia. Lancet **250**, 807 (1946).

71. BURNET, BEVERIDGE, BULL and CLARK: Investigations of an influenza epidemic in military camps in Victoria. Med. J. Austral. **2**, 371 (1942).

72. CHAMBERS and HENLE: Studies on the nature of the virus of influenza. I. The dispersion of the virus of influenza in tissue emulsions and in extraembryonic fluids of the chick. J. exp. Med. **77**, 251 (1943).

73. CHAMBERS, HENLE, LAUFFER and ANDERSON: Studies on the nature of the virus of influenza. II. The size of the infectious unit in influenza A. J. exp. Med. **77**, 265 (1943).

74. CHENG-J. WANG: The relation of infectious and hemagglutination titers to the adaptation of influenza virus to mouse. J. exp. Med. **88**, 515 (1948).

75. CHU and COOMBS: Modification of human red cells by virus action. Lancet **252**, I, 484 (1947).

76. CLARK and NAGLER: Haem-Agglutination by viruses. The range of susceptible cells with special reference to agglutination by vaccinia virus. Austr. J. exp. Biol. a. Med. Sci **21**, 103 (1943).

77. *Commission of acute respiratory diseases:* Haemagglutination by amniotic fluid from normal embryonated Hen's egg. Proc. Soc. exp. Biol. a. Med. **62**, 118 (1946).

78. COX, VAN DER SCHEER, CUSTON and BOHNEL: The purification and concentration of influenza-virus by means of alcohol precipitation. J. Imm. **56**, 149 (1947).

79. CROWLEY, THIPGEN and RICHARD: Isolation of influenza A virus from normal human contacts during an epidemic of influenza A. Proc. Soc. exp. Biol. a. Med. **57**, 354 (1944).

80. CUNHA, WEIL, D. BEARD, TAYLOR, SHARP and J. W. BEARD: Purification and characters of the Newcastle Disease virus (California strain). J. Imm. **55**, 69 (1947).

81. CURNEN and HORSFALL: Studies on pneumonia of mice (PVM). III. Haemagglutination by the virus; the occurrence of combination between the virus and tissue substance. J. exp. Med. **83**, 105 (1946).

82. — Properties of pneumonia virus of mice (PVM) in relation to its state. J. exp. Med. **85**, 39 (1947).

83. CURNEN, PICKELS and HORSFALL: Centrifugation studies on pneumonia virus of mice (PVM). The relative sizes of free and combined virus. J. exp. Med. **85**, 23 (1947).

84. DAVENPORT and HORSFALL: The associative reactions of pneumonia virus of mice (PVM) and influenza viruses: The effects of p_H and electrolytes upon virus-host cell combination. J. exp. Med. **88**, 621 (1948).

85. DAVIDSOHN and TOHARSKY: The production of bacteriogenic haemagglutination. J. inf. dis. **67**, 25 (1940).

86. — Bacteriogenic Haemagglutination II. J. Imm. **43**, 213 (1942).

87. Davies: A slide rule for determining chicken red cell agglutination titer. Science **106**, 273 (1947).
88. de Burgh, Howe and Bovarnick: Preparation from human red cells of a substance inhibiting virus haemagglutination. J. exp. Med. **87**, 1 (1948).
89. Delbrück: Interference between bacterial viruses. III. The mutual exclusion effect and the depressor effect. J. Bact. **50**, 151 (1945).
90. Dinter: Atypische Hühnerpest und Newcastle-Krankheit. Tierärztl. Umschau Nr. 15/16 (1947).
91. — Mündliche Mitteilung 1948.
92. — Über die haemagglutinative Aktivität des Rotlaufbakteriums. Tierärztl. Umschau **1948**, 143.
93. — Über den Haemagglutinationshemmungstest beim Rotlauf. Berl. und Münch. tierärztl. Wschr. Nr. 10, 113 (1948).
94. Dinter, Bakos und Angermair: Über den Haemagglutinationstest bei ägyptischer Hühnerpest. Berl. u. Münch. tierärztl. Wschr. Nr. 3, 32 (1948).
95. Doerr: Antikörper I. Teil. Immunitätsforschg. **1**, 223 (1947).
96. — Antikörper II. Teil. Immunitätsforschg. **4**, 17 (1949)
97. Doerr und Gold: Untersuchungen über das Virus der Hühnerpest. V. Mitteilung: Analyse der Septicaemie des infizierten Huhnes. Quantitative Verhältnisse der Virusadsorption in vitro. Z. Hyg. **113**, 645 (1932).
98. Doerr und Seidenberg: Untersuchungen über das Virus der Hühnerpest. VII. Mitteilung. Zur Virusadsorption in vitro. Z. Hyg. **114**, 269 (1932).
99. Dudgeon, Mellonby, Glover and Andrewes: Influenza A in Great Britain 1946/47. Brit. J. exp. Path. **29**, 132 (1948).
100. Dudgeon, Stuart-Harris, Glover, Andrewes and Bradley: Influenza B in 1945/46. Lancet **1946**, II, 627.
101. Eaton and Meiklejohn: Vaccination against influenza: A study in California during the epidemic of 1943/44. Am. J. Hyg. **42**, 28 (1945).
102. Eaton, Corey, van Herick and Meiklejohn: A comparison of various methods of demonstrating influenza virus in throat washings. Proc, Soc. exp. Biol. a. Med. **58**, 6 (1945).
103. Enders and Levens: "Diagnostic procedures for virus and rickettsial diseases." Am. Publ. Health Association New York 1948, 139.
104. Enders, Levens, Stockes, Maris and Berenberg: Attenuation of virulence with retention of antigenicity of mumps virus after passage in the embryonated egg. J. Imm. **54**, 283 (1946).
105. Evans and Curnen: Serological studies on infectious mononucleosis and other conditions with human erythrocytes modified by Newcastle disease virus. J. Imm. **58**, 323 (1948).
106. Fazekas de St. Groth: Destruction of influenza virus receptors in the mouse lung by an enzyme from V. cholerae. Austr. J. exp. Biol. a. Med. Sci **26**, 29 (1948).
107. — Regeneration of virus receptors in mouse lungs after artificial destruction. Austr. J. exp. Biol. a. Med. Sci **26**, 271 (1948).
108. — Modification of virus receptors by metaperiodat. I. The properties of JO_4-treated red-cells. Austr. J. exp. Biol. a. Med. Sci **27**, 65 (1949).
109. Fazekas de St. Groth and Graham: Modification of virus receptors by metaperiodat. II. Infection through modified receptors. Austr. J. exp. Biol. a. Med. Sci **27**, 83 (1949).
110. Feiner, Meyer and Steinberg: Bacterial lysis by lysozyme. J. Bact. **52**, 375 (1946).
111. Felsenfeld: The lecithinase activity of vibrio comma and the El Tor vibrio. J. Bact. **48**, 155 (1944).
112. Fenner: Studies in infectious ectromelia of mice. I. Immunization of mice against ectromelia with living vaccinia virus. Austr. J. exp. Biol. a. Med. Sci **25**, 257 (1947).
113. — Studies in infectious ectromelia in mice. II. Natural transmission: The portal of entry of the virus. Austr. J. exp. Biol. a. Med. Sci **25**, 275 (1947).

114. Fenner: Studies in infectious ectromelia in mice. III. Natural transmission: Elimination of the virus. Austr. J. exp. Biol. a. Med. Sci **25**, 327 (1947).
115. — The epizootic behaviour of mouse pox (infectious ectromelia). Brit. J. exp. Path. **29**, 69 (1948).
116. Fisher: The behaviour of Haemophilus pertussis in casein hydrolysate broth. Austr. J. exp. Biol. a. Med. Sci **26**, 299 (1948).
117. Flexner and Noguchi: Snake venom in relation to haemolysis, bacteriolysis and toxicity. J. exp. Med. **6**, 277 (1902).
118. Flick: Use of formalin-treated red cells for the study of influenza A virus hemagglutinating activity. Proc. Soc. exp. Biol. a. Med. **68**, 448 (1948).
119. Florman: "False positive" hemagglutination by allantoic fluids of embryonated eggs inoculated with unfiltered throat washings. J. Bact. **52**, 307 (1946).
120. — Hemagglutination with Newcastle disease virus (NDV). Proc. Soc. exp. Biol. a. Med. **64**, 458 (1947).
121. — Some alterations in chicken erythrocytes which follow treatment with influenza and Newcastle disease virus. J. Bact. **55**, 183 (1948).
122. Florman and Crawford: Use of agglutination-inhibition test in studying an epidemic of influenza. Am. J. Med. Sci **208**, 494 (1944).
123. Florman and Weiss: Storage of influenza virus for use in typing clinical cases. J. Bact. **49**, 507 (1945).
124. Francis: Apparent serological variation within a strain of influenza virus. Proc. Soc. exp. Biol. a. Med. **65**, 143 (1947).
125. — Dissociation of hemagglutinating and antibody-measuring capacities of influenza virus. J. exp. Med. **85**, 1 (1947).
126. Francis and Minuse: Influence of saliva upon hemagglutination by influenza virus. Proc. Soc. exp. Biol. a. Med. **69**, 291 (1948).
127. Francis and Salk: A simple procedure for the concentration and purification of influenza virus. Science **96**, 499 (1942).
128. Francis, Salk and Brace: The protective effect of vaccination against epidemic influenza B. J. A. M. A. **131**, 275 (1946).
129. Francis, Pearson, Salk and Brown: The status of immunity in human subjects artificially infected with influenza virus, type B. Am. J. publ. Health **34**, 317 (1944).
130. Friedenreich: Investigations into the Thomsen hemagglutination phenomena. Acta Path. et Microbiol. Scand. **5**, 59 (1928).
131. Friedenreich and Andersen: On the existence outside human blood of substances convertible into receptors T. Acta Path. et Microbiol. Scand. **6**, 236 (1939).
132. Friedewald: The immunological response to influenza virus infection as measured by the complement fixation test. J. exp. Med. **78**, 347 (1943).
133. — Qualitative differences in the antigenic composition of influenza A virus strains. J. exp. Med. **79**, 633 (1944).
134. — Adjuvants in immunization with influenza virus vaccines. J. exp. Med. **80**, 477 (1944).
135. — Enhancement of the immunizing capacity of influenza virus vaccines with adjuvants. Science **99**, 453 (1944).
136. Friedewald and Hook: Influenza virus infection in the hamster. J. exp. Med. **88**, 343 (1948).
137. Friedewald and Pickels: Size of infective particle and hemagglutinin of influenza virus as determined by centrifugal analysis. Proc. Soc. exp. Biol. a. Med. **52**, 261 (1943).
138. — Centrifugation and ultrafiltration studies on allantoic fluid preparations of influenza virus. J. exp. Med. **79**, 301 (1944).
139. Friedewald, Miller and Whatley: The nature of non-specific inhibition of virus hemagglutination. J. exp. Med. **86**, 65 (1947).
140. Fukuhara: Über haemaglutinierende Bakterien. Z. Immun.-Forsch. **2**, 313 (1909).

141. Gard and v. Magnus: Studies on interference in experimental influenza. II. Purification and centrifugation experiments. Ark. f. Kem., Min. u. Geol. 24 B, Nr. 8 (1947).
142. Gard, v. Magnus and Svedmyr: Physico-chemical aspects on inhibition and interference in experimental influenza. Proc. 4th intern. Congr. Microbiol. Copenhagen 1947, 301.
143. Ginsberg and Horsfall: Concurrent infection with influenza virus and mumps virus or pneumonia virus of mice (PVM) as bearing on the inhibition of virus multiplication by bacterial polysaccharides. J. exp. Med. 89, 37 (1949).
144. Ginsberg, Goebel and Horsfall: Inhibition of mumps virus by a polysaccharide. Proc. Soc. exp. Biol. a. Med. 66, 99 (1947).
145. — The inhibitory effect of polysaccharide on mumps virus multiplication. J. exp. Med. 87, 385 (1948).
146. — The effect of polysaccharides on the reaction between erythrocytes and viruses, with particular reference to mumps virus. J. exp. Med. 87, 411 (1948).
147. Glover and Andrewes: The antigenic structure of British strains of swine influenza virus. J. comp. Path. a. Therap. 53, 329 (1943).
148. Gordon: Demonstration of antigenic differences between different strains of influenza B. J. Imm. 44, 231 (1942).
149. — Adsorption of vaccinia by erythrocytes. N. Y. State Dep. Health Ann. Rep. Div. Lab. Res. p. 15 (1943).
150. Gottschalk and Lind: cit nach Fazekas de St. Groth (108) Hemagglutination of sheep cells by influenza viruses. Annual Report of the Walter and Eliza Hall Institute 1947-48, p. 26.
151. Green: Inhibition of multiplication of influenza virus by tannic acid. Proc. Soc. exp. Biol. a. Med. 67, 483 (1948).
152. Green and Woolley: Inhibition by certain polysaccharides of hemagglutination and of multiplication of influenza virus. J. exp. Med. 86, 55 (1947).
153. Green, Rasmussen and Smadel: Chemoprophylaxis of experimental influenza infections in eggs. Publ. Health Rep. 61, 1401 (1946).
154. Griffitts: Hemagglutination by bacterial suspensions with special reference to shigella alcalescens. Proc. Soc. exp. Biol. a. Med. 67, 358 (1948).
155. Gujot: Über die bakterielle Haemagglutination (Bacterio-Haemagglutination). Zbl. Bakter. Orig. I, 47, 640 (1908).
156. Hale and McKee: The value of influenza vaccination when done at the beginning of an epidemic. Am. J. Hyg. 42, 21 (1945).
157. — The intracranial toxicity of influenza virus for mice. Proc. Soc. exp. Biol. a. Med. 59, 81 (1945).
158. Hallauer: Beitrag zur Kenntnis der Erythrocytenveränderung nach Thomsen. Z. Immun-Forsch. 67, 15 (1930).
159. — Immunitätsstudien bei Hühnerpest. IV. Mitteilung: Über vererbte Immunität. Z. Hyg. 118, 605 (1936).
160. — Über den Virusnachweis mit dem Hirst-Test. Schweiz. Zschr. Path. u. Bakt. 9, 553 (1946).
161. — Beitrag zur Klassifikation der Hühnerpeststämme. Arch. ges. Virusforschg. 3, 356 (1947).
162. — Über Haemagglutination durch Virusarten. Bull, Schweiz. Acad. Med. Wiss. 3, 81 (1947/48).
163. — Die Haemagglutination durch murine Poliomyelitisstämme. Experentia (im Druck).
164. — Unveröffentlichte Versuche.
165. Hanig: Electrokinetic changes in human erythrocytes during adsorption and elution of PR 8 influenza virus. Proc. Soc. exp. Biol. a. Med. 68, 385 (1948).
166. Hardy and Horsfall: Reactions between influenza virus and a component of allantoic fluid. J. exp. Med. 88, 463 (1948).

167. Hare, Mackenzie, McClelland and Curl: Immunization of human beings with red cell eluate and other types of influenza vaccine. Brit. J. exp. Path. **28**, 141 (1947).
168. Hare, Morgan, Jackson and Stamatis: Immunization against influenza A. Canad. J. publ. Health **34**, 353 (1943).
169. Heinmets: Studies with the electron microscope on the interaction of red cells and influenza virus. J. Bact. **55**, 823 (1948).
170. Henle and Henle: Interference of inactive virus with the propagation of virus of influenza. Science **98**, 87 (1943).
171. — Neurological signs in mice following intracerebral inoculation of influemza viruses. Science **100**, 410 (1944).
172. — Interference between active and inactive viruses of influenza; incidental occurrence and artificial induction of phenomenon. Am. J. med. Sci **207**, 705 (1944).
173. — Interference between inactive and active viruses of influenza; factors influencing phenomenon. Am. J. Med. Sci **207**, 717 (1944).
174. — Interference between inactive and active viruses of influenza. III. Cross-interference between various related and unrelated viruses. Am. J. Med. Sci **210**, 362 (1945).
175. — Interference between inactive and active viruses of influenza. IV. The nature of the interfering agent. Am. J. Med. Sci **210**, 369 (1945).
176. — The toxicity of influenza viruses. Science **102**, 398 (1945).
177. — Effects of adjuvants on vaccination of human beings against influenza. Proc. Soc. exp. Biol. a. Med. **59**, 179 (1945).
178. — Studies on the toxicity of influenza viruses. I. The effect of intracerebral injection of influenza viruses. J. exp. Med. **84**, 623 (1946).
179. — Studies on the toxicity of influenza viruses. II. The effect of intraabdominal and intravenous injection of influenza viruses. J. exp. Med. **84**, 639 (1946).
180. — The effect of ultraviolet irradiation on various properties of influenza virus. J. exp. Med. **85**, 347 (1947).
181. Henle and McDougall: Mumps meningo-encephalitis. Isolation of virus from spinal fluid of a patient. Proc. Soc. exp. Biol. a. Med. **66**, 209 (1947).
182. Henle and Wiener: Complement-fixation antigens of influenza viruses type A and B. Proc. Soc. exp. Biol. a. Med. **57**, 176 (1944).
183. Henle, Henle and Harris: The serological differentiation of mumps complement-fixation antigens. Proc. Soc. exp. Biol. a. Med. **64**, 290 (1947).
184. Henle, Henle and Kirber: Interference between inactive and active viruses of influenza. V. Effect of irradiated virus on the host cells. Am. J. Med. Sci **214**, 529 (1947).
185. Henle, Henle and Rosenberg: The demonstration of one-step growth curves of influenza viruses through the blocking effect of irradiated virus on further infection. J. exp. Med. **86**, 423 (1947).
186. Henle, Henle and Stokes: Demonstration of the efficacy of vaccination against influenza type A by experimental infection of human beings. J. Imm. **46**, 163 (1943).
187. Henle, Henle, Groupé and Chambers: Studies on complement fixation with the viruses of influenza. J. Imm. **48**, 163 (1944).
188. Henle, Henle, Stokes and Maris: Experimental exposure of human subjects to viruses of influenza. J. Imm. **52**, 145 (1946).
189. Henle, Henle, Hampil, Maris and Stokes: Experiments on vaccination of human beings against epidemic influenza. J. Imm **53**, 75 (1946).
190. Herick, van and Eaton: An unidentified pleuropneumonialike organism isolated during passages in chick embryo. J. Bact. **50**, 47 (1945).
191. Hirst: The agglutination of red cells by allantoic fluid of chick embryo infected with influenza virus. Science **94**, 22 (1941).
192. — In vivo titration of influenza virus and of neutralizing antibodies in chick embryos. J. Imm. **45**, 285 (1942).

193. Hirst: Direct isolation of human influenza virus in chick embryos. J. Imm. **45,** 293 (1942).
194. — The quantitative determination of influenza virus and antibodies by means of red cell agglutination. J. exp. Med. **75,** 49 (1942).
195. — Adsorption of influenza hemagglutinins and virus by red blood cells. J. exp. Med. **76,** 195 (1942).
196. — Adsorption of influenza virus on cells of the respiratory tract. J. exp. Med. **78,** 99 (1943).
197. — Studies of antigenic differences among strains of influenza A by means of red cell agglutination. J. exp. Med. **78,** 407 (1943).
198. — Direct isolation of influenza virus in chick embryos. Proc. Soc. exp. Biol. a. Med. **58,** 155 (1945).
199. — Studies on the mechanism of adaptation on influenza virus to mice. J. exp. Med. **86,** 357 (1947).
200. — Comparison of influenza virus strains from three epidemics. J. exp. Med. **86,** 367 (1947).
201. — The nature of the virus receptors of red cells. I. Evidence on the chemical nature of the virus receptors of red cells and of the existence of a closely analogous substance in normal serum. J. exp. Med. **87,** 301 (1948).
202. — The nature of the virus receptors of red cells. II. The effect of partial heat inactivation of influenza virus on the destruction of red cell receptors and the use of inactivated virus in the measurement of serum inhibitor. J. exp. Med. **87,** 315 (1948).
203. Hirst and Pickels: A method for the titration on influenza hemagglutinins and influenza antibodies with the aid of a photoelectric densitometer. J. Imm. **45,** 273 (1942).
204. Hirst, Plummer and Friedewald: Human immunity following vaccination with formalinized influenza virus. Am. J. Hyg. **42,** 45 (1945).
205. Hirst, Richard and Friedewald: Studies in human immunization against influenza. Duration of immunity induced by inactive virus. J. exp. Med. **80,** 265 (1944).
206. Hirst, Richard and Whitman: A new method for concentrating influenza virus from allantoic fluid. Proc. Soc. exp. Biol. a. Med. **50,** 129 (1942).
207. Hirst, Richard, Whitman and Horsfall: Antibody response of human beings following vaccination with influenza viruses. J. exp. Med. **75,** 495 (1942).
208. Horsfall and McCarty: The modifying effects of certain substances of bacterial origin on the course of infection with pneumonia virus of mice (PVM). J. exp. Med. **85,** 623 (1947).
209. Hudson, Sigel and Markham: Antigenic relationship of British swine influenza strains to standard human and swine influenza viruses. J. exp. Med. **77,** 467 (1943).
210. Jones: Adaption of influenza virus to heat. Proc. Soc. exp. Biol. a. Med. **58,** 315 (1945).
211. Kalter and Chapman: Serological studies on influenza during a unic-month period. J. clin. Invest. **26,** 420 (1947).
212. Kalter, Chapman, Feeley and McDowell: An epidemic of influenza as due to an atypical strain. The relationship of this strain to other influenza viruses. J. Imm **59,** 147 (1948).
213. Kauffmann: On hemagglutination by Escherichia Coli. Acta path. **XXV,** 502 (1948).
214. Keogh and North: The hemagglutinin of hemophilus pertussis. I. Hemagglutinin as a protective antigen in experimental murine pertussis. Austr. J. exp. Biol. a. Med. Sci **26,** 315 (1948).
215. Keogh, North and Warburton: Hemagglutinins of the hemophilus group. Nature **160,** 63 (1947).
216. Kilham: Isolation of mumps virus from the blood of a patient. Proc. Soc. exp. Biol. a. Med. **69,** 99 (1948).

217. KLEIN and PEREZ: The reversal in vivo by Bal of inactivation of influenza A and Diplococcus pneumoniae by bichloride of mercury. J. Imm 60, 349 (1948).
218. KNIGHT: The stability of influenza virus in the presence of salts. J. exp. Med. 79, 285 (1944).
219. — A sedimentable component of allantoic fluid and its relationship to influenza viruses. J. exp. Med. 80, 83 (1944).
220. — The preparation of highly purified PR 8-influenza virus from infected mouse lung. Science 101, 231 (1945).
221. — The preparation of highly purified PR 8-influenza virus from infected mouse lung. J. exp. Med. 83, 11 (1946).
222. — Precipitin reactions of highly purified influenza viruses and related materials. J. exp. Med. 83, 281 (1946).
223. — The nucleic acid and carbohydrate of influenza virus. J. exp. Med. 85, 99 (1947).
224. KRAUS und LUDWIG: Über Bakteriohaemagglutinine und Antihaemagglutinine. Wien. klin. Wschr. S. 102 (1902).
225. LAMANNA: Hemagglutination by botulinal toxin. Proc. Soc. exp. Biol. a. Med. 69, 332 (1948).
226. LANNI and BEARD: Inhibition by egg-white of hemagglutination by swine influenza virus. Proc. Soc. exp. Biol. a. Med. 68, 312 (1948).
227. — The mechanism of egg-white inhibition of hemagglutination by swine influenza virus. Proc. Soc. exp. Biol. a. Med. 68, 442 (1948).
228. LAUFFER and MILLER: The sedimentation rate of the biological activities of influenza A virus. J. exp. Med. 80, 521 (1944).
229. LAUFFER and STANLEY: Biophysical properties of praparations of PR 8 influenza virus. J. exp. Med. 80, 531 (1944).
230. LAZARUS and WESTFALL: The incidence of neutralizing antibodies for type A and type B influenza virus in normal infants and children. J. Imm. 59, 159 (1948).
231. LENNETTE and KOPROWSKI: Interference between viruses in tissue culture. J. exp. Med. 83, 195 (1946).
232. LÉPINE, SAUTTER et REINIÉ: Technique de la reaction d'agglutination des hématies pour le diagnostic de la grippe. Ann. Inst. Pasteur 72, 523 (1946).
233. LEVENS and ENDERS: The hemagglutinative properties of amniotic fluid from embryonated eggs infected with mumps virus. Science 102, 117 (1945).
234. LEYMASTER and WARD: Direct isolation of mumps virus in chick embryos. Proc. Soc. exp. Biol. a .Med. 65, 346 (1947).
235. LIND: The nature of mumps virus action on red cells. Austr. J. exp. Biol. a. Med. Sci 26, 93 (1948).
236. LIND and MCARTHUR: The distribution of "T"-agglutinins in human sera. Austr. J. exp. Biol. a. Med. 25, 247 (1947).
237. LÖFFLER: Noch unveröffentlichte Versuche.
238. LOWELL and BUCKINGHAM: A comparison of the effect of various salt concentrations on the agglutination of red cells by influenza A virus and antibody. J. Imm. 58, 229 (1948).
239. LUSH: The chick red cell agglutination test with the viruses of Newcastle disease and fowl plage. J. comp. Path. a. Therap. 53, 157 (1943).
240. MAGILL and SUGG: Physical-chemical factors in agglutination of sheep erythrocytes by influenza virus. Proc. Soc. exp. Biol. a. Med. 66, 89 (1947).
241. — The reservibility of the O-D type of influenza virus variation. J. exp. Med. 87, 535 (1948).
242. MAGILL, PLUMMER, SMILLIE and SUGG: An evaluation of vaccination against influenza. Am. J. Hyg. 42, 94 (1945).
243. MAGNUS, VON: Studies on interference in experimental influenza. I. Biological observations. Ark. Kem., Min. u. Geol. 24B, Nr. 7 (1946).
244. — Studies on interference in experimental influenza. The formation of non-infective virus in chick embryos inoculated with undiluted egg passage virus. Proc. 4th intern. Congr. Microbiol. Copenhagen 1947, 300.

245. Mawson and Swan: Intranasal vaccination of humans with living attenuated influenza virus strains. Med. J. Austr. 1, 394 (1943).
246. McClelland and Hare: The adsorption of influenza virus by red cells and a new in vitro method for measuring antibodies for influenza virus. Canad. J. publ. Health 32, 530 (1941).
247. McCrea: Unspecific serum inhibition of influenza Hemagglutination. Austr. J. exp. Biol. a. Med. Sci 24, 283 (1946).
248. — Modification of red cell agglutinability by Cl. welchii toxins. Austr. J. exp. Biol. a. Med. Sci 25, 127 (1947).
249. — Mucins and mucoids in relation to influenza virus action. II. Isolation and characterization of the serum mucoid inhibitor of heated influenza virus. Austr. J. exp. Biol. a. Med. Sci 26, 355 (1948).
250. McKee and Hale: Reactivation of overneutralized mixtures of influenza virus and antibody. J. Imm. 54, 233 (1946).
251. McLean, D. Beard, Taylor, Sharp and J. W. Beard: The relation of antibody response in swine to dose of the swine influenza virus inactivated with formalin and with ultraviolet light. J. Imm. 51, 65 (1945).
252. — Antibody response of swine to repeated vaccination with formalin-inactivated, purified swine influenza virus. Proc. Soc. exp. Biol. a. Med. 60, 152 (1945).
253. — Antibody response of swine to vaccination with formolized swine influenza virus adsorbed on alum. Proc. Soc. exp. Biol. a. Med. 60, 358 (1945).
254. — The antibody response of swine to vaccination with inactivated swine influenza virus. Science N. Y. 101, 544 (1945).
255. McLean, D. Beard, Taylor, Sharp, J. W. Beard, Feller and Dingle: Influence of temperature of incubation on the increase of influenza virus B (Lee-strain) in the chorioallantoic fluid of chick embryos. J. Imm. 48, 305 (1944).
256. Miller: A study of conditions for the optimum production of PR 8 influenza virus in chick embryos. J. exp. Med. 79, 173 (1944).
257. — Influence of p_H and of certain other conditions on the stability of the infectivity and red cell agglutinating activity of influenza virus. J. exp. Med. 80, 507 (1944).
258. — Sedimentation, viscosity and electrophoretic studies on purified Lee influenza virus preparation. J. biol. Chem. 169, 745 (1947).
259. Miller and Stanley: Quantitative aspects of the red blood cell agglutination test for influenza virus. J. exp. Med. 79, 185 (1944).
260. Miller, Lauffer and Stanley: Electrophoretic studies on PR 8 influenza virus. J. exp. Med. 80, 549 (1944).
261. Mills and Dochez: Specific agglutination of murine erythrocytes by a pneumonitis virus in mice. Proc. Soc. exp. Biol. a. Med. 57, 140 (1944).
262. — Further observations on red cell agglutinating agent present in lungs of virus-infected mice. Proc. Soc. exp. Biol. a. Med. 60, 141 (1945).
263. Morgan, Enders and Wagley: A hemolysin associated with the mumps virus. J. exp. Med. 88, 503 (1948).
264. Nagler: Application of Hirst phenomenon to the titration of vaccinia virus and vaccinia immune serum. Med. J. Austr. 1, 281 (1942).
265. — Red cell agglutination by vaccinia virus. Application to a comparative study of vaccination with egg vaccine and standard calf lymph. Austr. J. exp. Biol. a. Med. Sci 22, 29 (1944).
266. North: A study of the immunological reactions of the variola and vaccinia grown in the developping egg. Austr. J. exp. Biol. a. Med. Sci 22, 105 (1944).
267. O'Connor: Hirst's hemagglutination phenomenon exhibited by rickettsia orientalis (syn. Tsutsugamushi). Med. J. Austr. 2, 459 (1945).
268. Panthier, Cateigne et Harmoun: Nouvelle technique d'étude du virus grippal; premières applications pratiques. Ann. Inst. Pasteur 75, 338 (1948).
269. Parodi, Lajmanovich y Mittelman: Adsorcion del virus de la influenza por los nucleos de los globulos rojos de pollo. Riv. Inst, Bacteriol "C. G. Malbran" 12, 312 (1944).

270. Pearson: Distribution of influenza virus type A in infected eggs and the survival of virus under certain conditions of storage. J. Bact. **48**, 369 (1944).

271. — Attempts to obtain specific agglutination of mixtures of collodion particles or bacterial cells with virus and antiviral serum. J. Imm. **49**, 117 (1944).

272. *Personnel of Naval Laboratory Research Unit No 1:* The use of the red blood cell agglutination test in the study of influenza. U. S. Naval Med. Bull. **41**, 114 (1943).

273. — Experimental human influenza. Am. J. Med. Sci **207**, 306 (1944).

274. Pickles: Effect of cholera filtrate on red cells demonstrated by incomplete Rh-antibodies. Nature **158**, 880 (1946).

275. Quilligan, Minuse and Francis: Homologous and heterologous antibody response of infants and children to multiple injections of a single strain of influenza virus. J. clin. Invest. **27**, 572 (1948).

276. Raettig: Untersuchungen zur Immunität und Serologie der Influenza. I. bis III. Mitteilung. Z. Bakter. I. Orig. **152**, 159, 186, 381 (1947).

277. Richard, Thigpen and Adams: Antibody response to strains of influenza and a swine influenza viruses in the serums of infants experiencing their first infection with influenza A. J. inf. Dis. **76**, 203 (1945).

278. Richard, Thipgen and Crowley: The isolation of influenza A virus by the intraallantoic inoculation of chick embryos with untreated throat washings. J. Imm. **49**, 263 (1944).

279. — Vaccination against influenza at the university of Minnesota. Am. J. Hyg. **42**, 12 (1945).

280. Rose and Geelhorn: Inactivation of influenza virus with sulfur and nitrogen mustards. Proc. Soc. exp. Biol. a. Med. **65**, 83 (1947).

281. Rosenthal: Agglutinating properties of Escherichia coli. Agglutination of erythrocytes, leucocytes, thrombocytes, spermatozoa, spores of molds and pollen by strains of E. coli. J. Bact. **45**, 545 (1943).

282. Salk: A simplified procedure for titrating hemagglutinating capacity of influenza virus and the corresponding antibody. J. Imm. **49**, 87 (1944).

283. — The immunizing effect of calcium phosphate adsorbed influenza virus. Science **101**, 122 (1945).

284. — Variation in influenza viruses. A study of heat stability of the red cell agglutinating factor. Proc. Soc. exp. Biol. a. Med. **63**, 134 (1946).

285. — Effect of formalin in increasing heat stability of influenza virus hemagglutinin. Proc. Soc. exp. Biol. a. Med. **63**, 140 (1946).

286. — Reactions to concentrated influenza virus vaccine. J. Imm. **58**, 369 (1948).

287. Salk, Menke and Francis: A clinical, epidemiological and immunological evaluation of vaccination against epidemic influenza. Am. J. Hyg. **42**, 57 (1945).

288. Salk, Pearson, Brown, Smyth and Francis: Immunization against influenza with observations during an epidemic of influenza A one year after vaccination. Am. J. Hyg. **42**, 307 (1945).

289. Sautter et Lépine: Etudes des causes d'erreur dans les réactions d'hémagglutination. Rôle du cuivre. Ann. Inst. Pasteur **74**, 261 (1948).

290. Sharp, Taylor, McLean, Beard, Beard, Feller and Dingle: Isolation and characterization of influenza virus B (Lee-strain). Science **98**, 307 (1943).

291. — Isolation and characterization of influenza B-virus (Lee-strain). J. Imm. **48**, 129 (1944).

292. Shrigley: Agglutination of staphylococcus aureus in the presence of chorioallantoic fluid from hen's eggs. Science **102**, 64 (1945).

293. Shwartzman: Association of the virus of lymphocytic choriomeningitis with erythrocytes of infected animals. J. Bact. **46**, 482 (1943).

294. — Recovery of the virus of lymphocytic chorio-meningitis from the erythrocytes of infected animals. J. Imm. **48**, 111 (1944).

295. Sigurdsson: Studies on influenza infections in the chick embryo. J. Imm. **48**, 39 (1944).

296. Skinner: The red cell agglutination test. Failure to demonstrate red cell agglutination by the foot- and mouth disease and vesicular stromatitis viruses. Rep. of the Research Institute Pirbright (1947).

297. Smith: Propagation of rabbit fibroma virus in the embryonated egg. Proc. Soc. exp. Biol. a. Med. **69**, 136 (1948).

298. Sugg and Magill: The serial passage of mixtures of different strains of influenza virus in embryonated eggs and in mice. J. Bact. **56**, 201 (1948).

299. Svec and Forster: Inhibition of the Hirst haemagglutination reaction by pneumococcal extract, normal serums and blood cell esterases. Proc. Soc. exp. Biol. a. Med. **66**, 20 (1947).

300. Svedmyr: Studies on a factor in normal allantoic fluid inhibiting influenza virus haemagglutination. Ark. Kem., Min. u. Geol. **24B**, Nr. 11 (1947).

301. — Studies on a factor in normal allantoic fluid inhibiting influenza virus. Proc. 4th intern. Congr. Microbiol. Copenhagen **1947**, 299.

302. — Studies on a factor in normal allantoic fluid inhibiting influenza virus haemagglutination. Occurrence, physico-chemical properties and mode of action. Brit. J. exp. Path. **29**, 295 (1948).

303. — Studies on a factor in normal allantoic fluid inhibiting influenza virus haemagglutination. Virus-inhibitor interaction. Brit. J. exp. Path. **29**, 309 (1948).

304. Schellner und Seyerl: Beiträge zum Rotlaufproblem. Tierärztl. Umschau **4**, 29 (1948).

305. Schwartz, Milstone, Bayliss and Decoursey: An influenza index on periodic determination of average antibody titers of population samples. J. Imm. **54**, 225 (1946).

306. Stanley: An evaluation of methods for the concentration and purification of influenza virus. J. exp. Med. **79**, 255 (1944).

307. — The size of influenza virus. J. exp. Med. **79**, 267 (1944).

308. — The preparation and properties of influenza virus vaccines concentrated and purified by differential centrifugation. J. exp. Med. **81**, 193 (1945).

309. — Biochemical studies on influenza virus. Chemical and engineering. New Am. chem. Soc. **24**, 750 (1946).

310. Stanley and Lauffer: Sedimentation constants of purified preparations of strains of influenza virus. J. physic. a. coll. chem. **51**, 148 (1947).

311. Stone: Inactivation of vaccinia and ectromelia virus haemagglutinins by lecithinase. Austr. J. exp. Biol. a. Med. Sci **24**, 191 (1946).

312. — Lipoid Haemagglutinins. Austr. J. exp. Biol. a. Med. Sci **24**, 197 (1946).

313. — Comparison of the action of vibrio cholera enzyme and of viruses on the red cell surface. Austr. J. exp. Biol. a. Med. Sci **25**, 137 (1947).

314. — Enzymic modification of the reaction between influenza virus and susceptible tissue cells. Nature **159**, 780 (1947).

315. — Prevention of virus infection with enzyme of V. cholerae. I. Studies with viruses of mumps-influenza group in chick embryo. Austr. J. exp. Biol. a. Med. Sci **26**, 49 (1948).

316. — Prevention of virus infection with enzyme of V. cholerae. II. Studies with influenza virus in mice. Austr. J. exp. Biol. a. Med. Sci **26**, 287 (1948).

317. — Tryptic inactivation of the receptor-destroying enzyme of V. cholerae and of the enzymic activity of influenza virus. Austr. J. exp. Biol. **27**, 229 (1949).

318. Stone and Burnet: The production of vaccinia haemagglutinin in rabbit skin. Austr. J. exp. Biol. a. Med. Sci **24**, 9 (1946).

319. Stuart-Harris: Observations on the agglutination of fowl red cells by influenza viruses. Brit. J. exp. Path. **24**, 33 (1943).

320. Stuart-Harris and Miller: Vagaries of the agglutination inhibition reaction with newly-isolated strains of influenza virus. Brit. J. exp. Path. **28**, 394 (1947).

321. Taylor, Sharp, McLean, Beard and Beard: Concentration and purification of influenza virus for the preparation of vaccines. J. Imm. **50**, 291 (1945).

322. Taylor, Sharp, Beard, Beard, Dingle and Feller: Isolation and characterization of influenza virus A (PR 8 strain). J. Imm. **47**, 261 (1943).

323. TAYLOR, SHARP, McLEAN, BEARD, BEARD, DINGLE and FELLER: Purification and character of the swine influenza virus. J. Imm. **48**, 361 (1944).

324. — Purification and character of the swine influenza virus. Science **98**, 587 (1943).

325. THIGPEN and CROWLEY: Isolation of influenza A by intraallantoic inoculation of untreated throat washings. Science **98**, 516 (1943).

326. TWYBLE and MASON: Hemagglutination by products of influenza virus using infected mouse-lung and chicken-embryo as the source of virus. J. Imm. **49**, 73 (1944).

327. *U. S. Army Medical Department:* A modified agglutination-inhibition test for the diagnosis of influenza. Bull. U. S. Army Med. Dep. **6**, 777 (1946).

328. VIEUCHANGE, REINIÉ et SAUTTER: Influence de la dilution du matériel d'inoculation sur la multiplication des virus grippaux dans l'oeuf de poule incubé (phénomene de l'optimum de concentration). C. R. Soc. Biol. **140**, 969 (1946).

329. VILCHES and HIRST: Interference between neurotropic and other unrelated viruses. J. Imm. **57**, 125 (1947).

330. VOLKERT and HORSFALL: The effect of sulfhydryl groups on pneumonia virus of mice (PVM). J. exp. Med. **86**, 383 (1947).

331. — Studies on a lung tissue component which combines with pneumonia virus of mice (PVM). J. exp. Med. **86**, 393 (1947).

332. WEIL, BEARD and BEARD: pH stability, response to antibiotics and factors influencing egg-culture of mumps virus. Proc. Soc. exp. Biol. a. Med. **68**, 308 (1948).

333. — Purification and sedimentation and electron micrographic characters of the mumps virus. Proc. Soc. exp. Biol. a. Med. **68**, 309 (1948).

334. WEINBERG et KEPINOW: Des Leuco-agglutinines. C. R. Soc. Biol. **18**, 880 (1921).

335. WELD and MITCHELL: Agglutination of rabbit leucocytes by staphylococcus aureus toxin. Proc. Soc. exp. Biol. a. Med. **49**, 370 (1942).

336. WELLER and ENDERS: Production of hemagglutinin by mumps and influenza A viruses in suspended cell tissue cultures. Proc. Soc. exp. Biol. a. Med. **69**, 124 (1948).

337. WELLER, CHEEVER and ENDERS: Immunologic reactions following the intradermal inoculation of influenza A and B vaccine. Proc. Soc. exp. Biol. a. Med. **67**, 96 (1948).

338. WENNER: Isolation of LCM virus in an effort to adapt poliomyelitis virus to rodents. J. inf. dis. **83**, 155 (1948).

339. WHEELER and NUNGESTER: Effect of mucin on influenza virus infection in hamsters. Science **96**, 92 (1942).

340. — The effect of atropin sulfate on the course of influenza virus infection. Science **100**, 523 (1944).

341. WHITMAN: Factors influencing the red cell agglutination-inhibitive reaction in influenza and their application to the diagnostic test. J. Imm. **56**, 167 (1947).

342. WIENER, HENLE and HENLE: Studies on the complement fixation antigens of influenza viruses types A and B. J. exp. Med. **83**, 259 (1946).

343. WIENHAUS: Zur Biochemie des Phasins. Biochem. Z. **18**, 228 (1909).

344. WILLIAMS and WYCKOFF: Electron shadow-micrography of virus particles Proc. Soc. exp. Biol. a. Med. **58**, 265 (1945).

345. WOOLLEY: Purification of an influenza virus substrate, and demonstration of its competitive antagonism to apple pectin. J. exp. Med. **89**, 11 (1949).

346. WOOLLEY and GREEN: Inhibition of hemagglutination and of multiplication of influenza virus by certain polysaccharides. J. Bact. **54**, 63 (1947).

347. ZIEGLER and HORSFALL: Interference between the influenza viruses. I. The effect of active virus upon the multiplication of influenza viruses in the chick embryo. J. exp. Med. **79**, 361 (1944).

348. ZIEGLER, LAVIN and HORSFALL: Interference between influenza viruses. II. The effect of virus rendered non infective by ultraviolet radiation upon the multiplication of influenza viruses in the chick embryo. J. exp. Med. **79**, 379 (1944).

Nachtrag bei der Korrektur.

[1] *Neurotrope Virusarten* (vgl. S. 189). Der Befund von Hallauer, wonach bestimmte *murine Poliomyelitisstämme* (SK und MM) Hammelerythrocyten agglutinieren, wurde von Olitzky und Yager (Proc. Soc. exp. Biol. a. Med. 71, 719 [1949]) und Verlinde und de Baan (Ann. Inst. Pasteur 77, 632 [1949]) bestätigt und von den erstgenannten Autoren dahin erweitert, daß für das *Mengo-Encephalomyelitis-* und das *Encephalomyocarditisvirus* ebenfalls Hammelerythrocyten-agglutinierende Eigenschaften nachgewiesen werden konnten. Der Befund von Verlinde und de Baan, wonach auch das *japanische Encephalitisvirus* Hammelerythrocyten agglutiniert, konnte von Olitzky und Yager sowie Hallauer nicht verifiziert werden. Nach Lahelle und Horsfall (Proc. Soc. exp. Biol. a. Med. 71, 713 [1949]) agglutiniert der *Stamm GD VII* des *Theilerschen Mäuseencephalomyelitisvirus* Menschenerythrocyten der Gruppe 0 bei + 4⁰ C, jedoch nicht bei 23⁰ und 37⁰ C; mit dem *Stamm FA* desselben Virus konnte eine solche Agglutination nicht beobachtet werden.

[2] *Vaccinevirus.* Wie Collier (Doc. Neerland. et Indones. de Morb. Trop. 1, No. 1, 81 [1949]) zeigte, besitzt die vom indischen Büffel gewonnene *Dermovaccine* gegenüber Hühnererythrocyten stark haemagglutinierende Eigenschaften (vgl. S. 177). Das haemagglutinierende Agens erwies sich bei der Lagerung als sehr stabil, kochbeständig und von hoher Resistenz gegenüber Formol (vgl. S. 145) und dürfte daher mit dem S-Antigen bzw. der S. S.-Substanz des Vaccinevirus identisch sein.

[3] *Normal-Haemagglutinine* (vgl. S. 198). Ein Haemagglutinin für Hühnererythrocyten fanden Francis und Minuse (Proc. Soc. exp. Biol. a. Med. 69, 291 [1948]) sowie Seltsam, Lanni und Beard (J. Imm. 63, 261 [1949]) im *menschlichen Speichel* bzw. im *Sekret der sublingualen und maxillaren Speicheldrüse.* Olitzky und Yager (Proc. Soc. exp. Biol. a. Med. 71, 719 [1949]) vermochten in Suspensionen von *Normal-Mäusegehirn* auch Agglutinine gegen Hunde-, Katzen-, Meerschweinchen- und Hamstererythrocyten nachzuweisen und ergänzten hierdurch frühere Beobachtungen von Hallauer (164).

[4] *Verschiedenartigkeit des haemagglutinationshemmenden und antiinfektiösen Antikörpers* (vgl. S. 187/88). Nach Walker und Horsfall (J. exp. Med. 91, 65 [1950]) ist die agglutinationshemmende und virusneutralisierende Wirkung von Influenzaimmunserum auf zwei distinkte Antikörperqualitäten zu beziehen. Denselben Befund erhob Collier (Doc. Neerland et Indones. de Morb. trop. 1, No. 2, 1 [1949]) für Immunsera gegen Vaccinevirus, und Hallauer (163) für Sera gegen den Poliomyelitisvirusstamm SK.

[5] *Inhibitoren.* Nach Francis und Minuse (Proc. Soc. exp. Biol. a. Med. 69, 291 [1948]) bewirken menschliche Speichelproben eine deutliche, wenn auch ungleich starke Hemmung der Haemagglutination gegen Influenzavirus. Der nachgewiesene Hemmstoff findet sich, wie Seltsam, Lanni und Beard (J. Imm. 63, 261 [1949]), zeigten, besonders reichlich im mukösen Sekret der sublingualen und submaxillaren Speicheldrüse und besitzt die Eigenschaften des von Francis beschriebenen Inhibitors (vgl. S. 173). Davenport und Horsfall (J. exp. Med. 91, 53 [1950]) überprüften ihren früher erhobenen Befund, wonach der gewebliche Hemmstoff bzw. Receptor des *Mäusepneumonievirus* nur in der Lunge empfänglicher Tierspecies vorhanden ist (vgl. S. 175), mit einer verfeinerten Methode und gelangten hierbei zu abweichenden Ergebnissen. Nahezu sämtliche Organgewebe von Mäusen und Kaninchen vermochten das Haemagglutinin in reversibler Art zu binden und zu hemmen; denselben Effekt zeigten kommerzielle Adsorbentien (Celitpräparate), mit welchen ebenfalls eine Adsorption und künstlich induzierte Elution des Haemagglutinins des Pneumonie- und Influenzavirus erreicht werden konnte. Die Bindung des Haemagglutinins an gewebliche Komponenten erscheint daher als unspezifischer Adsorptionsvorgang und nötigt nicht zur Annahme spezifischer Zellreceptoren, wie solche u. a. neuerdings von Collier (Ant. van Leeuwenhoek 15, 53 [1949]) im Lungengewebe von Säugetierspecies für Vaccinevirus postuliert werden.

[6] *Cholerakulturfiltrat (RDE).* Wie Verlinde und de Baan (Ann. Inst. Pasteur 77, 632 [1949]) feststellten, bewirkt die intraperitoneale Verabreichung von RDE bei weißen Mäusen einen bemerkenswerten Schutz gegen die intraperitoneale Infektion mit dem Hammelerythrocyten-agglutinierenden Stamm SK. Der Mechanismus dieser Schutzwirkung wurde im Sinne der australischen Autoren interpretiert.

Die Elektronenmikroskopie in der Virusforschung.

Von

HELMUT RUSKA, Berlin-Dahlem.

I. Die Bedeutung der Elektronenmikroskopie für die Virusforschung.

Als Kriterien des neueren Virusbegriffes, die seit den Arbeiten von LÖFFLER und FROSCH, sowie von BEIJERINCK (1898) entwickelt wurden, galten die *Filtrierbarkeit* des Virus durch bakteriendichte Filter, die unzureichende oder völlig mangelnde *Sichtbarkeit* im Lichtmikroskop und schließlich die fehlende *Züchtbarkeit* außerhalb einer empfänglichen Wirtszelle. Durch die Ultrafiltration (1915) ist das Virus mit Filtern aus der Flüssigkeit abtrennbar, durch die Elektronenmikroskopie (1938)[1] ist es sichtbar geworden. Mit Hilfe der neuen Verfahren konnten Größe, Gestalt und Struktur vieler Virusarten ermittelt

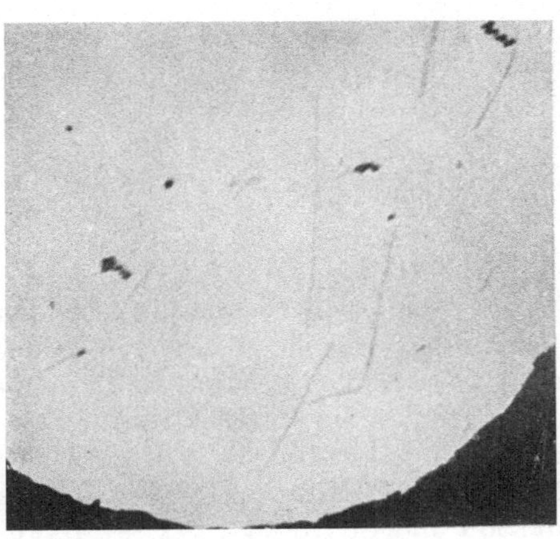

Abb. 1. Tabak Mosaikvirus (Einzelfäden) und kolloide Goldteilchen. 20000 : 1, Nach KAUSCHE, PFANKUCH und H. RUSKA (1939).

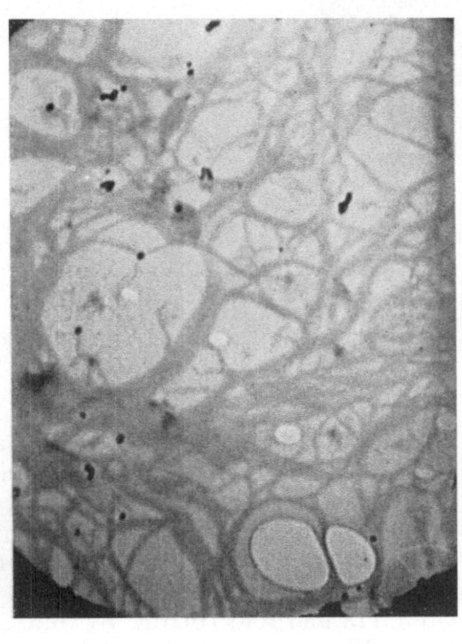

Abb. 2. Kartoffel X-Virus (Gelstruktur) und kolloide Goldteilchen. 25000 : 1. Nach KAUSCHE, PFANKUCH und H. RUSKA (1939).

werden. Die Kriterien der Filtrierbarkeit und der Unsichtbarkeit lassen sich seitdem nur noch auf die älteren Untersuchungsmethoden der Laboratoriumspraxis beziehen und haben ihre frühere Bedeutung verloren.

---

[1] Im September 1938 wurden auf der 95. Versammlung der Gesellschaft Deutscher Naturforscher und Ärzte die ersten Bilder des Tabak Mosaikvirus von KAUSCHE, PFANKUCH und H. RUSKA gezeigt. Vgl. v. BORRIES und E. RUSKA (1938/4).

Die Darstellung von Virus durch die Elektronenmikroskopie, die mit der
Abbildung makromolekularer Virusproteine (Abb. 1 und 2) begann, beschränkte
aber bald auch den Wert des letzten Kriteriums. Es zeigte sich, daß die größten
zum Virus gerechneten und nur in einer Wirtszelle züchtbaren Kontagien morphologisch wie bekannte Mikroorganismen gebaut sind. Behielten aber die Besonderheiten der Züchtung in der bisherigen Weise weiterhin ihre Geltung zur
Definition des Virusbegriffes, so würde man nach den neuen morphologischen
Erkenntnissen den Fehler begehen, die natürliche systematische Zusammen-

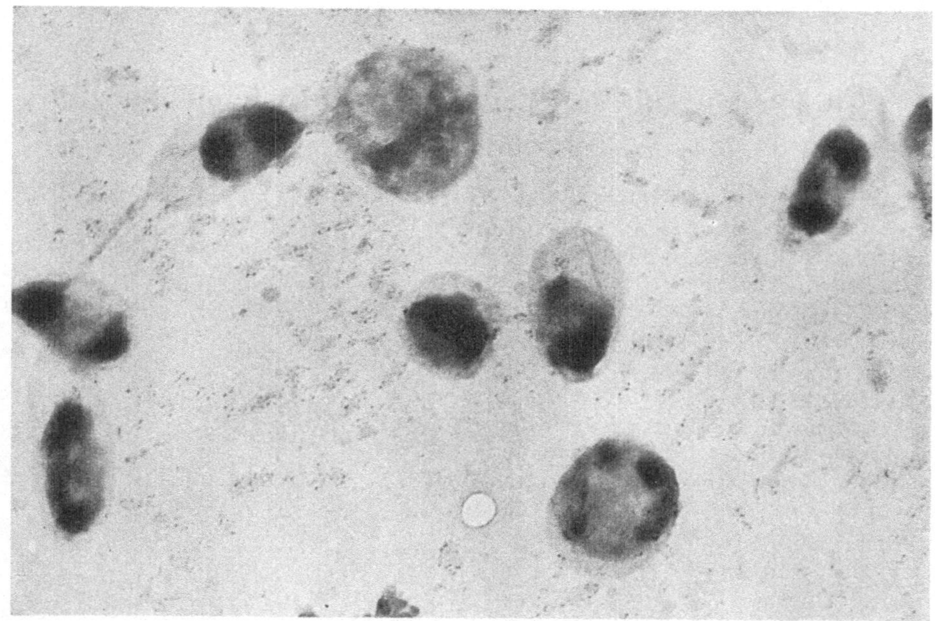

Abb. 3.   Fleckfieber-Rickettsien (R. Prowazeki), aus Läusedarm isoliert. 28000:1.   Nach Eyer
und H. Ruska (1944).

gehörigkeit nahe verwandter Organismen zu zerstören; denn die nur auf lebenden
Nährböden züchtbaren Mikroorganismen fielen dem Virus zu. Um diesen Fehler
zu vermeiden, haben die meisten Autoren keine Bedenken gehabt, die morphologische Zugehörigkeit als entscheidend anzusehen, als sie im Elektronenmikroskop den bakteriellen Bau der Rickettsien erkannten (Abb. 3) und die Übereinstimmung der großen Virusformen mit filtrierbaren Mikroorganismen fanden
(Abb. 113 und 109) (Plotz, Smadel, Anderson und Chambers; Weiss; Babudieri; Eyer und H. Ruska; Weyer, Friedrich-Freksa und Bergold; Weiss,
Rake, Rake, Hamre und Groupé; H. Ruska und Poppe). In klarer Weise
hatte Doerr 1938 (dieses Handbuch, Band I, S. 12), noch bevor die ersten
elektronenmikroskopischen Virusabbildungen bekannt wurden, vorausgesehen
und betont, daß sich das Virusgebiet „nicht nur durch Feststellung neuer Infektionsstoffe erweitern, sondern auch durch schärfere Bestimmung der bereits
als spezifisch erkannten Kontagien verkleinern kann".

Die Elektronenmikroskopie brachte die volle Bestätigung für die im Vorwort
zu diesem Handbuch von den Herausgebern betonte Auffassung, daß die Virusarten

— entgegen einer immer noch nicht allseits verlassenen Auffassung[1] — *keine* biologische Zusammengehörigleit zeigen. Sie erwiesen sich teils als *makromolekulare Infektionsstoffe*, teils als allerkleinste *Organismen*, teils als Gebilde, für die vorerst nur der unbestimmte Ausdruck *Virus* zur Verfügung steht. Virus ist also kein Begriff der biologischen Systematik, sondern eine Kollektivbezeichnung für verschiedenartige pathogene Agentien. Sowohl die ursprüngliche Auffassung von BEIJERINCK, daß es sich beim Virus um etwas anderes als Mikroben handle, als auch diejenige von LÖFFLER und FROSCH, daß sublichtmikroskopische Mikroorganismen vorlägen, erhielten durch die neueren Forschungsergebnisse grundsätzlich ihre Berechtigung. Das historische Verdienst von LÖFFLER und FROSCH erscheint daher heute wieder in anderem Licht als in jenen Jahren, die ganz unter dem Eindruck der STANLEYschen Entdeckung der Virus*proteine* standen und in denen bei manchen Forschern die Tendenz herrschte, den einfachen Proteincharakter *allen* Virusarten zuzuschreiben. Zwar trifft nach den heutigen Erkenntnissen die mikrobielle Natur nicht speziell für das Maul- und Klauenseuchevirus zu, das LÖFFLER und FROSCH zur Annahme unsichtbarer Mikroben führte, doch schloß man andererseits bis zum Aufkommen der Elektronenmikroskopie noch Formen der kleinsten Mikroben in den Sammelbegriff „Virus" ein. 50 Jahre nach den Anfängen der Virusforschung und 10 Jahre nach dem Beginn elektronenmikroskopischer Arbeiten sind alle Kriterien, die sich auf methodische Besonderheiten stützen und zeitweilig als grundsätzliche Grenzen galten, hinfällig geworden. Geblieben ist nur die Abgrenzung des Virus als eine Gruppe von vermehrungsfähigen, übertragbaren, in Zellen parasitierenden Agentien, *die eine einfachere Bauweise zeigen, als sie selbständig lebensfähige Organismen besitzen* (H. RUSKA 1949).

Die vor der Einführung der Elektronenmikroskopie gültige Fassung des Virusbegriffs scheidet nur die auf unbelebten Nährböden züchtbaren Agentien als filtrierbare Mikroben aus. Heute erscheint es richtig, den Einschnitt nicht mehr nach der Züchtbarkeit im zellfreien Nährmedium, sondern nach der morphologischen Struktur zu bestimmen. Dieser Gesichtspunkt führt zu einer anderen Lage der Begrenzung. Außerdem wird als Virus unabhängig von jeder Festlegung dasjenige bezeichnet, was — nach der strengen Definition inkonsequenterweise — immer noch den tatsächlichen Umfang der infektiösen Agentien in Lehr- und Handbüchern des Virusgebietes ausmacht.

Zum Studium der Virusmorphologie ist die Elektronenmikroskopie eine unerläßliche Methode, geworden. Kein Virus kann ohne ihre Hilfe ausreichend beschrieben werden, und keine Methode hat — neben der chemischen Forschung — bisher unsere Vorstellungen über das Wesen des Virus stärker beeinflußt als die hochauflösende Mikroskopie mit schnellen Elektronen.

1. Die Elektronenmikroskopie in der Morphologie.

Der Dimensionsbereich der Virusarten zwischen etwa 10 mμ und 500 mμ ist zur Erschließung der Protoplasmastrukturen auch außerhalb der Virusforschung von größtem biologischem Interesse. Der vielseitige architektonische Aufbau des Virus, der sich vom Makromolekül bis zur Mikrobenzelle erstreckt, fordert Rechenschaft über Lage und Ausdehnung des Virus innerhalb der morphologischen Größenskala. Außerdem sind auch die Zelle des Makroorganismus

---

[1] E. BUCHWALD 1947: „Viren sind eine biologisch einheitliche Gruppe kleinster Krankheitserreger ..."

einerseits, sowie die Proteinmoleküle der Immunkörper andererseits Gegenstand der Virusforschung. Das Auflösungsvermögen des Elektronenmikroskops von nur wenigen $m\mu$ genügt, um selbst bei kleinen Virusarten außer ihrer Größe noch Form und Strukturwerte zu erkennen.

Tabelle 1.

Objekte	Größenordnungen		Dispersionsgrad und Auflösungsgrenzen
Zellen	$10\,\mu$	10^{-3} cm	grobe Dispersionen
Bakterien			
Rickettsien..............	$1\,\mu$	10^{-4} cm	
Filtrierbare Mikroorganismen			Auflösungsgrenze des Licht-mikroskops
Größere Virusarten........	$100\,m\mu$	10^{-5} cm	
Bakteriophagen			kolloide Dispersionen
Virusproteine.............	$10\,m\mu$	10^{-6} cm	
Makromoleküle unter Berücksichtigung der Länge von Fadenmolekülen	$1\,m\mu$	10^{-7} cm	Auflösungsgrenze des Elektronenmikroskops
Moleküle			molekulare Dispersionen
Atome	$1\,Å$	10^{-8} cm	

Nach Tabelle 1, welche die Größenordnungen in Zehnerpotenzen abgestuft wiedergibt, überdecken die Virusdimensionen sowohl diejenigen der größten Moleküle als auch die der kleinsten Mikroben. Nach ihrem Dispersionsgrad gehören die Virusarten in den Bereich der Kolloide, der durch die Elektronenmikroskopie einer optischen Abbildung zugänglich geworden ist. Erst langsam wird erkennbar, daß sich vom Größenbereich der Strukturchemie bis zur lichtmikroskopischen Morphologie die Formen der lebenden Substanz kontinuierlich fortsetzen und von wesentlicher Bedeutung sind. Außer der Bakteriologie und Virusforschung vertiefte auch die Histologie ihre Einsichten in feinere Strukturen. Die an normalen Zell- und Gewebebestandteilen gewonnenen Ergebnisse werden zum Ausgangspunkt für die Fortentwicklung der Histo- und Cytopathologie der Virusinfektionen. Bisher verborgene Feinheiten von Protoplasmastrukturen (Wolpers und H. Ruska; Menke), Zellmembranen, Abb. 4 (H. Ruska 1941/7; Wolpers 1941/1), Chloroplasten, Abb. 5 (Kausche und H. Ruska 1940/3; Menke; Algera, Beijer, van Iterson, Karstens und Thung), Mitochondrien (Claude und Fullam 1945) und Chromosomen (Elvers) wurden der morphologischen Analyse zugänglich. Die Querstreifung der Fibrinfaser (H. Ruska und Wolpers; Wolpers 1947), des Kollagens, Abb. 6 (Hall, Jakus und Schmitt 1942; Wolpers 1944), des Trichocystenschafts (Jakus) und glatter Muskelfasern (Hall, Jakus und Schmitt 1945) wurde entdeckt und der Feinbau der quergestreiften Muskulatur, Abb. 7, weiter aufgeklärt (Hall, Jakus und Schmitt 1946). Zusammenhänge zwischen chemischer und morphologischer Struktur werden diskutierbar und das physiologische Geschehen wird im Hinblick auf den mikromorphologischen Bau in zunehmendem Maße verständlich.

Zunächst hat die Elektronenmikroskopie die isolierten Kontagien der Viruskrankheiten analysiert. Bei der besonders engen Verbindung, die das Virus zur Wirtszelle eingeht, ist aber auch jeder Fortschritt in der Analyse von Zellstrukturen, Membranen und ganzen Zellen für die Virusforschung von Bedeutung. Anfänge hierzu liegen in der Aufdeckung von

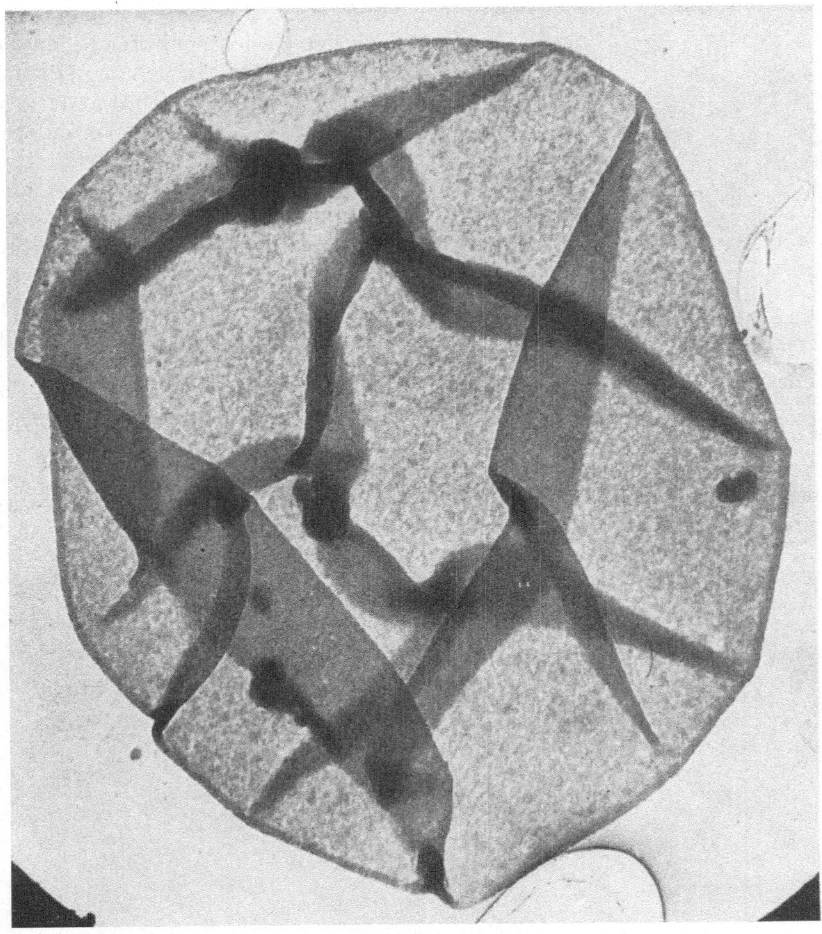

Abb. 4. Osmiumfixierte, durch osmotische Hämolyse gewonnene Erythrocytenmembran aus menschlichem Blut. 19000 : 1. Nach WOLPERS (1941).

Beziehungen des Tabak Mosaikvirus zum Chloroplasten der erkrankten Zelle durch KAUSCHE und H. RUSKA (1940/7), in den Untersuchungen zur bakteriophagen Lyse von H. RUSKA (1942/12), KOTTMANN (1942) und EDWARDS und WYCKOFF (1947), sowie in der Klärung der Struktur von Einschlußkörpern durch RAKE (1947). Besonderer Wert wird aber der Abbildung von Zellen aus Kulturen und von Gewebeschnitten zukommen, die von PORTER, CLAUDE und FULLAM (1945) und von CLAUDE und FULLAM (1946) begonnen wurde. Hierdurch wird es in Zukunft möglich sein, Stadier der Virusentwicklung und Virusreproduktion in der Wirtszelle zu erkennen[1].

Die Betrachtung der Morphologie der Kontagien führt zwangsläufig zu einer Aufstellung von Gruppen verwandter Virusformen innerhalb des verschiedenartigen Gesamtmaterials. Neben dem künstlichen System der Viruseinteilung nach Tropismen hat H. Ruska (1943/15) eine natürliche Gruppenbildung der Virusarten nach morphologischen Gesichtspunkten begonnen und ständig weiter ausgebaut (1949). Der Tropismus und die Art der pathologisch-anatomischen Veränderungen, die z. B. von Gildemeister, Haagen und Waldmann zur Grundlage der Einteilung der Viruskrankheiten gewählt worden waren, sind weitgehend unabhängig vom morphologischen Charakter des Virus. Dagegen finden sich Anhaltspunkte dafür, daß der morphologische Charakter des Virus für die Wege einer sinnvollen Vaccination und Chemotherapie entscheidend ist. Aus dem Streben nach Aufspaltung des komplexen Virusbegriffs und aus der Zusammenfassung eindeutig verwandter Formen zu kleineren Gruppen ergeben sich praktische Folgerungen (Liebermeister, 1949).

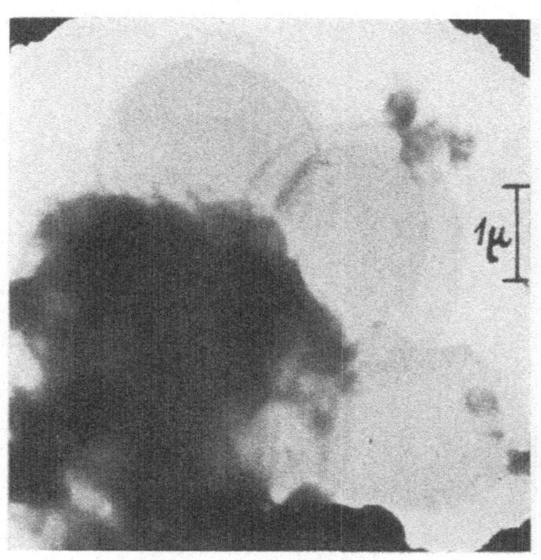

Abb. 5. Rand eines Chloroplasten von Nicotiana tabacum mit großen und kleineren mehrfach übereinanderliegenden Scheiben bzw. Blasen. 10000 : 1. Nach Kausche und H. Ruska (1940).

Abb. 6. Kollagene Fasern aus Rindersehne, osmiumfixiert. 42000 : 1. Nach Wolpers (1944).

[1] Anm. bei der Korr.: Siehe die neuen Versuche von Wirth und Athanasiu (1949) über die Abbildung von Epithelzellen, die mit Vaccinevirus infiziert sind.

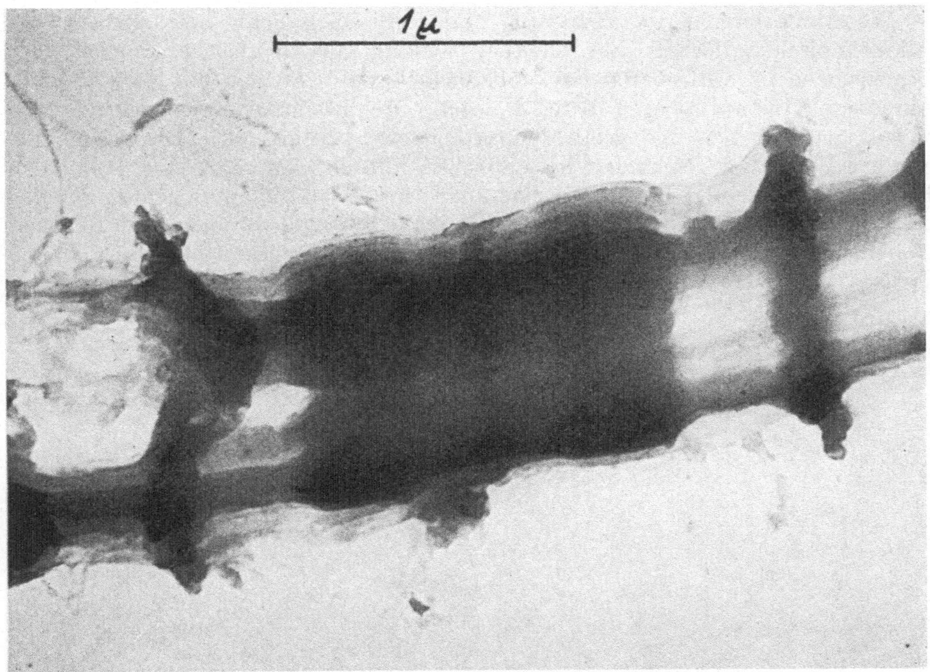

Abb. 7. Quergestreifte Muskelfasern von Skelettmuskeln der Maus. Flemming-Fixation. 40000 : 1.
Nach WOLPERS (1944).

2. Das Verhältnis der Elektronenmikroskopie zu anderen Methoden der Virusforschung.

Das Elektronenmikroskop vermag linear etwa 100mal kleinere Einzelheiten aufzulösen als das Lichtmikroskop. Bei der Besprechung der physikalischen Grundlagen des Elektronenmikroskops sind weitere Unterschiede aufzuzeigen. Vorher soll eine Gegenüberstellung zu anderen optischen und nichtoptischen Methoden erfolgen, die ebenfalls Aussagen über die im elektronenoptischen Bild enthaltenen Daten der Größe, Form und Struktur zu machen gestatten.

a) Licht- und ultraviolettmikroskopische Methoden.

α) Hellfeldmikroskopie.

Es liegt im Wesen der optischen Abbildung, daß keine kleineren Objekteinzelheiten getrennt dargestellt werden können, als etwa der halben Wellenlänge λ der angewandten Strahlung entspricht. Außer von der Wellenlänge ist das Auflösungsvermögen eines optischen Systems abhängig vom Öffnungswinkel 2α des vom Objekt auf das Objektiv treffenden Strahlenbündels und vom Brechungsindex n des zwischen Objekt und Objektiv befindlichen optischen Mediums. Der kleinste Abstand δ zweier Objekteinzelheiten, die noch getrennt werden können, ist nach ABBE $\delta = \dfrac{\lambda}{2\,\mathrm{n}\cdot\sin\alpha}$. Der absolute Wert von δ ist für eine mittlere Wellenlänge des sichtbaren Lichts von 560 mμ, den Berechnungsindex 1,4 und den optimalen Öffnungswinkel 2α von 180⁰ (sin $\alpha = 1$) bestenfalls 0,0002 mm = 0,2 μ = 200 mμ. Bei Verwendung von ultraviolettem Licht mit 257 mμ Wellenlänge können etwa 100 mμ als auflösbar angegeben werden. (Näheres siehe ELFORD, dieses Handbuch, Band I, S. 181).

Die Abmessungen des Virus sind zum Teil schon licht- bzw. ultraviolett-mikroskopisch auflösbar. Zur Sichtbarmachung müssen aber mindestens zwei Dimensionen im auflösbaren Größenbereich liegen. Partikel, die nur mit einer Abmessung im auflösbaren Bereich liegen, wie beispielsweise einzelne Tabak Mosaikvirusstäbchen und große Bakteriophagen, bleiben unsichtbar. Für eine Strukturbewertung reicht das lichtoptische Auflösungsvermögen aber in keinem Fall aus, da die feinere Form erst erkennbar wird, wenn die untersuchten Partikel mindestens die fünffache Größe des auflösbaren Intervalls besitzen. Für die elektronenmikroskopische Auflösung von etwa 1 mμ liegen dagegen auch die kleinsten Viruspartikel von 10 mμ oberhalb der Strukturbewertungsgrenze. Wo der

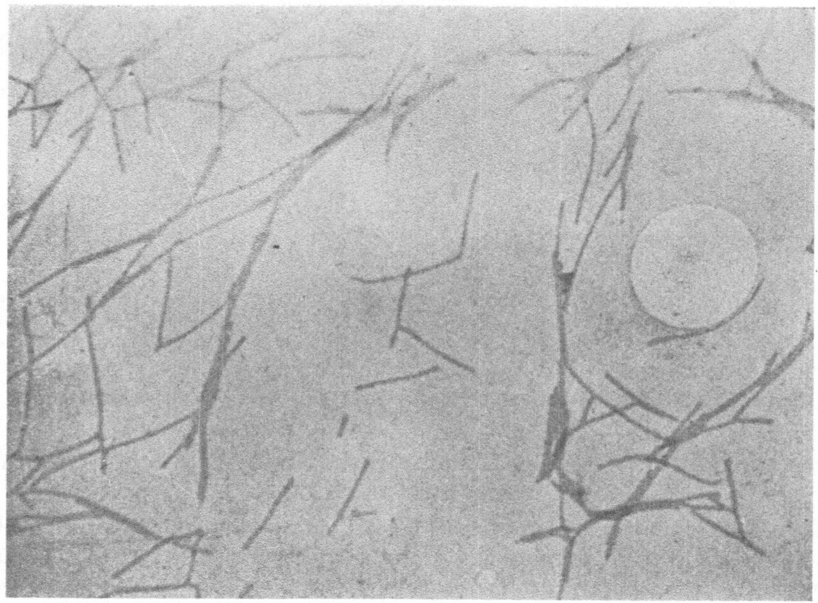

Abb. 8. Tabak Mosaikvirus. 39000 : 1. Nach Kausche und H. Ruska, aus H. Ruska (1942).

Kontrast organischer Partikel so gering ist, daß die Auflösung wesentlich beeinträchtigt wird, kann die chemische Behandlung der Objekte mit Metallsalzen oder die Schrägbedampfung mit atomarem Metall den Kontrast so verstärken, daß das Auflösungsvermögen nur wenig hinter dem optimalen Wert für Metallkolloide zurückbleibt. Als Beispiel für die unzureichende lichtmikroskopische Strukturbewertung sei erwähnt, daß die Elementarkörper der Vaccine, der Ektromelie der Maus, des Molluscum contagiosum und andere im Lichtmikroskop als Punkte erkennbar sind, jedoch ihre Quaderform (H. Ruska 1939) erst im Elektronenmikroskop sichtbar gemacht wurde. Nachträglich lassen sich allerdings auf den U.-V.-Aufnahmen von Barnard (siehe z. B. dieses Handbuch, Band I, S. 189) auch Quaderformen erkennen.

Wichtiger als das erhöhte Auflösungsvermögen des U.-V.-Mikroskops ist die spezifische Absorption geworden, die das U.-V.-Licht durch Zellkerne und chemisch kernähnliche Strukturen erfährt und deren Lokalisation z. B. in Bakterien (Piekarski u. a.) ermöglicht.

Ein Vorteil der lichtmikroskopischen Verfahren liegt in den vielseitigen Färbemethoden, welche die Virusteilchen durch Farbaufnahme stärker oder andersfarbig als Gewebebestandteile hervortreten lassen. Vielfach wird dabei zugleich ihre Größe, z. B. durch Metallanlagerung, heraufgesetzt (siehe M. Kaiser, dieses Handbuch, Band I, S. 252). Die Elektronenmikroskopie kennt als „Färbemethoden" die Steigerung des Bildkontrastes durch Bindung chemischer Stoffe großer Dichte oder durch Metallbedampfung (siehe Abschnitt IV, 4, d, β). Nachdem durch die Hochgeschwindigkeitsmikrotome von Fullam und Gessler (1946) die elektronenmikroskopische Untersuchung von Gewebeschnitten und von

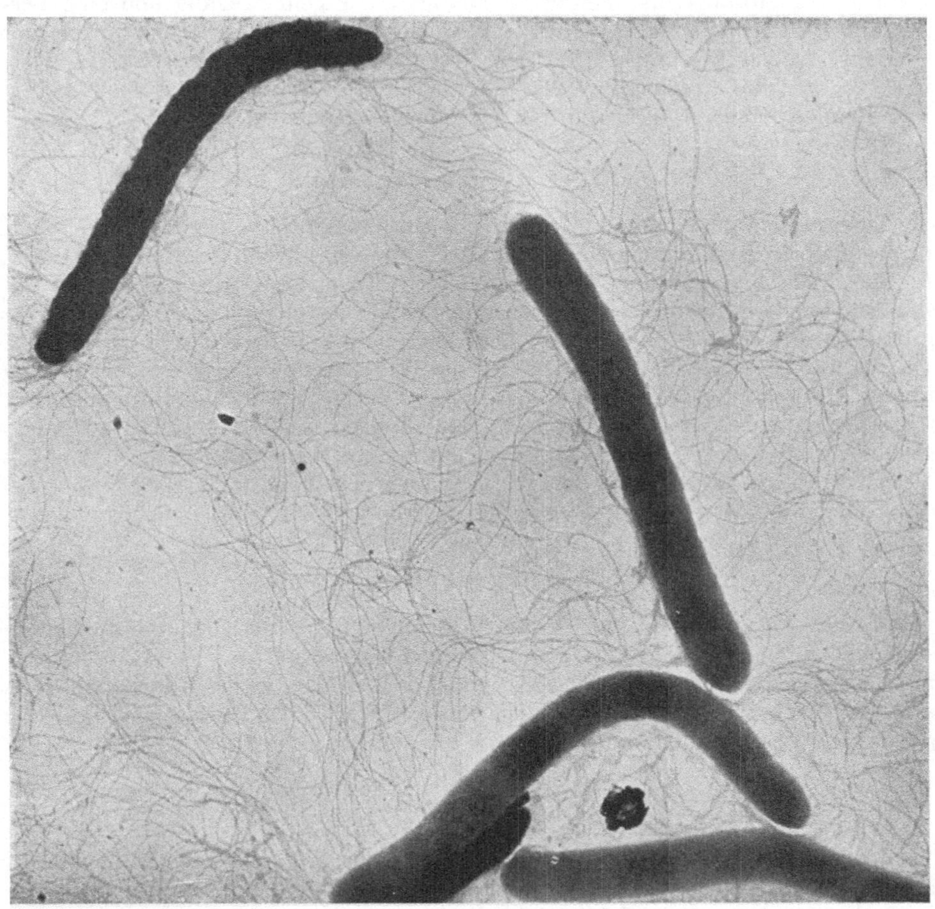

Abb. 9. Bacterium proteus. 10000 : 1. Nach H. Ruska, aus v. Borries (1949).

darin enthaltenen Einschlußkörpern grundsätzlich möglich geworden ist, bleibt als besonderer Vorteil der Lichtmikroskopie die Möglichkeit der Lebendbeobachtung größerer Virusformen und kleinster Mikroorganismen.

β) Dunkelfeld- und Ultramikroskopie.

Die Betrachtung im licht- und U.-V.-mikroskopischen Dunkelfeld und im Ultramikroskop läßt noch Virusteilchen als leuchtende Punkte erkennen, die unterhalb des lichtmikroskopischen Auflösungsvermögens liegen. Die Teilchen werden durch die Beugung des eingestrahlten Lichts zu selbstleuchtenden

Punkten. So sind beispielsweise Bakteriophagen von SCHLESINGER im Dunkelfeld beobachtet worden. Außerdem konnten filtrierbare Mikroorganismen im Dunkelfeld besonders eingehend untersucht werden. Bemerkenswerterweise sind einzelne Tabak Mosaikvirusstäbchen mit dem gleichen Verfahren bisher nicht nachgewiesen worden, obwohl sie dicker sind als die so häufig im Dunkelfeld studierten Bakteriengeißeln (vgl. Abb. 8 und 9). Die Differenz des Brechungsindex gegen das umgebende Medium scheint danach bei Geißeln größer zu sein als beim Tabak Mosaikvirus. Ferner begünstigt die Eigenbewegung und eine Verzopfung, vielleicht auch ein stärkerer Quellungszustand, die Sichtbarkeit der Geißeln. Lichtmikroskopisch können besonders im Dunkelfeld manche Entwicklungsvorgänge verfolgt werden, die im Elektronenmikroskop nur aus Serien von Momentbildern aufgetrockneter Virusteilchen erschließbar sind.

<h3 style="text-align:center">γ) Fluoreszenzmikroskopie.</h3>

Eine weitere Möglichkeit, Virus als leuchtende Punkte nachzuweisen, liegt in der Fluoreszenzmikroskopie (siehe M. HAITINGER, dieses Handbuch, Band I. S. 231). Die Viruspartikel, aber auch andere Teile der Präparate, werden mit fluoreszierenden Farbstoffen beladen und bei starker ultravioletter Beleuchtung schon mit verhältnismäßig geringer Vergrößerung durch die Fluoreszenz sichtbar. Da die Bindung der Fluorochrome an das Virus weniger spezifisch ist als die häufig sehr typische elektronenmikroskopische Virusstruktur, bleibt die Spezifität fluoreszenzmikroskopischer Virusnachweise mitunter fraglich (Maul- und Klauenseuche-Virus, Virus maligner Tumoren und andere). Der besondere Vorzug des Verfahrens liegt darin, mit wenig giftigen Fluorochromen (z. B. Akridinorange) die Zelle und das Virus vital anfärben zu können.

<h3 style="text-align:center">δ) Strömungsdoppelbrechung.</h3>

Stärkere Ungleichheit der Virusdimensionen (Anisodiametrie) läßt sich in strömenden Suspensionen mit polarisiertem Licht an der optischen Anisotropie mikroskopisch nachweisen. (TAKAHASHI und RAWLINS sowie andere Autoren, siehe W.M.STANLEY, dieses Handbuch, Band I, S. 511). Im Elektronenmikroskop ist dagegen die Anisodiametrie jedes Einzelteilchens direkt sichtbar.

<h3 style="text-align:center">b) Die Röntgenographie.</h3>

Viruspartikel, deren Anordnung oder Struktur *periodisch* ist, können röntgenographisch erfolgreich untersucht werden (WYCKOFF und COREY). Aus der *Lage* der Röntgenreflexe kann man direkt auf die Größe der Translationen, welche die Struktur mit sich selbst zur Deckung bringen würde, und daraus auf die Elementarzelle schließen. Die *Schärfe* oder *Unschärfe* der Reflexe läßt erkennen, ob jeweils eine große Anzahl von Elementarzellen aneinander gereiht ist oder ob in einer oder mehreren Richtungen nur wenige Perioden vorhanden sind. Diese Kenntnis und die der Elementarzelle läßt gewisse Schlüsse auf die Größe und Gestalt der Makromoleküle zu. So kann z. B. bei scharfen Reflexen jede Modellvorstellung ausgeschlossen werden, die als Volumen der Teilchen einen Wert ergibt, der zwar größer als das Volumen der Elementarzelle ist, jedoch nicht so groß, daß eine Periodizität innerhalb des Teilchens Platz hätte. Schlüsse auf die Gestalt der Teilchen sind im allgemeinen nur möglich, wenn man Aussagen über die *Intensitäten* der beobachteten Reflexe hinzunimmt, beispielsweise solche, die auf das Vorhandensein von Symmetrieelementen der Struktur schließen lassen. Daraus kann sich dann eine Aussage über die Symmetrie der Teilchen ergeben. Eine genaue Kenntnis der Intensitäten der Röntgenreflexe, von denen bei gut ausgebildeten Kristallen Hunderte zu beobachten sind, kann für präzise Aussagen über die Teilchengestalt ausgewertet werden.

Das Röntgenverfahren gibt bei geeignet gebautem Virus wesentlich feinere Aufschlüsse als die Elektronenmikroskopie. So wurde beispielsweise bei den Stäbchen des Tabak Mosaikvirus die Breite von 15,2 mμ ermittelt und eine Periodizität in Richtung der dreizählig symmetrischen Längsachse von 6,8 mμ (BERNAL und FANKUCHEN)[1]. Beim Tabak Nekrosevirus ließ sich zeigen, daß die nahezu kugelförmigen Virusteilchen keine kubische Symmetrie besitzen (CROWFOOT und SCHMITT). Außerdem können die Molekülanordnungen im Kristallgitter röntgenographisch zuverlässiger bestimmt werden als aus der Abbildung von Oberflächenabdrucken im Elektronenmikroskop, bei der Trocknungseffekte zu einer Verschiebung in der Molekülanordnung führen können (MARKHAM, SMITH und WYCKOFF 1948). Die Röntgenographie vermag für einen großen Teil des makromolekularen Virus unsere Kenntnisse entscheidend zu präzisieren und in kleinere Dimensionen fortzusetzen.

Die Anwendung der Elektronenbeugung wird bei organischen Substanzen dadurch erschwert, daß sie bei längerer Bestrahlung in Kohlenstoff übergehen (KÖNIG 1946/2). Die Viruspartikel behalten dabei ihre elektronenmikroskopische Form, erfahren aber einen Umbau der inneren Struktur zu einem lockeren, graphitähnlichen Kohlenstoffgerüst und zeigen Beugungsgitter, die nicht mehr für das untersuchte Ausgangsmaterial kennzeichnend sind. (Weitere Elektronenschäden siehe Abschnitt V, 1). Sehr gut läßt sich die Elektronenbeugung zur Identifizierung anorganischer Kristalle in organischen Objekten verwenden (KÖNIG und WINKLER, 1948).

c) Nicht-optische Verfahren.

α) Ultrazentrifugierung.

Die Bestimmung der Diffusions- und die der Sedimentationskonstante in der analytischen Ultrazentrifuge waren vor der Ermöglichung elektronenmikroskopischer Untersuchungen die bestmöglichen Verfahren für die Molekulargewichts- und Größenbestimmung eines Virus. SCHRAMM und BERGOLD (1947) bezeichnen sie auch heute noch als die sichersten Methoden, obwohl das Molekulargewicht sogar für jedes einzelne individuell abgebildete Virusteilchen über die Größenbestimmung im Elektronenmikroskop zu erhalten ist. PFANKUCH und H. RUSKA (1947) haben die Sedimentationskonstanten s des Tabak Mosaikvirus zu den verschiedenen Längen der Stäbchen in Beziehung gesetzt und für die Werte von s = 150· 10⁻¹³, bzw. 196· 10⁻¹³ und 214· 10⁻¹³ Stäbchenlängen von 50 bis 100 mμ, bzw. 250 bis 300 mμ und 500 bis 550 mμ gefunden. SHARP, TAYLOR, MCLEAN, BEARD und BEARD (1944) wiesen darauf hin, daß die Ergebnisse der analytischen Ultrazentrifuge und der Elektronenmikroskopie sich in wertvoller Weise ergänzen. Bei den verschiedenen Influenzavirusstämmen fanden sie in den Ergebnissen beider Methoden gute Übereinstimmung in bezug auf die Größenverteilung, weniger gute hinsichtlich der mittleren Teilchengröße, doch lagen hier Unstimmigkeiten in der Vergrößerungsangabe vor. Wichtig erscheint ihnen die Feststellung, daß sich bei ihren Untersuchungen unscharfe Sedimentations-

---

[1] Neuere Deutungen der Röntgendiagramme siehe bei KÄTHE DORNBERGER-SCHIFF (1949), der ich für die Durchsicht dieses Abschnittes zu Dank verpflichtet bin.

bande des Virus auf den Mangel an Monodispersität der Virusteilchen zurückführen ließen, daß die Teilchen sich aber ähneln und nicht der Art nach verschieden sind.

Die Analyse in der Ultrazentrifuge wie auch die schon früher erwähnte Untersuchung der Strömungsdoppelbrechung erfordern weitgehend gereinigte und angereicherte Präparate. Auch für die Elektronenmikroskopie ist in der Regel eine Reinigung und Anreicherung des Virus notwendig. Während sich in der Ultrazentrifuge aber erst einige mg Virus analysieren lassen, genügen für eine elektronenmikroskopische Aufnahme schon einige Tausendstel γ ($1\gamma = 10^{-6}$ g).

β) Ultrafiltration.

Die Ultrafiltration hat den großen Vorteil, mit wenig und verhältnismäßig unreinem Material arbeiten zu können. Zur ersten Orientierung liefert sie zweifellos mit dem geringsten experimentellen und apparativen Aufwand angenäherte Ergebnisse über die Virusgröße (vgl. Elford, dieses Handbuch, Band I, S. 128).

γ) Viskositätsmessung.

Für die Bestimmung des Formfaktors läßt sich die Viskositätsmessung zu Hilfe nehmen (Frampton und Neurath sowie Lauffer). Ihre quantitative Auswertung (Polson) ist aber schwierig und unsicher.

δ) Statistische Ultramikrometrie.

Als jüngste Methode der indirekten Größenbestimmung ist die statistische Ultramikrometrie zu nennen. Unter dieser Bezeichnung faßt Zimmer alle Verfahren zusammen, die durch Ionisierung mit Wellenstrahlen, α-Teilchen oder schnellen Neutronen eine Abtötung oder Inaktivierung der Virusteilchen herbeiführen und aus der Strahlendosis und dem Prozentsatz der erfolgten „Treffer“ die Berechnung des „Treffbereichs“ zulassen (vgl. Timoféeff-Ressovsky und Zimmer 1947). Die so gefundene Größe kann mit der Virusgröße identifiziert werden, wenn sie mit der auf andere Weise ermittelten Abmessung übereinstimmt. Abweichungen kommen bei größeren Virusformen vor und erfordern besondere Deutungen. Für sich allein und ohne genaue Kenntnis der physikalischen Voraussetzungen liefert die statistische Ultramikrometrie keine völlig gesicherten Ergebnisse über die Virusgröße. Wollmann und Lacassagne fanden mit Röntgenstrahlen bei kleinen Bakteriophagen Durchmesserwerte, die mit den durch andere Methoden ermittelten Abmessungen übereinstimmten. Bei der Untersuchung großer Phagen waren die ermittelten Werte deutlich zu klein. Sehr wahrscheinlich ist bei den großen Phagen der empfindliche Treffbereich auf eine Teilstruktur des ganzen Phagen beschränkt (Holweck; H. Ruska und Menze). Bonét-Maury hat bei statistisch-ultramikrometrischen Messungen mit α-Strahlen an Vaccine die bekannte Virusgröße gefunden, obwohl auch hier anzunehmen wäre, daß der Treffbereich kleiner ist als das sehr komplex aufgebaute Virus. Er vermutet deshalb, daß eine Energieleitung von außerhalb des Treffbereichs liegenden Bezirken zum Treffbereich stattfindet. Messungen von Bonét-Maury am Maul- und Klauenseuche-Virus ergaben größere Werte als die bisher durch Ultrafiltration festgestellten. Hierin ist möglicherweise ein Hinweis dafür zu sehen, daß der Treffbereich mit α-Strahlen hier ebenso wie beim Vaccinevirus zu groß gefunden wurde. Weitere Einzelheiten und ausführliche tabellarische Übersichten gibt Bonét-Maury (1948).

Lea (1946) fand mit Salaman gute Übereinstimmung der auf verschiedenen Wegen festgestellten Abmessungen bei dem kleinen Phagen S-13, aber mangelhafte Übereinstimmung bei größeren. Er schließt daraus — ohne auf elektronenmikroskopische Phagenbilder einzugehen —, daß kleine Phagen Makromoleküle sind, große jedoch sehr primitive Organismen mit 10 bis 20 Genen. Ebenso wie bei Phagen haben die genannten Autoren bei Vaccine mit Gamma-, Röntgen- und α-Strahlen Untersuchungen durchgeführt und im Gegensatz zu Bonét-Maury für die verschiedenen Strahlen Treffbereiche von 31, 41 und 70 mμ, also auch mit α-Strahlen kleinere Werte als die Virusgröße gefunden. Die Bedeutung der statistischen Ultramikrometrie liegt demnach weniger in der Ermittlung der Virusgröße als in der Feststellung eines Treffbereiches („target"), in dem die Ionisation wirksam wird und dessen spezielle biologische Struktur, Größe und Lokalisation von Fall zu Fall zu klären ist.

Pollard und Forro haben neuerdings an T_1-Phagen die Treffertheorie durch Bestrahlung mit Deuteronen geprüft. Der gemessene Treffbereich war größer als der wahre Treffbereich, da auch eine Inaktivierung des Phagen eintritt, wenn der Treffbereich von etwa 28 ± 4 mμ nicht unmittelbar getroffen wird.

Rückblickend läßt sich feststellen, daß eine Anzahl der besprochenen Verfahren zur Bestimmung von Größe, Form und Struktur der Virusteilchen Vorteile gegenüber der Elektronenmikroskopie besitzt und sie wesentlich ergänzt. Die Vorteile liegen für die größten Virusformen vor allem in der Möglichkeit der Lebendbeobachtung, für Virusarten jeder Größe in der Feststellbarkeit der Teilchengewichte und der Abmessungen ionisierungsempfindlicher Teilstrukturen, schließlich für die kristallisierbaren Virusproteine in der Möglichkeit der genauen Vermessung von Kristallformen, Elementarzellen, intramolekularen Perioden usw. Die Elektronenmikroskopie erlaubt demgegenüber die genaue Erkennung von einzelnen Viruspartikeln jeder Art, die Abbildung von Reaktionszuständen zwischen dem Virus und Kolloiden oder anderen Reaktionspartnern und dergleichen mehr.

II. Die Grundlagen der Mikroskopie mit Elektronenstrahlen.

Soweit es für verständnisvolles Mikroskopieren und Auswerten der Ergebnisse notwendig ist, muß der Mikroskopiker die physikalischen Erscheinungen kennen, die ihm seine Arbeit ermöglichen. Nur mit dieser Kenntnis sind optimale Ergebnisse zu erzielen. Für ein weitergehendes Eindringen in technische und theoretische Gebiete der Elektronenmikroskopie sei auf die größeren Gesamtdarstellungen von v. Ardenne (1940), v. Borries und E. Ruska (1940), Zworykin, Morton, Ramberg, Hillier und Vance (1945), Cosslett (1947) und besonders v. Borries (1949) verwiesen.

1. Die Elektronenstrahlen.

In den Metallen gibt es unter den Elektronen der äußeren Atomhüllen frei bewegliche, die sich ähnlich verhalten wie die Moleküle oder Atome eines Gases. Erhitzt man ein Metall, so nimmt die Bewegung der Elektronen so stark zu, daß ein Teil von ihnen aus der Oberfläche austritt. Erfolgt der Austritt in einen evakuierten Raum und gelangen die austretenden Elektronen unter den Einfluß eines elektrischen Feldes, das sie von der Austrittsstelle in den leeren Raum hinein beschleunigt, so bildet sich ein Elektronenstrahl. Der Strahl besteht aus den frei im Vakuum fliegenden kleinsten Bausteinen der Materie, welche die negative elektrische Elementarladung $e = 1,6 \times 10^{-19}$ Amp. sec. tragen und die Ruhemasse $m_0 = 0,91 \times 10^{-27}$ g' besitzen.

Abb. 10 zeigt ein evakuiertes Glasgefäß G mit einer eingeschmolzenen Elektronenquelle, der Kathode K, die durch den Heizakkumulator H zum Glühen gebracht werden kann. An der Kathode liegt außerdem die Beschleunigungsspannung B und ihr gegenüber auf Erdpotential die mit einer Durchtrittsöffnung versehene Anode A. Bei glühender Kathode entsteht ein Elektronenbündel (Kathodenstrahl), das sich zur Anode und jenseits der Anodenblende geradlinig fortsetzt. Der schmale, ausgeblendete Strahl fällt auf den Leuchtschirm S, der den Strahlquerschnitt als hellen Fleck sichtbar macht. Der Schirm ist mit einer Substanz (z. B. feinkristallinem Zinksulfid) überzogen, die beim Aufprall der Elektronen nicht wie die Metalle Röntgenstrahlen, sondern sichtbares Licht aussendet. Je höher die Beschleunigungsspannung gewählt wird, um so stärker werden die Elektronen zur Anode hin beschleunigt. Die Spannungen, welche in der Elektronenmikroskopie verwendet werden, liegen zwischen 30000 und mehreren 100000 Volt. Bei 80000 V erreichen die Elektronen etwa halbe Lichtgeschwindigkeit. Zu einem Elektronenmikroskop fehlen der Elektronenröhre von Abb. 10 die Elektronenlinsen, während Vakuumraum, Kathode, Beschleunigungsspannung und Leuchtschirm wie bei diesem vorhanden sind.

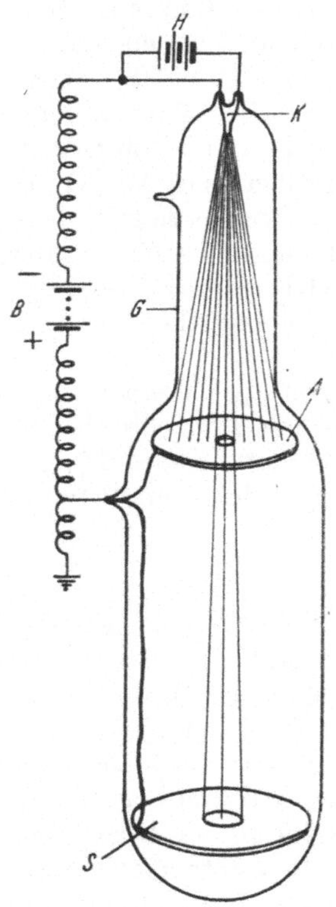

Abb. 10. Elektronenstrahlröhre. Nach v. BORRIES (1949). Erklärung siehe Text.

2. Die Elektronenlinsen.

Durch drehsymmetrische, zeitlich konstante, elektrische und magnetische Felder kann der Elektronenstrahl analog dem Lichtstrahl gebündelt und zu Abbildungszwecken verwendet werden.

a) Elektrische Felder und Linsen.

Innerhalb eines elektrischen Feldes werden die Elektronen in der Richtung zur positiven Elektrode beschleunigt (Abb. 11). Sofern die durch die Beschleunigungsspannung erzeugte Teilchengeschwindigkeit sich der Lichtgeschwindigkeit nicht zu sehr nähert, ist der Zuwachs der Bewegungsenergie eines geladenen Teilchens gleich dem Produkt aus Ladung e und Beschleunigungsspannung U. Ist die Anfangsgeschwindigkeit der Teilchen $v_a = 0$, wie die der Elektronen beim Austritt aus der Kathode, so ist die kinetische Energie

$$\frac{m_0}{2} v^2 = 10^7 \, e \times U \text{ Wattsec und}$$

$$v = 5,95 \times 10^7 \sqrt{U} \left(v \text{ gemessen in } \frac{cm}{sec}, \; U \text{ in Volt} \right).$$

Die Wirkung eines elektrischen Feldes auf einen Elektronenstrahl läßt sich durch die Zerlegung des Feldes in Schichten, welche durch die senkrecht zu den

Kraftlinien verlaufenden Äquipotentialflächen begrenzt sind, darstellen. Jede Schicht, die beliebig dünn gedacht werden kann, wirkt als brechendes Medium.

Bildet ein in das elektrische Feld eintretender Elektronenstrahl den Winkel α mit dem Lot auf die Äquipotentialfläche, so wird der Winkel β des gebrochenen Strahls kleiner als α, wenn die Feldschicht beschleunigend wirkt (Abb. 12), und

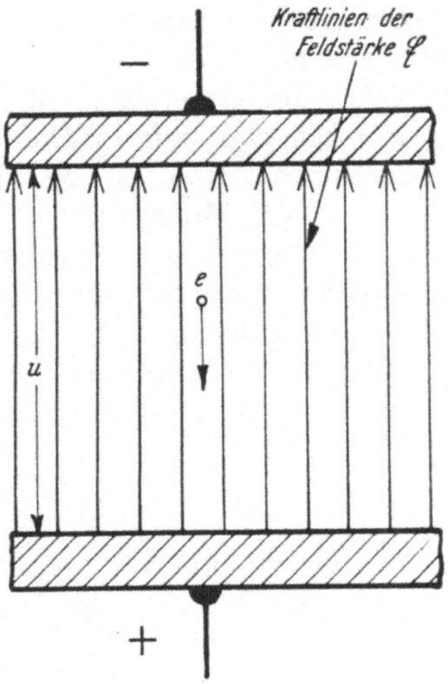

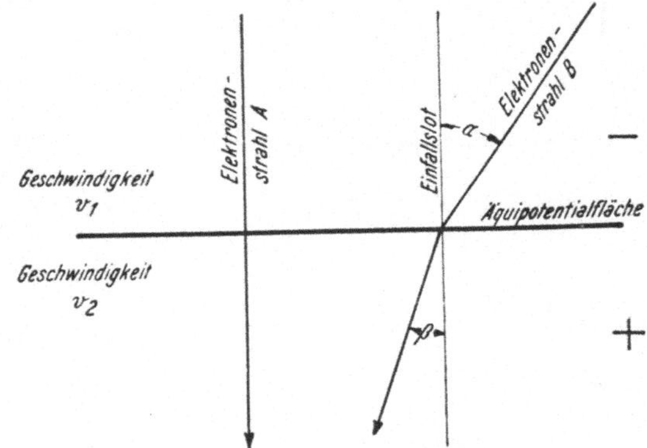

Abb. 11. Elektron e im homogenen elektrischen Kraftfeld; $u =$ Spannung zwischen den Elektroden.

Abb. 12. Verlauf zweier Elektronenbahnen durch eine Äquipotentialfläche.

größer als α, wenn sie verzögernd wirkt. Die Elektronen werden zum Lot hin gebrochen, wenn sie ihre Geschwindigkeit v erhöhen, bzw. vom Lot weg gebrochen, wenn sie dieselbe verringern. Es gilt dann wie in der Lichtoptik, daß der Brechungsindex

$$n = \frac{\sin \alpha}{\sin \beta} = \frac{v_2}{v_1}$$

ist. Je höher die Elektronengeschwindigkeit ist, um so geringer ist unter sonst gleichen Bedingungen die Ablenkbarkeit der Strahlen, d. h. um so „steifer" sind die Elektronenbahnen.

Elektrische Linsen für hochvergrößernde Mikroskope (Abb. 13 und 14) wirken infolge ihrer Rotationssymmetrie so, daß der Achsenstrahl alle Potentialflächen unbeeinflußt senkrecht durchläuft (vgl. Abb. 12 A), während die Randstrahlen durch die nach außen ansteigende Brechkraft insgesamt nach der Achse zu gelenkt werden. Die von KNOLL 1929 zur Strahlenkonzentrierung vorgeschlagene grundsätzliche Anordnung einer elek-

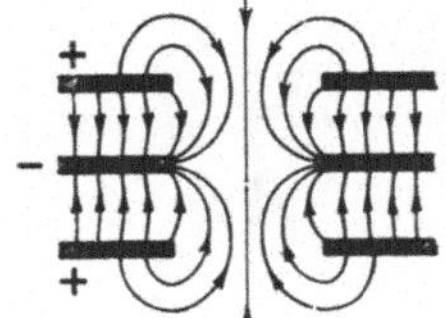

Abb. 13. Feldlinien einer elektrischen Lochblendenlinse. Nach v. BORRIES (1949).

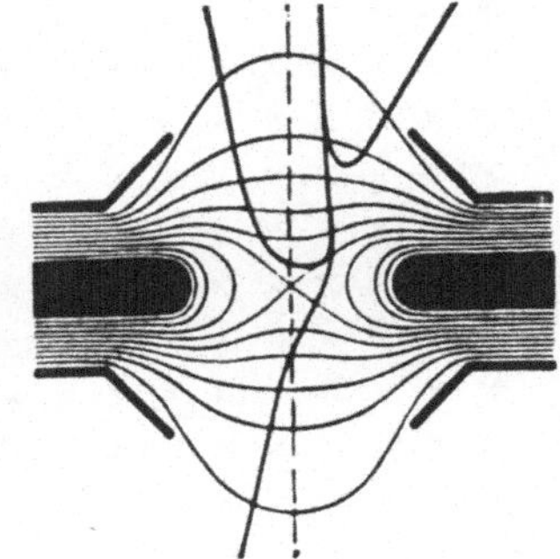

Abb. 14. Schnitt durch die Äquipotentialflächen einer elektrischen Lochblendenlinse. Nach RECKNAGEL (1937). Die Elektronen der eingezeichneten Bahn werden je nach ihrer Geschwindigkeit von der Linse reflektiert oder gebrochen.

trischen Linse besteht aus drei koaxialen Lochblenden (Abb. 13). Die mittlere weist in der Regel die gleiche Spannung auf wie die Kathode, während sich die äußeren wie die Anode auf Erdpotential befinden. Die gleiche Anordnung liegt den von Mahl für die elektrostatischen Übermikroskope verwendeten Linsen zugrunde. Die durch die Linse tretenden Elektronen werden bis zum Durchtritt durch die negative Mittelelektrode verzögert und dann um den gleichen Betrag beschleunigt. Die Äquipotentialflächen einer solchen Linse (Abb. 14) sind teils nach außen, teils nach innen gekrümmt. Die nach außen gewölbten Flächen wirken — bei innenliegender negativer Elektrode — zerstreuend, die nach innen gewölbten sammelnd. Da die nach innen gekrümmten Flächen mit der Sammelwirkung in der Mitte der Linse liegen, wo die Elektronen die geringste Geschwindigkeit haben, während die außen wirksame Zerstreuung nur an den schnelleren „steifen" Elektronen angreifen kann, überwiegt im ganzen die Sammelwirkung.

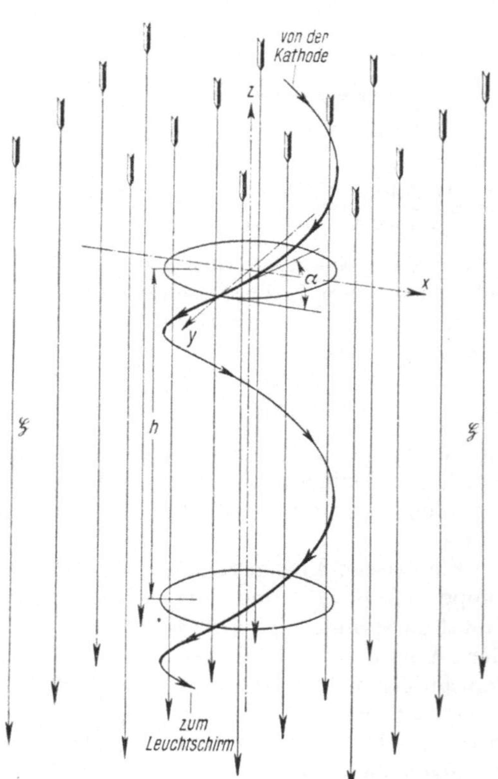

Abb. 15. Ablenkung des Elektrons im magnetischen Feld. a) senkrecht zu den Kraftlinien gesehen. b) in Richtung der in die Papierebene eintretenden Kraftlinien gesehen.

Abb. 16. Schraubenbahn des Elektrons im homogenen Magnetfeld der Feldstärke $\mathfrak{H}$. Nach v. Borries (1949).

b) Magnetische Felder und Linsen.

Elektronen, die sich in einem magnetischen Feld bewegen, lassen sich als Stromfäden betrachten. Nach den Gesetzen der Elektrodynamik erzeugt der Stromfaden um sich ein magnetisches Feld, das seinerseits mit einem von außen

wirksamen Magnetfeld in Wechselwirkungen treten kann. Läuft das Elektron
in Richtung oder entgegengesetzt den Kraftlinien, so tritt keine Beeinflussung
ein. Läuft es senkrecht zu den Kraftlinien wie in Abb. 15 a, so tritt senkrecht
zu diesen und zur Elektronenbahn eine Ablenkung ein. E sei ein aus der Papierebene heraustretender Elektronenstrahl. Als Stromfaden betrachtet, bildet er um
sich ein magnetisches Feld, dessen Kraftlinien kreisförmig im angegebenen Sinne
um E laufen. Der Elektronenstrahl wird dann nach der Seite hin abgelenkt,
wo das äußere Feld und das Feld des Strahls im gleichen Sinne verlaufen. Wenn
das äußere Magnetfeld so groß ist, daß der Strahl im Feld bleibt, so beschreibt
er eine Kreisbahn (Abb. 15 b). Unter einem beliebigen Winkel in das homogene
Kraftfeld eintretende Elektronen verlaufen in Spiralen (Abb. 16). Die Linsenwirkung ist nun dadurch möglich, daß die Steighöhe der Spiralen unabhängig
von dem Winkel ist, mit dem die Elektronen in das magnetische Feld eintreten.
Die von einem Objektpunkt in verschiedenen Winkeln ausgehenden Strahlen
können infolgedessen wieder im Bildpunkt vereinigt werden.

Im Magnetfeld erfolgt die Kraftwirkung auf das Elektron senkrecht zu seiner
Bewegung. Die Elektronengeschwindigkeit bleibt infolgedessen im magnetischen
Feld im Gegensatz zum elektrischen Feld konstant. Außerdem kann das Magnetfeld nur auf bewegte Ladungsträger einwirken, während das elektrische Feld
auch ruhende Ladungsträger zu bewegen vermag.

Magnetische Linsenfelder müssen so beschaffen sein, daß sie den Achsenstrahl
unbeeinflußt lassen und die achsenfernen Strahlen in zunehmendem Maße zur
Achse ablenken. Sie entstehen durch zwei koaxiale ferromagnetische Zylinder oder
Lochscheiben, die durch eine
Stromspule oder Permanentmagnete zu magnetischen Polen
erregt werden. Bei hochver-

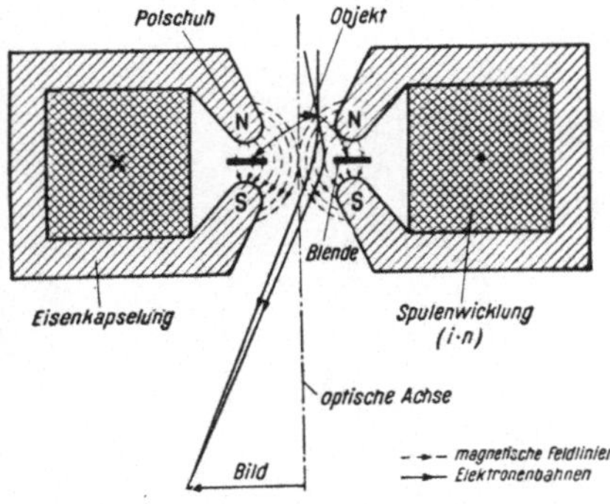

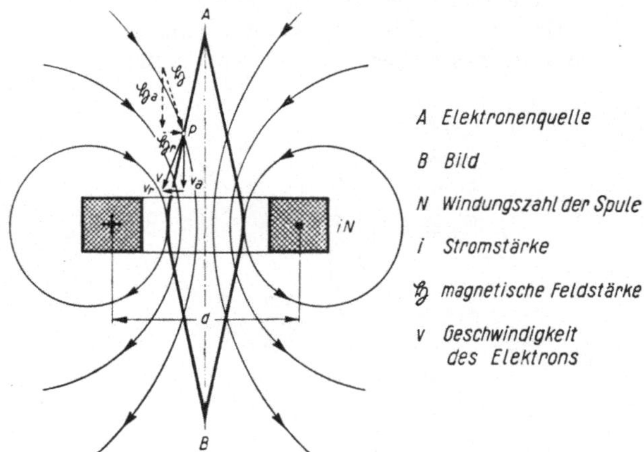

Abb. 17. Polschuhlinse. Nach v. Borries
und E. Ruska (1932/2).

Abb. 18. Stromdurchflossene Ringspule als magnetische
Elektronenlinse. Die spiraligen Elektronenbahnen sind
im Bild in die Zeichenebene zurückgedreht dargestellt.
Nach v. Borries und E. Ruska (1936).

größernden Linsen sind die feldbildenden Eisenteile in Form magnetischer
Polschuhe ausgebildet. Abb. 17 zeigt eine Polschuhlinse nach dem Vorschlag
von v. Borries und E. Ruska (1932). Das Magnetfeld wird durch eine Stromspule erzeugt und durch die mit Polschuhen versehene Eisenkapselung auf
engstem Raum zusammengedrängt.

Die Linseneinwirkung kommt nach Abb. 18 in folgender Weise zustande:
Die vom Punkt A mit einer bestimmten Geschwindigkeit kommenden Elektronen erfahren in P eine Ablenkung in die Zeichenebene hinein. Es entsteht
die auf der ursprünglichen Flugrichtung der Elektronen und auf den Kraft-

linien des Linsenfeldes senkrechte Spiralbewegung. Senkrecht zur tangentialen Spiralbewegung und zu den Kraftlinien des Linsenfeldes findet dann die für die Linsenwirkung entscheidende weitere Ablenkung in der Richtung zur Achse statt. Durch sie wird erst die Konzentration des Strahls herbeigeführt. Durch die Spiraldrehung der Strahlen kommt es zu einer entsprechenden Drehung der Bildlage gegen das Objekt, welche die Lichtoptik und auch die Elektronenoptik elektrostatischer Linsen nicht kennt. Änderung der Stromdurchflutung der Spule führt zur Änderung des magnetischen Linsenfeldes und damit der Brechkraft. Die magnetischen Linsen haben keine konstante Brennweite und können ähnlich der Augenlinse auf verschiedene Abstände des zu betrachtenden Objekts eingestellt werden.

3. Die Elektronenstrahlerzeuger.

Die auf Abb. 10 gezeigte Anordnung zur Erzeugung von Elektronenstrahlen ist für elektronenoptische und besonders elektronenmikroskopische Zwecke nur unvollkommen geeignet. Sie liefert einen Strahl, der zu wenig intensiv und zu wenig scharf gebündelt ist, um als „Mikroskopierlampe" zu dienen. Man bringt daher zwischen den haarnadelförmigen Glühkathodendraht und die Anode einen Zylinder, bzw. eine Blende, die elektrisch negativ gegen die Kathode geladen ist (Abb. 19). Dadurch entsteht ein Richtstrahler, der im Zusammen-

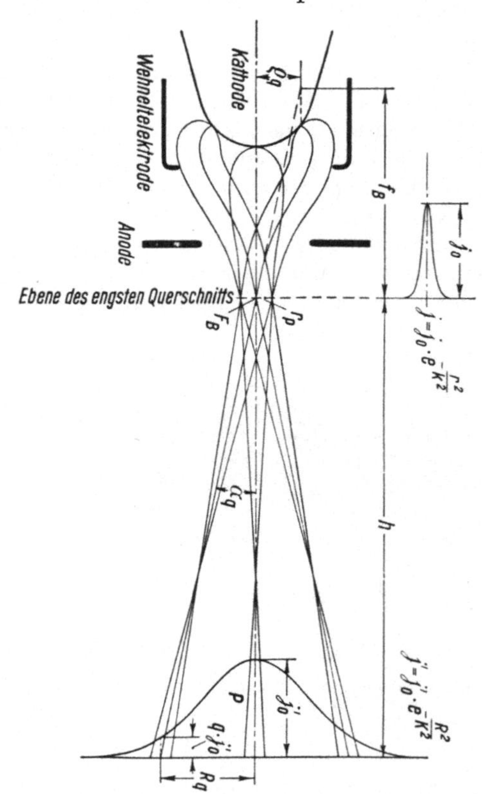

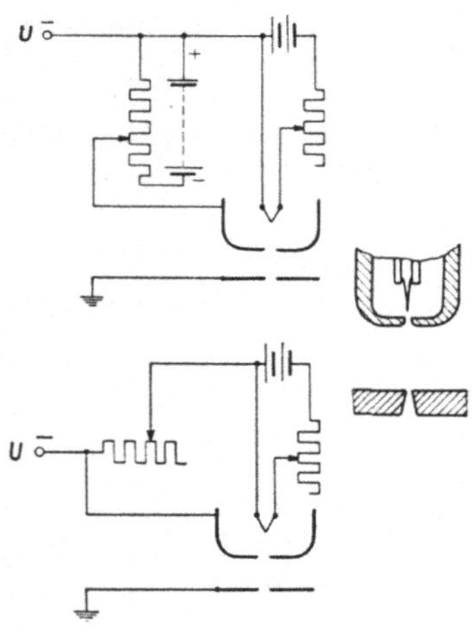

Abb. 19. Glühkathoden-Richtstrahler als Elektronenquelle. Rechts: Schematischer Schnitt. Oben: Betriebsschaltung mit Anodenbatterie zur Erzeugung der Steuerspannung. Unten: Betriebsschaltung mit regelbarem Kathodenwiderstand zur Erzeugung der Steuerspannung. Nach v. Borries (1949).

Abb. 20. Elementarbündel der Elektronen im Richtstrahler. j = Stromdichte im engsten Querschnitt; j' = Stromdichte in Entfernung h vom engsten Querschnitt; f_B = Brennweite der Beschleunigungslinse; F_B = Brennpunkt. Die Stromdichte ist über den engsten Querschnitt und in Annäherung auch über den Strahlöffnungswinkel α nach einem Gaußbuckel verteilt. Nach v. Borries (1949).

wirken des stark positiven Potentials der Anode mit dem schwach negativen des Steuerzylinders die zunächst nach allen Seiten ausgehenden Elektronen bündelt. Der Richtstrahler läßt sich zudem durch Änderung des Potentials am Steuerzylinder regulieren. Bei genügend starkem negativen Potential treten keine Elektronen durch die Steuerblende. Senkt man das Sperrpotential, so treten Elektronen aus und bilden zwischen Steuerzylinder und Anode einen kleinsten Strahlquerschnitt. Dieser vergrößert sich bei weiterer Senkung der Vorspannung irisblendenartig. Außer der Querschnittgröße wachsen die Flächendichte der Strahlung und der Öffnungswinkel (Abb. 20). Alle Erscheinungen tragen dazu bei, den austretenden Strahlstrom auf das notwendige Maß von 3 bis 50 Mikroampère (μA) zu vergrößern. Der Kathodendraht aus Wolfram muß, um optimal zu emittieren, auf 2300^0 bis 2400^0 C geheizt werden. Der engste Strahlquerschnitt wird für die optischen Erörterungn als Ausgangsfläche der Strahlung betrachtet, von der aus sich die Strahlen praktisch geradlinig fortpflanzen.

Statt der Glühkathoden können auch kalte Kathoden verwendet werden, bei welchen z. B. aufprallende positive Gasionen die Emission von Elektronen aus der getroffenen Kathodenfläche bewirken.

4. Die Elektronen-Indikatoren.

Um Elektronenbilder sichtbar zu machen und festzuhalten, werden Leuchtschirme und photographische Platten in den Vakuumraum der Elektronenröhre gebracht.

Die Leuchtschirme geben 3 bis 10 % der eingestrahlten Elektronenleistung als Lichtleistung zurück. Sie müssen so hell und so feinkörnig sein, daß noch eine Betrachtung bei drei- bis sechsfacher Vergrößerung möglich ist, um die Bilder scharfzustellen. Eine besonders hohe Leuchtschirmvergrößerung ist notwendig, wenn das Elektronenmikroskop nur verhältnismäßig geringe (einige tausendfache) Endvergrößerungen liefert. Die photographischen Platten setzen die Elektronenleistung in chemische Leistung um.

Da die elektronenmikroskopischen Präparate nicht als Dauerpräparate aufzubewahren sind, werden von jedem brauchbaren Objekt Aufnahmen gemacht. Erst über die Aufnahmen oder deren lichtoptische Vergrößerung sind die Ergebnisse auszuwerten. Agfa-Normal oder Spektral-Blau-Platten (hart und ultrahart) haben sich für die meisten Zwecke bewährt. Will man zur Erhaltung großer Objektfelder elektronenmikroskopisch niedrig vergrößerte Aufnahmen nachträglich höher vergrößern oder erfordert die Elektronenempfindlichkeit der Objekte das gleiche Vorgehen, um bei schwacher Bestrahlung arbeiten zu können, so werden besonders feinkörnige Emulsionen, wie Agfa-Mikrat-Platten, verwendet. Sie lassen über 50fache lineare Vergrößerungen zu (v. BORRIES 1942, KOPP und MÖLLENSTEDT 1946).

5. Der Strahlengang im Elektronenmikroskop.

Von den zahlreichen, zum Teil im Strahlengang und in der Art der Bildgewinnung grundsätzlich verschiedenartigen Elektronenmikroskopen werden hier nur jene betrachtet, die in Analogie zum Lichtmikroskop gebaut sind. Es sind Durchstrahlungs-Fokussierungsmikroskope, d. h. das Objekt wird durchstrahlt und das Bild durch Fokussierung der im Objekt gestreuten Strahlen gewonnen. In anglo-amerikanischen Ländern werden diese Elektronenmikroskope, ihrer größten praktischen Bedeutung entsprechend, als die „konventionellen" Elektronenmikroskope bezeichnet. In Deutschland ist für *alle* hochauflösenden Elektronenmikroskope die Bezeichnung „Übermikroskope" eingeführt worden. In bezug auf die Linsenfelder sind magnetische und elektrostatische Geräte zu unterscheiden.

Den Strahlengang eines magnetischen Gerätes zeigt Abb. 21. Die vom engsten Strahlenquerschnitt d_{min} vor der Kathode mit der Apertur α_s ausgehenden Elektronen werden durch den Kondensor gebündelt und treffen die Objektpunkte mit der doppelten Bestrahlungsapertur $2\alpha_k$. Im Objekt werden sie auf den Winkelbereich $2\alpha_0$ gestreut, der wesentlich kleiner ist als die doppelte Objektivapertur $2\alpha_{Ob}$, und durch das Objektiv zum Zwischenbild gesammelt. Vom Zwischenbild, das bei den meisten Geräten auf einem Leuchtschirm sichtbar wird, entwirft das Projektiv das Endbild auf einem zweiten Leuchtschirm oder auf der photographischen Schicht.

Das Magnetfeld des Kondensors kann durch Steigerung der Stromdurchflutung verstärkt und somit seine Brechkraft erhöht werden. Damit wächst der Winkel $2\alpha_k$ des bestrahlenden Bündels, woraus ein Ansteigen der Stromdichte im Objekt und der Bildhelligkeit folgt.

Die praktisch verwendete Bestrahlungsapertur α_k liegt zwischen 10^{-4} und 10^{-3}, sie ist also um 3 bis 4 Größenordnungen kleiner als in der Lichtmikroskopie. Verzicht auf die Kondensorlinse ist vor allem mit Verzicht auf eine höhere Stromdichte im Objekt (Bildhelligkeit) verbunden (Abb. 22).

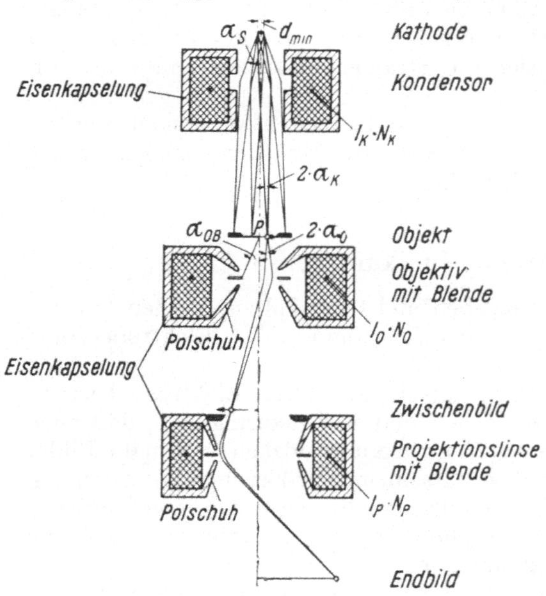

Abb. 21. Strahlengang des magnetischen Elektronenmikroskops. Nach v. Borries und E. Ruska (1939). Erklärung siehe Text.

Liegt als Objekt eine poröse Struktur vor, die ohne Objektträgerfolie über die Objektblende gelegt ist, wie etwa eine Diatomeenschale oder Schmetterlingsschuppe, so treten die Strahlen unbeeinflußt durch die im Strahlverlauf liegenden Objektlücken hindurch. In der bestrahlten Substanz werden die Elektronen gestreut und zu einem kleinen Anteil absorbiert. Je größer die Massendicke (Dicke mal Dichte) des Objektanteils ist, um so größer sind auch Streuung und Absorption. Die gestreuten Strahlen werden wegen der Linsenfehler zum größten Teil vom Objektiv nicht mehr zum Bildpunkt fokussiert. Sie fehlen also ebenso wie der absorbierte Anteil im Bild des Objekts. Dadurch erscheinen die substantiellen Objektanteile dunkler als die Lücken, und zwar um so dunkler, je größer die Massendicke des Objekts ist. Strahlen, die nicht so weit abgestreut werden, daß sie von der Objektivblende abgefangen werden, aber doch so, daß sie nicht mehr fokussiert werden, bewirken eine diffuse Helligkeit des Bilduntergrundes. Für die Erhaltung kontrastreicher Bilder ist es daher günstig, mit Blenden im Objektiv zu arbeiten, die so eng sind, daß sie möglichst viel gestreute Elektronen abfangen. Die Blendendurchmesser betragen z. B. 0,03 bis 0,07 mm. Die Lage der Blende befindet sich etwa dort, wo aus den unabgelenkt durchtretenden und den bisher noch nicht erwähnten, an Strukturrändern gebeugten Strahlen das sogenannte primäre Beugungsbild des Objekts erzeugt wird. Man versteht darunter die in der bildseitigen Brennebene entstehende Fokussierung, in der alle von *beliebigen* Objektteilen *in gleicher Richtung* abgebeugten oder gestreuten Strahlen vereinigt werden, während in der Bildebene alle vom *gleichen* Objektteil in *beliebiger* Richtung ausgehenden Strahlen zusammenkommen. Die Scharfstellung des Objekts durch das Objektiv erfolgt entweder durch Änderung des gegenseitigen Abstands wie in der Lichtoptik, oder durch Änderung der Brennweite der magnetischen Objektivlinse.

Die Bildebene des Objektivs liegt auf dem Zwischenbildleuchtschirm, bzw. etwas darunter. Die Elektronenbilder sind in sehr weiten Bereichen oberhalb und unterhalb der exakten Bildebene scharf, da die Aperturen der Strahlenbündel, die sich in den Bildpunkten vereinigen, außerordentlich klein sind. Sie liegen bei der Objektivstufe um 10^{-5}, d. h. bei Verlagerung der Leuchtschirmebene um 1 cm tritt erst eine Unschärfe von 0,1 μ auf. Aus dem gleichen Grund kann das Leuchtschirmbild des Endbildes, auf dem die Scharfstellung für die Aufnahme erfolgt, höher liegen als die photographische Platte, ohne daß die Bildschärfe merkbar leidet. Die Vergrößerung des Zwischenbildes beträgt bei magnetischen Elektronenmikroskopen das 100- bis 300fache. Sie ist also mitunter höher als bei den besten Immersionsobjektiven der Lichtoptik, aber das elektronenmikroskopische Bild ist besser aufgelöst als das lichtmikroskopische.

Die Projektivlinse kann daher in zweiter Stufe 300fache und höhere Vergrößerungen des Zwischenbildes auf dem Endbildschirm entwerfen, ohne daß die „förderliche" Vergrößerung überschritten wird. Als Endvergrößerung ergibt sich z. B. $150 \times 300 = 45\,000$. Zur Darstellung der kleinsten Auflösung von 1 bis 2 mμ muß so weit vergrößert werden, bis diese Strecke etwa 0,3 mm groß erscheint. Diese „förderliche" Vergrößerung liegt erst bei 150 000 bis 300 000 und kann leicht durch Nachvergrößerung der Originalaufnahme erreicht werden. Die Anfertigung von Nachvergrößerungen der elektronenmikroskopischen Aufnahme ist in den meisten Fällen ratsam. Man erhält bei niedriger Originalvergrößerung große Bildfelder und belastet das Objekt mit geringerer Elektronendichte. Man wird jedoch die elektronenmikroskopische Vergrößerung nicht so weit senken, daß die Feinheiten des Objekts infolge der Körnigkeit von Leuchtschirm und Photoplatte verlorengehen.

Der Endbildleuchtschirm ist auf einer Klappe angebracht, welche die darunter liegende Photoplatte zur Belichtung freigibt (Abb. 37). Wie schon erwähnt, sind Unterschiede in der Lage zwischen der Einstellung des Leuchtschirms und der Platte ohne Bedeutung. Im Endbild treffen sich die Strahlen nur mit Aperturen von etwa 10^{-7}. Bei 10 cm Lageunterschied ergeben sich erst Unschärfen von 10 mμ! Der Bildrand wird durch eine vor dem Projektivlinsenfeld liegende

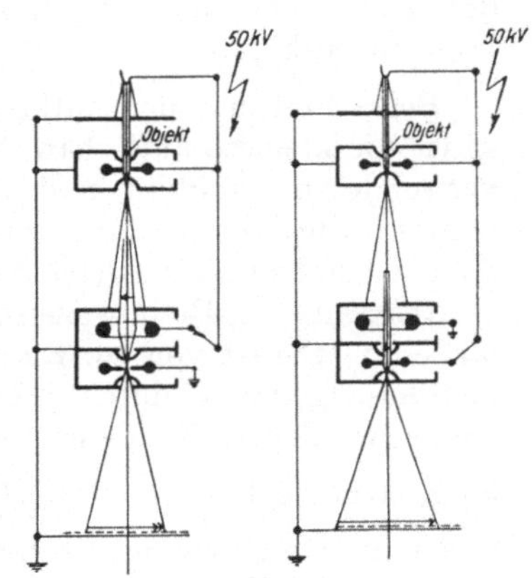

Abb. 22. Strahlengang des elektrostatischen Elektronenmikroskops. Nach H. MAHL (1940). Erklärung siehe Text.

Blende gebildet. Aus dem Produkt der Vergrößerung der Objektblende im Zwischenbild und der Projektivblende im Endbild ergibt sich die Gesamtvergrößerung des Endbildes. Bei den neuen Siemens-Übermikroskopen (E. RUSKA 1949) wird die Vergrößerung in einfacher Weise dadurch bestimmt, daß das Objekt mit Hilfe einer Mikrometerschraube so lange verschoben wird, bis ein ausgewählter Bildpunkt auf dem Endbildleuchtschirm um 6 cm $= 60\,000\,\mu$

verschoben ist. Die Vergrößerung ergibt sich dann durch Division von 60000 durch den an der Mikrometerschraube abgelesenen, in μ gemessenen Wert. Vgl. ferner Froula (1947), Farrant und Hodge (1948) sowie Mahl (1948).

In elektrostatischen Geräten ist der Strahlengang analog. Abb. 22 zeigt eine sogenannte Zweipolschaltung, bei der an Kathode und Linsenelektrode das gleiche Potential liegt. Dadurch ändert sich bei Spannungsschwankungen nicht nur die Beschleunigung der Elektronen, sondern auch die Linsenspannung. Beide Erscheinungen kompensieren sich gegenseitig in ihrer Wirkung auf den Strahlverlauf bis zu Spannungsschwankungen von 1%, während die magnetischen Geräte für Beschleunigungsspannung und Spulenstrom eine Konstanz von $^1/_{10}{}^0/_{00}$ benötigen. Das Projektiv besteht aus zwei Linsen, die wahlweise in den Strahlengang eingeschaltet werden können und etwa 1000fache, bzw. 10000fache Gesamtvergrößerung ergeben. Bei magnetischen Geräten kann die Vergrößerung durch das Projektiv mit Hilfe der Änderung der Stromdurchflutung in engen Grenzen und durch Auswechseln der Projektivpolschuhe oder Drehen eines im Projektiv angeordneten Revolvers mit Polschuhsystemen (Abb. 34) beliebig variiert werden. Besonders weitgehende Veränderungen des Abbildungsmaßstabes lassen Mikroskope zu, die mit dreistufiger Vergrößerung arbeiten, wobei zur stetigen Vergrößerungsregelung die Brechkraft der Mittellinse variiert wird.

6. Die Eigenschaften des elektronenmikroskopischen Bildes.

In den vorstehenden Abschnitten wurden schon zwei besondere Bildeigenschaften erwähnt: das im Vergleich zur Lichtmikroskopie erhöhte Auflösungsvermögen und die Proportionalität zwischen durchstrahlter Massendicke und Bildkontrast. Außerdem ist die elektronenmikroskopische Tiefenschärfe wesentlich größer und das Dunkelfeldbild bei gleicher Bildschärfe dunkler als in der Lichtmikroskopie.

Betrachtet man das Auflösungsvermögen nach der Abbeschen Formel (S. 227), so ist die kleinste auflösbare Strecke δ direkt proportional der Wellenlänge der verwendeten Strahlung und umgekehrt proportional der Apertur des Objektivs. Obwohl die Elektronenstrahlen Korpuskularstrahlen sind und andere Eigenschaften haben als die elektromagnetische Wellenstrahlung (harte γ-Strahlung, sichtbares Licht, Radiowellen), so kommt doch auch ihnen eine definierte Wellenlänge zu. Sie ist von de Broglie aus theoretischen Erwägungen gefolgert und unabhängig davon durch Davisson und Germer aus der Beugung der Elektronenstrahlen an Kristallgittern experimentell nachgewiesen worden. Die Wellenlänge λ beträgt $\dfrac{h}{m \cdot v}$, wobei h das Plancksche Wirkungsquantum, m die Masse und v die Geschwindigkeit der Elektronen ist. Für Elektronenstrahlen zwischen 50 und 100 kV Beschleunigungsspannung ergeben sich Wellenlängen, die etwa 100000mal kleiner sind als die des sichtbaren Lichtes. Infolge der 1000mal kleineren Apertur der Elektronenlinsen bleibt aber nur der Faktor 100 zugunsten eines besseren Auflösungsvermögens. In diesem Umfange ist das Auflösungsvermögen auch tatsächlich gesteigert worden, statt 5000 Gitterstriche pro mm lassen sich 500000 trennen.

In der Lichtmikroskopie wird das Objekt durch seine Farbe (Absorption), durch die vom Einbettungsmedium verschiedene Brechung und durch die Beugung an feinen Strukturen sichtbar. In der Elektronenmikroskopie steht an erster Stelle die Streuung, dann folgen Absorption und Beugung. Den verschiedenen lichtmikroskopischen Farben entsprechen in der Elektronenmikroskopie die verschiedenen Elektronengeschwindigkeiten (Beschleunigungsspannungen), mit

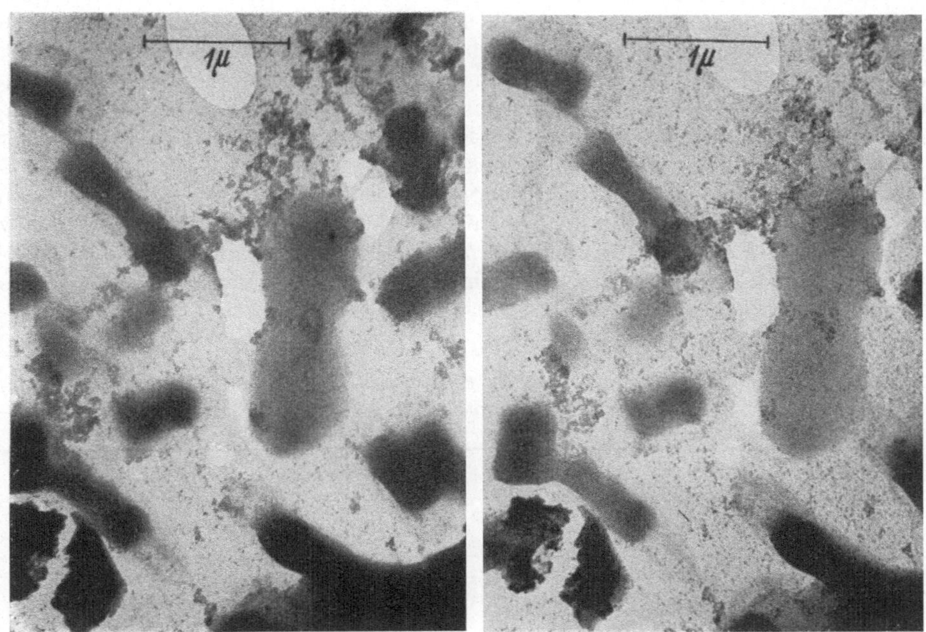

Abb. 23. Einfluß der Strahlspannung auf das elektronenmikroskopische Bild. Links: 38 *kV*; Rechts 89 *kV*. Bei der höheren Spannung sieht man die kolloiden Goldteilchen durch die Colibakterien hindurch. Nach v. BORRIES und E. RUSKA (1940).

welchen mikroskopiert wird. Strahlen mit hoher Beschleunigungsspannung, d. h. kürzerer Wellenlänge, sind durchdringender. Vergleichsweise dicke Objekte

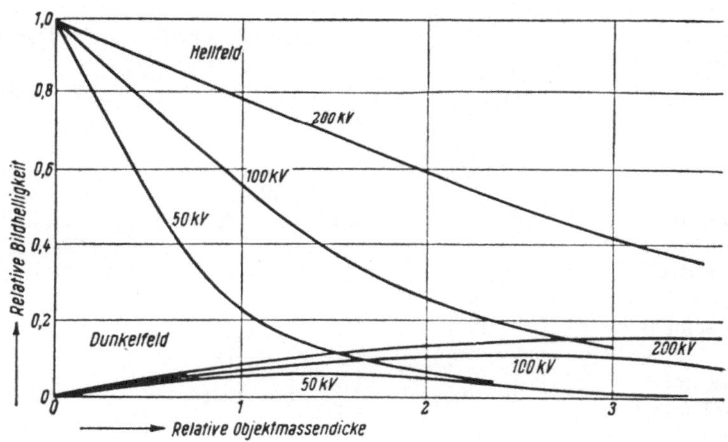

Abb. 24. Gradationskurven der Bildhelligkeit über der Objektmassendicke. Nach v. BORRIES und E. RUSKA (1940).

wird man bei hohen, dünne bei niedrigen Spannungen untersuchen, um günstige Kontraste zu erhalten. Strukturunterschiede dickerer Objektstellen zeigen sich oft erst bei höheren Spannungen (Abb. 23). Die Elektronenbilder sind in dieser Hinsicht mit Röntgenbildern vergleichbar (Tab. 2). Die Gradationskurven haben etwa den in Abb. 24 dargestellten Verlauf. Beim Durchtritt durch das Objekt verlieren die Elektronen zum Teil an Geschwindigkeit (eventuell bis zur völligen Absorption). Das ursprüngliche „monochromatische Licht" wird durch ungleichen Geschwindigkeitsverlust der Elektronen „polychromatisch". Da die Linsen nicht auf „Farbfehler" korrigierbar sind, müssen merkliche Geschwindigkeitsverluste der Strahlen vermieden werden. Die Beugung der Elektronenstrahlen an Strukturrändern kann als Saumbildung um die Strukturen sichtbar gemacht werden (Hillier 1940, Boersch 1943/44, E. Ruska 1943). Die Elektronenstreuung hat im lichtmikroskopischen Bild keine Analogie. Man kann sich aber Objekte vorstellen, die lichtmikroskopisch ausschließlich oder ganz vorwiegend durch ihre Trübung gegenüber dem Einbettungsmedium sichtbar werden. Ihre Einwirkung auf das Licht wäre derjenigen der Objekte der Elektronenmikroskopie auf die Elektronenstrahlen vergleichbar.

Tab. 2. *Eigenschaften des Durchstrahlungsbildes mit Licht-, Röntgen- und Elektronenstrahlen.*

	Lichtbild	Röntgenbild	Elektronenbild
Wellenlänge	800—200 mμ	1—0,1 mμ	etwa 0,005 mμ
Durchdringungsvermögen	gut	sehr gut	sehr gering
Bildarten	Schattenbild (Absorption) optisches Bild (Farbe, Brechung, Beugung)	Schattenbild (Absorption) —	Schattenbild (Absorption) optisches Bild (Streuung)
Kontrastentstehung innerhalb gleicher Schichtdicken	Brechungsexponent	Atomgewicht	Dichte
Kontrasterhöhung durch 1. Veränderung des Objekts 2. Verkleinerung der Wellenlänge 3. andere Mittel	Färbung ultraviolette Strahlen Dunkelfeld	Kontrastmittelfüllung „härtere" Strahlen —	Imprägnation, Bindung schwerer Atome. „schnellere" Elektronen. Metallbedampfung
Tiefenschärfe	um 1 μ	> 1 dm	> 1 μ
Förderliche Vergrößerung	bis 2000 : 1	um 1 : 1	> 100 000 : 1

Infolge der Abbildung der Massendicke läßt sich bei bekannter Gradationskurve der photographischen Platte die relative Dicke eines Objekts aus der Schwärzung der photographischen Schicht feststellen, doch muß außerdem noch die Größe der Objektivblende berücksichtigt werden.

Durch Einlegen einer zentralen Ringblende in den Strahlengang des Kondensors oder durch schräge Beleuchtung lassen sich Dunkelfeldbilder erzielen (Abb. 25). Schon eine geringe Kippung des Kondensors genügt, um den Strahl so schräg durch das Objekt in das Objektiv fallen zu lassen, daß er durch die Objektivblende abgefangen wird. Es gelangt also kein direktes Licht durch das Projektiv zum Endbild, wodurch der Bilduntergrund dunkel bleibt. Nur Strahlen, die an Objekt- rändern oder stärker hervor- tretenden inneren Struktu- ren abgestreut oder gebeugt werden, gelangen in das Endbild. Das elektronen- mikroskopische Dunkel- feldbild zeigt im Gegensatz zum lichtmikroskopischen bestenfalls die gleiche Auf- lösung wie das Hellfeld- bild. Während nämlich im Hellfeldbild die wirksame Apertur kleiner ist als die durch die Objektivblende gegebene und dadurch der Öffnungsfehler des Objek- tivs die Bildschärfe nicht sehr beeinträchtigt, wird im Dunkelfeldbild die ganze Objektivblende von Strahlen durchsetzt, die nur unvollkommen zum Bild fokussiert werden können. Durch extrem kleine Objek- tivblenden lassen sich die Dunkelfeldbilder unter Ver- lust an Bildhelligkeit bis zur Auflösung der Hellfeld- bilder verbessern. (Hell- und Dunkelfeldbilder von Tabak-Mosaikvirus siehe bei HALL 1948.)

Die Dicke eines Objekts wird lichtmikroskopisch durch zeitlich aufeinander- folgende Scharfstellung ver-

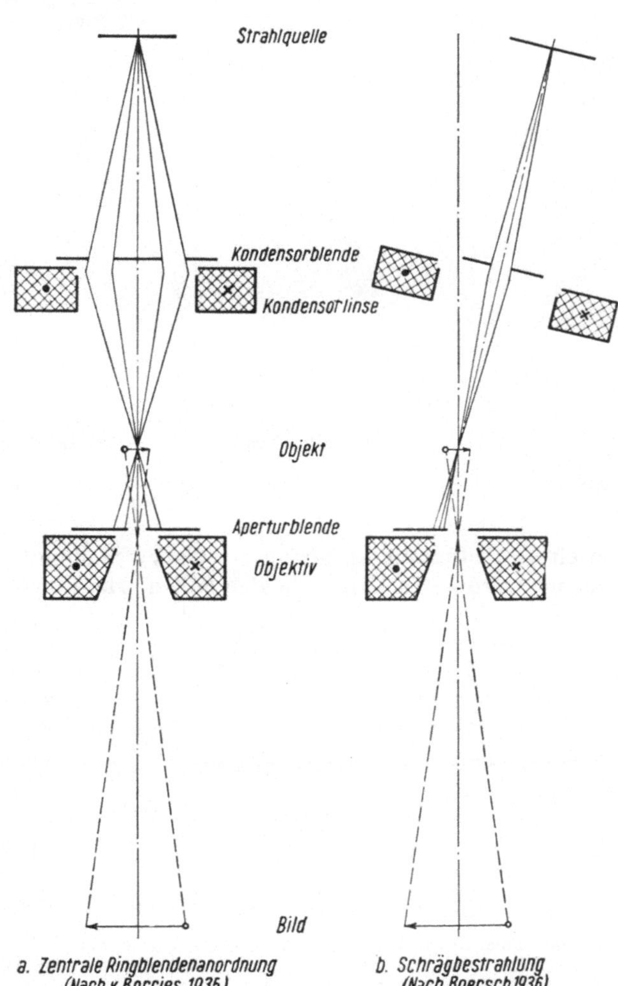

Abb. 25. Strahlengang bei Dunkelfeldbildern.

schiedener Objektebenen und Ausmessung der Objekttiefe mit der Mikrometer- schraube bestimmt. Dieses Vorgehen ist in der Elektronenmikroskopie nicht möglich, weil die Tiefenschärfe der Abbildung so groß ist, daß Objektebenen, die bis zum tausendfachen Wert des Auflösungsvermögens auseinanderliegen, zugleich scharf abgebildet werden. Man mißt daher die Tiefe aus stereo- skopischen Bildpaaren oder aus der „Schattenlänge" der Objektpartikel, die man durch schräg einfallende Metallbedampfung der Präparate vor der mikroskopischen Beobachtung erhalten kann (siehe Abschnitt V, 2b).

Die stereoskopischen Bildpaare werden so hergestellt, daß entweder das

Objekt im Strahlengang zwischen zwei Aufnahmen um eine in der Objekt-
fläche liegende Achse gekippt wird oder daß der Bestrahlungsapparat bei der

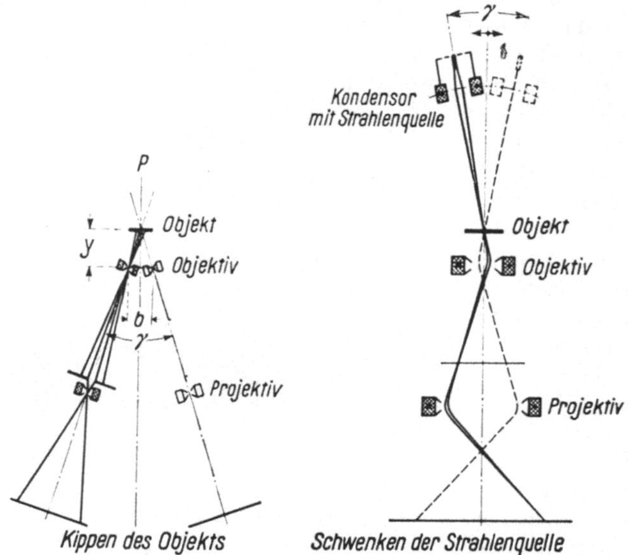

Abb. 26. Strahlengang bei der Aufnahme stereoskopischer Bildpaare. γ = Parallaxenwinkel. Nach
H. O. MÜLLER (1942).

zweiten Aufnahme gegenüber seiner Stellung bei der ersten Aufnahme
geneigt wird (Abb. 26). In beiden Fällen erhält man Bildpaare, die aus verschiedenen Richtungen betrachtet
sind. Im ersten Fall wird das Bild-
paar von der Projektivseite aus
gesehen, im zweiten von der
Kondensorseite aus. Das erste
Verfahren gibt die besseren Bilder,
erfordert aber eine besondere
Einrichtung, um die Objektebene
nach beiden Seiten um 3° bis 8°
kippen zu können. Bei schräger
Bedampfung des Objekts wird
die Tiefe aus einem einzigen Bild
bestimmt. Bei bekanntem Be-

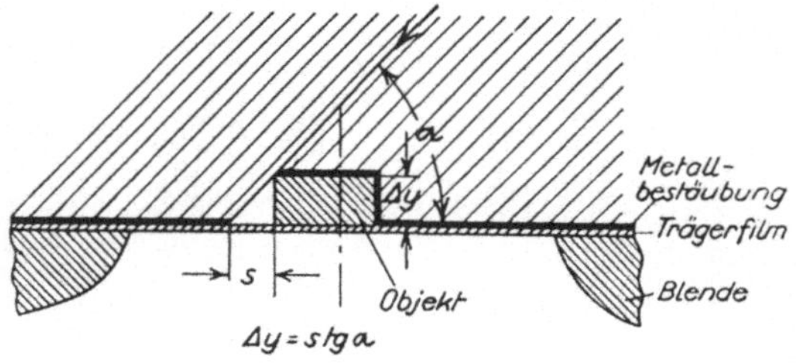

Abb. 27. Messung der Objektdicke aus der Schattenlänge
nach Schrägbedampfung des Objekts. Nach H.O. MÜLLER
(1942).
s = Schattenlänge, Δy = Objekttiefe, α = Bestäubungswinkel.

dampfungswinkel α ist die Objektdicke Δy gleich der Schattenlänge S × tg α
(Abb. 27). (Siehe u. a. v. ARDENNE 1940, EITEL und GOTTHARDT 1940, MAHL 1940,
GOTTHARDT 1942, H. O. MÜLLER 1942, MARTON 1944, KINDER 1946.)

7. Die physikalischen Grenzen der Elektronenmikroskopie.

Obwohl seit der Einführung der Elektronenstrahlen in die Mikroskopie das
Auflösungsvermögen um den Faktor 100 verbessert wurde, ist noch nicht das
ganze Gebiet der Mikromorphologie mikroskopierbar. Es besteht für molekulare
Dimensionen (vgl. Tab. 1) immer noch der Wunsch, die Grenzen der Mikroskopie
weiter hinauszuschieben und außerdem unter Bedingungen zu mikroskopieren,
welche die Objekte möglichst schonend der Einwirkung der Elektronenstrahlen

aussetzen. v. Borries (1949) hat die einzelnen Faktoren, welche die Elektronenmikroskopie begrenzen, eingehend untersucht und ist dabei zu Feststellungen gekommen, welche für die Beurteilung des Verfahrens, die Auswahl optimaler Geräte für die Aufgaben der Virusforschung und die Wahl des Photomaterials von Bedeutung sind. Die Kenntnis der physikalischen Grenzen gestattet, die äußersten Möglichkeiten zu erschöpfen und keine falschen Anforderungen zu stellen.

a) Optische Grenzen.

Der Abstand zweier noch auflösbarer Objektpartikel, der im Lichtmikroskop etwa von der Größenordnung der Lichtwellenlänge ist, beträgt beim Elektronenmikroskop noch das 200- bis 400fache der Elektronenwellenlänge. Ursachen dieser Begrenzung sind die Linsenfehler, insbesondere der Öffnungsfehler des Objektivs, und einige technische Umstände. Als Linsenfehler sind die geometrischen Bildfehler, der Beugungsfehler, der chromatische Fehler und der Symmetriefehler des Linsenfeldes zu nennen. Außer diesen können fremde variable Magnetfelder und mechanische Schwingungen zwischen Objekt und Objektiv die optische Leistung beeinträchtigen.

Für die visuelle Beobachtung des hochvergrößerten Elektronenbildes ist es notwendig, eine ausreichende Helligkeit des Leuchtschirms zu erzielen. Wie erwähnt, beträgt die Lichtausbeute der besten Leuchtstoffe nur wenige Prozent der eingestrahlten Energie, so daß die Elektronenleistungsdichte im Endbild eine beträchtliche Höhe aufweisen muß. Bedenkt man ferner, daß bei nur 10000facher Vergrößerung die Stromdichte im Objekt schon hundertmillionenmal höher liegt als im Endbild und daß die Objektivapertur kleiner als 1 : 1000 ist, so wird es verständlich, daß für höchste Vergrößerungen nur Elektronenstrahlerzeuger mit größter Leuchtdichte[1] verwendet werden können. Diese ergeben bei gut zentriertem Mikroskop eine genügende Helligkeit des Endbildes.

b) Durch Objekt und Elektronenindikatoren bedingte Grenzen.

Die Wechselwirkungen zwischen Strahl und Objekt bewirken einerseits Änderungen von Richtung, Intensität und Geschwindigkeit der Elektronen, andererseits Erwärmung und Ionisierung des Objekts. Die Streuung der Elektronen an den einzelnen Objektelementen ist die wesentliche Ursache der begrenzten Möglichkeit, dickere Objektschichten zu untersuchen. Zur Bildentstehung tragen fast nur diejenigen Elektronen bei, welche ungestreut das Objekt durchdringen. Infolge der Streuung und einer teilweisen Absorption erscheinen die massendicken[2] Objektteile dunkel, die massenfreien, bzw. -armen hell. Schwermetallkolloide von wenigen mμ Durchmesser sind für 100-kV-Elektronen praktisch schon „undurchstrahlbar", während organische Substanzen erst im Größenbereich der Bakterien die Durchdringungsgrenze erreichen. Für Objekte geringer Massendicke ist die Kontrastgrenze von Bedeutung. Sie stellt den Minimalwert der Objektdickenänderung dar, welcher im Endbild als Kontraständerung erkannt werden kann, und liegt bei Objektsubstanzen der Dichte 1 für geeignete photographische Platten bei 3 bis 5 mμ, für die Leuchtschirmbeobachtung noch wesentlich ungünstiger. Je niedriger die Beschleunigungsspannung der Elek-

---

[1] Leuchtdichte = Richtstrahlwert = Strahlstrom, der von der Einheit der strahlenden Fläche in die Einheit des Raumwinkels fließt.

[2] Massendicke = Dicke × Dichte.

tronen ist, um so besser wird der Kontrast, doch ist es aus hier nicht in allen Einzelheiten zu erörternden Gründen unzweckmäßig, mit der Beschleunigungsspannung unter 30 kV zu gehen.

α) Leistungsgrenze der Objektbelastung.

Gegen Temperaturerhöhung empfindliche Objekte können nicht beliebig stark bestrahlt werden. Um aber einen Objektpunkt im Leuchtschirmbild sichtbar zu machen, wird eine bestimmte Leistungsdichte des Elektronenstrahls im Objekt benötigt. Der absorbierte Anteil der aufgenommenen Leistung steigert die Objekttemperatur so lange, bis die mit der Objekttemperatur ansteigende Abstrahlung und Ableitung von Energie den Betrag der absorbierten Energie erreicht. Die Objekttemperatur wird um so größer, je geringer die Beschleunigungsspannung der Elektronen ist, weil langsame Elektronen stärker absorbiert werden als schnelle. v. Borries und Glaser haben die Temperaturen von Objektträgerfolien für verschiedene Durchmesser der Objektblenden berechnet und fanden, daß unter praktisch verwirklichten Bedingungen Temperaturen zwischen 100° und 300° C auftreten (Abb. 28). In den gegenüber der Trägerfolie massendicken Objekten selbst muß mit noch höheren Temperaturen gerechnet

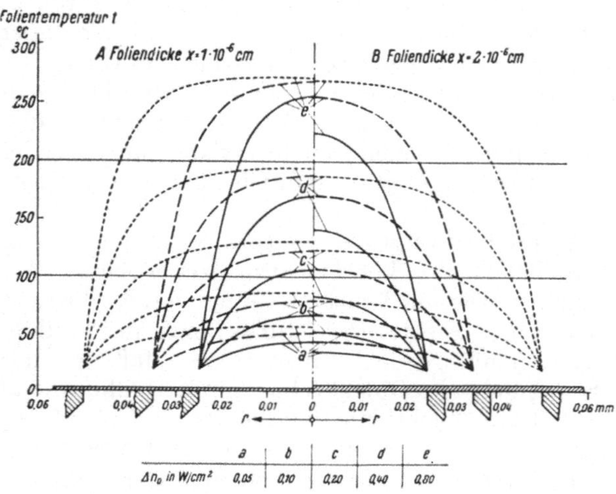

Abb. 28. Filmtemperatur t in Abhängigkeit vom Radius r für drei Werte des Blendenradius, fünf Werte der aufgenommenen Leistungsdichte Δn_0 und zwei Werte der Filmdicke x. Nach v. Borries und Glaser (1944).

werden. Die Erwärmung macht eine längere Beobachtung überlebenden Materials bei der üblichen elektronenmikroskopischen Untersuchungstechnik infolge der Leistungsgrenze der Objektbelastung unmöglich. Da die Objekte im Vakuum bereits ausgetrocknet sind, kommen sie durch die Erhitzung nicht zum Kochen und zeigen infolgedessen auch nicht die Schädigungen, welche z. B. an gekochten Bakterien nachzuweisen sind. Die weitaus meisten sichtbaren Strukturen bleiben vielmehr auch dann erhalten, wenn so stark bestrahlt wird, daß die organische Substanz in Kohlenstoff übergeht.

Eine wesentliche Herabsetzung der Temperatur kann nach v. Ardenne (1948) durch Vorschalten einer Feinblende von wenigen μ Durchmesser und durch Kühlung mit Wasserstoff von 1 Torr erreicht werden. Dadurch wird nur ein kleiner Teil des Objekts von der erwärmenden Strahlung getroffen und zudem ein wesentlicher Betrag der Wärme abgeführt. Statt der Feinblende läßt sich auch ein zweilinsiger Kondensor verwenden, mit dem der kleine Strahlquerschnitt auf optischem Wege zu erhalten ist.

Die unmittelbare Beobachtung von Lebensvorgängen im Elektronenmikroskop ist außerdem durch das fehlende feuchte Milieu unmöglich. Würde man die Objekte in eine dünne Feuchtigkeitsschicht einschließen, so wäre schon bei weniger als $1\,\mu$ Dicke die Durchdringungsgrenze für Elektronenstrahlen von

100 kV Beschleunigungsspannung erreicht, und die Brownsche Molekularbewegung der Objekte würde bei den verwendeten hohen Vergrößerungen so rasch erscheinen, daß eine Beobachtung kaum möglich wäre. Die Aufrechterhaltung eines feuchten Gasraums in der Objektebene ist andererseits dadurch begrenzt, daß höhere Gasdrucke, auch wenn sie nur im Objektbereich aufrechterhalten werden, eine scharfe Abbildung durch Streuung der Elektronen an den Gasmolekülen unmöglich machen (E. Ruska 1942).

β) Dosisgrenze der Objektbelastung.

Setzt man die bei einer Aufnahme dem Objekt zugeführte Strahlendosis dadurch herab, daß man es nur so lange den Strahlen aussetzt, wie zur Erhaltung einer photographischen Aufnahme notwendig ist, so genügt die Aufwendung der Bildpunktarbeit (Kraft × Weg) an Stelle der für die visuelle Beobachtung erforderlichen Bildpunktleistung (Kraft × Weg : Zeit). Erstere stellt die Mindestmenge an Elektronenenergie dar, welche durch den Objektpunkt hindurchtreten muß, um eine genügende Schwärzung der photographischen Schicht zu erzielen. v. Ardenne und Friedrich-Freksa haben gezeigt, daß es möglich ist, bei Bestrahlung mit dieser Mindestdosis die Sporen von Bakterien mit geringer Vergrößerung im Elektronenmikroskop aufzunehmen, ohne daß alle durch die eintretende Ionisierung abgetötet werden. Bakterien und Virus sind aber wesentlich ionisierungsempfindlicher als Sporen, so daß auch für die Elektronenmikroskopie getrockneter, überlebender Objekte so lange wenig Aussicht besteht, wie keine Photoschichten entscheidend höherer Empfindlichkeit zur Verfügung stehen (v. Borries 1948). Die Verwendung feinstkörniger photographischer Schichten, die 50fache und höhere optische Nachvergrößerung zulassen und daher niedrige elektronenmikroskopische Originalvergrößerungen ermöglichen, scheitert bisher daran, daß die aufzuwendende Bildpunktarbeit für diese Platten ungünstiger liegt als für Platten geringerer Vergrößerungsfähigkeit. Der Vorteil, bei niedriger elektronenmikroskopischer Originalvergrößerung mit geringer Strahlstromdichte im Objekt arbeiten zu können, wirkt sich also für die Fixierung des Bildes auf der photographischen Schicht bisher nicht aus.

8. Die historische Entwicklung.

Die Beschäftigung mit der Elektrizitätsleitung in verdünnten Gasen reicht über 100 Jahre zurück. Hittorf fand 1869 die geradlinige Fortpflanzung der von der Kathode ausgehenden Glimmstrahlung und ihre magnetische Ablenkbarkeit. Er beobachtete bereits die schraubenförmige Bewegung unter der Einwirkung magnetischer Felder. 1881 berechnete Riecke, daß unter einem kleinen Winkel zur Feldrichtung in ein homogenes Feld eintretende Kathodenstrahlen Spiralbahnen durchlaufen, deren Ganghöhe vom Einfallswinkel unabhängig ist. Wiechert benutzte 1898 das Feld einer langen Spule zur Bündelung von Elektronenstrahlen. Die Verwendung kurzer Spulen ist 1902 durch Lenard eingeführt und seit dieser Zeit zur Herstellung eines kleinen Schreibflecks in Braunschen Röhren vielfach verwendet worden. Die Bündelung mittels elektrischer Felder geht auf Wehnelt (1906) zurück. Er benutzte einen die Kathode umgebenden Zylinder zur Regelung des Elektronenstrahls, indem er an den Zylinder eine variable, gegenüber der Kathode negative Spannung legte, durch die der Strahlstrom gesteuert wurde.

Die theoretischen Voraussetzungen einer Elektronenoptik erkannten de Broglie (1924) und Busch (1926/27). de Broglie stellte die These auf, daß jeder Korpuskularstrahlung eine Wellenlänge zuzuordnen ist, und Busch erkannte die Konzentrationswirkung *kurzer* Felder als *Linsen*wirkung. Beugungserscheinungen der Elektronenstrahlung waren schon vor der Erkenntnis de Broglies (1922) von Davisson und Germer beobachtet worden. Knoll und E. Ruska haben zuerst daran gedacht, mittels Korpuskularstrahlung und Linsen Mikroskope zu bauen. Mit den Anfängen der Elektronenoptik wurden 1932 die ersten Elektronenmikroskope von Knoll und E. Ruska einerseits und von Brüche und Johannsen andererseits entwickelt. Das mit magnetischen Linsenfeldern arbeitende Instrument von Knoll und E. Ruska diente als Durchstrahlungs- und Emissionsmikroskop, d. h. zur Betrachtung durchstrahlter Blenden und der ausstrahlenden Elektronenquelle. Es war der unmittelbare Vorläufer des ersten hochvergrößernden Übermikroskops von E. Ruska (1933), das mit Polschuhlinsen nach v. Borries und E. Ruska ausgerüstet wurde, und außerdem der Vorläufer der ersten industriell von Siemens & Halske hergestellten Übermikroskope nach E. Ruska und v. Borries (1938). Das Emissionsgerät von Brüche und Johannsen war durch die Verwendung elektrostatischer Linsenfelder, nicht aber nach Strahlengang, Elektronengeschwindigkeit und technischer Ausbildung der Vorläufer des elektrostatischen Übermikroskops, das Mahl (1939) bei der Allgemeinen Elektrizitätsgesellschaft entwickelt hat.

Im Ausland begann Marton 1934 (damals Belgien), vor allem auf den Arbeiten von E. Ruska fußend, mit der eigenen Entwicklung am Mikroskop und seiner biologischen Anwendung. Ihm folgten die aus Tab. 3 ersichtlichen Autoren und v. Ardenne.

III. Die Elektronenmikroskope.

Für das physikalische Verständnis der Elektronenmikroskopie konnte in den vorhergehenden Abschnitten auf eine Erörterung des mechanischen Aufbaus der Geräte, der Vakuumanlagen und der elektrischen Versorgung verzichtet werden. Für das praktische Mikroskopieren ist jedoch eine Kenntnis der wichtigsten Teile der Gesamtapparatur unentbehrlich. Es ist indessen nicht Aufgabe dieses Handbuchs, alle Geräte in ihren Einzelheiten zu beschreiben. Einige grundsätzliche Ausführungen und eine Übersicht der im Handel befindlichen Geräte müssen genügen.

1. Der Gesamtaufbau.

Der Strahlengang verläuft bei Instrumenten hoher Leistung meist senkrecht von dem oben befindlichen Strahlerzeuger zu dem in Tischhöhe liegenden Leuchtschirm (Abb. 21, 22, 29, 30). Bei senkrechter Bauweise tragen sich die zum Teil recht schweren metallenen Bauteile gegenseitig ohne zusätzliche Stützvorrichtungen. Die säulenartige Mikroskopierröhre ruht auf einem Tisch, an dem das Bild von einer bis drei Seiten beobachtet werden kann. Von der vierten Seite werden Spannungen, Spulenströme, Kühlwasser- und Vakuumleitungen zugeführt. In dem Tisch oder einem auf seiner Rückseite mit ihm verbundenen Stativ können mechanische Vorrichtungen zum Öffnen der Mikroskopröhre, ferner Hochvakuumpumpe, sowie auch Teile elektrischer Anlagen untergebracht werden. Tisch oder Stativ tragen ferner die zur Bedienung des Elektronenmikroskops erforderlichen Schalter und Regelungen, die möglichst alle vom Beobachter im Sitzen bedienbar sein sollen. Die hochspannungführenden Teile (Kathode und Steuerzylinder mit zugehörigem Heizstromkreis und Regelschaltungen) sind innerhalb einer Wanne am Kopf des Stativs berührungssicher untergebracht.

Bei den meisten Gerätetypen wird die Hochspannungsanlage und die Stromquelle für die Magnetlinsen, sowie die Vorvakuumpumpe getrennt vom Mikroskop aufgestellt; bei zwei verbreiteten amerikanischen Geräten (RCA) sind die Hochspannungs- und Linsenstromquellen im Mikroskopstativ angebracht.

2. Die Mikroskopröhre.

Die notwendige Justierung des Bestrahlungsapparates zur Achse der Abbildungslinsen, das Einbringen und Bewegen des Objekts, die Bedienung der Aufnahmevorrichtung, das gelegentlich erforderliche Auswechseln von Kathode, Linsen, Blenden und Leuchtschirmen machen das einfache Elektronenstrahlrohr von Abb. 10 zu einem komplizierten, aus zahlreichen Teilen bestehenden Baukörper. Die einzelnen Teile sind zentrisch und vakuumdicht mit gefetteten Schliffen oder Gummidichtungen aufeinandergesetzt. Wo Bewegungen der Röhrenteile gegeneinander erforderlich sind, wirken Gummimanschetten oder Wellrohre als Teile der Vakuumwand. Die Bewegung einzelner Teile innerhalb des Mikroskops erfolgt bei Drehbewegungen meist über Schliffe oder gummigedichtete Stangen, bei Schiebebewegungen über Wellrohre oder Gummimanschetten.

Die meist als Glühkathode ausgebildete Elektronenquelle und die Steuerhülse (vgl. Abb. 19) werden durch ein Isolierrohr getragen. Der haarnadelförmige Glühdraht, welcher je nach der Güte des Vakuums und der Höhe der Kathodenheizung nach einiger Zeit durchbrennt und deshalb leicht auswechselbar sein muß, sitzt zentriert zur Öffnung der Steuerhülse (Wehneltzylinder) (Abb. 31). Heizung und Wehneltspannung sind regulierbar. Bei manchen Konstruktionen kann der Beleuchtungsapparat gegen die Mikroskopröhre verschoben und um das Objekt gekippt werden. Andere justieren die Kathode durch geringe Verschiebung innerhalb der Steuerhülse. Bei höheren Spannungen als 100 kV werden die Elektronen mehrstufig beschleunigt.

Die Kondensorlinse sammelt die divergierenden Strahlen auf dem Objekt. Meistens wird das ganze Objektfeld beleuchtet; oder es ist möglich, nur den für die jeweilige Abbildung benötigten Teil zu bestrahlen. Das zuletzt genannte Verfahren schont das Objekt, erfordert aber ein zweilinsiges Kondensorsystem. Da elektrostatische Linsen nur mit größerem Aufwand zu regulieren sind, wird heute bei elektrostatischen Elektronenmikroskopen entweder auf den Kondensor verzichtet (Abb. 22) oder ein magnetischer Kondensor verwendet. Die Justierung des Kondensors erfolgt entweder gemeinsam mit dem Strahlerzeuger oder unabhängig davon.

Vor die Objektivlinse wird das Objekt in das Vakuum eingeschleust. Die Abbildung 32 zeigt eine Schleusenkonstruktion, deren Einzelheiten in der Bildunterschrift erklärt sind. Bei kleinem Gesamtvakuumraum der Röhre und bei guten Pumpsystemen kann auf Schleusen verzichtet werden. Das Objekt muß möglichst nahe an die Objektivlinse herangebracht werden, damit kurze Brennweiten auszunützen sind. Magnetische und elektrostatische Objektivlinsen werden deshalb meist einseitig ausgebildet (Abb. 33). Das Objekt muß außerdem in der Objektebene beweglich sein, damit man seine Gesamtfläche durch Verschieben auf dem Endbildschirm abbilden kann. Dabei genügen Bewegungen von $\pm$ 0,3 mm. Meist wird das Objekt dem Objektiv fest aufgepreßt, damit keine von außen kommenden mechanischen Erschütterungen durch gegenseitige Bewegungen von Objekt und Objektiv die ruhige Lage des Bildes stören. Bei nicht regelbaren elektrostatischen Linsen ist zudem eine

Bewegung des Objekts in der optischen Achse erforderlich, um die Bildschärfe
einstellen zu können, während bei magnetischen Linsen die axiale Objektlage

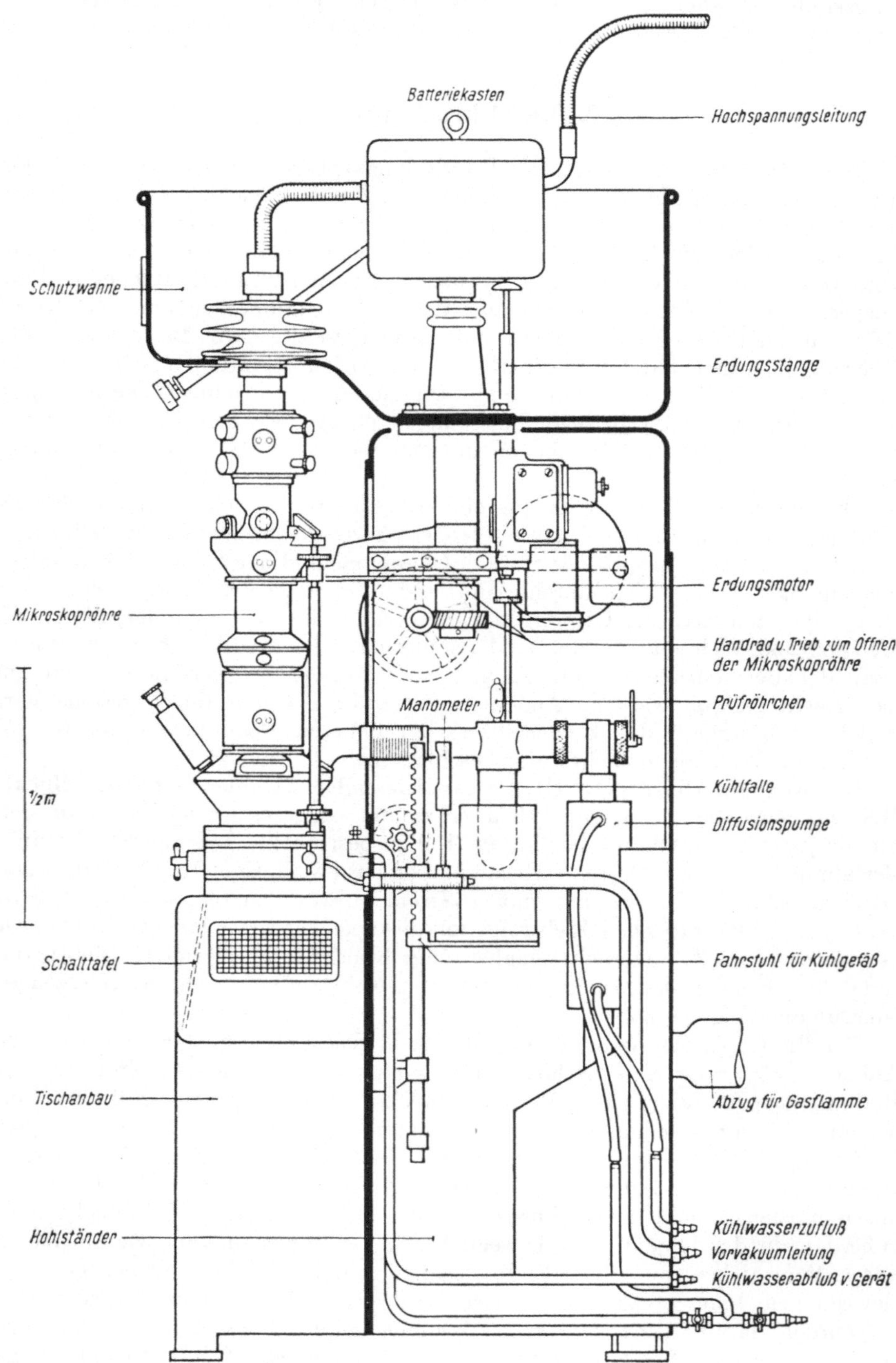

Abb. 29.   Elektromagnetisches Siemens-Übermikroskop nach E. Ruska und v. Borries (1939).
Aufriß von Röhre und Tischanbau sowie Schnitt durch das Stativ.

mcist fixiert ist und die Scharfstellung des Bildes durch Änderung der Objektivbrennweite erfolgt.

Der Tubus zwischen Objektiv und Projektiv trägt in der Regel Einblickfenster für einen über dem Projektiv liegenden Zwischenbildleuchtschirm. Durch ein Prisma oder einen Spiegel kann der Leuchtschirm auch im Sitzen beobachtet werden. Er erleichtert die Übersicht über das ganze Präparat, das im Zwischenbild in 100- bis 300facher Vergrößerung sichtbar ist. Außerdem dient das Zwischenbild dazu, in der ersten Vergrößerungsstufe den Abbildungsmaßstab zu bestimmen. Er ergibt sich aus der Division des Objektblendendurchmessers im Zwischenbild durch den tatsächlichen Durchmesser der Objektblende. Der Zwischenbildleuchtschirm kann wegklappbar sein, um in Verbindung mit einem Polschuh-Revolverprojektiv bei der Aufnahme von Beugungsdiagrammen den abgebeugten Strahlen den Durchtritt zum Endbildleuchtschirm freizugeben.

Der Wechsel der Gesamtvergrößerung wird durch Brennweitenänderung der Projektivlinsen erreicht. Bei magnetischen Mikroskopen können nach Öffnen der Röhre die Polschuhe ausgetauscht werden, bei neueren Konstruktionen ist ein Revolverprojektiv vorhanden (Abb. 34). Bei dreistufiger Abbildung kann eine kontinuierliche Regelung des Abbildungsmaßstabes durch Brennweitenregelung der Mittellinse ermöglicht werden. Elektrostatische Mikroskope haben mitunter zwei übereinanderliegende, wahlweise einschaltbare Projektive (vgl. Abb. 22).

Der Tubus zwischen Projektiv und Endbild ist wiederum mit Einblicköffnungen versehen. Sie sind meist so breit, daß das Endbild binokular betrachtet werden kann

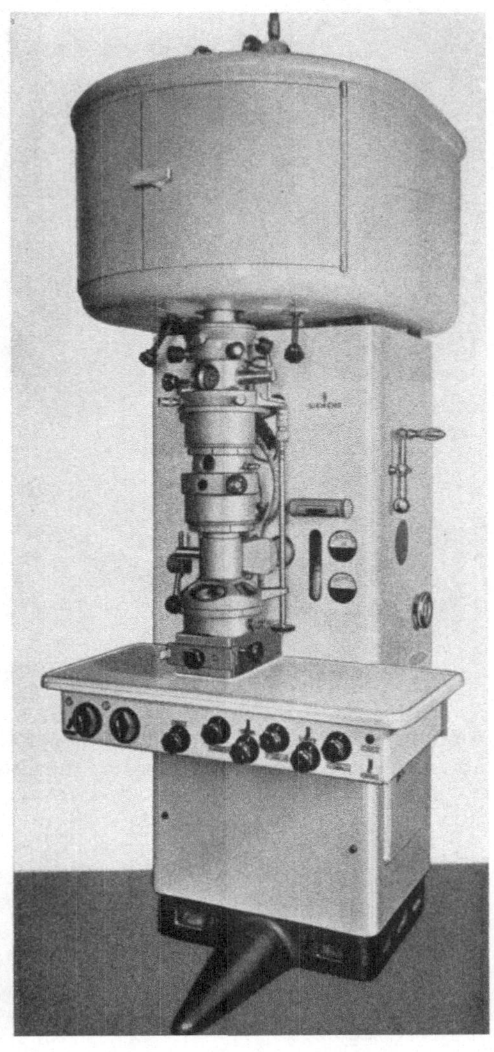

Abb. 30. Elektromagnetisches Siemens-Übermikroskop nach E. RUSKA und v. BORRIES (1949).

(Abb. 35). Die Scharfstellung des Bildes läßt sich durch vorschaltbare monokulare oder binokulare Lupen erleichtern und verbessern. Die Kamera zur Aufnahme der Photoplatten ist häufig als Schleuse ausgebildet und kann eine oder mehrere Platten sowie Filme aufnehmen (Abb. 36 und 37).

Die Baulänge der Mikroskopröhre ist bei gleicher Zahl der Vergrößerungsstufen (in der Regel 2) von der Art der Linsen, der verwendeten Strahlspannung und der geforderten Endvergrößerung abhängig. Je kürzer die Brennweiten der Linsen sind, um so kürzer kann auch die Abbildungslänge (Strahlengang zwischen Objekt und Endbild) sein. Die elektromagnetischen Linsen haben nach Abb. 38 kürzere Minimalbrennweiten als elektrostatische, und die Größe der Brenn-

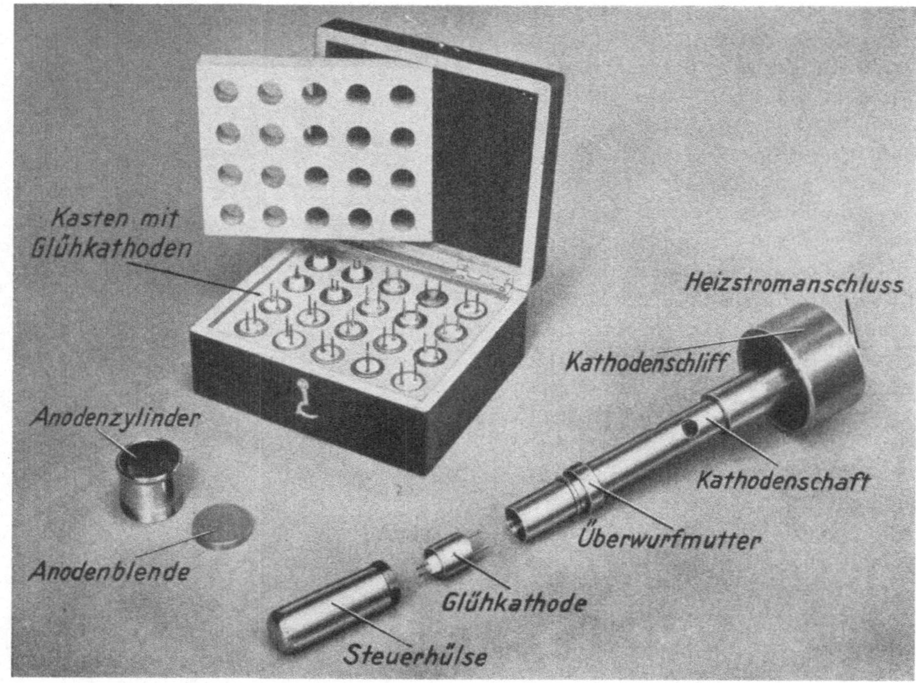

Abb. 31. Einzelteile des Elektronenstrahlrohrs von Abb. 30.

weiten wächst bei magnetischen Linsenfeldern langsamer mit der Strahlspannung als bei elektrischen. Da außerdem die Linsenfehler und das Auflösungsvermögen bei magnetischen Linsen bisher günstiger liegen als bei elektrischen, werden Hochleistungsgeräte bevorzugt mit elektromagnetischen Linsen ausgerüstet. Die Brennweiten sind bei den Objektiven größer als bei Projektiven, bei magnetischen Objektiven wegen der einseitigen Ausbildung der Polschuhe, bei elektrischen deshalb, weil das Objekt im Gegensatz zum Zwischenbild wegen

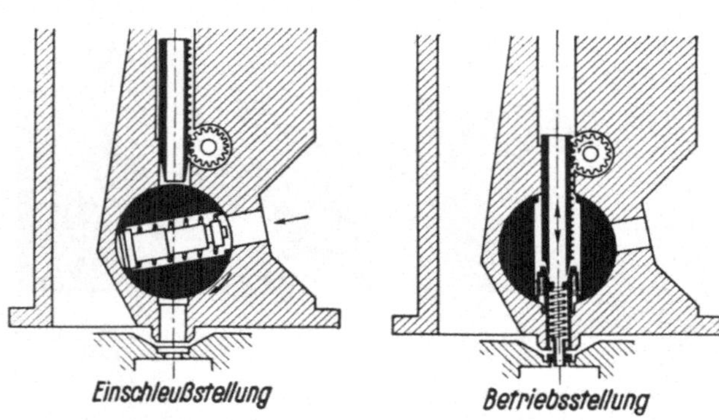

Abb. 32. Schnitt durch die Objektschleuse eines älteren Siemens-Übermikroskops.

der Gefahr der Entladung nicht beliebig nahe an die negative mittlere Linsenelektrode herangebracht werden kann. Aus Abb. 39 ist die Abhängigkeit der Abbildungslänge elektrischer und magnetischer Geräte von der Strahlspannung ersichtlich.

3. Die elektrischen Anlagen.

Zur Erzeugung der Beschleunigungsspannung dienen entweder an das 50periodische Licht- oder Kraftnetz angeschlossene Hochspannungsgleichrichter mit Glühventilen, entsprechend dem Schaltschema von Abb. 40, oder im Rundfunkfrequenzgebiet arbeitende Anordnungen, bestehend aus Generator, Transformator und Gleichrichter. Letztere können im Stativ des Elektronenmikroskops untergebracht werden.Die Höhe der Spannung kann fixiert, z. B. 50000 Volt, oder in mehreren Stufen, z. B. zwischen 40000 und 100000 Volt regelbar sein. Die Variabilität der Betriebsspannung macht die Geräte vielseitiger verwendbar, weil bei niedriger Spannung dünne Objekte noch gute Kontraste geben und bei hoher Spannung dickere Objekte besser durchstrahlt werden (Abb. 41). Für magnetische Elektronenmikroskope sind Spannungsanlagen höherer Konstanz erforderlich als für elektrostatische (vgl. S. 242).

Der von der Kathode ausgehende Strahlstrom wird durch die Vorspannung eingestellt und an einem in der Rückleitung vom Mikroskop zum Hochspannungserzeuger liegenden Meßinstrument laufend gemessen. Die Messung dient der ständigen Kontrolle von Höhe und Stetigkeit der Emission. Die negative Vorspannung wird durch eine

Abb. 33. Objektivspule des Elektronenmikroskops von Abb. 30 mit ausgebautem Polschuhsystem.

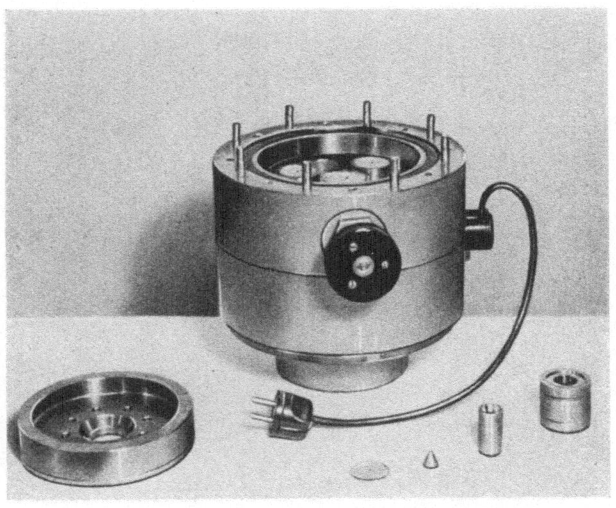

Abb. 34. Polschuhrevolverprojektiv eines Siemens-Übermikroskops; durch Drehen an dem Handrad können vier verschiedene Polschuhlinsen in den Strahlengang gebracht werden.

Anodenbatterie geliefert oder durch eine Stufenschaltung mit Hilfe von zwischen Kathode und Wehneltzylinder liegenden hochohmigen Widerständen erreicht (Abb. 19). Über eine isolierende Welle wird die Vorspannung reguliert.

Den Heizstrom für die Glühkathode liefert ein auf Kathodenpotential liegender Akkumulator (vgl. Batteriekasten, Abb. 29) oder bei niedrigeren Spannungen

ein in den Hochspannungserzeuger
eingebauter Heiztransformator. Im
ersteren Falle ist der Heizstrom über
eine isolierende Welle regulierbar, und
das Meßinstrument befindet sich auf
der Hochspannungsseite.

Während bei elektrostatischen Ge-

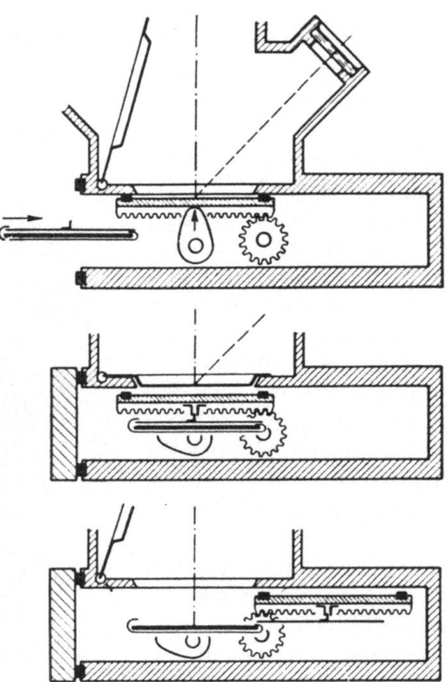

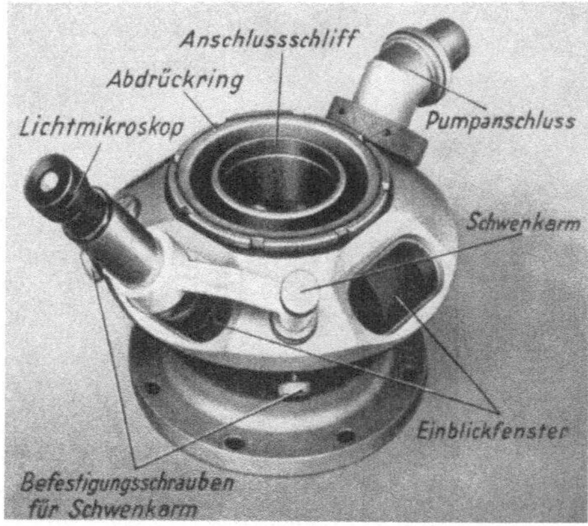

Abb. 35. Projektivtubus und Beobachtungs-
mikroskop des in Abb. 30 dargestellten Geräts.

Abb. 36. Schematischer Schnitt durch die
Plattenschleuse des in Abb. 30 dargestellten
Geräts. Oben: Der Schleusenraum ist gegen
die Mikroskopröhre verschlossen, nach außen
offen. Die Photoplatte wird in verschlossener
Kassette eingeschoben. Mitte: Der Schleusenraum ist gegen die Mikroskopröhre offen, nach
außen verschlossen. Unten: Wie in der Mitte,
jedoch Photokassette geöffnet.

Abb. 37. Ansicht zu Abb. 36.

räten die Hochspannungsanlage zugleich das Linsenpotential liefert (vgl. Abb. 22), ist bei elektromagnetischen Geräten eine zusätzliche, sehr konstante Stromquelle zur Erregung der magnetischen Spulen erforderlich. Am einfachsten

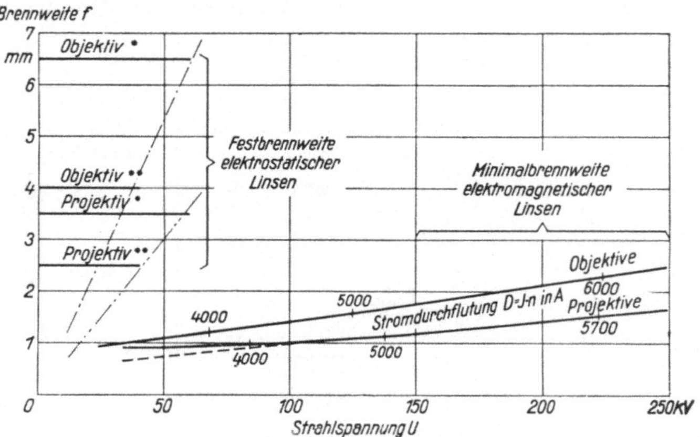

Abb. 38. Brennweiten elektrostatischer und elektromagnetischer Linsen in Abhängigkeit von der Strahlspannung. Nach E. Ruska (1942).

wird eine Akkumulatorenbatterie mit einer Leistung von etwa 150 bis 300 Watt und einem netzanschlußfähigen Ladeaggregat aufgestellt, doch können auch röhrengeregelte Netzanschlußgeräte Verwendung finden. Zwischen Stromquelle einerseits und Kondensor, Objektiv und Projektivlinse andererseits liegen Widerstände zur Regelung der Spulenströme. Die Regelung muß besonders beim Objektiv, mit dem die Scharfstellung des Endbildes erfolgt, stetig vor sich gehen und ist daher meist in eine Grob- und Feinregelung unterteilt.

An elektrischen Einrichtungen sind ferner die Heizung der Quecksilber- oder Öldiffusionspumpe sowie der Antrieb für eine rotierende Ölpumpe, bzw. der Antrieb einer Molekularluftpumpe zu nennen. Außerdem sind meist Beleuchtungen der Meßinstrumente, Warnlampen für die Hochspannung und Schalter für helle und abgedunkelte Raumbeleuchtung am Mikroskop vorgesehen.

4. Die Vakuumanlagen und Kühlwasserleitungen.

Die Mikroskopröhre liegt dauernd an einer oder zwei Hochvakuumpumpen. Bei Quecksilberdampfpumpen sind sogenannte

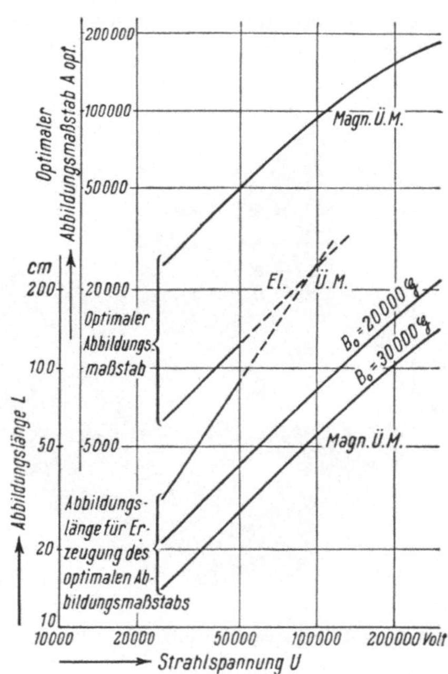

Abb. 39. Optimaler Abbildungsmaßstab und Abbildungslänge bei zweistufig vergrößernden Übermikroskopen in Abhängigkeit von der Strahlspannung. Nach v. Borries (1949).

Kühlfallen, die am besten mit flüssiger Luft gefüllt werden, zwischen Mikroskop und Pumpe geschaltet. Dadurch können keine Quecksilberdämpfe in die Mikroskopröhre eindringen. Die Pumpen sind gegen das Mikroskop durch

Hähne abschaltbar. Die Vakuumpumpe kann wahlweise mit der Mikroskopröhre, der Photokassette oder dem Vorvakuum der Hochvakuumpumpe verbunden werden. Je leistungsfähiger die Pumpen sind, um so rascher können
die Geräte, bzw. Teile davon, wie etwa die Photokassette, nach einer Luft-
füllung wieder in Betrieb genommen werden.

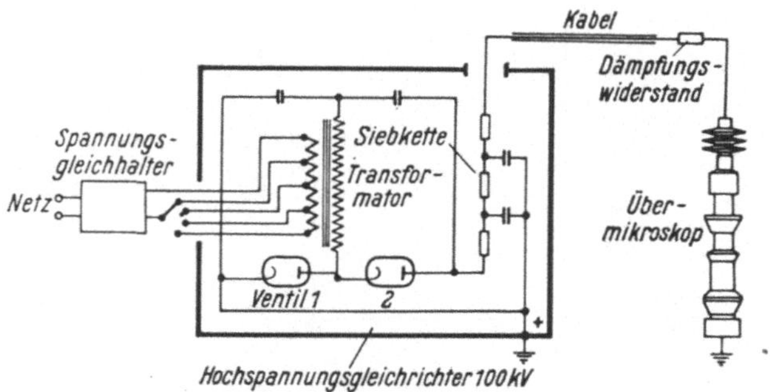

Abb. 40. Schaltbild eines Hochspannungserzeugers.

Die Hochvakuumpumpen werden vom Kühlwasser umströmt. Ebenso
müssen die Spulen der elektromagnetischen Objektive und Projektive von
Elektronenmikroskopen für hohe Strahlspannungen zum Abführen der in den
Spulen entstehenden Wärme gekühlt werden. Bei elektrischen und magneto-
statischen Linsen ist eine Kühlung nicht erforderlich.

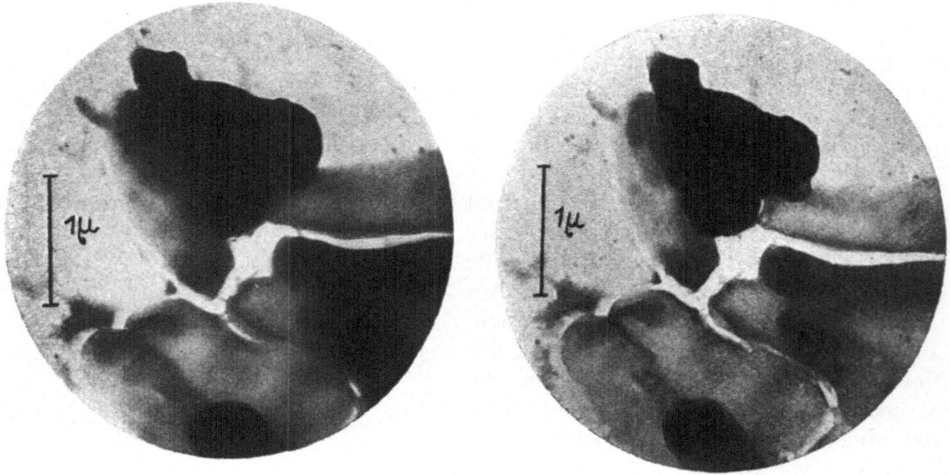

Abb. 41. Durchstrahlungsbild von Tuberkelbazillen bei 60000 V und 185000 V Strahlspannung.
Nach Müller und E. Ruska (1941).

5. Die im Handel befindlichen Elektronenmikroskope.

Die Darstellung der Elektronenmikroskope wurde auf die für die Virusforschung in Frage kommenden Durchstrahlungs-Fokussierungsmikroskope hoher
Auflösung beschränkt. Zu dieser Gruppe gehören alle serienmäßig von der Industrie gefertigten Elektronenmikroskope.

Die Mindestforderungen für das Auflösungsvermögen werden von allen in Tabelle 3 aufgeführten Elektronenmikroskopen der verschiedenen Firmen erfüllt. Hinsichtlich der Betriebsspannung und ihrer Variabilität, der Linsenfehler und des Auflösungsvermögens, sowie in der baulichen Ausführung bestehen jedoch beträchtliche Unterschiede. Für die ausschließliche Untersuchung von Viruspartikeln genügen niedrige Beschleunigungsspannungen. Schon bei Bakterien macht sich jedoch mitunter eine niedrige Spannung als Beschränkung in der Durchstrahlbarkeit der Zellen bemerkbar. Noch mehr dürfte dies für die Untersuchung von Einschlußkörpern oder ganzen Zellen zutreffen. Das Auflösungsvermögen soll möglichst so hoch sein, daß die größeren Virusteilchen noch innerhalb der Strukturbewertungsgrenze liegen. Wer bei kleinsten Virusformen nicht von der Lupenbetrachtung des Endbildes auf den Leuchtschirm abhängig sein möchte, sollte die maximale Endvergrößerung nicht unter 15 000 : 1 wählen. Kurze Baulängen der Mikroskopierröhre sind in der Bedienung bei sonst gleicher Leistung den großen Baulängen vorzuziehen. Leicht auswechselbare Projektive (Revolver) zum bequemen Wechsel des Vergrößerungsmaßstabes oder dreistufige Vergrößerung mit regelbarer Mittelstufe und Photokassetten mit mehreren Photoplatten erleichtern das Mikroskopieren wesentlich.

In Tabelle 3 sind nur die handelsüblichen, nicht aber die Sonderanfertigungen von einzelnen Elektronenmikroskopen, wie z. B. die Geräte von v. ARDENNE, aufgeführt, obwohl mit ihnen eine Reihe von Virusuntersuchungen durchgeführt worden ist. Wegen der konstruktiven Besonderheiten muß auf die im Schriftenverzeichnis angeführte zusammenfassende Literatur verwiesen werden. Die Tabelle bringt die Elektronenmikroskope in der Reihenfolge ihrer Entstehung. Ein elektrostatisches Gerät und den Schnitt durch eine elektrostatische Linse zeigen die Abb. 42 und 43.

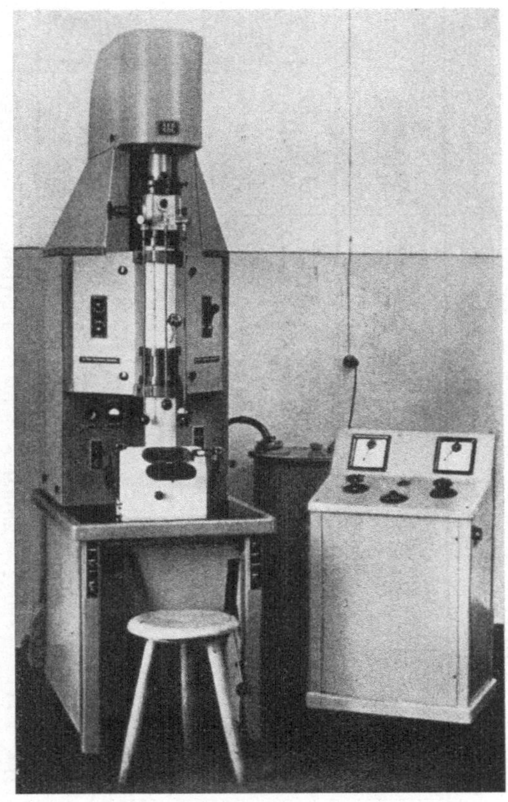

Abb. 42. Elektrostatisches Übermikroskop der Allgemeinen Elektrizitäts-Gesellschaft. Nach H. MAHL(1941), siehe auch BRÜCHE (1949).

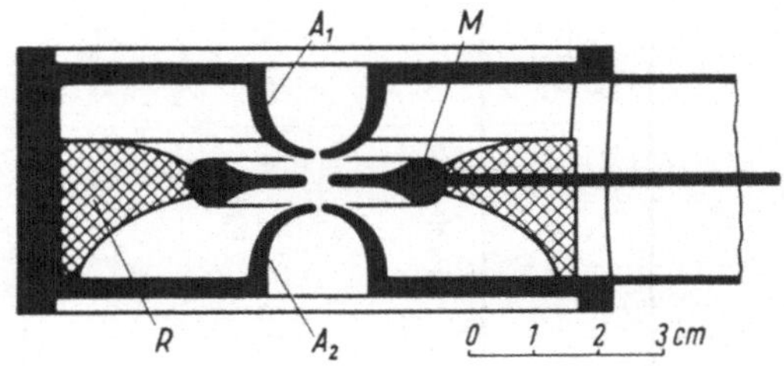

Abb. 43. Elektrische Hochspannungslinse kurzer Brennweite. Nach H. MAHL (1939).

Tabelle 3.

I. Geräte mit magnetischen Linsen.

Firma	Konstruktion	Modell und Jahr	Betriebsspannung in kV	Abbildungsstufen	Höhe des Geräts in cm	Länge des Strahlenganges in cm	Abbildung des Objekts auf der Platte bei höchster Strahlspannung minimal	maximal	Auflösungsvermögen in mμ	Aufnahmematerial	Aufnahmeeinrichtung	Bemerkungen
Siemens & Halske, Berlin	Ernst Ruska und v. Borries	1939—44	60—100	2	240	100	1000:1	40000:1	2	Platten 6,5×9 cm²	Schleuse für eine Platte	Lupenbetrachtung des Leuchtschirms
	Ernst Ruska	ÜM 100 1949	40—100	3	216	82	25:1	100000:1	2	wie oben und Normalfilm	wie oben oder Wechseleinrichtung für 12 Platten oder 40 Normalfilmbilder	
	,,	ÜM 60 1949	40—60	2	147	52	300:1	30000:1	3			ohne Kondensor
Radio-Corporation of America, Princeton, New Jersey	Hillier und Vance	EMB 1940	60	2	213	117	800:1	25000:1	5	Platten 2×10 Zoll² für 5 Aufnahmen 2×2 Zoll²	Schleuse	
	Smith und Piccard	EMU 1944	50	2	190,5	84	100:1	20000:1	2	,,	ohne Schleuse	Lupenbetrachtung des Leuchtschirms
	,,	EMC 1944	30	2	76,2	32	500:1	5000:1	<20	Plattenkassette 2×2 Zoll²	,,	ohne Kondensor, geneigter Strahlengang
Metropolitan Vickers Electrical Co., Manchester	Haine	EM 2 1947	50	2	213	124		10000:1				
	,,	EM 3 1949	25—100	3	210	90	600:1	40000:1	5	Platten 3,25×3,25 Zoll²	2 Platten ohne Schleuse	2fach vergrößernde Lupe im Vakuum

Schönander, Stockholm	Siegbahn	1946	30—85	2	~120	108		30000:1	2	Platte 8,2×8,2 cm²	Wechseleinrichtung für 8 Platten	horizontaler Strahlengang
Philips Eindhoven	Le Poole	1949 (ältere Modelle seit 1947)	40—100	3	118	66 bis zum Aufnahmefilm 42	1000:1	15000:1	2—5	Normalfilm 2,4×3,6 cm²	Kassette für 36 Aufnahmen	Schirmbild 60000:1, geneigter Strahlengang

II. Geräte mit magnetischen und elektrischen Linsen.

Trüb, Täuber & Co., Zürich	Induni	1949 (ältere Modelle seit 1946)	35—50	2	232	138	1500:1	12000:1	5	Platten 9×12 oder 6×9 cm² oder 10 m Film 8 cm breit	Wechseleinrichtung für 2 Platten mit und ohne Schleuse oder Filmkassette	Magnetischer Kondensor, magnetische Hilfslinse zur Scharfstellung des elektrostatischen Objektivs, elektromagnetisches Projektiv

III. Geräte mit elektrischen Linsen.

Allgemeine Elektrizitätsgesellschaft und Süddeutsche Laboratorien-GmbH., Mosbach (AEG-Zeiss)	Mahl, Brüche u. Gölz	EM 8 1949 (ältere Modelle seit 1941)	50	2	~250	90	1500:1	15000:1	< 4	Platten 6×9 cm²	Wechseleinrichtung für 24 Platten	ohne Kondensor. Elektrische Brennweitenregelung des Objektivs. Stigmator. Lupenbetrachtung des Leuchtschirms
Compagnie Générale de Télégraphie Sans Fil, Paris	Grivet und Bruck	1949 (ältere Modelle seit 1947)	30—65	2	252	~105	1900:1	14000:1	4	36-mm-Rollfilm 2,4×3,6 cm². 36 Aufnahmen oder 60-mm-Rollfilm 6×6 cm². 20 Aufnahmen	Film-kassetten	ohne Kondensor

IV. Die Präparationsmethoden.

Die Massendicke der Präparate für die Durchstrahlungsmikroskopie muß zwischen der Durchdringungsgrenze und der Kontrastgrenze liegen. Isolierte Viruspartikel erfüllen diese Bedingung immer. Bei Untersuchungen von Gewebeschnitten oder Zellen einerseits und Eiweißmolekülen, wie etwa Immunstoffen aus der Reihe der Globuline andererseits, machen sich die genannten Grenzen jedoch störend bemerkbar. Das Vakuum und die hohe Objekttemperatur entziehen dem Objekt die flüchtigen Stoffe, vor allem das Wasser. Zur Abbildung der durchstrahlbaren Massendicke ist es meist erforderlich, die Präparate ohne die natürlichen eiweiß- und salzhaltigen Milieuflüssigkeiten aufzutrocknen. Deshalb kann auch die schonendste Präparation bereits einen erheblichen Eingriff in das natürliche Gefüge bedeuten. Hinzu kommt der Beschuß mit Elektronen und dessen ionisierende und erwärmende Wirkung, wodurch sich die organische Substanz in Graphit verwandelt (König). Solange jedoch chemische Umwandlungen unsichtbar bleiben, sind sie für die rein morphologische Betrachtung ohne Bedeutung. Erst wenn wir über Mikroskope verfügen würden, deren Auflösungsvermögen bis in atomare

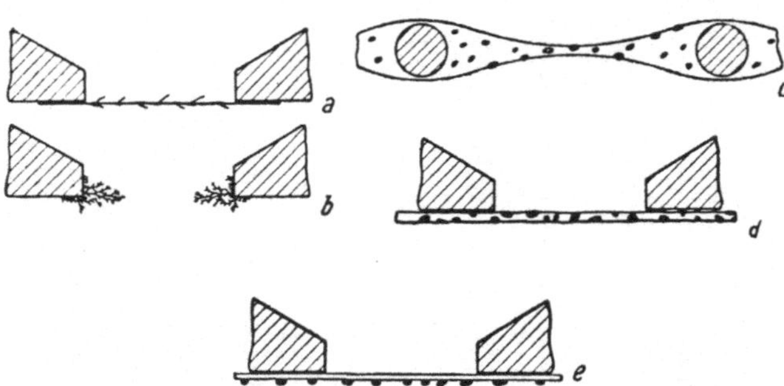

Abb. 44. Schematischer Schnitt durch elektronenmikroskopische Präparate auf Netz bzw. Blenden. a) den Strahlengang frei überquerendes Objekt, b) in den Strahlengang hineinragendes Objekt, c) und d) im Trägerfilm eingebettete Objekte, e) am Trägerfilm haftende Objekte. Nach H. Ruska (1939).

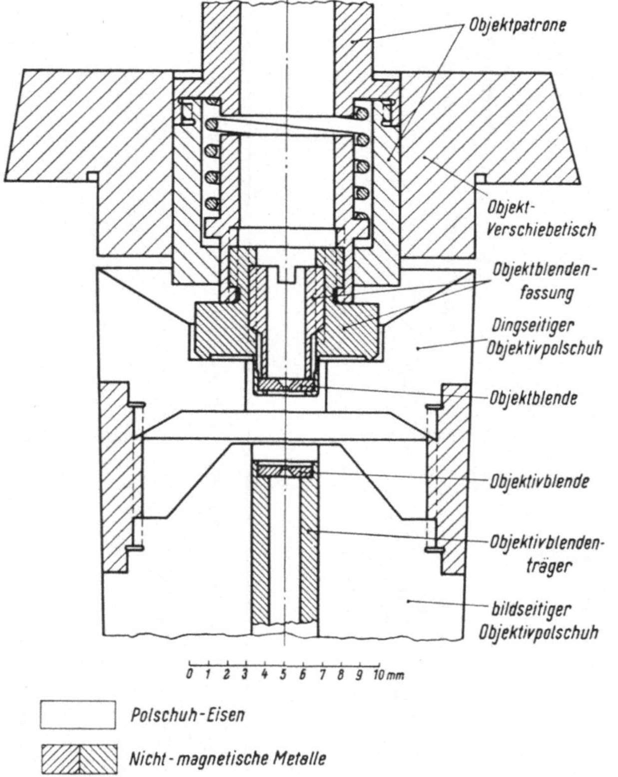

Abb. 45. Objektivpolschuhe und Objekthalterung im Siemens-Übermikroskop.

Dimensionen reichte, wäre jede chemische Änderung des Objekts auch morphologisch erkennbar. In den vergleichsweise immer noch groben Dimensionen der Elektronenmikroskopie kommt es, um Fehldeutungen zu vermeiden, darauf an, sichtbar werdende Änderungen der Struktur während der Betrachtung nicht zu übersehen. Typische Objektveränderungen können Hinweise auf die stoffliche Zugehörigkeit jener morphologischen Bezirke geben, die unter dem Einfluß der Elektronenbestrahlung in charakteristischer Weise zersetzt werden. Man muß sich aus den angeführten Gründen stets Rechenschaft darüber geben, was die Präparation und die Bestrahlung an der natürlichen Struktur geändert haben könnten. (Vgl. auch KÖNIG und WINKLER.)

1. Die Objekthalterung.

Da alle Substanzen in dickerer Schicht für Elektronen undurchlässig sind, müssen als Objektträger extrem dünne Schichten verwendet werden. Die Ausdehnung solcher Schichten quer zur Strahlung darf nur gering sein, da die dünnen Trägerschichten bei größerer Ausdehnung die auftretende Wärme schlecht ableiten und unter der Einwirkung der Elektronenstrahlen zerreißen. Es ist jedoch nicht bei allen Objekten notwendig, Trägerfilme zu verwenden.

a) Objektnetze und Objektblenden.

Die zu untersuchenden Gegenstände liegen auf Objektnetzen oder Objektblenden. Sie können entweder freischwebend darauf angebracht sein, in Trägerfilme eingebettet werden oder auf der Filmfläche haften. Eine schematische Darstellung der drei Möglichkeiten ist in Abb. 44 wiedergegeben. Die Objektnetze sind möglichst feinmaschig (200 Maschen pro Zoll, etwa 8 pro mm), haben die gleiche Größe von 2,5 bis 5 mm Durchmesser wie die Blenden und können zur Stabilisierung und Einebnung leicht gehämmert und auf beiden Flächen etwas abgeschliffen werden. Die Trägerfilme erhalten dadurch eine bessere Auflagefläche. Statt gewebter Netze werden auch auf galvanischem Wege hergestellte Gitter verwendet, die starr und eben sind. Netze und Blenden bestehen aus Material, das gegen Wärme und chemische Einflüsse möglichst widerstandsfähig ist und die Wärme ableitet. Die Blenden haben etwa 0,5 mm Dicke und in Richtung des Strahlenganges eine feine zentrale Bohrung von 0,1 bis 0,05 mm Durchmesser. Bei einer größeren Bohrung würde der Film bei der Bestrahlung besonders leicht zerreißen. Die Blenden können am besten aus Gold-Platin gefertigt werden, weil bei der Herstellung die Bearbeitungskosten gegenüber den Materialkosten weit überwiegen und Edelmetallblenden wiederholt benutzt werden können. Um einen langen, zu Reflexionserscheinungen und damit zu Bildstörungen Anlaß gebenden Bohrungskanal zu vermeiden, ist die Bohrung auf der Eintrittsseite des Strahls trichterförmig erweitert, wie es die Abbildungen 44 u. 45 im Schnitt des zentralen Teiles der Blende zeigen. Der Vorteil der Netze liegt in der größeren Anzahl der betrachtbaren Objektfelder und in den geringen Herstellungskosten. Nachteilig beim Gebrauch sind das niedrige Gewicht, die geringe Stabilität und der rauhe Netzrand. Durch das geringe Gewicht können die Netze bei jedem leichten Luftzug weggeweht werden. An dem rauhen Netzrand bleiben filmartig ausgebreitete Objekte beim Auffischen aus Flüssigkeiten leicht hängen oder werden verletzt. Außerdem ist die Wärmeableitung bei Netzen schlechter als bei Blenden. Die Nachteile der Netze lassen sich dadurch vermeiden, daß sie auf Ringe gespannt werden. Die Blenden besitzen Vorteile in der Handhabung und sind zur leichten Ermittlung der Vergrößerung in der ersten Abbildungsstufe des Elektronenmikroskopes besonders geeignet.

b) Reinigung der Objektblenden.

Bevor man die Objektblenden mit Filmen überzieht, werden sie sorgfältig gereinigt und lichtmikroskopisch auf ihre Brauchbarkeit geprüft. Zur Vergrößerungsbestimmung in der ersten Abbildungsstufe des Elektronenmikroskops kann dabei gleich der Blendendurchmesser gemessen werden. Da die Bohrungen nur 0,1 bis 0,05 mm betragen, ist eine grobmechanische Reinigung nicht möglich. Falls eine Säuberung wegen starker Verschmutzung wünschenswert erscheint, kann sie unter Ausnutzung des Wasserdrucks in nebenstehender Anordnung versucht werden. (Abb. 46.) Die bequemste, aber nicht die schonendste Reinigung, die besonders nach der Benutzung von Zaponlackfilmen mit darauf befindlichen organischen Objekten angewendet werden kann, ist das Ausglühen der Gold-Platin-Blenden auf einem Platinnetz in der Oxydationsflamme. Es darf dabei das Stadium der dunklen Rotglut nicht überschritten werden, da sonst die Blenden mit dem Netz verkleben oder sogar schmelzen. Objekte, die in der Flamme nicht verbrennen oder die in die Platinunterlage einbrennen, wie metallbedampfte Präparate, sollten auf nassem Wege in Schwefelsäure entfernt werden. Die Blenden werden dazu in einem Mikrokjedalkolben mit 50prozentiger Schwefelsäure und 1prozentigem Kupfersulfat auf einem Veraschungsgestell unter vorsichtiger Zugabe von etwas Wasserstoffsuperoxyd ausgekocht und anschließend in Wasser gespült.

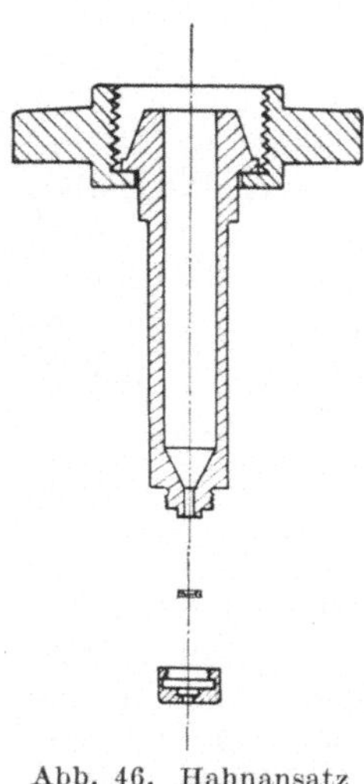

Abb. 46. Hahnansatz zur Durchspülung von Objektblenden.

Trotz schonender Behandlung ist es jedoch nicht zu vermeiden, daß die feinen Blendenbohrungen im Laufe der Zeit je nach Art der angewandten Trägerfilme und Objekte früher oder später ihre runde Form einbüßen. Die freie Öffnung wird kleiner und der Rand schließlich so rauh, daß die Filme bei der Bestrahlung einreißen. Damit werden die Blenden unbrauchbar und müssen von der weiteren Verwendung ausscheiden.

c) Objektträgerfassungen.

Objektnetze und Objektblenden werden nach der Beschickung mit Objekten nicht unmittelbar in die Mikroskopröhre gebracht, sondern in Fassungen eingeschraubt, die ihrerseits meist an einer Objektpatrone befestigt sind. Ein Ausführungsbeispiel für Wirkungsweise und Handhabung von Blenden und Blendenfassungen ist aus Abb. 45 und 47 ersichtlich. Ähnliche Konstruktionen sind mit gewissen Abwandlungen bei anderen Mikroskopen vorhanden. Die Objektpatrone oder die Blendenfassung selbst wird mit der Hand oder einem Spezialschlüssel in die Objektschleuse eingeschraubt und beweglich verankert. Sie muß zur Absuchung des Objektfeldes quer zur optischen Achse oder (bei elektrostatischen Objektiven) zur Scharfstellung des Objekts auch in der Achse verschiebbar sein. Da die Blende in einem Vorsprung der Fassung liegt oder selbst einen Vorsprung besitzt, ist es möglich, sie für kurze Brennweiten sehr nahe an das Feldmaximum des Objektivs heranzubringen. Die Objekte liegen in der Regel auf der objektivnahen Blendenseite. Nur wenn man befürchten muß, daß ein freischwebendes Objekt durch Adhäsion nicht genügend auf der Blende haftet und bei der Bestrahlung abspringen könnte oder wenn man bei geringer Vergrößerung größere Brennweiten verwendet, werden Objekt und Blende umgekehrt gelagert. Die Einklemmung zwischen Blende und Objektfassung kann auf dem in Abb. 47 dargestellten Gerät f durchgeführt werden. Eine besondere Halterung trägt den Schraubring b, auf dem die Objektblende a mit dem aufgelagerten Objekt sitzt. Darüber wird die kappenförmige Objektfassung c festgeschraubt. Der Stift g dient dazu, die Fassung zu öffnen und festgeklemmte Blenden herauszudrücken. d ist die Objektpatrone, die mit dem Schlüssel e in die Objektschleuse, Abb. 32, eingeschraubt wird, h trägt die Blende bei der Herstellung der Kollodiumfilme, i und k dienen zur Aufnahme von Objektträgerfassungen bzw. von Objektblenden.

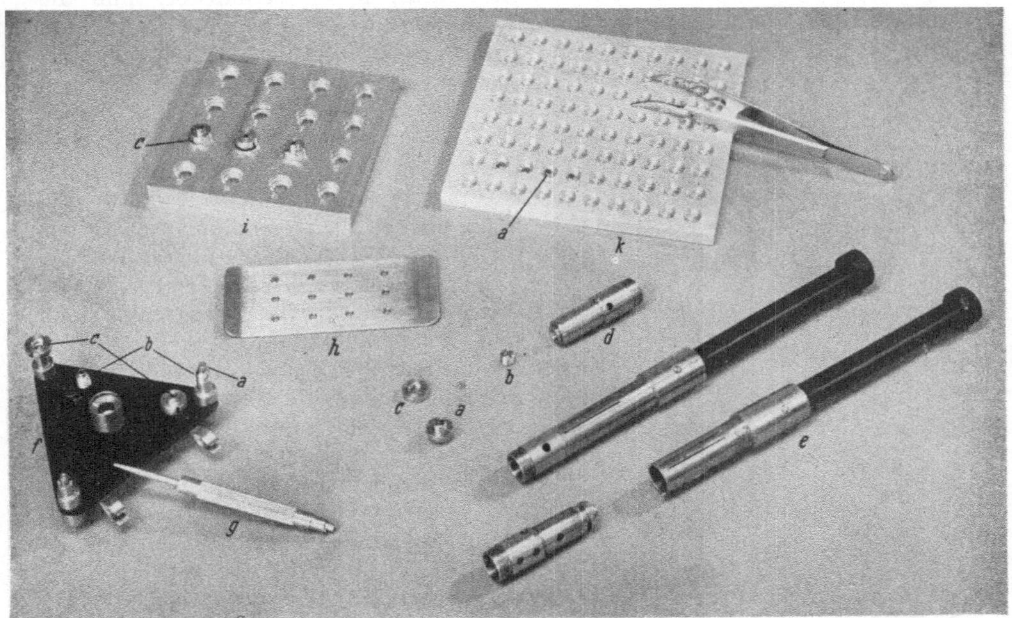

Abb. 47. Hilfsgeräte zur Handhabung und Halterung der Objektblenden. Erklärung siehe Text. (Ältere Siemens-Ausführung.)

2. Die Objektträgerfilme.

Für die Untersuchung von Bakterien, Virusteilchen und ähnlichen Objekten werden Trägerfilme aus verschiedenen Materialien hergestellt. Um sichere Aussage über die Objekte machen zu können, muß der Trägerfilm so dünn sein, daß er im Elektronenstrahl durchsichtig erscheint wie der Objektträger mit dem Deckglas im Lichtmikroskop. Er soll ferner einheitliche Dicke besitzen und frei von jeder elektronenmikroskopisch sichtbaren Eigenstruktur sein. Beim Aufbringen der Objekte und bei der Durchstrahlung mit Elektronen darf er nicht zerreißen, und die aufgebrachten Objekte müssen auf ihm festhaften. Schließlich soll der Film bei Elektronenbeugungsuntersuchungen keine störenden Eigeninterferenzen geben.

a) Die Herstellung von Kollodiumfilmen.

Filme, die den genannten Anforderungen genügen, werden am einfachsten aus Kollodium oder Agfa-Zaponlack Z 116 hergestellt. Die Fertigung dünner Kollodiumschichten wurde von W. Trenktrog angegeben und in den „Physikalischen Kunstgriffen" von E. v. Angerer beschrieben. F. Kirchner verwendete sie für Elektroneninterferenzversuche. Schon von Trenktrog wird Agfa-Zaponlack Z 116 empfohlen. Marton benutzte als erster für die Elektronenmikroskopie Zaponlackfilme. Ihre Herstellung wurde von H. Ruska (1939/4) in der heute üblichen Form ausgearbeitet. Man gibt einen Tropfen von in Amylacetat gelöstem Kollodium auf die Oberfläche eines mit Amylacetat gesättigten Wasserbades (1,5 g Kollodium in 100 g Amylacetat oder im Volumenverhältnis 1 : 2 mit Amylacetat verdünnten Zaponlack). Ingelmann und Siegbahn (1944) betonen, daß mit Zaponlack dünnere und stabilere Filme herzustellen sind als mit reiner Kollodiumlösung. Das entspricht auch unserer Erfahrung. Da das Amylacetat leicht verdunstet, muß die Verdünnung der Gebrauchslösung in gewissen Zeitabständen wieder auf das richtige Maß gebracht werden. Ein Tropfen von etwa 0,01 bis 0,05 g, dessen Größe durch Abfließenlassen an Glasstäben von verschiedenem Durchmesser verändert oder mit Pipetten genau gemessen werden kann, genügt zur Herstellung eines Films für zahlreiche Blenden. Auf der Wasseroberfläche breitet sich der Tropfen infolge seiner geringen Oberflächenspannung rasch und gleichmäßig dünn aus. Nach dem Abdunsten des Amylacetats schwimmt auf dem Wasser die Kollodiumhaut. Bei der Herstellung ist darauf zu achten, daß das Wasser, auf dem sich die Kollodiumlösung ausbreitet, völlig unbewegt, staubfrei und frei von kleinen Gasperlen ist. Beim Aufbringen darf der Tropfen, um heftige Wellenbewegungen zu vermeiden, nicht auf die Wasseroberfläche fallen, sondern muß mit dem Glasstab der Wasseroberfläche genähert werden und schließlich, indem er sie berührt, ohne jede Erschütterung abfließen. So entsteht auf einem hinreichend großen Wasserspiegel ein freischwimmender, etwa kreisförmig begrenzter Film. Beachtet man die beschriebenen Vorsichtsregeln nicht, so erhält man leicht eine unregelmäßige und zerrissene Begrenzung oder schon mit unbewaffnetem Auge sichtbare Löcher und

Abb. 48. Gerät zur Herstellung von Kollodiumfilmen. Nach H. Ruska (1940).

andere Fehler. Es hat sich als günstig erwiesen, das Wasser, auf das der Film gegossen werden soll, vorher mit Amylacetat zu sättigen, um die Diffusion des aufgetropften Amylacetats in das Wasser der Filmgießschale zu vermeiden. Sofort nach der völligen Ausbreitung des Tropfens erkennt man die Farben dünner Plättchen auf der Wasseroberfläche. Der fertige Film selbst zeigt nach dem Abdunsten des Amylacetats infolge der Schichtdickenabnahme keine Farben mehr, ist aber an einem deutlichen Seidenglanz erkennbar. Die Dicke der dünnsten Filme wird auf etwa 10 mμ berechnet. Sie sind dünner als die mittlere Länge der sie zusammensetzenden Kollodiummoleküle.

Eine zur Herstellung guter Kollodium- oder Zaponlackfilme geeignete Anordnung zeigt Abb. 48. Da ein Film nur einmal mit einem Objekt beschickt werden kann, stellt man sich die Zaponlackfilme in einem Filmgießgerät selbst her und bringt sie auf die Objektträgerblende auf. Destilliertes Wasser, das mit einem Überschuß von Amylacetat versetzt, durchgeschüttelt und klar abgesetzt ist, wird aus dem Scheidetrichter in das darunter befindliche flache Wasserbad abgelassen. Vorher wird eine rechteckige Metallplatte mit zwei aufgebogenen Rändern von der ungefähren Größe lichtmikroskopischer Objektträger (Abb. 47h), die eine größere Zahl von Objektblenden in entsprechenden Vertiefungen trägt, in die Mitte des Wasserbades gelegt. Um den Zaponlacktropfen auf die Wasseroberfläche bringen zu können, muß man den Scheidetrichter kurzfristig abnehmen. Die Anordnung ermöglicht staubfreies Arbeiten. Nach der Fertigstellung des Films, den man 10 Minuten oder länger erhärten läßt, wird der Inhalt des Wasserbades durch den am Boden befindlichen Hahn abgelassen. Dabei senkt sich der Film auf die Objektblenden und haftet vollkommen fest, nachdem der Blendenträger auf Filtrierpapier, an der Luft oder in einem Exsikkator getrocknet worden ist.

Die Kollodium- und Zaponlackfilme sind für die meisten Untersuchungen brauchbar. Hat man die Objekte in einem Suspensionsmittel aufgeschwemmt, in dem Kollodium löslich ist, so werden die unbelegten Filme in das Elektronenmikroskop eingeschleust, kurz mit Elektronen bestrahlt und nach dieser „Härtung" mit Objekten beschickt. Die Bestrahlung macht aus dem Kollodium ein in Amylacetat, Alkohol und dgl. unlösliches Kohlenstoffgerüst.

Runde Löcher im Film (z. B. Abb. 8), die sich bei der Bestrahlung häufig vergrößern, können schon bei der Filmherstellung entstanden sein. Sie verziehen sich zu ovalen Formen, wenn der Film zusätzlich bei der Bestrahlung reißt (Abb. 23). Die Einrisse können durch zu schnell ansteigende Bestrahlung, besonders mit niedrig beschleunigten Elektronen, durch zu dichte Belegung des Films mit zahlreichen Objekten oder durch zu große Objekte entstehen.

b) Die Berechnung der Dicke von Kollodiumfilmen.

Die Berechnung der Filmstärke x cm erfolgt unter der Voraussetzung einheitlicher Dicke nach folgender Überlegung:

Die auf die Wasseroberfläche gebrachte Lösungsmenge m cm^3 breitet sich zu einer Scheibe von D cm Durchmesser und h cm Höhe aus. Die Lösung besteht aus $n_K = 1{,}5$ g Kollodium vom spezifischen Gewicht $\gamma_K = 1{,}634$ und aus $n_A = 100$ g Amylacetat vom spezifischen Gewicht $\gamma_A = 0{,}88$. Es verhält sich x zu h wie das Volumen des Kollodiums V_K zum Volumen der Kollodiumlösung $m = V_A + V_K$. Setzt man für V_K und V_A die Gewichte G_A und G_K, dividiert durch die spezifischen Gewichte γ_A und γ_K ein, so ist:

$$\frac{x}{h} = \frac{\dfrac{G_K}{\gamma_K}}{\dfrac{G_K}{\gamma_K} + \dfrac{G_A}{\gamma_A}} = \frac{1}{1 + \dfrac{G_A}{G_K} \cdot \dfrac{\gamma_K}{\gamma_A}} \quad \text{für} \quad \frac{G_K}{G_A} = p = 0{,}015$$

ist $\dfrac{x}{h} = \dfrac{1}{1 + \dfrac{\gamma_K}{p \cdot \gamma_A}}$ da $h = \dfrac{4}{\pi} \cdot \dfrac{m}{D^2}$ ist und 1 im Nenner gegen $\dfrac{\gamma_K}{p \cdot \gamma_A}$

vernachlässigt werden kann, so ergibt sich für

$$x = \frac{4}{\pi} \cdot \frac{m}{D^2} \cdot p \cdot \frac{\gamma_A}{\gamma_K} \, \text{cm}$$

und daraus nach Einsetzen der Zahlenwerte für eine 1,5 %ige Lösung:

$$x = \frac{m}{100\,D^2}\ cm.$$

Für eine Tröpfchengröße von m = 0,04 ccm und einen Filmdurchmesser von D = 20 cm wird die Filmdicke 10 mμ.

c) Aluminiumoxydfilme.

Trägerfilme aus Aluminiumoxyd werden nach Hass und Kehler (1941), v. Ardenne und Friedrich-Freksa (1941) durch Aufdampfen einer dünnen Aluminiumschicht im Vakuum auf eine geschliffene Glasplatte vorbereitet. Nachträglich wird durch genau dosierte anodische Oxydation der Aluminiumoberfläche in einer organischen Säure oder in Ammoniumtartrat die Oxydschicht hergestellt. Die Oxydationszeit wird so gewählt, daß die gewünschte Filmdicke von 20 bis 40 mμ erreicht wird. Nach E. Semmler (1942) werden die Aluminiumschichten bei ca. 50 Volt in schwach alkalischem (p$_H$ 8,8) 2,5 %igen Ammoniumborat oxydiert. Sie zeigen dann weniger Eigenstruktur als bei höheren p$_H$-Werten. Die Oxydschicht wird nach der Fertigstellung in Quadrate von etwa 5 mm Kantenlänge geritzt und durch Einbringen der Glasplatte in gesättigte Sublimatlösung abgelöst. Die zum Gebrauch als Trägerfilme passend großen Stücke werden aus der Sublimatlösung für die Dauer einiger Minuten in verdünnte Salzsäure gebracht, in destilliertes Wasser übertragen und dann mit der auf einer Haltevorrichtung unter Wasser liegenden Objektblende so aufgefischt, daß sie die Blendenbohrung faltenlos überdecken. Die Filme aus Aluminiumoxyd sind stabiler als jene aus Kollodium und lassen sich daher über größere Objektblendenbohrungen spannen. Sie eignen sich bei größeren Einzelobjekten besonders für stereoskopische Aufnahmen, weil man nicht mit dem Einreißen des Films und mit unerwünschten Lageveränderungen des Objekts zu rechnen braucht, welche die Stereoskopie unmöglich machen. Wegen ihrer Beständigkeit gegen hohe Temperaturen (mehrere Stunden 500⁰ C) bei Anwesenheit von Sauerstoff sind die Aluminiumoxydfilme auch für eine Hitzebehandlung oder Mikroveraschung der Objekte geeignet. Die Erhitzung darf jedoch nicht über 600⁰ gehen, da bei 650⁰ bis 700⁰ eine sichtbare Kristallisation der vorher strukturlos erscheinenden Folie eintritt, die sie sowohl für Interferenzuntersuchungen als auch für die Übermikroskopie unbrauchbar macht (Hass und Kehler, 1941). Stärkste Elektronenbestrahlung führt ebenfalls zur Kristallisation. Die Anwendung des Aluminiumoxyds als Filmmaterial ist umständlicher und kostspieliger als die des Kollodiums.

d) Siliziumdioxyd- und Siliziummmonoxydfilme.

Gerould (1945) hat die Verwendung von Filmen empfohlen, die aus Quarz bestehen. Für ihre Herstellung werden Glasobjektträger der Lichtmikroskopie mit einem Polystyrolüberzug versehen, auf welchen Siliziumdioxyd im Vakuum aufgedampft wird. Die sorgfältig gereinigten und getrockneten Objektträger werden in eine 2- bis 5%ige Lösung von Polystyrol in Benzin getaucht, herausgenommen und getrocknet. Danach werden sie in 7 bis 10 cm Abstand über einem kegelförmig spiralig gewundenen Wolframglühdraht von 5 mm Länge befestigt, in dem sich etwa 0,8 bis 0,9 mg Quarzsplitter befinden. Die Anordnung wird unter einer Glocke auf 10^{-4} mm Hg oder höher evakuiert und der Glühdraht mit 10 bis 15 Volt, 25 bis 30 Ampère für die Dauer von 5 bis 15 Sekunden auf Weißglut erhitzt. Dabei verdampft der Quarz und schlägt sich zum

Teil auf dem mit Polystyrol überzogenen Glasobjektträger nieder. Von diesem wird die Quarzschicht, evtl. nach Aufteilung in kleinere Felder, abgelöst, indem man den Objektträger mit der Schichtseite nach oben in ein Gefäß mit Äthylbromid und 10% Benzin bringt. Nach etwa 30 Minuten löst sich der Quarzfilm ab und schwimmt nach oben. Er ist bei seitlicher Beleuchtung vor dunkelrotem Hintergrund gut sichtbar. Mit Hilfe eines feinmaschigen, nicht rostenden Stahlnetzes wird er in Äthylbromid mit einigen Prozent Isobutylalkohol übertragen. Nach dieser Spülung können geeignete Filmstücke auf Objektnetze oder -blenden aufgelegt werden.

Die Quarzfilme sind 10 bis 20 mμ dick, strukturlos (auch für Beugungsuntersuchungen verwendbar) und stabiler als Kollodiumfilme. Sie zeigen ein Minimum von Elektronenstreuung und elektrischer Aufladung. Die Objekte können auf ihnen durch kurzes Erhitzen fixiert werden.

KÖNIG (1944/48) weist darauf hin, daß bei der Verdampfung von Siliziumdioxyd Siliziumoxyd entsteht. Er verwendet Quarzglasscherben (Siliziumdioxyd) oder direkt Siliziummonoxyd. Dieses verdampft im Hochvakuum auf dem Wolframglühdraht, ohne zu zerfallen. Es wird in gewöhnlicher Zimmerluft nur spurenweise zu Siliziumdioxyd oxydiert. Solche Folien zeigen bei 40000facher Vergrößerung keinerlei Struktur. Sie sind auch als Träger für Hochtemperaturversuche im Elektronenmikroskop verwendbar, eignen sich aber nicht wie die durch Elektronenstrahlen in Graphit umgewandelten Kollodiumhäutchen (KÖNIG 1946) für Temperaturen über 1000°, dafür jedoch im Gegensatz zu den Graphithäutchen für Erhitzungsversuche bei Anwesenheit von Sauerstoff. Sie sind temperaturbeständiger als Aluminiumoxydschichten.

Die Herstellung der Filme geht besonders schnell und einfach: Mit Kollodium belegte Blenden werden im Vakuum mit Siliziummonoxyd bedampft, die Kollodiumhaut also gewissermaßen für höhere Temperaturen stabilisiert. Bei Temperaturen über 200° C wird dann die Kollodiumfolie vernichtet, und eine teils aus Siliziummonoxyd, teils aus Siliziumdioxyd bestehende temperatur- und säurebeständige Folie bleibt zurück. Auf diese Weise kann man in einem Arbeitsgang beliebig viele Blenden mit dieser Trägerhaut versehen, ohne für jede Blende einzeln eine Haut „fischen" zu müssen. Beugungsbilder und übermikroskopische Abbildungen von auf diesen Trägerfolien präparierten Objekten sind von der gleichen Güte wie die mit Kollodiumfolien hergestellten.

e) Silikatglasfilme.

MÖLLENSTEDT (1947) und ACKERMANN (1947) verwenden Silikatglas als Trägerfilme. Das glühende Ende eines zugeschmolzenen Jenaer Glasrohres wird in einem hoch geheizten Ofen mit Preßluft aus einer Flasche mit feinregulierbarem Ventil vorsichtig aufgeblasen. Man erhält dadurch dünnste Filme, die keine Farben dünner Plättchen mehr zeigen und sich glatt über Blenden oder Trägernetze spannen. Sie sind temperatur- und säurebeständig und besitzen keine stärkeren, bei Beugungsuntersuchungen störenden Eigeninterferenzen.

f) Berylliumfilme.

COSSLET (1948) empfiehlt die Verwendung von Berylliumfilmen, die schon von RÜDIGER (1942) als Abdruckfilme und von HAST (1947) als Trägerfolien verwendet worden waren. Er benutzt das Beryllium für zwei verschiedene Methoden.

1. Das Objekt wird in der üblichen Weise auf einen Kollodiumfilm gebracht.

der auf einem Trägernetz liegt. Dieses wird in einer Metallbedampfungsapparatur in 10 cm Abstand von 0,3 mg verdampfendem Beryllium (Sdp. 1500° C) befestigt. Die Bedampfungsschicht erreicht etwa 2,5 mμ Dicke. Das Netz wird dann auf den Objekthalter des Elektronenmikroskops übertragen, zur Lösung des Kollodiumfilms 20 Minuten in Amylacetat oder Aceton gebracht und auf Filtrierpapier vorsichtig getrocknet. Bei diesem Verfahren wird also während der Herstellung des endgültigen Trägerfilms zugleich das Objekt mit Beryllium bedampft.

2. Man überzieht einen Glasobjektträger mit einer Kollodiumschicht und dampft das Beryllium in der gewünschten Dicke auf. Die Oberfläche wird mit einer feinen Nadel in Quadrate geteilt und der Glasobjektträger in Amylacetat oder Aceton getaucht. Die quadratischen Stücke des Berylliumfilms steigen an die Oberfläche und werden durch Herausfischen oder durch Ablassen und Wegsaugen des Lösungsmittels auf Objektnetze übertragen, im Objekthalter befestigt und getrocknet.

Die klarsten Bilder liefert die erste Methode, da sie den Kontrast der Objekte erhöht (Bildbeispiele bei Cosslet: Tomaten Aucuba Mosaikvirus, Rüben Gelbmosaikvirus, Schweinepocken).

3. Die Beschickung der Filme mit Objekten.

Je nach Art des Untersuchungsmaterials und der Aufgabenstellung werden die Objekte in verschiedener Weise auf den Film gebracht. Spezielle Methoden sind für Teilchenzählungen entwickelt worden.

a) Tropfenmethode.

Aus Wasser oder flüchtigen organischen Lösungsmitteln werden die suspendierten Objekte mit einer Platinöse, wie sie in der Bakteriologie gebräuchlich ist, oder mit dünnen Glasstäben, sowie mit Kapillaren vorsichtig als Tropfen

Abb. 49. Schematischer Schnitt durch Objektblenden mit aufgebrachtem Flüssigkeitstropfen, in dem die Objekte suspendiert sind.

aufgebracht und getrocknet (Abb. 49). Dabei darf mit dem Draht oder Glas der über die Bohrung freigespannte Teil des Films nicht berührt werden. Die Verwendung einer Platinöse hat den Vorteil, daß sie beim Arbeiten mit Suspensionen, die keimfrei bleiben müssen, und beim Umgang mit infektiösem Material durch Ausglühen in der Flamme leicht sterilisiert werden kann. Nachteilig ist aber, daß verbrannte Präparatreste am Platindraht haften bleiben und, wenn sie nicht entfernt werden, in ein anderes Präparat als störende Fremdbestandteile verschleppt werden können.

Liegen Untersuchungsobjekte vor, die im Lichtmikroskop noch sichtbar sind, so kann die Dichte der Lagerung auf dem Film während der Eintrocknung des Tropfens beobachtet werden. Wenn die Lagerung der Teile zu dicht werden sollte, kann mit der Ecke eines Stückchens Filtrierpapier der Tropfen teilweise weggesaugt werden, bis die gewünschte Objektverteilung erreicht ist. Da Kollodiumfilme mit Wasser schlecht benetzbar sind, können nach völligem Wegsaugen des Tropfens mitunter alle Objekte vom Film entfernt sein. Oft kommt es beim Verdunsten der letzten Spuren des Suspensionsmittels zur Zusammenlagerung vorher getrennt schwebender Teilchen, doch glückt es immer bei der Anfertigung mehrerer Präparate, eine genügende Anzahl der Objekte auch in getrennter Lagerung zu erhalten.

Die flache Form der Objektträgerblende macht es möglich, lichtmikroskopisch auch mit starken Objektiven und selbst mit der Immersionslinse den beschickten Trägerfilm zu betrachten und zu photographieren. Immersionsöl kann mit Xylol sorgfältig vom Film entfernt werden. Bei der durchfallenden Beleuchtung lassen sich hohe Kondensoraperturen jedoch wegen des Bohrkanals der Objektträgerblende nicht voll ausnutzen. Handelt es sich um Teilchen, die lichtmikroskopisch unsichtbar sind, so kann über die Brauchbarkeit der Präparate oder einer zur Untersuchung angefertigten Suspension erst die elektronenmikroskopische Beobachtung entscheiden.

Bakterien und Virusteilchen in wässeriger Aufschwemmung neigen häufig dazu, sich beim Eintrocknen zusammenzulagern. Manchmal läßt sich durch chemische Zusätze, die die Oberflächenspannung herabsetzen, die Neigung zur Aggregation verhindern. In organischen Suspensionsmitteln macht sie sich weniger bemerkbar. Die Zusammenballung kann außerdem mechanisch oder durch Gefriertrocknung aufgehoben werden.

α) Der Objektträger-Vibrator.

Der Objektträgervibrator von v. ARDENNE (1940/41) (Abb. 50) ermöglicht die Befestigung von Objektblenden, die parallel zum Objektträgerfilm in horizontale Schwingungen versetzt werden. Die im aufzutrocknenden Tropfen suspendierten Teilchen werden durch die Vibration, bei welcher der Suspensionstropfen 200mal in der Sekunde die Geschwindigkeitsskala zwischen 0 und 12 m/sec durchlaufen kann, in so starke Eigenbewegung versetzt, daß sie nicht aggregieren. Die Eigenbewegung der Teilchen wird naturgemäß um so größer, je größer die Differenz zwischen dem spezifischen Gewicht der suspendierten Teilchen und dem des Suspensionsmittels ist. Daher bewährt sich der Objektträgervibrator besser bei dem von v. ARDENNE

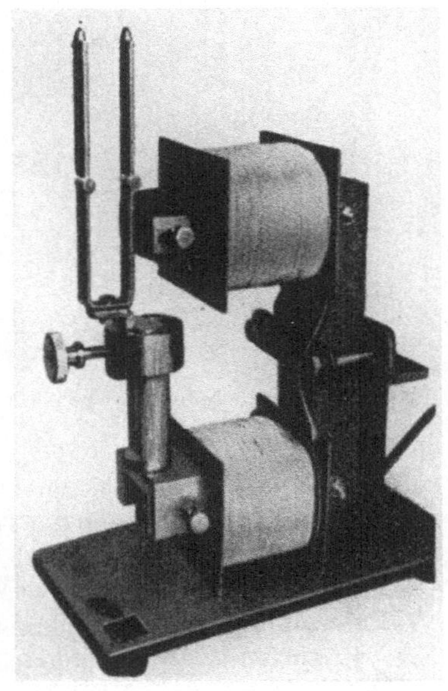

Abb. 50. Gesamtansicht eines Objektträgervibrators für Anschluß an 50-Perioden-Lichtnetz. Nach v. ARDENNE (1940).

als Beispiel gewählten keramischen Pulver als bei organischen Makromolekülen, Virusproteinen und dergleichen. Der auf den Trägerfilm aufgebrachte Tropfen muß möglichst klein sein, damit er bei der Vibration nicht abgeschleudert wird, oder man muß durch fest aufgekittete Randleisten das Abschleudern verhindern (vgl. Abb. 3b bei v. ARDENNE). Die maximale Vibration eignet sich für die Herstellung von Suspensionen und Emulsionen.

β) Die Gefrier-Vakuumtrockenkammer.

Die Häufigkeit und Größe der bei der Eintrocknung entstehenden Aggregate ist von der Eintrocknungstemperatur abhängig. Bei höherer Trockentemperatur muß die Aggregatbildung abnehmen, weil die Auftrocknungszeit verkürzt ist und die BROWNsche Molekularbewegung der Teilchen zunimmt. Wie weit sich ein solcher Einfluß praktisch auswirkt, scheint bisher nicht geprüft worden zu sein. Dagegen ist der entgegengesetzte Weg mit Erfolg beschritten worden.

Man kann durch Einfrieren des Tropfens die Bewegung der suspendierten Teilchen verhindern und im gefrorenen Zustand im Vakuum eintrocknen, so daß sich die Teilchen bei genügender Verdünnung nicht berühren und keine Aggregatbildung eintritt. Von dieser Überlegung ausgehend, haben nach mündlicher Mitteilung Friedrich-Freksa und Schramm Versuche zur Verhinderung der Aggregation gemacht.

Für ihre Durchführung eignet sich das in Abb. 51 gezeigte Gerät. Ein durch eine planparallele Glasplatte auf einem Randschliff vakuumdicht verschließbares Metallgefäß mit seitlichem Pumphahn und Manometer dient als Trockenkammer. Am oberen Rand befindet sich ein Kanal, durch den mittels eines Anschlußstückes, wie es bei Gefriermikrotomen üblich ist, unter Druck Kohlensäure aus der Stahlflasche geblasen werden kann. Der Gefrierkanal trägt auf einem

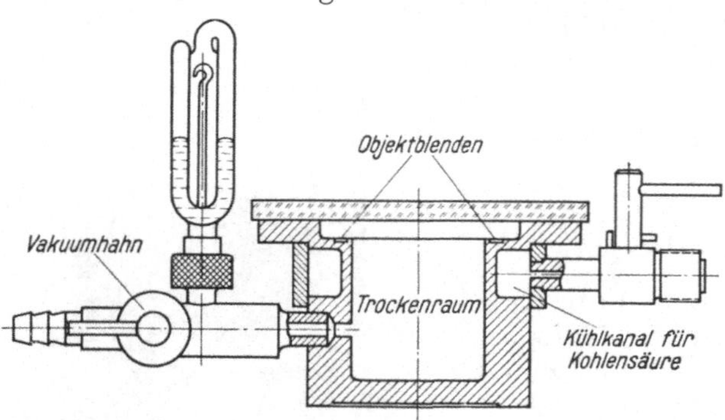

Abb. 51. Gefrier-Vakuumtrockenkammer für elektronenmikroskopische Präparate (Siemens & Halske).

Absatz 24 numerierte Einsenkungen für die Aufnahme von Objektträgerblenden. Vor der Benutzung wird das Gerät mit dem Trockenmittel (z. B. Phosphorpentoxyd) versehen und der Trockenraum verschlossen. Sodann wird Kohlensäure durch den Kühlkanal geleitet, bis die Kammer gut durchgefroren ist. In diesem Zustand wird der Trockenraum geöffnet. Entweder können dann die mit einem Tropfen beschickten Objektträgerblenden eingesetzt werden, oder die Tropfen werden auf die in der Kammer bereits vorgekühlten Blenden aufgebracht. Nach dem Verschließen wird erneut gefroren und mittels Wasserstrahl- oder Ölpumpe evakuiert. Wenn das Manometer Vakuum anzeigt und nach Verschluß des Vakuumhahns der Druck konstant bleibt, kann die Kammer abgenommen und zur Einsparung der Kohlensäure in einen Kühlschrank gestellt werden. Je nach Größe der aufgebrachten Tropfen sind diese in 5 bis 20 Minuten getrocknet. Die Tropfen werden am besten klein gehalten und exzentrisch auf die Blenden gesetzt, da erfahrungsgemäß bei großen, zentral sitzenden Tropfen die Trägerfilme beim Gefrieren einreißen können. Bei gut vorgekühlter Kammer gefrieren die Tropfen sofort beim Aufsetzen der Blenden in die numerierten Einsenkungen. Durch die Glasplatte kann beobachtet werden, ob die Tropfen in der verschlossenen Kammer flüssig, gefroren oder weggetrocknet sind. Das Dichtungsfett unter der Glasplatte und am Vakuumhahn muß auch bei Temperaturen unter 0° C gut abschließen. Von der aggregationsverhindernden Wirkung dieser Methode kann man sich leicht mit gröberen Objekten am Lichtmikroskop überzeugen.

Die Gefriertrocknung kann empfindliche Objekte schädigen. Meier und H. Ruska (1943, unveröffentlicht) haben bei manchen Bakterienstämmen Zerstörung der Geißeln beobachtet, während das Tabak-Mosaikvirus erhalten blieb. Die Gefriervakuumtrocknung eignet sich außer zur Verhinderung der Aggregatbildung auch zur besseren Beurteilung aller Versuche, bei denen eine vor der Trocknung in der Flüssigkeit eingetretene Aggregation festgestellt werden soll, z. B. bei Agglutinationsversuchen mit Bakterien oder Virus, zur Verfolgung der Präzipitatbildung und für ähnliche serologische Fragestellungen.

b) Niederschlagsmethode.

Bei der meist angewendeten Tropfenmethode ist der Tropfen erheblich größer als der Blendendurchmesser oder die einzelne Maschenlücke des Objektnetzes. Es ist daher nicht zu bestimmen, welche Schichtdicke oder welches Flüssigkeitsvolumen über einem gegebenen Objektfeld aufgetrocknet ist und ob die Verteilung verschiedenartiger Teilchen derjenigen in der aufgetropften Flüssigkeit entspricht oder eine zufällige Auswahl darstellt. Werden dagegen Tröpfchen auf die Objektträger gebracht, die kleiner sind als der Blendendurchmesser oder möglichst kleiner als ein Gesichtsfeld im Elektronenmikroskop (einige μ bis Bruchteile davon), so wird die durchschnittliche Teilchenverteilung in der Flüssigkeit sichtbar, und es ergibt sich die Möglichkeit, das aufgebrachte Flüssigkeitsvolumen zu bestimmen. Über die Ermittlung des Volumens mehrerer Tröpfchen läßt sich die Zahl der Teilchen in einer beliebigen Flüssigkeitsmenge feststellen. (RIEDEL und H. RUSKA, 1941).

Zur Erzeugung kleinster Tröpfchen wird die Flüssigkeit durch feinzerstäubende Düsen vernebelt, die zur Ausscheidung der größeren Tröpfchen gegen eine Prallplatte blasen. Stehen Flüssigkeitsmengen von einigen Kubikzentimetern zur Verfügung, so kann der in Abb. 56 gezeigte Zerstäuber verwendet werden. Für Mengen unter 1 cm³ hat sich das in Abb. 52 dargestellte eindüsige Prinzip als Mikrovernebler sehr be-

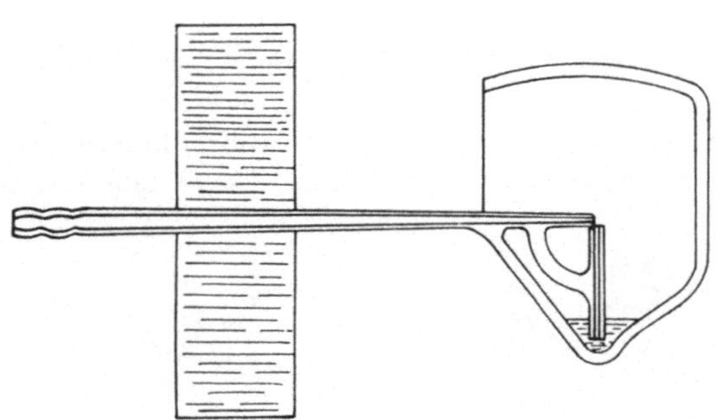

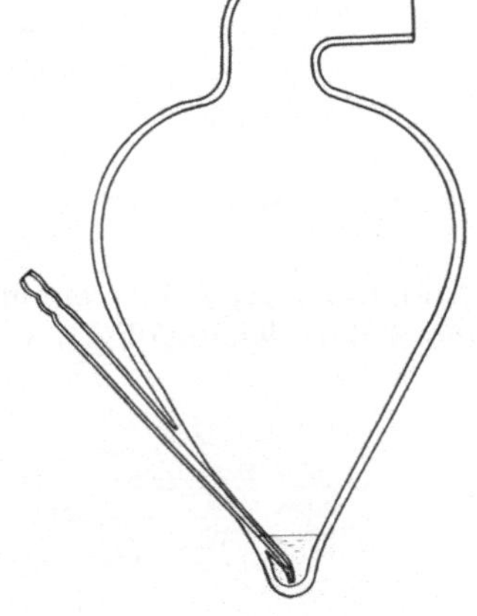

Abb. 52. Mikroverneblungsdüse für den Versuchsaufbau zum elektrischen Kernfäller von Abb. 58.

Abb. 53. Durchströmungsvernebler für den Versuchsaufbau zum elektrischen Kernfäller von Abb. 58

währt. Beide Anordnungen arbeiten so, daß größere Tropfen wieder in die Zerstäubungsflüssigkeit zurückfließen. Mitunter kann auch dadurch vernebelt werden, daß der Luftstrom durch die Flüssigkeit hindurchgeleitet wird (Abb. 53). Die an der Wasseroberfläche austretenden Gasblasen zerspritzen beim Platzen feinster Tröpfchen der Flüssigkeitslamelle und reißen diese im Gasstrom mit sich. Etwa entstehende Schaumblasen zerreißen beim Hochsteigen infolge des zunehmenden Querschnittes der Zerstäubungsbirne. Der Wirkungsgrad ist schlechter als beim Düsenprinzip.

Infolge des LENARD-Effekts sind die Wassertröpfchen polarisiert, d. h. ihre Oberfläche trägt eine negative, ihr Inneres eine positive Ladung. Gelöste Stoffe vermindern die Wirkung oder können die Ladungsverteilung umkehren. Die Aufladung hält die Tröpfchen in der Schwebe, ermöglicht aber auch, sie durch ein elektrisches Feld auf die Trägerfolien niederzuschlagen. Gleiches läßt sich mit einem thermischen Feld erreichen, doch ist das elektrische Feld zur Niederschlagung von Tröpfchen geeigneter, da vor dem Niederschlagen keine Erwärmung eintritt, die die Tröpfchen vorzeitig zum Verdampfen bringt.

α) Der elektrische Kernfäller.

Schnitt und Ansicht eines im Laboratorium für Übermikroskopie der Siemens & Halske A. G. von Riedel entwickelten elektrischen Kernfällers zeigen die Abb. 54 und 55. Das elektrische Feld, das von der mit Tröpfchen beladenen Luft durchströmt wird, bilden zwei konzentrische Zylinderflächen a und b

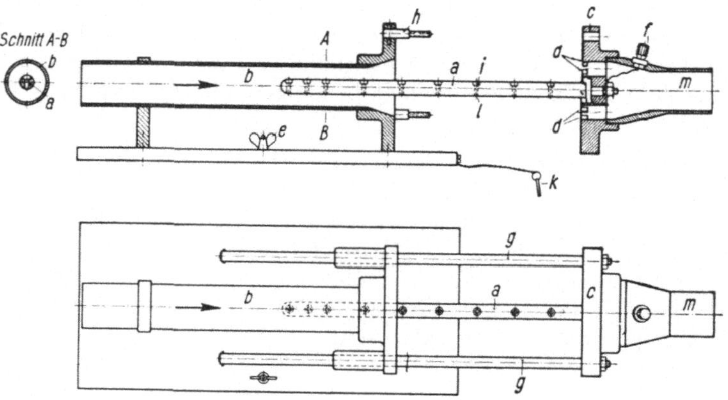

Abb. 54. Schnitt und Aufsicht des geöffneten elektrischen Kernfällers. Nach Riedel.

(Abb. 54). Gegenüber einem Plattenkondensator hat dieser Röhrenkondensator neben der gleichmäßigen Ladungsverteilung den Vorteil, daß lange Isolierungs-

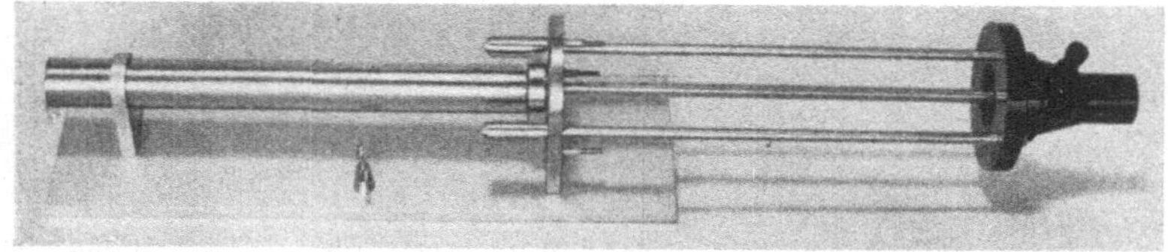

Abb. 55. Ansicht des geöffneten elektrischen Kernfällers. Nach Riedel (Siemens & Halske).

flächen, die einen Plattenkondensator zur Luftdurchführung seitlich begrenzen müßten, entbehrlich werden. Die nicht zu umgehende, isolierende Verbindung c zwischen den beiden Zylindern a und b des Röhrenkondensators, die außerdem für die Zentrierung erforderlich ist, befindet sich an demjenigen Ende des Kondensators, an dem die meisten Tröpfchen bereits ausgefallen sind. Die Durchströmungsrichtung ist in Abb. 54 durch den Pfeil gekennzeichnet. Weiter ist der Kriechweg zwischen den beiden Zylindern a und b längs des Verbindungsstückes c durch tiefe Rillen d wesentlich vergrößert. Durch diese Konstruktionsbesonderheiten kann mit dem Kernfäller eine Spannung bis zu 5000 Volt gehalten werden. Die erforderliche Gleichspannung wird einem dazugehörigen Aggregat entnommen und an die Klemmen e (Erde) und f (+ oder — Spannung) geführt. Der Innenzylinder a des Kernfällers ist ausziehbar, damit man auf ihn die mit Filmen versehenen Objektblenden auflegen kann. Die beiderseitige Führung g außerhalb des Kondensators gewährleistet ein erschütterungsfreies Aus- und

Einschieben. Die Paßschrauben h stellen beim Schließen jeweils die Zentrierung der beiden Zylinder a und b wieder her. Um das Feld durch die eingebrachten Objektträger nicht unnötig zu stören, sind Bohrungen i an der Oberseite des Zylinders a angebracht, in die die Objektblenden genau hineinpassen und aus denen sie mit Hilfe eines Stiftes k durch eine feine Bohrung l von unten herausgehoben werden können. Um quantitative Auswertungen zu ermöglichen, sind die Bohrungen in gleichen Abständen über den ganzen Zylinder verteilt. Die erste Bohrung liegt direkt hinter dem zur Vermeidung von Spitzenentladungen und Turbulenzströmungen abgerundeten vorderen Ende des Zylinders a. Zur Vermeidung von Druckänderungen bleibt der Querschnitt der Luftführung im Kernfäller und innerhalb des Verbindungsstückes c gleich. Die Länge des Kernfällers ist wegen der erforderlichen Zentrierung statisch begrenzt.

β) Der Versuchsaufbau.

Die notwendigen Zusatzteile zum Gebrauch des Kernfällers für die Luftzuführung und für die Verneblung des Untersuchungsmaterials lassen sich aus Glas herstellen. Druckpumpe oder Druckluftflasche und Rotameter werden in der handelsüblichen Ausführung verwendet. Abb. 56 zeigt eine Zusammenstellung aller zu den Versuchen benötigten Teile.

Virussuspensionen und dergleichen werden durch das Anblasen der Düsen folgendermaßen in künstliche Aerosole überführt. Durch die Druckluftpumpe 1

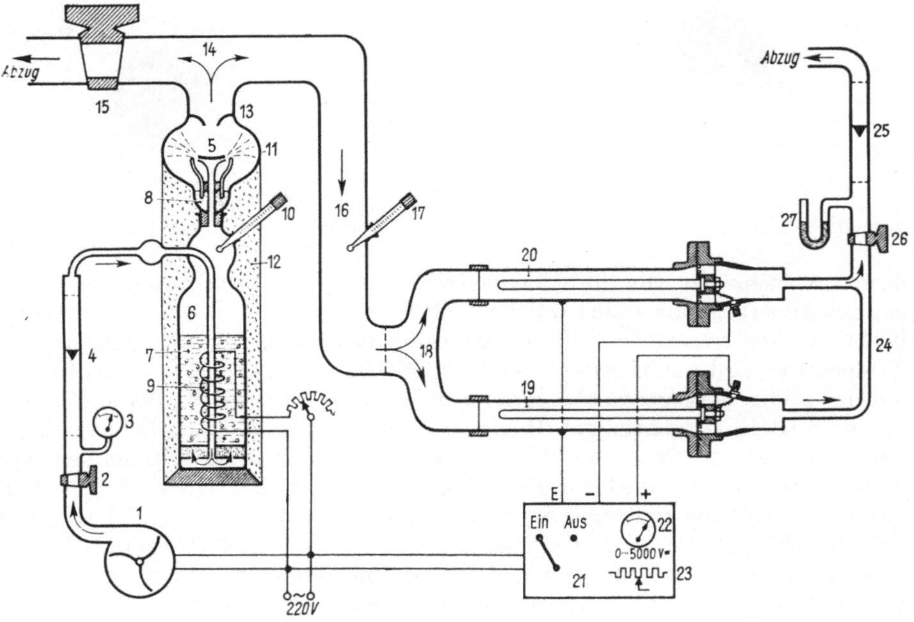

Abb. 56. Versuchsaufbau zum elektrischen Kernfäller. (Nach RIEDEL).

oder eine Preßluftflasche wird möglichst saubere Luft in die Apparatur geblasen. Mit dem Hahn 2 werden nach Manometer 3 ca. 2 atü eingestellt. Das weiter von der Luft durchströmte Rotameter 4 zeigt die sich ergebende Luftmenge an, die nicht beliebig geregelt werden kann, da sie vom Widerstand der Zerstäubungsdüsen 5 abhängig ist. Nun gelangt die Luft in die Waschflasche 6, deren Waschflüssigkeit 7 gleich dem Lösungsmittel der zu untersuchenden Substanz 8 ist und

mit einer Heizschlange 9 aufgeheizt werden kann. In den oberen Luftraum der
Waschflasche ist das Thermometer 10 eingepaßt und der dem Wiesbadener
Doppelinhalator ähnelnde Glaszerstäuber 11 eingelassen. Waschflasche und Zerstäuber befinden sich in einer wärmeisolierenden Umhüllung 12. Beim Austritt
aus dem Zerstäuber passiert die Luft den Tropfenabschneider 13 und tritt in
das T-Rohr 14 ein, dessen Querschnitt gleich dem der folgenden Rohre bzw. dem
doppelten Querschnitt des Kernfällers ist. Da für den Luftdurchgang durch die
Kernfäller sehr wenig Luft benötigt wird, läßt man aus Hahn 15 die überflüssige
Luft in den Abzug entweichen und regelt damit grob den Luftdurchgang durch
die Kernfäller (Vorsicht bei infektiösem Material). In dem Verweilrohr 16,

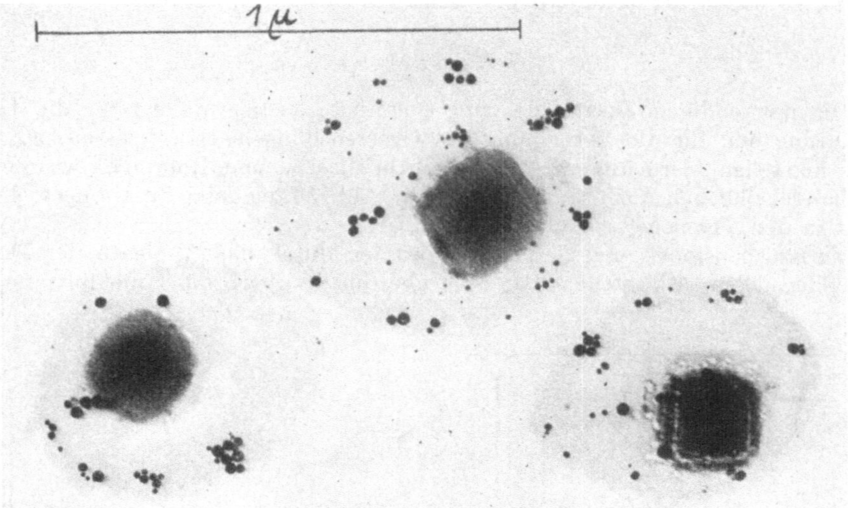

Abb. 57. 1 %iges Goldsol mit 1 % NaCl und 1 % LiCl vernebelt. Nach Riedel und H. Ruska.

dessen verschieden gewählte Länge bei gleicher Luftgeschwindigkeit in der
ganzen Apparatur die Dauer des Verweilens bedingt, nimmt die mit dem künstlichen Aerosol versehene Luft die erforderliche homogene Verteilung an. Ihre
Temperatur wird am Thermometer 17 abgelesen. Im T-Rohr 18 verteilt sie sich
auf die beiden entgegengesetzt geladenen Kernfäller 19 und 20, deren Spannung
von dem Gleichstromaggregat 21 bis zu mindestens 5000 Volt geliefert und nach
dem Voltmeter 22 mit dem Widerstand 23 eingestellt wird. Hinter den Kernfällern wird die Luft wieder in dem Rohr 24 vereinigt und, ehe sie in den Abzug
ausbläst, zur Mengenmessung durch das Rotameter 25 geführt, vor dem sich der
Feineinstellungshahn 26 und ein kleines Glasmanometer 27 befindet. Mit dem
Hahn 26 werden Druck und Menge des ausströmenden Aerosols reguliert. Abb. 57
zeigt mit der beschriebenen Anordnung erhaltene Tröpfchen im elektronenmikroskopischen Bild.

Die durch den Zerstäuber 5 strömende Luft wird erwärmt und mit dem Lösungsmittel gesättigt, um eine stärkere Verdampfung der Nebeltröpfchen, deren
Dampfdruck infolge der starken Oberflächenkrümmung erhöht ist, auf dem Wege
zum Kernfäller nach Möglichkeit zu verhindern. Bei längeren Versuchen führt
die Verdampfung der Nebeltröpfchen zu einer Zunahme der Konzentration der
Teilchen in der zerstäubten Flüssigkeit, da ein Teil der Tröpfchen immer wieder
nach 8 zurückfließt.

Die gleichzeitige Benutzung der entgegengesetzt geladenen Kernfäller erlaubt die Erfassung sowohl der negativ als auch der positiv geladenen Teilchen bei einem einzigen Versuch. Wenn schädliche Stoffe untersucht werden, ist die gesamte Apparatur luftdicht zu halten und nach jedem Versuch gründlich zu säubern.

Eine neuere, vereinfachte, für qualitative und quantitative Untersuchungen an Virussuspensionen und dergleichen ausreichende Anordnung zeigt Abb. 58. Aus der Preßluftflasche 1 wird durch ein Reduzierventil Luft oder Stickstoff zur Sättigung mit dem Suspensionsmittel durch die heizbare Waschflasche 2 gepreßt, von da durch den eindüsigen Mikrovernebler 3 (Abb. 52) in den Verweil-

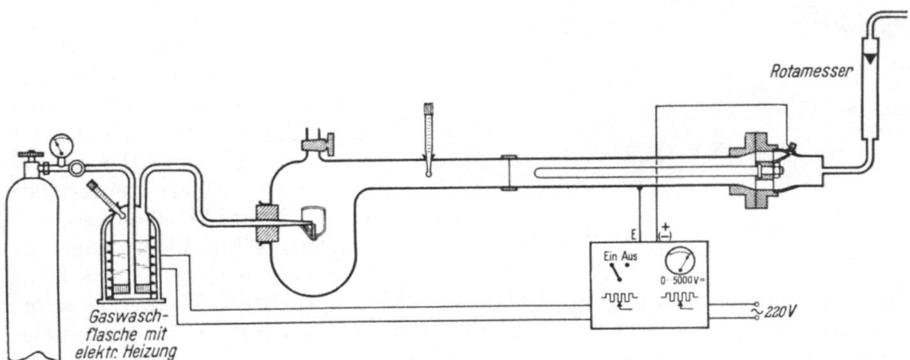

Abb. 58. Vereinfachter Versuchsaufbau mit elektrischem Kernfäller.

raum 4 und über den elektrischen Kernfäller 5 zum Rotameter 6. Der leicht auswechselbare, das Untersuchungsmaterial aufnehmende Zerstäuber wird zur Herausnahme durch eine Öffnung im Verweilraum eingesetzt, die im Gegensatz zur schematischen Zeichnung aus praktischen Gründen senkrecht zur Zeichenebene liegt.

Für das Arbeiten mit vernebelten Suspensionen seien einige Daten als Anhaltspunkte gegeben, die je nach den Anforderungen variiert werden müssen. Temperatur der Waschflasche ca. 5^0 C über Zimmertemperatur, erforderliche Versuchsmenge der Suspension 0,3 bis 1 cm³, Spannung am Kernfäller 3000 bis 5000 Volt. Versuchsdauer 3 bis 10 Minuten, durchgepreßte Luftmenge 1 bis 5 Liter pro Minute.

Wie aus dem aufgetrockneten Tropfen (Abb. 57) auf den Durchmesser des Nebeltropfens bei seiner Entstehung zu schließen ist und wie die Berechnung der Teilchenzahl einer Suspension für quantitative Versuche durchgeführt wird, folgt in Abschnitt V, 4c und d.

c) Oberflächenmethoden.

Die bisher beschriebenen Verfahren eignen sich zur qualitativen und quantitativen Untersuchung von mehr oder weniger gereinigten Bakterien und Virussuspensionen. Beim Aufbringen solcher Suspensionen auf den Trägerfilm geht der Zusammenhang von Bakterien, filtrierbaren Mikroorganismen oder Phagen, falls diese von festen Kulturoberflächen gewonnen wurden, verloren. Dies kann durch die folgenden Verfahren vermieden werden.

α) Das Abklatschverfahren.

Auf den mit Film bedeckten Blenden erfolgt die Herstellung der Präparate mit Hilfe von Mikromanipulator und Lichtmikroskop. Die mit dem Kollodiumfilm versehenen Objektträgerblenden werden mit der Filmseite nach unten in den auf Abb. 59 wiedergegebenen Halter gespannt, der seinerseits in dem Operationsstativ eines Mikromanipulators befestigt wird. Der Objektträgerfilm, dessen Größe das Gesichtsfeld des Lichtmikroskops nicht ausfüllt, wird nun genau horizontal in die optische Achse gebracht und mit dem Feintrieb des Operationsstativs vorsichtig so lange gesenkt, bis der Film die Oberfläche des darunter befindlichen Untersuchungsmaterials soeben berührt. Die Kulturoberfläche ist vorher in der Objektebene des Mikroskops scharf eingestellt worden. Beim Abheben des aufgetupften Films bleibt dann die oberflächliche Organismenschicht

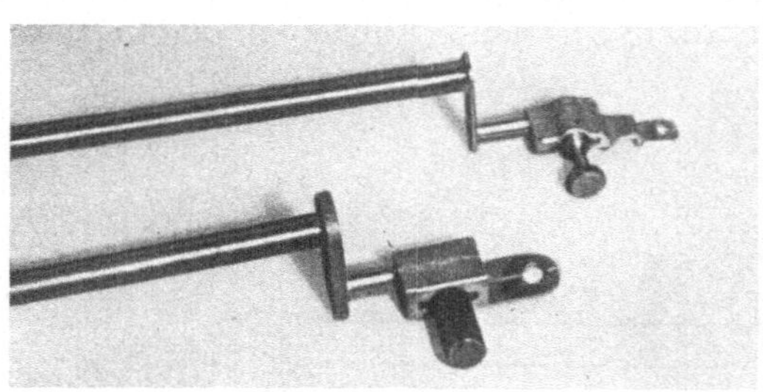

Abb. 59. Objektblendenhalter für das Abklatschverfahren.

an diesem hängen. Das von H. Ruska (19) angegebene Verfahren erfordert einige Übung und Erfahrung in der Auswahl von Nährboden mit geeigneter Feuchtigkeit. Kottmann hat damit eine sorgfältige Untersuchung über Phagen und Bakteriolyse auf sterilen Flecken durchgeführt (Abb. 103).

β) Das Abziehverfahren.

Ein größerer Tropfen von Kollodiumlösung oder Zaponlack wird beim Abziehverfahren unmittelbar auf den mit Bakterien oder ähnlichen Objekten bewachsenen, gut feuchten Nährboden gegeben. Auf diesem breitet er sich wie bei der Anfertigung eines Lackabdrucks bei leichter Bewegung und Neigung der Kulturschale sehr dünn aus. Es kann vorteilhaft sein, die Kollodiumlösung dünner anzuwenden als bei der Herstellung von Trägerfilmen. Nach dem Trocknen des Films wird der Nährboden durch Anblasen mit warmer Luft oder im Heizschrank zusätzlich getrocknet, damit die oberflächliche Bakterienschicht am Film fixiert wird. Dann wird der Nährboden aus der Kulturschale herausgenommen, in etwa 5 mm große Quadrate (Würfel) zerschnitten und zum Ablösen der Filme in warmes Wasser gebracht. Die Filme werden mit der Platinöse gefischt, zum Waschen nochmals in frisches Wasser übertragen und dann mit der Objektblende aufgenommen. Man erhält durch dieses Verfahren gute Präparate von der Lagerung der Zellen auf der Oberfläche nicht zu dick bewachsener Kulturen. Bei ganz jungen Bakterienrasen haften die Keime zu fest im Nährboden, bei sehr alten wird der Filmbelag leicht zu dick und undurchsichtig. Die Abziehpräparate geben außerdem bemerkenswerte Aufschlüsse über die sehr wechselnde Schrumpfung der Bakterien beim Trocknen. Man sieht mitunter die Zellen in einem hellen Hof liegen, der dem Abdruck des frischen Bakteriums bei der Filmherstellung entspricht. Die Differenz zwischen dem Hof und dem Bakterium zeigt das Ausmaß der Schrumpfung (Abb. 60). Lichtmikroskopisch läßt sich die Schrumpfung im Vakuum in einer Kammer nach Abb. 61 verfolgen (H. Ruska 1941). Das Abziehverfahren wurde 1943/44 im Laboratorium des Verfassers entwickelt. Hillier und Baker (1946) haben den gleichen Weg beschritten und außerdem durch eine leichte Bedampfung mit Gold die Bakterien so plastisch dargestellt, daß eine feine Faltenbildung der Zellmembran sichtbar wird. EDWARDS und WYCKOFF (1947) benutzten das Abziehverfahren ebenfalls in Kombination mit der Metallbedampfung (Abschnitt IV, 4c, β).

Weitere Variationen der Abziehverfahren siehe bei BRIEGER, CROWE und COSSLETT (1947).

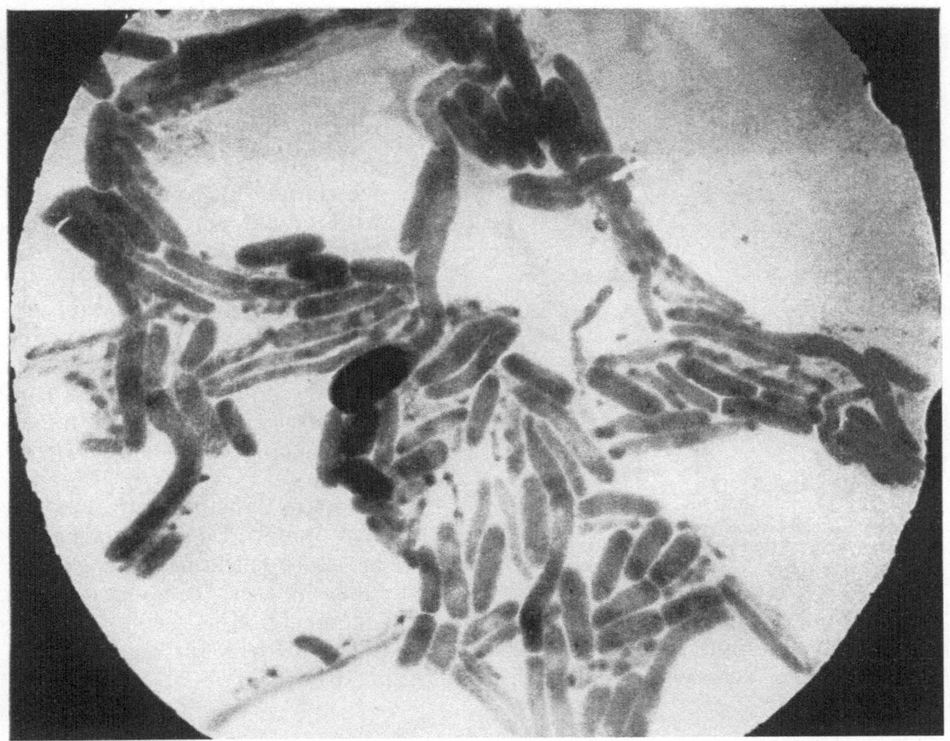

Abb. 60. Nach dem Abziehverfahren hergestelltes Präparat einer Proteus-Agarkultur. Man beachte die Wuchsformen. Nach BISS (1944) unveröffentlicht.

γ) Das Filmbewuchsverfahren.

HILLIER, KNAYSI und BAKER (1948) ließen die zu untersuchenden Bakterien oder Bakteriophagen direkt auf dem Trägerfilm wachsen. Sie beschränkten dadurch die möglichen Schädigungen der Objekte auf die Eintrocknung, die bereits vor dem Einschleusen in das Elektronenmikroskop erfolgt. Ihr Verfahren wird wie folgt durchgeführt:

Die dünne Agarschicht einer Petrischale wird zweimal mit sterilem destillierten Wasser gewaschen und dann mit einer Wasserschicht überdeckt (Abb. 62a). Auf diese wird ein Tropfen Kollodiumlösung (0,5 bis 1,0 % in Amylacetat) gebracht, der sich zu einem Film ausbreitet (b). Nach dem Verdunsten des Amylacetats wird das Wasser zwischen Film und Agar abgesaugt (c) und die

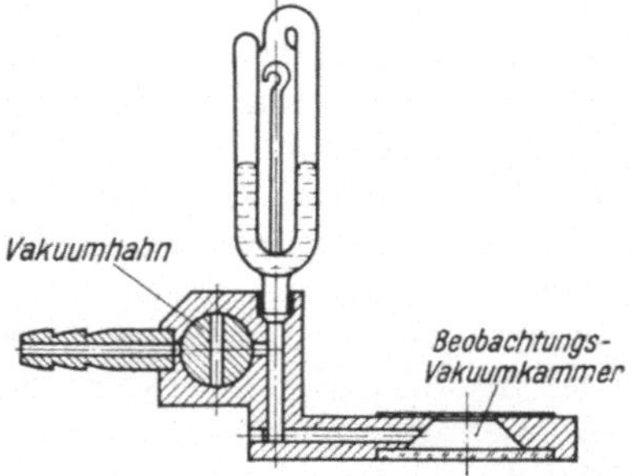

Abb. 61. Vakuumkammer für lichtmikroskopische Beobachtungen.

Membran mit einem oder mehreren Tröpfchen einer Bakteriensuspension beimpft (d). Wenn die Tropfen infolge ihrer hohen Oberflächenspannung die Membran schlecht benetzen, kann durch einen Zusatz von 0,05% Pepton, das durch tryptische Verdauung hergestellt ist, die Oberflächenspannung herabgesetzt werden. Nach dem Antrocknen wird die Kultur verschlossen (e) und 1 bis 3 Stunden in den Brutschrank gestellt. Das Wachstum kann lichtmikroskopisch kontrolliert werden (f). Wenn es das erwünschte Ausmaß erreicht hat, wird ein ausgewähltes Agarstück von etwa 10 mm Durchmesser herausgeschnitten (g), aus der Petri-

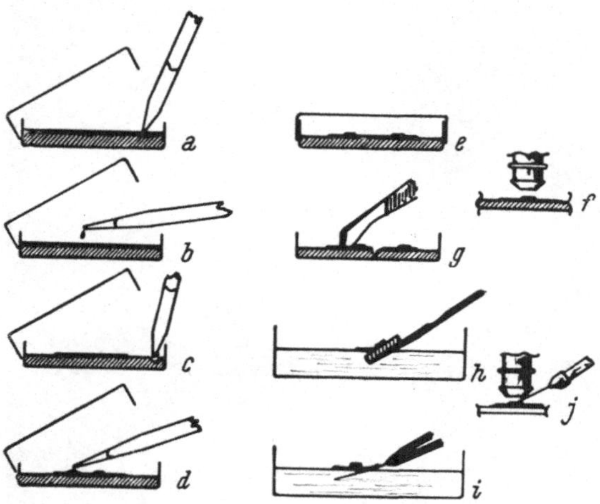

schale genommen und so unter die Oberfläche von destilliertem Wasser gebracht, daß sich der Film mit der Kultur ablöst und auf der Wasseroberfläche schwimmt (h). Aus dem Wasser wird er auf ein großes Objektträgernetz gebracht (i) und mit dem Netz auf Filtrierpapier getrocknet. Nach Auswahl geeigneter Stellen (j) werden aus dem Objektträgernetz Stücke herausgestanzt, die für das Einschleusen in das Elektronenmikroskop die geeignete Größe besitzen.

Abb. 62. Durchführung des Filmbewuchsverfahrens. Nach HILLIER, KNAYSI und BAKER (1948). Erklärung siehe Text.

Beim Arbeiten mit flüssigen Kulturen wird der Film zunächst in einer Petrischale hergestellt, die mit sterilem Wasser gefüllt ist. Dann wird das Wasser abgesaugt und an seiner Stelle Nährflüssigkeit unter den Film gebracht. Die Beimpfung der Filmoberfläche und das weitere Vorgehen entspricht dem für Agarkulturen beschriebenen Verfahren. (Vgl. auch BABUDIERI 1941/43).

δ) Die Abdruckverfahren.

Statt abgehobener Schichten mit anhaftenden Objekten können auch plastische Abdruckfilme des Oberflächenreliefs untersucht werden. Mit Hilfe dieses Verfahrens erhält man Bilder des Objekts, ohne die Objekte selbst den Einflüssen der Bestrahlung im Elektronenmikroskop aussetzen zu müssen. Die Reliefuntersuchung wurde von MAHL (1940 bis 1945) zunächst für undurchstrahlbare, metallische Objekte in die Elektronenmikroskopie eingeführt, dann aber auch für biologische Objekte verwendet. Die Zahl der für spezielle Zwecke entwickelten Methoden ist so groß, daß hier nur auf das Grundsätzliche hingewiesen werden kann, zumal sich nur ein Teil der Verfahren für biologische Objekte eignet. Die Abdrücke geben, wenn sie direkt vom Objekt hergestellt werden, das negative Relief wieder. Ein positives Relief wird über ein Hilfsobjekt gewonnen, dem zunächst das negative Relief aufgeprägt wird und von dem dann das Positiv als durchstrahlbarer Abdruckfilm zu erhalten ist. Werden die Abdrücke als Lackhäute hergestellt (MAHL), so folgt nur eine Seite der Haut den Feinheiten des Reliefs, während sich die andere nahezu eben ausbildet. Bedampft man die Oberfläche mit einer dünnen Quarz- oder Metallschicht (MAHL), die zur Untersuchung abgelöst wird, so folgt auch die nicht anliegende Fläche des Abdrucks der Form der Unterlage. Ihre Dicke wechselt je nach dem Winkel, unter dem die Atomar-, bzw. Molekularstrahlen auftreffen (Abb. 63). Die Dickenunterschiede bewirken zusammen mit der verschiedenen Lage der durchstrahlten Flächen den plastischen Eindruck des Elektronenbildes.

KRAUSE und MAHL (1943) haben Lackabdrucke von Knochen, Haut, roten Blutkörperchen und Leberzellausstrichen hergestellt. Das eingetrocknete lichtmikroskopische Ausstrichpräparat wird mit stark verdünntem Zaponlack über-

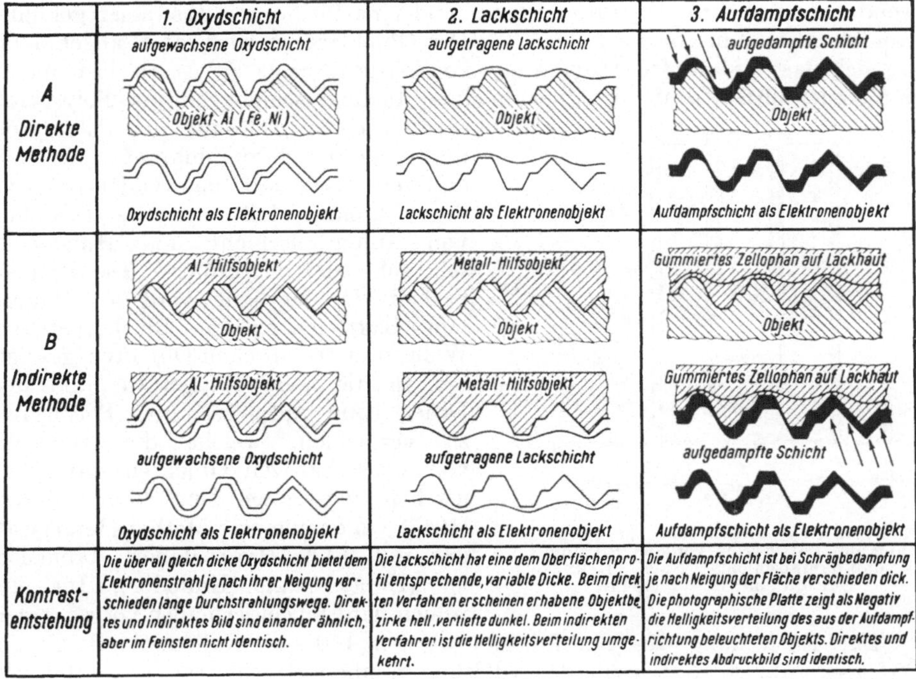

Abb. 63. Schematische Darstellung der Abdruckverfahren. Nach v. BORRIES (1949).

zogen und getrocknet. Durch verdünnte Flußsäure wird die Lackschicht mit dem Objekt vom Glas abgelöst und anschließend die organische Substanz durch Verdauungsfermente entfernt (negatives Relief). Auf diese Weise können z. B. an roten Blutkörperchen adsorbierte Virusarten (Influenza, Bartonellen) leicht dargestellt werden. Läßt man spezifisch wirksame Verdauungsfermente (Lipasen, Nukleasen, usw.) vor oder nach der Herstellung des Lackabdrucks einwirken, so werden die herausgelösten Strukturteile durch das Relief abgebildet.

KÖNIG (1946) hat die häufig bestehende Schwierigkeit, den Abdruck von seiner Unterlage abzulösen, dadurch umgangen, daß er nach dem Vorgehen von WOLF (1939) zunächst eine dicke Lackmatrize herstellte, die sich mit Hilfe eines aufgeklebten Zellophanstreifens leicht von der Unterlage abziehen läßt. Anschließend wird die Matrizenoberfläche mit Gold oder Aluminium bedampft. Der Lack läßt sich in Aceton wieder auflösen, wodurch die Aufdampfschicht (positives Relief) frei wird. Vom gleichen Objekt können auf diese Weise beliebig viele Abdrücke hergestellt werden.

Statt Lack verwenden BOWLING BARNES, BURTON und SCOTT (1945), sowie

Gerould (1947) bei 160^0 aufgepreßtes Polystyrol oder schon unter 100^0 schmelzendes Isobutyl-Metacrylat. Der Abdruck wird mechanisch entfernt und mit Siliziumdioxyd bedampft, wie es bei der Herstellung von Trägerfilmen aus Quarz beschrieben wurde (Abschnitt IV, 2 d). Geeignete Teile des Polystyrol-Quarzabdrucks werden so ausgestanzt, daß sie einen etwas kleineren Durchmesser haben als die Objektnetze oder -blenden, auf die sie gelegt werden. Das Polystyrol läßt sich danach über erwärmtem Äthylbromid, dessen Dampf sich am Objektnetz und dem aufgelegten Abdruck kondensiert, auflösen. Man verwendet dazu einen Äthylbromidsieder und eine kleine, mit Eiswasser gekühlte Kondensationseinrichtung (Abb. 64). Der Objektträger wird dort angebracht, wo sich das verdampfte Äthylbromid kondensiert, etwas Polystyrol löst und in den Sieder zurücktropft. Da die aufgedampfte Quarzschicht mit dem Polystyrol-

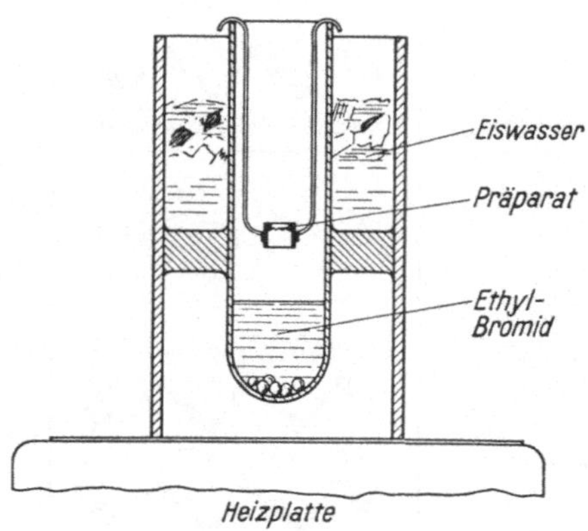

Abb. 64. Äthylbromidsieder zur Entfernung des Polystyrols von der Siliziumoxydaufdampfschicht. Nach Barnes, Burton und Scott (1945).

block vor der Ablösung des Polystyrols durch den Äthylbromiddampf in der endgültigen Lage auf das Objektnetz zu liegen kommt, läßt sich die Trennung von Aufdampfschicht und Hilfsobjekt sehr schonend durchführen. Die Brauchbarkeit der Methode wird an Bildern von Bakterien, roten Blutkörperchen, Wolle und technischen Objekten gezeigt.

Um die hohe Temperatur und den Druck beim Aufpressen des Polystyrols zu vermeiden, welche die Abdrücke von wasserhaltigen Objekten unmöglich machen, verwenden Brown und Jones (1947) monomeres Methyl-Metacrylat, das mit 0,1 % Hydrochinon stabilisiert ist (Handelsname Kallodoc). Im einzelnen wird folgendermaßen vorgegangen:

Zur Entfernung des Stabilisators werden 100 cm³ Kallodoc mit 20 cm³ 20%iger wäßriger Natriumhydroxydlösung in einem Scheidetrichter geschüttelt. Der braune Bodensatz wird abgelassen. Dann wird das Methyl-Metacrylat in einem Scheidetrichter mit 20 cm³ destilliertem Wasser gewaschen und nach Entfernung der Bodenschicht in einem Becherglas über 5 g wasserfreiem Kalziumchlorid getrocknet. Die entstabilisierte Flüssigkeit hält sich einige Monate, wenn sie kühl und dunkel aufbewahrt wird.

Etwa 5 cm³ der destabilisierten Flüssigkeit werden ungefähr eine halbe Stunde lang mit 3 mg Benzoylperoxyd erhitzt, wobei sie je nach der Dauer der Erhitzung mehr oder weniger viskös werden. Je zäher die Flüssigkeit wird, um so schneller polymerisiert sie bei der Herstellung der Matrize, aber um so schlechter dringt sie in die Feinheiten des Oberflächenreliefs ein.

Bei Temperaturen, die mit Rücksicht auf das Objekt zwischen 60^0 und niedriger Zimmertemperatur gewählt werden können, wird das Objekt mit der polymerisierenden Flüssigkeit übergossen und nach deren Erhärtung mit einer Rasierklinge abgelöst. Der Abdruck wird dann nach Heidenreich und Peck (1943) mit Siliziumoxyd bedampft und die Aufdampfschicht in Toluen abgelöst.

Weitere Verfahren siehe bei Zworykin und Ramberg (1942), Heidenreich und Matheson (1944), Schaefer (1942/43), Schaefer und Harker (1942) und Seeliger (1946).

4. Die Objektvorbereitung.

Die wenigsten Objekte können ohne besondere Vorbereitung nach einer der beschriebenen Methoden auf die Trägerfilme gebracht werden. Einfach ist die Präparation in der Bakteriologie und bei der Untersuchung bereits rein dargestellter Suspensionen von makromolekularen Virusarten. Angaben über die Objektvorbereitung können aber nicht auf das Virus allein beschränkt werden, da auch Zellbestandteile, ganze Zellen und Gewebe für die Pathologie der Virusinfektionen von Interesse sind. Die bereits beschriebenen Abdruckverfahren sind für die Untersuchung von Ausstrichen und Oberflächen jeder Art geeignet, die nicht selbst in das Elektronenmikroskop gebracht werden. Die folgenden Angaben beziehen sich auf Methoden, welche die direkte Untersuchung der Objekte betreffen.

a) Verfahren zur Untersuchung von Zellen und Geweben.

Zur Durchstrahlung von Geweben sind zerzupfte oder geschnittene Präparate von weniger als 0,5 μ Dicke erforderlich. Wenn wir von Vorschlägen von v. ARDENNE (1939), RICHARDS, ANDERSON und HANCE (1942) und SJÖSTRAND (1943) absehen, deren Erfolge unbefriedigend geblieben sind, so haben vor allem POHLMAN und WOLPERS (1944), FULLAM und GESSLER (1946) sowie CLAUDE und FULLAM (1946) wertvolle Verfahren angegeben.

α) Schnitt-Schall-Verfahren.

WOLPERS (1944) hat ein Schnitt-Schall-Verfahren entwickelt, bei dem zunächst möglichst dünne Gefrierschnitte hergestellt werden, die dann in Wasser suspendiert und magnetostriktiv mit etwa 10 kHz beschallt werden. Bei der

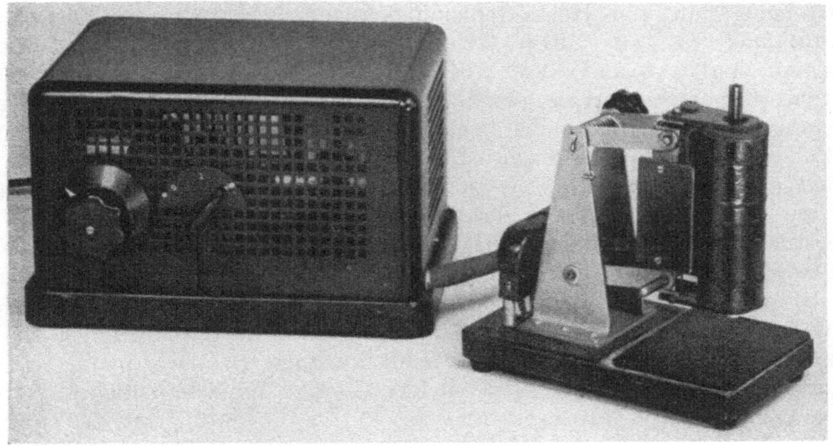

Abb. 65. Magnetostriktives Beschallungsgerät zur Herstellung elektronenmikroskopischer Präparate. Unter der Spule (rechts im Bild) sieht das Ende des Magnetstabes hervor, der die Schwingungen auf die Flüssigkeit überträgt.

Beschallung lösen sich aus den Schnitten die faserigen Bestandteile in feinster Form heraus, so daß sie der übermikroskopischen Untersuchung ausgezeichnet zugänglich sind. Auf diese Weise sind kollagene und elastische Fasern, glatte und quergestreifte Muskulatur, sowie Nervenfasern präpariert worden. Vergleichende Präparationsuntersuchungen von Pohlmann und Wolpers ergaben, daß die magnetostriktive Beschallung (7,5 kHz) der Faserschnittsuspensionen günstiger ist als die hochfrequente (350 kHz) piezoelektrische. Bei der piezoelektrischen Beschallung kommt es zu einer Aufwicklung der Fasern, wenn diese durch den Strahlungsdruck von den Schwingungsknoten des Wassers in die -bäuche transportiert werden. Bei der magnetostriktiven Beschallung bleibt dieser Effekt aus, da der Abstand von Schwingungsknoten zu Schwingungsbauch etwa 200 mm gegen 1 mm bei der piezoelektrischen Behandlung beträgt.

Zum Schneiden läßt sich jedes gute Gefriermikrotom der histologischen Technik verwenden. Für die Beschallung wurde bei Siemens und Halske ein kleines, an das Lichtnetz anschließbares Gerät entwickelt (Abb. 65). Seine Leistung reicht aus, um rote Blutkörperchen zu haemolysieren, Bakterien und Virus abzutöten und Schnittsuspensionen zu zerfasern. Der magnetostriktive Schallgeber besteht aus einem aus Nickelblech aufgerollten Stab, an dessen unterem Ende ein massiver Metallkopf in ein eng angepaßtes kleines Gefäß mit der zu beschallenden Suspension eintaucht. Er wird durch eine Spule erregt, die im Anodenkreis zweier handelsüblicher Rundfunkröhren (EL 12) eingeschaltet ist. Durch entsprechende Abstimmung und Rückkopplung auf dem Gitter der Röhren schwingt dieser Kreis mit der mechanischen Resonanzfrequenz des Nickelstabes (ca. 10 kHz).

β) Hochgeschwindigkeits-Mikrotome.

O'Brien und McKinley (1943) bauten ein Mikrotom, dessen Messer mit einer Geschwindigkeit von 40 m und mehr pro Sekunde das Objekt durchschneidet. Durch die hohe Schnittgeschwindigkeit (10^{-4} sec. Kontakt zwischen Messer und Gewebe) reduzierten sie die Deformation des geschnittenen Blocks und des Schnitts auf ein Minimum. Ihre Erkenntnis, daß sich Schnitte um so dünner anfertigen lassen, je höher die Schnittgeschwindigkeit ist, wurde von Fullam und Gessler (1946) beim Bau von Hochgeschwindigkeitsmikrotomen weiter ausgewertet. Claude und Fullam (1946) erzielten ausgezeichnete Bilder von Leberschnitten. Schuster und Gray (1947) machten Angaben über die Gewebepräparation. Die hohe Schnittgeschwindigkeit wird durch ein Messer erzielt, das radial oder tangential einer rotierenden Scheibe an-, bzw. aufsitzt, die 40000 und mehr Umdrehungen pro Minute macht. Die Geschwindigkeit des Messers wurde von Fullam und Gessler auf 150 bis 450 m pro Sekunde erhöht. Sie erhielten damit Schnitte, die nach stereoskopischer Vermessung 0,5 bis 0,1 μ dick waren. Zur Einbettung von osmiumfixierten Geweben eignet sich Paraffin, das durch Xylol oder Isopentan aus dem Schnitt herausgelöst wird. Zum Auffangen der Schnitte sind besondere Vorrichtungen vorgesehen, die ähnlich wie beim elektrischen Kernfäller die elektrische Ladung der abfliegenden Schnitte benutzen, um sie auf einem entgegengesetzt geladenen Schirm anzulagern. Die Custon Scientific Instruments, Inc. Arlington, N. J., liefert Geräte, die Schnittdicken von nur 0,05 μ herzustellen gestatten.

γ) Schnittverfahren für Bakterien.

Baker und Pease (1949) geben ein Verfahren an, mit dem Schnitte durch Bakterienzellen hergestellt werden können. Die Bakterien werden eine Stunde in 5 % Formalin fixiert, in der Zentrifuge abgeschleudert und in ansteigenden Alkoholkonzentrationen im Zentrifugenröhrchen entwässert. Nach der Entwässerung wird eine 3%ige „Parlodion"-Lösung mit gleichen Teilen Äther-

Alkohol zugesetzt. Die Konzentration des „Parlodion" wird langsam auf 12%
gebracht und das Lösungsmittel abgedampft. Nach 24 Stunden wird mit Chloroform gehärtet, der Block aus dem Zentrifugenröhrchen herausgebrochen, in
Karbol-Xylol nochmals entwässert und in Paraffin (Temperatur 65° C) eingebettet. Das Schneiden erfolgt nach den Angaben von PEASE und BAKER (1948)
mit einem gewöhnlichen Mikrotom. Das Paraffin wird mit Benzol aus dem Schnitt
herausgelöst. Dann wird der Schnitt in 0,2%ige Kollodium-Amylacetatlösung
getaucht und auf eine Wasseroberfläche gebracht, auf welcher er sich mit dem
anhaftenden Kollodium zu einem Film ausbreitet, der elektronenmikroskopisch
untersucht werden kann.

LAURELL (1949) „schneidet" Bakterien, indem er Klatschpräparate von
einer jungen Kultur auf einem geschliffenen Glasobjektträger herstellt, mit
Osmiumtetroxyd fixiert, nach HAST (1947) mit Beryllium bedampft und den
Berylliumfilm abzieht. Beim Abziehen werden größtenteils nur die oberfläch-
lichen Schichten der bedampften Zellen abgerissen, so daß man auf diese Weise
Schnitte und bei Wiederholung des Verfahrens Schnittserien erhalten kann.

b) Anreicherung und Reinigung von Virussuspensionen.

Virussuspensionen sind ohne Schwierigkeiten zu untersuchen, wenn das
Dispersionsmittel flüchtig und frei von gelösten oder anderen virusfremden
Substanzen ist. Da die natürlich anfallenden virushaltigen Flüssigkeiten keine
reinen Virussuspensionen sind und meist sogar sehr komplizierte Gemenge darstellen, ist die Reinigung des Virus eine wesentliche Aufgabe
der Präparation. Im ungereinigten Zustand werden die Virusteilchen bei der Auftrocknung oft von Fremdsubstanzen, insbesondere niedermolekularen Stoffen, Eiweiß, Zelltrümmern oder
Zellen völlig überlagert. Häufig muß der Reinigung eine Anreicherung der zu untersuchenden Teilchen vorangehen. Bei
bereits bekannten charakteristischen Virusformen braucht für
die morphologische Untersuchung die Reinigung nicht vollkommen
zu sein. Für die Frage, wieweit Reinigung und Anreicherung
fortgeschritten sind, liefert die Betrachtung eines Tropfens
oder, wenn möglich, eines Niederschlagspräparats mit kleinsten
Tröpfchen die schärfste Kontrolle.

Je nach dem Ausgangsmaterial (Pflanzenpreßsäfte, Lysate,
Körperflüssigkeiten, Zellen, Organe oder Gewebe) kann eine Be-
freiung von eingeschlossenem Virus aus Zell- und Gewebebestandteilen erforderlich sein, um eine gute Ausbeute zu erhalten. Das
übliche Zerreiben von Organen mit Seesand bringt unnötige Fremdbestandteile in das aufzuarbeitende Material. In den meisten
Fällen genügt das Zerdrücken in einem Glasschliffmörser nach
Abb. 66, wie er von WEIGL zur Freilegung von Rickettsien aus
den Darmzellen der Laus benutzt wurde. Er ermöglicht steriles
Arbeiten. Der quere Handgriff und die einseitige Ausbildung des
Rezipienten verhindern ein Wegrollen beim horizontalen Auflagern.

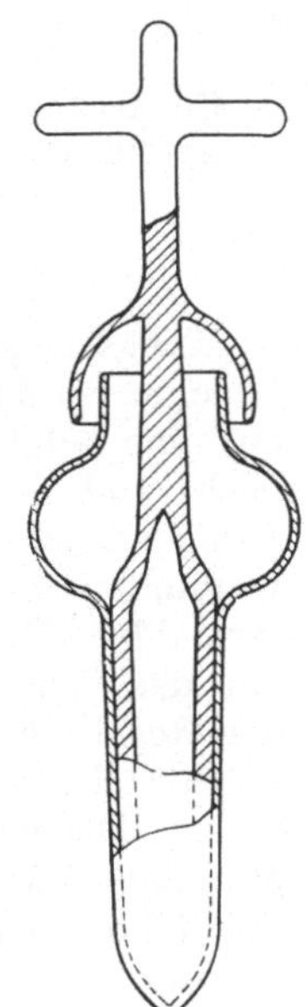

Abb. 66. Glas-
schliffmörser
nach WEIGL.

Der konische Schliff hat gegenüber zylindrischen Schliffen ähnlicher Konstruktion den Vorteil, daß er sich beim Auseinander-
nehmen nicht ansaugen kann. Er muß bis zur Spitze gut eingeschliffen sein.
Der Zerreibungseffekt ist von der Feinheit der Schliffflächen abhängig; denn
je feiner der Schliff ist, um so kleinere Objekte können zerrieben werden. Da
der Glasschliffmörser in jeder Größe leicht hergestellt werden kann, ist er
für kleinste Mengen brauchbar.

Versuche über die Wirkung eines feingeschliffenen Glasschliffmörsers an verschiedenen Objekten hatten die in Tab. 4 zusammengestellte Wirkung:

Tab. 4. *Aufschließung organischer Objekte im Glasschliffmörser.*

Mahlgut	Mahldauer 2—3 Min.	Mahldauer 10 Min.	Mahldauer 30 Min.	Kontrolle
Mäuseleber	Zerstörung der Zellen,Freilegen der Zellkerne	Beginnende Zerstörung der Zellkerne		Lichtmikroskop
Kaninchen-erythrozyten in isot. NaCl-Lösung	Fortschreitende Haemolyse, stark zerstörte Membranen neben intakten Zellen		starke Haemolyse	Elektronen-mikroskop
Kolibakterien	Geringfügige, nur langsam zunehmende Zerstörung einzelner Bakterien. Zellbruchstücke und Membranen.			Elektronen-mikroskop
Tabak Mosaik-virus	Keine nachweisbare Änderung der statistischen Längenverteilung gegenüber der Ausgangslösung (Zahl der vermessenen Stäbchen = 300)			Elektronen-mikroskop
Kollagene Fibrillen	Wesentlich schlechtere Aufteilung als beim Schnitt-Schall-Verfahren. Keine Zerstörung der Quer-streifung.			Elektronen-mikroskop

Tab. 4 zeigt, in welchen Zeiten und bis zu welchen Partikelgrößen organische Objekte bei der Zerreibung zerstört werden. Für die Isolierung von Strukturen aus dem Zellinnern ist die Tatsache bemerkenswert, daß Zellen verhältnismäßig leicht und sehr vollständig zerstört werden können, während stäbchenförmige Virusmoleküle nicht beeinträchtigt werden. Daher ist diese Methode zur Gewinnung von Virus besonders empfehlenswert. Zur Virusisolierung wird das zermahlene Rohmaterial durch Sedimentieren oder Zentrifugieren so vorgereinigt, daß das Virus in der überstehenden Flüssigkeit bleibt. Der Bodensatz wird in der Regel ein zweites Mal mit sterilem Wasser, Salz- oder Pufferlösungen aufgeschüttelt, erneut sedimentiert und der zweite Überstand mit dem ersten vereinigt. Die weitere Anreicherung und Reinigung erfolgt dann je nach Art des Materials durch Extraktion fettähnlicher Substanzen, durch abwechselnde Salzfällungen und Auflösungen des Virus oder durch Fällung der begleitenden Verunreinigungen und durch erneutes Zentrifugieren. Auch mehrmaliges Zentrifugieren allein kann zu befriedigender Reinigung führen. Nach der Vorreinigung wird ein oder mehrmals so hochtourig sedimentiert, daß das Virus in den Bodensatz kommt. Nach der letzten Aufnahme des Virussediments können Aggregate und dergleichen in einer Nachreinigung durch vorsichtiges Zentrifugieren entfernt werden, so daß danach der Überstand als Untersuchungsmaterial vorliegt. Häufig läßt sich die Aggregatbildung dadurch herabsetzen, daß das Virus nicht mit destilliertem Wasser, sondern mit hochverdünnten Salzlösungen präpariert wird. Einen Anhalt für die erforderlichen Sedimentationszeiten verschieden großer Partikel gibt Tab. 5.

Tab. 5. *Sedimentationszeiten.*

In Salzlösungen oder Bouillon suspendierte Objekte	Größter Durchmesser in mμ	Umdrehungszahlen pro Min.		
		3500 r ~ 8 cm	15000 r ~ 6 cm	30000 r ~ 6 cm
		Sedimentationszeiten in Min.		
Rote Blutkörperchen	7— 8000	10— 20		
Blutkörperchenmembranen	10000	20— 40	5— 10	
Blutplättchen	1000	30— 60	10— 20	
Bakterien	500— 1000	30— 60	10— 20	
Vaccine Virus	200— 300	60—120	30— 60	
Große Bakteriophagen	60— 120		60—120	10—20
Kleine Bakteriophagen	10— 20			30—60

Bei der Verarbeitung sehr großer Mengen und nicht zu kleiner Objekte (z. B. große Bakteriophagen) kommt die Verwendung von Cepa-Durchlaufzentrifugen (Fa. Padberg) in Frage, bei welchen das Zentrifugiergut zur Trennung mehrmals durch einen rotierenden Trennzylinder geschickt wird, aus dem die Bodenschicht und die überstehende Flüssigkeit getrennt ablaufen (40000 = 45000 Umdr./min.).

Alle Präparationsmaßnahmen sind mit Virusverlust verbunden, der durch Titerbestimmungen der Suspensionen in den verschiedenen Arbeitsgängen verfolgt werden kann. Es ist vorteilhaft, immer mit einem erheblichen Materialüberschuß zu arbeiten. Die für die morphologische Untersuchung benötigten Mengen sind jedoch verschwindend klein, so daß mit dem bloßen Auge nicht sichtbare Bodensätze schon ausreichen und kaum getrübte Suspensionen noch genügend Virus enthalten können. Die Viruszählung mit der Tröpfchenmethode erfordert dagegen hohe Viruskonzentrationen. Aus folgenden Überlegungen gehen die Zahlenwerte hervor:

1. Morphologische Untersuchung mit der Tropfenmethode.

Aufgetrocknete Schichthöhe = 1 mm. Geforderte Zahl der Virusteilchen pro Gesichtsfeld bei 10000facher Vergrößerung = 10. Größe des Bildfeldes etwa 5000 mm². Größe des dazugehörigen Objektfeldes = 0,00005 mm² und des darüber aufgetrockneten Volumens = 0,00005 mm³. Zahl der dazu erforderlichen Virusteilchen pro 1 cm³ = 200000000 = 2.10^8 (Vgl. v. BORRIES, 1949, Tab. 20).

2. Teilchenzählung mit der Niederschlagsmethode.

Durchschnittliches Volumen eines Tröpfchens = 1 μ^3. Durchschnittliche Zahl der Virusteilchen pro Tröpfchen = 1. Zahl der erforderlichen Virusteilchen pro 1 cm³ = 1000000000000 = 10^{12}. Für die Zählung von Virusteilchen mit der Niederschlagsmethode über die Vernebelung einer Virussuspension ist es also erforderlich, von Präparaten mit 10000fach höherer Viruskonzentration auszugehen als bei der einfachen morphologischen Untersuchung (Vgl. Abschnitt V, 4b und c).

c) Virusadsorption an rote Blutkörperchen.

Nach Haemolyse der Erythrocyten kann das an die durchsichtige Membran adsorbierte Virus (Influenza, New Castle Virus, Mumps, Hühnerpest) ausgezeichnet dargestellt und ausgezählt werden (Abb. 135, 136 und 137). DAWSON und ELFORD (1949/2) geben dafür folgende Technik an:

Hühnererythrozyten werden mit Saponin, entsprechend der Methode von Dounce und Lan (1943) haemolysiert und die lackfarbene Suspension vollkommen mit 0,9 % Kochsalzlösung, m/33 Phosphat pH 6,7 gewaschen. Nach einer bestimmten Zeit des Kontaktes mit dem Virus bei 0° werden die Zellen abzentrifugiert, der Bodensatz wird wieder in 0,9 % Kochsalzlösung aufgelöst, in 0,1 % Osmiumsäure fixiert, verschiedene Male mit destilliertem Wasser bei 500 Umdrehungen pro Minute je 5 Minuten gewaschen und die oben schwimmende Flüssigkeit abgegossen. Die Zellen werden dann direkt aus der Suspension in destilliertem Wasser auf gewöhnliche Objektblenden mit Kollodiumfilmen gebracht. Wenn die Lösung sich auf dem Film festsetzt, wird die Anzahl der Zellen unter dem Lichtmikroskop geprüft und entweder durch Abpipettieren der überschüssigen Suspension verringert oder soviel von dem Originalmaterial hinzugefügt, bis sich etwa sechs Zellen auf einer Strecke von 0,01 cm befinden. Die Probe wird dann durch direkte Übertragung, bzw. nach Bedampfung mit Gold oder Palladium untersucht. Wo die Zellmembranen in hoher Konzentration auftrocknen, tritt eine beachtliche Faltung der Membranen ein. Zum Zählen von Teilchen auf der Zellmembran ist es immer möglich, durch sorgsames Aufbringen ebene, gleiche Felder mit ganz geringer Faltung zu erhalten. Die Fläche, die man auszählt, wird auf der photographischen Platte durch ein Planimeter gemessen und die Zahl der Virusteilchen pro μ^2 der Zelle festgestellt. Nach Schäfer, Schramm und Traub (1949) genügen etwa 200 Virusteilchen je Erythrozyt, um eine deutliche Agglutination hervorzurufen.

Nach Flick (1948) bleibt das Adsorptionsvermögen der roten Blutkörperchen auch nach Fixierung mit 50 % Formalin und sorgfältiger Auswaschung des Formalinüberschusses erhalten.

d) Verfahren zur Kontraständerung.

Elektronen verschiedener Geschwindigkeit geben vom gleichen Objekt verschieden kontrastreiche Bilder (Abb. 23). Statt Erhöhung der Strahlspannung zur besseren Durchstrahlung dicker Objekte können auch die Objekte durch Extraktion bestimmter Substanzen, sowie durch teilweisen fermentativen Abbau (Dawson und McFarlane 1947) oder durch Veraschung (Draper und Hodge 1944) durchstrahlbarer gemacht werden. Umgekehrt läßt sich bei sehr dünnen, kontrastarmen Objekten der Kontrast außer durch Herabsetzung der Strahlspannung auch durch Kontraststeigerung des Objekts erhöhen.

α) Färbemethoden.

Als „Farben" eignen sich Stoffe, die in möglichst spezifischer Weise von einem Objekt oder dessen Teilstrukturen gebunden werden und seinen Kontrast durch Bindung größerer Substanzmengen oder durch die hohe Dichte des gebundenen Stoffes um mindestens 10 % erhöhen. An erster Stelle kommen Metallverbindungen in Frage. Wolpers und H. Ruska (1939) haben häufig Osmiumtetroxyd gebraucht. Auch von Babudieri (1948) wird Osmiumsäure als bestes Fixierungsmittel empfohlen. Jakus (1945) verwendete Phosphorwolframsäure. Gemeinsam mit Kausche wurden Kontrasterhöhungen bei der Behandlung von Tabak Mosaikvirus mit Gold- und Silbersalzen gesehen. Für Influenza- und Pferdeencephalomyelitisvirus wird von Sharp und Mitarbeitern Kalziumchlorid zur Kontrasterhöhung empfohlen. Systematische Untersuchungen hat Krabbe 1944 begonnen, aber nicht mehr veröffentlicht. Er hat schwere Atome, wie Jod und Metalle, in Bakterienfarbstoffe eingebaut und mit solchen

Verbindungen Kontraststeigerungen erzielt. Die bisher entwickelten Röntgenkontrastmittel, wie Jodtetragnost, sind wenig geeignet, weil sie nicht genügend
an den Objekten haften und diese unspezifisch schwärzen.

Die Färbung ist trotz ihrer großen Bedeutung noch sehr wenig systematisch
bearbeitet worden. Es wäre denkbar, daß Röntgenkontrastmittel durch kleine
Modifikationen in der Färbe-, Fixierungs- und Waschtechnik zum Haften gebracht werden können. Als Ziel muß aber nicht nur das Haften der Kontrastmittel angestrebt werden, sondern auch die Bindung an bestimmte, chemisch
definierbare Substrate, wie beispielsweise an Lipoide, einfache Eiweißstoffe oder
Nukleoproteine. Dadurch könnte die Analyse der Stoffverteilung im sublicht-
mikroskopischen Größenbereich wesentliche Fortschritte machen. Die Ableh-
nung von Färbe- und Fixierungsmitteln (SJÖSTRAND) ist nicht grundsätzlich
berechtigt.

Liegen statt komplexer Gebilde, wie z. B. Zellen, Zellbestandteilen und großen
Virusformen nur einfache makromolekulare Substanzen vor, so kann der Einbau
schwerer Atome unmittelbar an der untersuchten Substanz erfolgen. Ein vier-
fach mit Jodbenzoyl substituiertes Glykogen war nach E. HUSEMANN und
H. RUSKA (1940) kontrastreicher als das nichtsubstituierte Ausgangsmaterial.

β) Metallbedampfung (Schattenwurftechnik).

Eine andere Methode zur Kontraststeigerung ist die Metallbedampfung des
Objekts im Vakuum. Hierzu wird der mit Objekten belegte Trägerfilm in Richtung von etwa 10^0 bis 45^0 mit atomarem Metall bedampft, bzw. beschattet
(Abb. 68). Man erhält durch die vom Metall nicht getroffenen Schattenseiten
des Objekts und durch den Schatten auf der Trägerfolie sehr plastische Bilder,
deren Kontrast in unspezifischer Weise mit der Dicke der Bedampfungsssschicht
zunimmt. Das Verfahren wurde in der Durchstrahlungsmikroskopie zuerst von
H. O. MÜLLER (1942) zur Vermessung der Tiefenausdehnung aus der Schattenlänge
der Objekte verwendet und dann von WILLIAMS und WYCKOFF (1944, 45, 46)
in verbesserter Form in die Untersuchungstechnik eingeführt.

Während der Metallbedampfung muß das Vakuum in der Apparatur (Abb. 67)
so gut sein, daß eine geradlinige Ausbreitung des Metalldampfes im Raum zwischen der Bestäubungsquelle und dem Objekt gewährleistet ist. Wenn ein
merklicher Anteil der Metallatome auf ihrem Wege an Luftmolekülen von der
ursprünglichen Richtung abgestreut wird, ist die Begrenzung des Objektschattens
verwaschen, so daß eine Höhenmessung in demselben Maße ungenau wird.
Eine weitere Fehlerquelle für eine quantitative Auswertung liegt in der räumlichen Ausdehnung der Bestäubungsquelle. Beides wird in der Apparatur (Abb. 67)
auf ein erträgliches Mindestmaß herabgedrückt. Der Vakuumraum wird durch
eine Metallgrundplatte M und eine Glasglocke G begrenzt. Die Glocke ist mittels
Planschliff unter Verwendung von Hochvakuumfett gegen die Grundplatte ge-
dichtet und kann über den Vakuumstutzen V mit einer Quecksilberdiffusions-
pumpe ausgepumpt werden. Da die ins Vakuum eingebauten Teile nicht nach
den in der Röhrentechnik üblichen Verfahren ausgeheizt werden können (Zerstörung des Objekts), müssen die inneren Oberflächen der Apparatur vor Beginn
der Evakuierung gründlich gesäubert werden. Als Bestäubungsmittel werden
Beryllium, Aluminium, Chrom, Nickel, Kupfer, Silber, Antimon, Platin, Gold,
Uran u. a. Metalle, aber auch Siliziumdioxyd und Siliziummonoxyd verwendet.
Ein Tantalstreifen, in den eine wannenförmige Vertiefung W eingedrückt ist,
dient zur Aufnahme des Verdampfungsmaterials und zugleich als elektrischer
Widerstand zu dessen Erhitzung; auch Spiralen aus Wolframdraht können ver-

wendet werden. Der Heizstrom beträgt etwa 15 bis 20 A bei einer Spannung von 10 bis 15 Volt. Das zu bestäubende Objekt O ist zum bequemeren Wechsel an einem leicht herausnehmbaren, mit konischem Fettschliff versehenen Metalleinsatz K im oberen Teil der Glocke befestigt und kann gegen die Strahlachse um einen beliebigen Winkel gekippt werden. Damit ein Auftreffen gestreuter Metalldampfteilchen von der Seite vermieden wird, ist das Objekt von einem

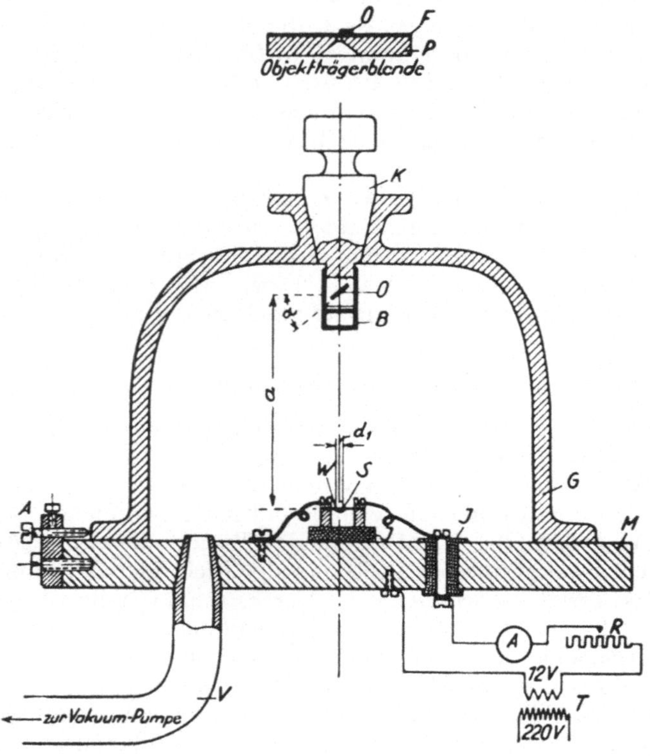

Abb. 67. Apparatur für die Schrägbedampfung elektronenmikroskopischer Präparate. Nach H. O. Müller (1942). a Abstand: Objektdampfquelle, α Bestäubungswinkel, B Doppelblende, d_1 Durchmesser der Dampfquelle, F Objektträgerfilm, I Isolation, R Widerstand, S Silber, T Transformator.

Blendenzylinder umgeben, der zur eindeutigen Festlegung der Bestrahlungsrichtung als Doppelblende ausgeführt ist. Durch Anschläge A ist die Lage der Glocke G so festgelegt, daß die zu bestäubende Objektstelle in einer Richtung mit den beiden Blenden und der Silberdampfquelle liegt. Das Objekt, das mit Metall bedampft oder dessen Höhe gemessen werden soll, wird auf eine Trägerblende oder ein Netz gebracht, damit es anschließend im Übermikroskop betrachtet werden kann. Über die Bohrung der Platinblende P erstreckt sich der Trägerfilm, dem das Objekt O aufliegt. Da sich bei genauen Ausmessungen die Lage des Objekts weder bei der Bestäubung noch im Elektronenstrahl des Übermikroskops verändern darf, wird als Material für den Trägerfilm mit Vorteil das gegenüber dem Kollodiumfilm mechanisch festere Aluminiumoxyd oder Siliziumdioxyd verwendet.

Die Menge des verdampften Materials beträgt einige Milligramm. Außer von der verdampften Menge ist die Dicke der Bedampfungsschicht vom Abstand zwischen Dampfstrahlquelle und Objekt abhängig. Um ein Objekt plastisch erscheinen zu lassen, genügen schon sehr geringfügige Schrägbedampfungen, die noch keine ganz geschlossene Schicht erzeugen (Abb. 68). Gut „beschattete" Objekte tragen Schichten von etwa 0,8 mμ. Zur Herstellung von Abdruckfilmen werden Schichten einer mittleren Dicke von 1 mμ oder mehr verwendet; Matrizen zur Gewinnung positiver Abdruckreliefs müssen noch wesentlich stärker sein. Gelegentlich tritt die Bestäubung zufällig während des Mikroskopierens im Elektronenmikroskop selbst auf (Abb. 68). Das aufgedampfte Material muß eine gleichmäßige Schicht bilden und darf auf dem Objekt nicht zu sichtbaren Kriställchen zusammenwachsen. KAUSCHE (1947) glaubt, daß für die örtliche Dicke der Aufdampfschicht nicht nur die Form des Objekts, sondern auch eine unterschiedliche Affinität zwischen Objektsubstanz und aufgedampfter Substanz maßgebend ist, so daß z. B. bei Bedampfung mit Metalllegierungen mit einer unterschiedlichen Adsorption verschiedener Objektstrukturen zu rechnen wäre.

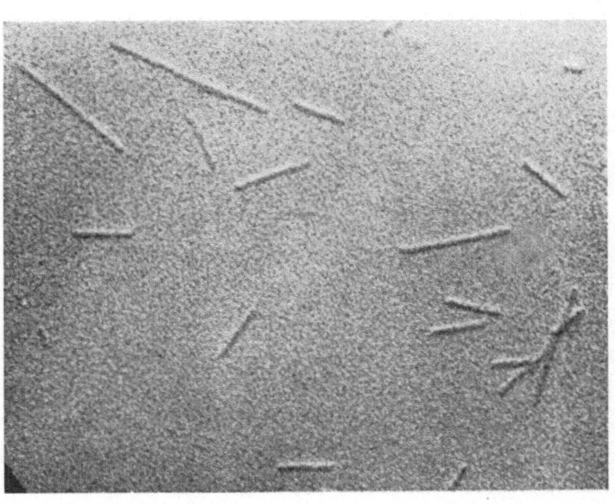

Abb. 68. Während der Untersuchung im Elektronenmikroskop zufällig bestäubte Tabak Mosaikvirus-Stäbchen. Nach KAUSCHE und H. RUSKA (1941) unveröffentlicht. 25 000 : 1.

Die Beschattungstechnik wird außerordentlich häufig angewendet. Sie kann plastische Einzelheiten sichtbar machen, die im einfachen Durchstrahlungsbild nicht erkennbar sind. Es darf aber nicht übersehen werden, daß sie auch Dichteunterschiede des Objekts, die sich nicht im Relief ausprägen, überdecken kann. Siehe die Bilder bei HILLIER, MUDD und SMITH (1949). Es sollte deshalb neben dem Beschattungsbild auch immer das nicht beschattete Objekt aufgenommen werden. Da Apparaturen zur Metallbedampfung auch für die Abdruckverfahren notwendig sind, gehören sie zu den wichtigsten Hilfseinrichtungen der Elektronenmikroskopie.

Die Metallbedampfungsbilder wirken besonders plastisch, wenn sie, wie es in der Regel geschieht, als Negative wiedergegeben werden. Die massedicken, stark bedampften, im positiven Bild dunklen Teile erscheinen dadurch hell, die Trägerfolie und die Schatten dunkel.

V. Die Auswertung von Beobachtungen und Aufnahmen.

Während der Herstellung der elektronenmikroskopischen Aufnahmen ist sorgfältig darauf zu achten, ob bei der Elektronenbestrahlung Formveränderungen des Objekts auftreten. Sind diese ausgeschlossen oder erkannt, so besteht häufig die Aufgabe, über die bildmäßige Wiedergabe und Beschreibung des Objekts hinaus noch zahlenmäßige Aussagen über Größen, Größenverteilungen und Häufigkeiten zu machen. Hierzu ist es notwendig, die Meßgenauigkeit des Elektronenmikroskops, die Grenzen der Gestaltwiedergabe, sowie die Art des Messens auf der photographischen Platte und auf dem Stereobild zu kennen. Außerdem müssen die einfachen Regeln der Statistik bekannt sein. Für das Zählen der Virusteilchen in Flüssigkeiten und in der Luft müssen schließlich besondere Verfahren angewendet werden.

1. Die Objektschädigungen durch Elektronenstrahlen.

Schon bei den ersten Versuchen, biologische Objekte im Elektronenmikroskop abzubilden, wurden Schädigungen beobachtet (E. RUSKA 1934, MARTON 1934/36). Später haben zahlreiche Autoren Veränderungen der Objekte beschrieben (LEMBKE und H. RUSKA 1940, RIEDEL und H. RUSKA 1941, H. RUSKA 1942, MARTON, DAS GUPTA und MARTON 1946, COSSLETT 1947, MAHL 1947, KINDER 1947, KÖNIG 1948, ECKARD 1948, WATSON 1948, HILLIER 1948 und zahlreiche andere). Hierdurch wurde bei vielen Biologen eine starke Skepsis gegenüber den Ergebnissen der Elektronenmikroskopie ausgelöst. Dieses war berechtigt in der Absicht, die Aufmerksamkeit immer wieder auf die Schädigungen zu lenken, die nicht nur durch die Bestrahlung, sondern auch schon durch die Präparation entstehen. Mit der fortschreitenden Entwicklung wurden durch die Analyse der Schädigungen wichtige Aufschlüsse über die Wechselwirkungen zwischen Elektronenstrahlen und Objekt, sowie über die chemische Beschaffenheit von Teilstrukturen erhalten. Von besonderem Interesse sind in diesem Zusammenhang die Untersuchungen über strahlenempfindliche Bakterienkörnchen von KÖNIG und WINKLER (1948). Sie konnten zeigen, daß die kugeligen, zunächst sehr schattendichten und bei starker Bestrahlung verdampfenden Einschlüsse mit den Volutinkörnchen identisch sind und aus dem Kalziumsalz einer Nukleinsäure bestehen. Die Verkohlung der organischen Substanz beginnt schon bei 200⁰ C, bleibt aber unsichtbar. Bei 400⁰ C verdampfen in Diphtheriebakterien eingeschlossene Tellurnadeln (CLAUBERG, II, Nährböden), bei 600⁰ C verdampft das Kalziumphosphat der Volutinkörnchen, das aus seiner organischen Bindung an Nukleinsäure schon bei niedrigerer Temperatur abgespalten wird[1]. Sehr kurze Elektronenbestrahlung, aber auch Verkohlen des Bakterienplasmas im Vakuum bei 300⁰ C macht die Grenzschichten säurefester Bakterien für Volutin-Lösungsmittel permeabel; andererseits kann etwas längere Bestrahlung eine Lösung der Tellurnadeln in Bromwasser und der Kalziumphosphatkörnchen in Salzsäure unmöglich machen. KÖNIG und WINKLER (1948) zeigten ferner, daß zwischen Kollodiumhäuten eingebettetes Kupfer nach Elektronenbestrahlung nicht mit Salzsäure herauslösbar ist. Sie vermuten, „daß oberflächenaktive Gruppen, beispielsweise OH-Gruppen, für die Durchlässigkeit der Haut verantwortlich zu machen sind und daß diese Gruppen, die bei nicht völlig veresterter Zellulose in reichem Maße vorhanden sind, durch die Elektronen abgespalten werden. Es ist nicht ausgeschlossen, daß oberflächenaktive Gruppen in ähnlicher Weise die Permeabilität des Bakterienplasmas steuern und daß eine Art Molekülzertrümmerung durch Elektronen das erste Stadium einer Veränderung des Präparates im Übermikroskop darstellt".

Die Verkohlung im Elektronenstrahl verläuft formbeständig, während bei der rein thermischen Verkohlung im Vakuum mitunter ein sichtbarer Zerfall eintritt. Nach KÖNIG und WINKLER beruhen die Veränderungen der Objekte im Elektronenmikroskop „zunächst weniger auf einer Wärmeentwicklung als vielmehr auf der ionisierenden Wirkung der Elektronen, die photochemische Prozesse verursacht, welche zu einem allmählichen Abbau des ursprünglichen Moleküls führen."

Befinden sich im Vakuumraum des Elektronenmikroskops vom Dichtungsfett ausgehende Kohlenwasserstoffdämpfe, die sich am Objekt kondensieren, so entstehen um das Objekt feine, verkohlende Hüllen (KÖNIG 1948). Zur Prüfung

---

[1] RUESS (1944) hat aus der Verpuffungstemperatur verschiedener Graphitoxyde bei etwa 40000facher Vergrößerung im Siemens-Übermikroskop (Typ 1939/45) die Objekttemperaturen zu 200 bis 300⁰ C bestimmt.

dieses Effekts kann NaCl pro Analyse auf dem Film aufgetrocknet werden. Verdampft es bei starker Bestrahlung, ohne eine äußere Kristallhülle zurückzulassen, so ist das Vakuum von störenden Kohlenwasserstoffen frei; meist bildet sich jedoch eine solche Hülle bei längerem Verweilen des Präparats im Mikroskop.

Weitere Artefakte hat MANDLE (1947) an goldbedampften Tabak Mosaikvirus-Präparaten festgestellt. Die Stäbchen erscheinen nach längerer intensiver Bestrahlung segmentiert und teilweise zerstört, und der vorher gleichmäßige Untergrund wird körnig. Der Effekt ist auf ein Schmelzen des Goldbelags zurückzuführen, bei dem sich kleine Tröpfchen, bzw. größere Goldkriställchen bilden. H. O. MÜLLER (1942) hat dieselbe Erscheinung in gröberem Ausmaß nach Silberbedampfung beobachtet. Die bedampften Präparate dürfen deshalb nur kurze Zeit der Bestrahlung ausgesetzt werden.

2. Die Vermessung der Objekte.

Für die Auswertung einer Aufnahme muß zunächst der Abbildungsmaßstab bekannt sein. Seine Bestimmung setzt sich aus vier Messungen zusammen. Die Durchmesser der Objektblendenbohrung O und die Projektivblende P werden im Lichtmikroskop bestimmt, die Bilder beider Blenden O_B und P_B auf dem Leuchtschirm, bzw. auf der photographischen Platte. Der Gesamtabbildungsmaßstab $\dfrac{O_B}{O} \times \dfrac{P_B}{P}$, der noch eine Korrektur erfährt, wenn der Zwischenbildleuchtschirm nicht am Ort der Bildebene vor dem Projektiv liegt, kann auf $\pm 5\,\%$, bei Anwendung besonderer Sorgfalt auf etwa $\pm 2\,\%$ ermittelt werden. Die lichtmikroskopisch auszumessenden Blenden müssen genau kreisförmig und in axialer Richtung wenig ausgedehnt sein. Außerdem muß die Projektivblende auf mindestens drei Ecken der photographischen Platte abgebildet werden, um den Kreisdurchmesser zu bestimmen. Bei sehr genauen Untersuchungen muß ferner die Verzeichnung des Endbildes berücksichtigt werden, derzufolge die Vergrößerung von der Bildmitte zum Rand hin um einige Prozente zunimmt (siehe auch MATHESON und HEIDENREICH 1945, sowie Abschnitt 11, 5).

a) Messen und Meßgenauigkeit in der Bildebene.

Die Ausmessung der Teilchengröße erfolgt am genauesten auf der Originalplatte mit schwach vergrößerndem Mikroskop und Okularmikrometer. Durch Kopie oder Vergrößerung können zusätzliche Ungenauigkeiten entstehen. Auf der überbelichteten Kopie sind alle Teilchen größer als auf einer unterbelichteten Kopie der gleichen Platte. Zur Vereinfachung der Größenbestimmung aller zu vermessenden Teilchen zahlreicher Aufnahmen ist es zweckmäßig, lichtoptische Nachvergrößerungen der Originalaufnahmen in einheitlichem Maßstabe herzustellen und auf diesem mit dem Millimetermaßstab zu vermessen. Es muß dabei aber auf normale Belichtung der Vergrößerungen geachtet werden. Auch die Abmessung der Originalplatte selbst liefert nur dann den richtigen Wert, wenn diese richtig exponiert ist: die gleichen Teilchen erscheinen auf einer überexponierten Platte kleiner als auf einer bei gleichem Abbildungsmaßstab aufgenommenen unterexponierten (v. BORRIES). — Häufig beobachtet man helle und gelegentlich auch dunkle Säume um die abgebildeten Partikel. In solchen Fällen ist die wahre Begrenzung des Teilchens dort zu suchen, wo die große Elektronenintensität des ersten hellen Saumes zur kleinen Intensität im Gebiete des Teilchenschattens übergeht.

Die Genauigkeit, mit der bei der Ausmessung die Größe des kleinsten Teilchens ermittelt werden kann, hängt außer vom Auflösungsvermögen des Geräts auch noch von dem durch die Objektmassendicke hervorgerufenen Elektronendichtekontrast ab. Nur bei ausreichendem Kontrast kann mit der Genauigkeit des Auflösungsvermögens selbst gemessen werden. Von Objekten, die etwas unterhalb des Auflösungsvermögens liegen, kann noch die Anwesenheit erkannt werden, wenn sie, wie Schwermetallkolloide, genügendes Streuvermögen haben. Ihre Größe kann aus den auf der photographischen Platte erzeugten hellen Flecken abgeschätzt werden, die je nach Teilchengewicht verschiedene Lichtdurchlässigkeit zeigen. (Ultra-Elektronenmikroskopie.) Bei genügender Erfahrung in der Deutung übermikroskopischer Bilder lassen sich die Größenabmessungen

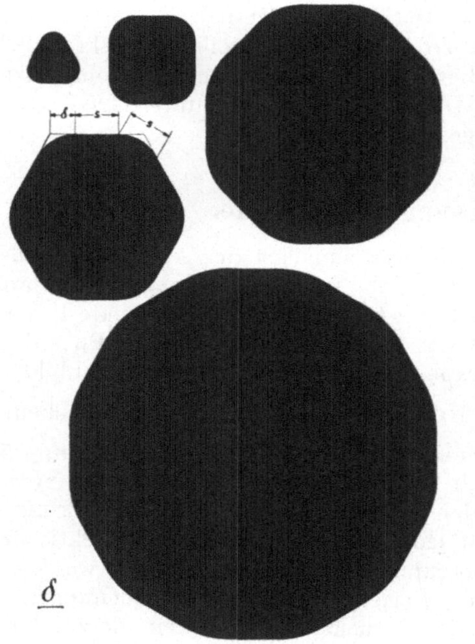

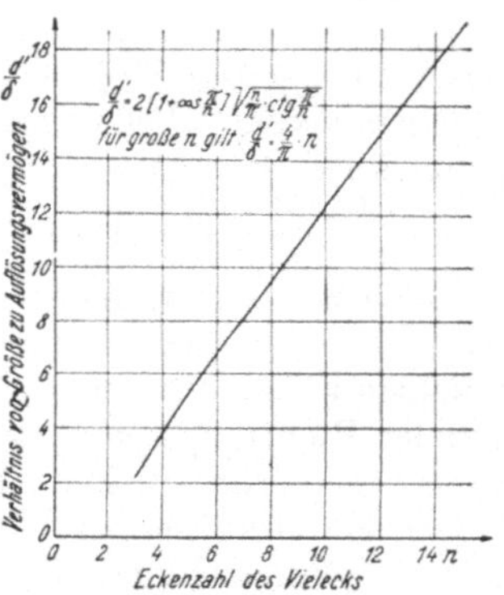

Abb. 69. Figuren verschiedener Vielecke, erzeugt mit gleichem Auflösungsvermögen δ. Je höher die Eckenzahl ist, um so größer müssen die Vielecke sein, wenn sie gleich ähnlich dargestellt werden sollen. Die Länge der Sehne des Abrundungsbogens ist hier jeweils gleich dem gradlinig dargestellten Kantenteil. Die Abrundungskreise berühren die Kante im Abstand δ von der Ecke. Nach v. Borries und Kausche (1940).

Abb. 70. Größe regelmäßiger Vielecke (gemessen als Durchmesser d' des flächengleichen Kreises) in Vielfachen des Auflösungsvermögens δ, abhängig von der Eckenzahl. Nach v. Borries und Kausche (1940).

eines Objekts ausreichender Massendicke aus Messung und Schätzung des Orts der wahren Teilchenbegrenzung etwa mit ± δ/2 ermitteln, wenn δ das von der betreffenden Aufnahme gelieferte Auflösungsvermögen ist.

Soll ein scharfkantig begrenzter Körper vermessen werden, so müssen die geradlinigen Kanten größer als das Auflösungsvermögen sein. Vielecke sind nur dann vermeßbar, wenn die geradlinig erscheinenden Kanten mindestens ebenso lang sind wie die Sehne des Abrundungsbogens, der an Stelle scharfer Ecken im Bild erscheint, Abb. 69 und 70.

Ein Vieleck wird nach v. Borries und Kausche (1940) erst dann erkennbar, wenn sein Durchmesser ungefähr so groß ist wie das mit der Eckenzahl multiplizierte Auflösungsvermögen der Aufnahme. Seeliger (1947/48) gibt für die Erkennbarkeit polygonaler Begrenzungen folgende Beziehung an

$$s = \delta \left[1 + \cos (\pi/n)\right].$$

worin s die Seitenlänge eines regelmäßigen n-Ecks und δ die Punktauflösung bedeutet (vgl. ferner HILLIER 1941). Ähnliche Erwägungen gelten auch für die Lichtmikroskopie, so daß man aus dem Fehlen der Quaderform des Vaccine-virus im Phasenkontrastbild nicht den Schluß ziehen kann, diese Form sei ein Artefakt der elektronenmikroskopischen Untersuchungstechnik (BAKER 1948).

b) Messen und Meßgenauigkeit in Richtung des Strahlenganges.

Wegen der großen Tiefenschärfe der Übermikroskope ist es nicht möglich, einzelne Ebenen des Objekts für sich scharf zu stellen und wie in der Lichtmikroskopie aus einer meßbaren Verschiebung des Objekts in Richtung des Strahlenganges die Objekttiefe auszumessen. Dafür sind andere, in der Lichtmikroskopie nicht oder wenig angewendete Verfahren zur Bestimmung der Tiefenausdehnung der Objekte entwickelt worden.

α) Messung aus dem Einzelbild.

Nur sehr ungenau kann die in Strahlrichtung liegende Dimension aus der Größe des Schwärzungskontrastes geschätzt werden. Bei kristallinen Objekten macht die Elektronenbeugung am Kristallgitter die Abschätzung unmöglich.

Einige Autoren haben die Dicke der Objekte dadurch gemessen, daß sie den Trägerfilm durch starke Bestrahlung so zum Einreißen brachten, daß sich das Objekt 90° um eine horizontale Achse drehte.

Mit größerer Genauigkeit läßt sich die Tiefenabmessung aus einem Einzelbild nach dem Verfahren von H. O. MÜLLER (1942) feststellen. Es beruht auf der Schattenbildung einer schräg auf die Trägerfolie treffenden Metallbedampfung und dient im wesentlichen für die Vermessung kompakter, einzeln auf einem Objektträger liegender Objekte. Wie aus der Abb. 27 hervorgeht, kann die Tiefe Δy eines Objekts aus der Schattenlänge s und dem Bedampfungswinkel α nach der Beziehung $\Delta y = s \cdot tg\alpha$ sehr einfach bestimmt werden. Zweckmäßig

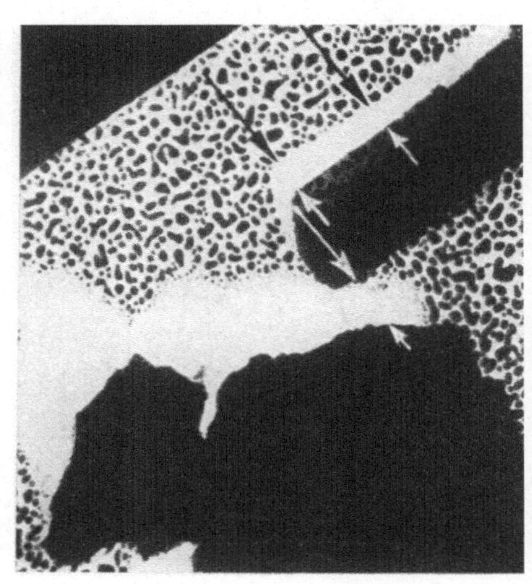

Abb. 71. Schmirgelkörner mit Silber unter 45° bestäubt. Nach H. O. MÜLLER (1942).

wählt man bei dickeren Objekten $\alpha = 45°$, damit man für $tg\alpha = 1$ mit der Ausmessung der Schattenlänge direkt die Objekttiefe erhält. Bei flachen Objekten wird mit wesentlich spitzerem Winkel beschattet, um die Messung aus einem langen Schatten möglichst genau durchführen zu können. Zum Aufbringen des Metallbelags dient das Bedampfungsverfahren (Abschnitt IV, 4 c β).

In Abb. 71 ist ein Präparat (Schmirgel), das zur Beurteilung der Korndimensionen mit dem Schrägbestäubungsverfahren (45°) behandelt worden ist, wiedergegeben. H. O. MÜLLER hat zugleich stereoskopische Aufnahmen angefertigt, um eine Nachprüfung der aus der Schattenausmessung erhaltenen Tiefenwerte mit Hilfe des Stereokomparators zu ermöglichen. Die Richtung des Schattens ist leicht festzustellen, seine Länge wird am besten an einer Objektstelle gemessen, die sich in der Schattenkontur eindeutig abzeichnet. Bei scharfkantigen Kristallen sind solche Punkte besonders leicht aufzufinden. In den nicht abgeschatteten Gebieten des Films ist der Silberbelag während der Betrachtung im Elektronenmikroskop zu kleinen Tröpfchen zusammengelaufen, ohne daß aber dadurch eine Aufhebung der Schattengrenze eingetreten ist.

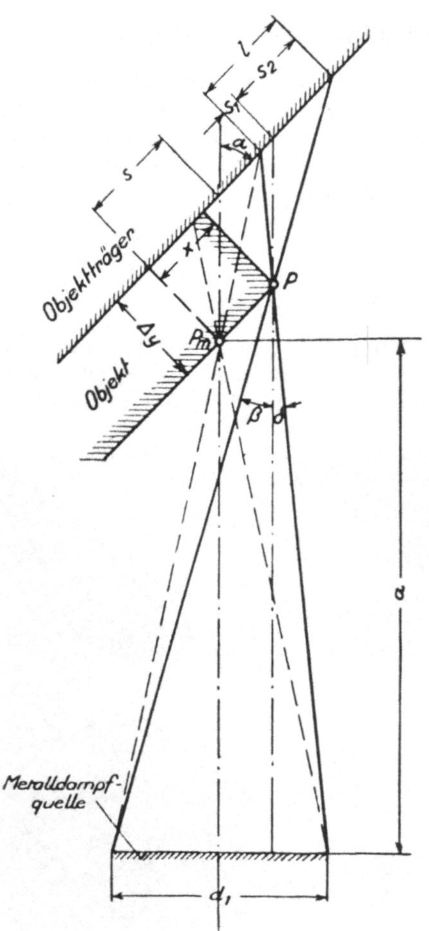

Abb. 72. Schatten- und Halbschattenlänge eines Objekts bei Schrägbestäubung. Nach H. O. Müller (1942).

Bei den in neueren Arbeiten verwendeten Bedampfungsstoffen tritt dieser Effekt nicht mehr oder nur in geringerem Umfange ein. Als Folge der räumlichen Ausdehnung der Bedampfungsquelle ist als Übergang vom „Schatten" zur vollen Versilberung ein „Halbschatten"-Gebiet entstanden, in dem infolge der zum „Kernschatten" hin abnehmenden Silberdeckung nicht mehr genügend Material zur Bildung von Tröpfchen vorhanden ist, so daß sich die Grenze des Halbschattens in einer plötzlichen Änderung der Tröpfchengröße markiert. Aus diesem Grunde ließen sich auch solche Präparate noch auswerten, bei denen eine über das normale Maß hinausgehende Silberbestäubung, verbunden mit starker Elektronenbestrahlung im Übermikroskop zu verhältnismäßig sehr großen Silbertropfen geführt hat. In dem Gebiet abnehmender Silberbestäubung wird die Tröpfchengröße mit sichtbarem Übergang kleiner. Die wahre Länge des Halbschattens, die bei der genauen Ausmessung der Schattenlänge berücksichtigt werden muß, ist schwer zu bestimmen, weil auch im Kernschattengebiet Silberdampfteilchen liegen, die durch Streuung von ihrer ursprünglichen Richtung abgekommen sind.

Da der Halbschatten bei der Ausmessung der Schattenlänge ins Gewicht fällt, wird rechnerisch die Länge des Halbschattens nach der in Abb. 72 aufgezeichneten geometrischen Anordnung bestimmt. In der Skizze bedeutet d_1 den Durchmesser der Metalldampfquelle, bzw. der vorgeschalteten Blende in seiner Ausdehnung senkrecht zur Bestrahlungsachse, die durch den Punkt p_m des um den Winkel α gegen die Achse geneigten Objekts verläuft. Da sich die Halbschattenlänge l mit dem Abstand x des betrachteten Punktes p von der Achse, mit der Tiefe y des Objekts über dem Objektträger und mit dem Neigungswinkel α ändert, wurde die Größe von l für den allgemeinen Fall bestimmt. Nach einfachen trigonometrischen Beziehungen ergibt sich:

$$l = s_1 + s_2 = \Delta y \,[\mathrm{ctg}\,(\alpha - \beta) - \mathrm{ctg}(\alpha + \delta)];$$

darin ist:

$$\operatorname{tg}\beta = \frac{\dfrac{d_1}{2} + x \sin\alpha}{a + x \cos\alpha} \qquad \operatorname{tg}\delta = \frac{\dfrac{d_1}{2} - x \sin\alpha}{a + x'\cos\alpha}$$

Für $x = 0$ wird $\beta = \alpha$.

Bei der Untersuchung der Genauigkeit seiner Methode fand H. O. Müller, daß für die Abmessungen $\alpha = 45^0$, $d_1 = 3$ mm und $a = 48$ mm die außeraxiale Lage des Punktes P unberücksichtigt bleiben kann. Auch die Änderung an der Länge des Halbschattens durch Lage- und Größenänderung der abdampfenden Silberoberfläche bleibt unter $\pm 1\%$, während die Unschärfe der Schattenbegrenzung $\pm 2,5\%$ beträgt.

Für die praktische Ermittlung der Tiefe eines Objektes ergibt sich folgender Arbeitsgang: Wie in Abb. 71 angedeutet ist, wird die zu einem Objekt gehörende Schattenlänge auf dem hochvergrößerten elektronenmikroskopischen Bild bis zu der Grenze ausgemessen, an der die gleichmäßige Silberbelegung der Folie (kenntlich an einheitlicher Tröpfchengröße) beginnt. Nach Abzug des halben Halbschattenwertes, der rund 6% der Schattenlänge beträgt, ergibt sich dann die wahre Schattenlänge S. Daraus läßt sich die Objekttiefe $y = \dfrac{S}{V}$ berechnen, wenn V die Vergrößerung der vermessenen Aufnahme ist. Da die Vergrößerung mit ± 2 bis $\pm 5\%$ bestimmt wird, kann die Teilchenausdehnung in Richtung des Strahlenganges mit etwa der gleichen Genauigkeit bestimmt werden wie die Ausdehnung in der Bildebene.

β) Messung aus einem stereoskopischen Bildpaar.

Die Herstellung stereoskopischer Bildpaare durch Neigung des Objekts gegen die Strahlachse oder durch Neigung des Bestrahlungsteils gegen den Abbildungsteil des Mikroskops ist in Abschnitt II, 6 beschrieben. Für das Verständnis der nicht ganz einfachen Stereophotogrammetrie muß auf die Arbeit von H. O. Müller (1942) verwiesen werden. Die plastische Darstellung der Objekte erfolgt fast ausschließlich durch die Schattenwurftechnik. Die Differenzen zwischen beiden Methoden der Tiefenausmessung betragen nur wenige Prozent. Bei der Vermessung von porösen Objektstrukturen wie Schmetterlingsschuppen oder Diatomeenschalen, deren räumliche Bildung nicht allein im Oberflächenrelief zu erfassen ist, bleibt die Stereo-

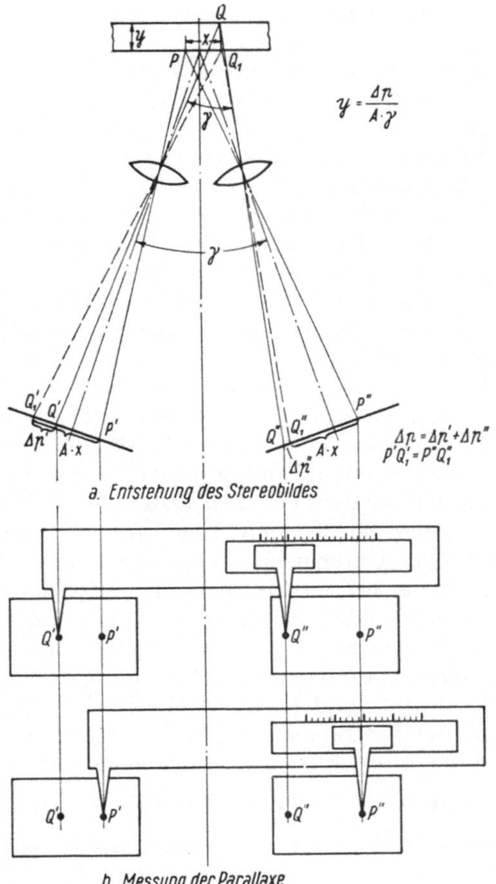

Abb. 73. Entstehung des Stereobildes und Messung der Parallaxe. Nach v. Borries (1949).

photometrie unentbehrlich. Man vergleiche ferner die Arbeiten von v. Ardenne 1940, Eitel und Gotthard 1940, Mahl 1940, Gotthard 1942, Marton 1944, Heidenreich und Matheson 1944, und Kinder 1947, sowie Abb. 73.

3. Die statistische Auswertung.

Die praktisch vorliegenden Systeme von kolloiden Teilchen, Viruspartikeln und Bakterien sind zumeist nicht monodispers. Vielmehr trifft man größere und kleinere Teilchen an, die mit einer gewissen Häufigkeit um einen bestimmten Betrag von einem am häufigsten vorkommenden Wert abweichen. Den Aufteilungsgrad solcher Systeme beschreibt man dann vollständig, wenn man die Häufigkeitsverteilung der Teilchen ermittelt und in Tabellen- und Kurvenform angibt. Zu diesem Zweck teilt man alle untersuchten Teilchen in Größenklassen ein und zählt aus, wieviel Teilchen zur Klasse der kleinsten Teilchen, zur Klasse der nächst größeren usw. gehören. Den gesamten von den Teilchen überstrichenen Größenbereich teilt man für genaue Untersuchungen in so viele Klassen ein, daß die einheitliche Größe der Einzelklasse, d. h. die Differenz zwischen unterer und oberer Grenze jeder Klasse nicht wesentlich größer als das Auflösungsvermögen der auszuwertenden Aufnahme ist. Man wird jedoch die Klassen größer wählen, wenn die Teilchengröße um ein sehr hohes Vielfaches des Auflösungsvermögens schwankt, wenn geringe Genauigkeit gefordert wird oder keine genügende Zahl von Teilchen für die Ausmessung zur Verfügung steht. In dem zuletzt genannten Fall würden auf die einzelnen, zu klein gewählten Klassen nur einige wenige, statistisch stark streuende Werte kommen, die das Ergebnis fälschen könnten. Erfahrungsgemäß erhält man bereits eine gut beurteilbare Kurve, wenn zehn genügend besetzte Klassen gebildet werden. Der grundsätzliche Charakter der Größenverteilungskurve eines kolloiden Systems und einer eingipfligen Häufigkeitskurve läßt sich bereits angeben, wenn etwa 100 bis 200 Teilchen ausgemessen werden. Nach v. Borries und Kausche weicht die dabei sich ergebende Kurve nur wenig von derjenigen ab, die sich bei der Auszählung von etwa 2000 Teilchen ergibt. Riedel und H. Ruska (1941), Hanson und Daniel (1947) u. a. haben diese Beobachtung bestätigt. Man kann damit rechnen, daß durch die Präparationstechnik eine selektive Bevorzugung besonders großer oder besonders kleiner Partikel nicht eintritt, obwohl bei dem üblichen Prozeß der Objektfilmbeschickung nicht alle in Tropfen vorhandenen Kolloide, sondern nur die zufällig über der Blendenbohrung liegenden der Untersuchung zugänglich sind. Die untersuchbaren Teilchen sind also trotz der bei der Eintrocknung auftretenden Anreicherung am Tropfenrand nicht im einen oder im anderen Sinne ausgelesen.

In den meisten Fällen ergeben sich eingipflige Häufigkeitsverteilungen, bei älteren Lösungen stäbchenförmiger Virusmoleküle auch mehrgipflige (z. B. Pfankuch und H. Ruska 1947). Die einfachen Kurven nähern sich mit größer werdender Teilchenzahl mehr und mehr der Gaußschen Verteilung. Diese Systeme kann man charakterisieren durch den Mittelwert M der Teilchengröße, die mittlere Abweichung σ und die Variabilität $\frac{\sigma}{M}$, die ein Maß für die Annäherung an den monodispersen Zustand, also für die Einheitlichkeit der Teilchengröße ist. Die mittlere Abweichung $\sigma = \pm\sqrt{\dfrac{\Sigma \cdot (\mathrm{pa}^2)}{n}}$ erhält man aus der Summe aller quadrierten Differenzen der Varianten vom Mittelwert $\Sigma (\mathrm{pa}^2)$ und der Anzahl der Varianten n. Wenn sich bei einer Untersuchung mehrgipflige Kurven oder solche mit stark unsymmetrischem Charakter ergeben, so gibt dies einen Hinweis, daß beim Entstehungsprozeß der Teilchen Besonderheiten vorgelegen haben. Um diese zu suchen, empfiehlt es sich, die Untersuchung mit einer größeren Anzahl von Einzelteilchen und möglichst mit mehr als 10 Größenklassen zu wiederholen, um den Charakter der Kurve genauer kennenzulernen.

In der Regel bestimmt man zunächst die Häufigkeitsverteilung der Linearabmessungen des betreffenden Systems. Aus dieser kann man die Verteilung der Oberflächen und Volumina auf folgende Weise berechnen: Man multipliziert jeweils die Oberfläche (das Volumen) des Einzelteils jeder Klasse mit der zugehörigen Klassenhäufigkeit und gewinnt so die von jeder Klasse gelieferte Oberfläche (das Volumen). Durch Addition der Oberflächen (Volumina) aller Klassen erhält man die Gesamtoberfläche O (das Gesamtvolumen V) des untersuchten Systems. Die Bestimmung des mittleren Teilchenvolumens ist für eines der Verfahren der Teilchenzählung wichtig (vgl. Abschnitt V, 4). Durch Division von O/V kann man die relative Oberfläche ermitteln, die natürlich um so größer ist, je feiner das Material aufgeteilt ist. Kennt man das spezifische Gewicht γ, so ergibt sich aus O/V, wenn man diese Größe in $\dfrac{1}{\mu}$ rechnet, durch Division mit γ unmittelbar die von einem Gramm des betreffenden Stoffes entwickelte Oberfläche in $\dfrac{m^2}{g}$.

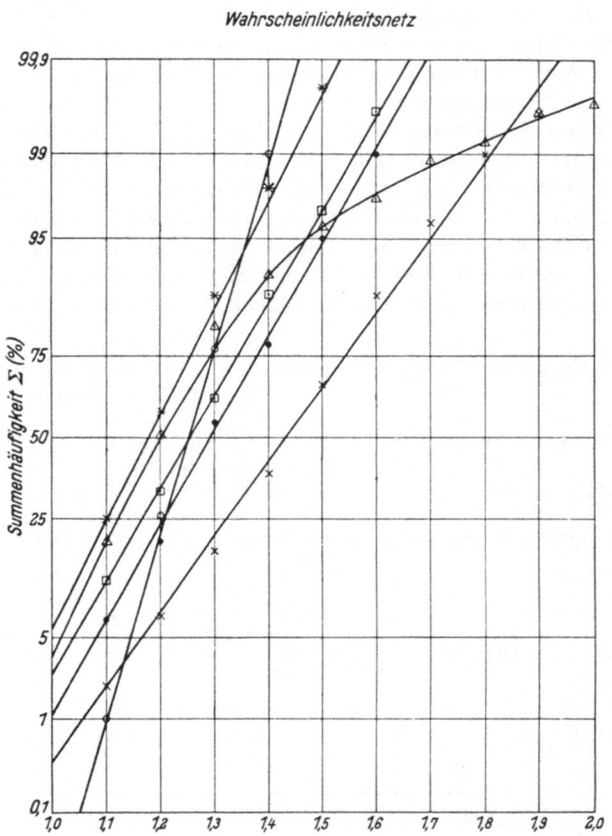

Abb. 74. Verteilung des Längen-Breitenverhältnisses von verschiedenen quaderförmigen Virusarten und von Staphylokokken auf dem Wahrscheinlichkeitsnetz. Die Abszisse gibt den Quotienten an. Nach H. Ruska und Kausche (1943).

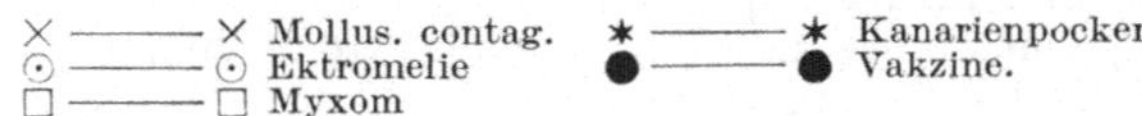

$$\frac{O}{G}\left[\frac{m^2}{g}\right] = \frac{O}{V}\left[\frac{1}{\mu}\right]\frac{1}{\gamma}\left[\frac{cm^2}{g}\right].$$

Diese Größe ist für das physikalisch-chemische Reaktionsvermögen (u. a. auch in der Ultrazentrifuge) bedeutsam.

Für biologische Fragestellungen nach Wachstum und Vermehrungsweise kann die Untersuchung des Längen- und Breitenverhältnisses der untersuchten Teilchen wichtig sein. Es läßt sich besonders bequem und genau ermitteln, weil Fehler der Vergrößerungsbestimmung, die zu zusätzlicher Streuung der Meßwerte führen, hierbei wegfallen (H. Ruska und Kausche 1943).

Die Häufigkeitsverteilung zeichnet man kurvenmäßig oder als Treppenpolygon auf. Die Kurvenpunkte geben die Klassenmitten wieder, die Polygone die Klassenbreiten. Berechnet man die Oberflächen oder Volumina aus den Linearabmessungen, für die man eine Einteilung in gleiche Klassen gewählt hat, so haben die daraus hervorgehenden Klassen für Oberfläche und Volumen nicht mehr gleiche Größen. Bei der Auftragung der Häufigkeitskurven muß man daher, wenn man als Abszisse die Oberflächengröße (die Volumengröße) einträgt, bei jeder Klasse die auf sie fallende Oberfläche (Volumen) durch ihre Klassengröße teilen. — Man kann die Ergebnisse der Häufigkeitsermittlung aber auch so darstellen, daß man über dem Ende jeder Klasse die bis dahin erreichte Summe aller Linearabmessungen (Oberflächen, Volumina) aufträgt. Diese Darstellungen bezeichnet man als Summenkurven, bzw. Summenpolygone. Die Summenkurven werden auf einem Wahrscheinlichkeitsnetz zu Geraden, wenn Gaußverteilungen vorliegen (Abb. 74). Der Schnittpunkt mit der Abszisse 50 stimmt mit dem Arithmetischen Mittel überein, falls die Verteilung symmetrisch ist (vgl. z. B. H. Ruska und Kausche 1943). Eine besondere Apparatur für die Vermessung und die statistische Auswertung elektronenmikroskopischer Aufnahmen ist von Hanson und Daniel 1947 entwickelt worden.

In Tab. 6 sind statistische Meßergebnisse an Bakterien und Virusarten nach v. Borries (1949) chronologisch zusammengestellt.

Tab. 6. *Ergebnisse der statistischen Auswertung von übermikroskopischen Aufnahmen verschiedener Objekte.*

Die Werte für H, M, σ, $\frac{\sigma}{M}$, $\frac{O}{V}$ und $\frac{O}{G}$ sind, soweit sie nicht in den betreffenden Arbeiten schon angegeben waren, aus den publizierten Häufigkeitskurven berechnet. In den Zeilen 11, 14, 17, 20, 23 und 26 stehen die Werte für M und σ nicht in mμ, sondern als reine Zahlen.

1	2	3	4	5	6	7	8	9	10	11
Nr.	Objekt	Teilchengestalt	Zahl der gem. Teilchen	Häufigster Wert H	Arithmet. Mittelwert	Mittlere (quadrat.) Streuung	Variabilität	Relative Oberfläche	Oberflächenentwicklung	Autor
			n	H	M	σ	$\frac{\sigma}{M}$	$\frac{O}{V}$	$\frac{O}{G}$	
			1	mμ	mμ	mμ	1	$\frac{1}{\text{m}\mu}$	$\frac{\text{m}^2}{\text{g}}$	
1	Goldkolloid ...	kugelig	969	30	28,5	6,9	0,24	191	10	v. Borries u. Kausche 1940
2	Goldkolloidgemisch......	,,	670	16	18,1	8,7	0,48	207	11	
3	kl. Partner von 2........	,,	609	16	15,9	4,8	0,31	326	17	
4	gr. Partner von 2	,,	61	40	39,5	9,7	0,24	136	7	
5	Tabak Mosaikvirus	Stäbchen	81	188	188	45,7	0,24	277	208	Trurnit u. Friedrich-Freksa 1940
6	Tomaten Mosaikvirus ..	,,	116	138	141	49,1	0,35	280	210	
7	Tabak Mosaikvirus ..	,,	58	280	(232)	(71,3)	(0,31)	275	207	Stanley u. Anderson 1941
8	Tabak-Mosaikvirus ..	,,	310	320	292	90,9	0,31	273	205	Kausche, Pfankuch u. H.Ruska, 1941

Fortsetzung auf Seite 301

Fortsetzung von Seite 300

1	2	3	4	5	6	7	8	9	10	11
Nr.	Objekt	Teilchengestalt	Zahl der gem. Teilchen	Häufigster Wert H	Arithmet. Mittelwert	Mittlere (quadrat.) Streuung	Variabilität	Relative Oberfläche	Oberflächenentwicklung	Autor
			n	H	M	σ	$\dfrac{\sigma}{M}$	$\dfrac{O}{V}$	$\dfrac{O}{G}$	
			1	$m\mu$	$m\mu$	$m\mu$	1	$\dfrac{1}{m\mu}$	$\dfrac{m^2}{g}$	
9	Molluscum contagiosum — Länge	Quader	150	H = M	255	61,3	0,24	28	21	
10	Breite	„		„	178	41,8	0,18			
11	Verh.	„		„	1,45	0,143	0,10			
12	Ektromelie — Länge	„	90	„	232	24,6	0,11	32	24	
13	Breite	„		„	172	9,6	0,06			
14	Verh.	„		„	1,34	0,024	0,02			
15	Kaninchen- Länge	„	81	„	287	20,7	0,07	24	18	
16	chen- Breite	„		„	233	19,0	0,08			
17	Myxom Verh.	„		„	1,22	0,122	0,10			H. Ruska u. Kausche 1943
18	Kanarien- Länge	„	75	„	311	26,1	0,08	22	17	
19	rien- Breite	„		„	263	22,7	0,09			
20	Pocken Verh.	„		„	1,14	0,108	0,09			
21	Vaccine — Länge	„	70	„	262	28,2	0,11	26	20	
22	cine Breite	„		„	209	23,9	0,11			
23	Verh.	„		„	1,25	0,124	0,10			
24	Staphylokokken — Länge	Rotationsellipsoid	320	„	608	120	0,20	11	8	
25	lokok- Breite			„	500	73,4	0,15			
26	ken Verh.			„	1,22	0,137	0,11			
27	Varicellen	kugelig	140	H ≈ M	143,5	26,2	0,18			H. Ruska 1943
28		„	130	„	144	16,5	0,11			
29	Bazillus Megatherium — Länge		32	5000	4400	1100	0,25	3,3	2,5	Dubin u. Sharp 1944
30	therium Breite		42	1500	1400	154	0,11			
31	Influenza-V.A.	kugelig	197	H = M	101	16,6	0,16	57	43	Sharp, Taylor usw., 1944
32	Influenza-V.B.	„	198	„	123	11,7	0,10	48	36	
33	Schweineinfluenza.....	„	574	„	96,5	11,7	0,12	61	46	
34	Tabak-Mosaik-virus — einf.	Stäbchen	765	H = M	271	51,9	0,19	274	206	Oster u. Stanley 1946
35	virus dopp.	„	35	„	552	52,4	0,10	271	203	
36	Tabak-Mosaik-virus — einf.	„	1278	„	288	52,3	0,18	274	206	
37	virus dopp.	„	60	„	596	35,2	0,06	270	203	Sigurgeirsson u. Stanley 1947
38	frisch dreif.	„	11	„	893	62,5	0,07	269	202	
39	Tabak-Mosaik-virus — einf.	„	669	„	290	55	0,19	274	206	
40	virus dopp.	„	142	„	603	48	0,08	270	203	
41	1 Tag alt dreif.	„	15	„	906	41	0,05	269	202	
42	Polyedervir. — Länge	„	150	340	350	23	0,07	50	44	Bergold 1947
43	Seidenraupe — Dicke	„	150	85	88	11	0,13			
44	Polyedervir. — Länge	„	138	430	415	81	0,20	29	26	
45	Schwammspinner — Dicke	„	138	170	160	39	0,25			

Weitere Statistiken bei Pfankuch und H. Ruska (1947), H. Ruska (1948/20), Dawson und Elford (1949/2).

4. Die Bestimmung der Teilchenzahl in dispersen Systemen.

Für die Teilchenzahlbestimmung muß zu einer auf der Aufnahme feststellbaren Partikelzahl das Flüssigkeits- oder Gasvolumen bekannt sein, aus dem die gezählten Partikel stammen. Die lichtmikroskopische Bestimmung von Teilchenzahlen erfolgt in Zählkammern definierter Höhe über einem in der Objektebene liegenden Zählnetz. Durch die Netzabstände und die Höhe der Zählkammer ist das Volumen, aus dem die Teilchenzählung erfolgt, gegeben. Dieser Weg ist in der Elektronenmikroskopie nicht gangbar. In der Objektebene (auf dem Trägerfilm) lassen sich zwar mit Abdruckfilmen Zählnetze anbringen, doch fehlt die Möglichkeit, eine definierte Schichthöhe so eintrocknen zu lassen, daß die Teilchenverteilung gleichmäßig wird. Die Untersuchung innerhalb einer Flüssigkeitsschicht definierter Dicke ist aber wegen des Vakuums, der Durchdringungsgrenze und der Brownschen Molekularbewegung nicht möglich. Man muß daher die Teilchen zur Zählung auf dem Objektfilm auftrocknen und das Volumen, aus dem die Teilchen stammen, auf andere Weise bestimmen.

a) Bestimmung aus Filmen mit eingebetteten Objekten.

Es wäre möglich, über die Herstellung von Objektträgerfilmen mit *eingebetteten* Objekten eine Teilchenzählmethode zu entwickeln. Hierzu muß nur das Volumen m cm³ des Tropfens, der die Teilchen enthält und aus dem der Film hergestellt ist, sowie der Durchmesser D cm des Films bekannt sein, von dem ein Teilstück untersucht wird. Unter der Annahme gleichmäßiger Filmdicke und Teilchenverteilung wäre dann bei z Teilchen pro 1 μ^2 im Bild die Teilchenzahl Z in 1 cm³

$$ Z = z \, \frac{\pi \, D^2}{4 \, m} \cdot 10^8 . $$

Dieser Weg ist trotz seiner Einfachheit bisher nicht beschritten worden. Er wäre für die Viruszählung nur durch die Schwierigkeiten beschränkt, die in der Herstellung von Filmen mit eingebetteten Objekten aus wäßriger Lösung liegen.

b) Bestimmung aus mittlerem Teilchengewicht und Teilchenkonzentration.

Ein zweites Verfahren, die Teilchenzahl im Volumen zu bestimmen, bedient sich der *Größenverteilungskurve*. Mit ihrer Hilfe wird das mittlere Teilchengewicht G bestimmt, woraus sich bei *bekannter Teilchenkonzentration* K die Teilchenzahl $Z = \dfrac{K}{G}$ ermitteln läßt. Riedel und H. Ruska (1941) fanden mit dieser Methode dieselben Werte wie mit den von ihnen entwickelten Verfahren der Teilchenbestimmung aus Suspensionen *unbekannter* Teilchenkonzentrationen. Am einfachsten läßt sich die Bestimmung bei isodiametrischen Teilchen durchführen, dann bei Plättchen und Stäbchen gleicher Dicke. Sind die Teilchen in allen drei Richtungen anisodiametrisch, so sind mühsame stereoskopische Messungen notwendig.

c) Bestimmung über die aerodisperse Zerteilung einer Suspension.

Um eine unmittelbare Zählung aus Suspensionen *unbekannter Konzentration* zu ermöglichen, haben Riedel und H. Ruska (1941), sowie Haardick, Kausche und H. Ruska (1944) aus kolloiden Lösungen Tröpfchen erzeugt und auf den Objektträgerfilm aufgebracht, die kleiner waren als dessen freier Durchmesser.

Man kann dann aus der Zahl der Teilchen in den aufgetrockneten Tröpfchen unter Voraussetzungen, die noch besprochen werden, auf die Zahl der Teilchen im Sol schließen.

Die Verneblung der Suspensionen und die Niederschlagung des Nebels im elektrischen Kernfäller ist im Abschnitt IV, 3 b beschrieben. Die *Tröpfchen* trocknen meist als kreisrunde *Flecken* auf dem Trägerfilm auf. Für die Bestimmung der Teilchenzahl ist es notwendig, die Beziehung zwischen der ursprünglichen Tröpfchengröße und dem Fleckdurchmesser zu kennen. Je nach dem Dampfdruck des Lösungsmittels, der Temperatur und Feuchte des zurückzulegenden Luftweges und der zwischen Erzeugung und Niederschlagung der Tröpfchen vergehenden Zeit werden die Tröpfchen in verschiedenem Ausmaß kleiner. Wegen der stark gekrümmten Oberfläche der kleinsten Tröpfchen ist der Dampfdruck des Lösungsmittels etwas höher als normal (bei 10 mμ großen Tröpfchen 10 %). Man sorgt deshalb dafür, daß sich die Tröpfchen bis zum abscheidenden elektrischen Feld in einer mit Dampf des Lösungsmittels nahezu gesättigten Atmosphäre befinden und daß die Schwebezeit recht kurz ist.

Man kann ferner, um die Versuchsbedingungen besser zu gestalten, künstlich den Dampfdruck des Lösungs-

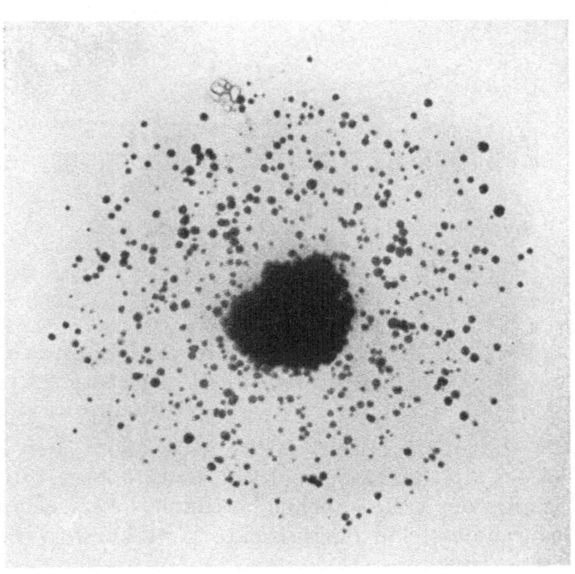

Abb. 75. 1 %iges Goldsol ohne Salzzusatz vernebelt, 64 000 : 1. Nach RIEDEL und H. RUSKA (1941).

mittels durch Beigabe bestimmter Substanzen erniedrigen. NaCl, MgCl, LiCl und Co(NO$_3$)$_2$ eignen sich wegen ihrer hygroskopischen Eigenschaften dafür besonders gut. Man erreicht mit dem Salzzusatz, daß die Tröpfchengröße innerhalb des Versuchsfeldes, wie die weiter unten gegebene Darstellung zeigt, wenigstens annähernd erhalten bleibt. Ein Vergleich der Abb. 75 und 57, die aufgetrocknete Tröpfchen eines 1 %igen Goldsols ohne Salz und mit LiCl + NaCl zeigen, läßt den Einfluß der Salze sofort erkennen. Die ohne Salz aufgetrockneten Flecke zeigen sehr viel mehr Gold pro Flächeneinheit als die mit Salz aufgetrockneten. Das Salz verhindert außerdem eine Zusammenballung des Goldes in der Mitte des aufgetrockneten Tröpfchens. Man gewinnt den Eindruck, daß in salzhaltigen, kleinsten Tröpfchen die Kolloidteilchen infolge der gegenseitigen elektrostatischen Abstoßung bevorzugt auf der Tröpfchenoberfläche liegen, und zwar besonders dann, wenn es nur sehr wenige sind.

Da bei der Bestimmung des Verhältnisses zwischen Tröpfchengröße und Fleckgröße die absolute Tröpfchengröße wegen der verschieden starken Abdampfung ungleich großer Tröpfchen berücksichtigt werden muß, haben RIEDEL und H. RUSKA für verschiedene Fleckdurchmesser die gesuchte Abhängigkeit bestimmt. Außerdem wurden verschiedene Salzzusätze berücksichtigt. Praktisch wird dabei so vorgegangen, daß ein seiner Konzentration nach bekanntes Goldsol ohne Salz und dann mit Salzzusätzen zerstäubt und niedergeschlagen wird. Aus der Menge des Goldsols in einem bestimmten Fleck läßt sich dann jeweils die zugehörige Tröpfchengröße errechnen, da das Verhältnis des Goldvolumens zum Tröpfchenvolumen bekannt ist.

Zur Aufstellung der Kurven von Abb. 76, die das Verhältnis der Tröpfchengröße zur Fleckgröße in Abhängigkeit vom Fleckdurchmesser angeben, wurden für jeden Meßpunkt aus Flecken der angegebenen Größe die Teilchenvolumina und daraus die Tröpfchenvolumina des 1%igen Sols bestimmt. Sind solche Kurven gewonnen, so läßt sich, wenn die gleichen Versuchsbedingungen streng eingehalten werden, die Teilchenzahl eines seiner Konzentration nach unbekannten Sols dadurch bestimmen, daß man die in einem Fleck bestimmter Größe aufgetrocknete Teilchenzahl auf das der Kurve zugehörige Tröpfchenvolumen bezieht.

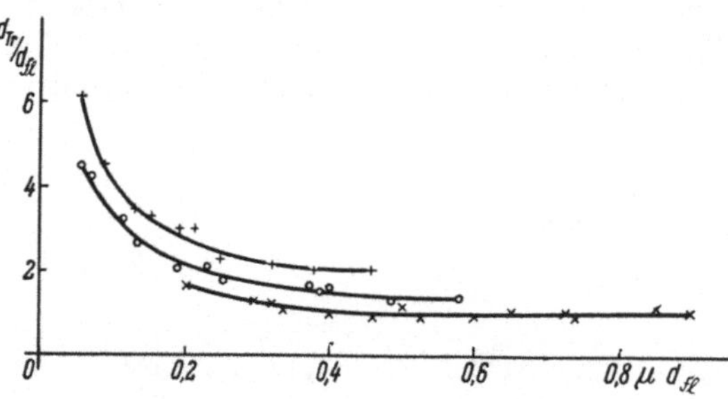

Abb. 76. Abhängigkeit des Quotienten des Tröpfchendurchmessers (d_{Tr}) zum Fleckdurchmesser (d_{Fl}) von der Größe des Fleckdurchmessers.

+ ——— + = salzfreies Sol,
○ ——— ○ = Sol mit 1% NaCl,
✕ ——— ✕ = Sol mit 1% NaCl + 1% LiCl.

α) Die Bestimmung der Teilchenzahl einer Suspension unter Zuhilfenahme der Beziehung zwischen Fleckgröße und Tröpfchengröße.

Nach Messung der Fleckgröße wird aus der zugehörigen Kurve, Abb. 76, die Tröpfchengröße und für dieses abgegrenzte Volumen die Teilchenzahl oder Länge der Virusstäbchen bestimmt. Wenn sich beim Reißen des Films der Fleckdurchmesser in einer Richtung verkürzt, wird mit dem größten Durchmesser gerechnet. Die gesuchte Zahl der Teilchen pro Volumeneinheit erhält man durch Umrechnung der Werte aus einem oder mehreren beliebigen Tröpfchen auf die Volumeneinheit. Wie Abb. 77 zeigt, liegen die Zählergebnisse verschiedener Einzeltröpfchen sehr gut beieinander. Jedes Tröpfchen allein gibt schon die richtige Größenordnung der Zahl der Goldteilchen im Goldsol an, und nur 10 bis 20 müssen ausgezählt werden, um sich dem wahren Wert auf

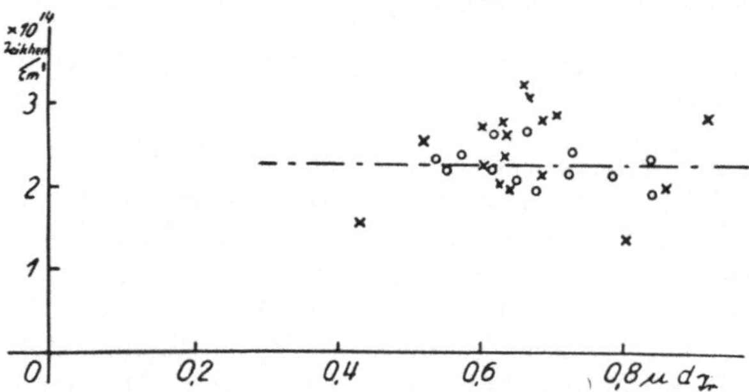

Abb. 77. Teilchenzahlen/ccm aus Flecken verschiedener Durchmesser, ermittelt über die Beziehung zwischen Fleckdurchmesser und Tröpfchendurchmesser (✕) und über den Mischkristall des NaCl + LiCl-Zusatzes (○).
— · — · — · — Teilchenzahl, ermittelt aus der Größenverteilung der Goldteilchen eines Sols bekannter Konzentration.

wenige Prozente zu nähern. Als Beispiel sind die Ergebnisse mit einer 1%igen Goldsollösung dargestellt, deren Teilchenzahl aus der Größenverteilung über die mittlere Teilchengröße zu $2{,}3 \cdot 10^{14}$ pro Kubikzentimeter bestimmt war. Die Methode ist sehr genau und zeigt die Zuverlässigkeit der Berechnung der Teilchenzahl über das Verhältnis von Fleckdurchmesser zu Tröpfchengröße. Gute Werte erreicht man besonders, wenn das stark hygroskopische Lithiumchlorid als verdampfungsvermindernder Zusatz verwendet wird.

β) **Die Bestimmung der Teilchenzahl einer Suspension unter Verwendung eines vermeßbar kristallisierenden Salzzusatzes.**

Da bei der Verwendung von LiCl zur Herabsetzung des Dampfdruckes der Tröpfchen keine vermeßbare Kristallisation des LiCl eintritt, muß für die vorliegende Aufgabe außer LiCl noch NaCl dem Goldsol zugesetzt oder nur NaCl geeigneter Konzentration verwendet werden. Es bilden sich dann Einzelkristalle mit quadratischen Grundflächen (Abb. 57), deren Seitenlängen nach Abb. 58 ein konstantes Verhältnis zu den Tröpfchendurchmessern haben. Der Tröpfchendurchmesser wurde dabei von Riedel und H. Ruska nicht aus den Kurven von Abb. 76 über den Fleckdurchmesser abgelesen, sondern aus dem Volumen der Goldteilchen über eine bekannte Goldkonzentration und das Tröpfchenvolumen errechnet.

Es bietet sich also die Möglichkeit, den Tröpfchendurchmesser aus der Seitenlänge der Grundfläche des Kristalls ohne Kenntnis der Abdampfungsgröße zu erhalten. Die Teilchenzahl der Ausgangslösung pro Volumeneinheit wird dann bestimmt, indem man aus einem beliebigen Fleck die

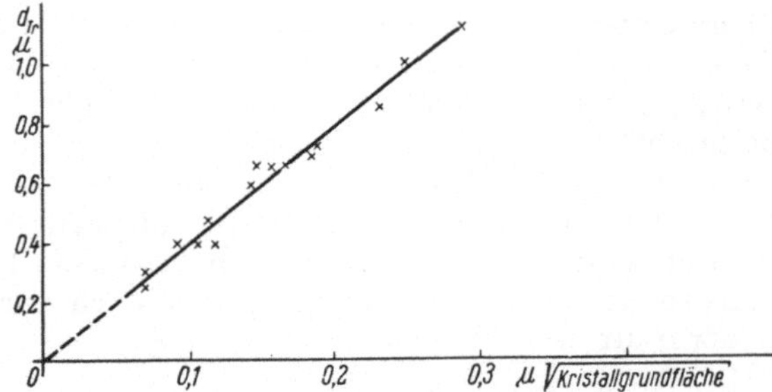

Abb. 78. Beziehung der Seitenlänge der Grundfläche des LiCl+NaCl-Mischkristalls zum Tröpfchendurchmesser.

Seitenlänge der Grundfläche des Mischkristalls ausmißt, in Abb. 78 den zugehörigen Tröpfchendurchmesser abliest und über das entsprechende Tröpfchenvolumen die Zahl der Teilchen des Flecks auf die Volumeneinheit bezieht.

Riedel und H. Ruska kamen auf diese Weise bei ihrer Versuchslösung wiederum zu der Zahl von 2,3 . 10^{14} Teilchen in 1 cm³. Wie die Einzelwerte, die in Abb. 77 ebenfalls eingetragen sind, zeigen, erhält man bei der Verwendung der Beziehung zwischen Kristallgröße und Tröpfchengröße noch weniger streuende Werte als unter Zuhilfenahme der Fleckgröße.

Ist die Teilchenzahl zur Auszählung zu hoch oder zu niedrig oder sind Salze oder andere Fremdsubstanzen im Sol vorhanden, so wird das Sol um einen bekannten Betrag eingedampft oder verdünnt, falls erforderlich und möglich, auch dialysiert und mit einem geeigneten Salzzusatz versehen. Bei jeder Suspension, deren Teilchen gezählt werden sollen, besteht zunächst die Aufgabe, für das Zählverfahren die geeigneten Versuchsbedingungen zu finden. Haardick, Kausche und H. Ruska (1944) haben in etwa 1%igen Tabak Mosaikvirus-Lösungen mit Zusatz von Co $(NO_3)_2$ die Virusstäbchen gezählt. Wie aus dem Beispiel mit Goldsol zu ersehen ist, müssen die Teilchenzahlen sehr hoch sein (10^{12} bis 10^{14} pro Kubikzentimeter), andernfalls erhält man „leere" Flecke. Bei der Zählung von Teilchen, die im Trägerfilm eingebettet sind, benutzt man nahezu die gleichen Konzentrationen. Sharp hat neuerdings (1949) eine Zählmethode entwickelt, bei der Virusteilchen in der Zelle einer Zentrifuge auf eine mit Kollodium überzogene Glasplatte geschleudert werden. Die Teilchen werden dann auf dem Kollodiumfilm im Elektronenmikroskop ausgezählt. Sie zeigen gleichmäßige Verteilung, und ihre Zahl stimmt gut mit indirekten Messungen überein. Ein Vorteil der neuen Methode ist darin zu erblicken, daß ganz wesentlich geringere Viruskonzentrationen verwendet werden können. Denn während bei der oben beschriebenen Methode Schichtdicken von der Größenordnung 1 μ auftrocknen, werden bei Sharp Schichtdicken von der Größenordnung 1 cm sedimentiert.

d) Bestimmung der Teilchenzahl in einem Gasvolumen.

Es kann auch die Aufgabe bestehen, nicht die Teilchen im Nebeltropfen, sondern die Nebeltröpfchen oder feste Teilchen (Bakterien, Virus) in einem bestimmten Gasvolumen zu zählen. Man bedient sich dazu der in Abschnitt IV, 3 b (Abb. 58) beschriebenen Anordnung und leitet den Gasstrom so langsam durch den Kondensator, daß praktisch alle Teilchen ausgeschleudert werden, d. h. daß auf der letzten Objektblende auf dem Innenzylinder des elektrischen Kernfällers keine Teilchen mehr nachweisbar sind. Das Volumen des durchgeleiteten Gases wird durch den Rotamesser bestimmt. Die niedergeschlagenen Teilchen werden auf allen Blenden, die am besten gleiche Bohrungsdurchmesser haben, ausgezählt und über der Oberfläche des Innenzylinders integriert. Daraus ergeben sich die Zahlen der positiv oder negativ geladenen Teilchen oder beide, falls mit zwei parallel geschalteten, entgegengesetzt geladenen Kernfällern gearbeitet wird. Das Verfahren kann besonders für Staubteilchenzählungen und meteorologische Untersuchungen, aber auch zur Bakterien- oder Viruszählung in der Luft wichtig sein.

VI. Ergebnisse der Virusforschung[1].

Bei einer Darstellung der elektronenmikroskopischen Morphologie der Virusarten ist eine Anordnung des Stoffes zu erstreben, bei der die infektiösen Elemente unabhängig von ihrem Tropismus und ihrer Pathogenität, ja auch unabhängig von der Frage des obligaten Zellparasitismus gruppiert werden. Neben der Größe des Virus ist in stärkerem Maße, als es früher möglich war, der Bau und die damit aufs engste verbundene chemische Zusammensetzung zu berücksichtigen.

Schon frühzeitig ließen sich mit Hilfe der elektronenmikroskopischen Abbildung gewisse Virusgruppen abgrenzen und in den Gruppen mehrere Virusarten zusammenfassen, die trotz verschiedener Tropismen und verschiedener Wirtsorganismen einander morphologisch nahestehen (H. Ruska 1943/15). Eine befriedigende Ordnung *aller* Virusarten und die Aufstellung eines natürlichen, auf phylogenetischen Beziehungen beruhenden Systems ist aber nicht durchführbar. Einerseits fehlt noch die morphologische Kenntnis sehr zahlreicher Virusformen, und andererseits sind wir über die Herkunft der weitaus meisten völlig im unklaren. Sicher ist nur, daß entsprechend der Heterogenität des Virusbegriffes die darunter zusammengefaßten infektiösen Agentien ganz verschiedenen Ursprung haben. So lassen sich die Rickettsien morphologisch von den Bakterien ableiten, einige große Virusarten dagegen von filtrierbaren Mikroorganismen. Für andere Formen werden Beziehungen zu autoreproduktiven Strukturen der Wirtszelle, wie Mitochondrien und Genen, diskutiert. Nicht ausgeschlossen erscheint es, daß die alte Vorstellung Guarnieris über Bezie-

---

[1] Dem Verfasser sind wahrscheinlich nicht alle elektronenmikroskopischen Virusarbeiten bekannt geworden. Auch von den zitierten Arbeiten konnten trotz der freundlichen Unterstützung zahlreicher in- und ausländischer Forscher einige nicht im Original eingesehen werden, wodurch Lücken in der Darstellung entstanden sein können. Die regelmäßige Überlassung von Sonderdrucken würde einem solchen Mangel für spätere Beiträge abhelfen.   *H. Ruska, Berlin-Dahlem, Faradayweg 16*

hungen zwischen Virus und Protozoen erneut geprüft werden muß, wenn umfassende elektronenmikroskopische Untersuchungen der verschiedenartigen Einschlußkörper vorliegen.

Die folgende Darstellung beginnt daher mit einigen Hinweisen auf Untersuchungen an Protozoen, Spirochaeten und Leptospiren. Ihr schließt sich eine Übersicht bakteriologischer Ergebnisse an, da die Virusforschung in mancher Hinsicht wieder näher an die Bakteriologie herangerückt ist, als es zur Zeit der Entdeckung der ersten makromolekularen Virusarten der Fall war. DOERR schrieb (1944 dieses Hb. 1, Erg. Bd.), „daß die Vorstellung der Viruseinheit als lebender Elementarorganismus, welcher durch chemisch-physikalische Methoden der Boden entzogen wurde, durch das Tor der Elektronenoptik bzw. der morphologischen Methodik wieder ihren Einzug gehalten hat." Seither sind nicht nur zahlreiche morphologische, sondern auch chemische Tatsachen erhoben worden, die BEARD (1945) und andere Autoren veranlaßten, sehr deutlich von einer Verallgemeinerung der Molekülhypothese abzurücken. Die Feststellung, daß bei den meisten tierpathogenen Virusarten und bei den Bakteriophagen keine Makromoleküle als Träger der Infektiösität vorliegen, stimmt aber noch nicht mit einem Nachweis lebender Organismen überein. Der Begriff „lebend" ist ohnehin zu umfassend, um naturwissenschaftlich alle die zahlreichen Abstufungen des Lebens, die schon in den Zellorganellen und in den Interzellularsubstanzen vorkommen, mit dem gleichen Wort zu bezeichnen. Bei den „großen Virusarten" kann auf Grund ihrer prinzipiellen morphologischen Übereinstimmung mit gewissen filtrierbaren Mikroben von „lebenden Organismen" im üblichen Sinne gesprochen werden. Ihre Darstellung folgt derjenigen der Bakterien. Daran schließen sich zusammengehörige Gruppen und einzelne warmblüterpathogene Virusarten in absteigender Größenskala. Ihnen folgen die Polyederviren der Raupen und schließlich die phytopathogenen Arten, die nicht nur nach den Wirtsorganismen, sondern auch in anderen Beziehungen eigene Gruppen darstellen.

Auch der umgekehrte Weg einer aufsteigenden Darstellung von den makromolekularen zu den zellulären Formen ist gangbar und früher (1.c.) beschritten worden. Keine der beiden Reihenfolgen erhebt aber den Anspruch, eine Bedeutung in phylogenetischer Hinsicht zu besitzen. Weder ist generell eine Entdifferenzierung von zelligen Parasiten bis zum Makromolekül erwiesen, noch kann von parasitierenden Makromolekülen aus eine geschlossene, aufsteigende Entwicklungslinie zur Zelle gefunden werden. Jede Darstellung geht von bekannten, gut analysierten Formen aus. Die Bakterien sind morphologisch und physiologisch, die makromolekularen Virusarten vor allem chemisch und strukturell der Erforschung bis ins feinste zugänglich geworden. Dazwischen gruppieren sich die methodisch schwerer erfaßbaren Virusarten, deren genauere Analyse erst in den letzten Jahren begonnen hat. Vom Tollwutvirus, dessen Züchtung VEERARAGHAVEN (1947) in zellfreiem Nährmedium gelungen zu sein scheint und das von ihm deshalb nicht mehr zum Virus im engeren Sinne gerechnet wird, sind dem Verfasser keine elektronenmikroskopischen Bilder bekannt geworden.

Die Erkennung einzelner Viruspartikel ließ die Frage entstehen, welche Bedeutung der Beobachtung zukommt, daß manche Virusarten eine sehr einheitliche Teilchengröße besitzen, anderen dagegen eine stark streuende (WYCKOFF 1946) zukommt. Das Problem ist einerseits mit der molekularen oder organismischen Natur der Teilchen, andererseits mit ihrem Duplikationsmechanismus verknüpft. Außer der Morphologie, der Einschlußkörperentwicklung und dem Wirtszellenaufbau kann die Elektronenmikroskopie die Reaktionsweise des Virus mit Antikörpern und anderen Kolloiden aufklären.

Seit den ersten Anfängen der elektronenmikroskopischen Virusforschung ist, abgesehen von Handbuchbeiträgen, eine Reihe von kürzeren Übersichten der Ergebnisse erschienen. Außer den bereits erwähnten Arbeiten seien noch folgende genannt:

H. Ruska, v. Borries und E. Ruska (1939), G. A. Kausche (1940, 1941), H. Ruska (1942, 1943), Anderson (1942), Stanley (1943,1947), Daneel (1944), Mudd und Anderson (1944), Mudd (1944), Stanley, Knight und Merre (1945).

1. Protozoen.

Die meisten einzelligen, tierischen Organismen sind zu dick, um sie mit den bisher angewandten Beschleunigungsspannungen der Elektronen nach dem Durchstrahlungsverfahren abbilden zu können. Es bleiben daher ihre Zellbestandteile und Innenstrukturen verborgen. Die Lichtmikroskopie bietet dagegen so gute Untersuchungsmöglichkeiten, daß zunächst wenig Anreiz zu elektronenmikroskopischen Forschungen bestand. Jedoch ist zu erwarten, daß mit der Fortentwicklung der präparativen Technik, z. B. durch fraktionierte Aufschließung der Zellen, die elektronenmikroskopische Untersuchung von Protozoen einsetzen wird, weil diese zum Studium von Protoplasmastrukturen besonders hochdifferenziertes Material liefern. Leicht zugänglich ist die Struktur der Bewegungsorganellen und der Trichocysten. H. Ruska (1939/2) hat eine feine, längsparallele Streifung an den Cilien von Colpoda duodenaria gefunden, und Jakus (1946) entdeckte die Querstreifung des Trichocystenschafts der Paramaecien. Sie hat im Hinblick auf die Querstreifung anderer Faserproteine für Fragen der Eiweißstruktur, besonders aber auch für die Erklärung des Streckungsmechanismus des Trichocystenschafts große Bedeutung.

Von den pathogenen Protozoen kennen wir durch Wolpers (1942/2) heranwachsende Schizonten von Plasmodium vivax im haemolysierten roten Blutkörperchen des Menschen. Zwar kann man weder den Kern noch das Pigment des Parasiten abgrenzen, erkennt aber deutlich, wie er von der Erythrocytenmembran eingehüllt wird und mit seinen Protoplasmaausläufern zum Teil bis zur Membran vorstößt und an ihr festhaftet. Bei manchen Formen sieht man eine große Vakuole und die Schüffnersche Tüpfelung. Emmel, Jakob und Gölz (1942) untersuchten Sporozoite von Plasmodium vivax und falciparum. Es fiel ihnen die sehr ungleiche Größe auf und die Unterscheidbarkeit eines schlanken und eines stumpfen Endes der mitunter deutlich längsgestreiften Sporozoite; die Längsstreifung führten sie auf kontraktile Elemente zurück. Nach Protoplasmaaustritt verbleibt ein kontrastarmer Periplast. Auf einer Flagellatenaufnahme von Emmel, Jakob und Gölz (Leishmania donovani) sind außer den dicken, stumpf endenden Geißeln noch Dichteunterschiede des Protoplasmas (Vakuolen?) zu sehen. Ähnliches beobachteten Wolpers und H. Ruska (unveröffentlicht) an Trypanosomen aus Rattenblut.

2. Spirochaeten und Leptospiren.

Die Morphologie der Spirochaeten birgt noch manche Unklarheiten. Durch die Elektronenmikroskopie wurden zwar einige neue Befunde gewonnen, aber die wesentlichen Fragen der Entwicklung, sowie des Vorkommens und der klinischen Bedeutung filtrierbarer Formen (Menze 1940) sind noch nicht gelöst. Ähnliches gilt von den Leptospiren. Viele der bekannt gewordenen Spirochaetenaufnahmen zeigen ungenügende Reinheit der Präparate und Objektschädigungen durch die Präparation, vor allem abnorme Streckung der Spiralform. Morton und ANDERSON (1942) haben Treponema pallidum aus Kulturstämmen, WILE, PICARD und KEARNY (1942) von syphilitischen Organen untersucht. Die Dichte des Spirochaetenkörpers ist oft ungleichmäßig. Einzelne Körnchen sind in ähnlicher Weise dicht und scharf begrenzt wie die Volutinkörnchen der Bakterien, doch scheint nicht bekannt, ob sie wie diese durch starke Elektronenbestrahlung zerstört werden. Ein zentraler Faden, der dem Plasmaleib die typische Form geben könnte, ist nicht sichtbar und wohl auch nicht vorhanden (MUDD, POLEVITZKY und ANDERSON 1942). Zarte plasmatische Brücken und lange einseitige Terminalfäden deuten nach den genannten Autoren und nach MAGERSTEDT (1944) auf eine vorausgegangene Querteilung hin. Weder LEVADITI (1943) noch MAGERSTEDT (1944) haben Terminalfäden an Treponema pallidum gesehen, wenn die Spirochaeten aus syphilitischen Organen oder Primäraffekten gewonnen waren; auch fehlen sie offenbar bei Treponema microdentium. Häufig wurden ohne erkennbare Gesetzmäßigkeit angeordnete, seitenständige Geißeln beobachtet, von denen mitunter mehrere an der gleichen Stelle, seitlich oder auch endständig zu entspringen scheinen (vgl. auch die Abb. bei ZWORYKIN und HILLIER 1944, O'BRIEN 1944, HILLIER und KURKIJAN 1944).

Die äußere Begrenzung des Spirochaetenkörpers soll nach LEVADITI aus Ektoplasma, nach MAGERSTEDT aus einer Hülle gebildet sein, die wahrscheinlich als Membran aufzufassen ist. Eindeutig geben MUDD, POLEVITZKY und ANDERSON an. daß die Spirochaeten eine Zellwand (Membran) besitzen. Nach JAKOB 1947 fehlt eine echte Membran. Undulierende Membrane. wie sie für manche Protozoen kennzeichnend sind, wurden elektronenmikroskopisch nur von BADUDIERI und BOCCIARELLI beobachtet.

Auf zahlreichen Spirochaetenabbildungen finden sich dichte, unregelmäßige, rundliche Körperchen am Ende oder in der Mitte der Zelle. Es sind die gleichen Gebilde, die schon im Lichtmikroskop gesehen und als Knospen, sporenähnliche Körperchen oder spirochaetogene Körnchen bezeichnet wurden. MUDD und Mitarbeiter vergleichen sie mit den reproduktiven Körnchen der Pleuropneumonie-Gruppe und schreiben ihnen die — wenn im einzelnen auch noch unbekannte — Bedeutung für eine neben der Querteilung vorkommende, asexuelle Vermehrung zu. Da der Durchmesser der „Knospen" größer ist als die Dicke der Spirochaeten (bis 400 gegen 100 bis 140 mµ), lassen sich die gelegentlich frei neben den Zellen liegenden „Knospen" nicht durch Filtration abtrennen. Durch diesen mißlichen Umstand ist die biologische Bedeutung der „Knospen" trotz des Fortschritts der morphologischen Methodik noch nicht gesichert. HAMPP, SCOTT und WYCKOFF 1948 gelang es, bei Borrelia vincentii in extrem großen, bis 5 µ messenden „Körnchen" Spirochaetenbruchstücke und begeißelte Spirochaeten nachzuweisen. Nach diesen Autoren sind die „Körnchen" eine sichergestellte Phase in der Entwicklung der Spirochaeten. (Vgl. auch BABUDIERI und BOCCIARELLI 1943, 1948; LOFGREN und SOULE 1945; JAKOB 1947.)

Leptospirenuntersuchungen sind an L. ictero-haemorhagiae, canicola, bataviae, biflexa u. a. durchgeführt worden. Morton und Anderson (1943) fanden im Gegensatz zu den Befunden an Spirochaeten keine Geißeln und keine Innenstrukturen, wohl aber eine zarte Membran, die nach Behandlung mit 0,59 mol. Bleiacetat als dunkle Konturlinie erscheint. Die Kontur kommt entweder durch Anlagerung des Schwermetallsalzes an die Membran oder durch die Aufnahme in den Spaltraum zwischen Membran und Zellplasma zustande. JAKOB (1947) fand in 1 bis 2 Monate alten Kulturen von L. canicola neben normal gewundenen Leptospiren gestreckte Exemplare, die am Ende birnenförmige Anschwellungen (1,5 × 0,7 µ) zeigten. Er bezeichnet sie wegen des spermienähnlichen Aussehens als S-Formen. Die Anschwellungen finden sich häufig auch ohne anhängende

Leptospirenzelle. Sie zeigen verschiedenartige Innenstrukturen (Abb. 79), eine zarte Membran und mitunter eine außerordentlich feine Geißel. An geplatzten Knospen läßt sich eine eng aufgerollte, kleine Leptospire erkennen. Ihre Bedeutung als besondere Reproduktionsform der Leptospiren ist nicht nur morphologisch, sondern auch tierexperimentell sehr wahrscheinlich gemacht worden. Die S-Form scheint nicht bei allen Leptospiren vorzukommen, auch nicht bei

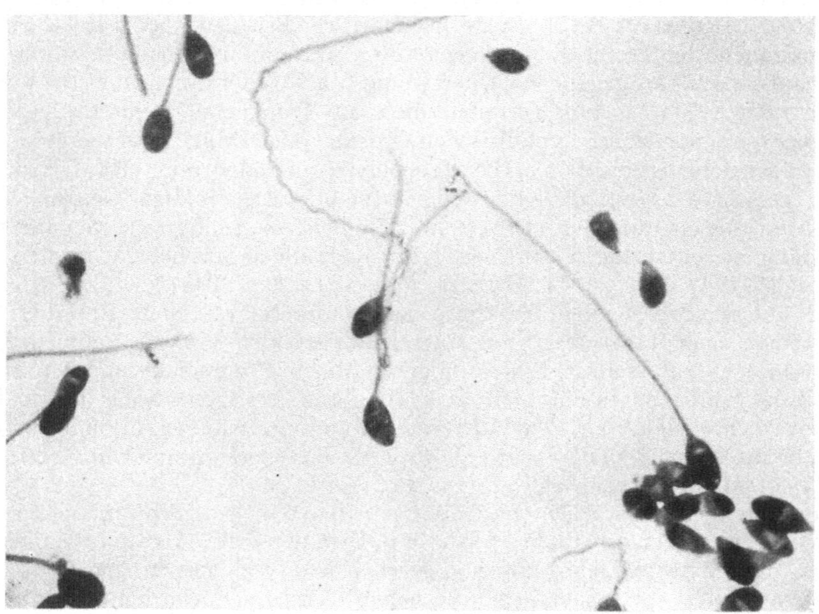

Abb. 79. Leptospira canicola, Stamm Schüffner-Gaethgens. 5 Wochen alte Kultur mit S-Formen und Normalformen, 6000 : 1. Nach Jakob (1947).

allen kultivierten Stämmen von L. canicola. Die Bedeutung der S-Formen der Leptospiren ist offensichtlich die gleiche wie die der Körnchen oder Knospen bei den Spirochaeten.

Neuerdings hat Jakob (1949) etwa 50 mμ große, runde Körperchen mit einem peitschenförmigen, etwa 1 mμ langen Fortsatz beschrieben und in Anlehnung an Gastinel und Mollinedo (1942) als „Leptospirogene" bezeichnet. Diese Leptospirogene (Abb. 80) konnten von ihm bei allen untersuchten Leptospiren (L. ictero-haemorrhagiae, grippotyphosa, pomona und canicola) als besondere Entwicklungsformen elektronenmikroskopisch nachgewiesen werden. Sie sind für die Biologie der Leptospiren sowie für die Klinik der Leptospirosen sicherlich von größter Bedeutung und zeigen, daß Fortpflanzungsformen, die kleiner sind als große Bakteriophagen, fähig sind, sich zu einem Leptospirenkörper zu entwickeln. Sollten ähnliche Gebilde bei Spirochaeten gefunden werden, so könnten diese „Spirochaetogene" eine entscheidende Stütze für die Auffassung von Menze (1940) abgeben, wonach ektodermale Luesmanifestationen durch filtrierbare, chemotherapeutisch bisher nicht angreifbare Formen hervorgerufen werden.

Auch Babudieri (1948) hat S-Formen abgebildet, hält es aber für fraglich, ob sie zur Entwicklung der Leptospiren gehören. Er beschreibt ferner Membran und Achsenfaden. Zu den S-Formen äußert er den Verdacht, daß es sich um Pilzfäden handeln könne. Ein solcher Verdacht kann, nachdem er einmal ausgesprochen ist, auch durch die Abb. 6 bei JAKOB (1947) nahegelegt werden, auf der man eine Verzweigung des S-Formenfadens sieht, sowie ein aus dem S-Formenkopf austretendes „Leptospirogen", das an ein „Spermatozoid" erinnert. Die aufgeworfene Frage bedarf offensichtlich weiterer Klärung.

VAN THIEL und VAN ITERSON (1947) stellen das Vorhandensein einer Membran bei Spirochaeten und Leptospiren in Abrede. Sie haben die JAKOBschen S-Formen

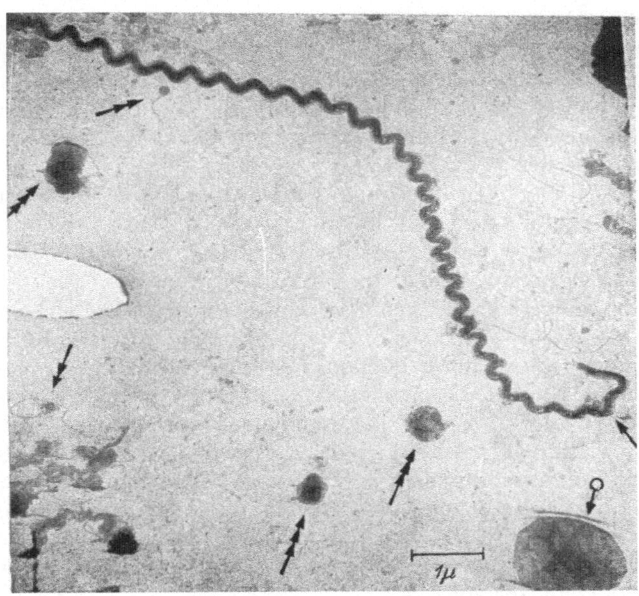

Abb. 80. Leptospira canicola Stamm M 468 Borg Petersen. 6 Monate alte Kultur mit sichtbarer Membran —→ an einer Normalform, Leptospirogene —→→, und größeren Körperchen —→→→, die als S-Formen gedeutet werden. Nach JAKOB (1949).

(bei L. biflexa) nicht beobachtet, wohl aber endständige Aufrollungen des Leptospirenkörpers, wie sie MUDD und Mitarbeiter bei T. pallidum gezeigt haben. In systematischer Hinsicht bezeichnen sie Leptospira ebenso wie Borrelia und Treponema als Subgenus der Spirochaeta.

3. Bakterien einschließlich Rickettsien.

Bisher fehlt trotz zahlreicher Einzeluntersuchungen eine lückenlose, systematische Kenntnis der elektronenmikroskopisch sichtbaren Morphologie aller pathogenen Bakterien. Die meisten elektronenmikroskopisch beobachteten Keime sind nicht aus dem erkrankten Organismus, sondern aus der künstlichen Kultur isoliert und untersucht worden. Für die Morphologie der Bakterienzellen ist dieses wahrscheinlich nicht ganz gleichgültig. Schon unter Laboratoriumsverhältnissen zeigen sich Zellveränderungen bei wechselnder Nährbodenbeschaffenheit und verschiedenem Kulturalter. Mit ebensolchen oder auch andersartigen Veränderungen müssen wir im Organismus rechnen, der je nach seiner Reaktion entweder eine gleichbleibende Nährbodenbeschaffenheit aufrecht erhält oder durch Entzündung und Immunitätsentwicklung sich auf andere Weise verändert als ein alternder künstlicher Nährboden.

Gleichgerichtete Einflüsse des Nährbodens machen sich bei verschiedenartigen Keimen auch in der gleichen Richtung morphologisch geltend. Sie können zu

morphologischen Unterschieden der Zellen der gleichen systematischen Art
führen, welche die Unterschiede gegenüber anderen, unter gleichen Bedingungen
gehaltenen Arten bei weitem übertreffen. Es ist deshalb die Darstellung einer
allgemein vergleichenden Morphologie von Membran, Kapsel, Schleimsubstanz,
Geißeln und Zellinhalt wichtiger als die Beschreibung der Besonderheiten einzelner
Keimarten. Schon hier sei vorweggenommen, daß unter der Zellmembran ein
flächenhaft ausgedehntes, beiderseits scharf begrenztes, die Zelle umhüllendes
Gebilde verstanden wird, das vom eingeschlossenen Inhalt chemisch und strukturell verschieden ist und sich leicht von der Unterlage ablösen läßt. Kapsel-
und Schleimsubstanzen liegen außerhalb der Zellmembran als Ausscheidungs-
produkte der Zellen. Der Zellinhalt ist innerhalb der Zellmembran von einer
besonderen Plasmahaut umhüllt, die strukturell und chemisch engste Beziehungen
zum Protoplasma besitzt.

Die alte, für die Virusforschung bedeutsame Frage nach dem Auftreten
filtrierbarer Formen von lichtmikroskopisch sichtbaren Keimen ist keinen
eingehenden Prüfungen unterzogen worden. Wohl aber läßt sich aus den vorliegenden Erfahrungen einerseits sagen, daß unter den vegetativen Keimen
gelegentlich besonders kleine Exemplare vorkommen, die in der Lage sein können,
ein Filter zu passieren, daß aber andererseits in Bakterienreinkulturen das Vorkommen von Formen, die typischen Viruselementarkörpern gleichen, bisher
nicht beobachtet wurde. Die engeren Beziehungen zwischen Virusforschung und
Bakteriologie, von welchen oben die Rede war, betreffen also nicht zusammen-
hängende Entwicklungszyklen von Bakterien und Virus, sondern die Gleichartigkeit allgemeiner Fragestellungen, wie etwa nach der Bauweise, nach der Lokalisation der für die Vererbungsvorgänge maßgebenden Substanzen, nach dem
Wachstums- und Teilungsmechanismus usw. Unmittelbarer sind die morphologischen Beziehungen zwischen Bakterien und Rickettsien, sowie bei den Erscheinungen der Bakteriophagie.

a) Zellmembran.

Alle vegetativen Bakterienzellen sind von einer zarten, etwa 10 mμ dicken
Membran eingeschlossen, für die sich im anglo-amerikanischen Schrifttum auch
der Ausdruck „Zellwand" findet. Es erübrigt sich daher eine Aufzählung einzelner
Keime, an welchen eine Membran nachgewiesen wurde. Die ersten elektronenmikroskopischen Beobachtungen von Bakterienmembranen sind durch Piekarski
und H. Ruska (1939/2) mitgeteilt worden. Genauere Untersuchungen, die zugleich eine scharfe Trennung der Begriffe Ektoplasma, Kapsel und Membran
zum Ziele hatten, führten Frühbrodt und H. Ruska (1940) durch. Wie unklar
die Vorstellungen bis dahin waren, geht aus der dort, sowie bei Jakob und Mahl
(1940) angeführten lichtmikroskopischen Literatur hervor. Auch mit Hilfe des
Elektronenmikroskops war die Beurteilung zunächst noch unsicher. Wenn die
Membran dem Zellinhalt dicht aufliegt, kann sie dem elektronenmikroskopischen
Nachweis entgehen. Häufig sieht man sie jedoch als schmalen, hellen Saum die
Zelle begrenzen, ein Zeichen dafür, daß der Inhalt geschrumpft ist (Abb. 99).
Ein Saum kann auch auf andere Weise um die Zelle hervorgerufen werden,
er muß aber, wenn er tatsächlich durch die Membran gebildet wird, bei intakten
Zellen eine Duplikatur der Membran sein. Daß dieses für Bakterien und Rickett-
sien zutrifft, ist an sich überkreuzenden Falten nachweisbar, von welchen eine
Falte der oberen und die andere der unteren Lage der Duplikatur angehört
(Abb. 82). Streng zu unterscheiden ist die Membran von Säumen, die dadurch
entstehen können, daß sich vor dem Auftrocknen der Zellen kleine Partikel

am Rand der Zelle ablagern, von denen sich beim Trocknen die schrumpfende Zelle samt der Membran zurückzieht. Meist schrumpft jedoch der Zellinhalt etwas stärker als die Membran, so daß diese sichtbar wird. Außer durch die Trocknung tritt eine Ablösung des Zellinhaltes von der Membran (Plasmolyse) auch durch andere Zellschädigungen ein, z. B. durch die Einwirkung von kolloidem Silber (Abb. 81). Der Nachweis einer osmotischen Plasmolyse, die ältere Autoren auf Grund lichtoptischer Beobachtungen angegeben hatten, ist dagegen A. TOSCH-

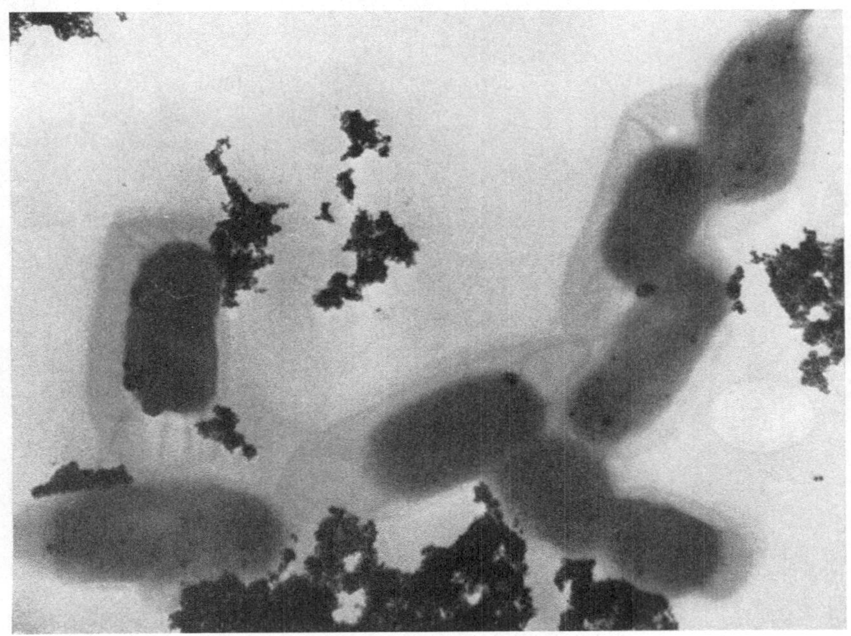

Abb. 81. Schrumpfung des Zellinhalts von Bacterium coli unter der Einwirkung von kolloidem Silber. Etwa 18 000 : 1.

KOFF (unveröffentlicht) im Laboratorium des Verfassers nicht gelungen. Die Zellmembran scheint dem Ionendurchtritt keinen wesentlichen Widerstand entgegenzusetzen. Durch feuchte Hitze (H. RUSKA 1941) oder durch hochfrequenten Schall (H. RUSKA 1947) kann die Membran zerreißen und vollständig vom Zellinhalt abgetrennt werden (Abb. 82). Auch MUDD und LACKMANN (1941), MUDD, POLEVITZKY, ANDERSON und CHAMBERS (1941), UMBREIT und ANDERSON (1942), MUDD und ANDERSON (1942), MUDD, POLEVITZKY, ANDERSON und KAST (1942) und MUDD, HEINMETZ und ANDERSON (1943), sowie zahlreiche weitere Autoren haben die Zellmembran beobachtet, deren Bedeutung vorher nicht immer richtig ausgelegt worden war.

An Milzbrandbazillen wurde von FRÜHBRODT und H. RUSKA (1940) beobachtet, daß sich die Zellmembran kappenartig am Zellende öffnet, um die Sporen austreten zu lassen: das gleiche sahen JAKOB und MAHL (1940) an Pseudotetanusbazillen. Sie zeigten ferner ebenso wie BRÜCHE und HAAGEN (1939) Zellmembranen an weiteren apathogenen Anaerobiern und an Gasbrandbazillen. Dadurch, daß diese Autoren die Bezeichnungen Membran und Kapsel in gleichem Sinne für Membran gebrauchen (vgl. auch JAKOB und MAHL 1948), obwohl beide schon von einigen älteren Autoren klar auseinandergehalten wurden[1], kamen sie zu der Auffassung, Kapseln an apathogenen Keimen nachgewiesen zu haben, die bisher als kapsellos galten. Auch der Ausdruck „Hülle" wird für

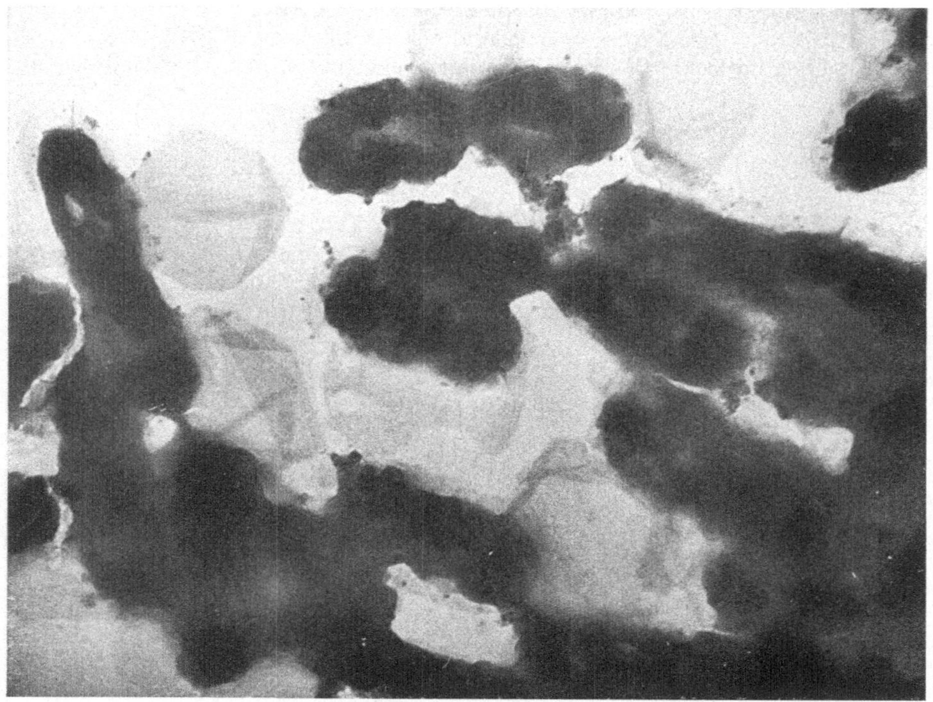

Abb. 82. Ablösung der Zellmembran von Bacterium coli unter der Einwirkung von feuchter Hitze. Etwa 18000 : 1.

die Zellmembran gebraucht. Van Iterson (1947) beschreibt, daß bei der Zellteilung von B. mycoides wie in Geweben höherer Pflanzen zunächst die Bildung einer Mittellamelle zu beobachten ist, die später mit der sich einziehenden äußeren Zellmembran an den Berührungsstellen verschmilzt. Der Zellmembran liegt innen der Zytoplasmakörper mehr oder weniger eng an. Seine Oberfläche wird von einer Plasmahaut, bzw. Grenzmembran überzogen (Abb. 83). Abgesehen von den Rickettsien und einigen L-Organismen, ist bei keinem Virus eine Zellmembran nachgewiesen, wohl aber sind bei manchen Virusarten besondere Oberflächenstrukturen oder Plasmahäute vorhanden.

Abb. 83. Kokken, die außer der Zellmembran eine dunkle Konturlinie als Begrenzung des Zytoplasmas erkennen lassen. (Auf der Reproduktion wesentlich schlechter erkennbar als auf dem Original.) 20000 : 1. Nach H. Ruska (1942).

---

[1] Vgl. z. B. E. Gotschlich in „Handbuch der pathogenen Mikroorganismen", Kolle, Kraus, Uhlenhuth I, 74ff (1929).

b) Kapsel und Schleimsubstanz.

Nach GOTSCHLICH (l. c.) sind Kapseln schleimige Hüllen, die im gefärbten lichtmikroskopischen Präparat den intensiv gefärbten Bakterienleib als hellen Hof umgeben. Es sind also ebenso wie der Bakterienschleim durch die Membran hindurch ausgeschiedene, mehr oder weniger schwer lösliche Substanzen. Ihre Begrenzung nach innen liegt dicht an der Membranoberfläche, während sie nach außen unregelmäßig auslaufen. Kapseln umgeben die einzelnen Bakterien; Schleimsubstanzen fließen dagegen infolge ihrer weniger festen Konsistenz oft von zahlreichen Bakterien zusammen. Beide werden in stärkerem Ausmaß nur von verhältnismäßig wenigen Bakterienarten gebildet und sind vom Kulturmedium abhängig oder treten nur im infizierten Organismus auf. Lichtmikroskopisch sieht man Kapseln am besten, wenn Bakterienausstriche mit Tusche, kolloidem Silber oder dergleichen gemacht werden. Die Kolloide liegen dann außerhalb der Kapsel, die

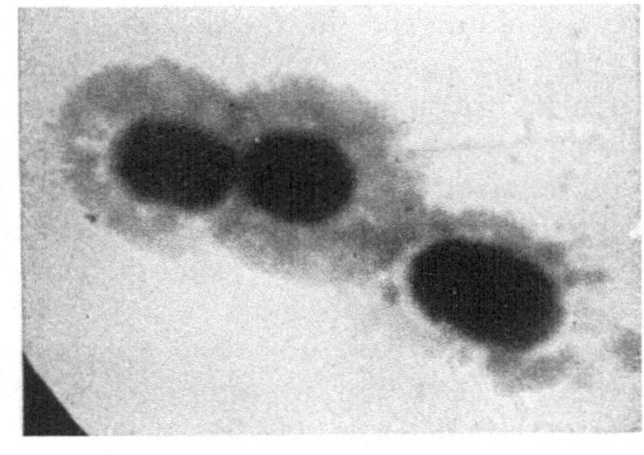

Abb. 84. Bacterium pneumoniae mit Kapselbildung. Aus Mäuseblut isoliert. 12 000 : 1. Nach FRÜHBRODT und H. RUSKA (1940).

vom Bakterienleib durch das schwächere Lichtbrechungsvermögen leicht zu unterscheiden ist. Irrtümer können aber, worauf FRÜHBRODT und H. RUSKA (1940) hingewiesen haben, auch mit diesem Verfahren vorkommen, wenn, wie es z. B. bei Milzbrandbazillen beobachtet wurde, durch Plasmaschrumpfung eine breite Lücke zwischen dem Zellinhalt (Zytoplasma) und der Zellmembran auftritt. Dann erscheint im Tuschepräparat der Plasmolysespalt als Kapsel.

Elektronenmikroskopisch haben die genannten Autoren Kapseln an Pneumoniebazillen aus Mäuseblut (Abb. 84) dargestellt. Ihre Bilder gleichen völlig denjenigen, die bei der Untersuchung der Kapselquellungsreaktion in vitro von MUDD, HEINMETZ und ANDERSON (1943/2) erhalten wurden. In Abb. 84 liegt demnach eine in vivo vor sich gegangene Quellungsreaktion vor. An manchen Zellen sieht man unter

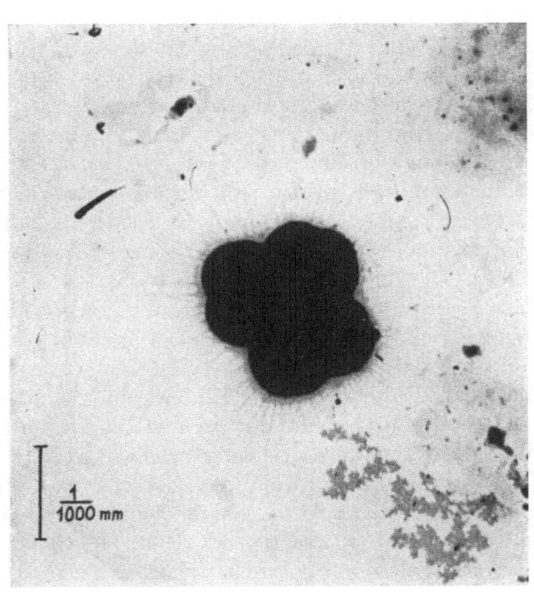

Abb. 85. Sarcine. Nach H. RUSKA (1939).

der Kapsel in scharfer Begrenzung die völlig intakte Membran. Beide lassen sich elektronenmikroskopisch sehr leicht erkennen und sicher unterscheiden. Mitunter zeigt sich eine radiäre Kapselstruktur, entsprechend der Ausscheidungsrichtung der Kapselsubstanz. MUDD und Mitarbeiter konnten zeigen, daß spezifische Antikörper und nicht spezifische Serumkomponenten in Spalten der Kapseln eindringen, gebunden werden und dadurch deren Dicken- und Dichtezunahme bewirken. Auch sie betrachten die Kapsel als ein der Zellwand außen dicht anliegendes, wäßriges Gel. Wood und Smith (1949) unterscheiden an Pneumokokken (Typ III) aus jungen Kulturen noch eine der festeren Kapsel aufsitzende lockere, voluminöse Schleimschicht, die bei einer Mutante des Typs III fehlte. Sie schützt vor der Phagocytose durch Leukocyten. Der schön ausgebildete „Strahlenkranz" der Sacrine auf Abb. 85 ist ebenfalls ein Ausscheidungsprodukt der Zellen.

Häufig sind neben den Bakterien kolloide Partikel zu sehen, die teils aus dem Kulturmedium stammen, teils aber auch wie Schleim und Kapsel als spezifische Ausscheidungen der Keime zu betrachten sind. Durch die Anfertigung von Klatschpräparaten, bei denen die Lagebeziehung zwischen den Bakterien und ihren Ausscheidungen im Gegensatz zu dem sonst üblichen Verfahren der Herstellung von Bakteriensuspensionen erhalten bleibt, ließ sich die Herkunft solcher kolloider Teilchen aus den Zellen nachweisen. Zugleich wurde an den verwendeten Kulturen von Chromobakterium prodigiosum gezeigt, daß der Farbstoff Prodigiosin am Zytoplasma und nicht an diesen Ausscheidungen haftet (H. Ruska 1941/6).

c) Geißeln.

Die beweglichen Bakterien besitzen als Fortbewegungsorgane 10 bis 20 mμ dicke Geißeln. Auf ihre leichte elektronenmikroskopische Darstellbarkeit haben Piekarski und H. Ruska (1939/1) hingewiesen. Die Geißeln sind ohne besondere Verfahren auf allen Aufnahmen beweglicher Bakterien zu sehen. Neu war der Umstand, daß die Geißeln meist wesentlich länger und zahlreicher sind, als nach lichtmikroskopischen Untersuchungen angenommen wurde (Abb. 9). Das lichtoptische Dunkelfeld erfaßt vielfach erst (oder ausschließlich ?) „Geißelzöpfe", und auch die Färbeverfahren sind unvollkommen. Kein Elektronenmikroskopiker wird daher der Vermutung von Rippel (1947) zustimmen, daß die Geißeln aus Fibrillen bestehen, die sich auflösen und mehrere Geißeln vortäuschen. Gegen diese Annahme spricht auch die Tatsache, daß beim Abziehverfahren freiliegende Einzelgeißeln in der üblichen Form zur Darstellung kommen (Hillier und Baker 1946). Die Theorie von Pijper (1946), der die Geißeln als schleimige Polysacharidfäden auffaßt, die sich durch die Bakterienbewegung von der Zelle ablösen, wird durch van Iterson (1947) abgelehnt. Vgl. auch Pijper (1949). Eine Beobachtung der Bakterien- und Geißelbewegung ist elektronenmikroskopisch nicht möglich. Daher geben die Bilder keinen sicheren Aufschluß über die Frage, welche der zahllosen Geißeln mit dem beobachteten Bakterium zusammenhängen. Auch über den genauen Ursprungsort der Geißeln (polar, subpolar, peritrich) und die Ursprungsstruktur (Membran, Zytoplasma, Basalkörnchen) lassen sich noch keine allgemeingültigen Aussagen machen. Vorläufig erscheint es noch zweifelhaft, ob die Angabe von Pietschmann (1942) richtig ist, wonach jede peritriche Begeißelung in Wirklichkeit subpolar ist. Die peritriche Begeißelung soll dadurch vorgetäuscht sein, daß an einem langen Bakterienfaden, der sich nicht in Einzelzellen getrennt hat, je Einzelzelle 4 Geißeln (an jedem Pol 2) subpolar entspringen. Auf der Abb. 1 von Piekarski und H. Ruska (1939/1) ist die Geißelzahl wesentlich größer, als nach der erwähnten Beziehung zu erwarten wäre, und bei van Iterson (1947) sind etwa 20 Geißeln zu sehen, die von einem einzigen Zellpol ausgehen. Man hat manchmal den Eindruck, als ob die Geißeln infolge der Zellschrumpfung beim Eintrocknen kurz vor dem Ansatz an die Zelle abgerissen seien. Möglicherweise ist die übliche Präparationstechnik zu wenig schonend, um die letzten Feinheiten zu erhalten. Einige Autoren neigen zu der Auffassung, daß die Geißeln an der Membran und nicht aus dem Zytoplasma entspringen. Gegen diese Annahme spricht schon stark der verschiedene Antigencharakter beider Gebilde, den MUDD und ANDERSON (1941) auch elektronenmikroskopisch bestätigt haben. RIPPEL (1947) vermutet daher mit Recht, daß die Geißeln aus einer ähnlichen Substanz bestehen wie die innere Plasmahaut. Beweisend für den zytoplasmatischen Ursprung erscheinen die Untersuchungen von VAN ITERSON (1947), der mit der Metallbedampfungstechnik an Pseudomonas fluorescens, Vibrio metschnikovii und Spirillum serpens gezeigt hat, daß sich das Relief der Geißeln unter der Zellmembran bis zum Zytoplasma verfolgen läßt. Bei einigen Vibrioarten wurden Basalgranula an der Ursprungsstelle beobachtet (MUDD und Mitarbeiter, VAN ITERSON), doch ist dies keine für alle Bakterien geltende Regel. Der Mechauismus der Geißelbewegungen ist noch rätselhaft. Man weiß nicht, ob die Geißeln ans kontraktiler Substanz bestehen und wie die im lichtoptischen Dunkelfelde sichtbare Koordination der Bewegung zustande kommt. Die Geißelsubstanz enthält etwa 16 % N, Spuren P, 1 % Kohlehydrat, ist also im wesentlichen ein einfaches Protein. Röntgendiagramme von ASTBURY und WEIBULL (1949) zeigten, daß das Geißelprotein zur Keratin-Myosin-Epidermis-Fibronogen-Gruppe gehört. Wahrscheinlich kann die Geißel als monomolekulares Haar oder monomolekularer Muskel aufgefaßt werden, der eigene Kontraktilität (α- und β-Form des Diagramms) besitzt. POLEVITZKY (1941) glaubt, daß die Geißeln eine röhrenförmige Struktur haben. JAKOB und MAHL (1948) äußern die Vermutung, daß die BROWNsche Molekularbewegung zusammen mit der gleichsinnigen elektrischen Ladung der Bakteriengeißeln und der dadurch bedingten gegenseitigen Abstoßung die Hauptursachen der Geißelbewegung sein könnte. Man darf aber nicht übersehen, daß äußere Reize die Richtung der Bakterienbewegung im positiven oder im negativen Sinne beeinflussen können und daß wohl noch andere Momente zu den von JAKOB und MAHL vermuteten hinzukommen.

d) Zellinhalt.

Das elektronenmikroskopische Bild des Zellinhaltes ist stark von den Kulturbedingungen abhängig, unter denen sich die Keime entwickelten. Eine Unterscheidung von Ektoplasma und Entoplasma, wie sie seit ZETTNOW (1896/97) üblich geworden ist, läßt sich nicht durchführen. Sie entbehrt einer weiteren Berechtigung, nachdem die Strukturen von Zellmembran und Kapsel, die vielfach als Ektoplasma gedeutet wurden, völlig einwandfrei zu erkennen sind. Eine Trennung des Bakterienzytoplasmas in eine äußere und innere Schicht erwies sich als undurchführbar.

Die wichtigste licht- und ultraviolettmikroskopische Feststellung über den Zellinhalt ist der Nachweis von thymonukleinsäurehaltigen Bezirken, die als Kernäquivalente angesprochen werden (STILLE; PIEKARSKI; NEUMANN; STAPP; BOIVIN; TULASNE; VENDRELY und MINCK u.a.). Es wurde daher versucht, diesen Befund auch elektronenmikroskopisch zu bestätigen. PIEKARSKI und H. RUSKA (1939/2) fanden zwar entgegen den Erwartungen elektronenmikroskopisch keine klar erkennbaren Kernäquivalente, jedoch eine charakteristische Massenverteilung innerhalb der Zelle, die den lichtmikroskopisch festgestellten Zelltypen entsprach. Der Zusammenhang zwischen der Verteilung der Nukleoide, dem Zelltyp und dem Kulturalter geht aus Abb. 86 hervor. Zu unterscheiden sind *primäre* Zellen aus jungen Kulturen mit *zwei* lichtmikroskopisch erkennbaren

Kernäquivalenten und sekundäre aus älteren Kulturen mit *einem* Kernäquivalent. Elektronenmikroskopisch erkennt man die Kernäquivalente nicht, wohl aber

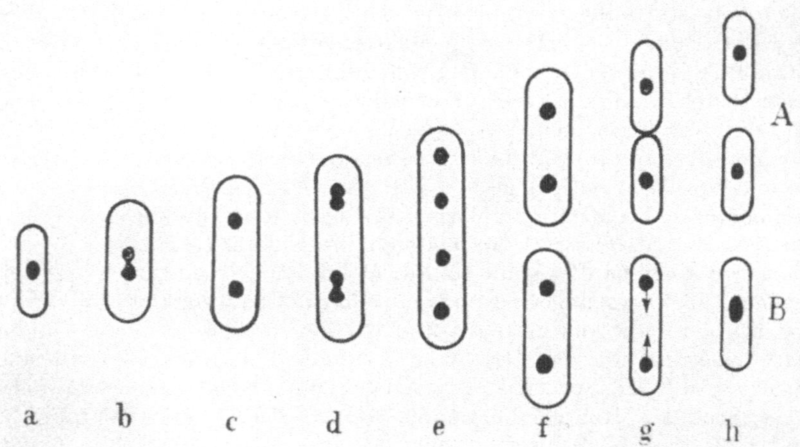

Abb. 86. Entwicklung einer sekundären Zelle (*a*) zur primären Zelle (*c*) und Teilung der primären Zelle (*c—f*) sowie Übergang der primären Zelle in sekundäre (*g—h*). Nach Piekarski.

die verschiedenen Zelltypen. Auch an Rickettsien läßt sich nach Eyer und H. Ruska (1944) eine Unterscheidung primärer und sekundärer Zellen durchführen. Außerdem gibt es zytologisch noch wenig geklärte, wahrscheinlich unter sich noch sehr verschiedene *tertiäre* Formen, deren genauere zytologische Analyse noch aussteht.

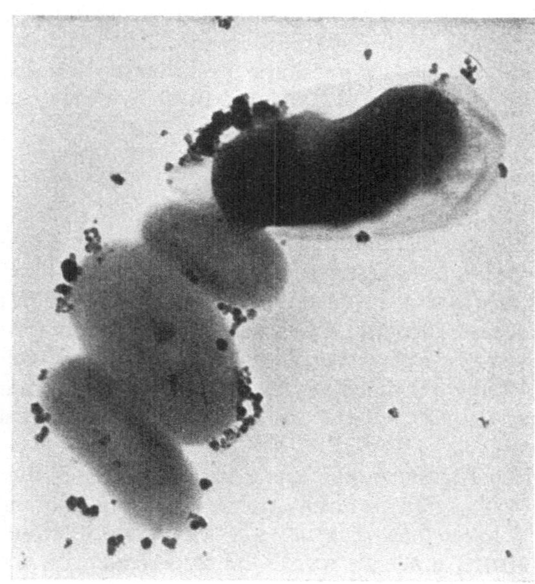

Abb. 87. Ruhrbakterien (Flexner). Typische sekundäre Zelle und drei atypische kleine Zellen mit Partikeln aus dem Agarnährboden, 23000 : 1. Nach H. Ruska (1942).

Abb. 9 und 87 zeigen typische primäre und sekundäre Zellen im elektronenmikroskopischen Bild, wie sie u. a. auch van Iterson (1947) sowie Rake und Oskay (1948) angaben. Bei den primären Formen ist der Zellinhalt meist nahezu gleichmäßig innerhalb der Membran verteilt, mitunter aber etwas geschrumpft, so daß an den Seiten und besonders an den Polen die Membran sichtbar wird. Mitunter ist die Dichte des Zytoplasmas an den Zellenden größer als in der Mitte. Bei den sekundären Zellen ist die größte Massendicke in der Zellmitte, das Zytoplasma hat

sich von einem oder beiden Zellpolen stark zurückgezogen, so daß man hier sehr gut die Membran erkennt (neuere Bilder metallbedampfter sekundärer Zellen von Donovania granulomatis siehe bei Rake und Oskay (1948). Wahrscheinlich hat an demjenigen Ende der Tochterzellen die vorausgehende Teilung stattgefunden, an dem die Retraktion des Zytoplasmas nur gering ist, da die Mutterzellen beiderseitige starke polare Zytoplasmaretraktion zeigen. Die Nukleoide bleiben unsichtbar. Es besteht keine wesentliche Massendifferenz zwischen den thymonukleinsäurehaltigen Bezirken und dem übrigen Zytoplasma.

KNAYSI und BAKER (1947) entwickelten eine elegante Methode zum elektronenoptischen Nachweis des Kernäquivalents. Sie ließen Sporen von Bac. mycoides in stickstofffreiem Medium keimen und erreichten durch diesen Nah-

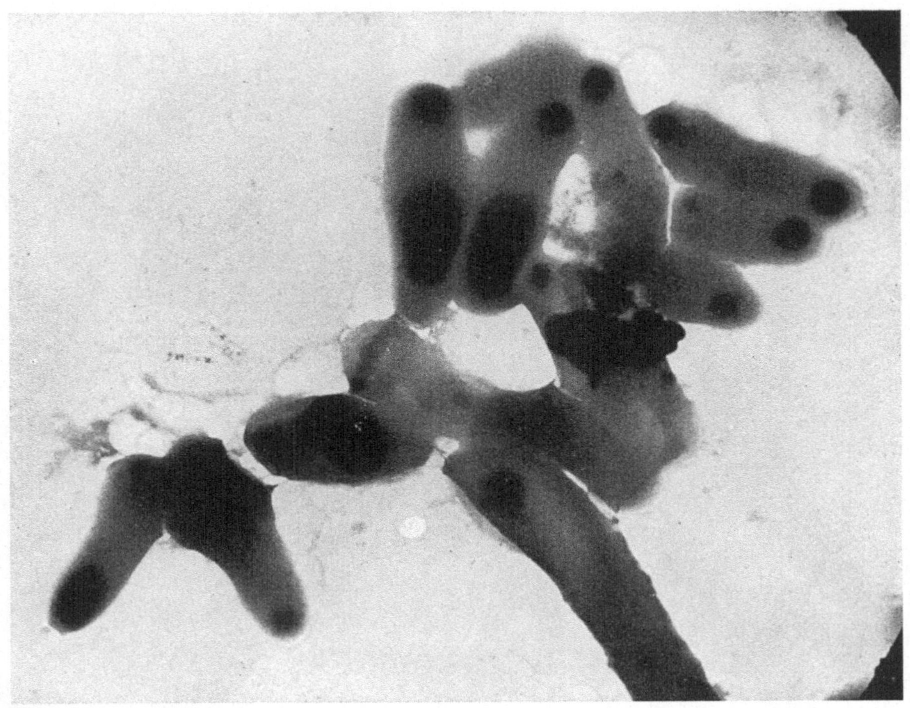

Abb. 88. Diphtheriebakterien mit Polkörpern. 18000 : 1. Nach H. RUSKA (1942).

rungsmangel, daß die Kernäquivalente als dichte Regionen im transparenten Zytoplasma erscheinen. Außerdem konnte die Entwicklung der Endospore und der Übergang der Kernäquivalente in die Sporen verfolgt werden. HILLIER, MUDD und SMITH (1949) wiesen die Kernäquivalente in B. coli mit Hilfe einer doppelten, besonders kontrastreich arbeitenden Objektivlinse kleinster Apertur als aufgehellte, leicht strukturierte Bezirke nach. BIELIG, KAUSCHE und HAARDICK (1949) berichteten über die Lagebeziehungen zwischen den Orten der UV-Absorption, der Feulgenreaktion, der Giemsafärbung und der Reduktion von Triphenyl-tetrazoliumchlorid.

Die von PIEKARSKI und H. RUSKA (1939/2), FRÜHBRODT und H. RUSKA (1940) u. a. in einzelnen, gut durchstrahlten Zellen gefundenen, scharf begrenzten Zelleinschlüsse (Abb. 88) wurden anfänglich als Nukleoide angesprochen, später jedoch als Volutingranula erkannt. Sie lassen sich durch starke Bestrahlung zerstören (Abb. 89) und gleichen darin den Polkörpern der Diphtheriebazillen (H. RUSKA 1942/12), sowie den Granulationen der Tuberkel- und Leprabazillen. MUDD, POLEVITZKY und ANDERSON (1942) beschreiben sie auch bei Vibrionen und Spirochaeten, Miller und Mitarbeiter (1948) bei Malleomyces mallei und pseudomallei. Die zuletzt genannten Autoren glauben, daß die Granula aus Lipoiden bestehen; sie fanden sie nach der Beschallung frei zwischen den Zelltrümmern verstreut. Knaysi und Mudd (1943) vermuten engere Beziehungen der Granula zu den Kernäquivalenten und nehmen an, daß dort, wo die Granula fehlen, die Kernsubstanz diffus in den Zellen verteilt sei. Lichtmikroskopisch lassen sich aber Kernäquivalente gerade in Zellen junger Kulturen nachweisen, die nur sehr selten Volutingranula zeigen. Nach Winkler und König (1948)

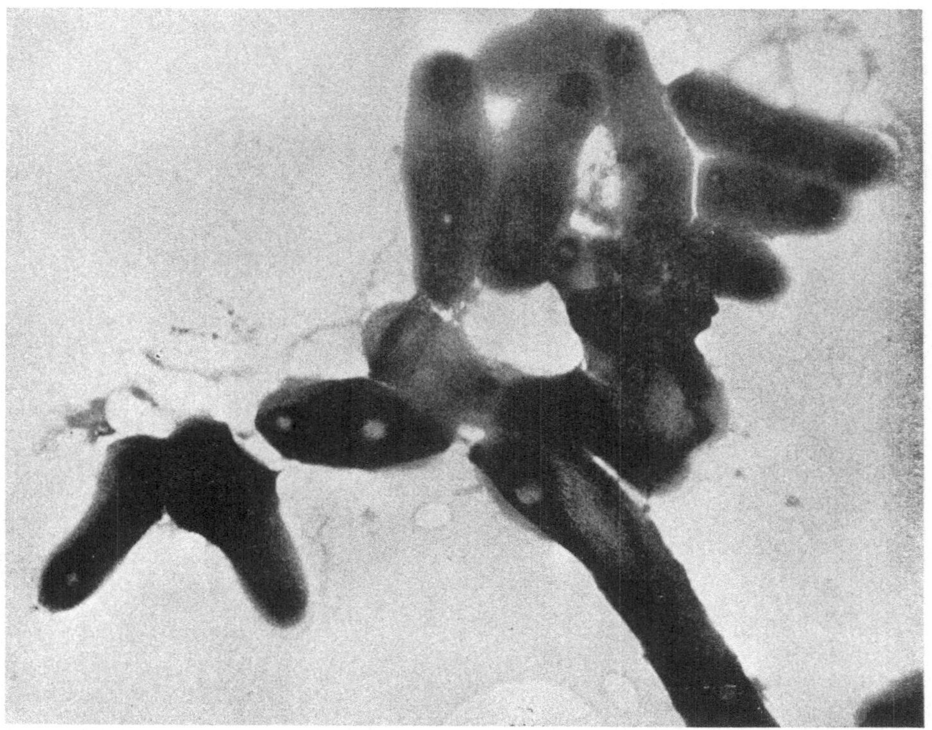

Abb. 89. Die gleiche Bakteriengruppe wie Abb. 88 nach Zerstörung der Volutingranula durch starke Elektronenbestrahlung. 18000 : 1.

bestehen die Granula wahrscheinlich aus dem Kalziumsalz einer Nukleinsäure und nicht, wie Lembke und H. Ruska (1940) nachgewiesen zu haben glaubten, aus ätherlöslichen Substanzen (Lipoiden). Wären die strahlenempfindlichen Substanzen Nukleoide, so müßten sie entgegen der Beobachtung in allen primären und sekundären Zellen nach starker Bestrahlung als helle, scharf begrenzte Bezirke sichtbar werden. Winkler und König klärten nicht nur die chemische Zusammensetzung der strahlenempfindlichen Granula auf, sondern bewiesen auch durch Löslichkeitsversuche, daß sie mit den Volutinkörnchen identisch sind. In verschiedenen Keimarten werden zu gleicher Zeit lichtoptisch die Volutinkörnchen und elektronenoptisch die strahlenempfindlichen Granula sichtbar und verschwinden beide zu Beginn der Sporenbildung. Schließlich gelang Winkler und König die färberisch lichtmikroskopische und die elektronenmikroskopische Abbildung an identischen Keimen. Die Beziehung zwischen den lichtoptisch nachweisbaren Kernäquivalenten und den Volutingranula besteht darin, daß es sich in einem Falle um Nukleoproteine, im anderen wahrscheinlich um nukleinsaures Kalzium handelt. Möglicherweise stellt die Nukleinsäure der Granula eine Reservesubstanz für die Bildung von Kernäquivalenten und Sporen dar. Es handelt sich aber trotz dieser chemischen Beziehungen um Gebilde ganz verschiedener zytologischer Bedeutung. Aus zahlreichen Abbildungen gewinnt man den Eindruck, daß die Volutingranula auf die Tochterzellen gesetzmäßig verteilt werden, indem z. B. in Kokken vor der Zellteilung zwei Granula auftreten. Dieses Verhalten begünstigt eine Verwechslung der Granula mit Kernäquivalenten. Nach den Reliefbildern zu urteilen, erheben sich die Volutin-

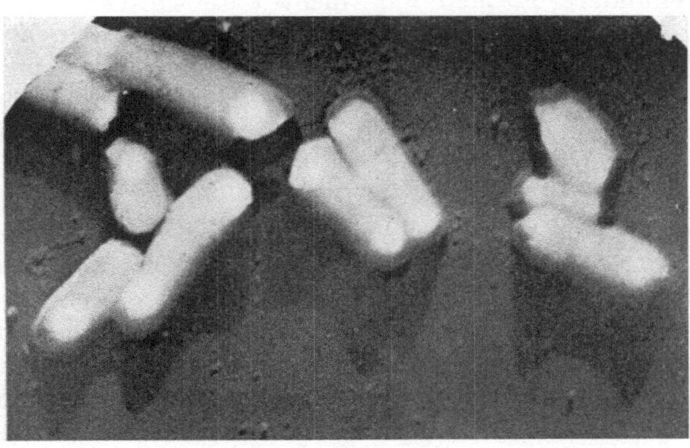

Abb. 90. Diphtheriebakterien mit Siliziumoxyd schräg bedampft. Etwa 10000 : 1. Nach KÖNIG und WINKLER (1948).

granula über das sonstige Niveau der flach aufgetrockneten Bakterienzelle (Abb. 90) und erscheinen nach der Herauslösung als Krater (Abb. 91).

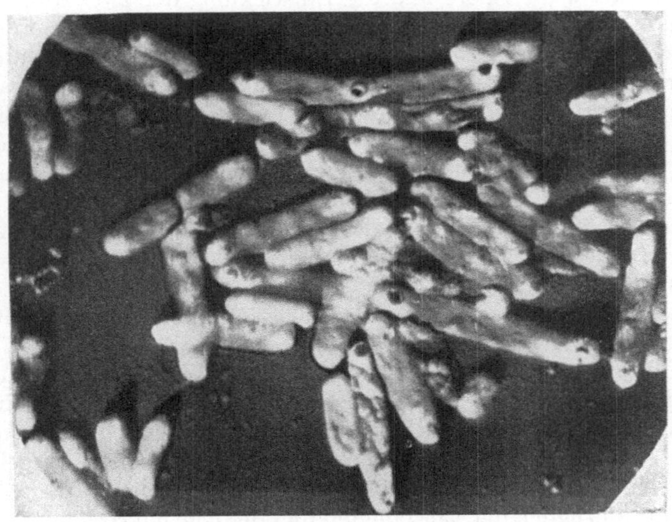

Abb. 91. Sporenbildende Stäbchen der Subtilisgruppe nach vorausgegangener Herauslösung der Volutinkörnchen durch Pikrinsäure mit Siliziumoxyd schräg bedampft. Etwa 5000 : 1. Nach KÖNIG und WINKLER (1948).

Hellere Aussparungen im Zytoplasma wurden in seltenen Fällen an Proteusbakterien beobachtet und sind als Vakuolen gedeutet worden (H. RUSKA (1942/12). Bei Tuberkelbazillen treten sie regelmäßig auf [LEMBKE und H. RUSKA (1940), WESSEL (1942)]. v. ARDENNE und AUGUSTIN (1941) beschreiben sie bei Mycobacterium leprae, MILLER, PANNELL, CRAVITZ, TANNER und INGALLS (1948) bei Malleomyces mallei und pseudomallei. Besonders bemerkenswert sind spiralige Anordnungen des Zytoplasmas, die z. B. UMBREIT und ANDERSON (1942) an

Thiobacillus thiooxydans, sowie Rake und Oskay (1948) in ganz gleicher Weise an Donovania granulamatis gezeigt haben.

Kokken zeigen die gleichen Bauprinzipien wie Stäbchen und enthalten mitunter, wie erwähnt, ebenfalls Volutingranula, doch lassen sich primäre und sekundäre Formen hier nicht unterscheiden.

An der Oberfläche des Zytoplasmas, d. h. an der Grenzfläche gegen die Membran oder den zwischen Membran und Zytoplasma liegenden Saftspalt, sowie an der Grenzfläche gegen die im Zytoplasma eingeschlossenen Vakuolen bildet sich eine besondere Struktur aus. Für äußere freie Protoplasmaoberflächen finden sich Bezeichnungen wie Plasmahaut, Plasmolemma, Pellicula oder Grenzmembran. Der Ausdruck „Grenzmembran" darf nicht zu Verwechslungen mit der Zellmembran oder Zellwand führen. Mitunter ist eine Grenzstruktur als feine Linie sichtbar (Abb. 83). Ihre Beschaffenheit ist wahrscheinlich für das grampositive oder -negative färberische Verhalten der Keime und für ihre Säurefestigkeit von größerer Bedeutung als die Beschaffenheit der Zellmembran (Winkler und König); wahrscheinlich sind Lipoide in ihr angereichert.

Systematische Untersuchungen über die Sporenreifung in den Zellen und über ihre Auskeimung zu neuen vegetativen Formen sind dem Verfasser nicht bekanntgeworden. Auf das Verhalten der Zellmembran beim Austritt der Sporen wurde schon hingewiesen, ebenso auf das Verschwinden der Volutingranula vor der Sporenreifung. Da spezifische elektronenmikroskopische Nachweismethoden für die Nukleoide der Bakterienzelle noch nicht angewendet wurden (z. B. fermentativer Abbau nach Boivin und Mitarbeitern), ist der Übergang von Kernsubstanz in die Spore und die Entwicklung der zwei Nukleoide in der primären Zelle, die aus der Spore entsteht, bisher nicht nach seinem elektronenmikroskopischen Erscheinungsbild bekannt. v. Ardenne und Friedrich-Freksa (1941) untersuchten Sporen unter einer besonderen biologischen Fragestellung. Es gelang ihnen bei schwacher Vergrößerung (500fach) und geringer Elektronenexposition, Sporen von Bacillus mesentericus vulgatus aufzunehmen und an einzelnen (8 %) nachträglich die erhaltengebliebene Keimfähigkeit nachzuweisen. Zu der beabsichtigten Lebendmikroskopie läßt sich das Verfahren bei dem heute gegebenen Stand der Technik aber nicht ausbauen (vgl. bei v. Borries 1949). Immerhin war dadurch gezeigt worden, daß die Elektronenschädigung in Grenzen zu halten ist.

Einige Besonderheiten sind noch bezüglich solcher Keime nachzutragen, die nicht zu den Eubakteriales gehören:

Bei den Rickettsien lassen sich, wie erwähnt, ebenso wie bei den Bakterien primäre und sekundäre Zellen nachweisen (Abb. 3 und 92). Ihre Entwicklung läuft nach Untersuchungen an R. prowazeki wahrscheinlich vom sogenannten Polstab über Doppelformen, unsymmetrische Polstäbe, bzw. Siegelringe zum Polstab und von den Kugelformen über symmetrische Polstäbe, bzw. Tönnchen zur Kugelform. Bezüglich der Einzelheiten und der lichtmikroskopischen Befunde muß auf die Arbeiten von Sikora (1942) und von Eyer und H. Ruska (1944) verwiesen werden. Gönnert (1947) beschrieb 1 μ bis mehrere μ große atypische runde und gestreckte, sogenannte R-Formen bei R. prowazeki, deren feinere Struktur elektronenmikroskopisch noch nicht untersucht ist. Vergleiche in Ergänzung hierzu Babudieri 1943 und Babudieri und Bocciarelli (1943/1). Die Übereinstimmung zwischen Rickettsien und Bakterien ist so weitgehend, daß man an elektronenmikroskopischen Bildern nicht entscheiden kann, welcher der beiden Organismen vorliegt, wenn die absolute Größe der Abbildung unbekannt ist. Plotz, Smadel, Anderson und Chambers (1943) haben außer R. prowazeki auch noch R. mooseri, R. rickettsii und R. burneti untersucht. Sie betonen ebenso wie WEISS (1943) und WEYER, FRIEDRICH-FREKSA und BERGOLD (1944) die Bakterienähnlichkeit und das Vorhandensein einer Zellmembran. SHEPARD

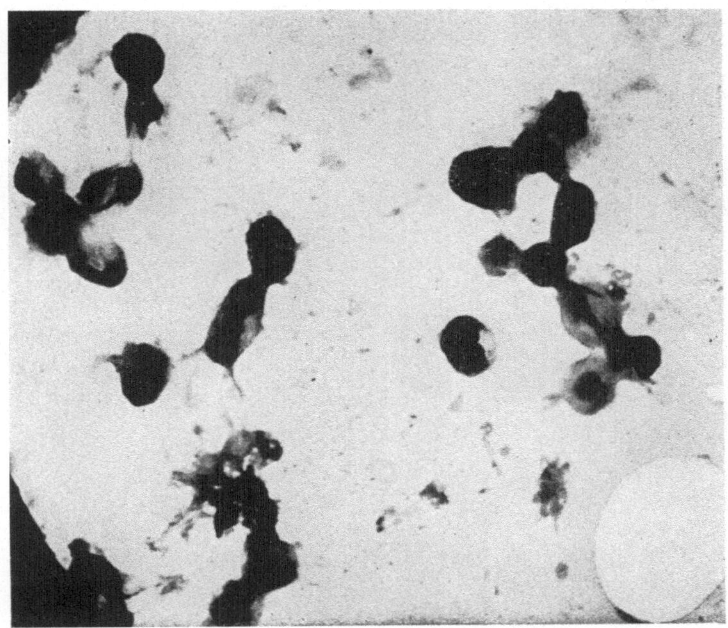

Abb. 92. Rickettsia quintana. Die Zellen sind kleiner als bei Rickettsia prowazeki und neigen stark zur Verklebung. Nach EYER und H. RUSKA (unveröffentlicht).

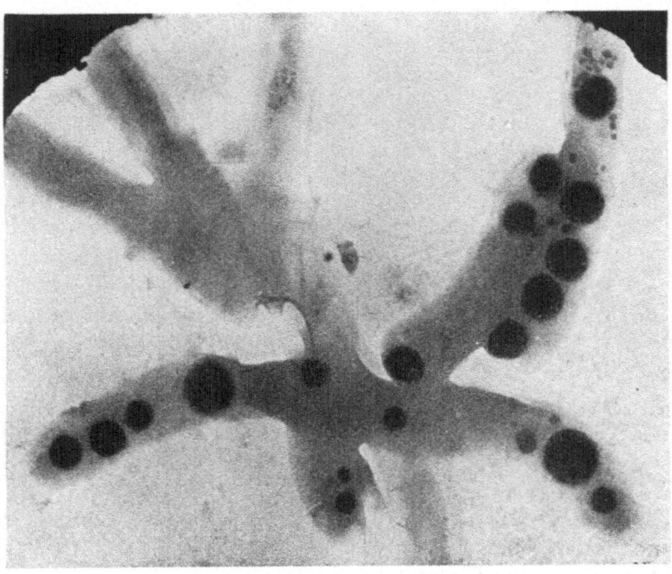

Abb. 93. Mycobacterium tuberculosis. Typus gallinaceus. Zahlreiche Volutingranula verschiedener Größe. 23 000 : 1. Nach LEMBKE und H. RUSKA (1940).

und WYCKOFF (1946) deuten dagegen die Membranen als Kapselsubstanzen, welche nicht die ganze Zelle umgeben. Sie zeigen an metallbedampften Präparaten, daß die Kapsel (bzw. Membransubstanz ?) in warmem Äther löslich und mit dem sogenannten löslichen Flecktyphus-Antigen identisch ist.

Von den Mycobakterien sind Tuberkulose- und Leprabazillen untersucht worden. Die klarsten Bilder erhält man vom Typus gallinaceus des Tuberkel-

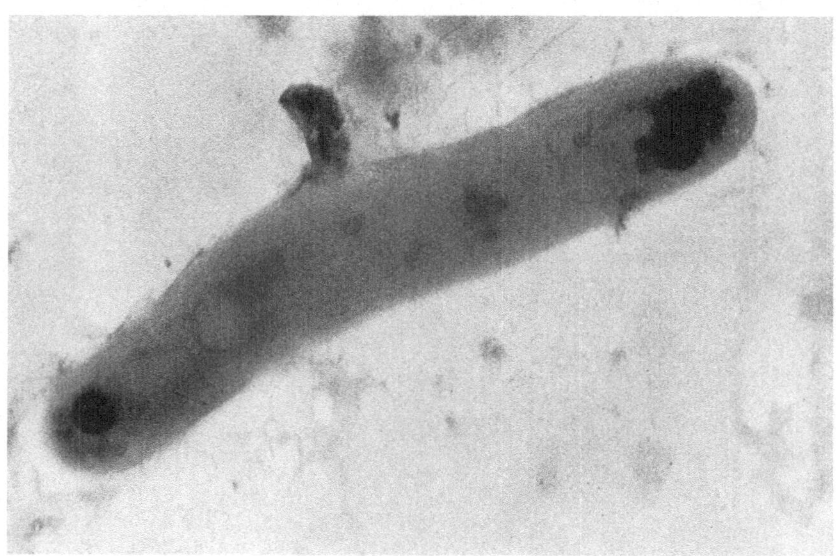

Abb. 94. Mycobacterium tuberculosis. Typus humanus. Volutingranula und Vakuolen. 30000 : 1. Nach Wessel (1942).

bazillus (Abb. 93). Sehr gut sind die schon von R. Koch und vielen späteren Untersuchern gesehenen „Sporen" oder Granula elektronenmikroskopisch zu erkennen. Da es sich hierbei nach Winkler und König um Volutin handelt, können die Granula nicht die lange umstrittene Bedeutung von keimfähigen und filtrierbaren Gliedern einer Entwicklungsreihe haben. Außer tiefschwarzen, durch die Elektronenstrahlen leicht zerstörbaren Granula sehr verschiedener Größe beobachtet man Vakuolen (Abb. 94), scharf begrenzte Bezirke, die heller als das umgebende Zytoplasma sind und beim Typus bovinus oft in zwei

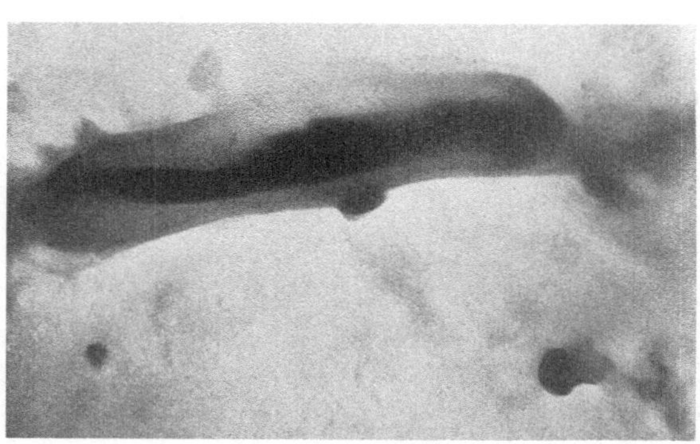

Abb. 95. Mycobacterium tuberculosis. Typus humanus. Alte Kultur, freiliegende, nicht aus Volutin bestehende Granula. 26000 : 1. Nach Wessel (1942).

Reihen nebeneinander liegen (bei Lembke und H. Ruska 1940, Abb. 12). Vakuolen und Volutin-Granulationen findet man auch gelegentlich bei anderen Bakterien, wie Pyocyaneus, Paratyphus und Proteus in der gleichen Zelle, während bei verschiedenen Kokken, Pneumobazillen und Diphtheriebakterien nur Volutingranula gefunden wurden. Als weitere rundliche Zellinhaltskörper beobachtet man an Typus gallinaceus scharf begrenzte Bezirke, deren Dichte geringer ist als die der Volutingranula, aber größer als die der Vakuolen. Innerhalb dieser Bezirke kann punktförmig (etwa 10 mμ Durchmesser) eine sehr dichte Substanz vorhanden sein. LEMBKE und H. RUSKA vermuteten, daß diese Gebilde Vorstufen der Volutingranula (Praegranula) sind, in welchen das Volutin aus einer übersättigten Lösung durch Entmischung tröpfchenförmig ausfällt. Gegen diese Deutung läßt sich allerdings einwenden, daß derartige „Praegranula" bei anderen Bakterien mit Volutinkörnern bisher nicht gesehen wurden. WESSEL (1942) beobachtete am Typus humanus, daß die dichten, großen Granula nicht alle strahlenempfindlich sind, und vermutete auf Grund seiner Aufnahmen, daß nur die strahlenresistenten Granula Sporencharakter haben können und als besondere Fortpflanzungsformen aus der Zelle ausgestoßen werden. Sie treten besonders in alten Kulturen auf (Abb. 95). v. ARDENNE und AUGUSTIN (1941) zeigten an einigen Aufnahmen weitgehende Analogien im Feinaufbau zwischen Mycobakterium leprae und tuberculosis. Über die zellphysiologische Bedeutung der verschiedenen Zytoplasmadifferenzierungen kann vorläufig nichts Bestimmtes ausgesagt werden. Eine wachsartige, dicke Hülle, wie sie wegen der Säureresistenz der Tuberkelbazillen vielfach angenommen wurde, läßt sich elektronenmikroskopisch nicht nachweisen, wohl aber eine sehr zarte Membran, wie auch von MUDD, POLEVITZKY und ANDERSON (1942) betont wird.

e) Verhalten gegenüber physikalischen und chemischen Einwirkungen.

Trockenes Erhitzen primärer Zellen führt zu keinen Strukturänderungen. Nach dem Erhitzen in wässeriger Suspension zeigt sich dagegen eine Koagulation

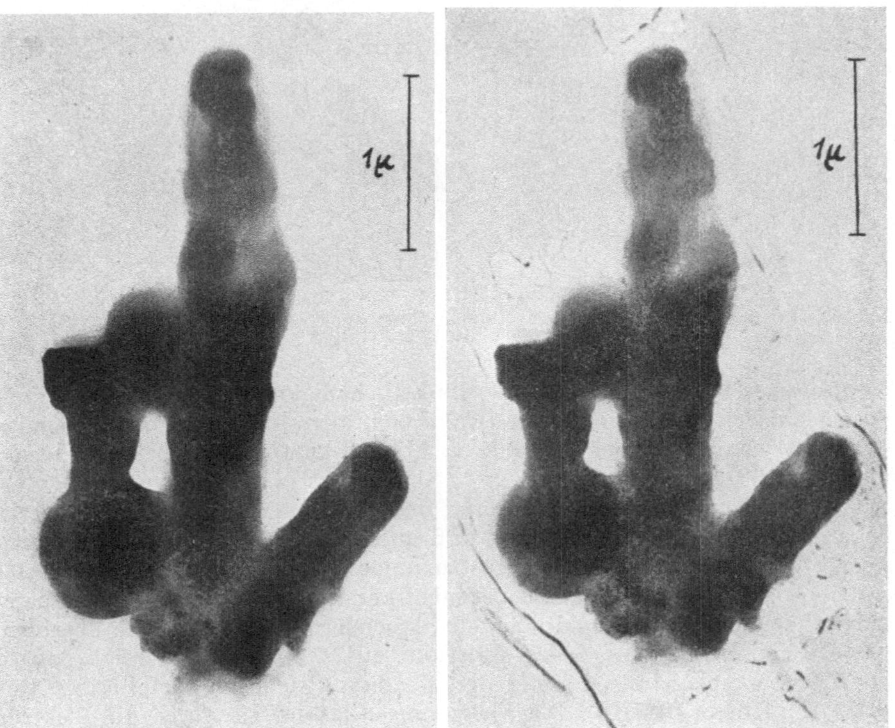

Abb. 96. Diphtheriebakterien vor und nach einer Erhitzung auf 420° C im Vakuumofen. Nach v. BORRIES (1948).

und teilweise Extraktion des Zellinhalts mit Ablösung von Zellmembranen
(Abb. 96 und 82). Ähnliche, aber geringere Störungen der Verbindung zwischen
Membran und Zytoplasma können sich bei Einwirkung tiefer Temperaturen
zeigen. Eine völlige Ablösung der Membran findet sich bei Behandlung der
Zellen mit hochfrequentem Schall. Wahrscheinlich sind Zellen, die ihre Membran
verloren haben oder starke Membrandefekte aufweisen, nicht mehr lebensfähig.
Die quantitativen Beziehungen zwischen den Absterbekurven und der Zahl
der durch Ultraschall sichtbar verletzten Zellen sind nicht untersucht, so daß
eine kritische Stellungnahme zu der Frage, ob die Zahl der abgestoßenen Zellen
mit derjenigen der verletzten parallel geht, noch nicht möglich ist. Eine Parallelität

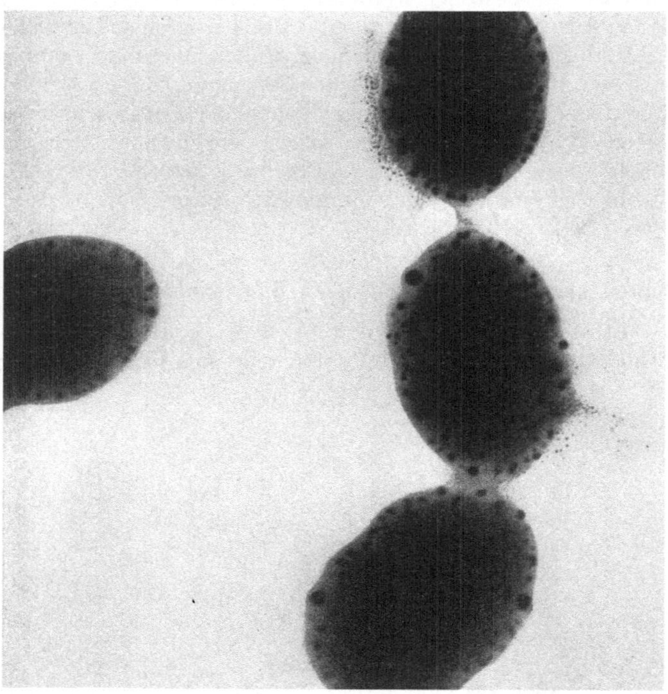

Abb. 97. Streptokokken nach einigen Minuten Einwirkung von 1⁰/₀₀ Sublimat. Etwa 50000:1.
Nach H. Ruska (1947).

würde wahrscheinlich machen, daß die Zellschädigung an vegetativen und nicht
an generativen Teilen der Zellen (Nukleoide) angreift, wie es für Schädigungen
vermutet worden ist, bei welchen die Absterbekurve dem Verlauf einer mono-
molekularen Reaktion gleicht.

Bei der Vergiftung mit Sublimat, dessen Bindung an die Bakterien lange
bekannt ist, läßt sich nachweisen, daß infolge der hohen Dichte des Hg's und
seiner Verbindungen eine Kontrastzunahme von Membran und Zytoplasma
eintritt. Diese Kontrastzunahme ist nicht nur Wirkung einer Eiweißverdichtung
durch die fällende Wirkung des Sublimats, sondern Folge der Quecksilberbindung
(Mudd und Anderson 1942, Gärtner 1943/1). Abb. 97 zeigt außerdem
kugelige Hg-Ablagerungen unter der Membran, vielleicht metallisches Quecksilber (H. Ruska 1947/19). Die Ablagerungen lassen sich durch die Elektronenbestrahlung zur Sublimation bringen, bevor die Volutingranula zerstört werden.
Die Bedingungen, unter welchen die Bakterien auf Sublimatzusatz in der zuletzt

beschriebenen Form reagieren, sind nicht näher bekannt, infolgedessen ist das Ergebnis nicht immer reproduzierbar. Offenbar ist der Zustand der Plasmahaut von Bedeutung, an der das Sublimat nach seinem Durchtritt durch die Zellmembran reagiert. Zu einer Kontraststeigerung der Objekte führen auch andere Metallverbindungen. Bei Osmiumtetroxyd wurde sie früh beobachtet, ferner bei Silbernitrat und Bleiacetat (MUDD und ANDERSON 1942). MORTON und ANDERSON (1941), sowie KÖNIG und WINKLER (1948) zeigten die Aufnahme von Tellur in Diphtheriebazillen (Abb. 98). All diesen Mitteln fehlt jedoch die Eignung für spezifische Färbungen, d. h. zu gesonderter Darstellung von Zellmembran, Plasmahaut, Nukleoiden oder anderen speziellen Strukturen des Zytoplasmas.

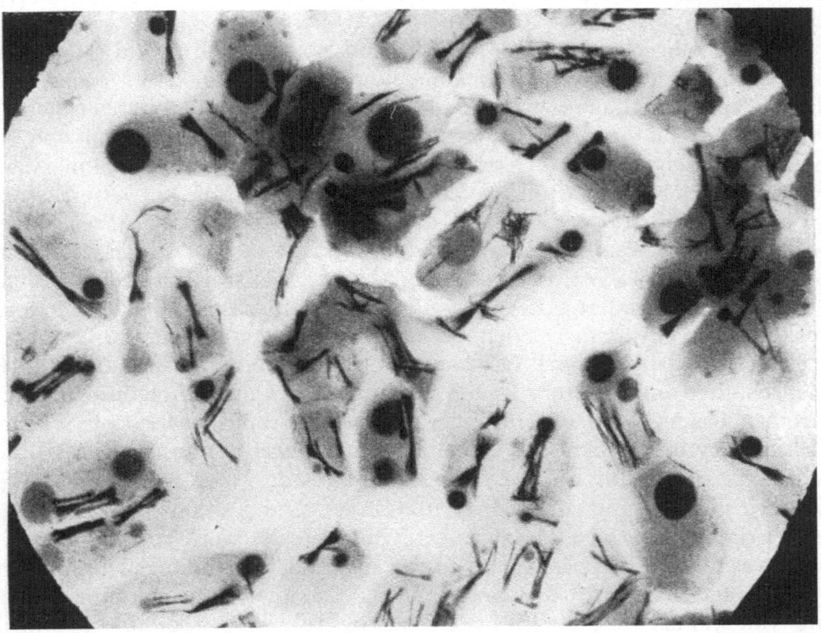

Abb. 98. Diphtheriebakterien von Claubergnährboden mit eingeschlossenen Tellurnadeln. 12 000 : 1. Nach KÖNIG und WINKLER (1948).

TROCH (1942) untersuchte die Einwirkung Peteosthor auf Tuberkelbazillen, einem Präparat von kolloidem Platin, Eosin „bläulich" und Thorium X, das zur Behandlung der Tuberkulose verwendet wird. Er fand schwere Destruktionen aller Zellbestandteile und nach 48 Stunden Einwirkung Verlust ihrer Entwicklungsfähigkeit. Kolloides Silber führt, wie erwähnt, nach mehrstündiger Einwirkung zu einer Retraktion des Plasmas von der Zellmembran (Plasmolyse). JAKOB und MAHL (1948) beobachteten bei einem Tetanusstamm X (nicht dagegen bei Stamm Zeißler) die Adsorption von Silber- und Goldkolloid in neutraler bis schwach saurer Lösung. B. coli, B. diphtheriae, B. pyozyaneum, B. prodigiosum, B. putrificus ver., Rauschbrandbakterien u. a. bleiben reaktionslos. B. Novy adsorbierte in schwach essigsaurer Kolloidlösung. Die Adsorption der negativ geladenen Metallkolloide fand besonders an den Geißeln statt, die eine den Zellmembranen entgegengesetzte elektrische Ladung tragen. Bei einem geißellosen Proteusstamm adsorbierten manche Membranen an der ganzen Oberfläche oder auch nur partiell, was als Ausdruck einer Ausscheidung elektropositiver Stoffwechselprodukte angesehen wird. Metallkolloide vermögen im Gegensatz zu den Metallionen nicht durch die Zellmembran hindurchzutreten.

Eingehendere Untersuchungen sind über die Wirkungsweise der Sulfonamide durchgeführt worden. Entsprechend unserer Aufgabe kann nur der enge Ausschnitt betrachtet werden, der sich mit den Änderungen der Zellstruktur befaßt. Vonkennel, Kimmig und Lembke (1943) fanden ein mit der Hemmung der Gasbildung einhergehendes, frühzeitiges Altern von Gonokokken, d h. frühzeitige Strukturänderungen, wie sie denjenigen entsprechen, die mit dem Altern der Kultur einhergehen. Liebermeister (1944) beobachtete an haemolytischen Staphylokokken erhebliche Vergrößerungen der sulfonamidgeschädigten Zellen, die auf ein fortgesetztes Wachstum bei gehemmter Teilungsfähigkeit hinweisen. Im gleichen Sinne sind die Fadenbildungen bei Typhusbakterien zu deuten, die Gärtner (1944) gezeigt hat. Außerdem ließen sich an den Aufnahmen Plasmolyse und mitunter ein so starker Plasmaschwund nachweisen, daß diese Zellen mit Sicherheit als abgestorben gelten können. Gärtner (1943) hat in seinen ausführlichen, zum Teil fluoreszenzmikroskopischen Untersuchungen auch elektronenmikroskopische Bilder gebracht, die ebenfalls das frühe Auftreten von Alterserscheinungen zeigen. Seine hochvergrößerten Aufnahmen geben meistens für den Ungeübten schwer beurteilbare Zellteilstücke wieder, die teilweise Elektronenschäden zeigen dürften. Man sieht auf seinen Abbildungen Auflockerung und Schwund von Zytoplasma, sowie Austritt von Zytoplasma durch die Membran (Plasmoptyse) (vgl. auch Gärtner 1944). Die im Elektronenmikroskop sichtbare Zahl strukturell veränderter Zellen ist identisch mit der Zahl, die nach fluoreszenzmikroskopischen Untersuchungen als abgestorben gelten können. Stickl und Gärtner (1943) betonen wie Vonkennel, Kimmig und Lembke die Ähnlichkeit der Zellschäden durch Alterung der Kultur und durch Sulfonamidzusätze. Sie beschreiben ferner die Ausbildung einer lockeren Protoplasmastruktur, Plasmolyse und Plasmoptyse. Die Art des Plasmaaustritts durch die geschädigte Membran soll charakteristisch für Sulfonamidwirkungen sein.

Untersuchungen mit einem Penicillinpräparat, das dem Verfasser 1944 von Levaditi dankenswerterweise zur Verfügung gestellt worden war, zeigten, daß Penicillinzusatz zu Bakteriensuspensionen ebenfalls zu Teilungshemmungen und zu Zytoplasmaschwund in den Zellen führt. Das gleiche zeigen eine kurze Mitteilung von Weiss (1943) und auch für weitere Antibiotika die ausführlichen Untersuchungen von Babudieri (1948) und Scanga (1948). Auch Lysozym führt zu einem Zerfall des Zellinhalts und schließlich zu dessen Auflösung (Babudieri 1944).

Kurze Zusammenfassungen der bakteriologischen Ergebnisse gibt Mudd 1948/49. Über das Problem des Bakterienzellkerns berichtet Piekarski (1949).

f) Serologische Reaktionen.

Die Abmessungen einzeln liegender molekularer Komponenten serologischer Reaktionen befinden sich an der Grenze oder unterhalb des elektronenmikroskopischen Auflösungsvermögens für organische Partikel. An den Reaktionsprodukten erkennt man aber die Anhäufung der molekularen Komponenten an, bzw. zwischen den reagierenden Oberflächen und erhält so Aufschlüsse über den Reaktionsmechanismus. Sowohl an Virusproteinen (v. Ardenne, Friedrich-Freksa und Schramm 1941, Anderson und Stanley 1941), als auch an Bakterien (Mudd und Anderson 1941, Mudd 1943) wurde wahrscheinlich gemacht, daß monomolekulare Filme auf den Oberflächen der Reaktionspartner die Präcipitation und Agglutination vermitteln. Auf die elektronenmikroskopische Untersuchung der Kapselquellungsreaktion von Pneumokokken und die des löslichen Rickettsienantigens wurde bereits oben hingewiesen (MUDD, HEINMETZ und ANDERSON 1943, SHEPARD und WYCKOFF 1946).

4. Bakteriophagen.

Im Anschluß an die Morphologie der Bakterien werden im Rahmen dieses Beitrags die bakterienpathogenen Virusarten behandelt. Wenn ihre Darstellung auch die Besprechung der Virusformen in absteigender Größenfolge unterbricht, da die *kleinsten* Phagenarten zu den kleinsten Virusformen rechnen, so lassen sie sich doch im Anschluß an ihre Wirtszelle am besten einordnen. Die gesonderte Darstellung ihrer Morphologie ist auch deshalb gerechtfertigt, weil sich bisher keine engeren Beziehungen zwischen Phagen (besonders den großen Formen) und irgendwelchen anderen, nicht bakterienpathogenen Virusarten gefunden haben. Die „Phagenähnlichkeit" (BANG 1946) des New Castle Virus (siehe S. 345, Abb. 118), soweit man überhaupt davon sprechen kann, gilt nur für wenige Formen, da das New Castle Virus im Gegensatz zu den Phagen äußerst polymorph ist. Die Ähnlichkeit der großen Phagen mit den Leptospirogenen JAKOBS beruht ebenfalls nicht auf einer näheren Wesensverwandtschaft, da es sich im einen Falle um Parasiten der Bakterien, im anderen um Fortpflanzungsformen der Leptospiren handelt.

Nach den Ergebnissen von NORTHROP (1938) konnten die Bakteriophagen als Phagenproteine zu den pflanzenpathogenen Virusproteinen gerechnet werden. Sie wären danach bei den makromolekularen Virusformen zu behandeln. Jedoch hat die Elektronenmikroskopie wichtige Gründe dafür aufgedeckt, wenigstens die in ihren feineren Einzelheiten sichtbar gewordenen, *großen* Bakteriophagen nicht mehr unter die Makromoleküle einzureihen. Eine Trennung der Phagen in kleine, molekulare und große, nichtmolekulare Formen durchzuführen, erscheint wegen der äußerst gleichartigen Wirkung der Phagen aller Größen zur Zeit nicht gerechtfertigt. Sie würde die Aufteilung der Agentien, die Gegenstand der Virusforschung sind, weiter komplizieren.

a) Größe und Form.

Indirekte Meßverfahren hatten ergeben, daß verschiedene Phagenstämme, auch wenn sie gegen die gleiche Bakterienart wirksam sind, sehr ungleiche Größen besitzen können, daß aber innerhalb desselben Stammes die Größe sehr gleichartig ist. Die Grenzen wurden etwa mit 10 und 120 mμ angegeben. Ruhrbakteriophagen, deren Wirkung D'HERELLE zur Entdeckung des Phänomens der bakteriophagen Lyse führte, waren auch die ersten Phagen, deren übermikroskopisches Bild (Abb. 99) von H. RUSKA (1941) gezeigt wurde. Die Form der Phagen erinnert an die eines Spermiums oder eines sporentragenden Tetanusbazillus. Der einzelne Phage besteht aus einem massendichten (dunklen) kugeligen bis eiförmigen Kopfteil und einem stäbchenförmigen Fortsatz. Der kugelige Teil kann eine Innenstruktur erkennen lassen und mehr oder weniger deutlich gedoppelt sein (diplokokkenförmig). Der Fortsatz zeigt vorwiegend einheitliche Länge, kann aber auch fehlen. Das Fortsatzende ist mitunter büschelartig verbreitert. Die großen Phagen messen am Kopfteil etwa 60 × 90 mμ und einschließlich des Fortsatzes etwa 250 mμ in der Länge. Die Stärke des Fortsatzes beträgt 15 bis 20 mμ und entspricht damit einerseits etwa der Dicke einer Bakteriengeißel, also einer Protoplasmadifferenzierung, andererseits der Breite des makromolekularen Tabak Mosaikvirusproteins. Die morphologische Gliederung der aufge-

fundenen Teilchen war so überraschend, daß ihre Identität mit den „Phagen"
zunächst zweifelhaft schien, obwohl sie einerseits nur in lytischen Bakterien-
kulturen, andererseits außer in Ruhrlysaten auch in solchen von Strepto-, Entero-
und Staphylokokken, sowie von Coli- und Proteusbakterien gefunden werden
konnten. Später wies dann H. Ruska (1942) auch kleinere und anders gestaltete
Phagen nach. Alle Formen wurden in Würdigung der Verdienste von F. d'Herelle,

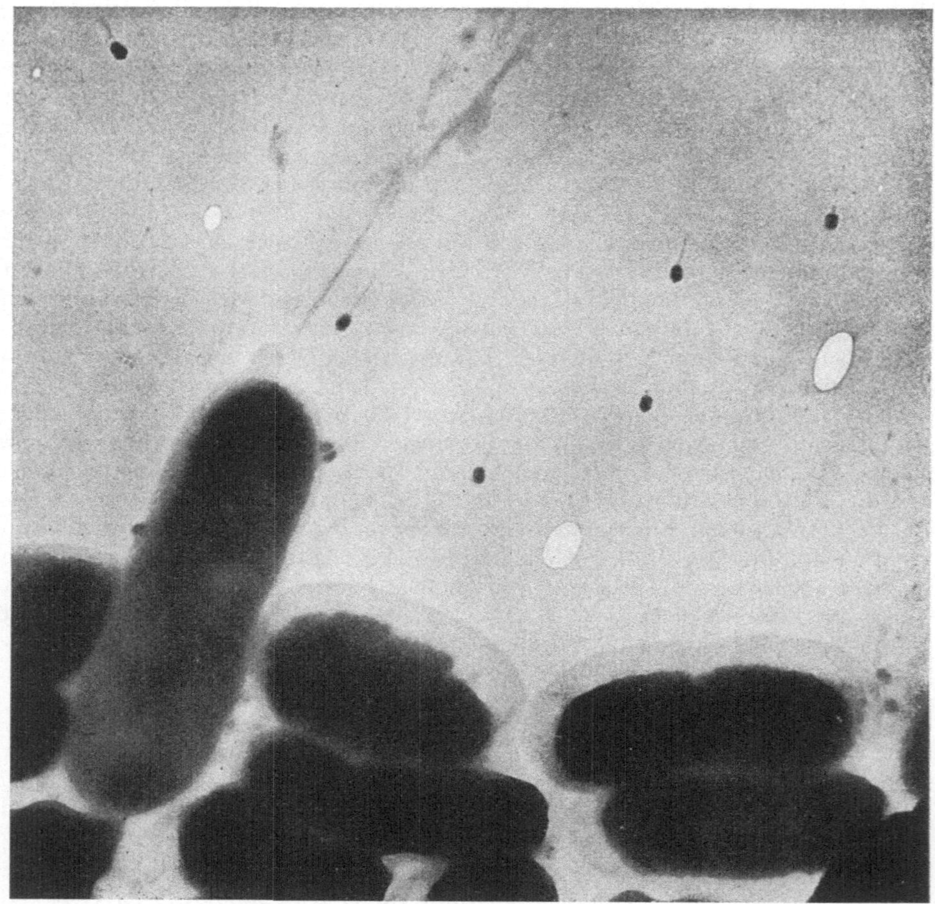

Abb. 99. Ruhrbakterien und keulenförmige Bakteriophagen. Etwa 21 000 : 1. Nach H. Ruska (1942).

dessen Arbeiten die intensive Phagenforschung ins Leben gerufen haben, als
„d'Herellen" bezeichnet. Nach der verschiedenen Gestalt sind kugelförmige
.(Abb. 100), keulenförmige (Abb. 99) und stäbchenförmige (Abb. 101) d'Herellen
zu unterscheiden. Ihr Volumen differiert fast um den Faktor 1000. Kleinere
kugelförmige Ruhrphagen (35 bis 45 mμ) zeigt Abb. 100, nahe an der Zellmembran
der Bakterien und frei aus der Folie liegend. Sie sind sehr kontrastarm. Es
erscheint nicht ausgeschlossen, daß sie wie die zuerst beschriebenen „Keulen-
formen" ebenfalls einen Fortsatz besitzen, der aber wegen seines zu geringen
Kontrastes beim einfachen Durchstrahlungsverfahren unter der elektronenmikroskopischen Nachweisbarkeitsgrenze liegt. Auch manche Enterokokken-

phagen erscheinen auf den ersten Blick kugelförmig. Sie haben einen Durchmesser von etwa 70 mμ, sind aber wesentlich kontrastreicher (Abb. 102). Bei genauer Betrachtung sieht man, daß sie einen sehr zarten Fortsatz besitzen. In Kolikulturen wurden außerdem „Stäbchenformen" mit Abmessungen von 35 × 140 mμ gefunden (Abb. 103). An den Enden tragen sie meist zwei punktförmige, nicht ganz symmetrische Verdichtungszentren von 25 mμ Durchmesser;

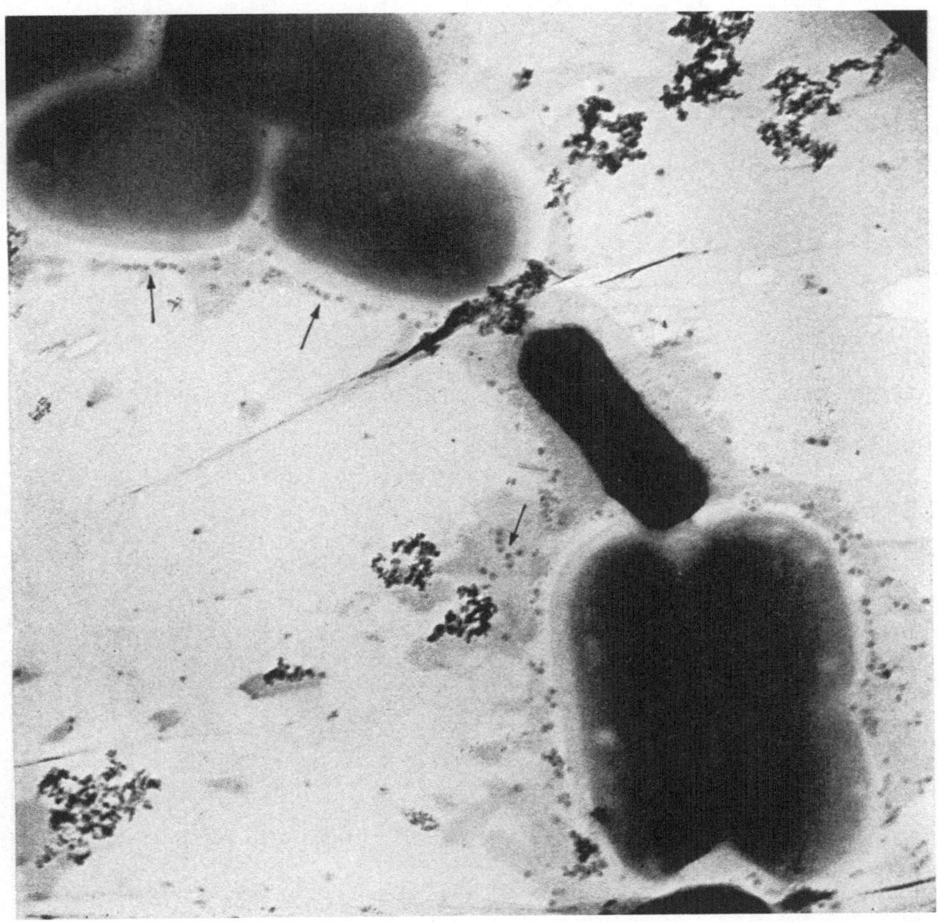

Abb. 100. Ruhrbakterien und kugelförmige Bakteriophagen. Etwa 20000 : 1. Nach H. RUSKA (1942).

sie erinnern dadurch an das lichtmikroskopische Bild von Diphtheriebazillen mit Polkörperchen. Phagenstämme, die diese Form zeigen, scheinen selten zu sein und wurden bisher in der anglo-amerikanischen Literatur nicht beschrieben. U. KOTTMANN (1942) hat sie eingehend untersucht, indem er nach dem Abklatschverfahren Präparate von „sterilen Flecken" herstellte. Abb. 103 zeigt die Stäbchen in charakteristischer Lagerung um die Zelle und einzeln oder zu mehreren auch freiliegend. Die anfängliche Vermutung, daß es Übergänge zwischen den verschiedenen Formen bei der Phagenvermehrung gäbe, wurde später (H. RUSKA 1943) wieder fallen gelassen. Neuerdings werden jedoch Entwicklungsformen diskutiert, die aber nicht mit den verschiedenartigen ausgereiften Endformen

übereinstimmen. Möglicherweise lagen bei den ersten Untersuchungen zum Teil
nicht ganz einheitliche Stämme (Mischphagen) vor. Edwards und Wyckoff

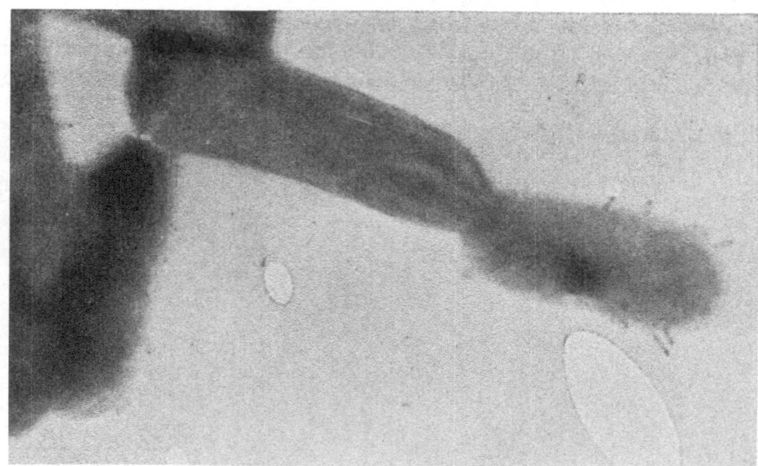

Abb. 101. Kolibakterien und stäbchenförmige Bakteriophagen. 20000 : 1. Nach H. Ruska (1942).

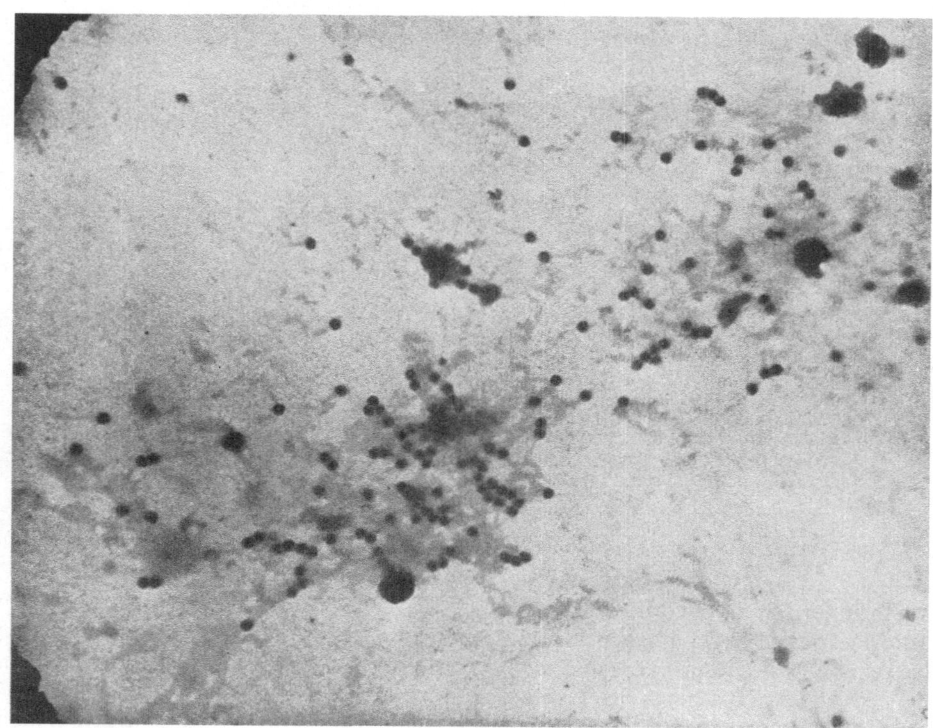

Abb. 102. Kugelförmige Phagen aus einer aufgelösten Enterokokkenkultur. 18000 : 1. Nach H. Ruska
(1942).

(1947), sowie Wyckoff (1947) stellten wie Kottmann 1942 Oberflächenbilder
von „sterilen Flecken" her, indem sie das Abziehverfahren benutzten. Die abge-

zogenen Filme, die teils das Relief der Oberfläche wiedergaben, teils haftengebliebene Bakterien und Phagen trugen, untersuchten sie nach zusätzlichen Metallbeschattungen.

Die Lagebeziehungen der d'Herellen zu den befallenen Bakterien und die

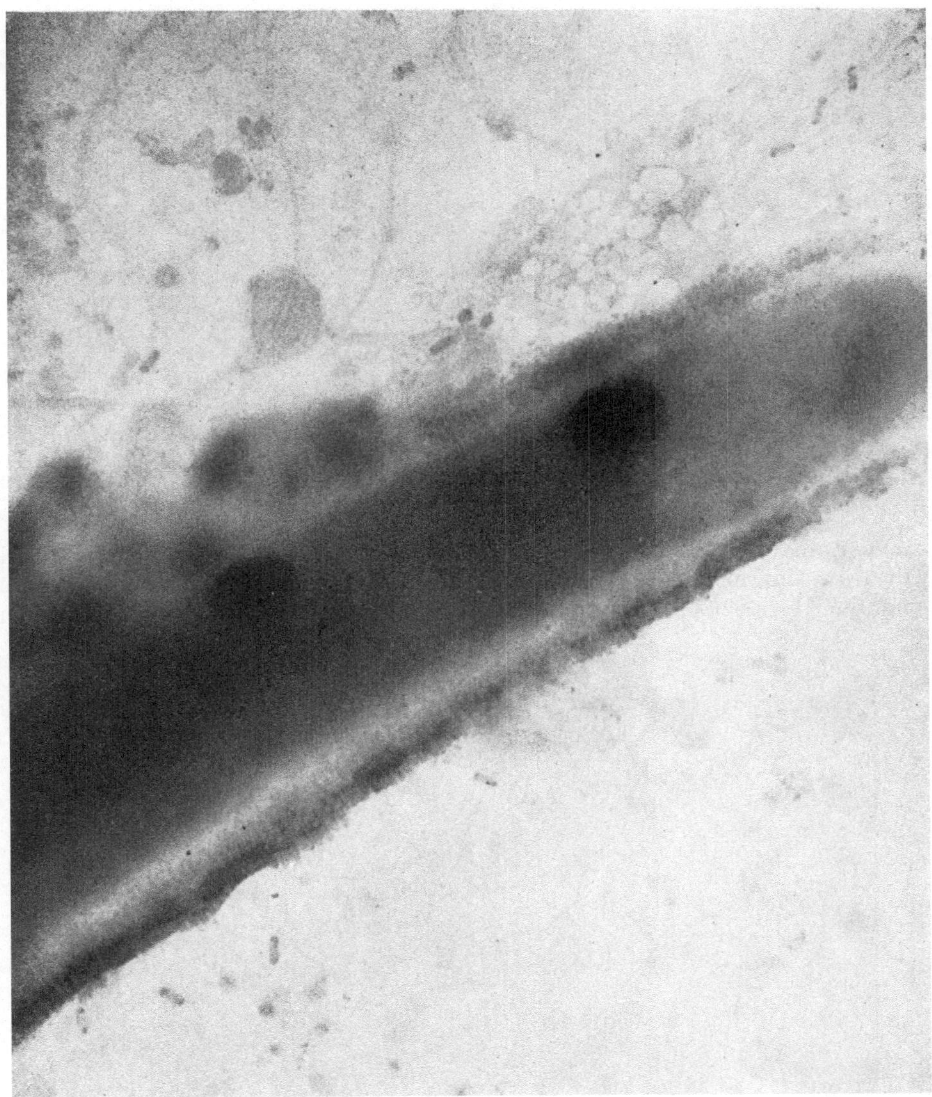

Abb. 103. Kolibacterium mit stäbchenförmigen Bakteriophagen. 32000 : 1. Nach KOTTMANN (1942).

ständige Bestätigung ihrer Formen bei verschiedenen Phagenstämmen, sowie die gleichen Befunde des AEG-Forschungsinstitutes, von LURIA und ANDERSON (1942), LURIA, DELBRÜCK und ANDERSON (1943), SHARP, TAYLOR, HOOK und BEARD (1946) ermöglichten es, die d'Herellen mit den Phagen zu identifizieren, wenn auch der elektronenmikroskopischen Abbildung keine völlige Reindarstellung des wirksamen Prinzips der bakteriophagen Lyse vorausgegangen war. Bedenken gegen die Identität der morphologisch nachgewiesenen Partikel mit den Phagen, die daraus abgeleitet werden konnten, daß Bloch (1940) eine Aufspaltbarkeit der Phagen in den 10000. bis 100000. Teil nachgewiesen zu haben glaubte, konnten durch H. Ruska (1945, erschienen 1948) zerstreut werden. Es gelang

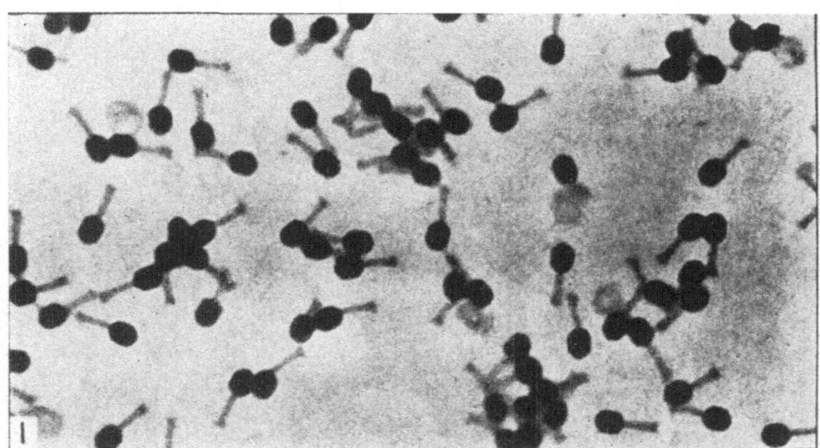

Abb. 104. Gereinigte, keulenförmige Kolibakteriophagen (T_2) mit 0,023 mol Kalziumchlorid präpariert 41000 : 1. Nach Hook, Beard, Taylor, Sharp und Beard (1946).

trotz Einhaltung der Blochschen Versuchsanordnung und Kontrolle durch Ultrafiltration nicht, die Aufspaltbarkeit der Phagen zu bestätigen. Außerdem konnten H. Ruska und Menze (1945, erschienen 1948) an sehr verschieden

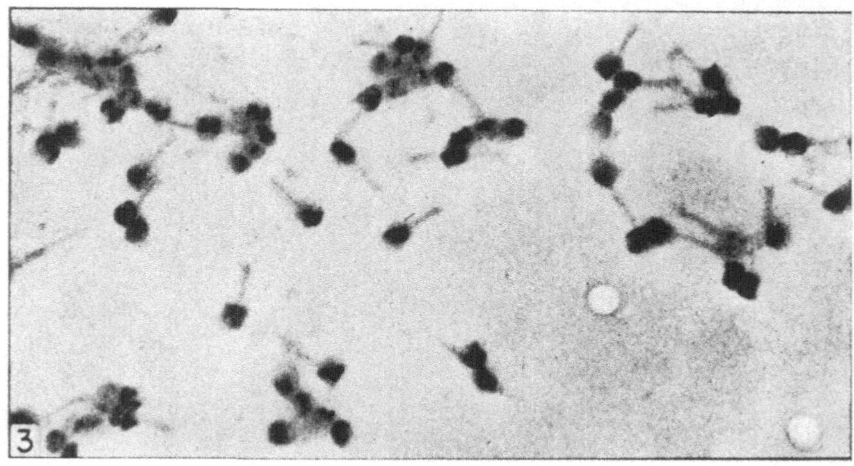

Abb. 105. Die gleichen Phagen wie in Abb. 104. Durch Dialyse durch destilliertes Wasser geschädigt. 41000 : 1. Nach Hook, Beard, Taylor, Sharp und Beard (1946).

großen Typhusbakteriophagen zeigen, daß die durch Filtrations-Endpunkt-bestimmung festgestellte Phagengröße mit der elektronenmikroskopisch sichtbaren Größe übereinstimmt. Nur die kleinsten Phagen konnten, obwohl sie mit 8 bis 10 mμ Durchmesser noch innerhalb der Sichtbarkeitsgrenze liegen, wegen der präparativen Schwierigkeiten und des Mangels an Kontrast nicht im Bild nach-

gewiesen werden. Die Form- und Massenunterschiede der vielerlei Phagen sind
so beträchtlich, daß man sie trotz ihrer großen Variabilität nicht wie D'HERELLE
(1942) als verschiedene Rassen ein und desselben Organismus auffassen kann,
zumal nach DELBRÜCK und Mitarbeitern bei den Variabilitätserscheinungen
stets der gleiche Formtyp beibehalten wird.

SHARP, HOOK, TAYLOR, BEARD und BEARD (1946) konnten die pH- und
Konzentrationsabhängigkeit der Sedimentationskonstanten mit der Aggregation
und Orientierung der Phagen im Schwerefeld in Übereinstimmung bringen. Die
gleichen Autoren (siehe HOOK und Mitarbeiter 1946) wiesen eine schwere Schädigung der Phagen in destilliertem Wasser nach, bei der insbesondere die Kopfteile
in Auflösung übergehen und schließlich völlig zerstört werden (Abb. 104 und 105).

b) Wirkungsmechanismus und Vermehrung.

Nach den zu Beginn der elektronenmikroskopischen Untersuchungen herrschenden Vorstellungen war zu erwarten, daß die Phagenteilchen an den Bakterien-
oberflächen zunächst haften, dann einzeln oder zu mehreren in die Zellen ein-

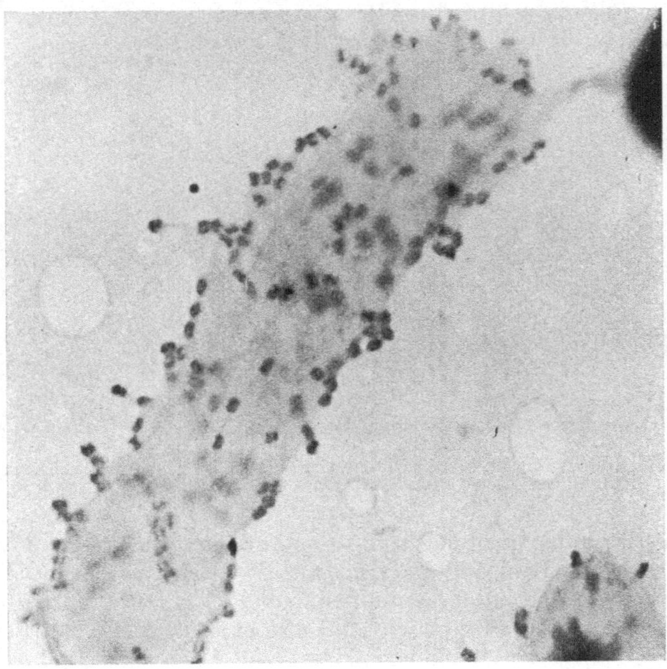

Abb. 106. Membran eines Proteusbakteriums von keulenförmigen Bakteriophagen besetzt. Die
Phagenfortsätze sind nach dem Zellinnern gerichtet. 29 000 : 1. Nach H. RUSKA (1941).

dringen und je nach dem Ausmaß ihrer Wirkung zu Änderungen der Wuchsform,
zu Mutantenbildung oder Zellauflösung führen. Die Phagenvermehrung wurde
im Zellinnern, möglicherweise im Zusammenhang mit der Bildung von Virus-
kristallen oder einer intrazellulären Viruskolonie gesucht, aus der die Phagen
nach lytischer Zerstörung der Bakterienmembran frei würden. Die Befunde,
welche H. RUSKA und KOTTMANN durch die übermikroskopische Abbildung
erhoben haben, weichen jedoch in zahlreichen Einzelheiten von den erwarteten
Ergebnissen ab. H. RUSKA kam zu der Auffassung, daß die Phagen *nicht* völlig

in das Innere der Bakterienzellen eindringen, sondern, wie es Abb. 106 zeigt, in bestimmter Orientierung außen an der Membran haften bleiben (Okkupationshypothese). Die Haftung wird, wie es an zahlreichen Abbildungen zu sehen ist, durch den Fortsatz, der wahrscheinlich in die Zellmembran eindringt und bis ins Zytoplasma reicht, oder durch einen Stäbchenpol vermittelt (vgl. auch Anderson 1948). Diese Teile sind also für die Haftung der Phagen an den Bakterien bestimmend, wobei nach Anderson (1945) und Delbrück (1948) noch die Anwesenheit eines oder mehrerer Cofaktoren (1 — Tryptophan, Ca^{++})

Abb. 107. Kolibakterien mit T_3-Phagen infiziert. Plastische Darstellung von regelmäßigen Aushöhlungen nach Chrombedampfung. 40000:1. Nach Wyckoff (1947).

erforderlich ist. Falls die Phagen nur von außen einwirken, muß angenommen werden, daß sie die Fähigkeit besitzen, aus den Bakterien eventuell vermittels des Fortsatzes Substanzen zu assimillieren. Bei vergleichenden Untersuchungen sieht man elektronenmikroskopisch zunächst eine Vergrößerung der befallenen Bakterien (vgl. auch den Befund von Hillier, Mudd und Smith (1949), dann zunehmende Substanzverarmung, bis nur noch Zellmembranen, Membranfetzen und ungeformte Zytoplasmareste übrigbleiben. Kernäquivalente oder Volutingranula sind nicht mehr aufzufinden. Vgl. dagegen die abweichenden Angaben von Bielig, Kausche und Haardick (1949). Aus chemischen Analysen kann gefolgert werden, daß die Nukleinsäure der Wirtszelle wahrscheinlich unter teilweiser Umwandlung von Hefe- in Thymo-Nukleinsäure in die Bakteriophagen übergeht. v. Borries (1949) hat darauf hingewiesen, daß in Abb. 106 nicht nur die am Membranrand liegenden Phagen, sondern auch die über der *Fläche* sichtbaren wegen ihrer übergangslosen, entweder scharfen oder verschwommenen Darstellung außerhalb der Membran liegen müssen. Die scharf umrissenen Phagen liegen objektivnahe und unscharf die umrissenen objektivfern, nicht jedoch wegen ungenügender Tiefenschärfe des Bildes, sondern wegen der Verwischung der Phagenkontur durch die dazwischenliegenden beiden Schichten der schlauchförmigen Bakterienmembran. Auch die Reliefbilder, die WYCKOFF (1947/48) von der Oberfläche phageninfizierter Kolikulturen erhalten hat, können dafür sprechen, daß die Phagenproduktion an Oberflächen der Bakterienzelle vor sich geht, obwohl WYCKOFF vermutet, daß nach Invasion eines einzelnen Phagen das Bakterienprotoplasma in Phagen verwandelt wird. Es ist gleich noch auf diese, vor allem von DELBRÜCK vertretene Auffassung zurückzukommen. Hier sei nur darauf hingewiesen, daß WYCKOFFS Befunde eine schöne Bestätigung und Erweiterung der KOTTMANNschen Ergebnisse sind, indem die von KOTTMANN gezeigte regelmäßige Phagenanordnung am Zellrand (Abb. 103) nunmehr auch von der Fläche aus gezeigt wird (Abb. 107).

Auch DELBRÜCK (1945/46/47) und seine Mitarbeiter versuchten, Phagen und Phagenvermehrung morphologisch im Zell*innern* nachzuweisen, ohne jedoch zu eindeutigen Befunden zu kommen. Sie stützen ihre „Penetrationshypothese“ auf die Ergebnisse ihrer serologischen und genetischen Versuche. Phagenantiserum vermag in einem Gemisch von Phagen und Bakterien die freien Phagen zu neutralisieren, nicht aber bereits infizierte Zellen vor der Lyse zu schützen. Außerdem gelingt es nicht, eine einzelne Bakterienzelle mit zwei völlig verschiedenartigen Phagen zu infizieren. Es wird deshalb angenommen, daß in der Regel nur ein Phage in die Zelle eindringt, dort vor der Einwirkung des Antiserums geschützt ist und die Zellmembran für weitere Phagen undurchdringbar macht (vgl. Befruchtungsvorgang). Der infizierte Phage ist aber für die Einwirkung des Antiserums auch dann unzugänglich, wenn der Fortsatz Angriffsort des Antiserums ist und der Phage, entsprechend unseren Beobachtungen, umgekehrt wie in einem Schema von DELBRÜCK (1945/46) bei freibleibendem Kopfteil mit dem Fortsatz in die Zellmembran eindringt. Der so entstehende Zustand könnte ebensogut wie das völlige Eindringen des ersten Phagen anderen den Zutritt unmöglich machen. Die häufig sichtbare Orientierung der Phagen mit dem Kopfteil zum Bakterium (Abb. 99) ist ein Effekt, der dadurch eintritt, daß bei der Auftrocknung der sich in Richtung auf die Bakterien zurückziehende Flüssigkeitstropfen die beliebig orientierten Phagen mit ihrem Kopfteil zur Bakterienzelle dreht. Der gleiche Effekt läßt sich im Modellversuch mit entsprechend geformten Glasstäbchen und fließendem Wasser leicht nachmachen. Er tritt auch bei abgetöteten Phagen auf und spricht somit nicht dafür, daß sich die Phagen aktiv, mit dem Kopf nach vorn gerichtet, auf die Bakterien zu bewegen oder gar mit dem Kopf in die Zelle eindringen. Eine weitere Beobachtung von DELBRÜCK (1945), daß nämlich bei Zusatz zweier verschiedenartiger UV-bestrahlter Phagensuspensionen (oder Wildform und r-Form) zu einer Bakterienkultur ein nicht infizierender Phage der einen Form die Neubildung des infizierenden Phagen der anderen Form beeinflussen kann und daß seine Einwirkung durch spezifisches Antiserum blockiert wird, läßt sich sogar besser verstehen, wenn nicht der ganze Phage, sondern nur der Fortsatz in die Membran eindringt. Vgl. ferner die Arbeiten von DELBRÜCK 1949/5 und 6, sowie BULBECCO 1949.

Die Penetrationshypothese ist aber durch morphologische Beobachtungen von WYCKOFF (1948/2) erneut zu stützen versucht worden. Man sieht auf den von ihm gewonnenen Aufnahmen die Abdrücke der Köpfe keulenförmiger Phagen in zerfallenden membranlosen Protoplasmamassen und zerplatzenden Zellen, aus welchen neben ihrer Dimension nach makromolekularen Kugeln und Fäden auch Phagen frei werden. Allerdings ist die Zahl der Phagen platzender Zellen verhältnismäßig gering gegenüber denjenigen, die auf Membranen oder Protoplasmaresten auftreten. Letztere stimmt bei einer Größenordnung von

100 mit der indirekt ermittelten Zahl der Phagen überein, die je Bakterienzelle gebildet werden. Besonders bemerkenswert ist das Auftreten von bläschenförmigen leeren Phagenköpfchen, die als unreife, in Entwicklung begriffene Phagen angesehen werden. Mit einer ähnlichen Untersuchungstechnik war auch schon Kottmann (1942) zur Vermutung eines Entwicklungsgangs gekommen. Es warten also hier noch wichtige Fragen der Phagenvermehrung auf eine Entscheidung. Die kontinuierliche Anlagerung von gleichen Teilbausteinen, wie sie bei der autoreproduktiven Vermehrung der pflanzenpathogenen Virusproteine erörtert wird, erscheint in Anbetracht der differenzierten Phagenstruktur unmöglich. Nach Luria (1947) kann damit gerechnet werden, daß sich bei großen Phagen 30 bis 50, bei kleinen weniger als 8 bis 10 Einheiten (Gene ?), die voneinander bis zu einem gewissen Grad unabhängig sind, in der Bakterienzelle reproduzieren und im ursprünglichen Mosaik oder in mutierter Form wieder zum Phagen zusammenfinden. Es ist dabei nicht notwendig, daß alle Einheiten von demselben Phagen stammen, vielmehr können sich die Einheiten aus verwandten, durch Bestrahlung inaktivierten Phagen gegenseitig vertreten. Die großen Phagen sind also bereits ein kompliziertes System verschiedener Duplikanten. Kombiniert man die Ergebnisse der Phagengenetik mit den morphologischen Befunden, so liegt die Annahme nahe, daß die Einheiten mittels des Phagenstiels aus dem Kopfteil in das Bakterium gelangen, sich darin reproduzieren und neue Phagen aufbauen. Danach würde sowohl die Okkupationshypothese als auch die Penetrationshypothese einen richtigen Kern enthalten.

Der morphologischen Gliederung keulenförmiger Phagen entspricht nach den festgestellten Lagebezeichnungen zwischen Phagen und befallenen Bakterien eine funktionelle Differenzierung des Fortsatzes als Haftorgan. Der streng gegliederte Aufbau ergab ein wesentliches Argument gegen die Auffassung, daß die Phagen fermentartige Proteinmoleküle seien. Das physikalisch-chemische Verhalten hatte zwar zur Molekülauffassung geführt, war dafür aber nicht beweisend. Weder ist eine Kristallisation des „Phagenproteins" gelungen, noch konnte nachgewiesen werden, daß die wirksamen Einzelteilchen selbst wie die Elementareinheiten der stäbchenförmigen Pflanzenviren eine durch das Röntgendiagramm erfaßbare Kristallstruktur besitzen. Infolgedessen hat durch die Elektronenmikroskopie und die Phagengenetik die Auffassung von F. d'Herelle, wonach die wirksamen Einheiten kleinste belebte Elemente sind, eine neue und, wie es scheint, endgültige Stütze erfahren. H. Ruska und C. Menze (1948) haben außerdem darauf aufmerksam gemacht, daß die zu geringe Phagengröße, die bei großen Phagen durch die statistische Ultramikrometrie von Wollmann und Lacassagne ermittelt wurde, darauf hindeutet, daß das Trefferzentrum ähnlich wie bei Bakterien und anderen Einzellern nicht mit dem ganzen bestrahlten Partikel identisch ist, sondern nur mit einer kleinen Partialstruktur. Hieraus folgt, daß der Phage nicht makromolekular ist, sondern eine organismische Struktur besitzt, die aus zahlreichen Molekülkomplexen besteht. In gleichem Sinne spricht die von Anderson (1945) gezeigte Abtrennbarkeit einer lytisch wirksamen Substanz des T_2-Phagen.

5. Filtrierbare Mikroorganismen und „große Virusarten".

Unter filtrierbaren Mikroorganismen werden die von Sabin und Klieneberger beschriebenen L-Organismen, die von Laidaw und Elford, sowie von Seiffert dargestellten Abwasserorganismen, ferner die Erreger der Pleuropneumonie der Rinder, der Agalaktie der Ziegen und Schafe, der Polyarthitis der Ratte, von einigen Pneumonieformen bei Hund, Ratte und Maus und andere

verstanden. Sie sind polymorph, zeigen schon im Lichtmikroskop Ring-, Kugel-
und Fadenformen, sowie reproduktionsfähige filtrierbare „Elementarkörper".
Letztere wachsen in zellfreien Kulturen auf geeigneten Nährböden bis zur Bildung
kleiner 10 bis 600 μ messender Kolonien heran, die ein dunkleres Zentrum und
eine durchscheinende Außenzone besitzen.

SABIN (1941), sowie KLIENEBERGER und SMILES (1942) sind auf Grund lichtmikroskopischer Untersuchungen zu der Auffassung gekommen, daß es kaum
möglich sei, die filtrierbaren Mikroorganismen unter die Schizomyceten einzureihen. SABIN schlägt als neue Klassenbezeichnung Paramycetes mit einer
parasitischen und einer saprophytischen Familie vor.

Zur Gruppe der „großen" nach GIEMSA färbbaren Virusarten rechnet GÖNNERT
das Virus des Lymphogranuloma inguinale (venereum), der Psittakose, des
Trachoms, der Schwimmbadkonjunktivitis und der von ihm gefundene Bronchopneumonie der Maus. RAKE
und Mitarbeiter (1946) sprechen von der Psittakose-
Lymphogranulom-Trachom-
Gruppe, zu der sie noch ein
menschenpathogenes Pneu-
monievirus (S. F.) und das
Virus der Katzenpneumonie
(Feline Pneumonitis, Baker)
rechnen. Sie bringen hierdurch die Zusammengehörigkeit in gleicher Weise zum
Ausdruck, wie mit der
späteren Wiederaufnahme
(RAKE 1947) des Begriffs
Chlamydozoaceae von LIP-
SCHÜTZ. EATON und Mit-
arbeiter (1945) fanden hier-
hergehörige Erreger atypischer menschlicher Pneu-

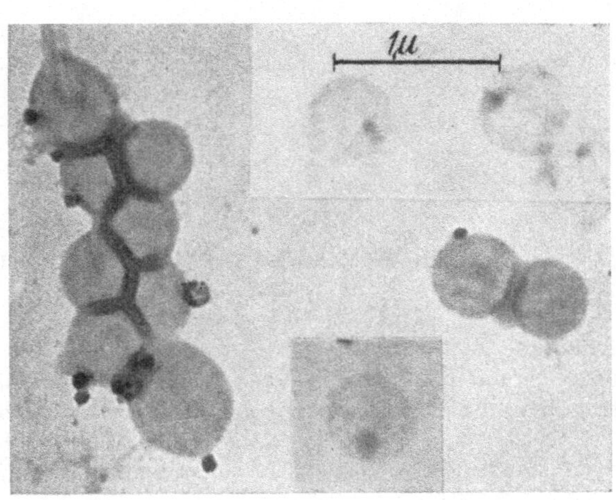

Abb. 108. SEIFEERTsche Mikroorganismen aus Serumbouillon.
Links: 5 Tage; oben und unten: 10 Wochen alte Kulturen.
Osmiumfixation. Nach H. RUSKA und POPPE (1947).

monien. Die Erreger wachsen nach bisherigen Beobachtungen nicht auf zellfreien Medien, sind ebenfalls polymorph und bilden aus den Elementarkörpern
intraplasmatische Initial- und Einschlußkörper, deren Entwicklung weitgehend
den Entwicklungsformen der filtrierbaren Mikroorganismen gleicht. (Siehe das
Schema bei KIKUTH und GÖNNERT (1949). H. RUSKA und POPPE (1947) wiesen
auf Grund elektronenmikroskopischer Befunde auf die Ähnlichkeit dieser beiden
Gruppen hin, die bisher nur mit Rücksicht auf die Züchtbarkeit voneinander
getrennt wurden, und schlugen als gemeinsamen Namen dieser neuaufgestellten
Klasse von *Mikroorganismen mit bläschenförmigen Elementarkörpern* die Bezeichnung Cysticetes (RUSKA, nov. class.) vor. Wie bei den Bakterien und Rickettsien besteht auch hier eine aus der Morphologie erschlossene Zusammengehörigkeit zwischen frei lebenden, saprophytischen, parasitischen und obligat zell-
parasitischen Formen (Virus).

Die einfachste Form der SEIFFERTschen Mikroorganismen ist nach H. RUSKA
und POPPE (1947/2) ein nahezu kreisrundes, fast völlig homogenes, unfixiert sehr
kontrastarmes, zwischen 400 und 700 mμ messendes Scheibchen. Bei engerer
Lagerung schieben sich die Scheibenränder übereinander oder stauchen sich
gegenseitig (Abb. 108). Die Dicke der Scheiben beträgt schätzungsweise 20
bis 30 mμ. In sehr alten Kulturen finden sich löcherige oder zerrissene, wahr-

scheinlich abgestorbene Exemplare (Abb. 108) mit einzelnen dunkleren Körnchen. Andere, oft in größeren Haufen liegende Elementarkörper sind kompakter und durchschnittlich kleiner als die zuerst beschriebenen. Im gleichen Präparat können auch verschieden kontrastreiche Elemente nebeneinander auftreten. In frischem, nicht ausgetrocknetem Zustand liegen wahrscheinlich nicht

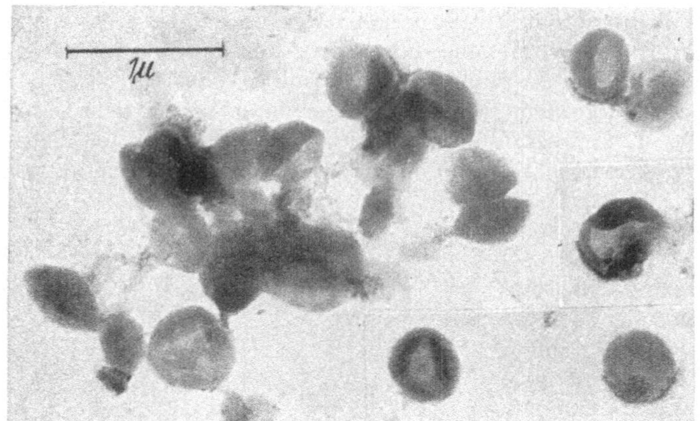

Abb. 109. Seiffertsche Mikroorganismen, 2 Tage alte Kultur. Osmiumfixation. Nach H. Ruska und Poppe (1947).

scheiben-, sondern bläschenförmige Gebilde vor, die erst nach ihrer Auftrocknung je nach der Größe ihres Wassergehaltes als dünne, runde Scheiben (kontrastarm) oder kokkenartig (kontrastreich) erscheinen.

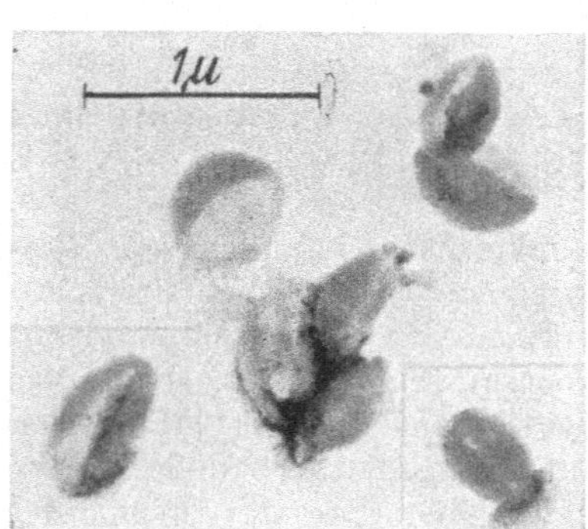

Abb. 110. Seiffertsche Mikroorganismen, 2 Tage alte Kultur. Osmiumfixation. Nach H. Ruska und Poppe (1947).

Weitere Formen unterscheiden sich von den bisher beschriebenen dadurch, daß sie in der Schwärzung, d. h. Dicke der durchstrahlten Substanz, nicht einheitlich sind. Außerdem ist der Umriß nicht immer kreisrund, sondern in Abhängigkeit von ihrer Lage im Strahlengang entweder kreisförmig, unsymmetrisch oval (wie der abnehmende Mond), oval oder zugespitzt oval. Die gleichmäßig runden Körperchen sind meistens außen am dicksten, so daß der Eindruck eines Ringes oder Siegelringes entsteht (Abb. 109). Es sind also abgeflachte Gebilde, die peripher einen Wulst und zentral eine Eindellung besitzen wie kernlose rote Blutkörperchen. Schon Tang und Mitarbeiter (1935) haben bei den Erregern der Lungenseuche diesen Vergleich gebraucht. Es wurden auch mehrere Eindellungen mit nach außen vorspringenden Falten beobachtet. Hierbei verläuft die stärkste Schwärzung quer über die Körperchen.

Sehr häufig ist die Eindellung von einer Seite aus so stark, daß durch Einstülpung Napf- oder Glockenformen entstehen (Abb. 110). Diese zeigen in der Seitenansicht die erwähnten verschiedenartigen asymmetrischen ovalen Umrisse. Die Einstülpungen können statt gleichmäßig rund auch schmal schlitzförmig oder dreieckig sein. Besonders deutlich werden die Napfformen, wenn die Invaginationsstelle zufällig ein zweites derartiges Gebilde teilweise in sich aufgenommen hat. Lichtmikroskopisch sind die Napfbildungen auch in feuchtem Zustand zu erkennen.

Bei den Erregern der Lungenseuche (Pleuropneumonia contagiosa) wurden Ring- und Invaginationsformen und außerdem zarte zusammengefaltete Bläschen mit eingeschlossenen Körnchen, sowie häufig membranartige kleine Bruchstücke gefunden. In mehrere Tage alten Kulturen der Lungenseuche-Erreger fanden sich außerdem fädige, ungegliederte bizarre Stränge von verschiedener Breite (Abb. 111), die schon Borrel 1910 abgebildet hat und die nach ihm fast alle Untersucher gesehen haben. Bei den SEIFFERTschen Mikroorganismen wurden keine ausgesprochenen Fadenbildungen gefunden, jedoch manchmal Protoplasmafortsätze (Abb. 112).

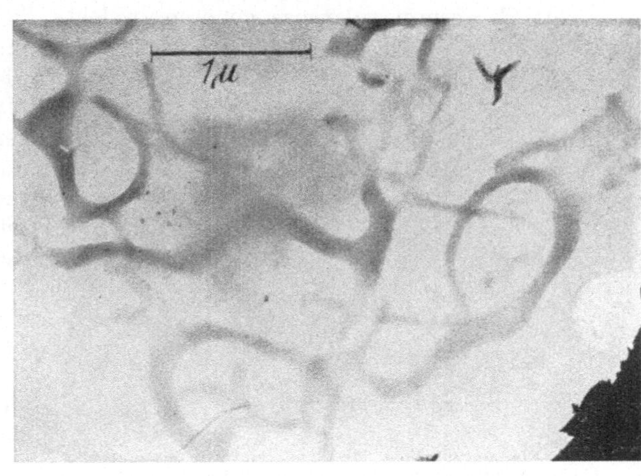

Abb. 111. Fädige Stränge aus einer Kultur von Lungenseuche-Erregern. Stamm Tessin, 6 Tage alte Kultur. Nach H. RUSKA und POPPE (1947).

Es scheint also möglich zu sein, daß die Fadenbildungen, in denen TANG und Mitarbeiter Protoplasmaströmung gesehen haben, unmittelbar von den runden Formen ausgehen. Die besonders von LEDINGHAM (1933) für die Erreger der Lungenseuche und von KLIENEBERGER für die L-Organismen betonte Plastizität besteht demnach auch bei den SEIFFERTschen Mikroorganismen (vgl. ferner WEISS 1944).

In einer Arbeit von SMITH, HILLIER und MUDD (1948) werden zwei aus der Cervix uteri isolierte L-Organismen beschrieben. L 4330 gleicht in seiner Morphologie und in seiner Empfindlichkeit gegen destilliertes Wasser (vgl. Abb. 6 der Originalarbeit mit Abb. 112) weitgehend den SEIFFERTschen Mikroorganismen. L 50 zeigt dagegen typische Bakterienformen

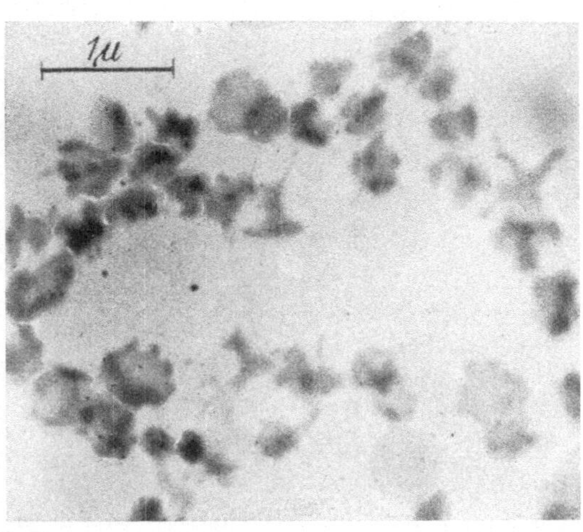

Abb. 112. SEIFFERTsche Mikroorganismen. Osmium-fixation. Nach H. RUSKA und POPPE (1947).

mit Zellmembran, Vakuolen und Volutingranulationen. Auch WEISS (1944) hat typisch bazilläre Formen eines L-Organismus beschrieben, der aus der Gelenkinfektion einer Ratte isoliert worden war. Von rein morphologischen Gesichtspunkten aus wären L-Organismen mit bazillären Formen nicht mit anderen in eine Gruppe zu stellen. Es ist aber auch möglich, daß Symbiosen

verschiedener Organismen vorliegen, von welchen die einen bazilläre Formen
bilden, die anderen Bläschen und pleomorphe Entwicklungsformen. Als weitere
Hypothese ergibt sich, daß zum mindesten ein Teil der L-Organismen pleomorphe Bakterien sind, die ähnliche Entwicklungsgänge wie die L-Organismen
zeigen. Zu einer solchen Gruppe würde nach Hesselbroock und Foshay (1945)
auch das Bacterium tularense gehören, welches diese Autoren zu den pleuro-
pneumonie-ähnlichen Orga-
nismen rechnen. Smith,
Mudd und Hillier (1948)
zeigen von Bacteroides funduliformis elektronenmikroskopische Bilder, die nicht
nur die übliche Zweiteilung
der Stäbchen zeigen, sondern
auch Vermehrungsformen, wie
sie von den L-Organismen
bekannt sind. Es bilden sich
große runde Körperchen in
der Mitte oder an den Enden
der Stäbchen. Diese werden
so groß wie die Länge der
Bakterien und senden Fäden

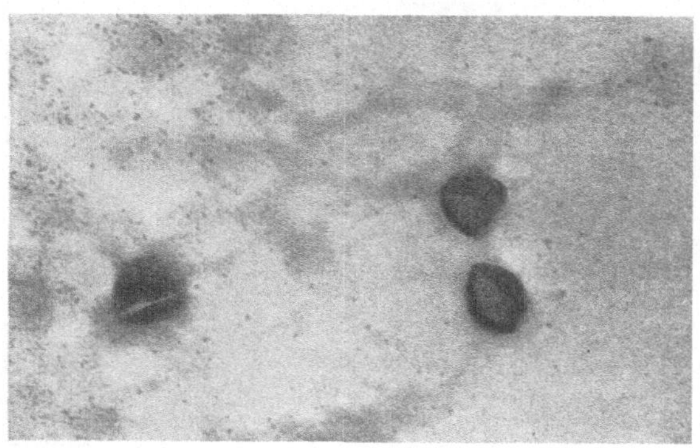

Abb. 113. Elementarkörper des Virus der Bronchopneumonie
der Maus. 20 000 : 1. Nach H. Ruska (1944).

aus, die sich zu neuen Zellen entwickeln. Außer normalen Kolonien sind unter
Penicillineinfluß L-Typ-Kolonien zu isolieren. Die L-Typ-Varianten können bei
B. funduliformis wieder in die normale Wuchsform übergehen, während andere

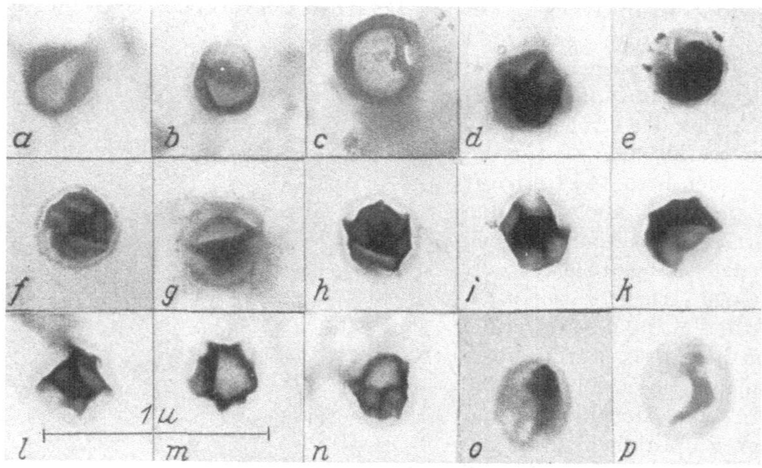

Abb. 114.   Verschiedene Elementarkörperformen der Bronchopneumonie der Maus aus mehreren
Präparaten. Nach H. Ruska (1944).

L-Organismen ihre Wuchsform beibehalten. Klieneberger ließ noch offen,
ob die L-Formen Symbionten der Bakterien sind. Durch Dienes (1945/46/47),
Smith, Mudd und Hillier (1948) scheint ein zusammengehöriger Entwicklungszyklus beider Formen erwiesen zu sein. Weitere Untersuchungen wären
nicht nur für die Systematik der Bakterien, der filtrierbaren Mikroorganismen
und der großen Virusarten, sondern auch für die Chemotherapie von Bedeutung,
da manche L-Varianten penicillinresistent sind.

H. Ruska (1944/18) hat das „Virus“ der Bronchopneumonie der Maus aus der erkrankten Lunge isoliert und damit die ersten „Elementarkörper“ (Elk.) einer der „großen Virusarten“ elektronenmikroskopisch untersucht. Sie gleichen morphologisch in vieler Hinsicht den Seiffertschen Mikroorganismen. Gegenüber vorher bekanntgewordenen Virusformen fiel außer der Größe und der starken Größenschwankung (230 bis 430 mμ) die ungewöhnliche Verschiedenartigkeit der einzelnen Elk. auf; typische Formen zeigt Abb. 113. Manche erscheinen wie Kaffeebohnen oder wie aufgebrochene Membranen keimender

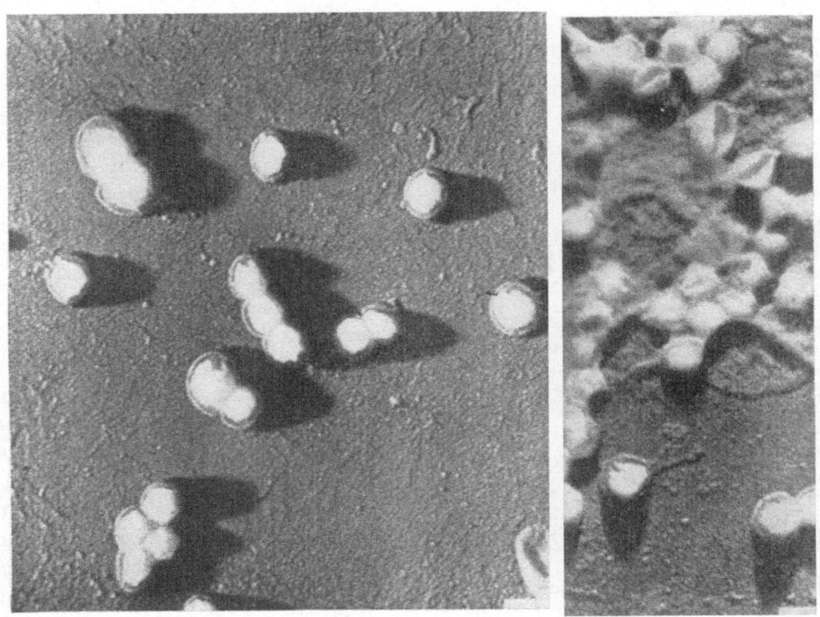

Abb. 115 und 116. Schräg mit Gold bedampfte Elementarkörper der Katzenpneumonie. 13000 : 1. Nach Hamre, Rake und Rake (1947).

Bakteriensporen. Der Formenreichtum ist auf Abb. 114 zu sehen. Der erste Eindruck erinnert an das lichtmikroskopische Bild von Erythrocyten eines technisch schlecht ausgeführten Blutausstrichs. Man sieht zentral und auch asymmetrisch liegende helle Zonen (a, b, c, g) neben ausgesprochenen Ringformen und kompakte oder zentral aufgehellte Stechapfelformen (h bis n). Außerdem fällt häufig eine mehr oder weniger blasse Randzone (e bis n) auf. Diese Randzone entsteht möglicherweise erst durch einen Schrumpfungsprozeß beim Eintrocknen innerhalb einer Plasmahaut. Mitunter finden sich auch kontrastarme Gebilde gleicher Größe, vielleicht mehr oder weniger „leere“ Plasmamembranen (o, p). Die Stechapfelformen entstehen wahrscheinlich ebenfalls erst als Folge der Eintrocknung. Außerdem ist mit morphologischen Veränderungen durch die Einwirkung des destillierten Wassers zu rechnen, das aber bei kurzer Einwirkungszeit die Infektiosität der Elementarkörper nicht aufhebt.

RAKE, Rake, Hamre und Groupé (1946) fanden beim Virus der Katzenpneumonie (feline pneumonitis, Baker), einem menschlichen Pneumonievirus (human pneumonitis S. F.), beim Lymphogranuloma inguinale (venerum) und bei der Psittakose ähnliche Bilder, wie sie H. Ruska schon für die beiden zuletzt genannten und das Trachomvirus vermutet hatte. Ihre Abbildungen von Elementarkörpern der Katzenpneumonie aus Dottersackkulturen gleichen zum Teil bis ins einzelne jenen des Erregers der Bronchopneumonie der Maus, und auch in der Deutung der Bilder besteht Übereinstimmung. Die Mehrzahl der Elk. mißt 440 bis 490 mμ. Außer einzelnen und doppelten Elk. fanden sich auch längere Ketten, wie sie bei der Bronchopneumonie der Maus, den Seiffertschen Mikroorganismen und den Erregern der Lungenseuche elektronenmikroskopisch nicht beobachtet worden sind.

Später isolierten Hamre, Rake und Rake (1947) die Elk. aus Allantois Flüssigkeit und stellten goldbedampfte Präparate (Abb. 115 und 116), sowie Oberflächenabdrücke her. Sie fanden dabei Formen, die sie ebenso wie H. Ruska und Poppe (1947) mit eingedrückten Gummibällen verglichen (vgl. Abb. 110 und 116). Der mittlere Durchmesser erschien mit 525 mμ noch größer als bei nicht bedampften Präparaten. Die aus der Schattenlänge gemessene Dicke betrug 175 bis 375 mμ. Die Autoren schließen auf das Vorhandensein einer Grenzmembran und einer gallertigen, wasserreichen Innensubstanz, die sich beim Trocknen von der Grenzmembran durch Schrumpfung ablöst. Mitunter erscheinen die Elk. verklebt oder in eine Grundsubstanz eingebettet.Rake (1947) zeigte außerdem zum erstenmal Initialkörper (IK.), die sich zu 0,5% unter den Elk. der Katzenpneumonie fanden. Die Durchmesser betrugen 700 bis 900 mμ. Man kann erkennen (Abb. 117), daß die Ik. eine einheitliche Grenzmembran besitzen und demnach als vergrößerte einzelne Elk. zu betrachten sind, die sich noch nicht in Elk. aufgeteilt haben. Kolonien von zahlreichen, gut abgegrenzten Elk.

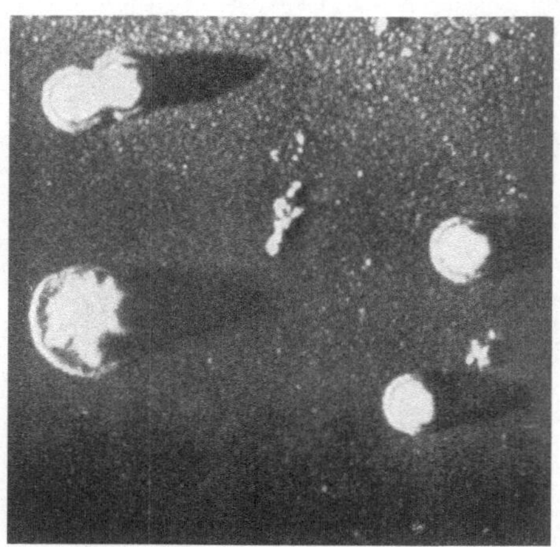

Abb. 117. Schräg mit Gold bedampfte Elementarkörper und ein Initialkörper des Erregers der Katzenpneumonie. 14000 : 1. Rake (1947).

wurden ebenfalls gefunden, und Rake vermutet, daß es sich dabei nicht um sekundäre Aggregate handelt, sondern um sogenannte Plaques (Einschlußkörper), die nach lichtmikroskopischen Befunden in eine gemeinsame Kapselsubstanz eingebettet sind. Die Kapselsubstanz ist jedoch, wie Rake und Jones (1942) in lichtoptischen Untersuchungen fanden, so empfindlich, daß bei der Präparation für das Elektronenmikroskop mit ihrer Zerstörung gerechnet werden muß.

6. Newcastle Virus.

Das Virus der atypischen Geflügelpest wurde von Bang (1946) und von Cuhna, Weil, Beard, Taylor, Sharp und Beard (1946) aus Eikulturen isoliert und elektronenmikroskopisch untersucht. Sowohl der Stamm B als auch der Stamm Kalifornien erwies sich als ein polymorphes Virus, das deshalb trotz der Angabe von Molekulargewichten (450 Millionen) nicht als makromolekulares Virus angesprochen werden kann. Es besteht aus 67% Protein, 27% Fettsubstanzen und Desoxypentose-Nukleinsäure (Cunha und Mitarbeiter). Die Identität der gefundenen Partikel (Abb. 118 und 119) mit dem Virus wurde von Bang durch die Agglutination mit spezifischem Immunserum elektronenmikroskopisch erwiesen. Die Viruspartikel sind rundlich bis oval und oft mit einem Fortsatz versehen (phagen-, bzw. spermienähnlich). Auch mehrere Fort-

sätze konnten nachgewiesen werden, so daß diese Partikel an Entwicklungsformen der filtrierbaren Mikroorganismen erinnern. Die Abmessungen des

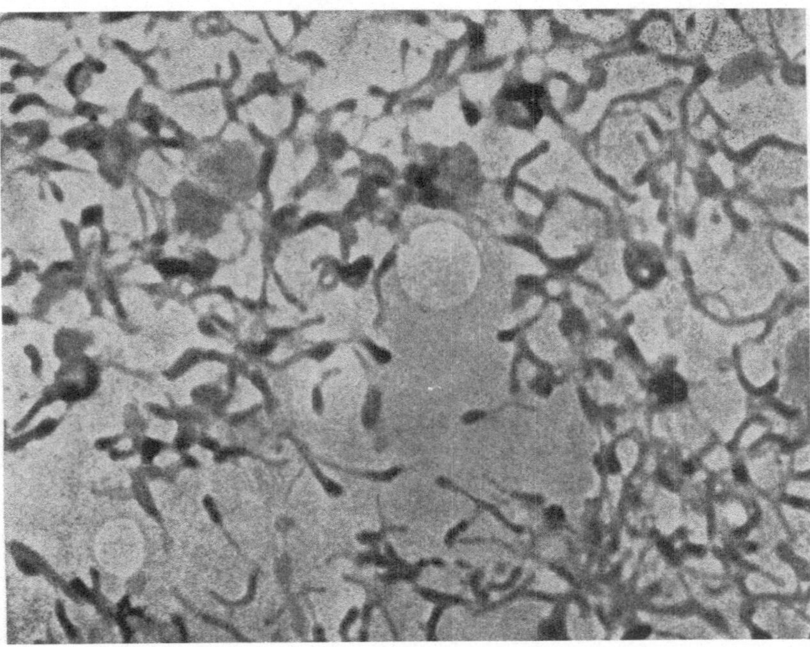

Abb. 118. Newcastle Virus aus Chorioallantoisflüssigkeit isoliert. 20000 : 1. Nach CUNHA, WEIL,
BEARD, TAYLOR, SHARP und BEARD (1946).

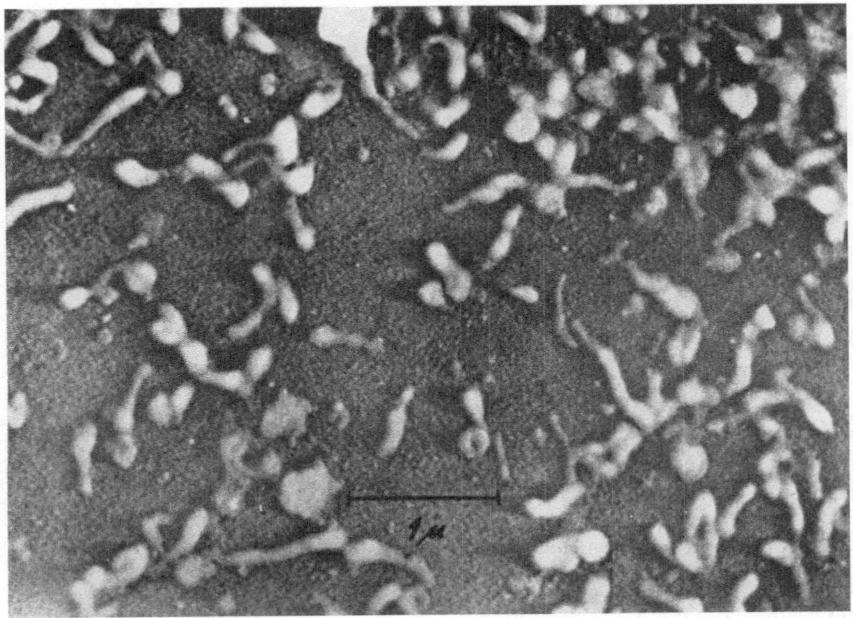

Abb. 119. Dasselbe wie Abb. 118 nach Schrägbedampfung mit Chrom. Nach CUNHA, WEIL, BEARD,
TAYLOR, SHARP und BEARD (1946).

Virus sind jedoch wesentlich kleiner als die der L-Organismen. Rundliche Formen sind meist 200 bis 500 mμ groß, die Fortsätze 20 bis 50 mμ dick, während äußerlich ähnliche Entwicklungsformen der L-Organismen oder L-Varianten von Bakterien die 10fachen Ausmaße haben können. In der Ultrazentrifuge finden sich zwei unscharfe Sedimentationsbande bei $S = 1800 \cdot 10^{-13}$ und $S = 1200 \cdot 10^{-13}$.

BABUDIERI hat dem Verfasser Aufnahmen von Newcastle Virus zur Verfügung gestellt, auf welchen rundliche Teilchen überwiegen, aber auch eine einzelne Fortsatzbildung zu sehen ist. Vgl. auch die Aufnahme von Newcastle Virusteilchen, die an Erythrocyten adsorbiert sind, bei DAWSON und ELFORD (1949/2) und die Arbeit von SCHÄFER, SCHRAMM und TRAUB (1949). Die zuletzt genannten Autoren fanden eine Sedimentationskonstante von 1340 S und ein mittleres Teilchengewicht von 800 Millionen. Den Unterschied zwischen den runden Formen aus salzfreier Lösung und den fädigen Formen aus salzhaltiger Lösung — den auch die älteren Autoren beobachtet haben — suchen sie durch die Annahme zu deuten, daß sich die fädigen Formen beim Auftrocknen der Präparate aus wäßrigen Suspensionen zu kugeligen Gebilden zusammenknäulen. Der Vorgang ist reversibel. Die Ursache für die Verhinderung der Knäuelbildung beim Auftrocknen aus Salzlösung ist noch ungeklärt, doch wird auf ähnliche Vorgänge bei fadenförmigen Nukleoproteinen hingewiesen, die in wäßrigen Lösungen eine Herabsetzung der Viskosität durch Verknäuelung zeigen.

Über Geflügelpest siehe die Fußnote im Abschnitt Influenza virus und Abb. 136.

7. Quaderförmige Virusarten.

Unter dieser Bezeichnung wurden von H. RUSKA (1941) mehrere Virusarten vereinigt, deren Elk. durch ihre gleichmäßige Pflaster- oder Ziegelsteinform ein gemeinsames Bauprinzip und somit ihre systematische Zusammengehörigkeit anzeigen. Es gehören ihr Viren mit verschiedenen Tropismen an. Nach

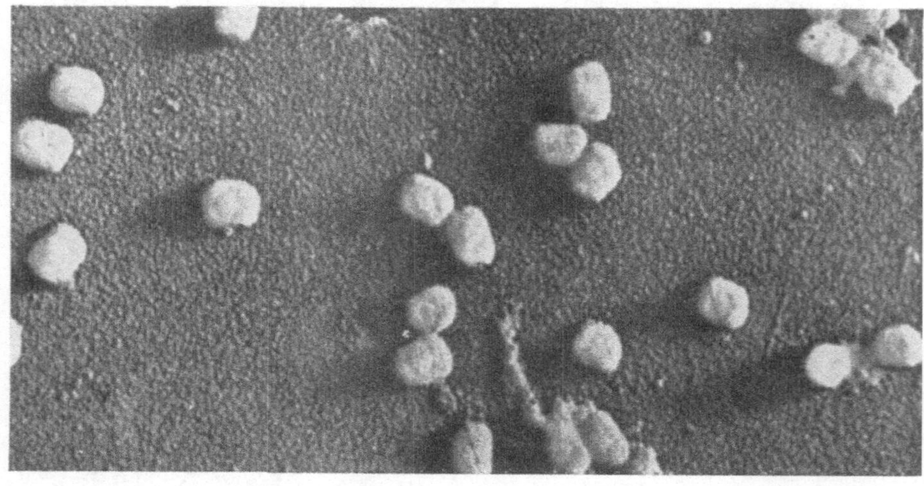

Abb. 120. Vaccine Virus. Schräg mit Gold bedampft. 25000:1. Nach SHARP, TAYLOR, HOOK und BEARD (1946).

bisher vorliegenden Untersuchungen von H. RUSKA, v. BORRIES und E. RUSKA (1939), H. RUSKA (1941 und 1943), GREEN, ANDERSON und SMADEL (1942), H. RUSKA und KAUSCHE (1943), GROUPÉ, OSKAY und RAKE (1946) und SHARP,

TAYLOR und BEARD (1946) sind das Vaccinevirus (Lapine, Abb. 120), das Virus des Molluscum contagiosum des Menschen (Abb. 121), das Virus der Ektromelie der Maus (Abb. 122), des Kaninchenmyxoms, das Kikuth-Gollubsche Kanarienvirus (Abb. 123) und das Virus der Geflügelpocken (Abb. 124) hierher zu rechnen. Auch die Erreger anderer Pockenkrankheiten der Tiere werden sich voraussichtlich als dazugehörig erweisen. Daß die Schafpocken quaderförmig sind, hat der Verfasser an einem Präparat von A. TOSCHKOFF (unver-

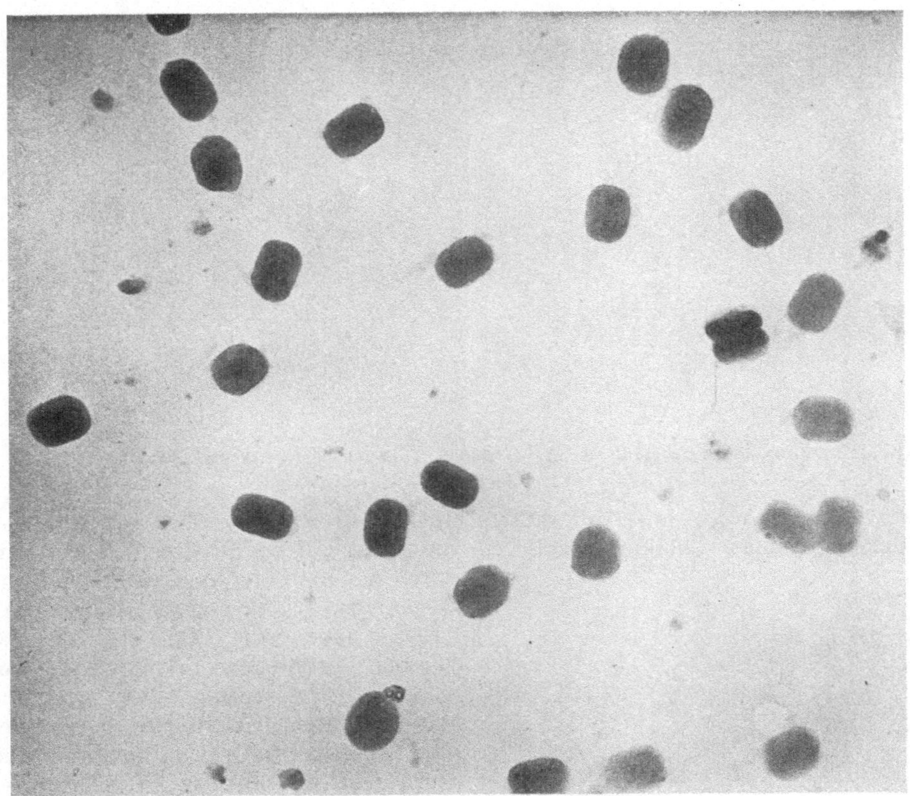

Abb. 121. Virus des Molluscum contagiosum. Annähernd 30 000 : 1. Nach H. RUSKA (1943/16).

öffentlicht 1943) gesehen. Mit dem Vaccine-Virus sind von KRAUSE (1938) die ersten Darstellungsversuche gemacht worden. Sie führten nicht zu der Erkennung der typischen Elk. (vgl. H. RUSKA, 1938). KRAUSE glaubte, Teilungsformen gefunden zu haben, die von keiner der späteren Untersuchungen bestätigt werden konnten. Die Quaderformen sind dagegen auch von GREEN, ANDERSON und SMADEL (1942) und von allen späteren Untersuchern als typisch erkannt worden.

a) Form, Größe und Struktur.

Auf den ohne Metallbedampfung hergestellten Durchstrahlungsbildern der Elk. war die Umrißform eines etwas gestreckten Spielwürfels zu erkennen. Es war aber nicht möglich, über die in Richtung des Strahlenganges liegende Abmessung und die Querschnittsform Aussagen zu machen. SHARP, TAYLOR,

Hook und J. W. Beard (1946) haben am Vaccine-Virus mit der Schattenwurftechnik die Höhe der Elk. des Vaccine-Virus bestimmt. Aus einzelnen, im Umriß
runden Elk., die einen besonders langen Schatten warfen (Abb. 120), haben sie
außerdem geschlossen, daß die Form der Elk. diejenige kurzer Säulenstümpfe

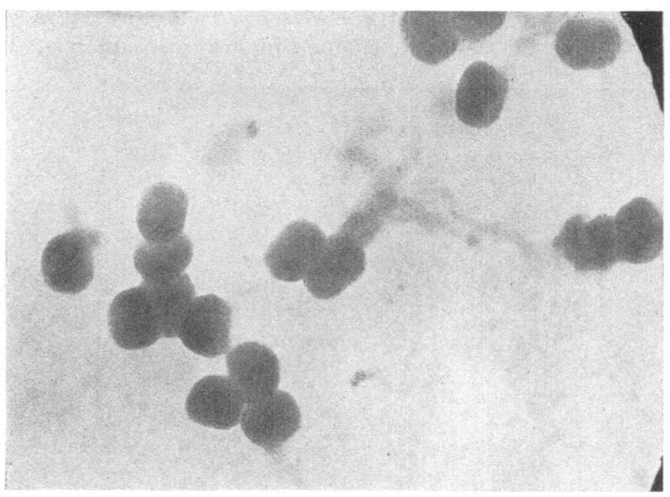

Abb. 122. Virus der Ektromelie der Maus. 32000 : 1. Nach H. Ruska und Kausche (1943).

sei. Nach Dawson und McFarlane (1948) scheinen aber doch in der Regel
backsteinförmige Gebilde vorzuliegen, die meist auf der größten Fläche, mitunter auch auf der längeren Seitenfläche liegen. H. Ruska und G. A. Kausche
(1943) haben die Elk. verschiedener hierher gehörender Virusarten einer
genaueren Vermessung unterzogen. Die Quader unterscheiden sich voneinander
durch bestimmte Größendifferenzen und durch ein verschiedenes Verhältnis der
langen (a) zur kurzen (b) Achse der Elk. Beim Virus des Moll. contag. fällt
die gestreckte Form besonders auf (Abb. 121). Sie liegt zwischen der Form
von Bakterien einerseits und der von Kokken andererseits, erhält aber durch
die abgerundeten Ecken der Teilchen und deren große Regelmäßigkeit ein von

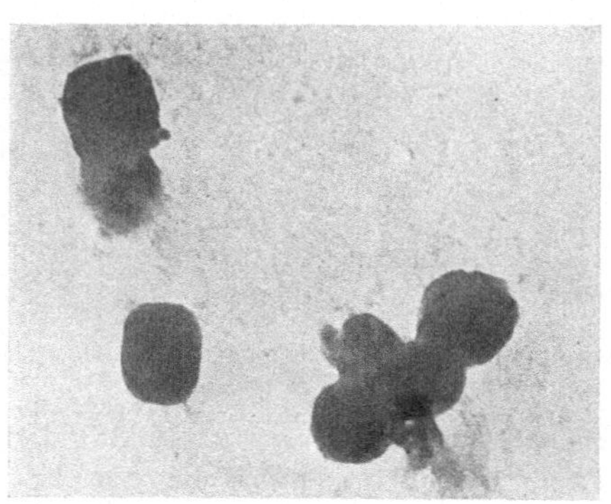

Abb. 123. Virus der Kanarienpocken. 32000 : 1.
Nach H. Ruska und Kausche (1943).

beiden abweichendes Aussehen. Zur Ermittlung der absoluten Größe der Teilchen
verschiedener Virusarten wurden die Längen (a) und Breiten (b) von jeweils n
Elk. ausgemessen, sowie das arithmetische Mittel und die mittlere Abweichung[1]
bestimmt.

Für die Elementarkörper der untersuchten Virusarten ergab sich:

---

[1] $\sigma = \pm \dfrac{\Sigma(p\,a^2)}{n}$, worin $\Sigma(p\,a^2) =$ die Summe aller quadrierten Differenzen der
Varianten vom Mittelwert und $n =$ Anzahl der Varianten bedeutet.

Tab. 7

Arithmetisches Mittel und mittlere Abweichung

	Zahl der Elk. n	lange Achse (mμ) a σ	kurze Achse (mμ) b σ	Achsen-verhältnis $Q = \dfrac{a}{b}$ σ	Größenangaben der Literatur
Mollus. contag.	150	255 ± 61,3	178 ± 41,8	1,45 ± 0,143	250 mμ [1]
Ektromelie	90	232 ± 24,6	172 ± 9,6	1,34 ± 0,024	100—150 mμ [2]
Myxoma cun. ..	81	287 ± 20,7	233 ± 19,0	1,22 ± 0,122	300 mμ [2]
Kanarienpocken	75	311 ± 26,1	263 ± 22,7	1,14 ± 0,108	125—175 mμ [2]
Vaccine	70	262 ± 28,2	209 ± 23,9	1,25 ± 0,124	125—200 mμ [2]

[1] Entnommen aus H. EYER, Handbuch der Viruskrankheiten. Jena, Fischer 1939.
[2] Entnommen aus E. HAAGEN, ebenda.

Die besondere Größe der Kanarienpocken (und der Geflügelpocken) wurde auch von GROUPÉ, OSKAY und RAKE (1946) festgestellt. Außerdem wurden die Werte in Klassengrößen von 20 mμ zusammengefaßt und die so gefundenen

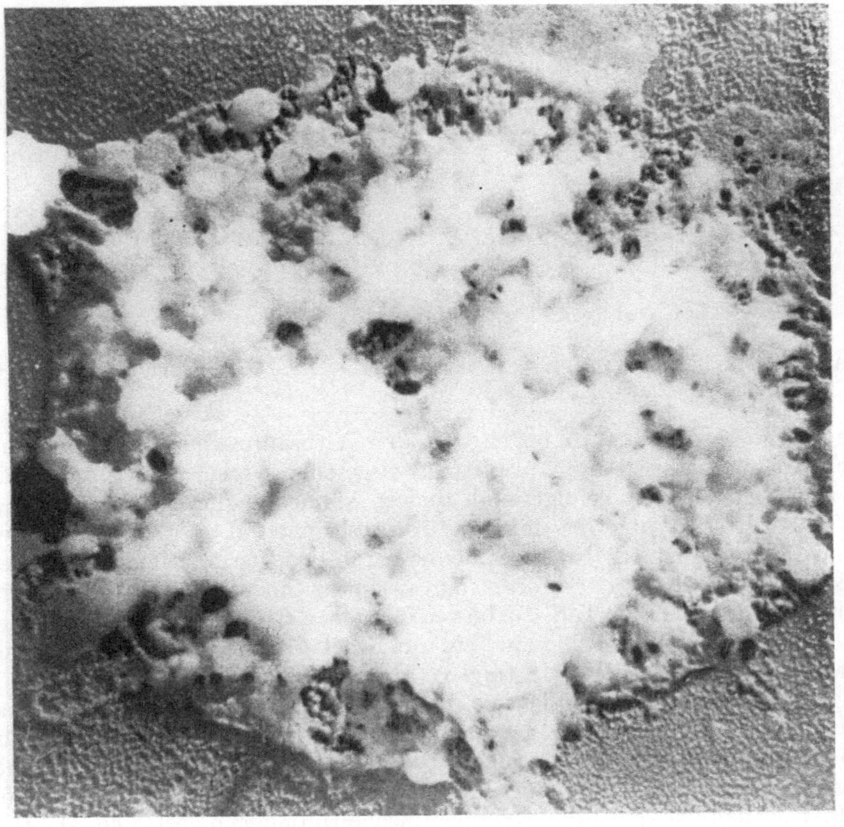

Abb. 124. Virus der Geflügelpocken, mit Gold schräg bedampft. Etwa 15000 : 1. Nach GROUPÉ und RAKE (1947).

Verteilungskurven mit der normalen Gausskurve verglichen, d. h. es wurden die Werte der Summen der einzelnen Größenklassen in Prozenten ausgedrückt und auf ein Wahrscheinlichkeitsnetz aufgetragen, wo sie bei Übereinstimmung mit der Gausskurve auf einer Geraden liegen müssen. Ein Teil der auf das Wahrscheinlichkeitsnetz aufgetragenen Werte der Elk. entsprach dieser Forderung, ein anderer Teil wich davon ab (bes. Vaccine), ohne in der Art der Abweichung einen regelmäßigen Gang zu zeigen. Einige Abweichungen rühren daher, daß

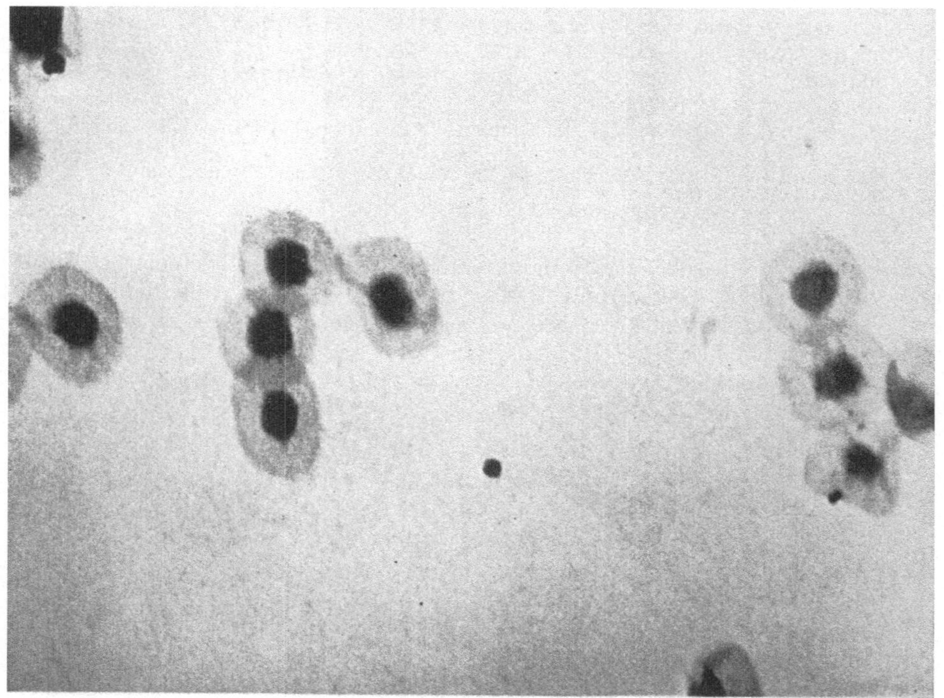

Abb. 125. Vaccine Virus mit Pepsin verdaut. 60000 : 1. Nach Dawson und McFarlane (1948).

die Vergrößerungsbestimmungen bei den Virusaufnahmen, die ursprünglich nicht für eine statistische Auswertung vorgesehen waren, nicht immer mit der gleichen Genauigkeit erfolgt sind, andere können dadurch bedingt sein, daß z. B. bei der Vaccine verschiedenes Ausgangsmaterial ausgewertet wurde.

Um die Achsenverhältnisse zahlenmäßig festzulegen, wurden bei den gleichen Elk. die Quotienten $Q = a/b$ ermittelt, hierfür arithmetisches Mittel und mittlere Abweichung bestimmt (Tab. 7), die verschiedenen Größenklassen zu den Klassenmitteln 1,05 — 1,15 — 1,25 usw. zusammengefaßt und diese ebenfalls auf ein Wahrscheinlichkeitsnetz aufgetragen (Abb. 74). Die graphische Darstellung zeigt, daß die Quotienten in sehr befriedigender Weise der normalen Häufigkeitsverteilung folgen und voneinander wesentlich verschieden, also für die einzelnen Virusarten typisch sind. Die gegenüber den Längen- und Breitenwerten bessere Lage der Meßpunkte auf den Geraden ist darauf zurückzuführen, daß Ungenauigkeiten der Vergrößerungsbestimmungen bei der Quotientenbildung ohne Einfluß bleiben. Die Dicke der Elk. läßt sich aus den metallbedampften Vaccinebildern von Sharp, Taylor, Hook und Beard mit 57 mμ

abschätzen. DAWSON und McFARLANE geben nach Gefriertrocknung 110 mμ an, nach Trocknung bei Zimmertemperatur in Übereinstimmung mit SHARP und Mitarbeitern die Hälfte dieses Wertes.

Ein Vergleich mit den Größenangaben der Literatur (Tab. 7) zeigt, daß die Ausmessung im Elektronenmikroskop z. T. größere Werte lieferte als die Untersuchungen mit der Zentrifuge und durch Filtration. Die Differenz ist um so bemerkenswerter, als diese Methoden den Durchmesser im feuchten (gequollenen)

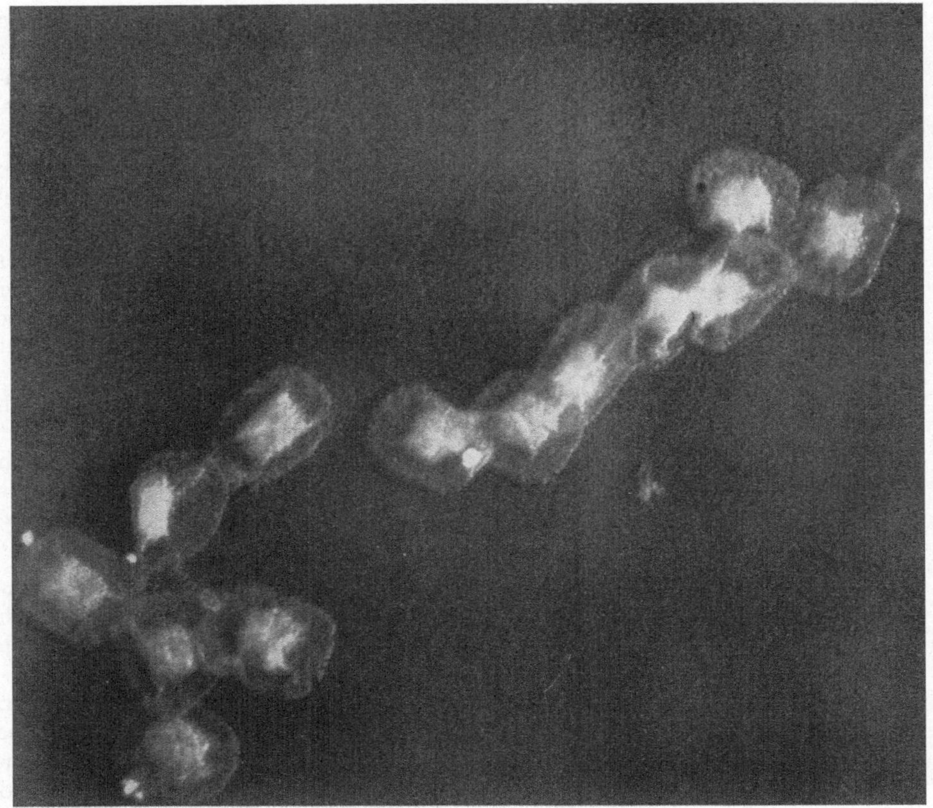

Abb. 126. Dasselbe wie Abb. 125, schräg mit Gold bedampft. 60000 : 1. Nach DAWSON und McFARLANE (1948).

Zustand ermitteln, die Abbildungen mit Elektronen aber von getrockneten Teilchen im Vakuum erfolgt. Größere als die in der Tabelle eingetragenen Literaturwerte wurden für Vaccine mit der Ultrazentrifuge 1938 von PICKELS und SMADEL (236 bis 252 mμ) und mittels der α-Strahlung des Radiums 1941 von BONÉT-MAURY und PERAULT (240 m$\mu \pm$ 20 mμ) ermittelt. Sie stimmen sehr gut mit der elektronenmikroskopischen Messung überein. LEVADITI (1943) und BONÉT-MAURY (1943) haben auf Grund der vom Verfasser angefertigten Elektronenbilder von Vaccine Elk. ihrer eigenen Präparate schon auf diesen Umstand hingewiesen. Sie stellten außerdem fest, daß die Vaccine-Elk. morphologisch den „normalen Korpuskeln" von SMADEL und WALL (1937) gleichen, die man bei der Überimpfung von gesundem Gehirn auf die Allantoismembran als unspezifisches Reaktionsprodukt erhält. Eine Bestätigung dieser Befunde wäre von großer Bedeutung für die Frage nach dem Ursprung und dem Wesen der quaderförmigen Elk.

Bezüglich der inneren Struktur haben Green, Anderson und Smadel (1942) darauf hingewiesen, daß die Vaccine-Elk. häufig eine oder mehrere zentrale Verdichtungen besitzen, die wie die fünf Punkte eines Spielwürfels angeordnet sein können. Auch auf den Aufnahmen der Elk. von Molluscum contagiosum haben H. Ruska und Kausche solche Verdichtungen geschen (Abb. 121). Besonders klar tritt die mittlere Verdichtung auf den plastischen Bildern von Sharp und Mitarbeitern hervor (Abb. 120). Sie zeigt sich als zentrale Erhebung in der Mitte der häufig leicht eingesunkenen größten Seitenfläche. Ein rundlicher Zentralkörper, der eine breite Seitenfläche überragt, ist allerdings mit der Annahme der zuletzt genannten Autoren schwer vereinbar, daß ein runder Querschnitt senkrecht zur längeren Achse der Elk. vorläge. Auf Bildern von Dawson und McFarlane sieht man Vaccine-Elk. in der Seitenansicht. Der Zentralkörper überragt nach beiden Seiten die Konturen wie der Saturn seinen Ring.

Dawson und McFarlane konnten außerdem zeigen, daß der Zentralkörper gegen Pepsinverdauung resistent ist, während $\frac{3}{4}$ der Elk.-Substanz in einer halben Stunde bei pH 3 und 37° C in Lösung gehen. Damit wurde die Angabe von Green und Mitarbeiter über das verschiedene chemische Verhalten beider Anteile bestätigt. Die Aufnahmen (Abb. 125 und 126) zeigen ferner, daß weitere Inhaltskörper nicht zu finden sind und daß somit die an den Ecken gelegentlich sichtbaren Verdichtungen entweder verdichtete oder andere aber ebenfalls mit Pepsin hydrolysierbare Virussubstanzen sind. Mit Desoxyribo-Nuklease kann Nukleinsäure herausgelöst werden, so daß bis auf die zentrale Dichte nichts am Aufbau der Elk. geändert wird. Es scheint danach berechtigt zu sein, den Innenkörper als Kernäquivalent zu betrachten.

Nach der Pepsinverdauung bleibt außer dem Zentralorgan eine zarte Plasmamembran übrig (s. Abb. 125), jedoch ist weder durch feuchte Hitze noch durch Ultraschall eine Zellmembran (Zellwand) wie bei Bakterien isoliert oder durch Indizien nachgewiesen worden. Dawson und McFarlane lehnen es im Gegensatz zu Anderson (1946) und Mudd (1949) ab, das Vaccine-Virus strukturell den Bakterien gleichzusetzen. Die Quaderform versuchen sie auf ein stabiles inneres Protein-Netzwerk (Stroma), bzw. auf einen teilweise parakristallinen Aufbau zurückzuführen. Ähnliche Erwägungen finden sich bei H. Ruska (1941) und H. Ruska und Kausche (1943).

b) Vermehrungsweise.

H. Ruska und Kausche (1943) versuchten, aus Elementarkörpervermessungen Schlüsse auf deren Vermehrung zu ziehen. Sie verglichen die an Elk. gewonnenen Ergebnisse mit Messungen an Staphylokokken, die infolge einer vor der Zellteilung eintretenden Streckung Unterschiede im Längen-Breiten-Verhältnis zeigen, die nicht der Gauss-Kurve folgen.

Vergleicht man die Variabilität der Längen und Breiten von Elk. durch die Bildung des Variabilitätskoeffizienten V, der das prozentuale Verhältnis der mittleren Abweichung zum arithmetischen Mittel ausdrückt $\left(V = \dfrac{100\,\sigma}{M}\right)$, so findet man, daß die Streuung der Längen und Breiten bei Elk. nicht wesentlich voneinander abweicht.

Tab. 8

Virus	Variabilitätskoeffizient V für	
	Länge a	Breite b
Mollus. contag.	24	18
Ektromelie	11	6
Myxoma cun.	7	8
Kanarienpocken	8	9
Vaccine	11	11

Hätte in jenem Zustand, in dem die Elk. isoliert wurden, eine Vermehrung derselben durch Streckung und anschließende Teilung stattgefunden, so müßte die Länge stärker variieren als die Breite. Das ist jedoch nicht nachzuweisen.

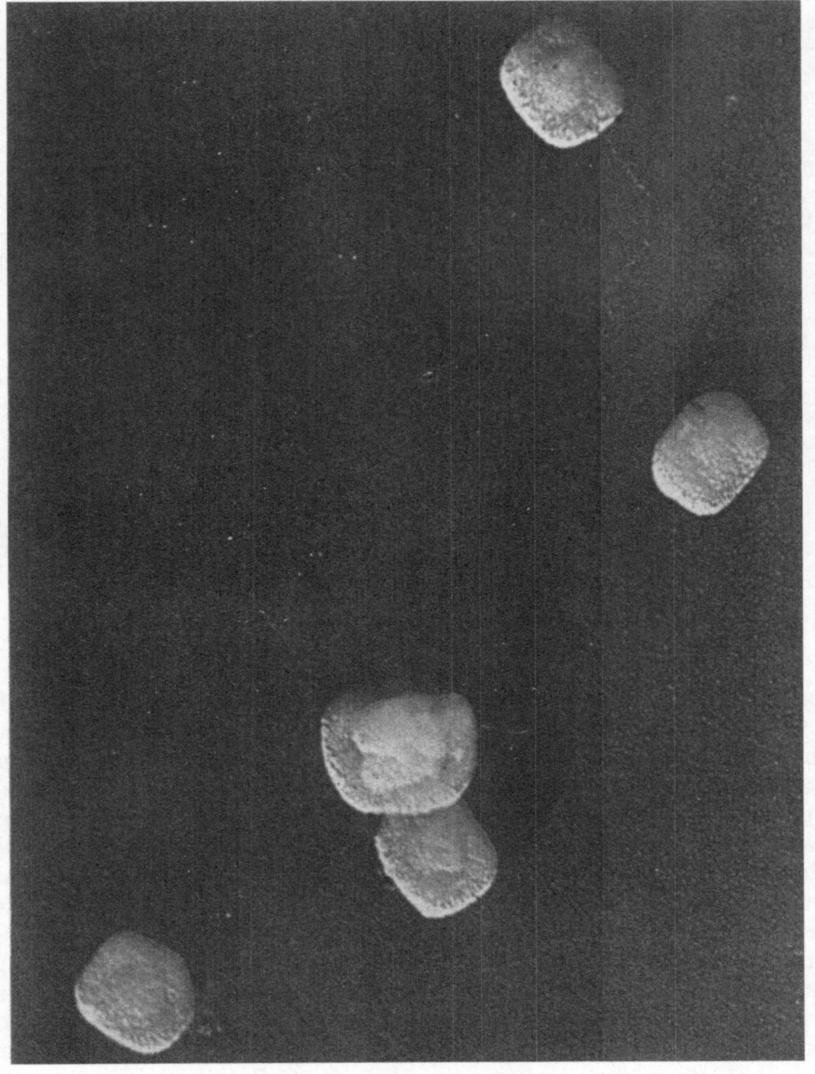

Abb. 127. Vaccine-Virus durch Salzfällung gereinigt und mit Gold schräg bedampft. 60 000 : 1. Nach DAWSON und MCFARLANE (1948).

Bei Staphylokokken ist dagegen die größere Ungleichheit bakterieller Zellen infolge Wachstum und Teilung durch die Ausmessung und die Wiedergabe der Meßwerte (n = 350) in den Kurven zu bestätigen Die Meßpunkte für das Molluscum contagiosum liegen sehr gut auf einer Geraden, während die Werte für die Breite der Kokken und noch mehr die Längenmaße eine deutliche Abweichung zeigen. Es treten mehr Zellen mit großer Länge, bzw. Breite auf, als der statistischen Verteilung einer bezüglich der Größe einheitlichen Population entsprechen würde. In der gleichen Art weichen auch die Quotienten nach der Seite höherer Werte stark von einer Geraden ab (Abb. 74). Der Zellstreckungs- und Teilungsvorgang läßt sich aus der Größenverteilungskurve deutlich ablesen.

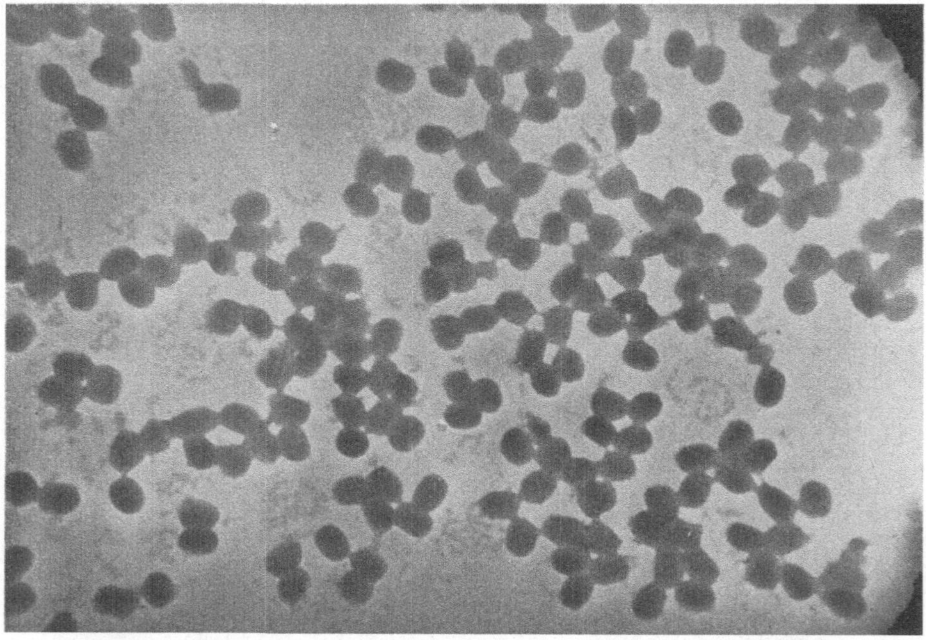

Abb. 128. Virus des Molluscum contagiosum. Eine Stunde gekocht. 21000 : 1. Nach H. Ruska und Kausche (1943).

Der Mangel an sicher deutbaren oder sich häufiger wiederholenden Teilungsformen hängt nach den genannten Autoren vielleicht damit zusammen, daß durch die Reindarstellung des Virus oder den Zeitpunkt der Virusgewinnung eine einseitige Auswahl stattgefunden hat. Es ist aber auch mit der Möglichkeit zu rechnen, daß die Vermehrung der Elk. entgegen der herrschenden Auffassung (Herzberg 1939 u. a.) nicht durch Teilung erfolgt, sondern daß sie ähnlich wie bei den „großen Virusarten" über die Bildung und den simultanen Zerfall der Einschlußkörper in mehrere Elk. vor sich geht. Auf einer dem Verfasser freundlicherweise von McFarlane überlassenen Aufnahme (Abb. 127) sieht man einen abnorm großen Elk., der an einen Initialkörper erinnert. Groupé, Oskay and Rake (1946) glauben allerdings, daß das kettenförmige Aneinanderliegen der Elk. an den Ecken oder die knospenartigen Auswüchse an diesen Stellen von Bedeutung für die Vermehrung seien. In einer späteren Arbeit zeigen Groupé und Rake (1947) größere Aggregate von Elk. der Geflügelpocken in einer Grundsubstanz eingebettet (Abb. 124). Sie nehmen an, daß es sich dabei, um Einschlußkörper (Bollinger-Körperchen) handelt, lassen aber auch die Möglichkeit offen, daß ein lösliches Antigen die Elk. umhüllt. Aneinanderlagerungen der Elk. und Strangbildungen zwischen ihnen sind von allen Untersuchern gesehen worden. H. RUSKA und KAUSCHE fanden sie beim Virus des Molluscum contagiosum besonders ausgeprägt nach Erhitzen der Virussuspension (Abb. 128). Es ist daher möglich, daß auch ohne besondere Behandlung die sichtbaren Verklebungen und Strangbildungen sekundärer Natur sind. Im ganzen gilt das Problem der Teilung oder Sprossung der quaderförmigen Elk., bzw. das „Ob" und „Wie" ihrer Entwicklung aus Einschlußkörpern als nicht endgültig gelöst.

c) Verhalten gegenüber physikalischen und chemischen Eingriffen.

Bringt man eine Elementarkörpersuspension zum Kochen (Abb. 128), so ist eine Verkleinerung der Elk. festzustellen. Nach Versuchen von H. RUSKA und KAUSCHE schrumpfen sie entsprechend der Übersicht in Tab. 9.

Tab. 9.

	n	Unbehandelt				Gekocht		
		a mμ	b mμ	Q	n	a mμ	b mμ	Q
Mollus. cont.	150	250	178	1,45	150	227	164	1,39
Ektr. inf.	90	232	172	1,34	30	191	154	1,23

Die Substanz der Elk. bleibt dabei homogen, während das Eiweiß des Bakterienzytoplasmas beim Erhitzen mehr oder weniger inhomogen flockig koaguliert und die Bakterienmembranen nach der Schrumpfung des Inhalts sich ablösen (H. RUSKA, 1941, Abb. 82). Die oberflächlichen Schichten der Elk. machen dagegen die Schrumpfung mit und neigen nach dem Kochen zur Verklebung mit anderen Elk. Es bilden sich zwischen ihnen fädige Verbindungsstränge, und

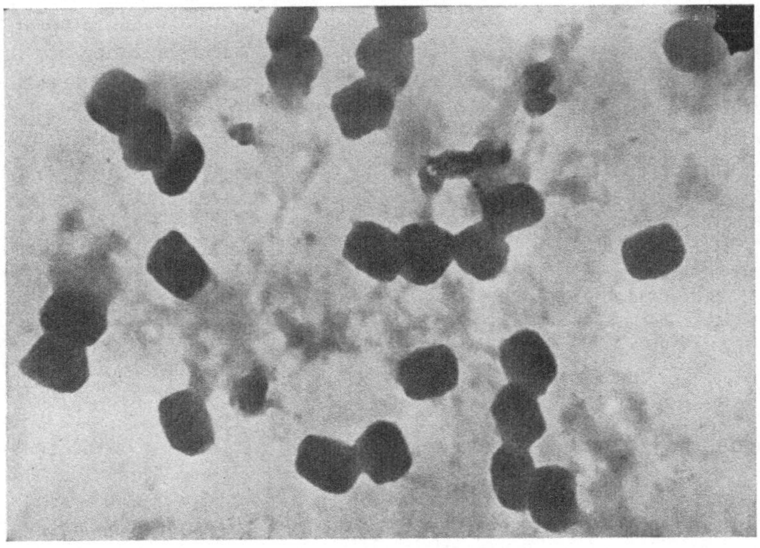

Abb. 129. Virus der Ektromelie der Maus, gefärbt mit Viktoriablau. 27 000 : 1. Nach H. RUSKA und KAUSCHE (1943).

Substanzteile werden herausgelöst, die neben den Elementarkörpern auftrocknen. Die *Grenzmembran* (Plasmahaut) der quaderförmigen Elk. ist nach ihrem physikalisch-chemischen Verhalten etwas anderes als die *Zell*membran, bzw. die Zellwand der Bakterien. Die Verkleinerung der Elk. ist mit einer Verschiebung des Achsenverhältnisses a/b im Sinne einer stärkeren Schrumpfung der langen Achse verbunden. H. Ruska und Kausche schließen daraus, daß die Elk. nicht nur nach der äußeren Form, sondern auch in bezug auf die innere Struktur polar gebaut sind. Zu ähnlichen Ergebnissen bezüglich des Längen-Breiten-Verhältnisses führte die Fixierung mit Osmiumtetroxyd und Sublimat, sowie die Färbung mit Viktoriablau (Abb. 129). Diese bewirkt aber keine Verkleinerung, sondern eine Vergrößerung der Elk., die der lichtmikroskopischen Sichtbarmachung dienlich ist. Die Schrumpfung durch Osmiumtetroxyd haben auch Dawson und McFarlane (1948) beobachtet; diese Autoren geben ferner an, daß gefriergetrocknete Elk. weniger geschrumpft sind als die bei Zimmertemperatur getrockneten. Die Schrumpfung findet in der zur aufgetrockneten Fläche senkrechten Richtung statt, indem sich vermutlich eine netzartige Innenstruktur ziehharmonikaartig zusammenfaltet.

Durch Ultraschall sind nach Beobachtungen des Verfassers mit Kausche die quaderförmigen Elk. wesentlich schwerer zu zerschlagen als die stäbchenförmigen Virusproteine der Pflanzen. Die entstehenden Bruchstücke sind unregelmäßig und nehmen keine Kugelform an; dieses spricht für eine verhältnismäßig starre, gallertartige Konsistenz. Es kommt auch nicht zu einer Ablösung der oberflächlichen Plasmahaut.

Bei der Behandlung mit Sublimat fand H. Ruska (1947/19) Körnchen an der Oberfläche adsorbiert, die möglicherweise aus reduziertem, metallischem Quecksilber bestehen, während Bakterien nach gleicher Vorbehandlung derartige Körnchen im Innern unter der Zellmembran zeigten.

8. Varicellen- und Zostervirus.

Das Virus der Varicellen und des Zoster ist von H. Ruska (1943), sowie von Farrant und O'Connor (1949), das Virus der Varicellen außerdem von Nagler und Rake (1948) dargestellt worden. Es ergab sich, daß die Differentialdiagnose zwischen Varicellen und Vaccine-Virus leicht zu stellen ist und daß in Brandblasen entsprechende Elk. fehlen. H. Ruska fand bei 270 vermessenen Varicellen-Elk. einen mittleren Durchmesser von etwa 145 mμ ($\sigma \sim 20$), bei 135 Zoster-Elk. den Wert 137 mμ ($\sigma \sim 20$). Das Zostervirus läßt sich morphologisch nicht von dem der Varicellen unterscheiden. Für die Differentialdiagnose dieser Virusarten gegenüber Vaccine macht H. Ruska folgende Angaben:

Tab. 10.

Morphologie	Vaccine	Varicellen
Form	quaderförmig	unregelmäßig, polygonal bis kugelförmig
Größe	260 × 210 mμ	150 mμ
Außenbegrenzung	meist scharf	häufig etwas wolkig oder mit kleinen Defekten an der Außenkontur
Innenstruktur	meist homogen, gelegentlich eine oder mehrere Verdichtungen	häufig eine zentrale runde Verdichtung von 50 mμ Durchmesser

Nagler und Rake geben als Durchmesser des Vaccine-Virus nach der Goldbedampfung, deren vergrößernde Wirkung bekannt ist, 302 × 244 mμ an und
für Varicellenvirus 177 × 210 mμ. Während in einem Fall der Durchmesser
des Varicellenvirus 63% desjenigen der Vaccine-Elk. beträgt, sind es im anderen
83%. Der Unterschied verringert sich etwas, wenn man berücksichtigt, daß
durch die Goldbedampfung das kleinere Virus verhältnismäßig stärker anwächst.
H. Ruska sah in den Elk. häufig eine zentrale Verdichtung von etwa 50 mμ
Durchmesser, die auf den mit der Schattenwurfmethode gewonnenen Bildern

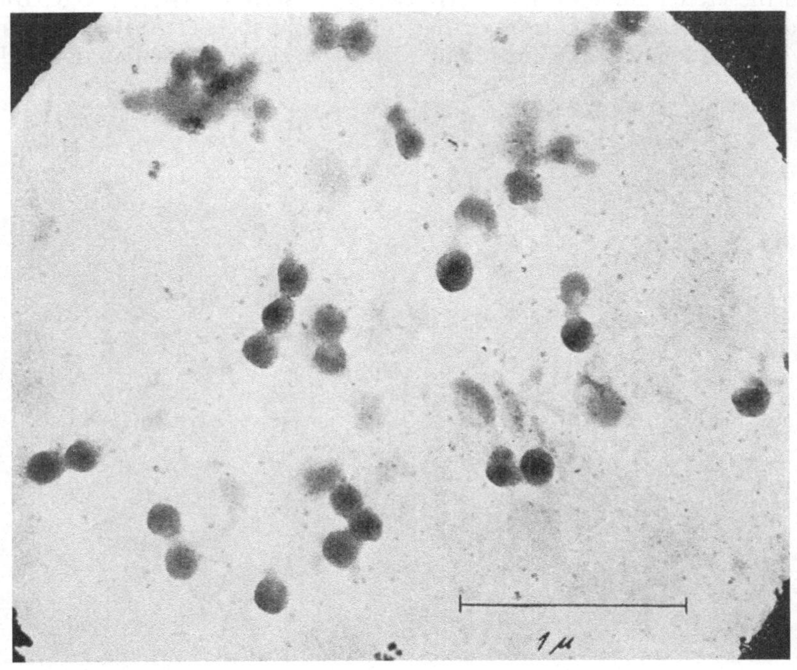

Abb. 130. Virus der Varicellen. 28000 : 1. Nach H. Ruska (1943).

nicht zu sehen ist, da der Zentralkörper nicht wie bei der Vaccine über die Oberfläche hinausragt. Farrant und O'Connor geben für Varicellen nach Platin-
bedampfung 240 ± 30 mμ, für Zoster 250 ± 80 mμ als Durchmesser an. Die
Höhe der aufgetrockneten Elk. beträgt nur 73 bis 80 mμ. Fixierung mit Formalin oder Osmiumsäure vermindert die bei der Trocknung auftretende Ab-
flachung und läßt die Elk. bei gleichbleibendem Volumen kleiner erscheinen
(180 × 110 mμ). Rekonvaleszentenserum agglutiniert und vergrößert die Elk.

Nagler und Rake sind der Ansicht, daß die Elk. der Varicellen vorherrschend
rechtwinklige Formen aufweisen, und rechnen sie daher mit zu den quaderförmigen
Virusarten. Der Unterschied zwischen langer und kurzer Achse der Elk. ist
allerdings besonders klein. H. Ruska trennt sie als nahestehende, aber besondere
Gruppe von den quaderförmigen Elk. ab, und zwar wegen ihrer rundlichen bis
polygonalen Formen (Abb. 130), ihrer minder scharfen Außenkontur, der geringeren
Größe, ihres anderen färberischen Verhaltens und der fehlenden Cytoplasmaein-
schlüsse in den Zellen des Wirtsorganismus. Auch nach den Ergebnissen von
Farrant und O'Connor scheint eine Sonderstellung gerechtfertigt. Mit großer
Wahrscheinlichkeit ist anzunehmen, daß der Zentralkörper hier ebenso wie bei
Vaccine resistent gegen Pepsinverdauung ist und Nukleinsäure enthält. Ent-

sprechende Versuche, wie sie Dawson und McFarlane am Vaccine Virus durchgeführt haben, dürften neben Untersuchungen über die Entwicklung der Varicellen-Elk. in der Zelle die Fragen nach der feineren Struktur (Zentralkörper, Grenzmembran), sowie nach der systematischen Stellung entscheiden.

9. Influenzavirus.

In einer Arbeit von Taylor, Sharp, Beard, Dingle und Feller (1943) werden die ersten Bilder von Influenzavirus A (Stamm PR 8) gezeigt. Die Präparation erfolgte in der Ultrazentrifuge mit anschließender Auflösung des Sediments in 0,023 mol. Kalziumchlorid. Dieses vermindert den in Ringerlösung

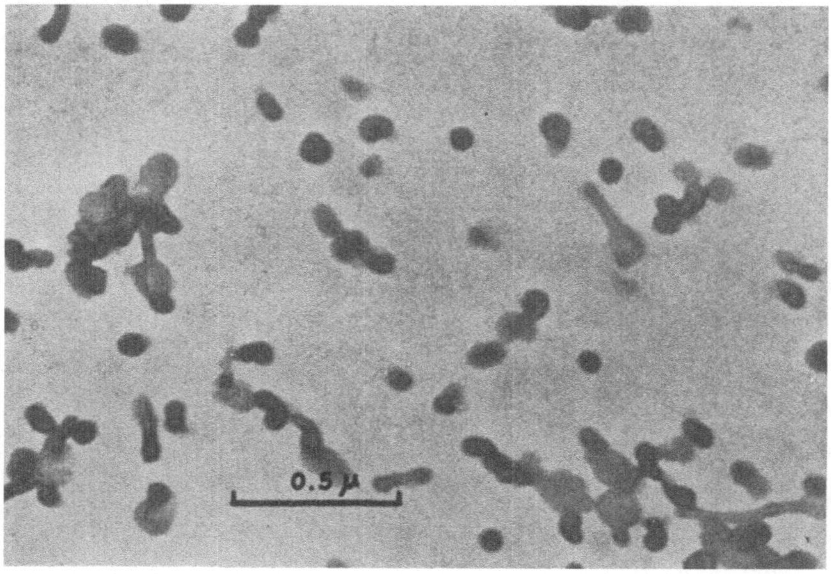

Abb. 131. Influenzavirus A mit der Ultrazentrifuge aus Chorioallantoisflüssigkeit aus Hühnerembryo konzentriert und mit 0,023 mol. Kalziumchlorid aufgetrocknet. Nach Taylor, Sharp, Beard, Beard, Dingle und Feller (1943).

und destilliertem Wasser durch teilweise Fällung eintretenden Virusverlust und verstärkt den Kontrast der Elektronenbilder (Abb. 131). Auch Agglutination roter Blutkörperchen durch Adsorption des Virus und anschließende Elution erwies sich als geeignet für die Gewinnung gereinigter Virussuspensionen. Durch Vergleich mit den 15 mμ breiten Stäbchen des Tabak Mosaikvirus wurde der mittlere Durchmesser der Influenza-A-Virus-Teilchen zunächst mit 77,6 mμ bestimmt. Die Teilchen weichen von der Kugelform häufig ab und lassen besonders bei Behandlung mit Kalziumchlorid z. T. eine zentrale Verdichtung erkennen; auch kurze Ketten aus mehreren runden Teilchen sind zu beobachten. Die Sedimentationskonstante S_{20} beträgt in guter Übereinstimmung mit der elektronenmikroskopisch vermessenen Virusgröße 742 . 10^{-13}.

Das Influenza-B-Virus (Stamm Lee, Abb. 132) ist mit 98 mμ Durchmesser ($S_{20} = 840 . 10^{-13}$) größer als das A-Virus (Sharp, Taylor, McLean, Beard, Beard, Feller und Dingle 1943 und 1944) und auch größer als das Virus der Schweineinfluenza (78,3 mμ, $S_{20} = 727 . 10^{-13}$, Abb. 133) nach Taylor, Sharp, McLean, Beard, Beard, Dingle und Feller, 1943 und 1944).

Der Unterschied in der Sedimentationskonstanten des gleichgroßen Schweine-
influenza- und des Influenza-A-Virus, der anfänglich noch größer gefunden
wurde als hier angegeben, konnte auf Differenzen des spezifischen Volumens

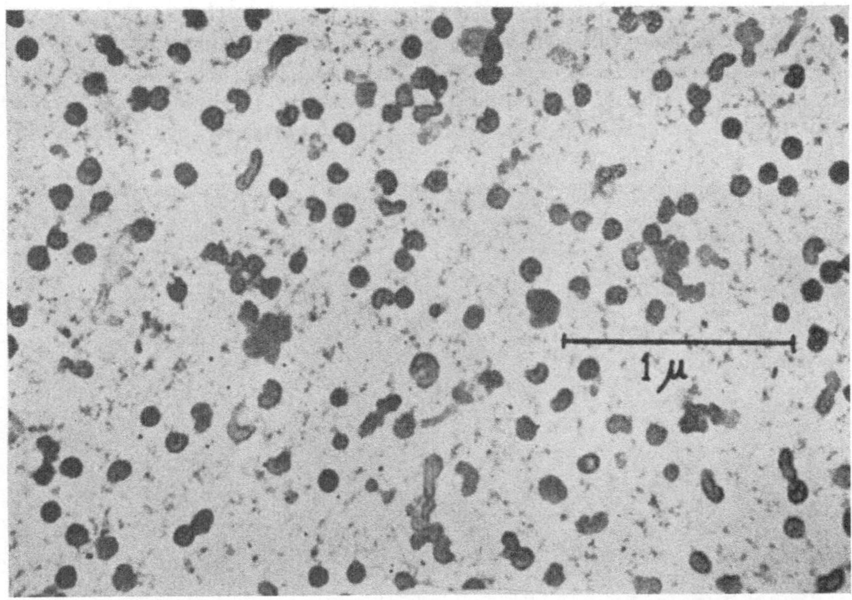

Abb. 132. Influenzavirus B (Stamm Lee) und Fremdbestandteile durch Adsorption und Elution
sowie Ultrazentrifugation konzentriert und mit Kalziumchlorid aufgetrocknet. Nach SHARP, TAYLOR
MCLEAN, BEARD, BEARD, FELLER und DINGLE (1944).

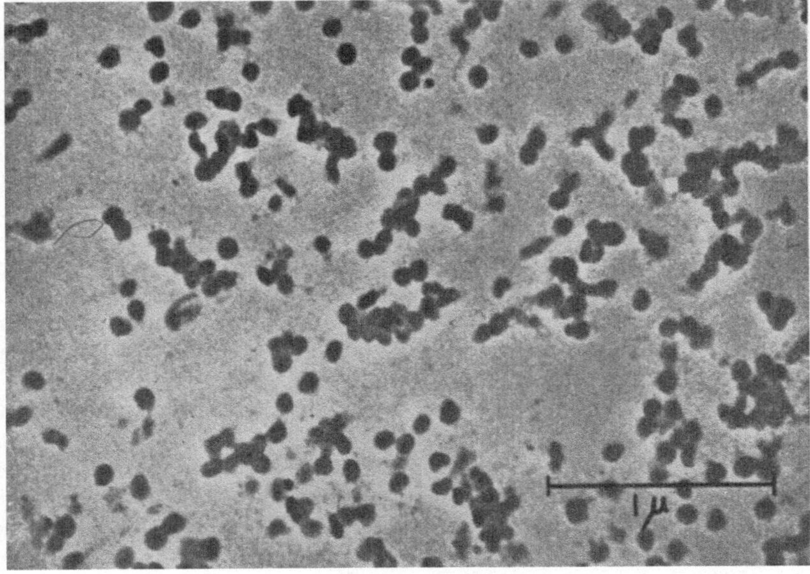

Abb. 133. Schweineinfluenzavirus durch Adsorption und Elution an rote Blutkörperchen sowie
durch Ultrazentrifugierung konzentriert und mit Kalziumchlorid aufgetrocknet. Nach TAYLOR-
SHARP, MCLEAN, BEARD, BEARD, DINGLE und FELLER (1944).

zurückgeführt werden. Die ursprünglich durch Vergleich mit dem Tabak-Mosaik-virus gemessenen Virusgrößen wurden in den späteren Arbeiten bei genauerer Bestimmung des elektronenmikroskopischen Maßstabs etwa 20% höher gefunden (101, 123, 96,5 für A, B, S). Der Wert des A-Virus entspricht der von Lauffer und Miller 1944 mit der Ultrazentrifuge bestimmten Größe.

Lauffer und Stanley (1944) geben die elektronenmikroskopisch gemessene Größe des A-Virus (PR 8) mit 110—115 mµ an, Chu, Dawson und Elford (1949) mit 90 ± 11,5 mµ für A und 103 ± 8 mµ für B. Bilder einer gemischten Vaccine der Stämme PR 8, Weiss und Lee finden sich bei Stanley (1946,

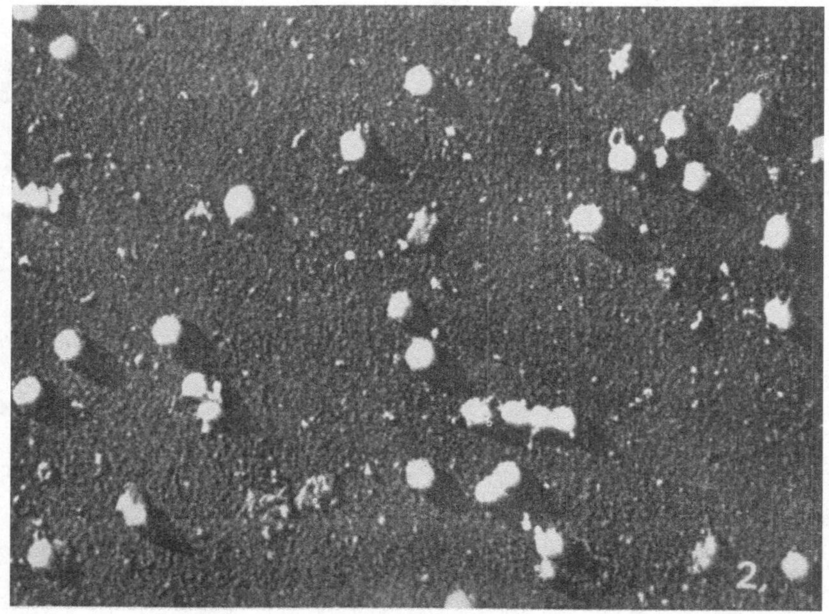

Abb. 134. Influenzamischvaccine mit Gold schräg bedampft. 35000 : 1. Nach Stanley (1946).

Abb. 134); Ergebnisse mit der Schattenwurftechnik sind außer durch die beiden zuletzt genannten Autoren auch von Williams und Wyckoff (1945) veröffentlicht worden.

Beard und Mitarbeiter (1944) betonen, daß die Influenza-Virus-Arten nach ihrer variablen Form und Größe, aber doch einheitlichen Morphologie nicht makromolekularer Natur sein können. Die chemische Analyse (60% Protein, 20% Fett, etwa 10% Kohlehydrate, davon ein Bruchteil aus Desoxypentose-Nukleinsäure) führte zu der gleichen Schlußfolgerung.

Die Autoren vergleichen das Influenzavirus mit kleinen Kokkobazillen, doch ist der Unterschied der Morphologie zweifellos wesentlich größer als zwischen Rickettsien und Bakterien. Neben den bisher beschriebenen kugeligen bis eiförmigen Virusteilchen fanden Mosley und Wyckoff (1946) besonders beim A-Virus (Stamm Weiss) lange, etwas segmentierte Fäden mit meist endständigen Knoten von mindestens der Größe der kugeligen Elk. Chu, Dawson und Elford (1949) haben diesen auffallenden Befund am PR 8 und Lee-Stamm bestätigt. Besonders häufig fanden sie Fäden bei den frisch isolierten Stämmen A Nederland I/49 und A Paris PL I/49, die erst über wenige Eipassagen geführt worden waren, seltener jedoch bei A Barret (1947) und B Blaskowic. Sie adsorbierten das Virus nach dem Vorgehen von Dawson und Elford (1949) an haemolysierte Hühnererythrozyten (Abb. 135) und bildeten es ebenso wie

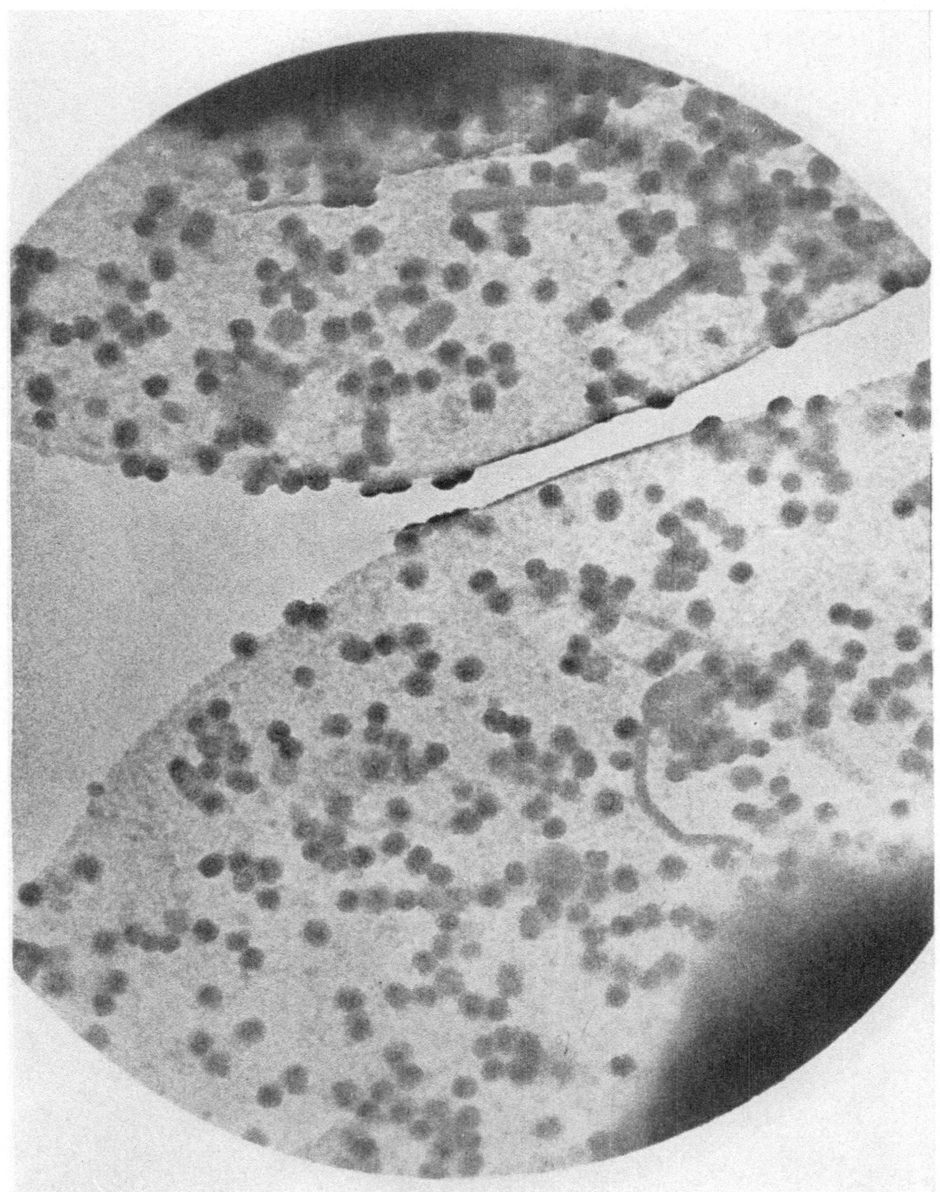

Abb. 135. Influenzavirus B an haemolysierte Erythrozyten adsorbiert. Etwa 30000 : 1. Nach Dawson und Elford (1949/2).
Abb. 135—137 mit freundlicher Genehmigung der Herausgeber des Journ. of Gen. Microbiol.

Heinmets (1948) mit den Erythrozytenmembranen ab. Die Fadenformen ließen sich auch im lichtmikroskopischen Dunkelfeld mit dem Kardioidkondensor (Zeiß) in unbehandelter Allantoisflüssigkeit nachweisen. Die Gliederung der

Fäden entspricht nicht der bekannten Kollagenperiode (Abb. 6), die Heinmets und Golub (1948) bei der Untersuchung normaler Allantoiszellen an fibrillären

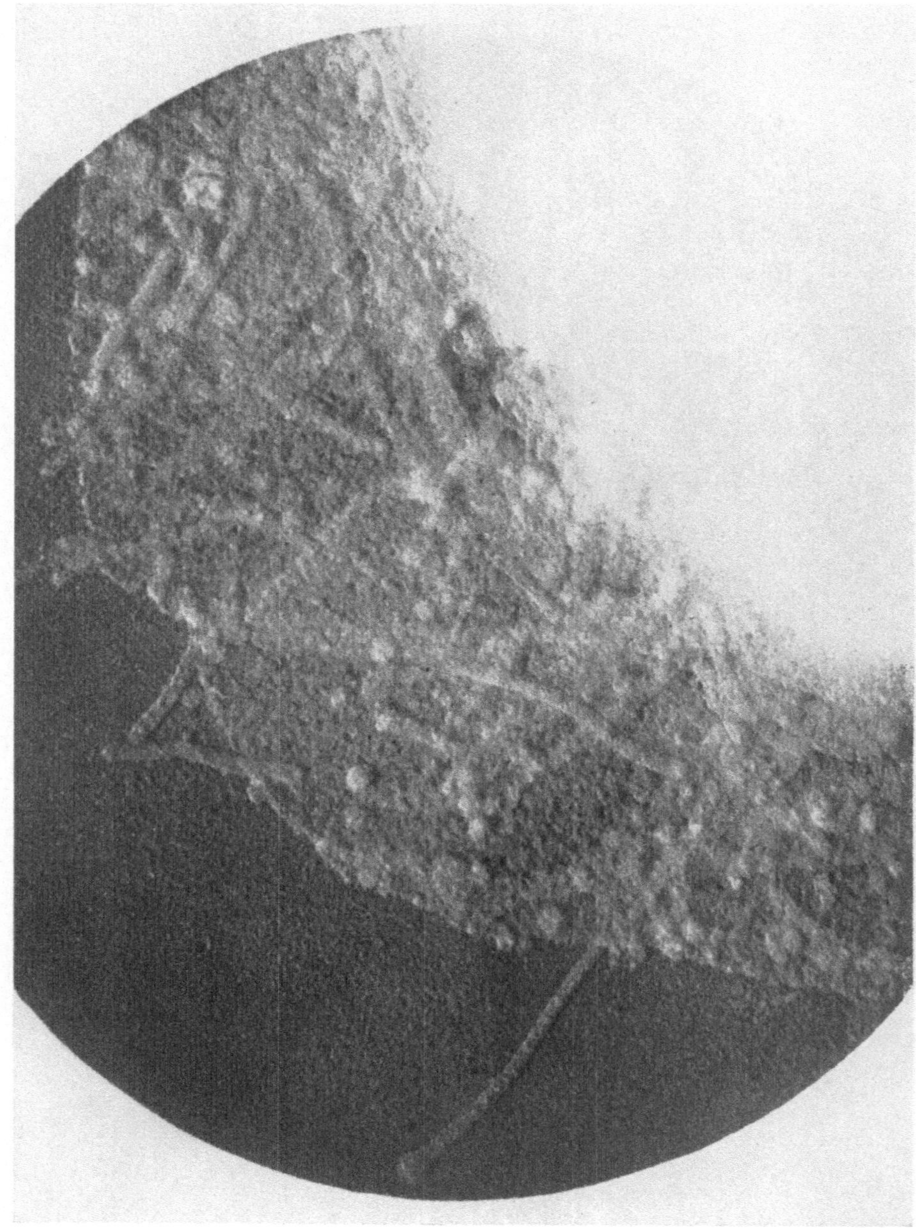

Abb. 136. Geflügelpestvirus nach Adsorption an haemolysierte Hühnererythrozyten mit Palladium schräg bedampft. Etwa 30000 : 1. Nach Dawson und Elford (1949/2).

Strukturen gefunden haben. Chu und Mitarbeiter halten es noch nicht für völlig gesichert, daß die Fadenformen zur Entwicklung des Influenza-Virus gehören, doch sprechen folgende Befunde dafür: gleiches Verhalten bei Adsorption und Elution mit roten Blutkörperchen, gleiche Agglutinierbarkeit mit homologem

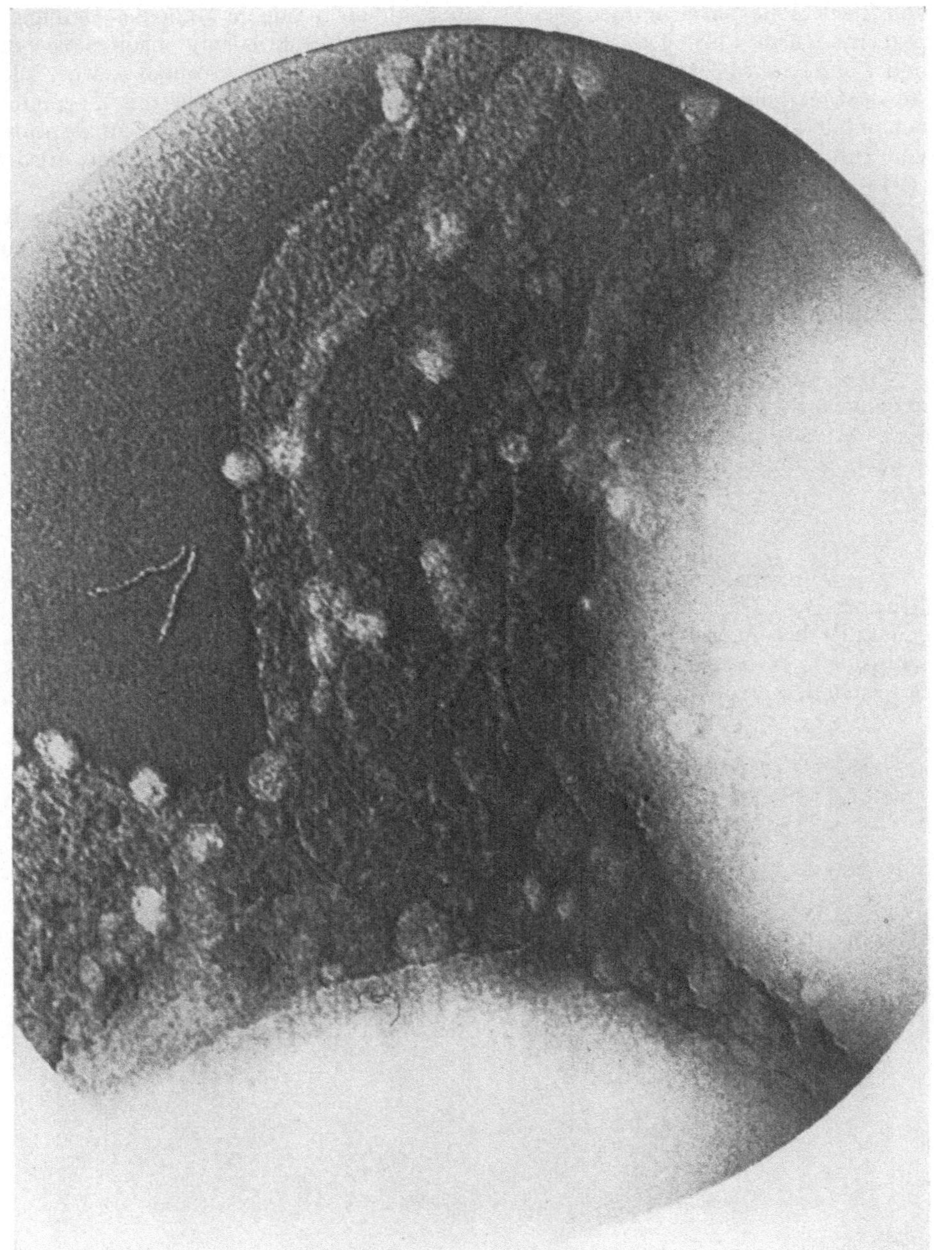

Abb. 137. Mumpsvirus an haemolysierte Hühnererythrozyten adsorbiert und mit Palladium schräg bedampft. Nach Dawson und Elford (1949/2).

Antiserum, Auftreten der Fadenformen bei der Impfung mit voraussichtlich fadenfreien Filtraten und bei zwischengeschalteter Mäusepassage, sowie schließlich gleiche (unveröffentlichte) elektronenmikroskopische Beobachtungen von LEPINE (Paris) und BABUDIERI (Rom). Siehe BABUDIERI und ARCHETTI (1949).

Dem fadenförmigen Influenza-A-Virus, das CHU, DAWSON und ELFORD (1949) von frisch isolierten Stämmen gezeigt haben, ähnelt in hohem Maße das Geflügelpestvirus (Abb. 136). Die Sedimentationskonstante S 20 beträgt nach SCHÄFER und SCHRAMM 340 S, ist aber knapp halb so groß wie beim Influenzavirus. Die Adsorption gelingt nach FLICK (1948) auch an Blutzellen, die mit Formalin behandelt sind. Über den Einfluß der Salzkonzentration auf die Adsorption von Influenza-A-Virus an Erythrocyten berichten FLICK, SANFORD und MUDD (1948).

Es liegt also nahe, anzunehmen, daß neben einer Zweiteilung der Elk. (ovale Formen) in Abhängigkeit von den Kulturbedingungen auch ein Auswachsen der Elk. zu Fäden mit anschließender Segmentierung vorkommt.

10. Mumps-Virus.

Über Mumpsvirus, das an Erythrozytenmembranen adsorbiert worden war, berichten DAWSON und ELFORD (1949/1 und 2, Abb. 137). Seine Größe liegt etwa bei 180 mμ, wobei besonders große Schwankungen des Teilchendurchmessers auffallen (100 bis 260 mμ).

11. Kaninchen-Papillomvirus.

Die Tumoren erzeugenden Virusarten gehören systematisch verschiedenen Gruppen an. Das Myxom-Virus wurde bereits bei den quaderförmigen Elk. genannt, das Agens der Milchdrüsentumoren der weißen Maus wird im Abschnitt VI, 15 beschrieben. Das Papillom-Virus wurde für die elektronenmikroskopische Untersuchung in 0,05 mol. Phosphatpuffer pH 6,5 von SHARP, TAYLOR,

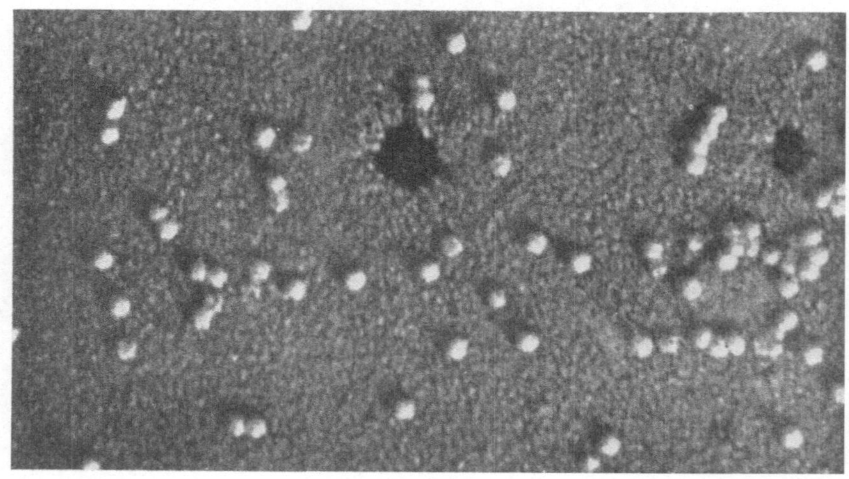

Abb. 138. Kaninchen-Papillomvirus mit Gold bedampft. 42 000 : 1. Nach SHARP, HOOK und BEARD (1946).

BEARD und BEARD (1942) in der Ultrazentrifuge isoliert. Die runden Elk. zeigten 44 mμ Durchmesser beim Vergleich mit 15 mμ breiten TM Virusstäbchen. Ähnlich wie beim Influenza-Virus dürfte der so gemessene Wert etwa 20% zu tief liegen, so daß die Größe mit 53 mμ angenommen werden kann. Die Teilchen sind weit gleichmäßiger als beim Influenzavirus und liegen gelegentlich zu Ketten aggregiert. Indirekte Größenbestimmungen ergeben 48,2 mμ. In einer späteren Untersuchung von SHARP, TAYLOR, HOOK und BEARD (1946) werden 66 mμ bei 58 Vol.% Wassergehalt angegeben. Mit der Schattentechnik aufgenommene Elk. erweisen sich als stark abgeflacht. Das Molekulargewicht der Teilchen beträgt 47 Millionen, doch dürfte die Frage noch offen sein, ob definierte Makromoleküle im chemischen Sinne vorliegen. Die gereinigten Suspensionen enthalten nur 1,5% in Fettlösungsmitteln extrahierbare Substanzen, 90% Proteine und 8,7% Thymonukleinsäure. Die Zusammensetzung ist also von dem fettreichen Influenzavirus (20%), dem Rous-Sarkom-Virus (40%) und dem Pferde-Enzephalomyelitisvirus (48%) recht verschieden. Ein chemischer Unterschied gegenüber den Influenza- und Pferde-Enzephalitisvirusarten kommt auch darin zum Ausdruck, daß mit Kalziumchlorid keine Kontrasterhöhung im elektronenmikroskopischen Bild eintritt (Abb. 138).

12. Pferde-Enzephalomyelitisvirus.

Aus Gewebeextrakten von 11 oder 12 Tage alten Hühnerembryonen haben SHARP, TAYLOR, BEARD und BEARD (1942) das Virus des westlichen Stamms der Pferde-Enzephalomyelitis durch Zentrifugen isoliert und elektronenmikroskopisch untersucht.

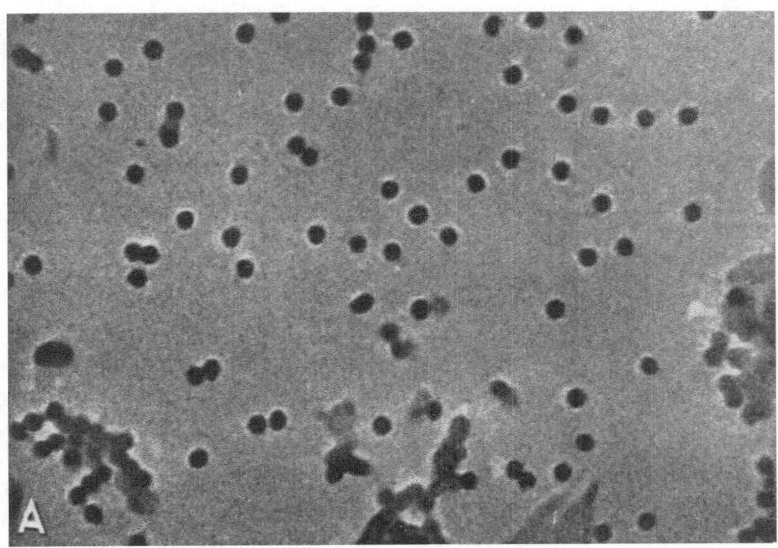

Abb. 139. Pferde-Enzephalomyelitisvirus östlicher Stamm, mit 0,023 mol Kalziumchlorid behandelt. 45000 : 1. SHARP, TAYLOR, BEARD und BEARD (1942).

Die runden bis ovalen etwa 40 mμ großen Elk. ergaben wenig kontrastreiche Bilder mit unscharfer Außenbegrenzung, besonders, wenn sie frisch isoliert und mit destilliertem Wasser gewaschen worden waren. Im Innern sah man eine dichtere Region, die annähernd $\frac{1}{3}$ des Teilchendurchmessers ausmachte. Am Virus des östlichen Stamms (TAYLOR und Mitarbeiter 1942) wurde nach Zusatz von 3 Teilen 0,25% $CaCl_2$ (ähnlich auch bei $MgCl_2$) zu einem Teil virushaltiger

Ringerlösung eine Kontrastzunahme, schärfere Begrenzung und Vergrößerung der Elk. auf etwa 47,5 mμ beobachtet (Abb. 139 und 140). Die Veränderung ist reversibel und wirkt sich nicht auf die Sedimentationskonstante der Teilchen aus. In einer späteren Mitteilung von Sharp und Mitarbeitern (1943) wurden folgende Werte angegeben:

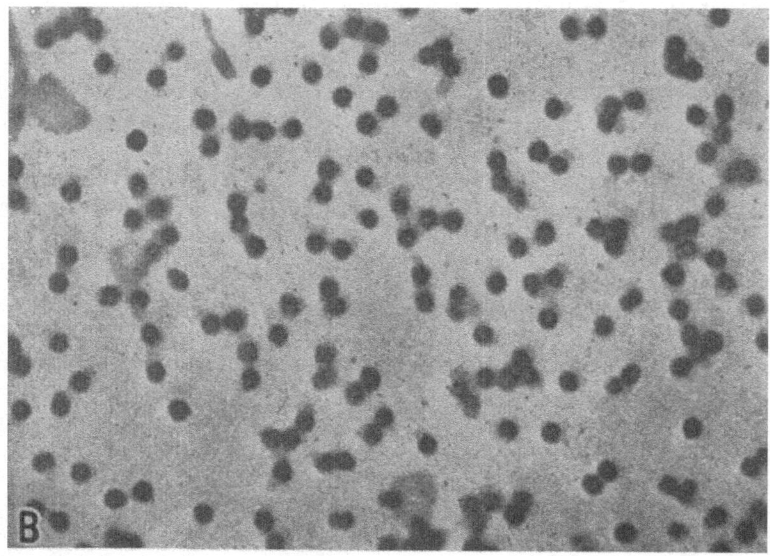

Abb. 140. Pferde-Enzephalomyelitisvirus, westlicher Stamm, behandelt mit 0,092 mol Kalziumchlorid. 45000 : 1. Nach Sharp, Taylor, Beard und Beard (1943).

Tab. 11.

	El.-mikr. bestimmter Durchmesser (mit 0,023 bis 0,092 mol. CaCl$_2$)		Berechneter Durchmesser	S_{20}	Spez. Vol.
	in mμ				
Westl. Stamm ...	39	(53)	56,8	$265 \cdot 10^{-13}$	0,864
Östl. Stamm	42	(47)	56,4	$237 \cdot 10^{-13}$	0,839

Wahrscheinlich kommen die höheren Durchmesserwerte der wirklichen Größe näher, da die Messung im Elektronenmikroskop durch Vergleich mit dem Tabak Mosaikvirus erfolgte, wodurch erfahrungsgemäß zu niedrige Werte erhalten werden (vgl. S. 360). Teilungs- oder Entwicklungsformen wurden nicht beobachtet. Die Verfasser isolierten außerdem eine makromolekulare Komponente aus dem Hühnerembryogewebe, die wesentlich kleiner war als das Virus (Abb. 141).

13. Poliomyelitis-Virus.

Tiselius und Gard (1942) bildeten zum erstenmal Poliomyelitis-Viruspräparate ab. Ihr Material stammte aus dem Nervensystem und dem Darm der Maus (Theiler-Virus), dem Darm von wilden, braunen Ratten und von

Menschen. Das Virus war nach dem Verfahren von Gard und Pedersen (1941) gereinigt, deren reinste Neuropräparate nach der Ultrazentrifugenanalyse aus einer einzigen Komponenten mit der Sedimentationskonstanten $S_{20} = 160$ bis $170 . 10^{-13}$ und der Diffusionskonstanten $D_{20} = 0{,}3 . 10^{-7}$ bestanden.[1] Unter

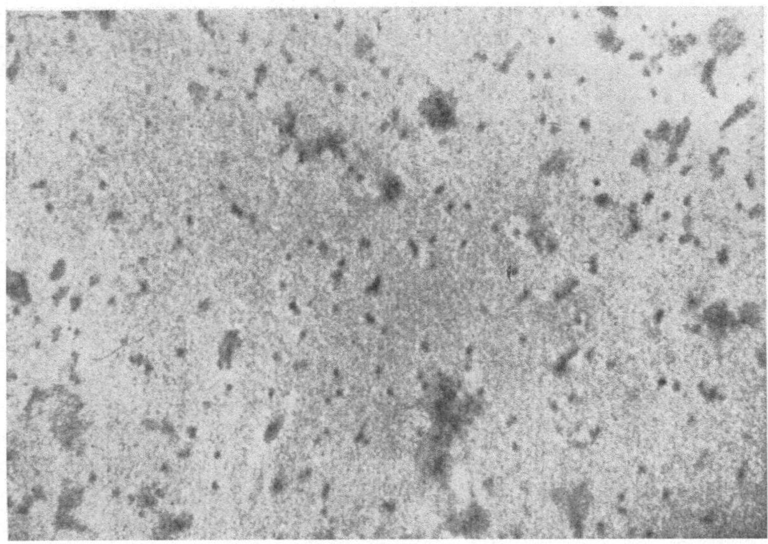

Abb. 141. Makromolekulare Komponente von normalem Hühnerembryogewebe. 45000 : 1. Nach SHARP, TAYLOR, BEARD und BEARD (1943).

gewissen Voraussetzungen wurde für Neurovirus auf ein Molekulargewicht von $50 . 10^6$ und Partikelabmessungen von 14 mμ Durchmesser und 640 mμ Länge geschlossen. Später gab Gard (1943/3) auf Grund der indirekten Methoden folgende Molekulargewichte und Partikelgrößen an:

Tab. 12.

	Molekular-gewicht	Partikel-durchmesser	Partikellänge
Neurovirus	$57 . 10^6$	12,5 mμ	580 mμ
Darmmaterial	$200 . 10^6$	12,2 mμ	2150 mμ

Abweichend vom Material aus Nervengewebe beträgt die Diffusionskonstante für Darmmaterial von Mensch, Maus und Ratte $D_{20} = 0{,}11 . 10^{-7}$. Die elektronenoptischen Abbildungen bestätigen die fadenförmige Gestalt des isolierten Materials (Abb. 142). Die kleinste Querdimension betrug nach Tiselius und Gard allerdings nur 5 mμ, was mit der beim Trocknen eingetretenen Schrumpfung infolge Verluste von Hydrationswasser in Zusammenhang gebracht wurde. Später führte Gard (1943/1) den auf manchen Präparaten unerwartet kleinen Durchmesser der Fäden wohl mit Recht auf eine sekundäre Streckung zurück und gab etwa 15 mμ als normale Fadendicke an. Die Vermessung der Faden-

---

[1] Vgl. auch Bourdillon (1943).

länge ergab unter Berücksichtigung der Streckung 3 bis 4 Häufigkeitsmaxima, die auf eine Ausbildung von ganzzahligen Vielfachen von 115 mμ hinweisen. Die Fäden sind z. T. ungewöhnlich lang (weit über 1 mμ). Durch Dehnung, Biegung oder Stauchung sind sie stärker deformierbar und bei der Aneinanderlagerung verschmelzen sie enger als beim Tabak-Mosaikvirus. Häufig bilden sich Verzweigungen und Netze aus. Je reiner sich die Präparate in physikalisch-chemischer Hinsicht erwiesen, um so stärker beherrschten die fädigen Strukturen das elektronenmikroskopische Bild und um so stärker war die Assoziationstendenz (Gard 1943/1). Die durchschnittliche Länge des humanen Neurovirus ist geringer als die des murinen. Mit einer an Sicherheit grenzenden Wahrscheinlichkeit wird angenommen, daß die Fäden mit der physikalisch-chemisch definierten Komponente der Präparate und mit dem Virus identisch sind.

Auffallenderweise finden sich die Fäden jedoch nicht nur in aktiven Viruspräparaten aus dem Nervensystem von kranken Mäusen und Menschen, sowie im Darm von jungen Mäusen und poliomyelitiskranken Menschen.

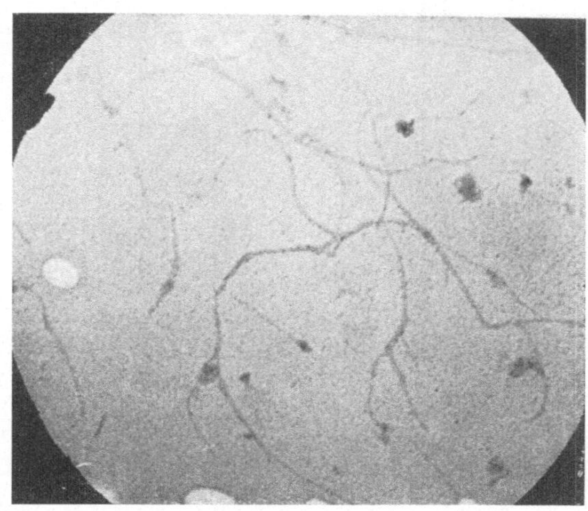

Abb. 142. Inaktives Theiler-Virus aus dem Darm von Ratten. 22000 : 1. Nach Gard (1943).

Gard konnte physikalisch-chemisch gleiches Material, das aber keine Virusaktivität besaß, aus dem Darm alter Mäuse, wilder, brauner Ratten und gesunder Menschen außerhalb der Zeit einer Poliomyelitisepidemie isolieren. Die gleiche Substanz wurde jedoch nicht im Kot von weißen Ratten, Meerschweinchen und Schweinen und in gesunden Gehirnen gefunden. Gard (1943/1) vermutet, daß die aktiven Formen neurotrope Varianten eines Darmvirus sind. Dafür spricht auch die Beobachtung stark wechselnder Virulenz muriner und humaner Stämme. Das inaktive Darmvirus verursacht von seiten des Darmes keinerlei Krankheitserscheinungen. Je Mäusehirn wurden nur etwa 3 γ Virus gefunden, während vom Darm mehrere Wochen lang täglich 10 bis 100 γ ausgeschieden werden. Die Isolierung führt auf chemischem Wege zu reineren, auf physikalischem Wege zu weniger veränderten Präparaten.

Abweichend von den Gardschen Befunden haben Loring, Marton und Schwerdt (1946) am Lansing-Stamm der Poliomyelitis sphärische Teilchen von 12 bis 34 (im Mittel 25) mμ Durchmesser gefunden. Die Partikel weisen eine Sedimentationskonstante von $S_{20} = 60 \cdot 10^{-13}$ auf. Loring hat seine Präparate zwei Wochen lang mit Kälte behandelt. In Präparaten, die nicht kältebehandelt worden waren, sedimentierte die aktive Komponente mit $S_{20} = 195 \cdot 10^{-13}$, also ähnlich wie das Material von Gard. So ist nach den Informationen des Verfassers zur Zeit die Frage offen, ob die Kälte eine Desaggregation der Fäden zu sphärischen Teilchen bewirkt, ohne daß die Virusaktivität verlorengeht, oder ob beim Lansing-Stamm keine Fäden vorliegen. Möglicherweise treten wie beim Influenza-Virus und beim Bushy-Stunt-Virus der Tomate sphärische und fädige Formen auf.

Kausche (1948) gibt für den Lansing-Stamm ebenfalls sphärische Teilchen an. Jungeblut und Bourdillon (1943) fanden beim SK-Stamm im Material aus Mäusehirn sphärische Teilchen von 25 bis 30 mμ, aber aus der Kultur des Virus in Mäusegewebe auch Fäden von etwa 20 mμ Dicke und 76 bis 5000 mμ Länge. Die Frage, ob die gefundenen Strukturen mit dem Virus identisch sind, lassen sie offen.

Das Virus der Teschener Krankheit der Schweine, das Gard den Poliomyelitis Virusarten von Mensch und Maus nahestellt, scheint noch nicht elektronenmikroskopisch untersucht zu sein.

Die chemische Analyse von menschlichem Darmmaterial ergab 16% Stickstoff und 0,6% Phosphor, sowie eine praktisch komplette U-V Absorption unter 2400 Å und ein relatives Maximum im Gebiet von 2600 bis 2700 Å. Danach ist die isolierte Substanz wahrscheinlich ein Nukleoprotein (Gard 1943/2).

Neben den Fäden finden sich abgerundete Körper von ziemlich einheitlichem Aussehen mit Durchmessern von 50 bis 100 mμ, die Gard als vorgebildete Zellstrukturen deutet. Außerdem sind kleinere Teilchen mit Durchmessern unter 10 mμ sichtbar, die wahrscheinlich ein der Normalkomponente aus Hirnmaterial entsprechendes molekulardisperses System darstellen. Bender und Kausche (1948) haben Gards Fadenformen für das Theilersche Virus bestätigen können (Fadenbreite 12 bis 15 mμ, Länge 120 bis 150 mμ oder ein Vielfaches dieser Länge). Es gelang den Autoren außerdem, in halbgesättigtem Ammonsulfat bei pH 7,0 lichtmikroskopisch sichtbare Kristalle nachzuweisen, die sich nach Fixierung durch Hitze, Pikrinsäure oder verdünnte Salpetersäure mit Viktoriablau 4 R anfärben ließen.

14. Virus der Maul- und Klauenseuche.

v. Ardenne und Pyl (1940) haben im Blaseninhalt der mit Maul- und Klauenseuche infizierten Meerschweinchen sehr kontrastarme Partikel von 20 bis 30 mμ Durchmesser nachgewiesen, die sie für das gesuchte Virus halten. Der nach den Ergebnissen der Ultrafiltration (8 bis 12, bzw. 20 mμ) etwas hohe Wert könnte durch sehr flaches Auftrocknen der Virusteilchen bedingt sein. Mit Hilfe der statistischen Ultramikrometrie hat Bonét-Maury (1948) ebenfalls einen Wert von etwa 30 mμ für den Durchmesser des dermotropen Maul- und Kleuenseuchevirus gefunden. Die Sedimentationskonstante beträgt nach Janssen (1941) $S_{28} = 17$ bis 18.10^{-13}, das Molekulargewicht 0,5 bis 1.10^6.

15. Das Agens des übertragbaren Milchdrüsentumors der Maus.

Nachdem Bittner in zahlreichen Untersuchungen gezeigt hatte, daß spontane Brustdrüsenkrebse der Maus durch die Milch und durch Organextrakte zellfrei übertragen werden können, versuchten Graff, Moose, Stanley, Randall und Haagensen (1947), sowie Passey, Dmochowski, Astbury und Reed (1947) die Reindarstellung und elektronenmikroskopische Abbildung des Krebsagens. Beide Autorengruppen fanden runde Teilchen uneinheitlicher Größe, deren Durchmesser mit 100, bzw. 20 mμ angegeben wird. In einer späteren Mitteilung berichteten Passey und Mitarbeiter (1948) über kleine Teilchen von 20 bis 35 mμ und größere bis zu 120 mμ Durchmesser, die bei 120000 bis 60000 g sedimentieren. Nicht nur in den Organen, sondern auch in der Milch der Mäusestämme, die den Spontantumor zeigten, fanden sie das Agens. Für die Identität der sichtbar gemachten Teilchen mit dem gesuchten Agens spricht, daß sie im Milchdrüsengewebe gesunder Stämme und in Methylcholanthrentumoren (Brustcarcinomen und Sarkomen) der Maus fehlten.

16. Polyeder Virus.

Die 1 bis 10 μ großen Polyeder, die in den Kernen fast aller Zellen polyeder-kranker Raupen bestimmter Schmetterlingsarten, sowie im Gewebesaft und im Blut frei vorkommen, erscheinen im Elektronenmikroskop als kaum durchstrahl-bare, kristallartige Körper (Bergold 1943). Die Tatsache, daß die Krankheit durch die Polyeder übertragen werden kann, veranlaßte ihre genauere physikalisch-chemische und elektronenmikroskopische Analyse. Es gelang Bergold und Mitarbeitern (1942/43), aus den Polyedern ein einheitliches Protein mit dem Molekulargewicht ~ 300000 und dem Moleculardurchmesser ~ 10 mμ zu isolieren. Unabhängig davon untersuchten Glaser und Stanley (1943), sowie Lauffer (1943) ein aus Lymphe isoliertes infektiöses Protein, das ebenfalls ein Molekulargewicht von etwa 300000 und Teilchendurchmesser von 10 mμ aufwies. Die gewonnenen Proteine wurden ursprünglich mit dem Virus identifiziert, doch zeigten neuere Untersuchungen von Bergold (1947/48), sowie von Bergold und Friedrich-Freksa (1947), daß zwischen Polyeder-Protein und eingeschlos-senem Virus scharf zu unterscheiden ist und daß das Virus, gleichgültig, ob aus Polyedern oder Lymphe isoliert, ein wesentlich höheres Teilchengewicht als 300000 hat.

Das Polyederprotein ist das erste genauer untersuchte Protein, das bei der Wechselwirkung zwischen Virus und infizierter Zelle entsteht. Als Ausgangs-material dienten Bergold Polyeder von Seidenraupen (Bombyx mori = Bm), Schwammspinner (Porthetria dispar = Pd) und Nonnenraupen (Lymantria monacha = Lm). Unter bestimmten Bedingungen und bei alkalischer Reaktion lösen sie sich im wesentlichen zu einem einheitlichen Protein auf. Die Molekular-gewichte (Mo) und Moleculardurchmesser (2 r_0) der Hauptkomponenten und der Spaltstücke der verschiedenen Polyederproteine sind in Tabelle 13 in mμ an-gegeben.

Tab. 13.

	Hauptkomponenten für nichthydratisierte Moleküle			Spaltkomponenten 1/6			Spaltkomponenten 1/18		
	Pd	Lm	Bm	Pd	Lm	Bm	Pd	Lm	Bm
Mo.........	276600	336000	378000	47250	—	60500	15360	18270	20230
2r_0........	8,64	9,22	9,59	4,79	—	5,21	3,30	3,49	3,62

Da die Hauptkomponenten im Gegensatz zu den Spaltstücken nicht wesentlich von der Kugelform abweichen und nahezu 10 mμ Durchmesser haben, dürften sie elektronenmikroskopisch gut darstellbar sein. Sie entsprechen in ihrer Größe dem von Glaser und Stanley aus Lymphe isolierten und abgebildeten Protein, sind aber serologisch von ihm verschieden. Die kleinsten Spaltstücke haben etwa das Molekulargewicht der Svedbergschen Proteineinheit von 17500 und die Größe der von O. Kratky an Bm-Polyedern röntgenographisch vermessenen Elementarzelle von 4,53 · 2,8 · 2,04 mμ. Das Polyederprotein zeigt serologische Beziehungen zum Polyedervirus, nicht aber zu den Proteinen der Lymphe.

Bezüglich der Herkunft des Polyederproteins erörtert Bergold (1947) drei Möglichkeiten:

1. Das Polyederprotein ist ein Bestandteil des Polyedervirus, der z. B. infolge nicht mehr verfügbarer Nukleinsäure nicht zum Viruspartikel aufgebaut werden kann.

2. Das Polyederprotein ist ein Abbauprodukt des Virusproteins, wofür die völlige Wasserunlöslichkeit sprechen könnte.

3. Das Polyederprotein ist ein Reaktionsprodukt der Wirtszelle, das um das Virus zum Polyeder auskristallisiert (v. Prowazek 1907).

Bergold glaubt, daß das Polyederprotein am ehesten mit dem löslichen Antigen anderer Virusarten, besonders des Pockenvirus, zu vergleichen ist, bei welchem man analog durch vorsichtige alkalische Spaltung ein dem Nukleoprotein serologisch verwandtes Protein erhält. Wesentliche Unterschiede dürften jedoch, wie die späteren Untersuchungen von Dawson und McFarlane (1948) vermuten lassen, darin bestehen, daß das lösliche Antigen des Pockenvirus mit dem Nukleoprotein zusammen in einer Grenzmembran eingehüllt ist und daß bei Verlust des löslichen Antigens kein aktives Virus zurückbleibt.

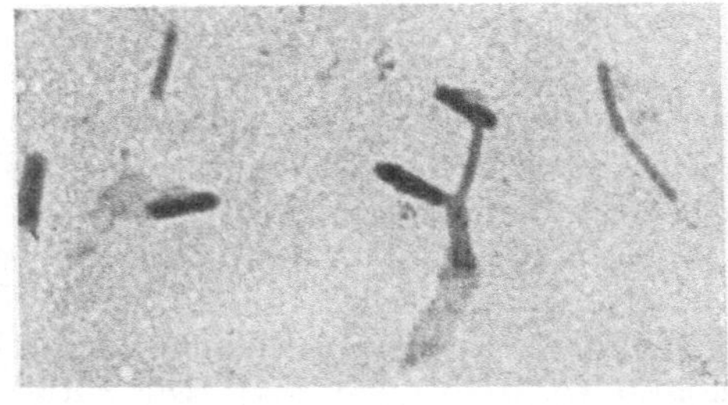

Abb. 143. Polyederviruspartikel von B. mori in Wasser. Doppel- und Einzelstäbchen. 20000:1. Nach Bergold (1948).

Die geringe Virusaktivität der Hauptkomponente des reinen Polyederproteins und die Beobachtung, daß bei der Ultrazentrifugierung die Aktivität in den oberen Schichten wesentlich schneller absinkt als der Gehalt an Polyederprotein, führten Bergold zu dem Schluß, daß das infektiöse Virus höher molekular sein mußte. Es gelang ihm, die gesuchte Komponente in der analytischen Ultrazentrifuge nachzuweisen und das Polyedervirus, das nur etwa 5% der Polyedersubstanz

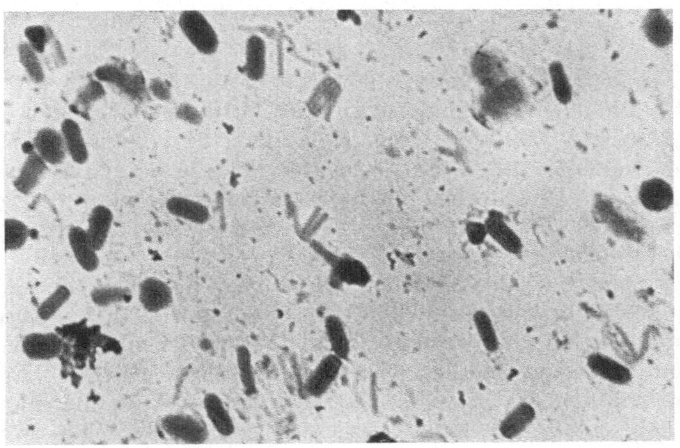

Abb. 144. Polyederviruspartikel von P. dispar in Wasser. Bündelförmige Virusaggregate in Auflösung zu Einzelstäbchen. 20000:1. Nach Bergold (1948).

ausmacht, zu isolieren. Die Bm-Viruslösungen ergeben eine einzige Sedimentationsgrenze und relativ breite Sedimentationskurven als Zeichen nicht ganz einheitlicher Partikel. Die Pd- und Lm-Lösungen zeigten drei deutlich trennbare Komponenten, deren Grund später mit Hilfe des Elektronenmikroskops geklärt werden konnte. Für nicht hydratisierte Teilchen ergaben die Partikelgewichte (Mo) und die indirekt ermittelten Durchmesser ($2\,r_0$) die Werte der folgenden Tabelle:

Tab. 14.

	Bm	Lm	Pd
Mo	916 000 000	1 136 000 000	2 146 000 000
$2\,r_0$	130,7	138,5	171,5

Bei Lm und Pd ist die Komponente mit der höchsten Konzentration zugrunde gelegt. Die Durchmesserwerte sind als Mittelwerte der von der Kugel abweichenden Formen aufzufassen.

Die elektronenmikroskopische Untersuchung zeigte beim Bm-Virus sehr einheitliche schlanke Stäbchen, beim Pd-Virus plumpere und uneinheitlichere Formen, wie es aus den Ergebnissen der Ultrazentrifugenanalyse erwartet werden konnte (Abb. 143 und 144). Die Mittelwerte und die aus den Abmessungen errechneten Partikelgewichte finden sich in Tabelle 15. Sie sind größer als es nach der Ultrazentrifugenanalyse zu erwarten war.

Tab. 15.

Virus	Anzahl der Teilchen	Mittelwerte in mμ		Partikel-gewicht
		Längen $\pm 3\,\sigma$	Breiten $\pm 3\,\sigma$	
Bm	150	349,74 $\pm$ 6,75	87,66 $\pm$ 2,55	1098 . 10^6
Pd	138	415,2 $\pm$ 20,82	160,00 $\pm$ 10,0	4510 . 10^6

Überraschend war die Bakterienähnlichkeit der Stäbchen, die durch eine gelegentlich vorhandene Einschnürung in der Mitte und durch fast endständige Verdichtungszentren verstärkt wird. Bergold (1947) betont jedoch, daß das Fehlen einer Membran gegen die Einreihung des Polyedervirus unter die Bakterien, bzw. Rickettsien sprechen dürfte, wofür sich später noch triftigere Gründe fanden. v. Borries (1949) weist darauf hin, daß die aus der ersten Arbeit von Bergold hervorgehende prozentual stärkere Streuung der Breite als die der Länge der Viruspartikel das Vorhandensein eines Längenwachstums, wie es bei Bakterien vorliegt, unwahrscheinlich mache. Bergold (1948) hat dann gefunden, daß die ursprünglich als Einheit aufgefaßten Virusteilchen in Wirklichkeit meist aus zwei und mehreren parallel gelagerten Stäbchen bestehen. Entsprechend den Ergebnissen mit der analytischen Ultrazentrifuge ist die Bündelung bei Bm selten (Abb. 143) und wenig ausgesprochen, insofern nur zwei Stäbchen nebeneinanderliegen. Bei Pd, wo verschiedene Sedimentationsbande auftreten, sind dagegen bis vier sich trennende Stäbchen in einem Bündel zu erkennen (Abb. 144). Eine Aufspaltung der Bündel ausschließlich in Einzelstäbchen ist bisher auf chemischem Wege weder durch Salzzusätze noch durch pH-Verschiebung gelungen. Nach der Ultrazentrifugenanalyse zerfallen zwar die Bm-Polyedervirusteilchen im sauren Bereich in etwa 1/4 und 1/8 Bruchstücke und im alkalischen in 1/12 und 1/24, aber diese Spaltung scheint nichts mit einer Stäbchentrennung zu tun zu haben. Elektronenmikroskopische Aufnahmen schwach saurer Viruslösungen von B. movi ließen stark verkrümmte Einzelstäbchen mit knotigen Abschnitten und veränderter Innenstruktur erkennen (Abb. 145). Die knotenförmigen Verdickungen erkennt man auf guten Aufnahmen auch am unbehandelten Virus. Sie liegen auf gleicher Höhe angeordnet, so daß eine Querstruktur entsteht. BERGOLD hält es für möglich, daß die Bündelung zu einer Längsspaltung beim Vermehrungsvorgang der Polyederviren in Beziehung zu setzen ist.

Beim Kochen schrumpfen die Virusaggregate unter teilweisem Substanzaustritt (Abb. 146); nach Behandlung mit 1% $HgCl_2$ zeigen sie zum Teil eine wenig dichte Hüllsubstanz (Abb. 147). Nach alkalischer Behandlung des Polyedervirus konnte BERGOLD in diesen Lücken nachweisen, in welchen die Virusstäbchen eingebettet waren (Abb. 148).

Über die Natur der Polyederviren kann noch nichts Endgültiges ausgesagt werden. Zwar spricht BERGOLD (1947) vielfach von ihrem „Molekular"-gewicht, doch schreibt er, daß die Empfindlichkeit gegen Glycerin, Alkohol, Äther und Einfrieren vielleicht eher für eine Organismennatur spreche. Polyederprotein wie Polyedervirus enthalten etwa 15% Stickstoff, das Virus aber erheblich mehr Phosphor (1,33 statt 0,04 bis 0,06%). Nach BERGOLD liegt wahrscheinlich der gesamte Phosphor des Virus als Thymonukleinsäure-Phosphor vor.

Dem Polyedervirus verwandt ist nach BERGOLD (1938) das Kapselvirus von Cacoecia murinana Hb. Im Zellplasma der erkrankten Raupen finden sich blasenförmige Gebilde von einigen bis 20 μ Durchmesser, in deren Innerem hell leuchtende Partikel vibrieren. Die Partikel (Kapseln) messen $360 \pm 40 \times$ $\times 230 \pm 55$ mμ (Abb. 149) und lassen in 0,02 mol. Na_2Co_3

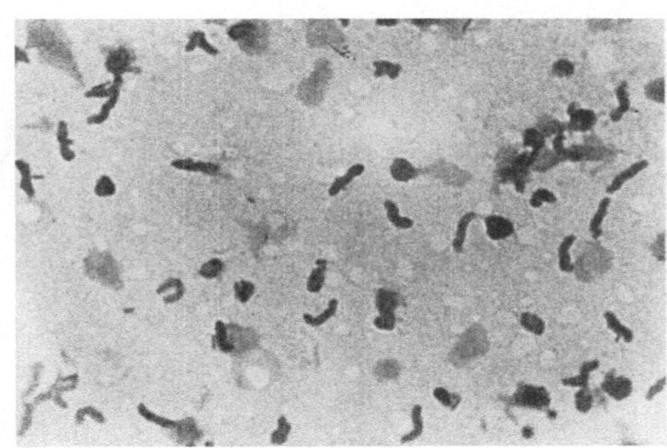

Abb. 145. Polyederviruseinzelstäbchen von B. mori, denaturiert in 0,05 m HCl. 20000 : 1. BERGOLD (1948).

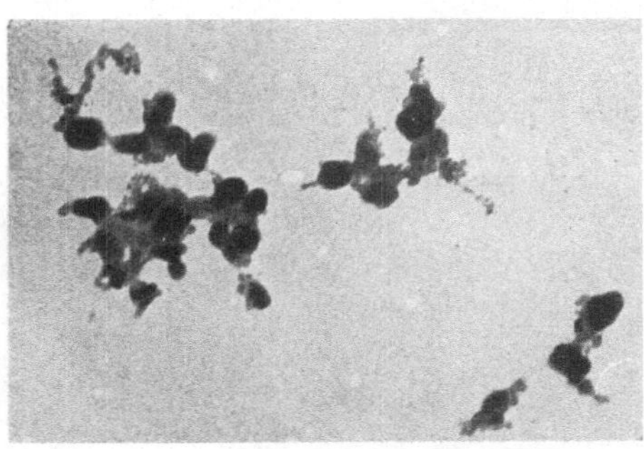

Abb. 146. Polyedviruspartikel von P. dispar, gekocht in Wasser. 20000 : 1. Nach BERGOLD (1948).

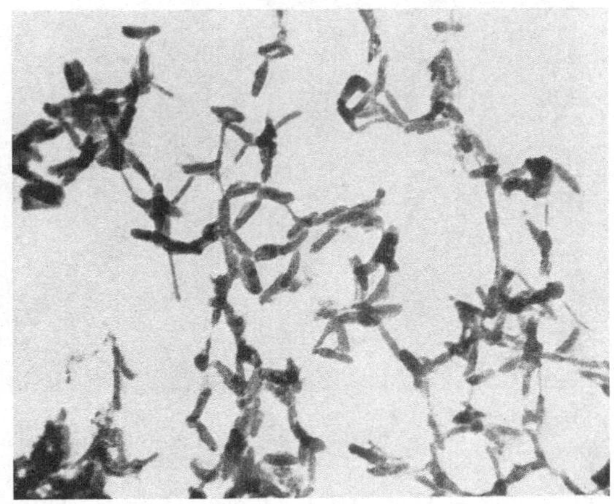

Abb. 147. Polyedviruspartikel von B. mori in 1%iger $HgCl_2$. 20000 : 1. Nach BERGOLD (1948).

stäbchenförmige Virusteilchen austreten, deren Bau den Polyedervirusteilchen gleicht (Abb. 151). Die Austrittsstellen aus der Kapsel markieren sich als stäbchenförmige Löcher, so daß die Kapseln ein kaffeebohnenartiges Aussehen annehmen (Abb. 150). In 0,04 mol. $Na_2Co_3 + 0,05$ mol. NaCl löst sich die Kapselsubstanz völlig auf. Das Kapselprotein besitzt das gleiche Molekulargewicht wie das Polyederprotein. Die Sedimentationskonstante der Virusstäbchen beträgt 1324 Svedberg-Einheiten, die Diffusionskonstante $D_{20}=0,278 \times 10^{-7}$, das aus diesen Daten als Molekulargewicht berechnete Teilchengewicht 460×10^6. Aus den elektronenmikroskopischen Abmessungen

Abb. 148. Teil eines P.-dispar-Polyeders mit Virusfehlstellen gelöst in schwacher Natriumkarbonatlösung. 26000 : 1. Nach BERGOLD (1947).

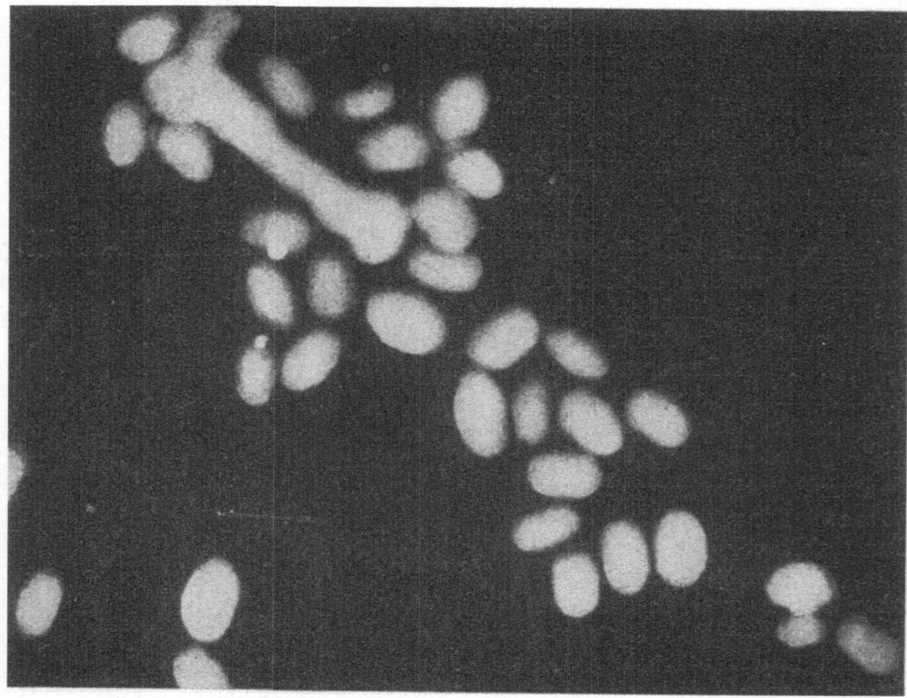

Abb. 149. Viruskapseln von Cacoccia murinana. 25000 : 1. Nach BERGOLD (1948).

von $262 \pm 37 \times 50{,}6 \pm 11$ mμ (Länge : Durchmesser 5,21) ergibt sich ein Teilchengewicht von 435×10^6.

Abb. 150. Viruskapseln von C. m., aus welchen die Virusstäbchen durch alkalische Behandlung herausgeschlüpft sind. 25000 : 1. Nach BERGOLD (1948). Vgl. Abb. 148.

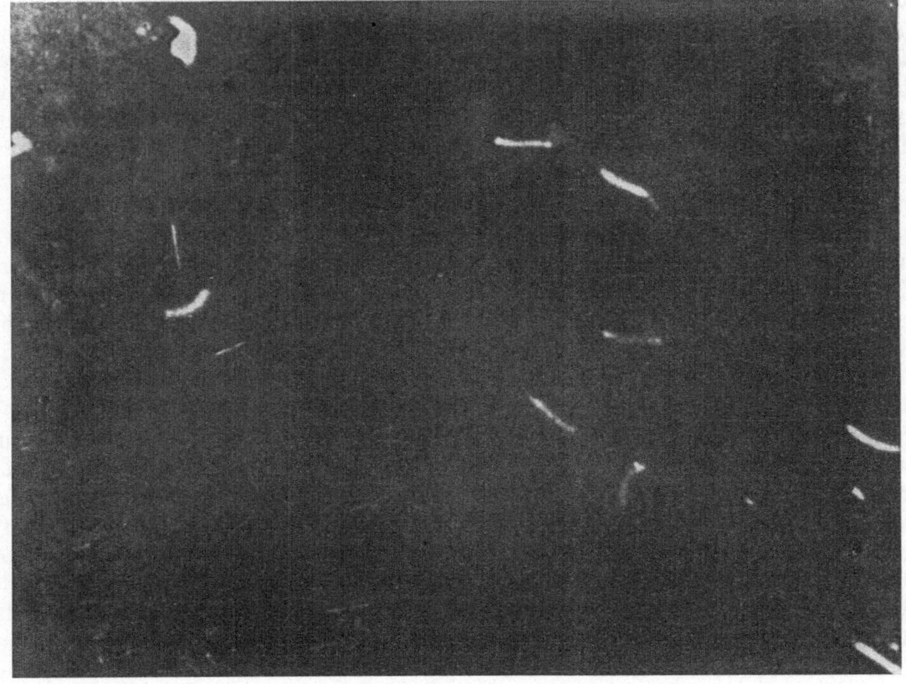

Abb. 151. Isolierte Kapselvirusstäbchen mit Querstrukturen. 25000 : 1. Nach BERGOLD (1948).

Bergold betont, daß die stäbchenförmigen Polyeder- und Kapselvirus-
arten durch die Hüllbildung und bündelförmige Ordnung eine höhere Organisa-
tionsstufe besitzen als die stäbchenförmigen Pflanzenvirusproteine. Er hält
es aber für unwahrscheinlich, daß engere Beziehungen des Kapselvirus zu den
Mikrosporidien bestehen, von welchen gewisse Coccosporidien bei Raupen
ähnliche Krankheitsbilder und Zellveränderungen hervorrufen wie das Kapselvirus.

17. Pflanzenpathogene Virusarten.

Die Virusproteine der Pflanzen unterscheiden sich physikalisch-chemisch
von fast allen bisher besprochenen Virusarten durch ihre Kristallisierbarkeit
und die an den Kristallen und an Gelen nachweisbaren Röntgeninterferenzen.
Chemisch sind sie reine Nukleoproteine. Morphologisch lassen sie sich in die
zwei Gruppen der stäbchenförmigen und kugelförmigen Virusproteine unterteilen.

a) Stäbchenförmige Virusproteine.

Zu dieser Gruppe gehören unter anderem das Tabak Mosaikvirus (Nicotiana
Virus 1 und 1 A, C, D), das Tomaten Mosaikvirus Dahlem, das Gurken Mosaikvirus 3 und 4, sowie das Kartoffel-X- und Y-Virus (Solanum Virus 1 und 2).
Die mit indirekten Methoden erschlossene Stäbchenform haben Kausche,
Pfankuch und H. Ruska (1939) am Tabak Mosaikvirus und Kartoffel-X-Virus
durch elektronenmikroskopische Abbildung bestätigt. Die Stäbchenbreite betrug
in Übereinstimmung mit den röntgenographischen Ergebnissen 15 mμ und erwies
sich als vollkommen einheitlich. Die Länge der Stäbchen war dagegen außerordentlich uneinheitlich (150 bis 1800 mμ), doch überwogen Abmessungen, die
von den genannten Autoren zunächst mit 300 mμ, später (1941) mit 320 mμ
angegeben wurden. Die größeren Längen ließen sich als Längsaggregate einer
Grundeinheit von 150 mμ, bzw. 300 mμ auffassen. Da frische Viruspräparate
in der Ultrazentrifuge eine einheitliche Sedimentationsbande zeigen, aus welcher
auf ein ebenfalls einheitliches Molekulargewicht geschlossen wurde, war die
Längenstreuung überraschend. Die spätere Diskussion beschäftigte sich eingehender mit der Frage, ob die Stäbchen erst bei der Präparation und elektronenmikroskopischen Abbildung durch Zerfall und Reaggregation uneinheitlich
werden, oder ob sie schon in der Wirtszelle verschieden sind. Durch stereoskopische Aufnahmen zeigten Melchers, Schramm, Trurnit und Friedrich-
Freksa (1940), daß alle untersuchten Stäbchen in der Filmebene lagen und
nicht durch winkliges Abstehen von der Folie in verschiedener perspektivischer
Verkürzung erschienen. Die präparative Erfahrung lehrte, daß die Längen des
Virus bei der Reindarstellung durch chemische Fällungsmethoden stärker modi-
fiziert werden als durch die Ultrazentrifugierung. Stanley und Anderson (1941)
fanden in schonend zentrifugierten Präparaten über 70% Tabak Mosaikvirus-
stäbchen von 280 ± 3,5% mμ Länge, für die sich bei einer Dichte von 1,33 ein
Molekulargewicht von 39,8 × 10^6 berechnen ließ. Dieser direkt ermittelte Wert
steht in bester Übereinstimmung mit der jüngst von Schramm und Bergold (1947)
aus der Messung von Sedimentations- und Diffusionskonstanten berechneten
Größe von 40,7 × 10^6. Einheitliche Stäbchenlängen zeigten sich auch bei der
Untersuchung frischer Preßsäfte (Melchers und Mitarbeiter 1941) und in den
Haarzellen virusinfizierter Pflanzen (Oster und Stanley 1946). Sigurgeirsson
und Stanley (1947) fanden ferner, daß in der Kälte (4° C) eine Längsaggregation
eintritt, die dazu führt, daß nach 20 Tagen mehr Stäbchen der doppelten als der

einfachen Länge von 280 mμ vorliegen. Die längsaggregierten Doppelstäbchen sind so stabil miteinander verbunden, daß sie bei Wärmeeinwirkung oder mechanischer Beanspruchung nicht an der Nahtstelle, sondern an beliebigen Orten

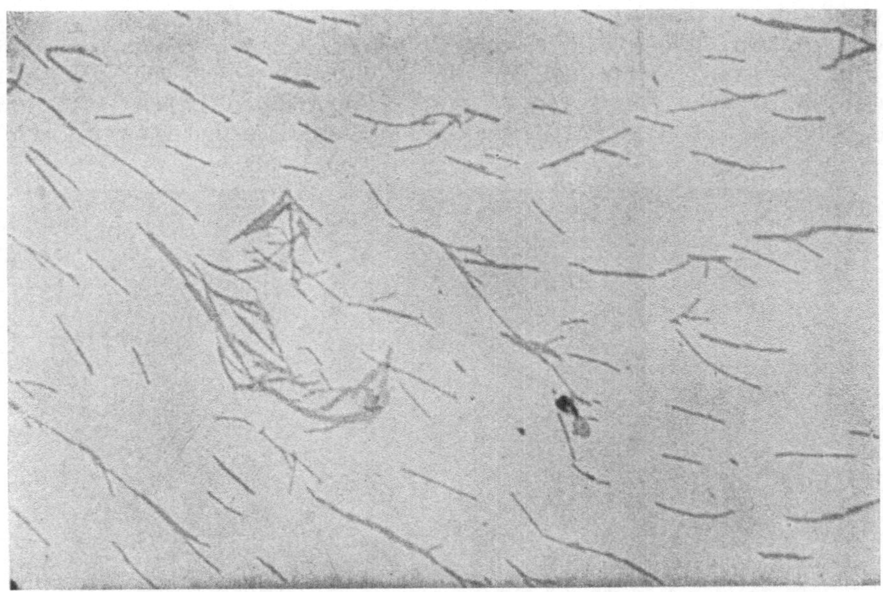

Abb. 152. Tabak Mosaikvirus. Uneinheitliche Stäbchenlängen. Nach Kausche aus H. Ruska (1943/15). 22000:1.

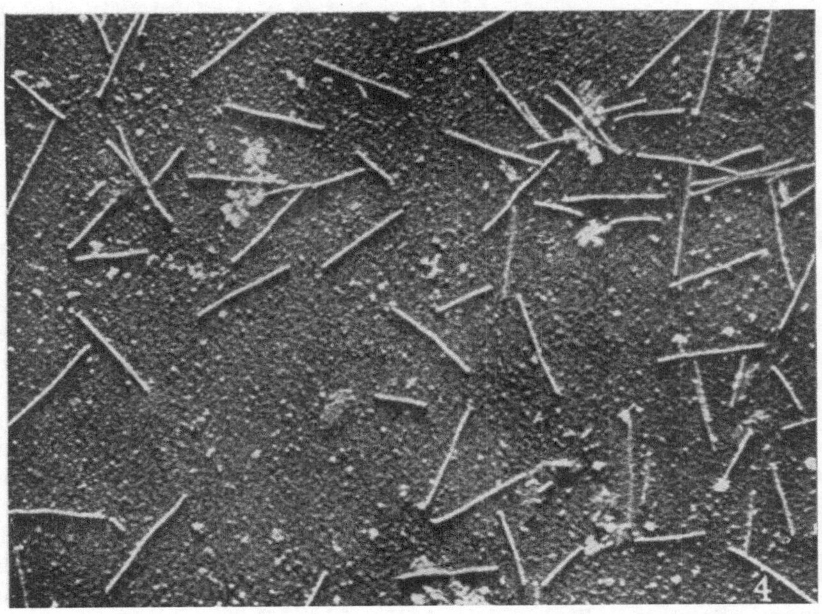

Abb. 153. Tabak Mosaikvirus aus Pflanzenpreßsaft, 20 Tage bei 4° C gehalten und mit dem 50fachen Volumen destillierten Wassers verdünnt. Mehr als die Hälfte der Stäbchen sind zur doppelten Normallänge aggregiert. Schrägbedampfung mit Gold. 23000:1. Nach Sigurgeirsson und Stanley (1947).

brechen. Nur bei der Durchschäumung einer Tabak Mosaikviruslösung fand H. Ruska (1948), daß ein sehr geringer Prozentsatz der Doppelstäbchen in die Normallänge zerfällt. Durch Abzentrifugieren der Normallängen machten Sigurgeirsson und Stanley (1947) in hohem Maße wahrscheinlich, daß nur die über 280 mμ messenden Stäbchen aktiv sind. Alle Befunde sprechen eindeutig für eine Größe der infektiösen Einheit des Tabak Mosaikvirus von 15 × 280 mμ (Abb. 152 bis 154). Mit diesem Wert lassen sich auch die Ergebnisse von Lauffer (1944) und Rawlins, Roberts und Utech (1946), von Pfankuch und H. Ruska (1947), sowie von Takahasi und Rawlins (1949)

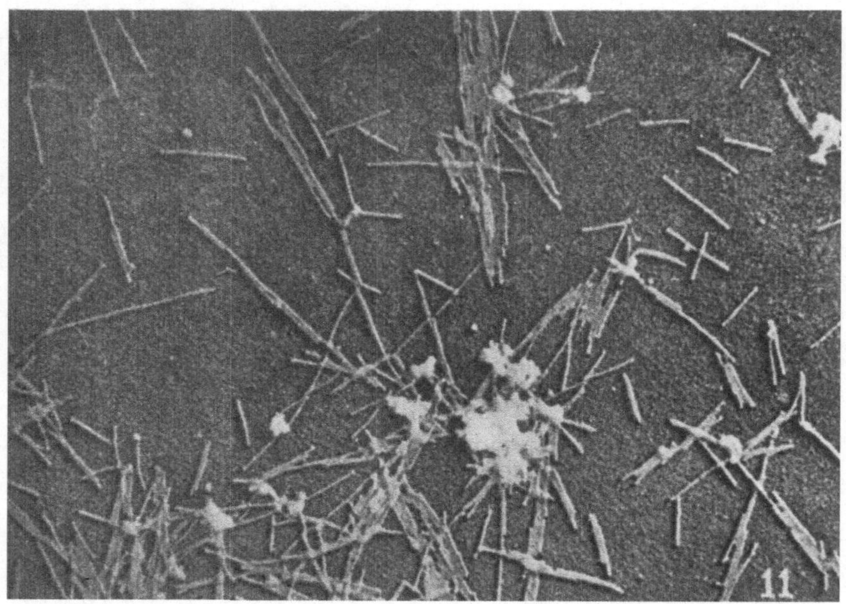

Abb. 154. Wie Abb. 153. Normallängen, Doppellängen und Mehrfachlängen der Virusstäbchen nach Säurebehandlung (pH 2,2). 23000 : 1. Nach Sigurgeirsson und Stanley (1947).

in Übereinstimmung bringen. (Vgl. dagegen die Bemerkung von Kausche 1948, welcher für das Theiler-Virus „in Analogie zu den neuesten Ergebnissen der experimentellen Forschung auf dem Gebiet der Tabak Mosaikviruskrankheit annimmt, daß kleine und kleinste Untereinheiten von der Größe der physikalischen Elementarzellen oder Identitätsperioden im Bereich von wenigen Angström-Einheiten sich aktiv fortbewegen und nach dem Matrizenschema neu gebildet werden".)

Die biologischen, serologischen, chemischen, physikalischen und morphologischen Eigenschaften des Tabak Mosaikvirus sind nach Gaw und Stanley (1947), sowie Gaw (1947) unabhängig von der Art der Wirtspflanze (Türkischer Tabak und Phlox). Für die aus Haarzellen isolierten Stäbchen der Tabak Mosaikvirusstämme aus Dahlem, Rothamsted und Princeton, die gleiche Krankheitssymptome hervorrufen, wiesen Oster, Knight und Stanley (1947) gleiche Teilchenabmessungen von 15,2 × 280 ± 8,6 mμ nach. Die älteren Angaben von Melchers, Schramm, Trurnit und Friedrich-Freksa (187,5 mμ) deuten daher auf Schädigungen, die wahrscheinlich durch die Ammonsulfatfällung und Wiederauflösung des Virus hervorgerufen wurden. Eine derartige Annahme erscheint besonders berechtigt, nachdem sich quantitative, serologische und biologische Unterschiede des Stammes Dahlem gegenüber dem von KAUSCHE und Mitarbeitern verwendeten Stamm nicht ergeben haben (FRIEDRICH-FREKSA, MELCHERS und SCHRAMM (1946). Für Stämme mit abweichenden Symptombildern (gelbes und grünes Aukuba Mosaikvirus, maskiertes Tabak Mosaikvirus u. a.) fanden KNIGHT und OSTER (1947) die gleichen Abmessungen von 280×15 mμ, ebenso TAKAHASI und RAWLINS (1947), während für das Gurken Mosaikvirus 3 und 4 etwa 300×13 mμ von KNIGHT und OSTER gefunden wurde. Es zeigen also nicht alle stäbchenförmigen Virusproteine der Pflanzen die gleichen Dimensionen. Durch die elektronenmikroskopische Kontrolle der indirekten Untersuchungen, die das Virusmaterial unter anderen physikalisch-chemischen Bedingungen prüfen, als sie bei der Abbildung mit Elektronen vorliegen, wurde in präparativer Hinsicht ein großes Erfahrungsmaterial gesammelt und eine vorzügliche Übereinstimmung der Resultate verschiedener Methoden erzielt. Zu den indirekten Methoden (Messung von Filtrationsendpunkt, Sedimentation, Diffusion, Viskosität, Röntgenbeugung) ist in den letzten Jahren die Größen- und Formbestimmung des Virus durch Messung der Lichtstreuung hinzugekommen (OSTER, DOTY und ZIMM (1947), OSTER (1948).

b) Kugelförmige Virusproteine.

Da nicht alle gereinigten, aus viruskranken Pflanzen gewonnenen Viruslösungen die Erscheinung der Strömungsdoppelbrechung zeigen, war bereits vor der Anwendung des Elektronenmikroskops bekannt, daß ein Teil der pflanzlichen Virusproteine kugelförmige oder annähernd kugelförmige Elementarteilchen hat. Hierher gehören unter anderem das Tabak Ringfleckvirus, das Rüben Gelbmosaikvirus, das Tabak Nekrosevirus, das südliche Bohnen Mosaikvirus und das Bushy Stunt Virus der Tomate. LAUFFER und STANLEY (1940) hatten in Übereinstimmung mit BAWDEN und PIRIE (1938), NEURATH und COOPER (1940) gezeigt, daß das Tomaten Bushy Stunt Virus aus etwa 26 mμ großen Teilchen besteht. Dieser Befund konnte von STANLEY und ANDERSON (1941) bei ihren ersten elektronenmikroskopischen Untersuchungen bestätigt werden. Sie fanden ferner etwa 20 mμ Durchmesser für die sphärischen Teilchen des Tabak Nekrosevirus. Durch Ultrafiltration (SMITH und MACCLEMENT 1940) und durch die statistische Ultrametrie (LEA 1940) waren Werte zwischen 13 und 20 mμ ermittelt worden. Die Sedimentationskonstante betrug nach PRICE und WYCKOFF (1939) 112×10^{-13}.

PRICE, WILLIAMS und WYCKOFF (1945) untersuchten metallbeschattete Präparate von Tomaten Bushy Stunt Virus und südlichem Bohnen Mosaikvirus. Infolge der dichten hexagonalen Anordnung der Teilchen konnten ihre Durchmesser mit großer Genauigkeit zu 25 mμ bestimmt werden. Für das Rüben Gelbmosaikvirus wurden von MARKHAM und SMITH (1946) elektronenmikroskopisch 22 mμ große Durchmesser gefunden.

c) Der Aufbau von Viruskristallen und -gelen.

Die elektronenmikroskopische Methodik erwies sich als wertvolle Ergänzung der Röntgenographie. An nadelförmigen Tabak Mosaikviruskristallen zeigten KAUSCHE und H. RUSKA (1939) den parakristallinen Aufbau aus parallel gelagerten stäbchenförmigen Makromolekülen. Da schon einzelliegende Längsaggregate der Moleküle die Stoßstellen der Teilstücke nicht erkennen lassen, waren quer zur Nadelachse keine Schichtebenen aufzufinden. Man sah aber an einzelnen Nadeln Unterbrechungen in den längsaggregierten Fäden, deren unregelmäßige Lage darauf hinwies, daß Schichtebenen fehlten. Dagegen ließ sich annehmen, daß die Moleküle Kettengitter bildeten. Alle Bilder zeigten, daß die Längen der Strukturelemente nach den Spitzen der Nadeln hin kürzer und unregelmäßiger wurden (Abb. 155). Da die Moleküllängen, wie spätere

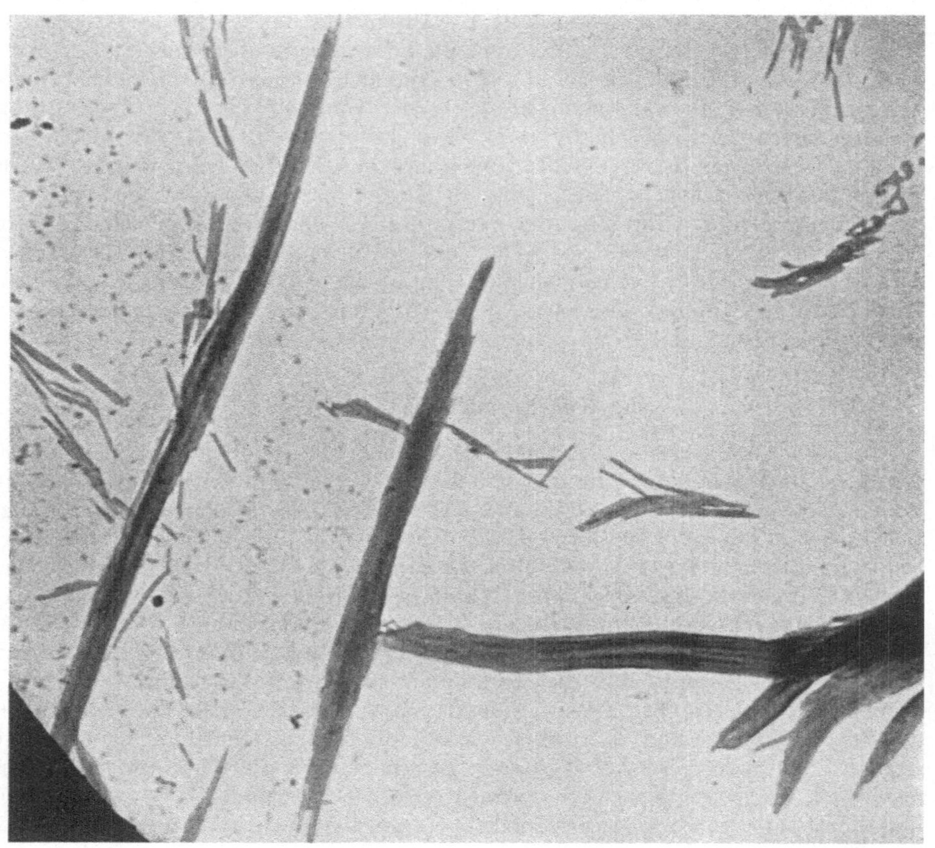

Abb. 155.   Tabak Mosaikviruskristalle, hergestellt durch Ammonsulfatfällung. 30 000 : 1.   Nach
KAUSCHE und H. RUSKA (1939).

Erfahrungen zeigten, bei der chemischen Virusdarstellung und Kristallfällung mit Ammonsulfat uneinheitlich werden, kann keine gleichmäßige Anordnung der Stäbchen quer zur Nadelachse erfolgen. In den hexagonalen Kristallplättchen, die in Haarzellen tabak-mosaikviruskranker Pflanzen nachgewiesen werden konnten und die KAUSCHE (1939) auch in vitro aus hochkonzentrierten Lösungen dargestellt hat, ist die Orientierung und genauere Anordnung der Moleküle aus elektronenmikroskopischen Untersuchungen nicht bekannt.

Die übliche Auftrocknung stäbchenförmiger Virusproteine auf die Objektträgerfolie führt bei höheren Konzentrationen zu Virusbildungen, wobei durch Längs- und Parallelaggregation Netzwerke aus Einzelmolekülen und breiten Strängen entstehen (KAUSCHE, PFANKUCH und H. RUSKA 1939, Abb. 2). WYCKOFF (1947) stellte ähnliche Präparate durch Gefriertrocknung und Schrägbedampfung mit Metall her. Außerdem wurden Auftrocknungen konzentrierter

Lösungen auf Glasplättchen mit Metall bedampft und zur Untersuchung vom
Glas abgelöst. Zum Teil entstanden zweidimensionale Blätter mit ein- oder
mehrschichtiger, bevorzugt paralleler Orientierung der Virusmoleküle. Wenn die
Konzentration des Virus mehr als 2 bis 4% beträgt, ist die Orientierung wahr-
scheinlich schon im Lösungszustand vorhanden, bei über 35% hat die Lösung
Gelkonsistenz.

Die geordnete Auftrocknung kugelförmiger Virusmoleküle untersuchten
PRICE, WILLIAMS und WYCKOFF (1945 und 1946) an Tomaten Bushy Stunt und
südlichem Bohnen Mosaikvirus (Abb. 156). Mikrotropfen der Virussuspensionen

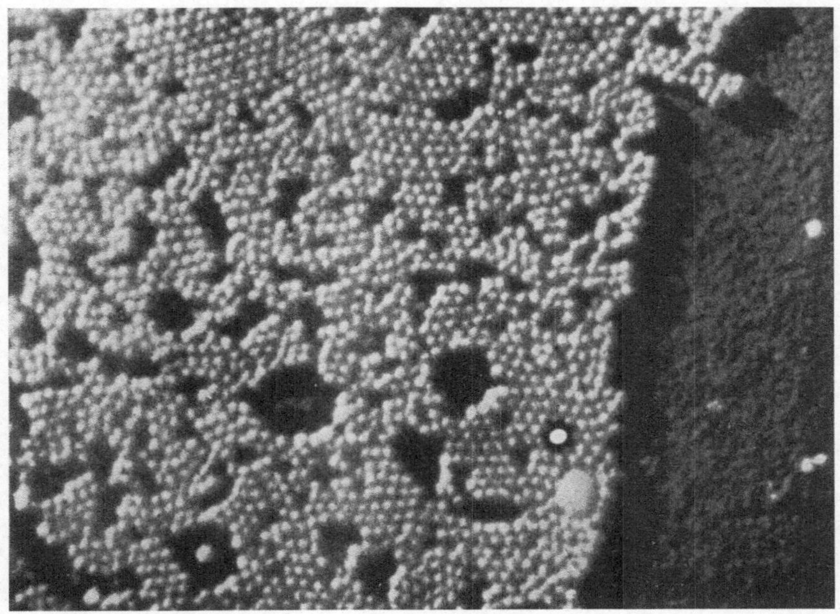

Abb. 156. Tomaten Bushy Stunt Virus. Schräg mit Gold bedampft. 62 000 : 1. Nach PRICE, WILLIAMS
und WYCKOFF (1946).

wurden nach 2 bis 5 Minuten mit der Pipette abgesaugt und die verbleibenden
Virusmoleküle mit Gold bedampft. Hierbei zeigte sich, daß ein- oder mehr-
schichtige Ablagerungen entstehen, in welchen die Moleküle größtenteils hexa-
gonal in der Ebene angeordnet sind. Die regelmäßigen Anordnungen können
als zweidimensionale Kristalle aufgefaßt werden. Außerdem fanden sich Anhäufungen ungeordneter Partikel. Am Rüben Gelbmosaikvirus fanden COSSLETT
und MARKHAM (1948) dem Diamantengitter entsprechende Lagerung der Moleküle. In den hexagonalen Ringen ist eine zentrale Lücke zu erkennen. Einen
weiteren Fortschritt brachte die Untersuchung der Oberflächen von Virus-
kristallen. PRICE und WYCKOFF (1946) zeigten die Lage der Moleküle an kleinen
Kristallen des Bohnen Mosaikvirus, die sie auf einen lichtmikroskopischen Glasobjektträger antrockneten, mit Gold bedampften, dann mit Kollodium abzogen
und mit der Aufdampfschicht ablösten. Am Tabak Nekrosevirus (Abb. 157)
machten MARKHAM, SMITH und WYCKOFF (1947 und 1948) körperzentriert
kubische Anordnung der Moleküle wahrscheinlich, während aus Röntgenreflexen
eine pseudokubische (trikline) Lagerung erschlossen worden war. Die Ursachen
dieser Differenz sind noch nicht völlig geklärt (verschiedene Kristallformen oder

Abb. 157. Tabak Nekrosevirus. Plastisches Bild der Kristalle. 68000 : 1. Nach Markham, Smith und Wyckoff (1948).

Molekülverlagerung bei der Auftrocknung der elektronenmikroskopischen Präparate).

Die Parallellagerung der Stäbchen und die hexagonale Packung der Kugeln ermöglichen eine genaue Bestimmung der Stäbchenbreite, bzw. der Kugelgröße.

Bei einer Diskussion mit DORNBERGER (unveröffentlicht) ist aufgefallen, daß auf den Elektronenbildern in den zitierten Arbeiten einzeln liegende Kugelmoleküle stets größere Durchmesser zeigen als die durch Gitterkräfte orientierten. Es kommt demnach durch die ordnenden Kräfte entweder zu einer gegenseitigen Zusammenpressung der Moleküle oder zu einer Ineinanderschiebung, wenn man annehmen kann, daß beispielsweise spiralige Strukturen eine partielle Ineinanderschiebung der orientierten Teilchen zulassen. Bei den stäbchenförmigen Molekülen würde die Annahme einer Spiralstruktur nach DORNBERGER nicht zu Widersprüchen mit den Ergebnissen der Röntgenographie führen und sogar einige Befunde besser deuten lassen als das bisher verwendete Modell sechskantiger Prismen. Auch die Molekülbruchstücke, die SCHRAMM (1943 und 47) gefunden hat und auf Quer- und Längsspaltungen des Prismas zurückführte, lassen sich vom Spiralmodell ableiten (DORNBERGER 1949).

d) Physikalisch-chemische und serologische Reaktionen.

Neben den gegenseitigen Lagebeziehungen der Virusmoleküle sind auch die zu kolloiden Reaktionspartnern geprüft worden. KAUSCHE (1938 und 1939) hatte gefunden, daß die in der Medizin zu differentialdiagnostischen Zwecken

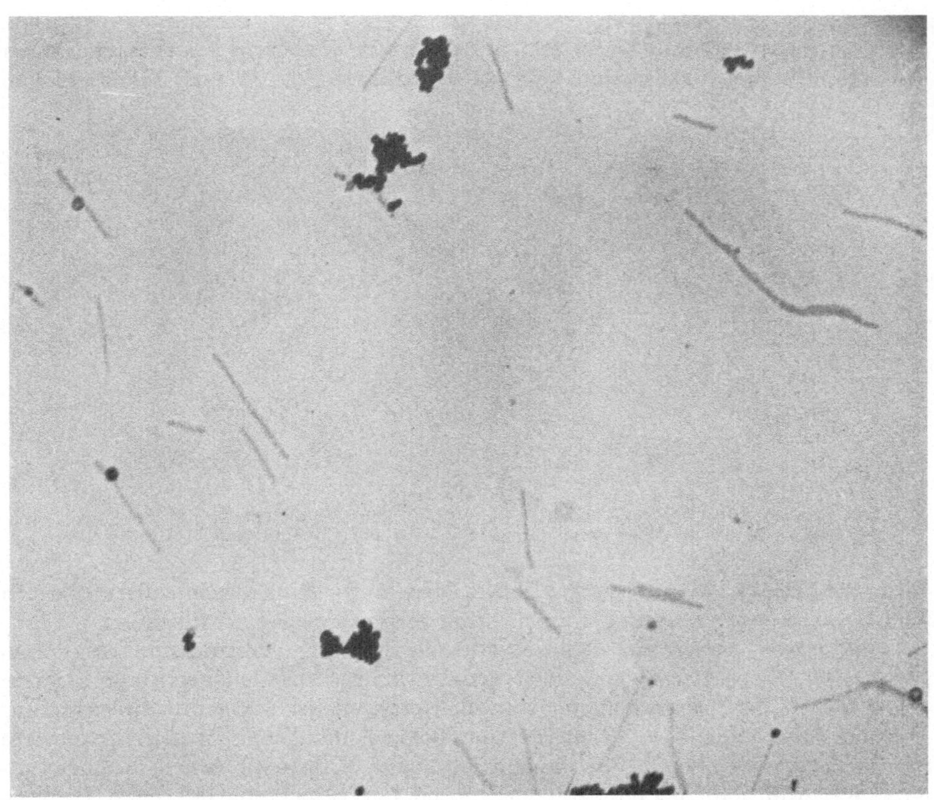

Abb. 158. Goldsol und Tabak Mosaikvirus als reaktionsloses Gemisch. 36000 : 1. Nach KAUSCHE und H. RUSKA (1939).

angewendete LANGEsche Goldsolreaktion des Liquor cerebrospinalis zum Nachweis und zur Unterscheidung verschiedener phytopathogener Virusproteine verwendbar ist. Es liegen ihr Adsorptionsvorgänge zwischen kolloidem Gold und Proteinmolekülen zugrunde, die von Kausche und H. Ruska (1939) sichtbar gemacht werden konnten. Die rote Goldsollösung ergibt durch Zusatz von Koch-

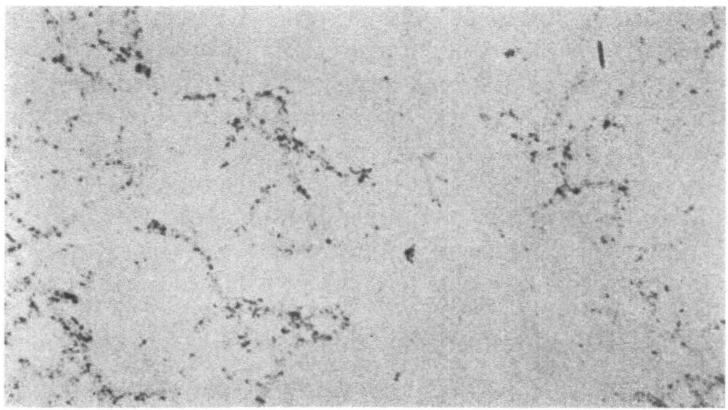

Abb. 159. Goldsol, HCl und Tabak Mosaikvirus. Virus-Goldbindung als sedimentierende Rotflockung. Besonders feines Goldkolloid. 60 000 : 1. Nach Kausche (1940).

salz eine makroskopisch blaue und durch Zusatz von Virus bei entsprechender Wasserstoffionenkonzentration eine rote Ausfällung. Setzt man Kochsalz und

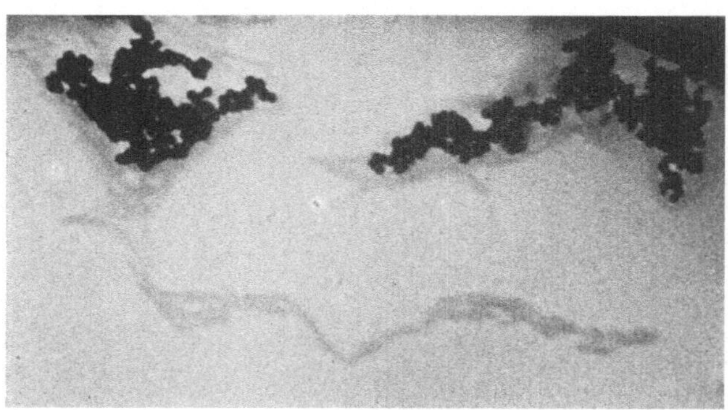

Abb. 160. Goldsol, 10 % NaCl und Tabak Mosaikvirus. Blaue „Schwebeflockung" des aggregierten Goldes zwischen Virusaggregaten. 30 000 : 1. Nach Kausche und H. Ruska (1939).

Virus zugleich zu, so entsteht bei Tabak Mosaikvirus in schwach saurem Bereich eine blaue Schwebeflockung und in stark saurem wieder Rotflockung.

Die reinen hochroten Goldsolhydrosole zeigten elektronenmikroskopisch gleichmäßig dispergierte, ganz vorwiegend primäre Einzelteilchen von kolloidem Gold. Bei 4% NaCl-Zusatz aggregieren die Einzelteilchen und sedimentieren unter Farbumschlag nach Blau. Gemische von Goldsol und Tabak Mosaikvirusprotein zeigten bei ausbleibender Sedimentation keine Beziehung beider Kolloide zueinander (Abb. 158), bei eintretender Rotflockung jedoch eine deutliche, ziemlich dichte Adsorption *einzelner* Goldteilchen an die Proteinmoleküle (Abb. 159). Das Wesen der Erhaltung hochroter Färbung trotz der völligen Sedimentation des Gemisches besteht darin, daß die einzelnen Goldteilchen getrennt voneinander an die TM-Einheiten gebunden werden. Bei gleichzeitiger Anwesenheit von NaCl aggregieren im entsprechenden pH-Bereich beide Reaktionspartner ge-

trennt für sich. Lockere Virusaggregate halten blaufärbende Goldaggregate
in Schwebe (Abb. 160). Eine Übersicht der Reaktionsmöglichkeiten (Abb. 161)

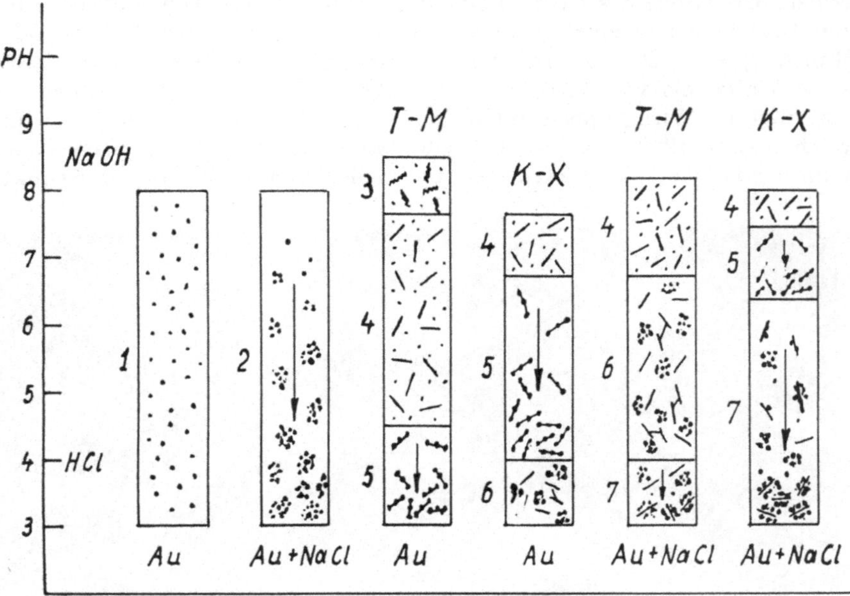

Abb. 161. Schematische Darstellung der Reaktion von kolloidem Gold und Tabak Mosaik- (TM) bzw.
Kartoffel-X-Virus (K-X) beim Fehlen und nach Zusatz von Kochsalz.

1. Monodisperses Goldsol.	5. Rotflockung.
2. Blauflockung.	6. Blaue Schwebeflockung.
3. Inaktives Virus in reaktionslosem Gemisch.	7. Violettflockung.
4. Reaktionsloses Gemisch.	

ist der Arbeit von KAUSCHE und H. RUSKA entnommen und nach KAUSCHE
(1940) ergänzt, um das unterschiedliche Verhalten von Kartoffel-X-Virus und

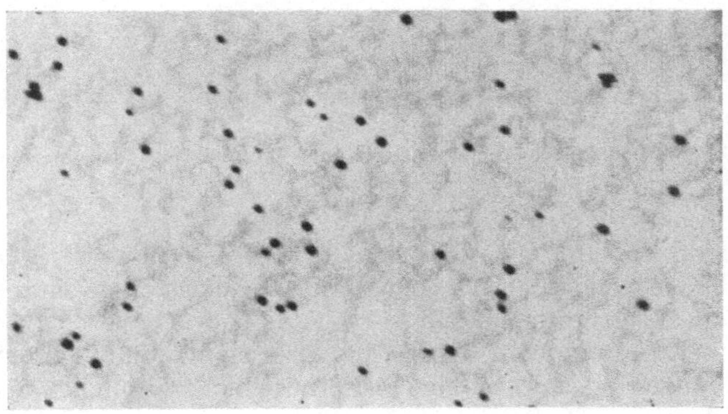

Abb. 162. Goldsol, n/500-HCl, Tabak Mosaikvirus und n/10-NaOH; Auflösung der Rotflockung
und Zerstörung der Tabak Mosaikvirusstäbchen durch Alkalisieren. 30000 : 1. Nach KAUSCHE und
H. RUSKA (1939).

Tabak Mosaikvirus zu zeigen. In beiden Fällen beruht die Rotfärbung auf einer
Umladung des einen der beiden elektronegativen Partner, und zwar des Virus,
so daß dadurch sein verschiedenes elektrophysikalisches Verhalten zum Ausdruck
kommt. Im alkalischen Bereich wird das Virus inaktiv und zerfällt (Abb. 162).

Die Gold-Virusbindung wird beim Tabak Mosaikvirus zwischen pH 4,3 und 4,9, beim Kartoffel-X-Virus erst zwischen pH 10,9 und 11,2 wieder gelöst, sie ist also reversibel. Die Größe des Goldkolloids (vgl. Abb. 158 mit Abb. 159) ist im geprüften Bereich bis zu einem achttausendfachen Teilchengewicht (3 bis 60 mμ Durchmesser) ohne Einfluß auf die grundsätzlichen Reaktionstypen. Bei geeignetem Verhältnis von Virus zu Gold geht alles Virus in die Fällung über, doch werden die Wirkgruppen durch das Gold nicht blockiert, das Virus bleibt aktiv (KAUSCHE 1940). KAUSCHE (1940) hat ferner gezeigt, daß das kolloide Gold *nicht* adsorbiert wird, wenn es erst in der Viruslösung durch Reduktion

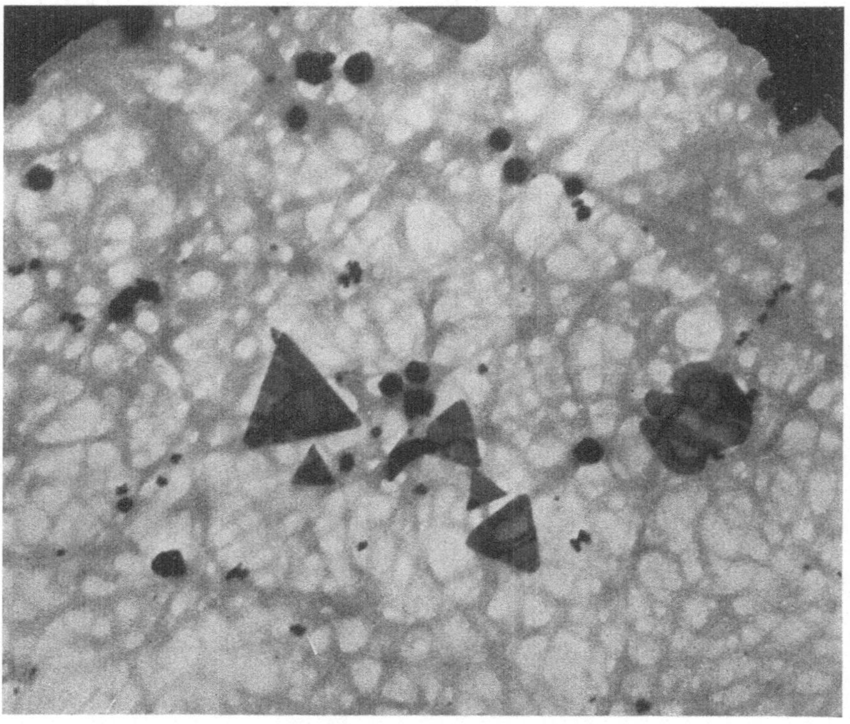

Abb. 163. AuCl₃ und Tabak Mosaikvirus. Gelbildung und teilweise Reduktion des Goldchlorids 32000 : 1. Nach KAUSCHE (1940).

von Goldchlorid entsteht. In den sich bildenden Virusgelen (Abb. 163) sieht man die kolloiden Goldteilchen neben Goldchloridkristallen ohne strenge Lagebeziehung zum Virusprotein liegen.

Bei der Behandlung von Tabak Mosaikvirus mit Sublimat ließ sich eine Adsorption reduzierten Quecksilbers in Form feinster Tröpfchen an den Proteinstäbchen nachweisen (H. RUSKA 1947). Ob und in welcher Form (kolloid oder molekular) eine Virus-Metallbindung eintritt, ist stark von den Versuchsbedingungen abhängig. Diffuse Kontrasterhöhungen der Virusmoleküle wurden vom Verfasser neben Abscheidungen von kolloidem Silber bei Behandlung mit Silbernitrat gesehen, ebenso bei Behandlung mit Kobaltnitrat (HAARDICK, KAUSCHE und H. RUSKA 1944) und mit Osmiumdämpfen (v. ARDENNE, FRIEDRICH-FRESKA und SCHRAMM 1941). Solche Untersuchungen sind für die Differenzierung zwar nahestehender, aber dennoch verschiedener Virusproteine

wesentlich, weil sich diese nicht nur durch unterschiedliche mittlere Stäbchenlängen, sondern auch an bildmäßig faßbaren Eigentümlichkeiten der physikalisch-chemischen Reaktion unterscheiden.

v. Ardenne, Friedrich-Freksa und Schramm (1941) untersuchten die
Präzipitinreaktion von Tabak Mosaikvirus mit Kaninchenantiserum. Da der
Reaktionspartner des Virus mit einem Molekulargewicht von 160000 als einzelnes Antikörperteilchen unter der Sichtbarkeitsgrenze bleibt, konnten nur die
präzipitierten und von Antikörpern umhüllten Virusteilchen dargestellt werden.
Die Antikörperhülle scheint sehr locker und ziemlich voluminös zu sein, wodurch
die Bilder der Präzipitate, wenn sie nicht mit Osmium fixiert sind, weich und
unscharf wirken. Die Verfasser ziehen Vergleiche zur O- und H-Agglutination
bei Bakterien. Ähnliche Versuche stellten Anderson und Stanley (1941) an.
Ihre Präzipitate sind infolge geringerer Vergrößerungen und vielleicht auch
wegen niedrigerer Strahlspannung kontrastreicher als die oben beschriebenen,
außerdem sind die unfixierten Einzelstäbchen durch Umhüllung mit dem
Antikörper größer und dicker. Anderson und Stanley vermuten daher, daß
letztere größenmäßig dem Antikörper der Kaninchen gegen Pneumokokken,
einem gestreckten Globulin von $3,7 \times 27,4$ mμ nahe stehen und sich radial
um die Virusstäbchen anordnen. Eine solche Auffassung würde sich mit theoretischen Vorstellungen über Struktur und Reaktionsweise der Antikörper von
Pauling (1940) decken. Antiseren gegen Tomaten Bushy Stunt Virus, latentes
Kartoffel Mosaik- und Tabak Ringfleckvirus gaben mit Tabak Mosaikvirus
weder mikroskopisch noch elektronenmikroskopisch sichtbare oder durch Änderung der Sedimentationskonstanten nachweisbare Reaktionen. Eine genauere
Klärung der Antigen-Antikörperreaktion wird von der Verwendung von Anti-
körper-Reinpräparaten erwartet. Schramm und Friedrich-Freksa (1941)
prüften außer Kaninchen- auch Schweineantiserum, das ein Molekulargewicht
von 930000 besitzt. Die Verdickung des Virusmoleküls durch die Bindung der
Schweineantikörper konnte elektronenmikroskopisch vermessen werden. Beide
Antikörper besetzen bei Antikörperüberschuß die Oberfläche der Viruspartikel
nahezu vollständig. Entsprechend ihrer verschiedenen Größe werden bis 600.
bzw. 55 Antikörpermoleküle am Tabak Mosaikvirusmolekül gebunden.

e) Die identische Reproduktion der Virusmoleküle.

Während bei den meisten tierpathogenen Virusarten und bei den Phagen
mit einem komplizierten Vermehrungsvorgang gerechnet werden muß, bei dem
verschiedenartige Stoffsysteme beteiligt sind, liegt bei den makromolekularen
Virusproteinen der Pflanzen das vergleichsweise einfache Grundproblem der
identischen Reproduktion eines makromolekularen Nukleoproteins vor. Das
Grundphänomen des organischen Lebens präsentiert sich hier in der einfachsten
Form, die wir kennen, aber auch noch in seiner vollen Rätselhaftigkeit. Friedrich-
Freksa (1940, 1948) versucht, die Kräfte, die bei der identischen Reproduktion
eines Virusmoleküls wirksam sind, mit jenen zu vergleichen, die die Paarung
der Chromosomen bewirken. In beiden Fällen handelt es sich um die Anziehung
identischer Strukturteile. Als wirksame Kräfte beansprucht Friedrich-Freksa
die elektrostatischen Ladungen der Oberfläche, Jordan (1941, 1944), von
gleichen Grundvorstellungen ausgehend, die quantenmechanische Resonanz-
anziehung thermisch rotierender Radikale der betreffenden Molekülkomplexe.
Beide Kräfte werden zwischen parallel orientierten Molekülen über weitreichende
Entfernungen wirksam. Unabhängig von der Diskussion um die wirksamen
Kräfte kann hier nur die Frage besprochen werden, welche morphologischen

Unterlagen für den Vorgang der Verdoppelung vorliegen. Keinesfalls können sie mit unseren morphologischen Kenntnissen über die mitotische Kernteilung und Chromosomenverdoppelung verglichen werden.

Kausche und H. Ruska (1940/4) fanden an Chloroplastentrümmern mosaikkranker Tabakpflanzen dem Virus gleichende Fäden in Verbindung mit Stromaresten und vermuten deshalb, daß die Virussynthese möglicherweise an die Chloroplasten gebunden ist, deren Veränderungen bekanntlich das sichtbare Krankheitsbild bestimmen. In gleicher Richtung könnte der Befund von Born, Lang, Schramm und Zimmer (1941) weisen, wonach gebundener radioaktiver Phosphor besonders reichlich im Chloroplastenrückstand und im Virusprotein auftritt. Williams und Steere (1949) werfen dagegen die Frage auf, ob die im Protoplasma nachweisbaren Kristallnadeln das geometrische Wachstumsmuster des Virus repräsentieren. Es ist also zunächst der Ort der Virusvermehrung innerhalb der Zelle noch unbekannt.

Außerdem wissen wir nicht, ob die identische Reproduktion unter dem Bilde eines Längenwachstums der Stäbchen und nachfolgender Querteilung oder — wie bei den Chromosomen des Zellkerns — durch Parallelanlagerung erfolgt. Zur ersten Auffassung neigen Stanley und Anderson (1941), obwohl die in frischen Preßsäften mosaikviruskranker Tabakpflanzen gefundene, außerordentlich geringe Streuung der Stäbchenlängen (s. Tab. 6) zunächst wenig für ein Längenwachstum spricht. Die Vorstellung von einer Neubildung der Stäbchen in Parallellage liegt dem Gedanken von Friedrich-Freksa zugrunde. Er macht die Annahme, daß die Verteilung der positiven Ladungen der Hexonbasen an sich verdoppelnden Eiweißmolekülen für die spezifische Struktur maßgebend ist. Die räumliche Anordnung der Ladungsverteilung wird durch die am Virusstäbchen seitenständigen Nukleinsäuren mit negativen Ladungen abgebildet, die nun ihrerseits die Anlagerung der positiven Eiweißbausteine in der spezifischen Anordnung bewirken usf. Auch für diese Vorstellung gibt die Elektronenmikroskopie keine bindenden Unterlagen. Die von Williams und Steere in der kurzen Zeit von 15 sec präparierten Kristallnadeln aus unverdünntem Pflanzensaft zeigen ungleiche Stäbchenlängen mit Verlagerungen der Stäbchenenden wie ammonsulfatgefällte Nadeln (Kausche und H. Ruska 1939/2) und haben die Länge mehrerer Grundeinheiten des Tabak Mosaikvirusstäbchens. Nur streng geordnete Stäbchenpakete mit quer durchgehenden Schichtebenen könnten für die Friedrich-Freksasche Auffassung sprechen.

In konzentrierten Lösungen zeigen die Virusstäbchen sowohl die Neigung zur parallelen als auch zur fadenförmigen Aggregation. Die Stäbchen besitzen ein Dipolmoment in Richtung der langen Achse, das möglicherweise außer für die End-zu-End-Aggregation auch für ein Längenwachstum von Bedeutung ist. Bei der Annahme, daß die Ladung der Stäbchenenden mit der Länge der Moleküle zunimmt, wird verständlich, warum die Stäbchen unter einer gewissen Minimallänge nicht mehr vermehrungsfähig sind, obwohl sie qualitativ alle erforderlichen Teilbausteine des aktiven Virus besitzen. Kurze Stäbchen (unter 250 mμ) hätten dann ein unzureichendes Potential der Stäbchenenden. Erfolgt die Anlagerung neuen Materials an der Stirnfläche der Stäbchen, so kann sich eine Änderung des Baues (Mutation des Virus) leicht durch die ganze Stäbchenlänge fortsetzen, während sie bei Reproduktion in Seit-zu-Seit-Lagerung eine isolierte Störstelle bleiben würde, die sich nicht auf die Teilbausteine der ganzen Stäbchenlänge auswirken könnte. Die neuerdings von Dornberger-Schiff postulierte Spiralstruktur der Virusstäbchen läßt nur Längenwachstum mit Querteilung, aber kein Breitenwachstum mit Längsteilung zu. Die in der zytologischen Literatur weit verbreitete Annahme, daß spiralige Chromosomen nach einer Verdoppelung des Spiralfadens seitlich auseinanderrücken können, beruht auf einem Irrtum. Ein derartiges Auseinanderrücken ist nur bei Zickzacklinien möglich, doppelte Spiralen müssen aber schraubenartig auseinandergedreht oder nach ihrer Streckung um ihre Achse rotiert werden, um sich trennen zu können. Es erscheint vorläufig nicht begründet, solche Vorgänge für die Reproduktion stäbchenförmiger Virusmoleküle zu fordern. Bei der Annahme von Längenwachstum und Querteilung der Stäbchen verschiebt sich die Analogie zwischen Virusfaden und Chromosomenfaden zu einer solchen zwischen Virusfaden und Chromomer eines Riesenchromosoms. Beide sind aus zahlreichen gleichen Teilbausteinen zusammengesetzt, während sich im Chromosomenfaden verschiedenartige Komplexe folgen.

Für die kugelförmigen Virusproteine sind hypothetische Vorstellungen über die identische Reproduktion und feinere Struktur weniger entwickelt, nicht zuletzt wohl darum, weil das morphologische Bild noch weniger Aufschlüsse erwarten läßt als das der Stäbchen. Beiden Formen müssen jedoch schließlich verwandte Strukturen zugrunde liegen, welche die gleichartige biologische Verhaltensweise der Virusmoleküle bedingen.

18. Die Systematik der Virusarten.

HOLMES (1948) hat unter Verwendung der binearen Nomenklatur in Bergey's Manual of Derminative Bacteriology ein System der Virusarten vorgeschlagen, das zunächst nach Wirtsorganismen (Bakterien, höheren Pflanzen, Tieren) trennt und dann nach Krankheitssymptomen weiter aufteilt. Bei diesem Vorgehen geraten z. B. die Erreger der Pocken und der Maul- und Klauenseuche in die gleiche Familie. Was man auf diese Weise erhält, sind keine systematischen Gruppen, sondern allenfalls Angehörige eines „Biotops". Niemand würde daran denken, pathogene Protozoen, Pilze oder Bakterien auf diese Weise zu systematisieren, und, nachdem die Morphologie der Virusarten elektronenmikroskopisch sowie röntgenographisch zugänglich geworden ist, sollte dieser Weg, der schon früher nur ein Behelf sein konnte, nicht mehr beschritten oder gar neu ausgebaut werden. Uns erscheint es (1943/15, 1949/22) ebenso wie FRIEDRICH-FREKSA (1950) richtiger, zunächst von morphologisch bekannten Virusarten auszugehen und diese mit ähnlichen in Familien einzuordnen, deren Zusammengehörigkeit zu Ordnungen und Klassen vorläufig noch offen bleiben muß. Auf diese Weise ergeben sich Gruppen, wie sie aus der Tab. 16 hervorgehen, die außerdem noch Angaben über die chemische Zusammensetzung, Molekulargewichte usf. enthält. Die morphologisch bekannten Arten lassen sich entsprechend den Anforderungen der internationalen Nomenklaturregeln beschreiben. Erst bei der Festlegung der Diagnose werden für morphologisch gleiche Arten die pathogenen Auswirkungen zur Unterscheidung herangezogen, ebenso sind diese bei der Erörterung der Beziehungen der verschiedenen Arten untereinander zu besprechen. Die durch verschiedene Virusarten hervorgerufenen gleichartigen oder ungleichartigen Krankheitssymptome können unserer Auffassung nach weder dazu dienen, größere Virusgruppen zusammenzufassen, noch einzelne Arten in weit auseinanderliegende Gruppen zu trennen. Erst wo morphologisch gleiche Virusformen vorliegen, können die durch sie hervorgerufenen ungleichen Krankheitsbilder zur Trennung nahestehender Virusarten dienen. Schon der Tropismuswechsel sollte von der Verwendung des Tropismus zur Systematisierung größerer Virusgruppen abhalten. Die Bedenken, die dagegen bestehen, aus der Kollektivbezeichnung „Virus" einen Begriff der biologischen Systematik zu machen, wurden am Anfang dieses Beitrags erörtert.

Tabelle 16. *Übersicht der elektronenmikroskopisch*

Kontagien	Morphologie		
	Größe (mμ)	Teilchen- u. Aggregationsf.	Innenbau
I. Makromolekulare Virusformen			
Tabak Ringfleckvirus (Nicotiana Virus 12)	19		
Rüben Gelbmosaikvirus (Brassica Virus)	22	Kugeln oder nahezu Kugeln	Oktaeder (Diamantgitter) Würfel (monoklin pseudokubisch)
Tabak Nekrosevirus (Nicotiana Virus 11)	24		Makromoleküle mit aperiodischer Struktur
Südl. Bohnen Mosaikvirus (Phaseolus Virus)	25		
Tomaten Bushy Stunt Virus (Lycopersicum Virus 4)	26		Dodekaeder (kubisch körperzentriert) achtseitige Prismen u. rhomboedrische Kristalle
und andere			
Tabak Mosaikvirus (Nicotiana Virus 1, 1 A, C, D)	15×280	Stäbchen mit dreizähliger Symmetrie, möglicherweise enge Spiralen. Plumpe Stäbchen	Gele, nadelförmige Parakristalle, beim TM-Virus auch hexagonale Plättchen
Gurken Mosaikvirus 3 und 4 (Cucumis Virus)	13×300		Makromoleküle mit linearer Periodizität
Kartoffel-X-Virus (Solanum Virus 1)	etwa 15×430		
Kartoffel-Y-Virus (Solanum Virus 2)			
Kartoffel Gelbzwergvirus (Solanum Virus 16)	etwa 50×200		
und andere		unbekannt	unbekannt
Maul- und Klauenseuchevirus	um 12	Kugeln	unbekannt
Milchfaktor aus Brustdrüsenkrebs der Maus	20 oder größer	Kugeln	unbekannt
Poliomyelitisvirus, Mensch	um 20 (bzw. 12× mehrere 100 ?)	Kugeln, Fäden stark variabler Länge	Fäden, wahrscheinlich periodisch
Poliomyelitisvirus, Maus Stamm Lansing Stamm Theiler	25 um 12× mehrere 100 ?	Kugeln Fäden stark variabler Länge / nadelförmige Parakristalle aus Fäden	
II. Organisierte Virusformen			
Kaninchen Papillomvirus	44	Kugeln	erscheint gleichmäßig dicht
Pferde Encephalomyelitisvirus Östlicher Stamm Westlicher Stamm	40—50	ungleichmäßige Kugeln	
Influenzavirus A Stamm B Stamm Schweineinfluenza	um 80 um 100 um 80	ungleichmäßige Kugeln und plumpe Fäden	zentral dichter
Geflügelpestvirus	um 100	ungleichmäßige Kugeln	
Varizellen- und Zostervirus	um 150	ungleichmäßige Kugeln	
Quaderförmiges Virus (Tesserulata) Kanarienpocken Kaninchenmyxom Vaccine Ektromelie der Maus Molluscum contagiosum weitere Pocken der Tiere	etwa 260×310 230×290 210×260 170×230 180×250	Quader, Seitenverhältnis der größten Fläche: 1 : 1,14 1 : 1,22 1 : 1,25 1 : 1,34 1 : 1,45	an Vaccine festgestellt: pepsinresistente innere Verdichtung innerhalb pepsinverdaulichem Protein

untersuchten Virusarten und einiger Mikroorganismen

	Physiologie			
Oberfläche	Stoffwechsel	Wachstum	Vermehrung	chem. Angaben
Grenzflächen von Makromolekülen mit oberflächlich liegenden, am Gesamtmolekül gebundenen Ribose-Nukleinsäuren	Nur bei Anwesenheit lebender Zellen züchtbar, in reinem Zustande keine Fermentsysteme und keine Atmung nachgewiesen	Apposition kleiner, aus 6—12 Aminosäuren bestehender Teilbausteine aus dem Plasma der Wirtszelle	Einzelheiten der Verdopplung und Abtrennung unbekannt	Nukleoproteide mit meist 14—17 %, Tabak Ringfleckvirus mit 40 % Ribose-Nukleinsäure (R. N. S.) Molekulargewichte $(3,4$ bis $8)\times10^6$
			wahrscheinlich Längenzunahme mit anschließender Querdurchtrennung ohne differenzierten inneren Teilungsmechanismus	Nukleoproteide mit meist etwa 6 %. R. N. S. Molekulargewichte T. M. V. 40×10^6 K. X V. um 30×10^6
	übertragbarer Milchfaktor bisher nicht außerhalb des Mäuseorganismus gezüchtet	wahrscheinlich wie oben	siehe oben	Molekulargewicht $(0,5—1,0)\times10^6$ wahrscheinlich reine Nukleoproteide, Fäden etwa 6 % N. S. Molekulargewichte der Kugelformen $(0,4$ bis $0,7)\times10^6$, der Fäden $(60$ bis $200)\times10^6$
wenig scharf begrenzt				8,7 % Desoxyribosenukleinsäure (D. N. S.), 1,5 % Fett (F) Molekulargewicht 25×10^6
wenig scharf begrenzt, besondere Außensubstanz	wie unter I.	unbekannte Vorgänge in der Wirtszelle, beim Influenzavirus wahrscheinlich Auswachsen von Kugeln zu Fäden, die sich möglicherweise wieder segmentieren		4,4 % R. N. S., 48,5 % F, Teilchengewicht um 25×10^6 1,3 % D. N. S., 23 % F. Toxische Komponente, Teilchengewicht um 400×10^6 wahrscheinlich ähnlich Influenza unbekannt, höchstwahrscheinlich Nukleinsäure vorhanden
scharf begrenzt; pepsinresistentes Plasmolemma	nur bei Anwesenheit lebender Zellen züchtbar; kein Gaswechsel, verschiedene Fermentsysteme	aus der Vermessung der Elementarkörper keine Streckung und Teilung nachweisbar. Wahrscheinlich Entstehung der Elementarkörper über Einschlußkörper, die wieder in Elementarkörper zerfallen		Vaccine: 5 % D. N. S., 5,7 % F., 0,005 % Cu, Spuren Flavine, Biotin, wahrscheinlich Katalase, Phosphatase u. Polynukleotidase, Teilchengewichte $(2000$ bis $5000)\times10^6$

Fortsetzung auf Seite 392

Fortsetzung von Seite 391

Kontagien	Morphologie		
	Größe (mμ)	Teilchen- u. Aggregationsf.	Innenbau
Atypische Geflügelpest (New Castle Virus) Mumpsvirus	um 190 um 180	Ellipsoide, Fäden, Spermienformen u. dgl. bes. ungleich große Kugeln (100 bis 260)	ungleiche Dichte verschiedener Strukturteile
Polyedervirus Seidenraupenvirus (Bombyx mori) Schwammspinnervirus (Porthetria dispar) Kapselvirus Virus von Cacoecia murinana Virus von Junonia coenia und andere	45 bzw. 90×350 40 bzw. 160×415 50×260 40×300	Stäbchen und Parallelaggregate mit abgerundeten Enden, z. T. in Polyeder eingebettet Stäbchen, in sporenähnliche Kapseln eingebettet	polnahe Verdichtung
Bakteriophagen (d'Herellata)	um 12—40 um 35×140 40—90×300	Kugeln Polstäbchen Keulen	gleichmäßig dicht polare Verdichtungen Kopfteil dichter als Fortsatz
III. Mikroorganismen Cysticeten (Cysticeta) Bronchopneumonievirus der Maus, Katzenpneumonievirus und andere große Virusarten	etwa 230—430	Kugeln und Bläschen, z. T. mit Einstülpungen oder Fortsatzentwicklung, protoplasmatische Stränge, starke Polymorphie	Bläschen zentral weniger dicht als peripher
Erreger der Pleuropneumonie des Rindes u. a. auf unbelebtem Nährboden züchtbare pathogene Formen	300—700 und 200× viele 100		
SEIFFERTsche Mikroorganismen u. a. apathogene Formen	300—700		
Spaltpilze (Bacteria) Rickettsien (Rickettsiales) R. prowazeki R. mooseri R. rickettsi R. quintana u. a.	um 300 bis 400×1400	Kugeln und Stäbchen	Zytoplasma wechselnder Dichte, wahrscheinlich Kernäquivalente
Bakterien (Eubacteriales) Mycobakterien (Mycobacteriales) usw.	etwa 200×500 bis 1000×9000	kugel- und stäbchenförmige Zellen, Sporen	Zytoplasma, Kernäquivalente, Volutinkörner, Vakuolen
Urtiere (Protozoa)	etwa 1000 bis über 107	Zellen, Cysten, Gameten usw.	Grundplasma mit Mitochondrien, Golgiapparat, Nahrungs- und Exkretionsvakuolen, Kernen, Centrosomen und anderen Organellen

| Oberfläche | Physiologie | | | |
	Stoffwechsel	Wachstum	Vermehrung	chem. Angaben
deformierbare Oberfl. aus reversiblem Formenwechsel erschließbar	wie unter I. Haemolysine nachgewiesen	unbekannte Vorgänge in der Wirtszelle		N. C. V.: Kleine Mengen Nukleinsäure, davon ein Teil D. N. S., 27 % F. Teilchengewichte um (300 bis 400) $\times 10^6$
scharf begrenzt	wie unter I	unbekannte Vorgänge in der Wirtszelle bzw. in Polyedern oder Kapseln	sehr wahrscheinlich Längsdurchtrennung der Stäbchen	P. V.: 14 % D. N. S. Teilchengewichte (900 bis 2100)$\times 10^6$ K. V. (C. m.) Teilchengewicht 435$\times 10^6$
besondere Substanzen der Oberflächenschicht und des Fortsatzes	wie unter I, wahrscheinlich abtrennbare Lysine	komplexe Vorgänge auf der Membran oder im Innern der in lebhaftem Stoffwechsel begriffenen Bakterien		T_2-Phage: 42 % D.N.S., 4 % R.N.S., 2,2 % F., Teilchengewichte versch. Phagen (0,4 bis 300)$\times 10^6$
Plasmolemma von wechselnder Formbeständigkeit	morphologisch ähnliche Formen teils nur in Wirtszellen züchtbar, teils auf unbelebten Nährböden; für letztere eigene Fermentsysteme unzweifelhaft	in der Wirtszelle / im Wirtsorganismus u. auf unbelebten Nährböden / auf unbelebten Nährböden	Entwicklung über Einschlußkörperbildung und deren Zerfall in Elementarkörper / vielerlei Vermehrungsweisen beschrieben: Sprossung, Teilung, Segmentierung, Gonidienbildung und anderes	beim Psittakosevirus D. N. S. nachgewiesen, z. T. toxische Komponenten, komplexe chemische Zusammensetzung, Teilchengewichte (1400 bis 8500)$\times 10^6$
Zellmembran	nur bei Anwesenheit lebender Zellen züchtbar, eigene Fermentsysteme, verschiedene Antigene	in der Wirtszelle	Längenwachstum, Teilung	vorwiegend R. N. S. (zirka 20 %), D. N. S. (zirka 4 %) in Kernäquivalenten, z. T. Toxine, komplexe chemische Zusammensetzung, Fermente usw., Teilchengewichte Rickettsiales um 12000$\times 10^6$ Eubacteriales um 200000$\times 10^6$
Zellmembran, außerdem z.T. Geißeln, Kapseln oder Schleimsubstanzen	auf unbelebten Nährböden züchtbar, äußerst vielfältige Fermentsysteme und Antigene; aerobe und anaerobe Formen	Assimilation gelöster Substanzen	Längenwachstum, Teilung, z. T. Sporenbildung, keine Gameten	
Pellikula mit Wimpern, Wimperplatten, Geißeln, Trichozysten und anderen Organellen	fast alle auf unbelebtem Nährboden züchtbar, eigene Fermentsysteme, Atmung	Assimilation auch ungelöster Substanzen durch Aufnahme ins Zellinnere	Teilung und verschiedenartige geschlechtliche Vermehrungsvorgänge	D. N. S. in Kernen, R. N. S. im Plasma, sehr komplexe chemische Zusammensetzung, Teilchengewichte in weiten Grenzen um $10^6 \times 10^6$

Literatur.[1]

Ackermann, I.: Silikatglas als Trägerfolie für elektronenoptische Anordnungen. Optik, **2**, 280—282 (1947).

Algera, L., J. J. Beijer, W. van Iterson, W. K. Karstens and T. H. Thung: Some data on the structure of the chloroplast, obtained by electron microscopy. Biochem. et Biophys. Acta **1**, 517—526 (1947).

Anderson, T. F.: (1) The nature of viruses. The Pennsylvania Gazette May 1942.

— (2) Electron microscopic study of bacteriophage. J. Bact. **45**, 303 (1943).

— (3) On a bacteriolytic substance associated with a purified bacterial virus. J. Cell. Comp. Physiol. **25**, 1—13 (1945).

— (4) The rôle of tryptophan in the adsorption of two bacterial viruses on their host, E. coli. J. Cell. Comp. Physiol. **25**, 17—26 (1945).

— (5) Morphological and chemical relations in viruses and bacteriophages. Cold Spring Harbor Symposia **11**, 1—13 (1946).

— (6) The activation of the bacterial virus T 4 by l-tryptophan. J. of Bact. **55**, 637—649 (1948).

Anderson, T. F. and W. M. Stanley: A study by means of the electron microscope of the reaction between tobacco mosaic virus and its antiserum. J. Biol. Chem. **139**, 339—344 (1941).

Angerer, v. E.: Technische Kunstgriffe bei physikalischen Untersuchungen. Braunschweig: Verlag Vieweg 1924 und 1936.

Ardenne, M. v.: (1) Die Keilschnittmethode, ein Weg zur Herstellung von Mikrotomschnitten mit weniger als 10^{-3} mm Stärke für elektronenmikroskopische Zwecke. Zschr. wiss. Mikrosk. **56**, 8—23 (1939).

— (2) Elektronen-Übermikroskopie. Springer-Verlag, Berlin, 1940.

— (3) Der Objektträger-Vibrator, ein neues Hilfsgerät der Übermikroskopie und Mikroskopie. Kolloid-Z. **93**, 158—163 (1940).

— (4) Stereo-Übermikroskopie mit dem Universal-Elektronenmikroskop. Naturw. **28**, 248—252 (1940).

— (5) Über ein Universal-Elektronenmikroskop für Hellfeld-, Dunkelfeld- und Stereobildbetrieb. Z. f. Physik **115**, 339—368 (1940).

— (6) Die Anwendung des Objektträger-Vibrators zur Herstellung von Emulsionen. Angewandte Chemie **54**, 144—146 (1941).

— (7) Über eine neue Einrichtung am Universal-Elektronenmikroskop zur fast vollständigen Vermeidung der thermischen Objektbelastung. Kolloid-Z. **111**, 22—30 (1948).

Ardenne, M. v. und H. Augustin: Elektronenmikroskopische Darstellung des Lepraerregers (Mycobacterium Leprae). Klin. Wschr. **20**, 753—755 (1941).

Ardenne, M. v. und H. Friedrich-Freksa: Die Auskeimung der Sporen von Bacillus vulgatus nach vorheriger Abbildung im 200-kV-Universal-Elektronenmikroskop. Naturw. **29**, 523—528 (1941).

Ardenne, M. v., H. Friedrich-Freksa und G. Schramm: Elektronenmikroskopische Untersuchung der Präcipitinreaktion von Tabak-Mosaikvirus mit Kaninchenantiserum. Arch. ges. Virusforsch. **2**, 80—86 (1941).

Ardenne, M. v. und G. Pyl: Versuche zur Abbildung des Maul- und Klauenseuchevirus mit dem Universal-Elektronenmikroskop. Naturw. **28**, 531—532 (1940).

Astbury, W. T. and C. Weibull: X-Ray diffraction of the structure of bacterial flagella. Nature **163**, 280 (1949).

Babudieri, B. (1): Lo sviluopo di rickettsia prowazeki, nell' embrione di pollo. Rendiconti, Dell' istituto superiore di sanita **6**, 292—297 (1943).

— (2) Note di tecnica per la microscopia electronica. II. Fissazione di preparati. Rendiconti, Dell' istituto superiore di sanita **6**, 1225—1229 (1948).

— (3) Un nuovo procedimento per l'allestimento di preparati batterici destinati all' osservazione col supermicroscopio elettronico. Rendiconti, Dell' istituto superiore di sanita **11**, 572—576 (1948).

---

[1] In das Verzeichnis wurden noch einige Arbeiten aufgenommen, die im Text nicht mehr berücksichtigt werden konnten.

BABUDIERI, (4) L'azione antibatterica di alcuni antibiotici, studiata col microscopio elettronico. Rendiconti, Dell' istituto superiore di sanita 11, 577—598 (1948).
— (5) Ricerche di microscopia elettronica. III. Studio del genere leptospira. Rendiconti, Dell' istituto superiore di sanita 11, 1046—1066 (1948).
— (6) Applicazioni della microscopia elettronica. Tecnica Italiana, Rivista D'ingegneria e scienze, 3. 3—7 (1948).
BABUDIERI, B. et I. Archetti: Isolamento del virus dell' attuale epidemia influenzale. Rendiconti, Dell' istituto superiore di sanita 12, 417—428 (1949).
BABUDIERI, B., G. BIETTI et D. BOCCIARELLI: La batteriolisi prodotta dal „lysozym" lacrimale studiata al supermicroscopio elettronico. Rendiconti, Dell' istituto superiore di sanita 7, 604—615 (1944).
BABUDIERI, B. et D. BOCCIARELLI (1): Ricerche di microscopia elettronica. 1. Studio morfologico di rickettsia prowazeki. Rendiconti, Dell' istituto superiore di sanita 6, 298—304 (1943).
— (2) Ricerche di microscopia elettronica. II. Studio morfologico del genere spironema. Rendiconti, Dell'istituto superiore di sanita 6, 305—314 (1943).
— (3) Electron microscope studies on relapsing fever spirochaetes. J. Hygiene 46, 438—439 (1948).
BAKER, R.: Phase-contrast microscopy of viruses. Nature 14, 251 (1948).
BAKER, R. F. and D. C. PEASE. Sectioning of the bacterial cell for the electronmicroscope. Nature 163, 282 (1949).
BANG, F. B.: (1) Filamentous forms of Newcastle virus. Proc. Soc. Biol. Med. 63, 5 (1946).
— (2) Newcastle virus: conversion of spherical forms to filamentous forms. Proc. Soc. exper. Biol. Med. 64, 135 (1947).
— (3) Formation of filamentous forms of Newcastle disease virus in hypertonic concentration of sodium chloride. Proc. Soc. Exp. Biol. Med. 71, 50—52 (1949).
BANG, F. B. and G. O. GEY: Electron microscopy of tissue cultures infected with the virus of eastern equine encephalomyelitis. Proc. Soc. Exp. Biol. Med. 71, 78—80 (1949).
BARNES, R. B., Ch. J. BURTON and R. G. SCOTT: Electronic microscopical replica technique for the study of organic surfaces. J. Appl. Phys. 16, 730—739 (1945).
BAWDEN, F. C. and N. W. PIRIE: (1) Brit. J. exper. Path. 19, 251 (1938).
— (2) The separation and properties of tobacco mosaik virus in different states of aggregation. Brit. J. exper. Path. 26, 294—312 (1945).
BAYLOR, M. B., M. D. APPLEMAN, O. H. SEARS and G. L. CLARK. Some morphological characteristics of nodule bacteria as shown by the electronmicroscope Il. J. Bacter 50, 249—256 (1945).
BAYLOR, M. R. B., J. M. SEVERENS and G. J. CLARK: Electron microscope studies of the bacteriophage of salmonella pullorum. J. Bakt. 47, 277—282 (1944).
BEARD, J. W.: The ultracentrifugal, chemical and electron micrographic characters of purified animal viruses. Proc. inst. med. Chicago 15 (1945).
BEARD, J. W., D. G. SHARP, A. R. TAYLOR, I. M. MCLEAN, D. BEARD, D. FELLER and J. H. DINGLE: Ultracentrifugal, chemical and electron microscopic identification of the influenza virus. Southern Med. J. 37, 313 (1944).
BENDER, A. und G. A. KAUSCHE: Zur Kenntnis der Kristallisierbarkeit eines tierpathogenen Virus aus der Poliomyelitisgruppe. Klin. Wschr. 26, 489—490 (1948).
BERGOLD, G.: (1) Über Polyederkrankheiten bei Insekten. Biol. Zbl. 63, 1—44 (1943).
— (2) Die Isolierung des Polyedervirus und die Natur der Polyeder. Z. Naturforsch. 2b, 122—143 (1947).
— (3) Bündelförmige Ordnung von Polyedervirus. Z. Naturforsch. 3b, 25—26 (1948).
— (4) Über die Kapselviruskrankheit. Z. Naturforsch. 3b, 338—342 (1948).
BERGOLD, G. und J. HENGSTENBERG: Ultrazentrifugenversuche mit Insektenvirus. Kolloid-Z. 98, 304 (1942).
BERGOLD, G. und H. FRIEDRICH-FREKSA: Zur Größe und Serologie des Bombyxmori Polyedervirus. Z. Naturforsch. 2b, 410—414 (1947).
BERGOLD, G. und G. SCHRAMM: Biologische Charakterisierung von Insektenviren. Biol. Zbl. 62, 105 (1942).

Bernal, J. D. and I. Fankuchen: Structure types of protein „crystals“ from virus-infected plants. Nature **139**, 923 (1937).

Black, L. M., V. M. Mosley and R. W. G. Wyckoff: Electron microscopy of potato yellow dwarf virus. Biochem. Biophys. A. **2**, 121—123 (1948).

Black, L. M., W. C. Price and R. W. G. Wyckoff: The electron micrography of plant virus-antibody mixtures. Proc. Soc. Exper. Biol. Med. **61**, 9—12 (1946).

Bloch, H.: Über einen Potenzierungseffekt bei Bakteriophagen. Arch. ges. Virusforschg. **1**, 560 (1940).

Boersch, H.: (1) Randbeugung von Elektronen. Physikal. Zschr. **44**, 32—38 (1943).

— (2) Über Ränder in elektronenmikroskopischen Abbildungen. Kolloid-Z. **106**, 169—174 (1944).

Bonét-Maury, P.: (1) Evaluation, par irradiation alpha, de la taille des ultravirus. J. Chem. Physique **39**, 116 (1942).

— (2) Evaluation, par irradiation alpha, de la taille du virus de la fièvre aphteuse. Ann. Inst. Pasteur **69**, 22—26 (1943).

— (3) La mesure des dimensions du virus vaccinal d'aprés la micrographie électronique. Monographies de l'Institut Alfred Fournier, Paris 5—6 (1934).

— (4) L'irradiation des Virus. Les ultravirus des maladies humaines 2me Édition, Maloine, Paris, 1948, 1—47.

Bonét-Maury, P. et Perault: Action du radon sur le virus vaccinal. Evaluation du diamètre des corpuscules. C. r. Soc. Biol. **135**, 1117—1119 (1941).

Borrel, du Jardin-Beaumetz, Jeantet et Jouan: Le microbe de la péripneumonie. Ann. Inst. Pasteur, Paris, **24**, 169 (1930).

Borries, B. v.: (1) Über die Intensitätsverhältnisse am Übermikroskop. I. Mitteilung: Die Schwärzung photographischer Platten durch Elektronenstrahlen. Physikal. Zschr. **43**, 190—204 (1942).

— (2) Über die Intensitätsverhältnisse am Übermikroskop. II. Mitteilung: Vergrößerungsfähigkeit, Körnigkeit und Auflösungsvermögen elektronengeschwärzter photographischer Platten. Zschr. f. angew. Photographie **4**, 42—58 (1942).

— (3) Über die Intensitätsverhältnisse am Übermikroskop. III. Mitteilung: Eignung und Empfindlichkeitsgrenzen photographischer Platten für übermikroskopische Aufnahmen. Zschr. f. Physik **119**, 498—521 (1942).

— (4) Über günstige Entwicklungsverfahren und Grenzauflösungsvermögen photographischer Platten für übermikroskopische Aufnahmen. IV. Mitteilung über die Intensitätsverhältnisse am Übermikroskop. Zschr. f. Physik **122**, 539—572 (1944).

— (5) Die energetischen Daten und Grenzen der Übermikroskopie. VI. Mitteilung über die Intensitätsverhältnisse am Übermikroskop. Optik **3**, 321—377; 389—422 (1948).

— (6) Der Minimalkontrast im übermikroskopischen Bild. VIII. Mitteilung über die Intensitätsverhältnisse am Übermikroskop. Optik **4**, 235—250 (1948).

— (7) Elektronenstreuung und Bildentstehung im Übermikroskop. VII. Mitteilung über die Intensitätsverhältnisse am Übermikroskop. Zschr. f. Naturforschung **4a**, 51—70 (1949).

— (8) Die Übermikroskopie, Einführung, Untersuchung ihrer Grenzen und Abriß ihrer Ergebnisse. Saenger, Berlin 1949.

Borries, B. v. und W. Glaser: Über die Temperaturerhöhung der Objekte im Übermikroskop. V. Mitteilung über die Intensitätsverhältnisse am Übermikroskop. Kolloid-Z. **106**, 123—128 (1944).

Borries, B. v. und G. A. Kausche: Übermikroskopische Bestimmung der Form und Größenverteilung von Goldkolloiden. Kolloid-Z. **90**, 132—141 (1940).

Borries, B. v., und H. Koppe: Unelastische Streuung und diskrete Geschwindigkeitsverluste schneller Elektronen. Naturw. **34**, 187 (1947).

Borries, B. v. und E. Ruska: (1) Das kurze Raumladungsfeld einer Hilfsentladung als Sammellinse für Kathodenstrahlen. Zschr. f. Physik **76**, 649—654 (1932).

— (2) Magnetische Sammellinse kurzer Feldlänge. DRP 680284 v. 17. 3. 1932, bek. gem. 3. 10. 1935, ausg. 26. 8. 1939.

— (3) Angewandte Elektronenoptik. VDI-Zschr. **80**, 989—994 und 1075—1083 (1936).

BORRIES, B. v. und E. RUSKA: (4) Vorläufige Mitteilung über Fortschritte im Bau und in der Leistung des Übermikroskopes. Wissensch. Veröff. d. Siemens-Werke 17, 99—106 (1938).
— (5) Das Übermikroskop als Fortsetzung des Lichtmikroskops. Verhandlungen der Ges. deutsch. Naturforsch. u. Ärzte 95, Versammlung zu Stuttgart vom 18. bis 21. September 1938, 72—77.
— (6) Aufbau und Leistung des Siemens-Übermikroskopes. Zschr. f. wissensch. Mikro. 56, 317—333 (1939).
— (7) Versuche, Rechnungen und Ergebnisse zur Frage des Auflösungsvermögens beim Übermikroskop. Z. techn. Phys. 20, 225—235 (1939).
— (8) Ein Übermikroskop für Forschungsinstitute. Naturw. 27, 577—582 (1939).
— (9) Mikroskopie hoher Auflösung mit schnellen Elektronen. Erg. exakt. Naturw. 19, 237—322 (1940).
— (10) Der Einfluß der Strahlspannung auf das übermikroskopische Bild. Z. Phys. 116, 249—256 (1940).
BORRIES, B. v., E. RUSKA und H. RUSKA: (1) Bakterien und Virus in übermikroskopischer Aufnahme (mit einer Einführung in die Technik der Übermikroskopie). Klin. Wschr. 17, 921—925 (1938).
— (2) Übermikroskopische Bakterienaufnahmen. Wissensch. Veröff. der Siemens-Werke 17, 107—111 (1938).
BOURDILLON, J.: Purification, sedimentation and serological reaction of the murine strain of SK poliomyelitis virus. Arch. Biochem. 3, 285—297 (1944).
BRAUN, A. C. und R. P. ELROD: Elektronenmikroskopische Untersuchung an Pseudomonas tumefaciens. J. Bacter (Am.) 52, 695—702 (1946).
BRIEGER, E. M., G. R. CROWE and V. E. COSSLETT: Electron microscopy of bacteria. Nature 160, 864 (1947).
BROWN, A. F. and W. M. JONES: A methyl methacrylate-silica replica technique for electron microscopy. Nature 159, 635 (1947).
BRÜCHE, E.: Neues vom Elektronenmikroskop. Die Umschau in Wissenschaft und Technik 13 (1949).
BRÜCHE, E. und E. HAAGEN: Ein neues, einfaches Übermikroskop und seine Anwendung in der Bakteriologie. Naturw. 27, 809—811 (1939).
BRÜCHE, E. und H. JOHANNSON: Elektronenoptik und Elektronenmikroskop. Naturw. 20, 353—358 (1932).
BUCHWALD, E.: Das Doppelbild von Licht und Stoff. Fachverlag Schiele und Schön, Berlin 1947.
BUSCH, H.: (1) Berechnung der Bahn von Kathodenstrahlen im axialsymmetrischen elektromagnetischen Felde. Ann. d. Physik 81, 974—993 (1926).
— (2) Über die Wirkungsweise der Konzentrierungsspule bei der Braunschen Röhre. Arch. f. Elektrotechnik 18, 583—594 (1927).
CHAMBERS, L. A., W. HENLE, M. A. LAUFFER and T. F. ANDERSON: Studies on the nature of the virus of influenza: II. The size of the infectious unit in influenza A. J. exper. Med. 77, 265—276 (1943).
CHU, C. M., I. M. DAWSON and W. J. ELFORD: Filamentous forms associated with newly isolated influenza Virus. The Lancet 9, 602 (1949).
CLAUDE, A. and E. F. FULLAM: (1) An electron microscope study of isolated mitochondria. Method and preliminary results. J. exper. Med. 81, 51—62 (1945).
— (2) The preparation of sections of guinea pig liver for electron microscopy. J. exper. Med. 83, 499 (1946).
COHEN, S. S.: New crystalline forms of tomato bushy stunt virus. Proc. Soc. Exper. Biol. Med. 51, 104—105 (1942).
COLLAN, F.: Purification of the MM poliomyelitis virus. Proc. Soc. Exper. Biol. Med. 67, 364, 366 (1948).
CONN, H. J. and R. P. ELROD: Concerning flagellation and motility. J. Bact. 54, 681—687 (1947).
COSSLETT, V. E.: (1) Electron microscopy and electron diffraction. From "Electronics and their application", Pilot Press 535—641 (1947).
— (2) Particle "growth" in the electron microscope. J. Appl. Phys. 18, 844—845 (1947).

Cosslett, V. E.: (3) Beryllium films as object supports in the electron microscopy of biologial specimens. Biochem. et Biophys. Acta 2, 239—245 (1948).

Cosslett, V. E. and R. Markham: Structure of turnip yellow mosaic virus crystals in the electron microscope. Nature 161, 250—252 (1948).

Cunha, R., M. L. Weil, D. Beard, A. R. Taylor, D. G. Sharp and J. W. Beard: Purification and characters of the Newcastle disease virus (California strain). J. Immun. 55, 69—89 (1946).

Danneel, R.: Wege und Ziele der Virusforschung. Arch. f. Kinderheilkunde 131, 42—53 (1944).

Davisson, C. and L. H. Germer: Diffraction of electrons by a crystal of nickel. Physical Review. 30 (2), 705—740 (1927).

Dawson, L. M. and W. J. Elford: (1) Electron microscope studies on the interaction of certain viruses with fowl red cell membranes. Nature 163, 63 (1949).

— (2) The investigation of influenza and related viruses in the electron microscope, by a new. technique. J. gen. Microbiol. 3, 298—311 (1949).

Dawson, I. M. and A. S. McFarlane: Structure of an animal virus. Nature 161, 464—466 (1948).

Delbrück, M.: (1) J. Bakt. 50, 151 (1945).

— (2) Experiments with bacterial viruses (Bacteriophages). The Harvey Lectures Series XLI, 161—187 (1945/46).

— (3) Über Bakteriophagen. Naturw. 34, 301—306 (1947).

— (4) Biochemical mutans of bacterial viruses. J. of Bact. 56, 1—16 (1948).

— (5) Génétique du bactériophage. Unités biologiques douées de continuité génétique. 91—104 (1949).

Delbrück, M. and W. T. Bailey: Induced mutations in bacterial viruses. J. Cold Spring Harbor symp. Quant. Biol. 11, 33—37 (1946).

Delbrück, M. und M. B. Delbrück: Vermehrungsmechanismus von Bakteriophagen. Naturwissenschaftliche Rundschau 7, 301—306 (1949).

D'Herelle, F.: Le critère de la vie. La presse Medicale 33, 1—8 (1942).

Dienes, L.: (1) Morphology and nature of the pleuropneumonia group of organisms. J. Bakt. 50, 441—458 (1945).

— (2) Complex reproductive processes in bacteria. Cold Spring Harbor Symposia Quant. Biol. 11, 51—59 (1946).

— (3) The morphology of the L_1 of Klieneberger and its relationship to streptobazillus moniliformis. J. Bact. 54, 231—237 (1947).

— (4) Isolation of an L type culture from a gram positive spore-bearing bacillus. Proc. Soc. Exper. Biol. Med. 71, 30—33 (1949).

Dornberger-Schiff, K.: Zur Deutung der Röntgendiagramme gewisser Eiweißstoffe. I. Annalen der Physik 5, 14—32 (1949).

Dounce, A. L. and T. H. Lan: Isolation and properties of chicken erythrocyte nuclei. Science 97, 584 (1943).

Draper, M. H. and Hodge: Sub-microscopic localization of minerals in skeletal muscle by internal micro-incineration. Nature 163, 576—577 (1949).

Dulbecco, R.: The number of particles of bacteriophage T_2 that can participate in intracellular growth. Genetics 34, 126—132 (1949).

Eaton, M., G. Meiklejohn, W. v. Herick and M. Corey: J. exper. Med. 82, 317 (1945).

Eckardt, A.: Über die Veränderung übermikroskopischer Präparate im Elektronenstrahl. Optik 3, 53—58 (1948).

Edwards, O. F. and R. W. G. Wyckoff: Electron micrographs of bacterial cultures infected with bacteriophage. Proc. Soc. exper. Biol. Med. 64, 16—19 (1947).

Eitel, W. und E. Gotthardt: Über die stereophotogrammetrische Dickenmessung kleinster Kristalle nach übermikroskopischen Aufnahmen. Naturw. 28, 367 (1940).

Elvers, I.: An electron microscopic study of chromosomes and cytoplasm in Lilium. Ark. Bot. 3 OB (4) 1942.

Emmel, L., E. Gölz und A. Jakob: Elektronenoptische Untersuchungen an Malariasporozoiten. Deutsch. Tropenmed. Zschr. 46, 573—575 (1942).

EMMEL, L., A. JAKOB und G. GÖLZ: Elektronenoptische Untersuchungen an Malaria-sporozoiten und Beobachtungen an Kulturformen von Leishmania donovani. Deutsch. Tropenmed. Zschr. **46**, 254—258 (1942).

ENDERLEIN, G.: Über die elektronenoptische Bestätigung der vergleichend-morphologischen und entwicklungsgeschichtlichen Forschungen bei Bakterien und deren Jugendstadien (Chondrit und Protit), dem sogenannten Virus. Arch. Entwicklungsgesch. Bakt. **2**, 8—13 (1943).

EYER, H. und H. RUSKA: Über den Feinbau der Fleckfieber-Rickettsie. Zschr. f. Hyg. u. Infektionskrankheiten **125**, 483—492 (1944).

FARRANT. J. L. and O'CONNER: Elementary bodies of varicella and herpes zoster. Nature **163**, 260—261 (1949).

FARRANT, J. L. and A. J. HODGE: An interferometric method for the calibration of electron microscope magnification. J. Appl. Phys. **19**, 840 (1948).

FLICK, J. A.: Use of formalin-treated red cells for the study of influenza A virus hemagglutinating activity. Proc. Soc. Exper. Biol. Med. **68**, 448—450 (1948).

FLICK, J. A., B. SANFORD and A. MUDD: The effect of salt concentration on the interaction of influenza A virus and erythrocytes. J. Immun. **61**, 65—77 (1949).

FRAMPTON, L. V.: The size and shape of the tobacco mosaic virus protein particle. Science **95**, 232—233 (1942).

FRAMPTON, C. and J. NEURATH: An estimate of the relative dimensions and diffusion constant of the tobacco mosaic virus protein. Science **87**, 468 (1938).

FRIEDRICH-FREKSA, H.: (1) Strukturforschung am Tabak-Mosaikvirus. Angew. Chem. A/**60**, 22 (1948).

— (2) Eine Modellvorstellung des Vorgangs der Selbstvermehrung. Angew. Chem. A. **60**, 23 (1948).

— (3) Die stammesgeschichtliche Stellung der Virusarten. In: Die Evolution der Organismen. 2. Auflage, 1950, Fischer, Jena, in Vorbereitung.

FRIEDRICH-FREKSA, H., M. v. ARDENNE und G. SCHRAMM: Elektronenmikroskopische Untersuchungen der Päcipitinreaktion von Tabak Mosaikvirus mit Kaninchen-antiserum. Arch. ges. Virusforsch. **2**, 80—86 (1941).

FROULA, H. C.: Determination of magnification in electron micrographs, J. Appl. Phys. **18**, 19—20 (1947).

FRÜHBRODT, E. und H. RUSKA: Untersuchungen über Bakterienstrukturen unter besonderer Berücksichtigung der Bakterienmembran und der Kapsel. Arch. f. Mikrobiol. **11**, 137—154 (1940).

FULLAM, E. F. and A. E. GESSLER: A high speed microtom for the electron microscope. J. Appl. Phys. **17**, (1946), Referat.

GARD, S.: (1) Purification of poliomyelitis viruses. Experiments on murine and human strains. Acta Med. Scandinavica 1943, Supplementum 143.

— (2) Übermikroskopische Beobachtungen an gereinigten Poliomyelitis-Virus-Präparaten. II. Ein Beitrag zur Frage der Epidemiologie der Kinderlähmung. Klin. Wschr. **22**, 315—318 (1943).

— (3) Übermikroskopische Beobachtungen an gereinigten Poliomyelitis-Virus-präparaten. III. Ein Vergleich mit den physikalisch-chemischen Versuchsergebnissen. Arch. ges. Virusforsch. **3**, 1—17 (1943).

— (4) Vergleichende physikalisch-chemische und übermikroskopische Studien an gereinigten Poliomyelitis-Viruspräparaten. Arkiv för Kemi, Mineralogi och Geologi **17**, 1—4 (1943).

GARD, S. and K. O. PEDERSEN: Science **94**, 493 (1941).

GARDNER, G. M. and R. S. WEISER: A bacteriophage for mycobacterium smegmatis. Proc. Soc. Exper. Biol. Med. **66**, 205—206 (1947).

GÄRTNER, K.: (1) Ein Beitrag zur Färbbarkeit der lebenden und toten Bakterienzelle. Zschr. f. Hyg. usw. **125**, 86—99 (1943).

— (2) Die Sulfonamidwirkung im Lichte der Fluoreszenz- und Elektronenmikroskopie. Z. Bakt. usw. I. **150**, 97—115 (1943).

— (3) Die sichtbare Beeinflussung der Krankheitserreger in vitro durch Sulfonamide. Deutsch. Med. Wschr. **70**, 336—338 (1944).

Gaw, H. Z.: A comparative study of the properties of two strains of tobacco mosaic virus prepared from the sap and from the leaf residues of diseased turkish tobacco plants. Arch. ges. Virusforsch. **3,** 347—355 (1947).

Gaw, H. Z. and W. M. Stanley: Comparative properties of purified preparation of two distinctive strains of tobacco mosaic virus obtained from diseased turkish tobacco and phlox plants. J. Biol. Chem. **167,** 765—772 (1947).

Gerould, C. H.: Preparation and uses of silica replicas in elctron microscopy. J. Appl. Phys. **18,** 273 and 333—343 (1947).

Gildemeister, E., E. Haagen und O. Waldmann: Handbuch der Viruskrankheiten. Band I und II, Fischer, Jena (1939).

Giuntini, J., P. Lepine, P. Nicolle und O. Croissant: Elektronenoptische Abbildungen einiger Bakteriophagen und die Festlegung ihrer Gestalt. Ann. Inst. Pasteur **73,** 579—582 (1947).

Glaser, R. W. and W. M. Stanley: Biochemical studies on the virus and the inclusion bodies of silkworm jaundice. J. exper. Med. **77,** 451—466 (1943).

Gollan, F. and J. F. Marvin: Electron-microscopy of the purified MM poliomyelitis virus. Proc. Soc. Exper. Biol. Med. **67,** 366—367 (1948).

Gönnert, R.: (1) Die Bronchopneumonie, eine neue Viruskrankheit der Maus. Z. Bakt. **147,** 161 (1941).

— (2) Über ein neues, dem Erreger des Lymphogranuloma inguinale ähnliches Mäusevirus. Klin. Wschr. **3,** 76—78 (1941).

— (3) Zur Morphologie der Fleckfieber-Rickettsie in der Laus. Zentralbl. f. Bakt. **152,** 203—209 (1947).

Gotthardt, E.: Zur räumlichen Ausmessung von Objekten mit dem Elektronenmikroskop. Zschr. f. Physik **118,** 714—717 (1942).

Graff, G., D. H. Moore, W. M. Stanley, H. T. Randall and C. D. Haagensen: 4th Intern. Cancer Research Congress, Sept. 2—7, 144 (1947).

Green, R. H., T. F. Anderson and J. E. Smadel: Morphological structure of the virus of vaccina. J. exper. Med. **75,** 651—656 (1942).

Groupé, V. and G. Rake: Studies on the morphology of the elementary bodies of fowl pox. J. Bakt. **53,** 449—454 (1947).

Groupé, V., S. Oskay and C. Rake: Electron micrographs of the elementary bodies of fowl pox and canary pox. Proc. Soc. Exper. Biol. Med. **63,** 477 (1946).

Haagen, E.: Die Bedeutung des Elektronenmikroskops für die experimentelle Virusforschung. Jahrbuch d. AEG.-Forschung. **7,** 88—90 (1940).

Haardick, H., G. A. Kausche und H. Ruska: Elektronenmikroskopische Bestimmung der Konzentration von Tabak Mosaikviruslösungen. Naturw. **32,** 226—228 (1944).

Hall, C. E.: Dark-field electron microscopy. I. Studies of crystalline substances in dark-field. J. Appl. Phys. **19,** 198—212 (1948).

Hall, C. E., M. A. Jakus and Schmitt: (1) Electron microscope observations of collagen. J. Am. Chem. Soc. **64,** 1234 (1942).

— (2) The structure of certain muscle fibrils as revealed by the use of elctron strains. J. Appl. Phys. **16,** 459—465 (1945).

— (3) An investigation of cross striations and myosin filaments in muscle. Biological Bullet. **90,** 32—50 (1946).

Hampp, E. G., D. B. Scott and R. W. G. Wyckoff: Morphologic characteristics of certain cultured strains of oral spirochetes and treponema pallidum as revealed by the electron microscope. J. of Bact. **56,** 755—769 (1948).

Hamre, D. H. Rake and G. Rake: Morphological and other characteristics of the agent of feline pneumonitis grown in the allantoic cavity of the chick embryo. J. exper. Med. **86,** 1—6 (1947).

Hanson, E. E. and J. H. Daniel: An instrument for measuring the particle diameters and constructing histograms from electron micrographs. J. Appl. Phys. **18** (1947).

Hass, G. und H. Kehler: Über eine temperaturbeständige und haltbare Trägerschicht für Elektroneninterferenzaufnahmen und übermikroskopische Untersuchungen. Kolloid-Z. **95,** 26—29 (1941).

HAST, N.: Structure of clay. Nature **159**, 354 und 370 (1947).

HEIDENREICH, R. D. and L. A. MATHESON: Electron-microscope determination of surface elevations and orientations. J. Appl. Phys. **15**, 423—435 (1944).

HEIDENREICH, R. D. and V. G. PECK: Fine structure of metallic surfaces with the electron microscope. J. Appl. Phys. **14**, 23—29 (1943).

HEINMETS, F.: Studies with the electron microscope on the interaction of red cells and influenza virus. J. Bacter. **55**, 823—831 (1948).

HEINMETS, F. and O. J. GOLLUB: J. Bakt. **56**, 509 (1948).

HERZBERG, K.: Handbuch der Viruskrankheiten. Jena, Fischer 1939.

HESSELBROCK, W. and L. FOSHAY: The morphology of bacterium tularense. J. Bakt. **49**, 209—231 (1945).

HILLIER, J.: (1) Fresnel-diffraction of electrons as a contour phenomenon in electron-super-microscope images. Phys. Review **58**, 842 (1940).

— (2) The electron microscope in the determination of particle size characteristics. Am. Soc. Test. Mat. Symposium on new methods for particle size determination in subsieve range 1941, 90—94.

— (3) On the investigation of specimen contamination in the electron microscope. J. Appl. Phys. **19**, 226—230 (1948).

HILLIER, J. and R. F. BAKER: The mounting of bacteria for electron microscope examination. J. Bakt. **52**, 411—416 (1946).

HILLIER, J., G. KNAYSI and R. F. BAKER: New preparation techniques for the electron microscopy of bacteria. J. Bakt. **56**, 569—576 (1948).

HILLIER, J. and A. KURKJIAN: On the artefacts produced by the use of distilled water as an intermediate medium in the mounting of bacterial specimens for the electron microscope. J. Appl. Phys. **16**, 264 (1944).

HILLIER, J., ST. MUDD and A. S. SMITH: Internal structure and nuclei in cells of escherichia coli as shown by improved electron microscopic techniques. J. Bacter. **57**, 319—338 (1949).

HITTORF: Über die Elektrizitätsleitung der Gase. Pogg. Ann. **136**, 8 u. 17 u. 197 (1869).

HOLMES, F. O.: The filterable viruses (Supplement No. 2, Ed. VI), Bergey's Manual of Determinative Bacteriology 1948.

HOOK, A. E., D. BEARD, A. R. TAYLOR, D. G. SHARP and J. W. BEARD: Isolation and characterisation of the T_2 bacteriophage of escherichia coli. J. Biol. Chem. **165**, 241—257 (1946).

HUSEMANN, E. und H. RUSKA: Die Sichtbarmachung von Molekülen des p-Jodbenzoyl-glykogens. Naturw. **28**, 534 (1940).

INGELMAN, B. and K. SIEGBAHN: An electron-microscopic study of dextran molecules. Ark. Kemi Min. Geol. **18 B** (1943).

ITERSON v. W.: Some electron-microscopical observations on bacterial cytology. Biochem. Biophys. Acta **1**, 527—548 (1947).

JAKOB, A.: (1) Einige neuere Ergebnisse in der Leptospirenforschung mit dem Elektronenmikroskop. Phys. Blätter **3**, 286 (1947).

— (2) Über die Morphologie der Leptospira canicola. Med. Klinik **42**, 22—25 (1947).

— (3) Neuere Untersuchungsergebnisse in der Spirochätenforschung mit dem Elektronenmikroskop. Ein Beitrag zur Morphologie der Spirochaeta Pallida. Klin. Wschr. **24/25**, 882—886 (1947).

— (4) Ein Beitrag zur Frage der Dauerformen (Körnchenstadium) bei den Leptospiren. Klin. Wschr. **27**, 346—366 (1949).

JAKOB, A. und E. GÖLZ: (1) Über den Feinbau der pathogenen Kultur-Leptospiren. Neue Phys. Blätter **1**, 14—15 (1946).

— (2) Über den Feinbau einer Kulturform der Spirochaeta Pallida. Neue Phys. Blätter **1**, 16 (1946).

JAKOB, A. und H. MAHL: (1) Anwendung des Übermikroskops in der Bakteriologie, insbesondere für Versuche der Kapseldarstellungen. Jahrbuch der AEG-Forsch. **7**, 77—87 (1940).

— (2) Strukturdarstellungen bei Bakterien, insbesondere die Kapseldarstellung bei Anaerobiern mit dem elektrostatischen Elektronenübermikroskop. Arch. exper. Zellforsch. **24**, 87—104 (1940).

Jakob, A. und H. Mahl: (3) Über die Adsorption von Metallkolloiden an Bakterien. Zschr. Naturforsch. **30**, 26—31 (1948).

Jakus, M. A.: (1) The structure and properties of the trichocysts of paramecium. J. exper. Zoology **100**, 457—485 (1945).

— (2) The structure of trichocysts of paramecium. J. Appl. Phys. **17** (1946), Referat.

Janssen, L. W.: Die Zentrifugierung und die Sedimentationskonstante des Maul- und Klauenseuchevirusproteins. Naturw. **29**, 102—103 (1941).

Johnson, F. H., N. Zworykin and G. Warren: A study of luminous bacterial cells and cytolysates with the electron microscope. J. Bact. **46**, 167—185 (1943).

Jungeblut, C. W. and J. Bourdillon: Electron micrography of murine poliomyelitis virus preparations. J. Am. Med. Ass. **123**, 399—402 (1943).

Kausche, G. A.: (1) Über Versuche zum Nachweis und zur Sichtbarmachung von pflanzlichem Virus. Mitt. d. Biol. Reichsanstalt f. Land- und Forstwirtschaft **59**, 15—23 (1939).

— (2) Ergebnisse und Probleme der experimentellen Virusforschung bei Pflanzen (mit übermikroskopischen Aufnahmen). Ber. d. Deutsch. Bot. Ges. **58**, 200—222 (1940).

— (3) Über den Mechanismus der Goldsolreaktion beim Protein des Tabak-Mosaik- und Kartoffel-X-Virus. Biolog. Zentralbl. **60**, 179—200 (1940).

— (4) Untersuchungen zum Problem der biologischen Charakterisierung rhyto-pathogener Virusproteine. Arch. ges. Virusforsch. **1**, 362—372 (1940).

— (5) Wesen und Leistung der Übermikroskopie für die Struktur- und Virusforschung. Nachrichtenblatt für den deutsch. Pflanzenschutzdienst **21**, 41—44 (1941).

— (6) Zur Methode der „Schrägbedampfung" in der Elektronenmikroskopie. Physikal. Blätter **3**, 285—286 (1947).

— (7) Zur Reindarstellung eines Poliomyelitisvirus (Lansingstamm). Verh.d.Deutsch. Ges. f. innere Med. **54.** Kongreß, 324—327 (1948).

Kausche, G. A., E. Pfankuch und H. Ruska: (1) Die Sichtbarmachung von pflanzlichem Virus im Übermikroskop. Naturw. **27**, 292—299 (1939).

— (2) Beobachtungen über Schall- und Ultraschalleinwirkungen am Protein des Tabak-Mosaikvirus. Naturw. **29**, 573—574 (1941).

Kausche, G. A. und H. Ruska: (1) Die Sichtbarmachung der Adsorption von Metall-kolloiden an Eiweißkörpern. I. Die Reaktion kolloides Gold/Tabak-Mosaikvirus. Kolloid-Z. **89**, 21—26 (1939).

— (2) Die Struktur der „kristallinen Aggregate" des Tabak-Mosaikvirusproteins. Biochem. Zschr. **303**, 221—230 (1939).

— (3) Zur Frage der Chloroplastenstruktur. Naturw. **28**, 302—304 (1940).

— (4) Über den Nachweis von Molekülen des Tabak-Mosaikvirus in den Chloro-plasten viruskranker Pflanzen. Naturw. **28**, 303 (1940).

Kikuth, W. und R. Gönnert: Zur Ätiologie der primären atypischen Pneumonie oder Viruspneumonie. Klin. Wschr. **27**, 185—188 (1949).

Kinder, E.: (1) Elektronenstrahlschäden an Kristallen. Naturw. **34**, 23 (1947).

— (2) Ein Stereo-Elektronenmikroskop. Naturw. **33**, 367 (1946).

Kirchner, F.: Ein Kathodenstrahl-Interferenzapparat für Demonstration und Strukturuntersuchungen. Phys. Zschr. **31**, 772—773 (1930).

Klieneberger, E. and J. Smiles: Some new observations on the developmental cycle of the organism of bovine pleuropneumonia and related microbes. J. Hyg. **42**, 110—123 (1942).

Knaysi, G.: (1) A morphological study of Streptococcus faecalis. J. Bacter. **42**, 575 (1941).

— (2) With and origin of bacterial flagella, electron microscope studies. Science **95**, 405—407 (1942).

— (3) Cytology of bacteria II. Botan. Rev. **15**, 106—151 (1949).

Knaysi, G. and R. F. Baker: Demonstration with the electron microscope of a nucleus in bacillus mycoides grown in a nitrogen-free medium. J. Bact. **53**, 539—553 (1947).

Knaysi, G. and J. Hillier: A study with the high-voltage electron microscope of the endospore and life cycle of bacillus mycoides. J. Bacter. **53**, 525—537 (1947).

KNAYSI, G. and S. MUDD: The internal structure of certain bacteria as revealed by the electron microscope. A contribution to the study of the bacterial nucleus. J. Bact. **45**, 349—359 (1943).

KNIGHT, C. A. and G. OSTER: The size of the particles of some strains of tobacco mosaic virus as shown by the electron microscope. Arch. Biochem. **15** ,289—294 (1947).

KNOLL, M. und E. RUSKA: Das Elektronenmikroskop. Zschr. f. Physik **78**, 318—339 (1932).

KÖNIG, H.: (1) Ein elektronenmikroskopisches Abdruckverfahren für biologische Objekte. Nachr. d. Akad. d. Wiss. Göttingen, Math.-Phys. Klasse 1946, 68—70.

— (2) Veränderung organischer Präparate im Elektronenmikroskop. Nachr. d. Akad. d. Wiss. Göttingen, Math.-Phys. Klasse 1946, 24—25.

— (3) Elektronenbeugungsversuche an Silizium und seinen Oxyden. Optik **3**, 419 (1948).

— (4) Veränderungen von Präparaten im Elektronenmikroskop. Zusammenfassung dieses Vortrages in: Bericht über die Tagung der Deutschen Physikalischen Gesellschaft in der britischen Zone e. V. vom 9. bis 11. September 1948 in Clausthal-Zellerfeld, von F. H. MÜLLER. Kolloid-Z. **111**, 63—65 (1948).

KÖNIG, H. und A. WINKLER: Über Einschlüsse in Bakterien und ihre Veränderung im Elektronenmikroskop. Naturw. **35**, 136—144 (1948).

KOPP, CH. und G. MÖLLENSTEDT: (1) Einstufige Elektronenmikroskopie unter Benutzung feinstzeichnender photographischer Emulsionen und· Leuchtschirme. Gött. Nachr. Math.-Phys. Kl. 79—82 (1946).

— (2) Eigenschaften äußerst feinzeichnender photographischer Emulsionen bei Elektronenbestrahlung. Optik I, 327—342 (1946).

KOTTMANN, U: Morphologische Befunde aus taches vierges von Colikulturen. Arch. f. ges. Virusforsch. **2**, 388—396 (1942).

KRAUSE, F.: Aufnahmen von Viren mit dem Elektronenmikroskop. Naturw. **26**, 122 (1938).

KRAUSE, F. und H. MAHL: Die übermikroskopische Oberflächenabbildung medizinisch-biologischer Objekte nach dem Abdruckverfahren. Kolloid-Z. **105**, 53—55 (1943).

KREGEL, E. A., J. W. APPLING, B. J. SCHEMA and G. R. SEARS: A technique for the electron microscopic examination of encepsulate bacteria. Science **94**, 592 (1941).

LAUFFER, M.: (1) The molecular weight and shape of tobacco mosaic virus protein. Science **87**, 469 (1938).

— (2) The viscosity of tobacco mosaic virus protein solutions. J. of Biol. Chem **126**, 443 (1938).

— (3) Proc. Soc. exper. Biol. Med. **52**, 330 (1943).

— (4) The size and shape of tobacco mosaic virus particles. J. Amer. Chem. Soc. **66**, 1188—1194 (1944).

LAUFFER, M. A. and G. L. MILLER: The sedimentation rate of the biological activities of influenza A virus. J. of exper. Med. **80**, 521—529 (1944).

LAUFFER, M. A. and W. M. STANLEY: (1) J. Biol. Chem. **135**, 463 (1940).

— (2) Biophysical properties of preparations of PR 8 Influenza virus. J. exper. Med. **80**, 531—548 (1944).

LAURELL, A. H. F.: A method of sectioning bacteria in situ for electronmicroscopical and cytochemical investigations. Nature **163**, 282 (1949).

LEA, D. E.: (1) Nature, **146**, 137 (1940).

— (2) The action of radiations on viruses and bacteria. Brit. Med. Bulletin **4**, 24—26 (1946).

LEDINGHAM, I. C. G.: The growth phases of pleuropneumonia and agalactia on liquid and solid media. J. Path. a. Bact. **37**, 393—410 (1933).

LEMBKE, A.: (1) Übermikroskopische Untersuchungen an Bakterien. (In: „Das Übermikroskop als Forschungsmittel"), Verlag Walter de Gruyter & Co., Berlin 1941, 33—47.

— (2) Untersuchungen an den Erregern der Tuberkulose. Zbl. Bakt. **159**, 239 (1947).

LEMBKE, A. und R. BÖNICKE: Experimentelle Störungen des Lebensvorganges und deren Wert für die Deutung des lebendigen Geschehens bei Einzellern. Zentralbl. Bakt. usw. **153**, 145—173 (1949).

LEMBKE, A. und H. RUSKA: Vergleichende mikroskopische und übermikroskopische Beobachtungen an den Erregern der Tuberkulose. Klin. Wschr. **19**, 217—220 (1940).

LENARD: Über Kathodenstrahlen. Nobel-Vortrag, de Gruyter, Berlin und Leipzig **11**, (1920).

LEVADITI, C.: Aspect et dimensions des corps élémentaires vaccinaux et des corpuscules normaux en lumière électronique. Monographies de l'Institut Alfred Fournier, Paris, 1—2 (1943).

— (2) Le Treponema pallidum en microscopie électronique. Monographies de l'Institut Alfred Fournier, Paris, 7—8 (1943).

— (3) Images électroniques en Microbiologie. (Neueste französische Literatur.) Librairie Maloine, Paris (1949).

LIEBERMEISTER, K.: (1) Übermikroskopische Untersuchungen zur Wirkungsweise von Cibazol auf hämolytische Staphylokokken. Deutsch. Med. Wo. **70**, 125—127 (1944).

— (2) Versuche mit Streptomycin, Penicillin und Sulfonamiden bei filtrierbaren Organismen der Pleuropneumoniagruppe. Klin. Wschr. **27**, 64—66 (1949).

LOFGREN, R. and M. H. SOULE: The structure of spirochaeta novyi as revealed by the electron microscope. J. Bact. **50**, 679—690 (1945).

LORING, H. L., L. MARTON and C. E. SCHWEDT: Electron microscopy of purified lansing virus. Proc. Soc. Exper. Biol. and Med. **62**, 291 (1946).

LURIA, S. E.: Reactivation of irradiated bacteriophage by transfer of self-reproducing units. Proc. nat. Academy of Science **33**, 255—264 (1947).

LURIA, S. E. and T. F. ANDERSON: The identification and characterisation of bacteriophages with the electron microscope. Proc. Nat. Acad. Sci. U. S. **28**, 127—130 (1942).

LURIA, S. E., M. DELBRÜCK and T. F. ANDERSON: Electron microscope studies of bacterial viruses. J. Bact. **46**, 75—76 (1943).

MAGERSTEDT, C.: Ein Beitrag zur Morphologie der Syphilisspirochäte. Arch. f. Dermatologie und Syphilis. **185**, 272—280 (1944).

MAHL, H.: (1) Über das elektrostatische Elektronenmikroskop hoher Auflösung. Z. techn. Phys. **20**, 316—317 (1939).

— (2) Ein plastisches Abdruckverfahren zur übermikroskopischen Untersuchung von Metalloberflächen. Metallwirtschaft **19**, 488—491 (1940).

— (3) Metallkundliche Untersuchungen mit dem elektrostatischen Übermikroskop. Zschr. f. techn. Physik **21**, 17—18 (1940).

— (4) Stereoskopische Aufnahme mit dem elektrostatischen Übermikroskop. Naturw. **28**, 264 (1940).

— (5) Über das elektrostatische Elektronenmikroskop und einige Anwendungen. Koll. Z. **91**, 105—117 (1940).

— (6) Über das plastische Abdruckverfahren zur übermikroskopischen Untersuchung von Oberflächen. Zschr. f. techn. Physik **22**, 33—38 (1941).

— (7) Die übermikroskopische Oberflächendarstellung mit dem Abdruckverfahren. Naturw. **30**, 207—217 (1942).

— (8) Fortschritte zur Oberflächenübermikroskopie nach dem Abdruckverfahren. Metallwirtschaft **22**, 9—12 (1943).

— (9) Die elektronenmikroskopische Untersuchung von Oberflächen. Erg. d. exakt. Naturw. **21**, 266—312 (1944/45).

— (10) Über die Deutung übermikroskopischer Elektronenbilder. Optik **2**, 106—113 (1947).

— (11) Längen- und Dickenmessungen im Elektronenmikroskop. ATM V, 1121—1127, T 81—83 (1948).

MANDLE, R. J.: Artifacts in gold shadowed electron micrographs due to electrons of high intensity. Soc. exper. Biol. and Med. **64**, 362—366 (1947).

MARKHAM, R. and K. M. SMITH: Nature, **158**, 300 (1946).

MARKHAM, R., K. M. SMITH and R. W. G. WYCKOFF: (1) Electron microscopy of tobacco necrosis virus crystals. Nature 159, 574 (1947).
— (2) Molecular arrangement in tobacco necrosis virus crystals. Nature 161, 760—761 (1948).
MARTON, C.: (1) A bibliography of electron microscopy. I. J. Appl. Phys. 14, 522—531 (1943).
— (2) A bibliography of electron microscopy. II. J. Appl. Phys. 15, 575—579 (1944).
— (3) A bibliography of electron microscopy. III. J. Appl. Phys. 16, 373—378 (1945).
MARTON, L.: (1) Electron microscopy of biological objects. Physic Review 46, 527—528 (1934).
— (2) La microscopie électronique des objects biologiques. Bull. de L'Acad. royale de Belgique (Classe des Science) 20, 439—446 (1934).
— (3) Bull. Acad. Belg., Classe des Science 23, 672 (1937).
— (4) The electron microscope: A new tool for bacteriological research. J. Bact. 41, 397—413 (1941).
— (5) The electron microscope in biology. Ann Rev. of Biochem. 12, 587—614 (1943).
— (6) Stereoscopy with the electron microscope. J. Appl. Phys. 15, 726—727 (1944).
MARTON, L., DAS GRUPTA and C. MARTON: Modification of specimens in electron microscopy. Science 104, 35 (1946).
MARTON, L. and L. I. SCHIFF: Determination of object thickness in electron microscopy. J. Appl. Phys. 12, 759—765 (1941).
MATHESON, L. A. and R. D. HEIDENREICH: Magnification calibration of the electron microscopes. J. Appl. Phys. 16, 263—266 (1945).
MELCHERS, G., G. SCHRAMM, H. TURNIT und H. FRIEDRICH-FREKSA: Die biologische, chemische und elektronenmikroskopische Untersuchung eines Mosaikvirus aus Tomaten. Biol. Zbl. 60, 524—556 (1940).
MELNICK, J. L.: Detection with the electron microscope of rod-shaped particles in stools of normal and poliomyelitis individuals. J. Immun. 48, 25—28 (1944).
MENKE, W.: Untersuchungen über den Feinbau des Protoplasmas mit dem Universal-Elektronenmikroskop. Protoplasma 35, 115—130 (1940).
MENZE, H.: Zur Pathogenese der Lues nervosa. Dermatologische Wschr. 111, 915—929 (1940).
MILLER, W., L. PANNELL, L. GRAVITZ, W. A. TANNER and M. S. INGALLS: Studies on certain biological characteristics of malleomyces mallei and malleomyces pseudomallei. J. of Bact. 55, 115—126 (1948).
MÖLLENSTEDT, G.: Silikatglas als haltbare, temperaturbeständige und säurefeste Trägerfolie für Elektroneninterferenzen und Elektronenmikroskopie. Optik 2, 276—277 (1947).
MORTON, H. E. and T. F. ANDERSON: (1) Electron microscopic studies of biological reactions. I. Reduction of potassium tellurite by corynebacterium diphtheriae. Proc. Soc. exper. Biol. and Med. 46, 272—276 (1941).
— (2) Electron microscopic studies of the reduction of potassium tellurite by corynebacterium diphtheriae. J. Bact. 41, 25 (1941).
— (3) Some morphological features of the Nichols strain of trepanoma pallidum as revealed by the electron microscope. Am. J. Syphilis, Gonorrhoea, Veneral Diseases 25, 565—573 (1942.)
— (4) Morphology of leptospira ictero-hemorrhagiae and L. canicola, as revealed by the electron microscope. J. Bact. 45, 143 (1943).
MOSLEY, V. M. and R. W. G. WYCKOFF: Electron micrography of the virus of influenza. Nature 157, 263 (1946).
MUDD, S.: (1) Changes in the bacterial cell brought about by the action of germicides and antibacterial substance as demonstrated by the electron microscope. Am. J. Public Health 33, 167—168 (1943).
— (2) Relationship to immunity. J. Am. Med. Ass. 126, 632—639 (1944).
— (3) Submicroscopic structure of the bacterial cell as shown by the electron microscope. Nature 161, 302—305 (1948).
— (4) Electron microscopic visualization of bacteria. Med. Ann. Dis. Columbia 28, Nr. 2 (1949).

Mudd S.: (5) Electron microscopy in relation to the medical sciences. Annals Internal Med. **31**, 570—581 (1949).

Mudd, S. and T. F. Anderson: (1) Demonstration by the electron microscope of the combination of antibodies with flagellar and somatic antigens. J. Immunol. **42**, 251—266 (1941).

— (2) Selective "Staining" for electron micrography. The effects of heavy metal salts on individual bacterial cells. J. exper. Med. **76**, 103—108 (1942).

— (3) Pathogenic bacteria. Rickettsiae and viruses as shown by the electron microscope. J. Am. Med. Ass. **126**, 561—571 (1944).

Mudd, S., F. Heinmets and T. F. Anderson: (1) Bacterial morphology as shown by the electron microscope. VI. Capsule, cell-wall and inner protoplasm of pneumococcus, type III. J. Bact. **46**, 205—211 (1943).

— (2) The pneumococcal capsula swelling reaction studied with the aid of the electron microscope. J. exper. Med. **78**, 327—332 (1943).

Mudd, S. and D. B. Lackman: Bacterial morphology as shown by the electron microscope. I. Structural differentiation within the streptococcal cell. J. Bact. **41**, 415—420 (1941).

Mudd, S., K. Polevitzky and T. F. Anderson: (1) Bacterial morphology as shown by the electron microscope. IV. Structural differentiation within the bacterial protoplasm. Arch. Path. **34**, 199—207 (1942).

— (2) Bacterial morphology as shown by the electron microscope. V. Treponema pallidum, T. macrodentium, and T. microdentium. J. Bact. **46**, 15—24 (1943).

Mudd, S., K. Polevitzky, T. F. Anderson and L. A. Chambers: Bacterial morphology as shown by the electron microscope. II. The bacterial cell membrane in the genus bacillus. J. Bact. **42**, 251—264 (1941).

Mudd, S., K. Polevitzky, T. F. Anderson and C. Kast: Bacterial morphology as shown by electron microscope. III. Cell wall and protoplasm in a strain of fusobacterium. J. Bact. **44**, 361—366 (1942).

Müller, H. O.: (1) Die Ausmessung der Tiefe übermikroskopischer Objekte. Kolloid-Z. **99**, 6—28 (1942).

— (2) Einrichtung zur genauen Exponierung übermikroskopischer Aufnahmen. Kolloid-Z. **109**, 152—156 (1944).

Müller, H. O. und E. Ruska: Ein Übermikroskop für 220 kV Strahlspannung. Kolloid-Z. **95**, 21—25 (1941).

Nagler, F. P. O. and G. Rake: The use of the electron microscope in diagnosis of variola, vaccinia and varicella. J. Bact. **55**, 45—51 (1948).

Neumann, F.: Untersuchungen zur Erforschung der Kernverhältnisse bei den Bakterien. Zbl. f. Bakt. **103**, 385 (1941).

Neurath, H. and G. R. Cooper: J. Biol. Chem. **135**, 455 (1940).

Northrop, J. H.: Concentration and purification of bacteriophage. J. gen. Physiol. (Am) **21**, 335 (1938).

O'Brien, H. C. and G. M. McKinley: New microtome and sectioning method for electron microscope. Science **98**, 455—456 (1943).

O'Brine, J.: Wonder of the ages. "Pic", August **1**, 10—12 (1944).

Oster, G.: (1) Studies on the sonic treatment of tobacco mosaic virus. J. gen. phys. **31**, 89 (1947).

— (2) The scattering of light and its applications to chemistry. Chem. Reviews **43**, 319—360 (1948).

Oster, G., P. M. Doty and B. H. Zimm: Light scattering studies of tobacco mosaic virus. J. of Am. Chem. Soc. **69**, 1193 (1947).

Oster, G., C. A. Knight and W. M. Stanley: Electron microscope studies of Dahlem, Rothamsted, and Princeton samples of tobacco mosaic virus. Arch. Biochem. **15**, 279—288 (1947).

Oster, G. and W. M. Stanley: An electron microscope study of the contents of hair cells from leaves diseased with tobacco mosaic virus. Brit. J. exper. Path. **27**, 261 (1946).

PASSEY, R. D., L. DMOCHOWSKI, W. T. ASTBURY and R. REED: Electron microscope studies of normal and malignant tissues of hight and lowbreast-cancer strains in mice. Nature **160**, 565 (1947).

PASSEY, R. D. DMOCHOWSKI, W. T. ASTBURY, R. REED and P. JOHNSON: Ultracentrifugation and electron microscope studies of tissues of inbred strains of mice. Nature **15**, 759 (1948).

PAULING, L. J.: Am. Chem. Soc. **62**, 2643 (1940).

PEASE, D. C. and R. F. BAKER: (1) Sectioning techniques for electron microscopy using a conventional microtome. Proc. Soc. Exp. Biol. Med. **67**, 470—474 (1948).

— (2) Preliminary investigations of chromosomes and genes with the electron microscope. Science **109**, 8—10 (1949).

PFANKUCH, E. und F. PIEKENBROCK: Zur Spaltung von Virusprotein der Tabak-Mosaikvirusgruppe. Naturw. **31**, 94 (1943).

PFANKUCH, E. und H. RUSKA: Versuche über Länge und Sedimentationskonstante des Tabak-Mosaikvirusmoleküls vor und nach seiner Beschallung. Zschr. Naturforsch. **2b**, 358—360 (1947).

PICKELS, E. G. and SMADEL J. E.: Ultracentrifugation studies on the elementary bodies of vaccine virus I. General methods and determination of particle size. J. exper. Med. **68**, 583—606 (1938).

PIEKARSKI, G.: (1) Zytologische Untersuchungen an Paratyphus- und Colibakterien. Arch. Mikrobiol. **8**, 428 (1937).

— (2) Zytologische Untersuchungen an Bakterien im ultravioletten Licht. Zentralbl. Bakt. **142**, 69 (1938).

— (3) Lichtoptische und übermikroskopische Untersuchungen zum Problem des Bakterienzellkerns. Zbl. Bakter. usw. I. Orig. **144**, 140 (1939).

— (4) Zum Problem des Bakterienzellkerns. Erg. Hyg. Bakt. Immun. u. Exper. Therapie. **26**, 333—364 (1949).

PIEKARSKI, G. und H. RUSKA: (1) Übermikroskopische Darstellung von Bakteriengeißeln. Klin. Wschr. **18**, 383—386 (1939).

— (2) Übermikroskopische Untersuchungen an Bakterien unter besonderer Berücksichtigung der sogenannten Nukleoide. Arch. Mikrobiol. **10**, 302—321 (1939).

PIETSCHMANN, K.: Über die Begeißelung der Bakterien. Arch. f. Mikrobiol. **12**, 377—472 (1942).

PIJPER, A.: (1) J. Pathol. and Bact. **58**, 325 (1946).

— (2) Evidence that amputation of bacterial flagella does not affect motility. Science **109**, 379 (1949).

PLOTZ, H., J. E. SMADEL, T. A. ANDERSON and L. A. CHAMBERS: Morphological structure of rickettsiae. J. exper. Med. **77**, 355 (1943).

POHLMANN, R. und C. WOLPERS: Über das Verhalten histologischer Suspensionen im Ultraschallfeld. Kolloid-Z. **109**, 106—112 (1944).

POLEVITZKY, K.: Pictures of bacterial forms taken with the electron microscope. J. Bact. **41**, 260 (1941).

POLLARD, E. C. and F. FORRO: Examination of the target theory by deuteron bombardment of T_1-phage. Science **109**, 375 (1949).

POLSON, A. G.: (1) Über die Berechnung der Gestalt von Proteinmolekülen. Kolloid-Z. **88**, 51 (1939).

— (2) Diffusion constants of the E. coli bacteriophages. Proc. Soc. Exper. Biol. Med. **67**, 294—296 (1948).

PORTER, K. R., A. CLAUDE and E. F. FULLAM: A study of tissue culture cells by electron microscopy. Methods and preliminary observations. J. exper. Med. **81**, 233 (1945).

PRICE, W. C., R. C. WILLIAMS and R. W. G. WYCKOFF: (1) Science **102**, 277 (1945).

— (2) Electron micrographs of crystalline plant viruses. Arch. Biochem. **9**, 175—185 (1946).

PRICE, W. C. and R. W. G. WYCKOFF: (1) Phytopathology **29**, 83 (1939).

— (2) Electron micrographs of molecules on the face of a crystal. Nature **157**, 764 (1946).

Prowazek, v. S.: Gelbsucht der Seidenraupen. Arch. Protistenkunde **10**, 358 (1907).

Rake, G.: The initial body and the plague form in the chamydozoaceae. J. Bact. **54**, 637—640 (1947).

Rake, G., H. Blank, L. L. Corriell, F. P. Nagler and S. T. F. McNair: The relationship of varicella and herpes zoster: Electron microscope studies. J. Bact. **56**, 293—303 (1948).

Rake, G. and H. P. Jones: Studies on lymphogranuloma venereum. I. Development of the agent in the yolk sac of the chicken embryo. J. exper. Med. **75**, 323—338 (1942).

Rake, G. and J. Oskay: Cultural characteristics of donovania granulomatis. J. Bakt. **55**, 667—675 (1948).

Rake, G., H. Rake, D. Hamre and V. Groupé: Electron micrographs of the Agent of Feline Pneumonitis. (Baker). Proc. Soc. exper. Biol. and Med. **63**, 489—491 (1946).

Rawlins, T. R.: (1) Recent evidence regarding the nature of viruses. Science **96**, 425—426 (1942).

— (2) Stream double refraction studies on the orientation of tobacco mosaic virus particles. Science **99**, 447—449 (1944).

Rawlins, T. E., C. Roberts and N. M. Utech: Am. J. Botany **33**, 356 (1946).

Recknagel A.: Elektronenspiegel und Elektronenlinse; Beiträge zur Elektronen-optik, Barth (1937).

Richards, A. G., T. F. Anderson and R. T. Hance: A microtome sectioning technique for electron microscopy, illustrated with sections of striated muscle. Proc. Soc. exper. Biol. Med. **51**, 148—152 (1942).

Riecke: Über die Bewegung eines elektrischen Teilchens in einem homogenen magnetischen Felde und das negative elektrische Glimmlicht. Wiedemanns Ann. **13**, 191—194 (1881).

Riedel, G.: Ein elektrischer Kernfäller zur Gewinnung übermikroskopischer Präparate. Kolloid-Z. **103**, 228—232 (1943).

Riedel, G. und H. Ruska: Übermikroskopische Bestimmung der Teilchenzahl eines Sols über dessen aerodispersen Zustand. Kolloid-Z. **96**, 86—96 (1941).

Rippel-Baldes, A.: Grundriß der Mikrobiologie. Berlin und Göttingen, Springer-Verlag 1947.

Rosenblatt, M. V., E. F. Fullam and A. E. Gessler: Studies of mycobacteria with the electron microscope. Am. Rev. Tuberculosis **46**, 587—598 (1942).

Rüdiger, O.: Über die Eignung von Beryllium zur Herstellung übermikroskopischer Abdruckfolien. Naturw. **30**, 279 (1942).

Ruess, G.: Zur Objekterwärmung im Siemens-Elektronenübermikroskop. Kolloid-Z. **109**, 149—152 (1944).

Ruska, E.: (1) Das Elektronenmikroskop als Übermikroskop. Forsch. u. Fortschr. **10**, 8 (1934).

— (2) Über die Fortschritte im Bau und in der Leistung des magnetischen Elektronenmikroskops. Zschr. Phys. **87**, 580—602 (1934).

— (3) Beitrag zur übermikroskopischen Abbildung bei höheren Drucken. Kolloid-Z. **100**, 212—219 (1942).

— (4) Über die Linsen hochauflösender Elektronenmikroskope. Arch. f. Elektrotechnik **36**, 431—454 (1942).

— (5) Zur Entstehung der Säume um übermikroskopisch abgebildete Partikel und über ihre Veränderung mit der optischen Einstellung. Kolloid-Z. **105**, 43—52 (1943).

— (6) Über den Bau und die Bemessung von Polschuhlinsen für hochauflösende Elektronenmikroskope. Arch. f. Elektrotechnik **38**, 102—130 (1944).

— (7) Über neue magnetische Durchstrahlungselektronenmikroskope im Strahlspannungsbereich von 40—220 kV. Teil I. Kolloid-Z. 1950 (im Druck).

Ruska, H.: (1) Bakterien und Virus in übermikroskopischer Aufnahme. Deutsch. med. Wschr. **64**, 33 (1938).

— (2) Übermikroskopische Bilder zu Strukturproblemen. Verhandl. deutsch. zool. Ges. 295—302 (1939).

Ruska, H.: (3) Übermikroskopische Darstellung organischer Struktur (vom Größenbereich der Zelle bis zum Ultravirus). Arch. exper. Zellforsch. **22**, 673—680 (1939).

— (4) Übermikroskopische Untersuchungstechnik. Naturw. **27**, 287—292 (1939).

— (5) Bedeutung und Ergebnisse der Übermikroskopie. Siemens-Zschr. **20**, 228—234 (1940).

— (6) Beobachtungen über geformte Stoffwechselprodukte bei Chromobacterium prodigiosum. Zschr. Hyg. usw. **123**, 289-293 (1941).

— (7) Über den Einfluß keimschädigender oder keimabtötender Maßnahmen auf die Bakterienstruktur. I. Mitteilung: Temperatureinwirkungen. Zschr. Hyg. usw. **123**, 294—301 (1941).

— (8) Über Grenzfragen aus dem Gebiet der Strukturforschung und Mikrobiologie. Deutsch. med. Wschr. **67**, 281—286 (1941).

— (9) Über ein neues, bei der bakteriophagen Lyse auftretendes Formelement. Naturw. **29**, 367—368 (1941)

— (10) Objektwahl, Objektvorbereitung und Deutung des Bildes in der Übermikroskopie (Buch: „Das Übermikroskop als Forschungsmittel"). Verlag Walter de Gruyter & Co., Berlin, 88—104 (1941).

— (11) Fragen der Virusforschung. Forsch. u. Fortschr. **17**, 363—365 (1941).

— (12) Morphologische Befunde bei der bakteriophagen Lyse. Arch. f. Virusforsch. **2,** 345—387 (1942).

— (13) Die Erweiterung des mikroskopischen Sehens und ihre Auswirkung auf Morphologie, Kolloidforschung und Mikrobiologie. Jahrbuch d. Auslandsamtes d. Deutsch. Dozentenschaft. 65—71 (1943).

— (14) Ergebnisse der Bakteriophagenforschung und ihre Deutung nach morphologischen Befunden. Erg. d. Hyg. Bakt. usw. **25**, 437—498 (1943).

— (15) Versuch einer Ordnung der Virusarten. Arch. ges. Virusforsch. **2**, 480—498 (1943).

— (16) Über das Virus der Varicellen und des Zoster. Klin. Wschr. **22**, 703—704 (1943).

— (17) Ziele und Erfolge⁻ der Übermikroskopie in der medizinischen Forschung. „Scienta" (Revue de Synthese Scientifique) **37**, 16—21 (1943).

— (18) Über die Elementarkörper des Virus der Bronchopneumonie der Maus. Klin. Wschr. **23**, 121—122 (1944).

— (19) Über die Bindung des Sublimats an Bakterien aus Virus. Arch. Path. Pharm. **204**, 576—584 (1947).

— (20) Der Einfluß der Zerschäumung auf Tabak-Mosaikviruslösungen. Kolloid-Z. **110**, 79—102 (1948).

— (21) Zur Frage der Potenzierung von Bakteriophagenlösungen durch Zerschäumen. Kolloid-Z. **110**, 175—177 (1948).

— (22) Virus. Vortrag Düsseldorf (1949), Handbuch der Biologie (1950).

Ruska, H., B. v. Borries und E. Ruska: Die Bedeutung der Übermikroskopie für die Virusforschung. Arch. ges. Virusforsch. **1**, 155—169 (1939).

Ruska, H. und G. A. Kausche: Über Form, Größenverteilung und Struktur einiger Viruselementarkörper. Zentralbl. Bakt. usw. **150**, 311—318 (1943).

Ruska, H. und C. Menze: Die Größe von Typhusbakteriophagen nach der Filtrationsendpunktbestimmung und dem elektronenmikroskopischen Bild. Kolloid-Z. **110**, 103—105 (1948).

Ruska, H. und K. Poppe: (1) Elektronenmikroskopische Untersuchungen zur Morphologie der Seiffertschen Mikroorganismen und des Erregers der Lungenseuche des Rindes. Z. Hyg. **127**, 201—215 (1947).

— (2) Morphologische Beziehungen zwischen filtrierbaren Mikroorganismen und großen Virusarten. Z. Naturforsch. **2 b**, 35—36 (1947).

Ruska, H. und C. Wolpers: Zur Struktur des Liquorfibrins. Klin. Wschr. **19**, 695 (1940).

Rybak, B., P. Lépine und C. Croissant: Biochemische Bedingungen für die Vermehrung eines Bakteriophagen. Nachweis mit dem Elektronenmikroskop. C. R. hebd. Séances Acad. Sci. **227**, 238—240 (1948).

Sabin, A. B.: The filterable microorganisms of the pleuropneumonia group. Bact. Revs. **5**, 1—6 (1941).

Saudek, E. C. und D. R. Colingwort: Ein Bakteriophage in der Streptomycinforschung. J. Bakt. **54**, 41 (1947).

Scanga, F.: (1) L'azione della streptomicina, del solfone e di alcuni disinfettanti sui germi, studiate al microscopio elettronico. Rendiconti, Dell'istituto superiore di sanita **11**, 970—979 (1948).

— (2) L'azione della streptomicina sui germi, osservata al microscopio elettronico con la tecnica delle ombre. Rendiconti, Dell'istituto superiore di sanita **11**, 980 bis 989 (1948).

Schaefer, V. J.: (1) New methods of preparing surface replicas for microscopic observation. Phys. Rev. **62**, 495 (1942).

— (2) Dry stripped replicas for the electron microscope. Science **97**, 188 (1943).

Schaefer, V. J. and D. Harker: Surface replicas for use in the electron microscope. J. Appl. Phys. **13**, 427—433 (1942).

Schäfer, W. und G. Schramm: Die Reindarstellung des Virus der klassischen Geflügelpest. Z. Naturforsch. **4b**, 123—124 (1949).

Schäfer, W., G. Schramm und E. Traub: Untersuchungen über das Virus der atypischen Geflügelpest. Zeitschr. Naturforsch. **4b**, 157—167 (1949).

Schlesinger, M.: Beobachtung und Zählung von Bakteriophagenteilchen im Dunkelfeld. — Die Form der Teilchen. Z. Hyg. **115**, 774 (1933).

Schramm, G.: (1) Über die Spaltung des Tabak-Mosaikvirus in niedermolekulare Proteine und die Rückbildung hochmolekularen Proteins aus den Spaltstücken. Naturw. **31**, 94—96 (1943).

— (2) Über die Spaltung des Tabak-Mosaikvirus und die Wiedervereinigung der Spaltstücke zu höhermolekularen Proteinen. I. Die Spaltungsreaktionen. Chem. Zbl. **118**, 1001 (1947).

— (3) Über die Spaltung des Tabak-Mosaikvirus und die Wiedervereinigung der Spaltstücke zu höhermolekularen Proteinen. II. Versuche zur Wiedervereinigung der Spaltstücke. Z. Naturforsch. **2b**, 249—257 (1947).

Schramm, G. und G. Bergold: Über das Molekulargewicht des Tabak-Mosaikvirus. Z. Naturforsch. **2b**, 108—112 (1947).

Schramm, G. und H. Friedrich-Freksa: Die Präcipitinreaktion des Tabak-Mosaikvirus mit Kaninchen- und Schweineantiserum. Hoppe-Seylers Zschr. f. physiol. Chem. **270**, 233—246 (1941).

Schultz, E. W., P. R. Thomassen and L. Marton: Electron microscopic observations on pseudomonas aeruginosa bacteriophage. Proc. Soc. Exp. Biol. Med. **68**, 451—455 (1948).

Schuster, M. C. and C. E. Gray: Preparation of tissue for high speed sectioning. J. Appl. Phys. **18**, 271 (1947).

Seeliger, R.: (1) Das „Prägeabdruckverfahren" zur übermikroskopischen Oberflächenabbildung. Neue physikal. Blätter **1**, 15—16 (1946).

— (2) Über den Zusammenhang zwischen den Grenzen der Punktauflösung und der Erkennbarkeit polygonaler Begrenzung kleiner Teilchen im Elektronenmikroskop. Phys. Blätter **3**, 287 (1947).

— (3) Punktauflösung und polygonal begrenzte Objekte im Elektronenmikroskop. Optik **3**, 315—319 (1948).

Semmler, E.: Über die Eigenstruktur von Aluminiumoxydfilmen für übermikroskopische Oberflächenuntersuchungen. Zschr. Metallkunde **34**, 229—231 (1942).

Shanahan, A. J. and F. W. Tanner: Further studies on the morphology of escherichia coli exposed to penicillin. J. of Bact. **55**, 535—544 (1948).

Sharp, D. G.: Enumeration of virus by electron micrography. Proc. Soc. exper. Biol. Med. **70**, 54—59 (1949).

Sharp, D. G., A. R. Taylor and J. W. Beard: Vaccinia virus. Electron microscope study. Proc. Soc. exper. Biol. Med. **61**, 259 (1946).

Sharp, D. G., A. R. Taylor, D. Beard and J. W. Beard: (1) Electron micrography of the western strain of equine encephalomyelitis virus. Proc. Soc. exper. Biol. and Med. **51**, 206—207 (1942).

SHARP D. G. A. R. TAYLOR, B. and J. W. BEARD: (2) Study of the papilloma virus protein with the electron microscope. Proc. Soc. exper. Biol. and Med. 50, 205—207 (1942).
— (3) Morphology of the eastern and western strains of the virus of equine encephalomyelitis. Arch. Path. 36, 167—176 (1943).
SHARP, D. G., A. G. TAYLOR, A. E. HOOK and J. W. BEARD: Rabbit papilloma and vaccinia viruses and T_2 bacteriophage of E. coli in "shadow" electron micrographs. Proc. Soc. exper. Biol. Med. 61, 259 (1946).
SHARP, D. G., A. E. HOOK, A. R. TAYLOR, D. BEARD and J. W. BEARD: Sedimentation characters and pH stability of the T_2 bacteriophage of escherichia coli. J. Biol. Chem. 165, 259—270 (1946).
SHARP, D. G., A. R. TAYLOR, I. W. McLEAN, D. BEARD and J. W. BEARD: Density and size of influenza virus A (PR 8 strain) in solution. Science 100, 151—153 (1944).
— (2) Sedimentation velocity and electron micrographic studies of influenza viruses A (PR 8 strain) and B (Lee strain) and the swine influenza virus. J. Biol. Chem. 156, 585—600 (1944).
SHARP, D. G., A. R. TAYLOR, I. W. McLEAN, J. D. BEARD, J. W. BEARD, A. E. FELLER and J. H. DINGLE: (1) Isolierung und Charakterisierung des Influenzavirus (Lee-Stamm). Science 98, 307—308 (1943).
— (2) Isolation and characterization of influenza virus B (Lee-Strain). J. Immun. 48, 129—153 (1944).
SHEPARD, C. D. and R. W. G. WYCKOFF: The nature of the soluble antigen from typhus rickettsiae. Public Health Reports 61, 761—767 (1946).
SIGURGEIRSSON, S. and W. M. STANLEY: Electron microscope studies on tobacco mosaic virus. Phytopath. 37, 26—38 (1947).
SIKORA, H.: Z. f. Hyg. 124, 250 (1943).
SJÖSTRAND, F.: (1) Electron microscopic examination of tissues. Nature 151, 725—726 (1943).
— (2) Fixierung und Präparation für elektronenmikroskopische Untersuchungen von Gewebe. Hygiea 105, 1207—1212 (1943).
— (3) Eine neue Methode zur Herstellung sehr dünner Objektschnitte für die elektronenmikroskopische Untersuchung von Geweben nebst einigen vorläufigen elektronenmikroskopischen Beobachtungen über den submikroskopischen Bau der Skelettmuskelfaser. Arkiv f. Zoologie 35A, 5 (1943).
SMADEL, J. E. and J. WALL: Elementary bodies of vaccinia from infected chorioallantoic membranes of developing chick embryos. J. exper. Med. 66, 325—336 (1937).
SMITH, W. E., J. HILLIER and S. MUDD: Electron micrograph studies of two strains of pleuropneumonialike (L) organisms of human derivation. J. Bact. 56, 589—601 (1948).
SMITH, K. M. and W. D. MacCLEMENT: Parasitology 32, 320 (1940).
SMITH, W. E., S. MUDD and J. HILLIER: L-Type variation and bacterial reproduction by large bodies as seen by electron micrographic studies of bacteriodes funduliformes. J. Bact. 56, 603—618 (1948).
STANLEY, W. M.: (1) Isolation of a crystalline protein possessing the properties of tobacco mosaic virus. Science 81, 644—645 (1935).
— (2) Viruses and the electron microscope. Chronica Botan. 7, 291—294 (1943).
— (3) The size of influenza virus. J. exper. Med. 79, 267—283 (1944).
— (4) The efficiency of different Sharples centrifuge bowls in the concentration of tobacco mosaic and influenza viruses. J. Immun. 53, 179—189 (1946).
— (5) Chemical studies on viruses. Chemical and engineering news, American Chemical Society 25, 3786 (1947).
STANLEY, W. M. and T. F. ANDERSON: A study of purified viruses with the electron microscope. J. Biol. Chem. 139, 325—338 (1941).
STANLEY, W. M., C. A. KNIGHT et L. J. DE MERRE: Les Virus, études biochimique et biophysique récentes. Actual. Med. Chir. 6, 10—81 (1945).
STAPP, C.: Der Pflanzenkrebs und sein Erreger Pseudomonas tumefaceus. XI. Mitteil. Zytologische Untersuchungen des bakteriellen Erregers. Zbl. Bakt. II 105, 1 (1942).

412 H. Ruska: Die Elektronenmikroskopie in der Virusforschung.

Steinhaus, E. A. and C. G. Thompson: Granulosis disease in the buckeye caterpillar, junonia counia Hübner. Science 110, 276 (1949).
Stickl, O. und K. Gärtner: Die Wirkungsweise der Sulfonamide und ihre chemotherapeutische Anwendung bei Ruhr. Zschr. Hyg. usw. 125, 226—264 (1943).
Stille, B.: Zytologische Untersuchungen an Bakterien mit Hilfe der Feulgenschen Nuclealreaktion. Arch. Mikrobiol. 8, 125 (1937).
Takahashi, W. and T. E. Rawlins: (1) Stream double refraction of preparation of crystalline tobacco mosaic protein. Science 85, 103 (1937).
— (2) An electron microscope study of mutation in tobacco mosaic virus. Phytopathologie 37, 73—76 (1947).
— (3) The effect of washing frozen mosaic leaves on the length of tobacco mosaic virus particles. J. Bact. 57, 131—134 (1949).
Tang, F. F., H. Wei, D. L. Whiter and J. Edgar: (1) An investigation of the causal agent of bovine pleuropneumonia. J. Path. Bact. 40, 391—406 (1935).
— (2) Further investigation on the causal agent of bovine pleuropneumonia. J. Path. Bact. 42, 45—51 (1936).
Taylor, A. R., D. G. Sharp, D. Beard and J. W. Beard: Electron micrography of the eastern strain of equine encephalomyelitis virus. Proc. Soc. exper. Biol. Med. 51, 332—334 (1942).
Taylor, A. R., D. G. Sharp, D. Beard, J. W. Beard, J. H. Dingle and A. E. Feller: Isolation and characterization of influenza virus A (PR 8 Strain). J. Immun. 47, 261—282 (1943).
Taylor, A. R., D. G. Sharp, I. W. McLean, D. Beard, J. W. Beard, J. H. Dingle and A. E. Feller: (1) Purification and characterization of the swine influenza virus. Science 98, 587—589 (1943).
— (2) Purification and characterization of the swine influenza virus. J. Immun. 48, 361—379 (1944).
Thiel, van, P. H. and W. v. Iterson: An electron microscopical study of leptospira biflexa. Proc. Kon. Med. Akad. v. Wetensch, Amsterdam, 50, 976, 979 (1947).
Timoféeff-Ressovsky, N. W. und K. G. Zimmer: Biophysik Band I. Das Trefferprinzip in der Biologie. Leipzig, Hirtzel 1947, XII, 317.
Tiselius, A. und S. Gard: Übermikroskopische Beobachtungen an Poliomyelitis-Viruspräparaten. Naturw. 30, 728—731 (1942).
Troch, P.: Übermikroskopische Beobachtungen an Tuberkelbazillen nach Einwirkung von Peteosthor. Zschr. Hyg. usw. 124, 513—518 (1942).
Trurnit, H. und H. Friedrich-Freksa: Die elektronenmikroskopischen Untersuchungen des „Tomaten Mosaikvirus Dahlem 1940". Biol. Ztrbl. 60, 546—556 (1940).
Umbreit, W. W. and T. F. Anderson: A study of thiobacillus thiooxidans with the electron microscope. J. Bact. 44, 317—320 (1942).
Veeraraghavan, N.: In-Vitro-Züchtung von Rabies-Virus. Nature 159, 782 (1947).
Vonkennel, J., J. Kimmig und A. Lembke: Gasanalytische und elektronenoptische Untersuchungen zur Wirkungsweise der Sulfonamide. Deutsch. med. Wschr. 69, 129—130 (1943).
Watson, J. H. L.: Specimen contamination in electron microscopes. J. Appl. Phys. 19, 110—111 (1948).
Weiss, L. J.: (1) Electron micrographs of rickettsiae of typhus fever. J. Immunol. 47, 353—357 (1943).
— (2) Electron micrographs of bacteria medicated with penicillin. Proc. Indiana Acad. Science 52, 27—29 (1943).
— (3) Electron micrographs of pleuropneumonia-like organisms. J. Bakt. 47, 523—527 (1944).
Wessel, E.: Übermikroskopische Beobachtungen an Tuberkelbazillen von Typus humanus. Zschr. f. Tuberkulose 88, 22—36 (1942).
Weyer, F., H. Friedrich-Freksa und G. Bergold: Die Beziehungen der Rickettsien zu Bakterien und Viren. Naturw. 32, 361—365 (1944).

WICHERT, E.: Experimentelle Untersuchungen über die Geschwindigkeit und die magnetische Ablenkbarkeit der Kathodenstrahlen. Ann. d. Phys. Chem. **69**, 739—766 (1899).

WILE, W. J. and KEARNEY: The morphology of treponema pallidum in the electron microscope; demonstration of flagella. J. Am. Med. Ass. **122**, 167—168 (1943).

WILE, W. J., R. G. PICARD and E. B. KEARNEY: The morphology of spirochaeta pallida in the electron microscope. J. Am. Med. Ass. **119**, 880—881 (1942).

WILLIAMS, C. und R. C. BACKUS: Die elektronenmikroskopische Struktur von beschatteten Filmen und Oberflächen. J. Appl. Phys. **20**, 98—106 (1949).

WILLIAMS, R. C. and R. L. STEERE: Electron micrographic observations of tobacco mosaic virus in crude, undiluted plant juice. Science **109**, 308 (1949).

WILLIAMS, R. C. and W. C. WYCKOFF: (1) Thickness of electron microscopic objects. J. Appl. Phys. **15**, 712—716 (1944).

— (2) Electron shadow micrography of the tobacco mosaic virus protein. Science **101**, 594—596 (1945).

— (3) Electron shadow micrography of virus particles. Proc. Soc. exper. Biol. med. **58**, 265—270 (1945).

— (4) Shadowed electron micrographs of bacteria. Proc. Soc. exper. Biol. Med. **59**, 265 (1945).

— (5) Application of metallic shadow-casting to microscopy. J. Appl. Phys. **17**, 23—33 (1946).

WINKLER, A. und H. KÖNIG: Zur Deutung elektronenoptischer Befunde an Bakterien. Zbl. Bakt. **152**, 9—15 (1948).

WIRTH, J. and P. ATHANASIU: Electron microscopy of cells from tissue cultures infected with vaccine virus. Proc. Soc. exper. Biol. Med. Vol. **70**, 59—61 (1949).

WIRTH, J., P. ATHANASIU, G. BARSKI und O. CROISSANT: Elektronenmikroskopische Untersuchungen der protoplasmatischen Einschlüsse in mit Vaccinevirus infizierten Reinkulturen von Nierenzellen. C. R. Hebd. Séances Acad. Sci. **225**, 827—829 (1947).

WOLF, J.: Über die Herstellung mikroskopischer Präparate der Oberflächen verschiedener Objekte mit Hilfe der Adhäsionsmethode. Zschr. Mikroskopie **56**, 181 (1939).

WOLLMAN, E. et A. LACASSAGNE: (1) Evalution de la taille relative des bakteriophages par leur radiosensibilité. C. r. Soc. Biol. Paris, **131**, 959 (1939).

— (2) Recherches sur le Phénomène de Twort-D'Hérelle. Ann. Inst. Pasteur **64**, 5—39 (1940).

WOLPERS, C.: (1) Zur Feinstruktur der Erythrocytenmembran. Naturw. **29**, 416—410 (1941).

— (2) Zur elektronenoptischen Darstellung der Malaria tertiana. Klin. Wschr. **21**, 1049—1054 (1942).

— (3) Die Querstreifung der kollagenen Bindegewebsfibrille. Virchows Archiv **312**, 292—302 (1944).

— (4) Die Fibrinquersteigerung. Klin. Wschr. **24/25**, 424—426 (1947).

WOLPERS, C. und H. RUSKA: Strukturuntersuchungen zur Blutgerinnung. Klin. Wschr. **18**, 1077—1081, 1111—1117 (1939).

WOOD, B. and M. R. SMITH: The inhibition of surface phagocytosis by the capsular "slime layer" of pneumococcus Type III. J. exper. Med. **90**, 85—96 (1949).

WYCKOFF, R. W. G.: (1) Some recent developments in the field of electron microscopy. Science **104**, 21—26 (1946).

— (2) Electron micrographs from concentrated solutions of the tobacco mosaic virus protein. Biochem. Biophys. acta. **1**, 139 (1947).

— (3) Symmetrical patterns of bacteriophage production. Proc. Soc. exper. Biol. Med. **66**, 42—44 (1947).

— (4) The electron microscopy of developing bacteriophage. II. Growth of T_4 in liquid culture. Biochem. Biophys. acta. **2**, 246—253 (1948).

— (5) Multiplication of bacteriophage. Nature **162**, 649—650 (1948).

— (6) Multiplication of the T 3 bacteriophage against E. coli, Proc. Soc. Exp. Biol. Med. **71**, 144—146 (1949).

414   H. Ruska: Die Elektronenmikroskopie in der Virusforschung.

Wyckoff, R. W. G. and R. B. Corey: X-Ray diffraction pattern of crystalline tobacco mosaic proteins. J. Biol. Chem. **116**, 51 (1936).

Zimmer, K. G.: Statistische Ultramikrometrie mit Röntgen-Alpha- und Neutronenstrahlung. Phys. Zschr. **44**, 233 (1943).

Zworykin, V. K. and J. Hillier: Electronic microscopy. The Scientific Monthly, Sept. 165—179 (1944).

Zworykin, V. K. and F. G. Ramberg: Surface studies with the electron microscope. J. Appl. Phys. **12**, 692—695 (1941).

Zworykin, V. K., G. A. Morton, E. G. Ramberg, J. Hillier and A. W. Vance: Electron optics and the electron microscope. New York, John Wiley u. Sons, Inc. 1945, S. 766. London: Chapman u. Hall, Limited.

Literaturnachtrag.

Anderson, T. F.: The reactions of bacterial viruses with their host cells. Bot. Rev. **15** 464—505 (1949).

Beard, J. W.: Purified animal viruses. J. Immun. **58**, 49—108 (1948).

Borries, B. v.: (1) Ein magnetostatisches Objektiv-Projektiv-System für das Elektronenmikroskop. Kolloid-Zschr. **114**, 164—167 (1949).

— (2) Die gegenwärtige Lage der Mikroskopie. Zschr. Ver. Deutsch. Ing. **92**, 240—248 (1950).

Boswell, F. W.: Electron microscope studies of virus elementary bodies. Brit. J. Exp. Path. **28**, 253—260 (1947).

Bretschneider, L. H.: (1) Anwendung und Ergebnisse der Elektronenmikroskopie. Mikroskopie **3**, 160—175 (1948).

— (2) A simple technique for the electron microscopy of cell and tissue sections. Proc. Kon. Ned. Akad. v. Wetensch., Amsterdam, **52**, 2—14 (1949).

— (3) L. H.: (3) La technique d'une étude histologique au microscope électronique. Journ. Cyto-embryol. belgo-néerland. Gand, 9—10 (1949).

Castaneda, M. Ruiz: Preparation and properties of purified rickettsial suspensions. J. Immun. **58**, 238—292 (1948).

Claude, A.: Electron microscope studies of cells by the method of replicas. J. Exp. Med. **89**, 425—430 (1948).

Danon, M., E. Guyénot, E. Kellenberger and J. Weigle: Electron micrograph of a chromosome of triton. Nature **165**, 33 (1950).

Evans, A. S. and J. L. Melnick: Electron microscope studies of the vesicle and spinal fluids from a case of herpes zoster. Proc. Soc. Exp. Biol. Med. **71**, 283—286 (1949).

Farrant, J. L. and A. J. Hodge: An interferometric method for the calibration of electron microscope magnification. J. Appl. Phys. **19**, 840—844 (1948).

Friedrich-Freksa, H.: Die stammesgeschichtliche Stellung der Virusarten; in: Die Evolution der Organismen. Herausgeg. v. Heberer, G. (1950). (Unveröffentlicht.)

Fulton, F.: Growth-cycle of influenza virus. Nature **164**, 189—190 (1949).

Giuntini, J., O. Croissant, P. Athanasiu et L. Reinié: Morphologie comparée au microscope électronique des corps élémentaires du virus vaccinal et des corpuscules de même taille extraits des cellules normales. Soc. Biol. **749** (1947).

Gollan, F.: Purification of the MM poliomyelitis virus. Proc. Soc. Exp. Biol. Med. **67**, 364—366 (1948).

Grün, L. und W. Tischer: Ein einfaches Gerät zur Herstellung von Abklatschpräparaten für elektronenmikroskopische Untersuchungen. Optik **6**, 129—132 (1950).

Haardick, H.: (1) Über die Sichtbarmachung der Bakteriophagenadsorption im Elektronenmikroskop. Zeitschr. f. Hygiene **130**, 428—435 (1949).

— (2) Über Wechselwirkungen zwischen Aufdampfschichten und mikroskopischen Objekten. Optik **5**, 549—554 (1949).

Hamre, D.: The effect of ultrasonic waves upon klebsiella pneumoniae, saccharomyces cerevisiae, miyagawanella felis, and influenza virus A. J. Bact. **57**, 279—295 (1949).

HAST, NILS: Production of extremely thin metal films by evaporation on to liquid surfaces. Nature **162**, 892 (1948).

HEDÉN, C. G. and R. W. G. WYCKOFF: The electron microscopy of heated bacteria. J. Bact. **58**, 153—160 (1949).

HERZBERG, K.: (1) Virus-Atlas. Transmare-Photo. Berlin (1947).

— (2) Neuere Ergebnisse und Auffassungen aus dem Virusgebiet. Verh. d. Deutsch. Ges. f. innere Med. 54. Kongreß, 201—222 (1948).

— (3) Über Viruskrankheiten in der Dermatologie. Arch. Derm. Syph. **188**, 526—549 (1949).

HILLIER, J.: Some remarks on the image contrast in electron microscopy and the two-component objective. J. Bact. **57**, 313—317 (1949).

HOUWINK, A. L. and M. v. ITERSON: Electron microscopical observations on bacterial cytology; II. A study on flagellation. Biochim. Biophys. Acta **5**, 10—44 (1950).

HOYLE, L.: (1) The growth cycle of influenza virus A. Brit. J. Exp. Path. **29**, 390—399 (1948).

— (2) Growth-cycle of influenza virus. Nature **164**, 1137—1138 (1949).

HUGHES, K. M.: A demonstration of the nature of polyhedra using alkaline solutions. J. Bact. **59**, 189—195 (1950).

KERBY, G. P., R. A. GOWDY, E. S. DILLON, M. L. DILLON, T. Z. CZÁKY, D. G. SHARP and J. W. BEARD: Purification, pH stability and sedimentation properties of the T$_7$ bacteriophage of escherichia coli. J. Immun. **63**, 93—107 (1949).

LANNI, F., D. G. SHARP, E. A. ECKERT, E. S. DILLON, D. BEARD and J. W. BEARD: The egg white inhibitor of influenza virus hemagglutination. J. Biol. chem. **179**, 1275—1287 (1949).

LEMBKE, A., M. KÖRNLEIN und H. FRAHM: Beiträge zum Brucelloseproblem; I. Mitteilung. Zbl. Bakt. I, **155**, 16—31 (1950).

LIEBERMEISTER, K. und E. ZEHENDER: Zur Morphologie und systematischen Stellung des Q-Fieber-Erregers. Klin. Wochenschr. **28**, 276—278 (1950).

McFARLANE, A. S., M. A., B. Sc., M. B., CH. B.: Electron microscopy of bacteria and viruses. Brit. Med. J. **2**, 1247 (1949).

MORTON, H. E. and J. OSKAY: Electron microscope studies of treponemes; II. The effect of penicillin on the Nichols strain of treponema pallidum. Am. J. Syph., Gonor. and Ven. Dis. **34**, 34—39 (1950).

PALAY, S. L. and A. CLAUDE: An electron microscope study of salivary gland chromosomes by the replica method. J. Exp. Med. **89**, 431—438 (1948).

PARMELEE, C. E., P. H. CARR and F. E. NELSON: Electron microscope studies of bacteriophage active against streptococcus lactis. J. Bact. **57**, 391—397 (1948).

PASSEY, R. D., L. DMOCHOWSKI. W. T. ASTBURY, R. REED and P. JOHNSON: Electron microscope studies of normal and malignant tissues of high- and low-breast-cancer strains of mice. Nature **165**, 107 (1950).

RHIAN, M., S. G. LENSEN and R. C. WILLIAMS: An electron microscope study of material from tissue of the central nervous system of poliomyelitic and normal mice and cotton rats. J. Immun. **62**, 487—504 (1949).

RIS, H. and J. P. FOX: The cytology of rickettsiae. J. Exp. med. **89**, 681—686 (1949).

SCHÄFER, W. und G. SCHRAMM: Über die Isolierung und Charakterisierung des Virus der klassischen Geflügelpest. Z. Naturforschg. **5b**, 91—102 (1950).

SCHLOSSBERGER, H., A. JAKOB und G. PIEKARSKI: Zur systematischen Stellung der Spirochäten. Naturwiss. **37**, 186—187 (1950).

SCHULTZ, E. W.: The present status of viruses and virus diseases. J. Amer. Med. Assoc. **138**, 1075—1079 (1948).

SPAETH, E.: Das Virusproblem. Wien. klin. Wschr. **58**, 118 (1946).

STEINHAUS, E. A.: (1) Nomenclature and classification of insect viruses. Bact. Rev. **13**, 203—223 (1949).

— (2) Polyhedrosis ("Wilt disease") of the alfalfa caterpillar. J. Econ. Ent. **41**, 859—865 (1949).

— (3) Principles of insect pathology. McGraw-Hill Book Co., Inc., New York (1949).

STEINHAUS, E. A., K. M. HUGHES and H. BLOCK: Demonstration of the granulosis virus of the variegated cutworm. J. Bact. **57**, 219—224 (1949).

Strauss, M. J., E. W. Shaw, H. Bunting and J. L. Melnick: "Crystalline" virus-like particles from skin papillomas characterized by intranuclear inclusion bodies. Proc. Soc. Exp. Biol. Med. 72, 46—50 (1949).

Theismann, H. und K.-H. Wallhäuser: Elektronenmikroskopische Untersuchungen an beschallten Bakterien. Naturwiss. 37, 185—186 (1950).

Weil, M. L., D. Beard, D. G. Sharp and J. W. Beard: (1) Purification, pH stability and culture of the mumps virus. J. Immun. 60, 561—582 (1948).

— (2) Purification and sedimentation and electron micrographic characters of the mumps virus. Proc. soc. Exp. Biol. Med. 68, 309 (1948).

Winkler, A.: Elektronenoptische Untersuchungen von Bakterien. Zbl. Bakt. I, 154, 189—194 (1949).

Wyckoff, R. W.: (1) Electron Microscopic study of viruses. J. Amer. Med. Assoc. 138, 1081—1083 (1948).

— (2) Electron microscopy. Interscience Publishers. New York and London (1949).

Vorträge aus der zweiten Tagung der Deutschen Gesellschaft für Elektronenmikroskopie in Bad Soden, April 1950

Bolt, W., O. Küchenhoff und T. Vogel: Elektronenoptische Studien über Grippevirus und das Bindungs-Lösungsphänomen an der Erythrozytenmembran.

Grün, L. und W. Hennessen: Einige Beobachtungen bei der bakteriophagen Lyse.

Grünholz, G. und H. Grünholz: Elektronenoptische Untersuchungen über Morphologie und Entwicklung pleuropneumonieartiger Mikroorganismen menschlicher Herkunft.

Haussmann, H. und H. Kehler: Über einige elektronenmikroskopische Beobachtungen an ultraschall- und wärmegeschädigten Bakterien.

Helmcke, J.-G.: Über Versuche mit hochpolymeren Einbettungsmitteln für licht- und elektronenmikroskopische Untersuchungen an biologischen Substanzen.

Helwig, G. und H. König: Neue Präparationsmethoden.

Kleinschmidt, A. und E. Kinder: Elektronenoptische Untersuchungen an Trypanosomen.

Liebermeister, K.: Neuere bakteriologische Präparationsverfahren in der Elektronenmikroskopie.

Mudd, S. und A. G. Smith: Elektronen- und lichtmikroskopische Kernstudien an Bakterien.

Nauck, E. G.: Elektronenoptische Darstellung von Bartonella muris.

Schäfer, W.: Elektronenmikroskopische Differenzierung von Geflügelviren.

Winkler, A.: Vergleichende licht- und elektronenoptische Abbildungen von Bakterien.

Zahn, H.: Die Morphologie der Proteine.

Nachtrag bei der Korrektur:

Angulo, J. J. and J. H. L. Watson: Electron microscopy study of chick embryo erythrocytes. Proc. Soc. Exp. Biol. Med. 71, 646—651 (1949).

Bergold, G. H.: The multiplication of insect viruses as organisms. Can. J. Research 28, 5—10 (1950).

Cherry, W. B. and D. W. Watson: The streptococcus lactis host-virus system. I. Factors influencing quantitative measurement of the virus. II. Characteristics of virus growth and the effect of electrolytes on virus adsorption. J. Bact. 58, 601—620 (1949).

Jung, S. F.: Strukturprobleme am roten Blutkörperchen. Naturw. 37, 229—233 (1950).

Piekarski, G.: Haben Bakterien einen Zellkern? Naturw. 37, 201—205 (1950).

Porter, K. R. and C. van Zandt Hawn: Sequences in the formation of clots from purified bovine fibrinogen and thrombin. A study with the electron microscope. J. Exp. Med. 90, 225—232 (1949).

Reagon, Lillie, Hickmann und Brueckner: Die infektiöse Anämie der Pferde. Am. J. Vet. Res. XI (39), 157 (1950).

RHOADES, R. P.: Low speed microtomy for the electron microscope. Proc. Soc. Exp. Biol. Med. **71**, 660—661 (1949).

ROBERTIS, E. de: An electron microscope analysis of nerves infected with the B Virus. J. Exp. Med. **90**, 291—295 (1949).

ROBERTIS, E. de and F. O. SCHMITT: An electron microscope study of nerves infected with human poliomyelitis virus. J. Exp. Med. **90**, 283—289 (1949).

SCHLOSSBERGER, H., A. JAKOB und G. PIEKARSKI: Zur systematischen Stellung der Spirochäten. Naturw. **37**, 186—187 (1950).

THEISMANN, H. und K. H. WALLHÄUSER: Elektronenmikroskopische Untersuchungen an beschallten Bakterien. Naturw. **37**, 185—186 (1950).

ZAHLER, S. A.: Nucleotide content of bacteriophage genetic units. Science **111**, 210 (1950).

Sachverzeichnis.

Abwasserorganismen 338.
Antikörper, haemagglutinationshemmender 178, 193, 198.
—, Verhältnis zum antiinfektiösen Antikörper 159, 187, 188, 220.
Auflicht, Myxom im 43.
—, Pustelausstriche im 43.
Auflösungsvermögen, der optischen Systeme 227.
Aureomycin 93.
—, against primary atypical pneumonia 93.
Avidität, haemagglutinierender Virusarten 179, 201.
Bakterien 191, 311.
—, elektronenmikroskopische Darstellung 311.
—, —, Geißeln 316.
—, —. —, Bewegung 316.
—. —. —, Ursprung 316.
—, —, Kapsel 315.
—, —, Plasmolyse 328.
—, —, Schleimsubstanz 315.
—, —, Serologische Reaktionen 328.
—, —, Sporen 322.
—, —, Verhalten gegen physikalische und chemische Agenzien 325.
—. —, —, hochfrequenter Schall 326.
—, —. —, Sublimat 326.
—, —, —, Silberkolloid 327.
—, —. —, Goldkolloid 327.
—, —, —, Sulfonamide 328.
—, —, Zellinhalt 317.
—, —, —, Ektoplasma 317.
—, —, —, Endoplasma 317.
—, —, —, Kernäquivalente 317.
—, —, —, Polkörper 319.
—, —, —, Vakuolen 321.
—, —, —, Volutingranula 319, 322.
—, —, Zellmembran 312, 325.
—, Haemagglutinine 191.

Bakteriophagen 189, 329.
—, elektronenoptische Darstellung 329.
—, —, Größe und Form 329.
—, —, Vermehrung 335.
—, —, Wirkungsmechanismus 335.

Bakteriophagen, elektronenoptische Darstellung, Okkupationshypothese 336.
—, —, —, Penetrationshypothese 337.
—, Fehlen des Haemagglutinins 189.
Bildpaare, elektronenoptische 246.
Bronchopneumonia, bacterial 91.
—, differentiation from virus pneumonia 91f.
Bronchopneumonievirus der Maus, elektronenmikroskopisch 339, 342, 343, 392.
Bulbushalter (Wild) 28.

Canine distemper virus 71.
Cellextrusions 38.
Cholerakulturfiltrat (siehe auch receptor destroying enzyme) 142, 170, 182, 192, 220.
Choriomeningitis, Virus der, Fehlen des Haemagglutinins 189.
Clostridium welchii, Kulturfiltrat 142, 182, 192.
Cold agglutination in primary atypical pneumonia 101.
—, cold agglutinin, characteristics 103f.
—, methods 101f.
—, in other diseases 102.
Cotton rat agents W 2 and W 3 137.
—, carried by Sigmodon hispidus 137.
Cysticetes 339.

Drierite 19.
Drying, of viruses 11.
—, pre-freezing 11.
—, after freezing (siehe Freeze-drying) 11.
—, —. blood plasma 12.
—. —. —, lipoidal constituents 12.
—, —. enzymatic changes 12.
—, —, gelatin 12.
—, —, mechanism 12.
Dunkelfeld 36.
—, Ektromelieknötchen im 34.
—, Leukocyten, virusbeladene, im 24.
—, Vaccineeffloreszenzen im 33.
Dunkelfeldmikroskopie 229.

Einschlußkörper, elektronenmikrosko-
pisch 225.
Ektromelievirus, elektronenmikroskopische Darstellung 301, 347, 348, 349,
353, 355, 390.
—, haemagglutinierende Eigenschaften
(siehe Haemagglutinine der Virus-
arten).
Elektrokinetisches Potential, virusvorbehandelter Erythrocyten 170.
Elektronenindikatoren 239.
Elektronenlinsen 234.
Elektronenmikroskope 250.
—, elektrische Anlagen 255.
—, Gesamtaufbau 250.
—, im Handel befindliche 258.
—, Kühlwasserleitungen 257.
—, Mikroskopröhre 251.
—, Vakuumanlagen 257.
Elektronenmikroskopie 221.
—, historische Entwicklung 249.
—, physikalische Grenzen 246.
—, Objektvorbereitung 283.
—, —, zur Untersuchung von Zellen und
Geweben 283.
—, —, —, Schnitt-Schall-Verfahren
283.
—, —, —, Hochgeschwindigkeitsmikro-
tome 284.
—, —, —, Schnittverfahren für Bakte-
rien 284.
—, —, Anreicherung und Reinigung von
Virussuspensionen 285.
—, —, Virusadsorption an Erythrocyten
287.
—, —, Verfahren zur Kontraständerung
288.
—, —, —, Färbemethoden 288.
—, —, —, Metallbedampfung (Schat-
tenwurftechnik) 289.
—, Präparationsmethoden 262.
—, —, Objekthalterung 263.
—, —, —, Objektnetze und Objekt-
blenden 263.
—, —, —, —, Reinigung der Objekt-
blenden 264.
—, —, —, Objektträgerfassungen 264.
—, —, Objektträgerfilme 265.
—, —, —, Kollodiumfilme 265.
—, —, —, —, Berechnung der Dicke
267.
—, —, —, Aluminiumoxydfilme 268.
—, —, —, Siliziumoxydfilme 268.
—, —, —, Silikatglasfilme 269.
—, —, —, Berylliumfilme 269.
—, —, Beschickung der Filme 270.
—, —, —, Tropfenmethode 270.

Elektronenmikroskopie, Präparations-
methoden, Beschickung der Filme,
Objektträgervibrator 271.
—, —, —, —, Gefriertrockenkammer
271.
—, —, —, Niederschlagsmethode 273.
—, —, —, —, elektrische Kernfäller
274.
—, —, —, —, Versuchsaufbau 275.
—, —, —, Oberflächenmethoden 277.
—, —, —, —, Abklatschverfahren 278.
—, —, —, —, Abziehverfahren 278.
—, —, —, —, Abdruckverfahren 280.
—, —, —, —, Filmbewuchsverfahren
279.
Elektronenmikroskopische Darstellung,
Objekte 306, 308.
—, Agens des Milchdrüsentumors der
Maus 369.
—, Bakterien (siehe daselbst) 311.
—, Bakteriophagen (siehe daselbst) 329.
—, filtrierbare Mikroorganismen 338.
—, Haemagglutination, virusbedingte
165, 171.
—, Influenzavirus 358.
—, Leptospiren 308.
—, Mumpsvirus 364.
—, Papillomvirus 364.
—, Pferdeencephalomyelitisvirus 365.
—, pflanzenpathogene Virusarten 376.
—, —, stäbchenförmige Virusproteine
376.
—, —, kugelförmige Virusproteine 379.
—, —, Viruskristalle und Virusgele
379.
—, —, physikochemische und serolo-
gische Reaktionen 383.
—, Poliomyelitisvirus 366.
—, Polyedervirus 370.
—, Protozoen 308.
—, quaderförmige Virusarten (siehe da-
selbst) 346.
—, Rickettsien 311.
—, Spirochaeten 308.
—, Varicellenvirus, Zostervirus 356.
—, Virus der Maul- und Klauenseuche
369.
Elektronenoptische Aufnahmen 291.
—, Objektschädigung durch Elektronen-
strahlen 292.
—, Vermessung der Objekte 293.
—, —, Messen in der Bildebene 293.
—, —, Messen in der Richtung des
Strahlenganges 293.
—, —, —, aus dem Einzelbild 295.
—, —, —, aus stereoskopischem Bild-
paar 297.

Elektronenoptische Aufnahmen, statistische Auswertung (siehe daselbst) 298.
—, —, Bestimmung der Teilchenzahl 302.
—, —, —, aus Filmen mit eingebetteten Objekten 302.
—, —, —, aus Teilchengewicht und Teilchenkonzentration 302.
—, —, —, aerodisperse Zerteilung einer Suspension 302.
—, —, —, in einem Gasvolumen 306.
Elektronenoptisches Bild, Eigenschaften 242.
—, Dunkelfeldbilder 245.
Elektronenstrahlen 233.
Elektronenstrahlerzeuger 238.
—, Leuchtdichte 247.
Elutionsphase, Virushaemagglutination 142, 165, 185.
Encephalitisvirus, japan., Haemagglutinin 220.
Encephalitisvirus, St. Louis, Fehlen des Haemagglutinins 189.
Encephalomyelitisvirus der Mäuse (siehe Mäuseencephalomyelitisvirus [Theiler]).
Encephalomyocarditisvirus, Haemagglutinin 220.
Enzymatische Reaktion, bei Virushaemagglutination 184, 185
Epilampen (Zeiß) 25.
Equines Encephalomyelitisvirus, elektronenmikroskopische Darstellung 365, 390.
—, Fehlen des Haemagglutinins 189.
Erythrocytenveränderung 168, 191.
—, durch bakterielle Agenzien 191, 192.
—, durch Virus 168.

Felder, und Linsen 234, 236.
—, elektrische 234.
—, magnetische 236.
Feline pneumonia (siehe Pneumonievirus der Katze) 134.
Fibromvirus der Kaninchen, Fehlen des Haemagglutinins 189.
Filtrierbare Mikroorganismen 338.
—, Abwasserorganismen 338.
—, Agalaktie der Ziegen 338.
—, Elementarkörper 239.
—, L-Organismen 338.
—, Pleuropneumonie der Rinder 338.
—, Polyarthritis der Ratten 338.
Fluoreszenzlicht, Beobachtungen 24.
Fluoreszenzmikroskopie 230.
Formol, Haemagglutinine und 146.
Francis-Inhibitor (siehe Seruminhibitor).

Freeze-drying, viruses 11, 12.
—, apparatus, for laboratory research 19.
—, —, for commercial production 18.
—, mechanism 12.
—, —, removing water vapor 16.
—, —, —, condensation 16.
—, —, —, dessicating substances 18.
—, —, —, pumping 19.
—, results 13.
—, —, rabbit brains 13.
—, —, vaccine 13.
—, —, canine distemper 14.
—, —, influenza virus 14.
—, —, tobacco mosaic virus 14.
—, —, hog cholera 14, 15.
—, —, cattle plague 15.
—, technic 15.
—, —, removing water vapor 16.
—, —, dessicating substances 18.
—, —, apparatus, for laboratory research 19.
—, —, —, Cryochem type 19, 21.
Freezing, of viruses 13.

Geflügelpestvirus, atypisches (siehe Newcastle disease virus).
—, klassisches 362, 364, 390.
—, —, elektronenmikroskopisch 362, 364, 390.
—, —, Haemagglutinin (siehe Haemagglutinine der Virusarten).
Geflügelpocken, Virus der 347, 349, 390.
—, —, elektronenmikroskopisch 347, 349, 390.
Gelbfiebervirus, Haemagglutinin 189.
Glucose, Poliomyelitisagglutinin und 176.
Glutathion, Mäusepneumonievirus und 176.
Gonadotropes Hormon, Inaktivierung durch Influenzavirus 182.
Grenzmembran 344, 356.
Grey Lung virus 136, 189.
—, Fehlen des Haemagglutinins 189.
—, pathogenic for mice and other rodents 136.
Guarnierikörperchen 32.

Haemagglutinine, der Virusarten 141.
—, antigenes Vermögen 187.
—, Beziehung zu geformten Viruselementen 157, 160.
—, Beziehung zur Infektiosität 149, 150, 156.
—, Beziehung zu Interferenzphänomenen 184, 186.
—, Bildung während Virusvermehrung 149.
—, chemische Natur 142, 149.

Haemagglutinine, Dissoziation, von anderen Virusfunktionen 141, 145, 147, 148, 188.
—, fermentative Eigenschaften 143, 181.
—, Inhibitoren 174.
—, Klassifikation 157.
—, Reaktion mit Erythrocyten 168.
—, Reaktion mit Wirtszellen des Virus 143, 184, 186.
—, Reaktionskinetik 142, 165.
—, Resistenz 144, 149, 188.
—, spezifische Hemmung durch Antikörper 178.
—, Wirkungsbreite 162.
—, wirtseigene 198.
Haemagglutination der Virusarten 141.
—, Anwendung 193.
—, Entdeckung 141.
—, Fermenthypothese 181.
—, haemagglutinierende Virusarten 144.
—, intravitale 187.
—, Mechanismus 144.
—, Phänomenologie 141.
—, Technik 194.
Haemagglutinationsspektrum 144, 162.
Haemagglutinationstest 193.
—, Anwendung 193.
—, spezielle Methoden 194.
—, Technik 194.
Haemagglutinationshemmungstest 178.
—, Analyse 178.
—, Anwendung 193.
—, spezielle Methoden 202.
—, Technik 198.
Haemagglutinierende Einheit 197.
Haemagglutinationstiter 197.
—, Infektiositätstiter und 150.
Haemagglutinationstypus, virusbedingter 144, 161.
Haemolysine, der Virusarten 161.
Hamster agent W. I. 137.
—, gray lung virus and 137.
Hellfeldmikroskopie 227.
Hornhautfärbung, für U.-O.-Betrachtung 28.
Hühnerpestvirus (siehe Geflügelpestvirus).

Influenza, antibody levels 79f.
—, —, heterologous strains 79.
—, —, in children 79.
—, —, persistance after vaccination 80.
—, —, resistance and 79.
—, —, respiratory secretions and 80.
—, group of viruses 47.
—, —, haemagglutination 48.
Influenza, group of viruses, haemagglutination, receptor in the erythrocytes 48.
—. —, range of size, identity 48.
—, —, viruses included in the group 48.
—, —, —, fowl plague 48.
—, —, —, influenza A 47, 53.
—, —, —, influenza B 47, 53.
—, —, —, —, differences between A and B 48.
—, —, —, —, similarities from A and B 48.
—, —, —, mumps 48.
—, —, —, Newcastle disease of fowls 48.
—, —, —, swine influenza 48.
—, immunity 66.
—, —, cellular 80.
—, —, circulating antibody and 80.
—, —, epidemiological evidence 66.
—, —, experiments in animals 67.
—, —, after disease 69.
—, —, —, by modified infection 70.
—, —, after infection 67.
—, —, —, ferrets 67.
—, —, —, mice 67.
—, —, after vaccination 68.
—, —, —, ferrets 68.
—, —, —, swine 68.
—, —, passive 77.
—, —, resistance in man 69.
—, —, —, in general population 69.
—, —, —, in human volunteers 69.
—, vaccination against (siehe auch vaccination of man) 71.
Influenzavirus, antigenic differences amongst strains 52.
—, activity in human convalescent sera 53.
—, —, varying breadth of activity 53.
—, cross immunity test in mice 58.
—, determined by neutralization with immune ferret sera 57.
—, immune rabbit sera 57.
—, ratio convalescent serum titre: acute serum titre 54.
—, significance, observed with human immune sera 55.
—, concentration 73, 74.
—, —, adsorption at red blood cells and elution 74, 75.
—, —, by freezing and thawing 73.
—, conditions for epidemics 47.
—, —, emergence of more virulent mutants 47.
—, elektronenmikroskopische Darstellung 30, 358, 390.
—, enzymic activity 58.

Influenzavirus, enzymic activity,
variations of different strains 58.
—, Haemagglutination (siehe Haemagglutinine der Virusarten).
—, inactivating procedures 69, 71, 72.
—, type A 53, 55, 62.
—, —, definition 53.
—, —, —, American swine influenza
(S. 15) as A virus 53, 62.
—, —, —, classification of strains 55.
—, —, —, inactivation by human convalescent serum with broad action 53.
—, —, —, —, by patients sera 53.
—, —, haemagglutination 49.
—, —, —, titre with fowl cells (F) 49.
—, —, —, —, with guinea pig cells (G)
49.
—, —, —, —, ratio F/G 49.
—, —, —, —, —, in amniotic fluid 49,
50.
—, —, —, —, —, in allantoic fluid 49,
50.
—, —, laboratory strains 60.
—, —, neurotropic strains 60.
—, —, O-Phase in mice 51.
—, —, —, immunizing power for mice
51.
—, —, —, passage experiments in mice
51.
—, —, —, pathogenicity for man 52, 60.
—, type B 51, 53.
—, —. definition 53.
—, —, growth in the allantoic cavity 52.
—, —, no O-D change 51.
—, —, vaccination against 74, 75.
—, —, —, effects 76.
—, —, —, inhalation of 75.
—, —, —, subcutaneous injection 75.
Influenzaviruses, variation in 47.
—, adaptation to laboratory passage 49,
59.
—, —, to the allantoic cavity 49.
—, —, —, O-D mutation 50.
—, —, to mice 51.
—, —, —, influenza A 51.
—, —, —, influenza B 52.
—, general principles 61.
—, significance to the epidemiology 62.
Inhibitoren, der Virushaemagglutinine
172, 174, 183, 220.
—, Inaktivierung durch Virus 183.
—, Mucopolysaccharide 172.
—, Polysaccharide bakterieller und
pflanzlicher Provenienz 177.
—, wirtseigene, unbekannter Natur 174,
175, 220.
Interferenzphänomen 143, 184, 186.

Kanarienpocken, Virus der 301, 347, 348,
353, 390.
—, —, elektronenmikroskopische Abbildung 301, 347, 348, 353, 390.
Kapselsubstanzen (bakterielle), als Inhibitoren 177, 186.
Klassifikation der Virusarten, elektronenmikroskopische 389.
Kompressor, für Bulbus 27.

Lebendmikroskopie 24.
Lebenszyklus, Vaccinevirus 37.
L-Organismen (SABIN und KLIENEBER-
GER) 338.
Lipoide, Haemagglutinine 157.
Lymphogranuloma venereum, Virus des,
elektronenmikroskopisch 339, 343.
Lyssavirus, elektronenmikroskopisch 307.
—, Haemagglutinin, fehlendes 189.

Maul- und Klauenseuchevirus, elektronenmikroskopisch 269, 390.
—, Haemagglutinin, fehlendes 189.
Mäuseencephalitis Theiler, elektronenmikroskopische Darstellung 366, 390.
—, Haemagglutinin 189, 220.
Mäusepneumonievirus (siehe Haemagglu-
tinine der Virusarten).
Mengo-Encephalomyelitisvirus, Haemag-
glutinin 200.
Mikrokultur, Vaccinevirus 39.
Mikroorganismen, sublichtmikroskopi-
sche 223.
Milchdrüsentumor der Mäuse, Agens des
369, 390.
—, —, elektronenmikroskopisch 369,
390.
Molluscum contagiosum, Virus des 301,
347, 349, 353, 355, 390.
—, —, elektronenmikroskopisch 301,
347, 349, 353, 355, 390.
Mononucleose-Sera, Reaktion mit virus-
vorbehandelten Erythrocyten 169.
Mucin 143, 172.
Mucinase 143, 183.
Mucopolysaccharid 143, 172.
Mumpsvirus, elektronenmikroskopische
Darstellung 364, 392.
—, Haemagglutinin (siehe Haemagglutinine der Virusarten).
Myxomvirus 301, 347, 349, 353, 390.
—, elektronenmikroskopisch 301, 347,
349, 353, 390.

Newcastle disease virus, elektronenmikroskopische Darstellung 344, 392.
—, Haemagglutinin (siehe Haemagglutinine der Virusarten).

Normalkomponente (KNIGHT) 158.
—, chemische Natur 158.
—, Glucosamingehalt 159, 173.
—, Hemmung von Virushaemagglutinin 154, 158, 173.
—, physikalische Eigenschaften 158.
—, serologische Beziehung zum Virushaemagglutinin 159.

Objektbelastung, Leistungsgrenze 248.
—, Dosisgrenze 249.
Objektträger, Reinigung 44.
O-D-Phase, bei Virushaemagglutination 50, 177.

Papillomvirus des Kaninchens 364, 390.
—, elektronenmikroskopisch 364, 390.
Paramycetes 339.
Pektin, als Inhibitor 177.
Penicillin 93.
—, against viruses of the psittacosis group 93.
Perjodat-Inaktivierung 148, 173.
—, von Inhibitoren 173, 174.
—, von Virus 173, 174.
—, von Zellreceptoren 143, 166, 172.
Peripneumonieerreger, elektronenmikroskopisch 338, 341, 392.
—, Haemagglutinin 144, 163.
Pferdeanaemievirus, Haemagglutinin 189.
Pflanzensamen, Haemagglutinine 142, 161, 191.
Phloxin 28.
Phänomen von HIRST 141
Phänomen von THOMSEN-FRIEDENREICH 142, 170, 182, 187, 192.
Phytagglutinine 142, 161, 191.
Phytopathogene Virusarten 376, 390.
—, elektronenmikroskopisch 376, 390.
Pleuropneumonia-like organisms 135, 144, 163, 338, 341, 392.
Pneumonia, primary atypical 93.
—, clinical characteristics 97.
—, distribution 94.
—, epidemiology 95.
—, etiology 109.
—, —, contaminating viruses 118.
—, —, transmission, to chick embryos 119.
—, —, —, to cotton rats 114.
—, —, —, to laboratory animals 112.
—, —, —, to man 109.
—, —, —, to mongoose 112.
—, pathology 99.
—, serological reactions 101.
—, —, agglutination (indifferent Streptococci) 106.

Pneumonia, serological reaction, cold agglutination 101.
—, —, —, characteristics 103.
—, —, complement fixation (non specific) 105.
—, —, neutralization with human sera 123.
—, —, relation of streptococcal agglutinins to other reactions 107.
Pneumonievirus, der Katze 134, 339, 343, 392.
—, der Maus (siehe Mäusepneumonievirus, Bronchopneumonie der Mäuse).
—, des Menschen 109, 339.
—, der Rinder (siehe Peripneumonieerreger).
Poliomyelitisvirus, elektronenmikroskopische Darstellung 366, 390.
—, Haemagglutination 189, 190, 220.
Polyarthritis der Ratte, Virus der 338.
—, —, elektronenmikroskopisch 338.
Polyedervirus 301, 370, 392.
—, elektronenmikroskopisch 301, 370, 392.
Potenzierungseffekt 167, 184.
Pseudopestvirus der Hühner (siehe Newcastle disease virus).

Quaderförmige Virusarten 346.
—, Einschlußkörper 354.
—, Form, Größe und Struktur 347.
—, Grenzmembran 356.
—, physikalische und chemische Eingriffe 355.
Quarzstäbe, lichtleitende 24.
Q-fever 93.
—, streptomycin against 93.

Rezeptoren, für Virushaemagglutinine 171.
—, chemische Natur 171.
—, fermentativer Abbau 182.
—, hemmende Wirkung 171.
—, Inaktivierung 182.
—, Isolierung 171.
—, Transformation, durch RDE 170, 172, 182.
—, —, durch Virus 168, 170, 182.
—, Vorkommen in Erythrocyten 143, 171.
—, —, in Gewebszellen des Wirtes 143, 185.
—, —, im Hühnerei 173.
—, —, im Schleimdrüsensekret 172.
—, —, im Serum 172.
Receptor destroying enzyme (RDE) 183, 185.
Receptor gradient 59, 168, 183, 185.

Reflektoren, künstliche 35.
Reticulierende Degeneration 31.
Rhesus-Antikörper 169, 170.
Rickettsien 144, 163, 221, 311, 392.
—, elektronenmikroskopische Darstellung 221, 311, 392.
—, Haemagglutinin 144, 163.
Roller tube cultures 1.
—, apparatus 3.
—, technic 2.
—, —, addition of virus 4.
—, —, embryonic extract 3.
—, —, incubation 3.
—, —, modifications 6.
—, —, morphological methods 4.
—, —, routine care of 3.
—, —, Simms solution 2
—, —, tissue 2.
—, virus studies 7.
Röntgenographie 230.

Senfgas, Einfluß auf Influenzavirus 147.
Spaltlampenmikroskopie 24.
Speichel, menschlicher 173, 220.
—, Gehalt an Haemagglutininen 220.
—, hemmende Wirkung auf Haemagglutination 173, 220.
Strahlengang im Elektronenmikroskop 239.
Streptococcal agglutinins, in primary atypical pneumonia 106.
—, relation to other serological reactions 107.
Strömungsdoppelbrechung 230.
Sulfonamides 93.
—, little value in virus pneumonias 93.
System, der Virusarten 224, 389.

Tanninsäure, als Hemmstoff 178.

Übermikroskopische Aufnahmen, statistische Auswertung 300.
—, Ektromelie 301.
—, Influenzavirus A 301.
—, Influenzavirus B 301.
—, Kanarienvirus 301.
—, Kaninchenmyxom 301.
—, Molluscum contagiosum 301.
—, Pocken 301.
—, Polyedervirus 301.
—, Schwammspinner 301.
—, Schweineinfluenza 301.
—, Tabakmosaikvirus 301.
—, Tomatenmosaikvirus 301.
—, Varicellen 301.
Ultrafiltration 232.
Ultramikrometrie, statistische 232.
Ultramikroskopie 229.

Ultraviolettes Licht, Haemagglutinine und 147.
Ultrazentrifugierung 231.
Ultropak (Heine) 25.
Universalopakilluminator (Reichert) 25.
U.-V.-Mikroskop 227.

Vaccination of man, with influenzavirus 71.
—, duration of effect 81.
—, effects 75, 76.
—, factors influencing immunity by 78.
—, influence of dosage 72, 73, 80.
—, relation to antibodies 71, 72, 73, 79, 82.
—, strain differences 78.
—, systemic reactions 82.
—, routes 70, 71, 81.
—, —, instillation in the nose 70.
—, —, intracutaneous 71, 81.
—, virus preparations 71, 72, 73, 82.
—, —, allantoic fluid 73.
—, —, —, concentration 73.
—, —, culture virus 71.
—, —, tissue from chick embryo 71.
Vaccinevirus, elektronenmikroskopische Darstellung 301, 347, 390.
—, Haemagglutinin (siehe Haemagglutinine der Virusarten).
Varicellen, Virus der 301, 356.
—, —, elektronenmikroskopisch 301, 356, 390.
Variolavirus (siehe Haemagglutinine der Virusarten).
Vergrößerung, förderliche, im Elektronenmikroskop 241.
—, Bestimmung 241.
Vertikalilluminator 25.
Vesicularstomatitis, Virus, fehlendes Haemagglutinin 189.
Virusarten 306.
—, System 306.
—, Rickettsien und 306.
—, Protozoen und 307.
—, Molekülhypothese 307.
Virusarten, große 338.
—, Bronchopneumonie der Maus 339.
—, Lymphogranuloma inguinale 339.
—, Pneumonie der Katzen 134, 339.
—, Psittakose 339.
—, Schwimmbadkonjunktivitis 339.
Virusbegriff, Definition 221.
Viruscysten 38.
Virusdifferenzierung, serologische 178, 181, 194, 201.
Virusdimensionen 224.

Virusforschung, elektronenoptische, Ergebnisse 306.
Viruspneumonia and pneumonitis viruses 87.
—, etiology 88.
—, —, experiments with known viruses 89.
—, historical 87.
—, in man 91.
—, —, differentiation from bacterial pneumonia 91.
—, —, epidemiologicaldifferentiation 92.
—, —, failure of response to chemotherapy 93.

Viruspneumonia in man, pathology, characteristic 93.
Virusproteine, makromolekulare 222.
Viskositätsmessung 232.
Vitalmikroskopie 23.

Wachstumstypen, des Virus (MERLING) 38.
Wasserblau 28.

Zoster, Virus des 356.
—, —, elektronenmikroskopisch 356, 390.